Cornelia Schmitt
März 72

ITERATIVE SOLUTION
OF NONLINEAR EQUATIONS
IN SEVERAL VARIABLES

Computer Science and Applied Mathematics
A SERIES OF MONOGRAPHS AND TEXTBOOKS

Editor
Werner Rheinboldt
University of Maryland

Hans P. Künzi, H. G. Tzschach, and C. A. Zehnder
NUMERICAL METHODS OF MATHEMATICAL OPTIMIZATION: WITH ALGOL AND FORTRAN PROGRAMS, 1968

Azriel Rosenfeld
PICTURE PROCESSING BY COMPUTER, 1969

James Ortega and Werner Rheinboldt
ITERATIVE SOLUTION OF NONLINEAR EQUATIONS IN SEVERAL VARIABLES, 1970

ITERATIVE SOLUTION OF NONLINEAR EQUATIONS IN SEVERAL VARIABLES

J. M. Ortega and W. C. Rheinboldt

UNIVERSITY OF MARYLAND
COLLEGE PARK, MARYLAND

 1970

ACADEMIC PRESS   New York and London

Copyright © 1970, by Academic Press, Inc.
ALL RIGHTS RESERVED
NO PART OF THIS BOOK MAY BE REPRODUCED IN ANY FORM,
BY PHOTOSTAT, MICROFILM, RETRIEVAL SYSTEM, OR ANY
OTHER MEANS, WITHOUT WRITTEN PERMISSION FROM
THE PUBLISHERS.

ACADEMIC PRESS, INC.
111 Fifth Avenue, New York, New York 10003

United Kingdom Edition published by
ACADEMIC PRESS, INC. (LONDON) LTD.
Berkeley Square House, London W1X 6BA

Library of Congress Catalog Card Number: 79-107564

PRINTED IN THE UNITED STATES OF AMERICA

*TO
SARA AND CORRIE*

CONTENTS

Preface xiii
Acknowledgments xvii
Glossary of Symbols xix

Introduction 1

Part I BACKGROUND MATERIAL

1. Sample Problems

1.1.	Two-Point Boundary Value Problems	9
	Notes and Remarks	12
	Exercises	13
1.2.	Elliptic Boundary Value Problems	14
	Notes and Remarks	16
1.3.	Integral Equations	18
	Notes and Remarks	20
	Exercises	20
1.4.	Minimization Problems	21
	Notes and Remarks	25
	Exercises	26
1.5.	Two-Dimensional Variational Problems	26
	Notes and Remarks	30
	Exercises	32

2. Linear Algebra

2.1.	A Review of Basic Matrix Theory	34
	Notes and Remarks	37
	Exercises	37
2.2.	Norms	38
	Notes and Remarks	44
	Exercises	44
2.3.	Inverses	45
	Notes and Remarks	50
	Exercises	51
2.4.	Partial Ordering and Nonnegative Matrices	51
	Notes and Remarks	57
	Exercises	57

3. Analysis

3.1.	Derivatives and Other Basic Concepts	59
	Notes and Remarks	65
	Exercises	66
3.2.	Mean-Value Theorems	68
	Notes and Remarks	73
	Exercises	74
3.3.	Second Derivatives	74
	Notes and Remarks	81
	Exercises	82
3.4.	Convex Functionals	82
	Notes and Remarks	88
	Exercises	89

Part II NONCONSTRUCTIVE EXISTENCE THEOREMS

4. Gradient Mappings and Minimization

4.1.	Minimizers, Critical Points, and Gradient Mappings	93
	Notes and Remarks	97
	Exercises	97
4.2.	Uniqueness Theorems	98
	Notes and Remarks	101
	Exercises	102
4.3.	Existence Theorems	104
	Notes and Remarks	108
	Exercises	109
4.4.	Applications	110
	Notes and Remarks	115
	Exercises	116

5. Contractions and the Continuation Property

5.1.	Contractions	119
	Notes and Remarks	124
	Exercises	125
5.2.	The Inverse and Implicit Function Theorems	125
	Notes and Remarks	131
	Exercises	132
5.3.	The Continuation Property	132
	Notes and Remarks	139
	Exercises	140
5.4.	Monotone Operators and Other Applications	141
	Notes and Remarks	145
	Exercises	145

6. The Degree of a Mapping

6.1.	Analytic Definition of the Degree	147
	Notes and Remarks	154
	Exercises	155
6.2.	Properties of the Degree	156
	Notes and Remarks	160
	Exercises	160
6.3.	Basic Existence Theorems	161
	Notes and Remarks	163
	Exercises	164
6.4.	Monotone and Coercive Mappings	165
	Notes and Remarks	167
	Exercises	168
6.5.	APPENDIX. Additional Analytic Results	169

Part III ITERATIVE METHODS

7. General Iterative Methods

7.1.	Newton's Method and Some of Its Variations	181
	Notes and Remarks	187
	Exercises	188
7.2.	Secant Methods	189
	Notes and Remarks	200
	Exercises	205
7.3.	Modification Methods	206
	Notes and Remarks	212
	Exercises	214
7.4.	Generalized Linear Methods	214
	Notes and Remarks	222
	Exercises	229

7.5.	Continuation Methods	230
	Notes and Remarks	234
	Exercises	235
7.6.	General Discussion of Iterative Methods	236
	Notes and Remarks	239

8. Minimization Methods

8.1.	Paraboloid Methods	240
	Notes and Remarks	242
	Exercises	243
8.2.	Descent Methods	243
	Notes and Remarks	247
	Exercises	248
8.3.	Steplength Algorithms	249
	Notes and Remarks	258
	Exercises	259
8.4.	Conjugate-Direction Methods	260
	Notes and Remarks	263
	Exercises	266
8.5.	The Gauss–Newton and Related Methods	267
	Notes and Remarks	269
	Exercises	271
8.6.	APPENDIX 1. Convergence of the Conjugate Gradient and the Davidon–Fletcher–Powell Algorithms for Quadratic Functionals	271
8.7.	APPENDIX 2. Search Methods for One-Dimensional Minimization	275

Part IV LOCAL CONVERGENCE

9. Rates of Convergence—General

9.1.	The Quotient Convergence Factors	281
	Notes and Remarks	286
	Exercises	286
9.2.	The Root Convergence Factors	287
	Notes and Remarks	293
	Exercises	294
9.3.	Relations between the R and Q Convergence Factors	295
	Exercises	298

10. One-Step Stationary Methods

10.1.	Basic Results	299
	Notes and Remarks	305
	Exercises	308

10.2.	Newton's Method and Some of Its Modifications	310
	Notes and Remarks	316
	Exercises	318
10.3.	Generalized Linear Iterations	320
	Notes and Remarks	331
	Exercises	332
10.4.	Continuation Methods	334
	Notes and Remarks	340
10.5.	APPENDIX. Comparison Theorems and Optimal ω for SOR Methods	341

11. Multistep Methods and Additional One-Step Methods

11.1.	Introduction and First Results	347
	Notes and Remarks	352
	Exercises	354
11.2.	Consistent Approximations	355
	Notes and Remarks	363
	Exercises	366
11.3.	The General Secant Method	369
	Notes and Remarks	378
	Exercises	379

Part V SEMILOCAL AND GLOBAL CONVERGENCE

12. Contractions and Nonlinear Majorants

12.1.	Some Generalizations of the Contraction Theorem	383
	Notes and Remarks	390
	Exercises	391
12.2.	Approximate Contractions and Sequences of Contractions	393
	Notes and Remarks	398
	Exercises	399
12.3.	Iterated Contractions and Nonexpansions	400
	Notes and Remarks	406
	Exercises	407
12.4.	Nonlinear Majorants	409
	Notes and Remarks	414
	Exercises	414
12.5.	More General Majorants	415
	Notes and Remarks	419
	Exercises	420
12.6.	Newton's Method and Related Iterations	421
	Notes and Remarks	428
	Exercises	430

13. Convergence under Partial Ordering

13.1.	Contractions under Partial Ordering	432
	Notes and Remarks	439
	Exercises	440
13.2.	Monotone Convergence	441
	Notes and Remarks	446
	Exercises	446
13.3.	Convexity and Newton's Method	447
	Notes and Remarks	454
	Exercises	454
13.4.	Newton–SOR Interactions	456
	Notes and Remarks	463
	Exercises	464
13.5.	M-Functions and Nonlinear SOR Processes	464
	Notes and Remarks	470
	Exercises	471

14. Convergence of Minimization Methods

14.1.	Introduction and Convergence of Sequences	473
	Notes and Remarks	478
	Exercises	478
14.2.	Steplength Analysis	479
	Notes and Remarks	493
	Exercises	493
14.3.	Gradient and Gradient-Related Methods	494
	Notes and Remarks	499
	Exercises	500
14.4.	Newton-Type Methods	501
	Notes and Remarks	506
	Exercises	507
14.5.	Conjugate-Direction Methods	509
	Notes and Remarks	512
	Exercises	513
14.6.	Univariate Relaxation and Related Processes	513
	Notes and Remarks	518
	Exercises	519

An Annotated List of Basic Reference Books   521

Bibliography   523

Author Index   559
Subject Index   566

PREFACE

This book is an outgrowth of research work undertaken by us and several of our Ph.D. students during the past five years; most of the material was also covered in graduate courses we offered during that time at the University of Maryland. Our aim is to present a survey of the basic theoretical results about nonlinear equations in n dimensions as well as an analysis of the major iterative methods for their numerical solution. Such a comprehensive presentation of this rapidly growing field appears to be needed and should benefit not only those working in the field but also those interested in, or in need of, information about specific results or techniques. At the same time we also hope to provide here a text for graduate numerical analysis courses in this area, and, in order to meet this aim, we have endeavored to make the main text as self-contained as possible, to prove all results in full detail, and to include a number of exercises throughout the book. In order to make the work useful as a reference source, we have supplemented each section with a set of "Notes and Remarks" in which literature citations are given, other related results are discussed, and various possible extensions of the results of the text are indicated. In addition, the book ends with a comprehensive bibliography of the field.

The main text presupposes that the reader has a preparation equivalent to the material covered in standard advanced multivariate-calculus and linear algebra courses. Some of this material is reviewed and collected in the form we need it in Chapters 2 and 3. Also some familiarity with the basic techniques for solving single equations in one unknown as well as systems of linear equations will be helpful but is not essential.

In particular, all of the background material we shall require from the theory of the numerical solution of linear equations is collected in Chapter 2.

By necessity, we have had to make various decisions limiting the scope of our material. We include no specific treatment of iterative methods for one-dimensional equations or for systems of linear equations since these are covered in the excellent monographs of Ostrowski [1966] and Traub [1964] for the former subject and those of Forsythe and Wasow [1960] and Varga [1962] for the latter. Except in a few places, we restrict ourselves to the problem of determining isolated, and not otherwise constrained, solutions of n real equations in n real unknowns. The problem of "solving" m equations in n unknowns is for $m > n$ a topic of approximation theory, while for $m < n$ it appears to have little independent interest. The introduction of a set of side conditions constraining the desired solution requires rather different techniques and is in certain settings a problem of nonlinear programming, while, in its general form, it is as yet little understood. Similarly, the case of nonisolated solutions is an almost completely open research area. We do not discuss iterative methods which require second or higher derivatives in their formulation because the analysis of such methods tends to be uninformative and cumbersome and, more importantly, because the evaluation of the kth derivative of a mapping of R^n into itself requires in general n^{k+1} functional evaluations. Consequently, we regard methods requiring more than the first derivative as numerically unattractive except, perhaps, for special problems. Finally, we do not include numerical examples. It appears that very little is gained if various methods are applied to incidental examples for which theoretical results already guarantee the absence of all difficulties. In order to provide meaningful insight into the numerical behavior of an iterative process, extensive numerical computations are needed in which not only the equation to be solved and its dimension are varied systematically, but also many different initial approximations are used in each case. To our knowledge no adequate computational effort of this type has so far been undertaken, and our own computational experiments leave more questions open than they answer. The influence of the variation of the equation, the dimension, and the initial data upon the outcome of the computation is still very little understood both from a practical as well as a theoretical viewpoint and, in particular, there are few results about the influence of the various types of computational error.

Perhaps the most important delineation of our material, however, results from our decision to restrict ourselves to finite dimensions throughout the main text in spite of the fact that many of the research results in this field are currently being presented in a more general setting. But, we have endeavored to present as many of the results as possible in such a way that the extension to, say, operators on Banach spaces is immediately clear. We believe this has the advantage that the book is meaningful and accessible both to readers with an extensive functional analysis background or none at all. In addition, in the "Notes and Remarks" we indicate these extensions to infinite dimensional spaces under the assumption that the reader is familiar with the necessary terminology. For a treatment of some of these topics in a functional analytic setting, the reader is referred to the books by Collatz [1964], Goldstein [1967], Kantorovich and Akilov [1959], Rall [1969], and Vainberg [1956].

ACKNOWLEDGMENTS

It is our pleasure to acknowledge with gratitude the support received for part of our own research, as well as that of our students, from the National Aeronautics and Space Administration under Grant NsG 398, the National Science Foundation under Grant GJ-231, and the U.S. Army Research Office Durham under Grant OOR-DA-AROD-31-124-G676. Moreover, we wish to thank the University of Maryland for providing the facilities and the atmosphere necessary and conducive for such work, and in particular Mrs. Dawn Shifflett for her excellent and painstaking work in typing the entire manuscript. We are also grateful to our students—John Avila, Ray Cavanagh, Steven Rauch, Robert Stepleman, and Robert Voigt—for reading and commenting on various parts of this material, and especially Jorge Moré, who read the entire manuscript. Finally, we wish to thank our long-suffering wives, to whom this book is dedicated, for the patience and understanding which made it possible.

GLOSSARY OF SYMBOLS

Spaces (see Section 2.1)

R^n	real n-dimensional space
C^n	complex n-dimensional space
$L(R^n, R^m), L(R^n)$	the linear space of linear operators from R^n to R^m, or R^n to R^n
$R^n \times R^m$	the product space
$(R^n)^m$	the m-fold product space $R^n \times \cdots \times R^n$

Vectors and Sets (see Chapter 2)

$e^1, \ldots, e^n$	the coordinate vectors of R^n
$x = (x_1, \ldots, x_n)^T$	a column vector with components x_i
x^T	the transpose of x
$\{x^1, \ldots, x^m\}$	a set of m vectors
$\{x^k\}$	a sequence of vectors
$\|\cdot\|$	an arbitrary norm on R^n
$\|\cdot\|_p$	the l_p-norm on R^n, $1 \leq p \leq \infty$
(x, y)	an inner product on R^n
$S(x, r)$	the open ball $\{y \in R^n \mid \|y - x\| < r\}$
$\bar{S}(x, r)$	the closed ball $\{y \in R^n \mid \|y - x\| \leq r\}$
$\bar{S}$	the closure of the set S
$\dot{S}$	the boundary of the set S
int(S)	the interior of the set S
$[x, y]$	the set $\{z \in R^n \mid z = tx + (1-t)y,\ t \in [0, 1]\}$
$x \leq y$	the partial ordering $x_i \leq y_i$, $i = 1, \ldots, n$
$\langle x, y \rangle$	the set $\{z \in R^n \mid x \leq z \leq y\}$
$\lvert x \rvert$	the vector with components $\lvert x_i \rvert$, $i = 1, \ldots, n$

Matrices (see Chapter 2)

$A = (a_{ij})$	an $n \times n$ matrix with elements a_{ij}
A^{-1}	the inverse of A

GLOSSARY OF SYMBOLS

A^T	the transpose of A
$\det A$	the determinant of A
A^+	a generalized inverse of A
A^k	the kth power of A
$\{A_k\}$	a sequence of matrices
$\rho(A)$	the spectral radius of A
$\|A\|$	an arbitrary norm of A
$\|A\|_p$	the l_p-norm of A, $1 \leq p \leq \infty$
$\operatorname{rank} A$	the rank of A
I	the identity matrix
$A \leq B$	the partial ordering $a_{ij} \leq b_{ij}$, $i, j = 1,\ldots, n$
$(a^1,\ldots, a^n)$	a matrix with columns $a^1,\ldots, a^n$
$\operatorname{diag}(a_1,\ldots, a_n)$	a diagonal matrix with elements $a_1,\ldots, a_n$

Functions and Derivatives (see Chapter 3)

$F: D \subset R^n \to R^m$	a mapping with domain D in R^n and range in R^m
$F(D)$ or FD	the set $\{y \in R^m \mid y = Fx, x \in D\}$
$F'(x), F''(x)$	the first or second (G or F) derivative of F at x
F^p	the pth power of F
F^{-1}	the inverse of F
F_U	the restriction of F to the set U
$\partial_i f$	the partial derivative with respect to the ith variable
$\partial_i F$	the vector partial derivative (see **5.2.2**)
$F(\cdot, y)$	the mapping derived by holding y fixed
$H_g(x)$	the Hessian matrix of g at x
Δ	the Laplacian

Miscellaneous

$\in$	element inclusion
$\subset$	set inclusion
$\cup, \cap$	union, intersection
$S_1 \sim S_2$	the difference of the sets S_1 and S_2
$\forall$	for all
Π, Σ	product, sum
δ_{ij}	Kronecker delta
$\operatorname{sgn} x$	the function $\operatorname{sgn} x = 1$, if $x \geq 0$, $x = -1$, if $x < 0$
$\lim_{t \to a-}$	the one-sided limit as $t \to a$ with $t < a$
∎	end of proof or definition
$\mathscr{I}$	a general iterative process
$C(\mathscr{I}, x^*)$	the set of sequences generated by $\mathscr{I}$ and converging to x^*
$R_p\{x^k\}, R_p(\mathscr{I}, x^*)$ $Q_p\{x^k\}, Q_p(\mathscr{I}, x^*)$ $O_R\{x^k\}, O_R(\mathscr{I}, x^*)$ $O_Q\{x^k\}, O_Q(\mathscr{I}, x^*)$	rate-of-convergence symbols (see Chapter 9)

INTRODUCTION

We denote by R^n the real n-dimensional linear space of all column vectors

$$x = \begin{pmatrix} x_1 \\ \vdots \\ x_n \end{pmatrix}.$$

However, except for some coordinate-dependent iterative processes, such as the SOR methods described in Section **7.4**, essentially all our discussions will be basis-independent, and thus R^n may also be regarded as an abstract real n-dimensional linear space. We shall use lower-case Latin letters, with or without superscript, for the vectors of R^n and denote the components of these vectors by subscripts; for example, x_i^k is the ith component of the vector x^k. For further related notational conventions about vectors and matrices, see Chapter 2.

The space R^n will usually be assumed to be equipped with some otherwise unspecified norm (see Section **2.2**), and whenever a result is dependent upon a particular norm this will be explicitly stated. The reader is assumed to be familiar with the basic topological concepts on R^n, such as open, closed, or compact sets, neighborhoods of a point, limits and Cauchy sequences, continuity and uniform continuity of functions, etc. If the reader is familiar with these concepts only in the usual Euclidean norm, it is important to note that they carry over immediately to arbitrary norms by means of the norm-equivalence theorem **2.2.1**, which implies that all topological considerations on R^n are norm-independent.

We denote the closure, boundary, and interior of a set $S \subset R^n$ by $\bar{S}$, $\dot{S}$, and int(S), respectively. Particularly important subsets of R^n are the open and closed balls (with respect to some norm $\|\cdot\|$)

$$S(x^0, r) = \{x \in R^n \mid \|x - x^0\| < r\}, \qquad \bar{S}(x^0, r) = \{x \in R^n \mid \|x - x^0\| \leq r\},$$

with center at x^0 and radius $r > 0$.

A function F (also called a mapping or operator) with domain D in R^n and range in R^m will be denoted by $F: D \subset R^n \to R^m$, or sometimes by $F: D \to Q$ if the dimensionalities of the sets D and Q are evident. For $m > 1$, the components of $F: D \subset R^n \to R^m$ are indicated by $f_1, \ldots, f_m$ and $Fx \in R^m$ is assumed to be represented by the column vector

$$Fx = \begin{pmatrix} f_1(x) \\ \vdots \\ f_m(x) \end{pmatrix}.$$

With these notational conventions, the problem we discuss is to find solutions of the system of equations

$$f_i(x_1, \ldots, x_n) = y_i, \qquad i = 1, \ldots, n, \tag{1}$$

or, more compactly,

$$Fx = y, \tag{2}$$

where $F: D \subset R^n \to R^n$ is a given operator and $y \in R^n$ a fixed vector. It is usually no restriction to absorb the vector y into F and consider only the equation

$$Fx = 0. \tag{3}$$

Before attempting to find a solution of (3), it is important to realize that such a problem may not even have a solution, or, alternatively, that there may be arbitrarily many of them. To illustrate this, consider the system

$$f_1(x_1, x_2) = 0, \qquad f_2(x_1, x_2) = 0. \tag{4}$$

Each of these two equations represents a (not necessarily continuous) curve in R^2, and the solutions of (4) are therefore the intersections of these two curves. If in the simple example

$$f_1(x_1, x_2) \equiv x_1^2 - x_2 + \alpha, \qquad f_2(x_1, x_2) \equiv -x_1 + x_2^2 + \alpha,$$

INTRODUCTION

the real parameter α varies between $+1$ and -1, the following cases arise (see Fig. **I.1**):

(a) $\alpha = 1$: no solution.
(b) $\alpha = \frac{1}{4}$: one solution, $x_1 = x_2 = \frac{1}{2}$.
(c) $\alpha = 0$: two solutions, $x_1 = x_2 = 0$; $x_1 = x_2 = 1$.
(d) $\alpha = -1$: four solutions, $x_1 = -1$, $x_2 = 0$; $x_1 = 0$, $x_2 = 1$; $x_1 = x_2 = \frac{1}{2}(1 \pm \sqrt{5})$.

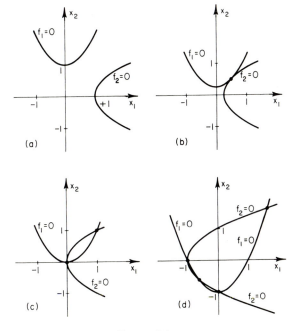

FIGURE I.1

Similarly, the system

$$f_1(x_1, x_2) \equiv \tfrac{1}{2}x_1[\sin(\tfrac{1}{2}\pi x_1)] - x_2 = 0, \qquad f_2(x_1, x_2) \equiv x_2^2 - x_1 + 1 = 0$$

has denumerably many solutions (Fig. **I.2**), while, for

$$f_1(x_1, x_2) \equiv x_1^2 - |x_2| = 0, \qquad f_2(x_1, x_2) \equiv x_1^2 - x_2 = 0,$$

there is even a continuum of solutions (Fig. **I.3**).

These examples indicate the need for discussing at least some of the major existence and uniqueness results for equations of the form (2)

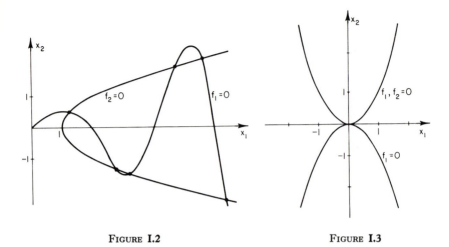

FIGURE I.2   FIGURE I.3

and (3). A comprehensive survey of existence theorems would be beyond the scope of this book, but, in Part II, most of the main approaches and results for finite-dimensional equations are covered.

Provided that the equation (3) indeed has solutions, our main concern will be the description and analysis of methods for approximating these solutions. In contrast to the case of linear systems of equations, direct methods for the solution of nonlinear equations are usually feasible only for small systems of a very special form. Consequently, our attention will be restricted to iterative methods. Probably the most basic iteration, and certainly the most central to our considerations, is Newton's method:

$$x^{k+1} = x^k - F'(x^k)^{-1} Fx^k, \qquad k = 0, 1, \dots . \tag{5}$$

Here, $F'(x)$ denotes the derivative or Jacobian matrix of F (see Section 3.1) and $F'(x)^{-1}$ its inverse. More generally, any iterative method consists of a procedure for generating a sequence $\{x^k\}$ of vectors starting from $p \geq 1$ given initial vectors $x^0, \dots, x^{-p+1}$, and Part III is devoted to a discussion of a large number of possible methods for the system (3).

There are three basic problems connected with the analysis of iterative processes. The first is to ascertain that the iterates will be well defined. For example, if the algorithm requires the evaluation of F at each x^k, it has to be guaranteed that the iterates remain in the domain of F; in the case of the Newton iteration (5), the derivative must also exist and be nonsingular at each x^k. It is, in general, impossible to find the exact set of all initial data for which a given process is well defined, and we restrict ourselves to giving conditions which guarantee that an iteration sequence is well defined for certain specific initial vectors.

The second, and most fundamental, problem concerns the convergence of the sequences generated by a process and the question of whether their limit points are, in fact, solutions of the equation. There are several types of such convergence results. The first, which we call a *local convergence theorem*, begins with the assumption that a particular solution x^* exists, and then asserts that there is a neighborhood U of x^* such that for all initial vectors in U the iterates generated by the process are well defined and converge to x^*. In Chapters 10 and 11, we will consider a variety of such results. The second type of convergence theorem, which we call *semilocal*, does not require knowledge of the existence of a solution, but states that, starting from initial vectors for which certain—usually stringent—conditions are satisfied, convergence to some (generally nearby) solution x^* is guaranteed. Moreover, theorems of this type usually include computable (at least in principle) estimates for the error $x^k - x^*$, a possibility not afforded by the local convergence theorems. Finally, the third and most elegant type of convergence result, the *global theorem*, asserts that starting anywhere in R^n, or at least in a large part of it, convergence to a solution is assured. Part V contains both of these latter two types of results.

The third basic problem concerns the economy of the entire operation, and, in particular, the question of how fast a given sequence will converge. Here, there are two approaches, which correspond to the local and semilocal convergence theorems. As mentioned above, the analysis which leads to the semilocal type of theorem frequently produces error estimates, and these, in turn, may sometimes be reinterpreted as estimates of the rate of convergence of the sequence. Unfortunately, however, these are usually overly pessimistic. The second approach deals with the behavior of the sequence $\{x^k\}$ when k is large, and hence when x^k is near the solution x^*. This behavior may then be determined, to a first approximation, by the properties of the iteration function near x^* and leads to so-called asymptotic rates of convergence. Results of this type are contained in Part IV in connection with the local convergence theorems.

As a final point, it may be useful to indicate here the general numbering scheme employed throughout the book. The chapters (numbered from 1 to 14, independent of their combination into parts) are each divided into several sections, denoted by decimal numbers of the type **12.1**, **12.2**, etc. At the end of each section, there is also a collection of Notes and Remarks and Exercises. In addition, some chapters have one or more appendices which cover some related material not within the main line of the book. All definitions and results are numbered consecutively within each section in the form **12.1.1**, **12.1.2**, etc. Definitions are identified as such, but none of the usual designations as lemma, proposition, or theorem are used. The only exception is that certain results generally known under a specific name are also given that title. Notes and Remarks, as well as Exercises, are attributed to specific sections by a self-explanatory code of the form **NR 12.1-3** or **E 10.3-4**. Equations are numbered by (1), (2), etc., consecutively in each section, and any reference to an equation in the same section is by this number only, while equations in another section are identified, for instance, by (10.1.3), meaning Eq. (3) of Section **10.1**.

Part I / BACKGROUND MATERIAL

This first part collects various background material from analysis and linear algebra which will be used frequently in later chapters. More specifically, we consider in Chapter 1 several mathematical problems for which the numerical solution requires solving a system of nonlinear equations. These problems are intended to give some idea of typical areas of numerical analysis where nonlinear systems of equations arise, but the list is by no means exhaustive, nor is each problem stated in its most general form. A number of these problems will be discussed in later sections of the book to illustrate existence and uniqueness results as well as convergence theorems for iterative methods.

Chapter 2 is devoted to linear algebra and, in particular, to basic results about eigenvalue problems, norms on n-dimensional spaces and their induced matrix norms, a variety of results about the invertibility of linear operators on R^n, and properties of R^n as a partially ordered linear space.

In Chapter 3, we present an introduction to n-dimensional calculus, and especially to the theory of Gateaux and Frechet derivatives and to their properties. In addition, a section on convex functionals is also included.

Other, more specialized results from linear algebra and analysis are presented in later chapters of the book where needed.

Chapter 1 / SAMPLE PROBLEMS

1.1. TWO-POINT BOUNDARY VALUE PROBLEMS

A wide variety of problems in many areas, such as trajectory calculation or the study of oscillatory systems, may be stated in terms of a boundary value problem for an ordinary differential equation. For example, in the investigation of the forced oscillations of a simple pendulum, one encounters problems of the type

$$u'' = c \sin u + g(t), \qquad 0 \leqslant t \leqslant 1, \quad u(0) = u(1) = 0. \tag{1}$$

More generally, we shall consider the problem

$$u'' = f(t, u), \qquad 0 \leqslant t \leqslant 1, \quad u(0) = \alpha, \quad u(1) = \beta, \tag{2}$$

of which (1) is a special case. If we assume that f is twice-continuously differentiable on the set

$$S = \{(t, y) \mid 0 \leqslant t \leqslant 1, \; -\infty < y < +\infty\}, \tag{3}$$

and that

$$f_y(t, y) \geqslant \eta > -\pi^2, \qquad \forall (t, y) \in S, \tag{4}$$

then it is known (see **NR 1.1-1**) that the problem (2) possesses a unique twice-continuously differentiable solution. In particular, these conditions are satisfied for the sample problem (1) if $|c| < \pi^2$.

In order to compute a numerical approximation to the solution u of (2), we consider first the following discrete analog. Let

$$t_j = jh, \quad h = 1/(n+1), \quad j = 0,\ldots, n+1,$$

be a uniform subdivision of the interval $[0, 1]$, and, at each point t_j, $j = 1,\ldots, n$, approximate $u''(t_j)$ by the second central difference quotient:

$$u''(t_j) \doteq (1/h^2)[u(t_{j+1}) - 2u(t_j) + u(t_{j-1})], \quad j = 1,\ldots, n. \tag{5}$$

When this approximation is used in (2), we find that the solution u satisfies

$$(1/h^2)[u(t_{j+1}) - 2u(t_j) + u(t_{j-1})] = f(t_j, u(t_j)) + r(t_j, h), \quad j = 1,\ldots, n, \tag{6}$$

at the grid points $t_1, \ldots, t_n$. Here, $r(t_j, h)$ are the errors introduced by the approximation (5), and it can be shown (**E 1.1-4**) that $\lim_{h \to 0} r(t_j, h) = 0$, provided u is suitably differentiable.

We now drop the error term in (6) and define approximations $x_1, \ldots, x_n$ to the values of the solution u at the grid points t_j, by requiring that the x_i satisfy the system of n equations

$$x_{j+1} - 2x_j + x_{j-1} = h^2 f(t_j, x_j), \quad j = 1,\ldots, n, \quad x_0 = \alpha, \quad x_{n+1} = \beta. \tag{7}$$

If we introduce the $n \times n$ matrix

$$A = \begin{pmatrix} 2 & -1 & & \bigcirc \\ -1 & 2 & \ddots & \\ & \ddots & \ddots & -1 \\ \bigcirc & & -1 & 2 \end{pmatrix}, \tag{8}$$

and the mapping $\phi: R^n \to R^n$ defined by

$$\phi x = h^2 \begin{pmatrix} f(t_1, x_1) - (\alpha/h^2) \\ f(t_2, x_2) \\ \vdots \\ f(t_{n-1}, x_{n-1}) \\ f(t_n, x_n) - (\beta/h^2) \end{pmatrix}, \tag{9}$$

then the system of equations (7) may be written in the compact form

$$Ax + \phi x = 0. \tag{10}$$

This system is an example of a particularly simple class of equations which will occur throughout the book.

1.1. TWO-POINT BOUNDARY VALUE PROBLEMS

1.1.1. Definition. A nonlinear mapping $\phi: D \subset R^n \to R^n$ is *diagonal* if, for $i = 1, \ldots, n$, the ith component φ_i of ϕ is a function of only the ith variable x_i. A mapping $F: D \subset R^n \to R^n$ is *almost linear* if F can be written in the form $F = A + \phi$, where A is an $n \times n$ matrix and ϕ is diagonal. ∎

We shall show in Section **4.4** that the system (7) has a unique solution whenever f is continuous and monotone increasing in y for each fixed t. For example, f may be of the form $f(t, y) = a(t) + e^y$, or $f(t, y) = a(t) + y^m$, where m is an odd, positive integer. Moreover, these existence results will allow f to satisfy the condition (4) provided that h is sufficiently small.

The same type of discretization may also be applied to more general equations. Consider, in place of (2), the problem

$$u'' = f(t, u, u'), \quad 0 \leq t \leq 1, \quad u(0) = \alpha, \quad u(1) = \beta. \tag{11}$$

Here, the first derivative u' on the right-hand side of (11) must be approximated as well, and considerations similar to those above lead, for example, to the system of approximating equations

$$x_{j+1} - 2x_j + x_{j-1} = h^2 f(t_j, x_j, (2h)^{-1}[x_{j+1} - x_{j-1}]), \quad j = 1, \ldots, n, \tag{12}$$

where, again, $x_0 = \alpha$, $x_{n+1} = \beta$. Note that this system is no longer almost linear.

Similarly, the same procedure may be used for the general implicit problem

$$g(t, u, u', u'') = 0, \quad 0 \leq t \leq 1, \quad u(0) = \alpha, \quad u(1) = \beta, \tag{13}$$

which would lead to the system of equations

$$g(t_j, x_j, (2h)^{-1}[x_{j+1} - x_{j-1}], h^{-2}[x_{j+1} - 2x_j + x_{j-1}]) = 0,$$
$$j = 1, \ldots, n, \quad x_0 = \alpha, \quad x_{n+1} = \beta. \tag{14}$$

For all of the above problems, we may also use more general approximations to the derivatives. For example, the grid points t_j need not be equally spaced (see **NR 1.1-3**) and higher-order approximations may be employed.

A quite different approach to the solution of two-point boundary value problems is the so-called *shooting method*. Consider a system of differential equations written in vector form,

$$u'' = f(t, u, u'), \quad 0 \leq t \leq 1, \quad u(0) = a, \quad u(1) = b. \tag{15}$$

Assume that, for any $x \in R^n$, the associated initial value problem

$$z'' = f(t, z, z'), \quad 0 \leqslant t \leqslant 1, \quad z(0) = a, \quad z'(0) = x, \qquad (16)$$

has a unique solution, which we denote by $z(t; x)$. Then we may define a mapping $F: R^n \to R^n$ by $Fx = z(1; x)$; that is, the value of F for given x is the solution of the corresponding initial value problem (16) evaluated at the end point 1. Therefore, if x^* is a solution of the system of equations $Fx = b$, then $u(t) = z(t; x^*)$ is the solution of the problem (15).

Since the evaluation of F for given x requires the solution of the initial value problem (16), the mapping F cannot, in general, be defined explicitly. Indeed, in practice, the solution of the initial value problem will be effected by a step-by-step numerical integration in order to obtain an approximation $\hat{z}(t; x)$ to $z(t, x)$. Therefore, instead of the system $Fx = b$ we are, in reality, attempting to solve the related system $\hat{F}x \equiv \hat{z}(1; x) = b$. In this case, the mapping $\hat{F}$ is defined explicitly, although in a complicated fashion, by means of the numerical integration scheme chosen.

The shooting method may also be used, with natural modifications, for certain control problems. Consider, for example, the problem

$$u' = f(t, u; x), \quad 0 \leqslant t \leqslant 1, \quad u(0; x) = a, \quad u(1; x) = b. \qquad (17)$$

Here, $x \in R^n$ is a "control vector," and it is desired to find an x^* such that $u(t; x^*)$ is a solution of (17). For example, u could represent the state vector of a missile, that is the position-velocity vector, and the problem is to select the parameter vector x, which may represent thrust profile, launch angle, etc., so that at time $t = 1$ the missile is at a specified point in space with a specified velocity.

As before, we may define a mapping $F: R^n \to R^n$ such that $Fx = z(1; x)$, where now z is the solution of the initial value problem

$$z' = f(t, z; x), \quad 0 \leqslant t \leqslant 1, \quad z(0; x) = a.$$

Then a solution of the system $Fx = b$ gives the desired control vector.

NOTES AND REMARKS

NR 1.1-1. The existence and uniqueness result quoted for (2) is due to Lees [1966]. For various other existence and uniqueness results for two-point boundary value problems, see Keller [1968] or Bailey, Shampine, and Waltman [1968] and the references therein.

NR 1.1-2. Whenever a differential equation is approximated by an analogous difference equation, there arises the problem of estimating the *discretization*

1.1. TWO-POINT BOUNDARY VALUE PROBLEMS

error. For example, let $u = u(t)$ and $x_j = x_j(h)$, $j = 1,..., n$, be the solutions of the boundary value problem (2) and of the approximating problem (7), respectively. Then the discretization error at each grid point t_j is $\epsilon_j = u(t_j) - x_j(h)$. Under the conditions that f is four times continuously differentiable and satisfies (4) and that h is sufficiently small, Lees [1966] has shown that this error satisfies

$$\max_{j=1,...,n} |u(t_j) - x_j(h)| \leq ch^2, \tag{18}$$

where c is a constant which does not depend on h. Hence, in this case, the solution of the discrete problem tends to that of the continuous problem as $h \to 0$ and at the rate given by (18). For further discussion of the discretization error for two-point boundary value problems, see Henrici [1962] and Keller [1968].

NR 1.1-3. An important property of the matrix A of (8) is its symmetry, and this property is preserved under appropriate formulation even if the subdivision points t_j of the approximation to u'' are not equally spaced. Consider, in fact, the problem

$$[p(t) u'(t)]' = f(t, u), \quad u(0) = \alpha, \quad u(1) = \beta, \tag{19}$$

where p is a positive, continuously differentiable function on $[0, 1]$, and let $0 = t_0 < t_1 < \cdots < t_n < t_{n+1} = 1$ be an arbitrary subdivision of $[0, 1]$. Set $h_i = t_{i+1} - t_i$, $i = 0,..., n$, and replace (19) by the approximating system

$$p_{i+(1/2)} \left(\frac{x_{i+1} - x_i}{h_i} \right) - p_{i-(1/2)} \left(\frac{x_i - x_{i-1}}{h_{i-1}} \right)$$
$$= \left(\frac{h_i + h_{i-1}}{2} \right) f(t_i, x_i), \quad i = 1,..., n, \tag{20}$$

where $p_{i \pm (1/2)} = p(t_i \pm \frac{1}{2} h_i)$ and $x_0 = \alpha$, $x_{n+1} = \beta$. Clearly, if $p(t) \equiv 1$ and $h_i = h$, $i = 0,..., n$, then (20) reduces to (7). Moreover, it is easy to see (**E 1.1-5**) that the coefficient matrix of the linear part of (20) is symmetric, whereas the symmetry is lost if the equations (20) are multiplied by $2(h_i + h_{i-1})^{-1}$ to obtain the form resulting from the direct approximation of $[p(t) u'(t)]'$.

NR 1.1-4. For further discussion and analysis of the shooting method, see Keller [1968].

EXERCISES

E 1.1-1. Show that the following two-point boundary value problems have the indicated solutions:

(a) $u'' = 2(u - \frac{1}{2} t + 1)^3$, $u(0) = u(1) = 0$; $u(t) = [1/(1 + t)] + \frac{1}{2} t - 1$.

(b) $u'' + a^2(u')^2 + 1 = 0, u(0) = u(1) = 0$;
$u(t) = (1/a^2) \ln\{\cos[a(t - \tfrac{1}{2})]/\cos(a/2)\}$, $-\pi < a < \pi$.
(c) $u'' = \tfrac{1}{2}u^3$, $u(0) = 1, u(1) = 2; u(t) = 2/(2-t)$.
(d) $u'' = \tfrac{1}{2}u^3 + 3u' - [3/(2-t)] + \tfrac{1}{2}$,
$u(0) = 0, u(1) = 1; u(t) = t/(2-t)$.
(e) $u'' = \tfrac{3}{2}u^2$, $u(0) = 4, u(1) = 1; u(t) = 4/(1+t)^2$.

E 1.1-2. Show that the general solution of $u'' = e^u$ is

$$u(t) = \ln\{\tfrac{1}{2}c^2/\cos^2[c(t+d)/2]\}.$$

E 1.1-3. Apply the discretizations of Section 1.1 to the boundary value problems of **E 1.1-1** and **E 1.1-2** and attempt to solve the discrete problems for $n = 2$. Compare your solutions with the exact solution.

E 1.1-4. Assume that $u: [0, 1] \to R^1$ is four times continuously differentiable on $[0, 1]$. Show that there is a constant c, independent of t, such that, for any $t \in (0, 1)$ and suitably small $h > 0$,

$$|h^{-2}[u(t+h) - 2u(t) + u(t-h)] - u''(t)| \leqslant ch^2.$$

E 1.1-5. Show that the coefficient matrix of the linear part of the equations (20) of **NR 1.1-3** is symmetric.

1.2. ELLIPTIC BOUNDARY VALUE PROBLEMS

The considerations of the previous section extend in a natural way to boundary value problems in more than one variable. Consider the two-dimensional analog of the two-point boundary value problem (1.1.2), namely, the Dirichlet problem

$$\Delta u \equiv u_{ss} + u_{tt} = f(s, t, u), \quad (s, t) \in \Omega, \quad u(s, t) = \varphi(s, t), \quad (s, t) \in \dot{\Omega}. \quad (1)$$

Here, Ω is a simply-connected, bounded open region in the plane, and φ is a given function defined on the boundary $\dot{\Omega}$ of Ω. It is known (see **NR 1.2-1**) that if $f: \Omega \times R^1 \to R^1$ is a continuously differentiable function which satisfies

$$f_u(s, t, u) \geqslant 0, \quad \forall (s, t) \in \Omega, \quad u \in R^1, \quad (2)$$

then, under mild conditions on Ω and φ, the problem (1) has a unique solution.

In order to obtain a discrete analog of (1) analogous to the system of equations (1.1.7), we assume, for simplicity, that the domain Ω is the

1.2. ELLIPTIC BOUNDARY VALUE PROBLEMS

unit square $(0, 1) \times (0, 1)$, and impose a uniform square mesh on Ω by defining the grid points

$$P_{ij} = (ih, jh), \quad h = 1/(m+1), \quad i,j = 0,\ldots, m+1; \quad (3)$$

this is depicted in Fig. 1.1 for $m = 2$. At each interior grid point P_{ij}, $i,j = 1,\ldots, m$, the partial derivatives $u_{ss}(P_{ij})$ and $u_{tt}(P_{ij})$ are now approximated by the central difference quotients corresponding to (1.1.5); that is,

$$u_{ss}(P_{ij}) \doteq h^{-2}[u(P_{i+1,j}) - 2u(P_{ij}) + u(P_{i-1,j})],$$
$$u_{tt}(P_{ij}) \doteq h^{-2}[u(P_{i,j+1}) - 2u(P_{ij}) + u(P_{i,j-1})], \quad i,j = 1,\ldots, m. \quad (4)$$

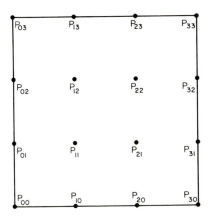

FIGURE 1.1

If we set $x_{ij} = u(P_{ij})$, $i,j = 0,\ldots, m+1$, and replace u_{ss} and u_{tt} in (1) by the approximations (4), we obtain the following discrete analog of (1):

$$4x_{ij} - x_{i-1,j} - x_{i+1,j} - x_{i,j+1} - x_{i,j-1} + h^2 f(ih, jh, x_{ij}) = 0, \quad i,j = 1,\ldots, m. \quad (5)$$

The values of x_{ij} at the boundary points are assumed to be given by the boundary conditions; that is,

$$x_{0,j} = \varphi(P_{0j}), \quad x_{m+1,j} = \varphi(P_{m+1,j}), \quad x_{j,0} = \varphi(P_{m0}), \quad x_{j,m+1} = \varphi(P_{j,m+1}), \quad (6)$$

for $j = 0,\ldots, m+1$. Therefore, (5) is a system of $n = m^2$ equations in the n unknowns x_{ij}, $i, j = 1,\ldots, m$.

In order to write the system (5) in matrix notation, we associate vectors $x \in R^n$ with the unknowns x_{ij} by the correspondence

$$x_1 = x_{11},\ldots, x_m = x_{m1}, \quad x_{m+1} = x_{12},\ldots, x_n = x_{mm},$$

and define the *block tridiagonal* matrix

$$A = \begin{pmatrix} B & -I & & & \bigcirc \\ -I & B & \cdot & & \\ & \cdot & \cdot & \cdot & \\ & & \cdot & \cdot & -I \\ \bigcirc & & & -I & B \end{pmatrix}, \tag{7}$$

where I is the $m \times m$ identity matrix and B is the $m \times m$ matrix

$$B = \begin{pmatrix} 4 & -1 & \cdots & 0 & 0 \\ -1 & 4 & \cdots & 0 & 0 \\ \vdots & \vdots & & \vdots & \vdots \\ 0 & 0 & \cdots & 4 & -1 \\ 0 & 0 & \cdots & -1 & 4 \end{pmatrix}. \tag{8}$$

The system (5) may then be written as

$$Ax + \phi x = b, \tag{9}$$

where the ith component φ_i of the nonlinear operator ϕ is defined by

$$\varphi_i(x) = h^2 f(kh, lh, x_i), \quad i = lm + k, \tag{10}$$

and $b = (b_1, ..., b_n)^T$ is a vector containing the boundary values.

Note that, as for the corresponding two-point boundary value problem, the matrix A of (7), (8) is again symmetric, and that the system is almost linear in the sense of Definition **1.1.1**. We shall show in Section **4.4** that the system of equations (5) has a unique solution provided that (2) holds.

NOTES AND REMARKS

NR 1.2-1. For a discussion of existence and uniqueness results for nonlinear elliptic boundary value problems, see, for example, Ladyzhenskaya and Ural'tseva [1964] or Bers, John, and Schechter [1964].

NR 1.2-2. For a review of the discretization-error problem for linear elliptic equations and, in particular, the classical theorem of Gerschgorin, see Forsythe and Wasow [1960]. For more recent results on the linear problem, see, for example, Hubbard [1966] and the references therein.

The discretization error for the nonlinear equation (1) and the discrete analog (5) has been studied by Bers [1953], who showed, under the condition (2) and certain smoothness assumptions on the solution u of (1), that the solution of (5) tends to u as $h \to 0$.

1.2. ELLIPTIC BOUNDARY VALUE PROBLEMS

NR 1.2-3. The type of approximation used in the text is easily extended, at least formally, to more general equations. Consider, for example, the general quasilinear equation

$$au_{ss} + 2bu_{st} + cu_{tt} = f(s, t, u, u_s, u_t), \tag{11}$$

where the coefficients a, b, and c may be functions of s, t, u, u_s, and u_t, but not of the second derivatives, such that the ellipticity condition

$$a > 0, \quad c > 0, \quad ac - b^2 > 0$$

is satisfied for all values of the arguments of a, b, and c. A famous example of (11) is the Plateau equation (see **1.5** and **NR 1.5-2**)

$$(1 + u_t^2) u_{ss} - 2u_s u_t u_{st} + (1 + u_s^2) u_{tt} = 0. \tag{12}$$

We discretize the equation (11) in a manner analogous to the discretization of (1). Assume again for simplicity that the domain Ω is the unit square $(0, 1) \times (0, 1)$ and that the solution u of (11) is to agree with a given function φ on the boundary of Ω. Imposing on Ω the same uniform mesh as before, approximating the first derivatives u_s and u_t by

$$u_s(P_{ij}) \doteq (2h)^{-1}[u(P_{i+1,j}) - u(P_{i-1,j})], \quad u_t(P_{ij}) \doteq (2h)^{-1}[u(P_{i,j+1}) - u(P_{i,j-1})],$$

the second derivatives u_{ss} and u_{tt} by (4) and u_{st} by

$$u_{st}(P_{ij}) \doteq (2h^2)^{-1}[u(P_{i+1,j+1}) - u(P_{i+1,j-1}) - u(P_{i-1,j+1}) + u(P_{i-1,j-1})],$$

and then again setting $x_{ij} = u(P_{ij})$, now gives the system of equations

$$a_{ij}[x_{i+1,j} - 2x_{ij} + x_{i-1,j}] + b_{ij}[x_{i+1,j+1} - x_{i+1,j-1} - x_{i-1,j+1} + x_{i-1,j-1}]$$
$$+ c_{ij}[x_{i,j+1} - 2x_{ij} + x_{i,j-1}]$$
$$= h^2 f(ih, jh, x_{ij}, (2h)^{-1}[x_{i+1,j} - x_{i-1,j}], (2h)^{-1}[x_{i,j+1} - x_{i,j-1}]) \tag{13}$$

where $i, j = 1, \ldots, m$. Here, the values of x_{ij} at the grid points of the boundary are again defined by (6), while

$$a_{ij} = a(ih, jh, x_{i,j}, (2h)^{-1}[x_{i+1,j} - x_{i-1,j}], (2h)^{-1}[x_{i,j+1} - x_{i,j-1}])$$

with identical definitions for b_{ij} and c_{ij}.

It is important to note, however, that the discretization (13) may be highly unsatisfactory unless the magnitude of b is suitably small relative to that of a and c. This problem already manifests itself for linear equations of the form (11), in which a, b, c, and f are functions only of s and t, and led Bramble and Hubbard [1962] to devise a much more complicated discretization for which they were able to prove that the solution of the discrete problem tends to that of the continuous problem. This work complemented earlier theoretical results of Motzkin and Wasow [1953] in which it was shown that such "good" discretizations must exist. The discretization of Bramble and Hubbard has been extended to nonlinear equations of the form (11) by Frank [1967], and also by Stepleman [1969], who points out a discrepancy in Frank's work. Both of these authors apply their results, in particular, to the Plateau equation (12). See also McAllister [1966a] for related work.

1.3. INTEGRAL EQUATIONS

From the previous two sections, it should be clear that discretization not only of differential equations, but also of other types of operator equations, such as integral equations or integrodifferential equations, will lead to systems of equations in n dimensions. We shall consider here only an integral equation of the form

$$u(s) = \psi(s) + \int_0^1 K(s, t, u(s), u(t))\, dt, \tag{1}$$

where ψ and K are given functions. Similar considerations will apply to more general equations involving, for example, derivatives of u, or to higher-dimensional problems involving unknown functions of two or more variables.

Equations of the form (1) arise in a variety of contexts. For example, in connection with a problem of radiative transfer, Ambartsumian and Chandrasekar were led to the so-called H-equation:

$$u(s) = 1 + \int_0^1 \frac{su(s)\,u(t)}{s+t} \varphi(t)\, dt, \tag{2}$$

where φ is a known function.

In order to discretize (1) we first choose a quadrature formula

$$\int_0^1 f(t)\, dt = \sum_{j=1}^n \gamma_j f(t_j) + r, \tag{3}$$

where $0 \leqslant t_1 < t_2 < \cdots < t_n \leqslant 1$ are the grid points of the formula, $\gamma_1, \ldots, \gamma_n$ are the weights, and r is the remainder or error term. Applying this quadrature formula to the integration in (1), dropping the remainder term, and setting $x_i = u(t_i)$, $i = 1,\ldots, n$, we are led to the discrete analog

$$x_i = \psi(t_i) + \sum_{j=1}^n \gamma_j K(t_i, t_j, x_i, x_j), \qquad i = 1,\ldots, n, \tag{4}$$

which is a system of n equations in the n unknowns $x_1, \ldots, x_n$.

1.3. INTEGRAL EQUATIONS

A special, but important, case of (1) is the Urysohn equation

$$u(s) = \psi(s) + \int_0^1 K(s, t, u(t))\, dt, \qquad (5)$$

in which $u(s)$ does not appear explicitly under the integral. Here, the discrete analog corresponding to (4) is simply

$$x_i = b_i + \sum_{j=1}^n \gamma_j K_{ij}(x_j), \qquad i = 1,\ldots, n, \qquad (6)$$

where we have set $b_i = \psi(t_i)$ and $K_{ij}(t) = K(t_i, t_j, t)$. An important special case of (5), in turn, is the Hammerstein equation

$$u(s) = \psi(s) + \int_0^1 H(s, t) f(t, u(t))\, dt, \qquad (7)$$

which arises frequently in connection with boundary value problems for differential equations. Indeed, by means of Green's function

$$H(s, t) = \begin{cases} -s(1-t), & \text{for } s \leq t, \\ -t(1-s), & \text{for } s \geq t, \end{cases} \qquad (8)$$

for the homogeneous problem

$$u''(t) = 0, \qquad u(0) = u(1) = 0,$$

the two-point boundary value problem

$$u''(t) = f(t, u), \qquad u(0) = \alpha, \qquad u(1) = \beta, \qquad (9)$$

of Section 1.1 may be put into the equivalent form (7) with $\psi(s) = \alpha + (\beta - \alpha)s$.

In general, the discrete analog of (7) corresponding to (4) is

$$x_i = \psi(t_i) + \sum_{j=1}^n \gamma_j b_{ij} f(t_j, x_j), \qquad i = 1,\ldots, n, \qquad (10)$$

where $b_{ij} = H(t_i, t_j)$. This may be written in matrix form as

$$x = b + B\phi x,$$

where $x \in R^n$, B is the $n \times n$ matrix with components b_{ij}, and

$$b = \begin{pmatrix} \psi(t_1) \\ \vdots \\ \psi(t_n) \end{pmatrix}, \qquad \phi x = \begin{pmatrix} \gamma_1 f(t_1, x_1) \\ \vdots \\ \gamma_n f(t_n, x_n) \end{pmatrix}.$$

For the special case where H is given by (8), the t_j are chosen as $t_j = jh$, $j = 1,\ldots, n$, $h = (n+1)^{-1}$, and the quadrature coefficients are taken as $\gamma_j = h$, $j = 1,\ldots, n$, we have

$$b_{ij} = \begin{cases} -ih(1-jh), & i \leq j, \\ -jh(1-ih), & j \leq i, \end{cases} \quad i, j = 1,\ldots, n, \qquad (11)$$

and $\varphi_i = hf(ih, x_i)$, $i = 1,\ldots, n$. Therefore, the system of equations (10) becomes, in this case,

$$x_i = \alpha + i(\beta - \alpha)h - h^2 \sum_{j=1}^{i} j(1-ih)f(jh, x_j) - h^2 \sum_{j=i+1}^{n} i(1-jh)f(jh, x_j),$$
$$i = 1,\ldots, n. \qquad (12)$$

It is easy to see **(E 1.3-2)** that $B = -hA^{-1}$, where A is the matrix defined by (1.1.8), and that (12) is equivalent to the system (1.1.7).

NOTES AND REMARKS

NR 1.3-1. For further discussion of nonlinear integral equations, see Anselone [1964] and Krasnoselskii [1956]. In particular, in one of the articles in the first reference, Moore [1964] gives numerical solutions of the H-equation (2).

NR 1.3-2. Note that (6) is a special case of equations of the general form

$$f_i(x) \equiv \sum_{j=1}^{n} \gamma_{ij} f_{ij}(x_j) = 0, \quad i = 1,\ldots, n, \qquad (13)$$

where each f_i is a linear combination of nonlinear functions f_{ij} of a single variable. We note that many nonlinear systems which arise from the discretization of continuous problems possess this property; this is easily seen, in particular, for the equations (1.1.7) and (1.2.5). Furthermore, if each f_{ij} is a polynomial, then (13) is, in turn, a special case of a general polynomial system of equations

$$f_i(x) = \sum_{j_1,\ldots,j_n=0}^{n} a_{j_1,\ldots,j_n} x_1^{j_1}, \ldots, x_n^{j_n} = 0, \quad i = 1,\ldots, n.$$

EXERCISES

E 1.3-1. Use any n-point quadrature formula (3) with

$$0 < t_1 < t_2 < \cdots < t_n < 1$$

1.4. MINIMIZATION PROBLEMS

to obtain a discrete analog of the H-equation (2). Show that the resulting system of equations has the solution

$$x_i = \left(1 \Big/ \prod_{j=1}^n t_j\right) \prod_{j=1}^n (t_j + t_i) \Big/ \prod_{j=1}^n (1 + \alpha_j t_i), \quad i = 1,\ldots, n,$$

where $\alpha_1, \ldots, \alpha_n$ are the nonnegative solutions of the equation

$$1 = 2 \sum_{j=1}^n [\gamma_j \varphi(t_j)/(1 - \alpha^2 t_j^2)]$$

(Chandrasekhar [1950]).

E 1.3-2. Let A be defined by (1.1.8) and B by (11). Show that $B = -hA^{-1}$ and that (1.1.7) is equivalent to (12).

1.4. MINIMIZATION PROBLEMS

In many different types of applications it is required to find a point x^*, called a minimizing point or *minimizer*, that minimizes a given functional $g: R^n \to R^1$; that is, for which $g(x^*) = \min\{g(x) \mid x \in R^n\}$. If g is differentiable, then it is known from the calculus (see **4.1.3** for a precise statement) that the partial derivatives of g must all vanish at x^*; that is, x^* is a solution of the system of equations

$$f_i(x) \equiv \frac{\partial}{\partial x_i} g(x) = 0, \quad i = 1,\ldots, n. \tag{1}$$

Therefore, the problem of locating minimizers leads naturally to the solution of systems of equations [see Section **4.1** for a more thorough discussion and, in particular, Theorem **4.1.4** for a sufficient condition that a solution of (1) is actually a minimizer].

One of the most ubiquitous minimization problems is that of *least-squares approximation*. This problem typically arises in the course of attempting to estimate certain parameters in a functional relationship by means of experimental data. Suppose, for example, that some quantity y is assumed to satisfy a relationship of the form $y(t) = f(t; x)$, where f is a known function of t and x, t is an independent variable (perhaps time), and x is an unknown n-dimensional parameter vector. For various values of t, say, $t_1, \ldots, t_m$, measurements y_i, $i = 1,\ldots, m$, perhaps subject to error, are made of $y(t_i)$ and it is desired to estimate the vector x. If these measurements were exact, then the parameter vector x would satisfy $f(t_i, x) = y_i$, $i = 1,\ldots, m$, which is a system of m equations in the n unknowns $x_1, \ldots, x_n$. However, in general, the y_i will be subject to error. Then a standard procedure is to take more measurements than the number of unknowns, so that $m > n$, and to find x so as to minimize the sum of the squares of the deviations $[y_i - f(t_i\,;x)]$; that is, we seek to minimize the functional $g: R^n \to R^1$ defined by

$$g(x) = \sum_{j=1}^{m} [y_j - f(t_j\,;x)]^2.$$

In this case, the system of equations (1)—the so-called normal equations of the least-squares problem—takes the form

$$\sum_{j=1}^{m} [y_j - f(t_j;x)] \frac{\partial f}{\partial x_i}(t_j;x) = 0, \qquad i = 1,\ldots, n.$$

Another source of minimization problems is the calculus of variations. In general terms, a variational problem has the form: Minimize a given functional J defined on some (infinite-dimensional) function space X over a certain subset of X. Our interest here will center on related finite-dimensional minimization problems which approximate the infinite-dimensional problem.

We begin by considering the following more concrete problem. Let $C^1[0, 1]$ denote the linear space of real functions which are continuously differentiable on $[0, 1]$, and define the mapping $J: C^1[0, 1] \to R^1$ by

$$Ju = \int_0^1 f(s, u(s), u'(s))\, ds, \qquad (2)$$

where $f: [0, 1] \times R^2 \to R^1$ is a given continuous function. Moreover, for fixed α and β, set

$$S = \{u \in C^1[0, 1] \mid u(0) = \alpha,\ u(1) = \beta\}. \qquad (3)$$

The variational problem we consider here then states:

$$\text{Find}\quad u^* \in S \quad \text{such that}\quad Ju^* = \inf_{u \in S} Ju. \qquad (4)$$

Under certain conditions on f, it is known that this problem has a unique solution (see **NR 1.4-3**).

It is known from the calculus of variations (see **NR 1.4-3**) that if f is sufficiently differentiable, then the solution of the minimization problem (4) must satisfy the so-called Euler equation

$$\frac{d}{dt}\left[\frac{\partial f}{\partial u'}(t, u, u')\right] - \frac{\partial f}{\partial u}(t, u, u') = 0, \qquad (5)$$

1.4. MINIMIZATION PROBLEMS

together with the boundary conditions $u(0) = \alpha$, $u(1) = \beta$. This is a two-point boundary value problem for u, and we could attempt to obtain an approximation to u by proceeding as in Section 1.1. However, there is no reason why we should not work directly with the given minimization problem rather than with its Euler equation. This we shall do by considering various replacements of the variational problem by finite-dimensional minimization problems.

A natural approach to this replacement by a finite-dimensional problem is the *Ritz method*. Let $u_1, \ldots, u_n$ be given functions in $C^1[0, 1]$ such that $u_i(0) = u_i(1) = 0$, $i = 1, \ldots, n$, and define an n-dimensional subspace of $C^1[0, 1]$ by

$$L_n = \left\{ u \in C^1[0, 1] \mid u = \sum_{i=1}^{n} c_i u_i, \quad c_i \in R^1, \quad i = 1, \ldots, n \right\}; \tag{6}$$

that is, L_n is the collection of all linear combinations of $u_1, \ldots, u_n$. For simplicity's sake, suppose the u_i are linearly independent (that is, if $\sum_{i=1}^{n} c_i u_i(s) = 0$ for all $s \in [0, 1]$, then $c_i = 0$, $i = 1, \ldots, n$). Next, introduce the set

$$S_n = \{ v \in C^1[0, 1] \mid v = u + \varphi, \ u \in L_n, \ \varphi(s) = \alpha + s(\beta - \alpha), \ s \in [0, 1] \}; \tag{7}$$

hence, any function $v \in S_n$ satisfies the boundary conditions $v(0) = \alpha$, $v(1) = \beta$. Finally, we define a functional

$$g: R^n \to R^1, \quad g(x) = J\left(\sum_{i=1}^{n} x_i u_i + \varphi \right), \tag{8}$$

where J is given by (2), and consider the minimization problem:

$$\text{Find} \quad x^* \in R^n \quad \text{such that} \quad g(x^*) = \inf_{x \in R^n} g(x). \tag{9}$$

Clearly, the problem (9) is equivalent to finding an element $u_n^* \in S_n$ such that $Ju^* = \inf\{Ju \mid u \in S_n\}$. The intent, of course, is that as $n \to \infty$ the functions $\{u_1, \ldots, u_n, \ldots\}$ span $C^1[0, 1]$ in some sense and u_n^* tends to the solution u^* of (4) (see **NR 1.4-4**).

The evaluation of g at some point $x \in R^n$ requires the integration indicated by (8) and (2). In practice, this integration will probably be done only approximately by means of a quadrature formula, and it is convenient, in this case, to redefine the functional g to embody this approximation. Let $s_1, \ldots, s_M$ be M points in $[0, 1]$ and let

$$\int_0^1 q(s)\, ds \doteq \sum_{j=1}^{M} \gamma_j q(s_j) \tag{10}$$

be a quadrature formula. We then define an approximation J_M to J by

$$J_M: C^1[0,1] \to R^1, \qquad J_M u = \sum_{j=1}^{M} \gamma_j f(s_j, u(s_j), u'(s_j)), \qquad (11)$$

and replace in problem (9) the functional (8) by the new functional

$$g: R^n \to R^1, \qquad g(x) = J_M\left(\sum_{i=1}^{n} x_i u_i + \varphi\right). \qquad (12)$$

This procedure shall be termed a *discrete Ritz method*. Note that for the functional (12), the system of equations (1) becomes

$$\sum_{j=1}^{M} \gamma_j \left[\frac{\partial f}{\partial u}(s_j, \psi(s_j), \psi'(s_j)) u_i(s_j) + \frac{\partial f}{\partial u'}(s_j, \psi(s_j), \psi'(s_j)) u_i'(s_j)\right] = 0, \quad i = 1,\ldots, n, \qquad (13)$$

where, of course, $\psi(s) = \sum_{i=1}^{n} x_i u_i(s) + \varphi(s)$.

Another approach to the replacement of the variational problem by a finite-dimensional problem is analogous to the previous discretizations of differential equations. That is, we can approximate the functional J of (2) by replacing the integration by a numerical quadrature and the derivatives of u by difference quotients. Let

$$t_j = jh, \qquad h = 1/(n+1), \qquad j = 0,\ldots, n+1, \qquad (14)$$

be a uniform subdivision of $[0, 1]$ and define an approximation to J by

$$J_n u = h \sum_{j=0}^{n} f(t_j + \tfrac{1}{2}h, u(t_j + \tfrac{1}{2}h), u'(t_j + \tfrac{1}{2}h)); \qquad (15)$$

that is, the integral of $f(s, u(s), u'(s))$ over the interval $[t_j, t_{j+1}]$ is approximated by h times the value of the integrand at the midpoint of the interval. We now make the further approximations

$$u(t_j + \tfrac{1}{2}h) \doteq \tfrac{1}{2}[u(t_j) + u(t_{j+1})], \qquad u'(t_j + \tfrac{1}{2}h) \doteq h^{-1}[u(t_{j+1}) - u(t_j)]. \qquad (16)$$

If we set $x_j = u(t_j)$, $j = 1,\ldots, n$, $x_0 = \alpha$, and $x_{n+1} = \beta$, then, using (16) in (15), we obtain the approximating functional $g: R^n \to R^1$ defined by

$$g(x) = h \sum_{j=0}^{n} f(t_j + \tfrac{1}{2}h, \tfrac{1}{2}(x_{j+1} + x_j), h^{-1}(x_{j+1} - x_j)). \qquad (17)$$

More generally, we may use one-sided approximations to the derivatives, or higher-order approximations, as well as different quadrature

1.4. MINIMIZATION PROBLEMS

formulas. Therefore, it will be convenient to consider the more general functional g defined by

$$g(x) = \sum_{j=0}^{M} \gamma_j f\left(s_j, \sum_{k=0}^{n+1} \alpha_{jk} x_k, \sum_{k=0}^{n+1} \beta_{jk} x_k\right), \tag{18}$$

where the s_j are given points in $[0, 1]$. Note that (17) is the special case of (18) in which

$$M = n, \quad s_j = (j + \tfrac{1}{2})h, \quad \gamma_j \equiv h, \quad \alpha_{jj} = \alpha_{j,j+1} = \tfrac{1}{2}, \quad \beta_{jj} = -1/h, \quad \beta_{j,j+1} = 1/h,$$

and $\alpha_{jk} = \beta_{jk} = 0$, otherwise. Another special case of (18) is the functional

$$g(x) = \gamma_0 f(0, x_0, h^{-1}[x_1 - x_0]) + \gamma_{n+1} f(1, x_{n+1}, h^{-1}[x_{n+1} - x_n])$$
$$+ \sum_{j=1}^{n} \gamma_j f(jh, x_j, (2h)^{-1}[x_{j+1} - x_{j-1}]), \tag{19}$$

which results from the approximation of the derivatives by central difference quotients at the interior grid points and one-sided differences at the end points.

We note that for (18) the equations (1) take the form

$$\sum_{j=0}^{M} \gamma_j \left[\frac{\partial f}{\partial u} \alpha_{ji} + \frac{\partial f}{\partial u'} \beta_{ji}\right] = 0, \quad i = 1, \ldots, n, \tag{20}$$

where the arguments of $\partial f/\partial u$ and $\partial f/\partial u'$ are those of f in (18).

NOTES AND REMARKS

NR 1.4-1. For further discussions of nonlinear least-squares problems, see, for example, Draper and Smith [1966].

NR 1.4-2. The nonlinear least-squares problem is the special case of the more general problem of minimizing a functional $g: R^n \to R^1$ defined by

$$g(x) = \varphi(Fx), \tag{21}$$

where $F: R^n \to R^m$, and $\varphi: R^m \to R^1$. For the least-squares problem discussed in the text, $Fx = (y_1 - f(t_1; x), \ldots, y_m - f(t_m; x))^T$ and $\varphi(z) = \sum_{i=1}^{m} z_i^2$. For the choice $\varphi(z) = \max_{1 \leq i \leq m} |z_i|$, (21) becomes the more difficult *uniform* (or *Chebyshev*) approximation problem. For results on this problem, whose specific treatment is outside the scope of this book, see, for example, Cheney [1966] and Meinardus [1964].

NR 1.4-3. The basic ideas and results of the calculus of variations are presented in Bliss [1925] or Akhieser [1955]. Many applications are discussed in, for example, Weinstock [1952].

NR 1.4-4. For current interesting work on the Ritz method in which so-called "spline approximations" are used as the basis functions, see Ciarlet, Schultz, and Varga [1967] and the references therein. There, results on the convergence of the approximate solutions to the true solution as the number of basis functions tends to infinity are also given. For earlier work on the Ritz method, see, for example, Mihklin [1957].

NR 1.4-5. Allen [1966] gives numerical results involving certain approximations of the form (18).

EXERCISES

E 1.4-1. For the three variational problems $Ju = \int_0^1 f(s, u(s), u'(s)) \, ds = \min$ with

(a) $f(s, u, u') = (1 + u')^{1/2}$
(b) $f(s, u, u') = (u')^2 (1 + u')^2$
(c) $f(s, u, u') = u[1 + (u')^2]^{1/2}$

determine the general, twice-continuously differentiable solutions of their Euler equation (5).

E 1.4-2. For the problem **E 1.4-1(b)**, consider the boundary conditions $u(0) = 0$, $u(1) = \frac{1}{4}$ and show that $u(s) = -\frac{1}{4}s$ is the unique solution of the Euler equation satisfying these boundary conditions, but that $Ju > Jv$ for the piecewise linear function

$$v(s) = \begin{cases} 0 & \text{for } 0 \leq s \leq \frac{3}{4} \\ -s + \frac{3}{4} & \text{for } \frac{3}{4} \leq s \leq 1. \end{cases}$$

E 1.4-3. For polynomials in s as trial functions and the trapezoidal rule as quadrature formula, set up the discrete Ritz method (12) for the problem

E 1.4-1(c). Solve the system (13) for small n and M and compare the solution with that of the continuous problem.

1.5. TWO-DIMENSIONAL VARIATIONAL PROBLEMS

In the previous section, we considered various discrete analogs of one-dimensional variational problems. These considerations extend in a natural way to higher-dimensional problems; for simplicity, we restrict ourselves to two dimensions.

1.5. TWO-DIMENSIONAL VARIATIONAL PROBLEMS

As in **1.2**, let Ω be a simply-connected, bounded, open set in R^2 with boundary $\dot{\Omega}$ and closure $\bar{\Omega}$, respectively. Let $C^1(\Omega)$ denote the class of functions which are continuously differentiable on Ω and continuous on $\bar{\Omega}$, and define the functional $J: C^1(\Omega) \to R^1$ by

$$Ju = \iint_\Omega f(s, t, u(s, t), u_s(s, t), u_t(s, t))\, ds\, dt, \tag{1}$$

where $f: \bar{\Omega} \times R^3 \to R^1$ is a given continuous function. Finally, introduce the set

$$S = \{u \in C^1(\Omega) \mid u(s, t) = \varphi(s, t),\ \forall (s, t) \in \dot{\Omega}\}, \tag{2}$$

where $\varphi: \dot{\Omega} \to R^1$ is also a given continuous function. As in **1.4**, we then consider the variational problem:

$$\text{Find}\quad u^* \in S \quad \text{such that}\quad Ju^* = \inf_{u \in S} Ju. \tag{3}$$

Under rather general conditions on f, φ, and Ω, it is known (see **NR 1.5-1**) that this problem has a unique solution.

For simplicity, we shall restrict ourselves in the sequel to problems of the form

$$Ju = \iint_\Omega f(u_s(s, t), u_t(s, t))\, ds\, dt; \tag{4}$$

that is, the integrand function depends on only u_s and u_t. An example of a concrete problem of this type is the *minimal surface* or *Plateau* problem, in which, if the mapping $\varphi: \dot{\Omega} \to R^1$ is interpreted as representing a curve in R^3, one is to find the surface of minimal area which passes through φ. This leads to a functional J of the form

$$Ju = \iint_\Omega [1 + u_s^2(s, t) + u_t^2(s, t)]^{1/2}\, ds\, dt. \tag{5}$$

Another example of a functional of the form (4) arises from magnetostatics (see **NR 1.5-3**); here, J is defined by

$$Ju = \iint_\Omega [u_s^2 + u_t^2 - (c - d) \ln(c + u_s^2 + u_t^2)]\, ds\, dt, \tag{6}$$

with certain constants $c > d > 0$.

The Ritz and discrete Ritz methods, described in **1.4** for the one-dimensional problem, extend formally in a natural way to higher dimensions (see **NR 1.5-4**), and we consider here only the finite-difference approach.

For simplicity, assume that, as in **1.2**, Ω is the unit square $(0, 1) \times (0, 1)$ and again impose on Ω a square mesh of width h as illustrated for $h = \frac{1}{3}$ in Fig. **1.1**. We denote by P_{ij} the grid points (ih, jh), $i, j = 0,..., m + 1$, $h = 1/(m + 1)$, and by Ω_{ij} the mesh square with vertices $P_{i-1,j-1}$, $P_{i,j-1}$, $P_{i-1,j}$, $P_{i,j}$. Then, perhaps the simplest approximation to the integral (4) arises from approximating it over Ω_{ij} by the area, h^2, of Ω_{ij} times the values of the integrand at some point $Q_{ij} \in \Omega_{ij}$; that is,

$$Ju \doteq h^2 \sum_{i,j=1}^{m+1} f(u_s(Q_{ij}), u_t(Q_{ij})). \tag{7}$$

We next need to specify Q_{ij} as well as finite-difference replacements for the derivatives. Consider first the choice $Q_{ij} = P_{ij}$ and the backward difference quotients

$$u_s(P_{ij}) \doteq (1/h)(x_{ij} - x_{i-1,j}), \qquad u_t(P_{ij}) \doteq (1/h)[x_{i,j} - x_{i,j-1}],$$

where, as in **1.2**, we have set $x_{ij} = u(P_{ij})$. The approximation (7) then becomes

$$Ju \doteq g(x) \equiv h^2 \sum_{i,j=1}^{m+1} f(h^{-1}[x_{ij} - x_{i-1,j}], h^{-1}[x_{ij} - x_{i,j-1}]), \tag{8}$$

and the variational problem (2)–(4) is replaced by the minimization of the functional $g: R^{m^2} \to R^1$ defined by (8). Here, the values $x_{0,j}$, $x_{m+1,j}$, $x_{j,0}$, and $x_{j,m+1}$, $j = 0,..., m + 1$, are assumed to be known by the boundary conditions.

We could, of course, obtain the analogous approximation

$$Ju \doteq g(x) \equiv h^2 \sum_{i,j=0}^{m} f(h^{-1}[x_{i+1,j} - x_{ij}], h^{-1}[x_{i,j+1} - x_{ij}]), \tag{9}$$

by taking the point Q_{ij} of (7) to be $P_{i-1,j-1}$ and by replacing the derivatives by forward difference quotients.

Both of the approximations (8) and (9) have a certain asymmetric character. One simple way of obtaining a more symmetric approximation is to average (8) and (9); that is, if the functionals g of (8) and (9) are denoted by g_B and g_F, respectively, then a new approximation is defined by

$$Ju \doteq g(x) \equiv \tfrac{1}{2}[g_F(x) + g_B(x)]. \tag{10}$$

This approximation also has a natural interpretation in terms of integration over triangles rather than squares (see **NR 1.5-5**).

1.5. TWO-DIMENSIONAL VARIATIONAL PROBLEMS

A perhaps more natural construction of a symmetric approximation proceeds as follows. Let Q_{ij} be the center of the square Ω_{ij} as shown in Fig. 1.2. In order to approximate the derivatives u_s and u_t at Q_{ij} in terms

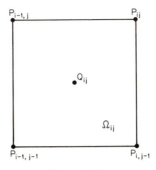

FIGURE 1.2

of the P_{kl}, we take averages of the difference quotients along the horizontal and vertical sides of Ω_{ij}; that is, we set

$$u_s(Q_{ij}) \doteq (2h)^{-1}[x_{ij} - x_{i-1,j} + x_{i,j-1} - x_{i-1,j-1}] \equiv b_{ij}(x),$$
$$u_t(Q_{ij}) \doteq (2h)^{-1}[x_{ij} - x_{i,j-1} + x_{i-1,j} - x_{i-1,j-1}] \equiv c_{ij}(x),$$

where again $x_{kl} = u(P_{kl})$. Then (7) becomes

$$g(x) = h^2 \sum_{i,j=1}^{m+1} f(b_{ij}(x), c_{ij}(x)). \qquad (11)$$

For the analysis of the functionals (8)–(11) it is again convenient to associate the unknowns x_{ij}, $i, j = 1,\ldots, m$, with the components $x_1, \ldots, x_n$ of a vector in R^n, and to denote the boundary values $x_{0,j}$, $x_{m+1\,j}$, $x_{j,0}$, $x_{j,m+1}$ by η_k, $k = 1,\ldots, N = 4(m+1)$. It is then clear that, for appropriate choices (see **E 1.5-1**) of the constants M, γ_i, α_{ij}, β_{ij}, $\hat{\alpha}_{ij}$, and $\hat{\beta}_{ij}$, each of the functionals (8), (9), (10), and (11) may be written in the general form

$$g(x) = \sum_{i=1}^{M} \gamma_i f\left(\sum_{j=1}^{n} \alpha_{ij} x_j + \sum_{j=1}^{N} \hat{\alpha}_{ij} \eta_j, \sum_{j=1}^{n} \beta_{ij} x_j + \sum_{j=1}^{N} \hat{\beta}_{ij} \eta_j\right). \qquad (12)$$

In **4.4**, we shall give conditions on the constants of (12), as well as on f, which ensure that (12) has a unique minimizer. These conditions will allow immediate application to the specific functionals (8)–(11); moreover, they allow consideration of more general domains and discretizations.

We conclude this section by considering a discretization which cannot be cast in the form (12). Note first that the two sample problems (5) and (6) may be written as

$$Ju = \iint_\Omega \psi(u_s^2(s, t) + u_t^2(s, t))\, ds\, dt, \tag{13}$$

where the mapping $\psi: R^1 \to R^1$ is given by

$$\psi(t) = (1 + t)^{1/2} \tag{14}$$

for (5), and by

$$\psi(t) = t - (c - d) \ln(c + t), \quad c > d > 0, \tag{15}$$

for (6). Again with Q_{ij} as the center of the square Ω_{ij} (Fig. **1.2**), we now approximate u_s^2 and u_t^2 simultaneously by

$$u_s^2(Q_{ij}) + u_t^2(Q_{ij}) \doteq \tau_{ij}(x) \equiv (2h^2)^{-1}\,[(x_{ij} - x_{i-1,j})^2 + (x_{ij} - x_{i,j-1})^2 \\ + (x_{i,j-1} - x_{i-1,j-1})^2 + (x_{i-1,j} - x_{i-1,j-1})^2]. \tag{16}$$

Then the discrete approximation to (13) is

$$Ju \doteq g(x) \equiv h^2 \sum_{i,j=1}^{m+1} \psi(\tau_{ij}(x)), \tag{17}$$

which is not of the form (12).

NOTES AND REMARKS

NR 1.5-1. An excellent discussion of existence and uniqueness theorems for the variational problems of this section is given in Ladyzhenskaya and Ural'tseva [1964]. A survey of classical results may be found in Rado [1951].

NR 1.5-2. As with the one-dimensional problem of **1.4**, associated with the problem (1)–(3) is the Euler equation

$$f_{pp}u_{ss} + 2f_{pq}u_{st} + f_{qq}u_{tt} + f_{pr}u_s + f_{qr}u_t + f_{rs} + f_{rt} - f_r = 0, \tag{18}$$

where $f = f(s, t, r, p, q)$, and the derivatives of (18) are the corresponding functions of s, t, u, u_s, and u_t; hence, (18) is a special case of the general quasi-linear elliptic equation (1.2.11). Note that the ellipticity condition becomes, in this case,

$$f_{pp}f_{qq} - f_{pq}^2 > 0, \quad f_{pp} > 0, \quad f_{qq} > 0, \tag{19}$$

which shows that, for each fixed s, t, and r, the function $f(s, t, r, \cdot, \cdot)$ is strictly convex (see **3.4**).

1.5. TWO-DIMENSIONAL VARIATIONAL PROBLEMS

Now, suppose that f has the form

$$f \equiv f(r, p, q) = \tfrac{1}{2}(p^2 + q^2) + \int \sigma(t)\, dt.$$

Then (18) reduces to the mildly nonlinear equation (1.2.1): $\Delta u = \sigma(u)$. In this case, many of the discretizations of this section reduce to those of 1.2 (see E 1.5-3).

On the other hand, consider the class of functionals given by (13) in which

$$f \equiv f(p, q) = \psi(p^2 + q^2). \tag{20}$$

Here, the Euler equation (18) becomes

$$[2\psi'' u_s^2 + \psi']\, u_{ss} + [2\psi'' u_t^2 + \psi']\, u_{tt} + 4\psi'' u_s u_t u_{st} = 0, \tag{21}$$

where ψ' and ψ'' are evaluated, of course, at $u_s^2 + u_t^2$. Equation (21) may also be written in the sometimes more convenient divergence form

$$\frac{\partial}{\partial s}(u_s \psi') + \frac{\partial}{\partial t}(u_t \psi') = 0.$$

For the minimal surface problem, we have $\psi(t) = (1 + t)^{1/2}$, so that (21) becomes

$$\tfrac{1}{2}[(1 + u_t^2)\, u_{ss} + (1 + u_s^2)\, u_{tt} - 2u_s u_t u_{st}](1 + u_s^2 + u_t^2)^{-3/2} = 0,$$

which is equivalent to the Plateau equation (1.2.12).

When f has the special form (20), the ellipticity condition (19) becomes

$$\psi'(t) > 0, \qquad \psi'(t) + 2t\psi''(t) > 0, \qquad \forall t \in [0, \infty); \tag{22}$$

these quantities are the eigenvalues of the matrix

$$\begin{pmatrix} f_{pp} & f_{pq} \\ f_{qp} & f_{qq} \end{pmatrix}.$$

For the function ψ of (15), we find that

$$\psi'(t) = (d + t)(c + t)^{-1}, \qquad \psi''(t) = (c - d)(c + t)^{-2},$$

and, since $c > d$, it follows that (22) is satisfied. Similarly, for the minimal surface problem we have $\psi(t) = (1 + t)^{1/2}$, so that

$$\psi'(t) = \tfrac{1}{2}(1 + t)^{-1/2}, \qquad \psi''(t) = -\tfrac{1}{4}(1 + t)^{-3/2},$$

and again (22) holds.

NR 1.5-3. The numerical solution of the problem (6) has recently been considered by Concus [1967a] using the "nonlinear discretization" (17).

NR 1.5-4. Although the Ritz method extends in a formal way to two- and higher-dimensional variational problems, there is a severe practical difficulty in generating suitable basis functions for general regions. For some recent results on obtaining such basis functions by bivariate Hermite and spline interpolation, see Birkhoff, Schultz, and Varga [1968].

NR 1.5-5. The approximation (8) has been studied by Schechter [1962], while (10) has been used for numerical computations by Greenspan [1965b]. Greenspan, however, derives (10) in a somewhat different fashion. With the notation of the text, let T_{ij} and S_{ij} be right triangles which divide Ω_{ij} as shown in Fig. 1.3, and approximate the integral over Ω_{ij} by $\tfrac{1}{2}h^2$ times the value of the

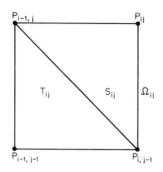

FIGURE 1.3

integrand at P_{ij} plus $\tfrac{1}{2}h^2$ times the integrand at $P_{i-1,j-1}$. Next, at P_{ij} and $P_{i-1,j-1}$ approximate the derivatives by backward and forward difference quotients, respectively. Then the resulting approximation to J is given by (10).

NR 1.5-6. For the more general problem (1), the difference approximations (8), (9), and (11) go over in a natural way to

$$Ju \doteq h^2 \sum_{i,j=1}^{m+1} f(P_{ij}, x_{ij}, h^{-1}[x_{ij} - x_{i-1,j}], h^{-1}[x_{ij} - x_{i,j-1}]),$$

$$Ju \doteq h^2 \sum_{i,j=0}^{m} f(P_{ij}, x_{ij}, h^{-1}[x_{i+1,j} - x_{ij}], h^{-1}[x_{i,j+1} - x_{ij}]),$$

$$Ju \doteq h^2 \sum_{i,j=1}^{m+1} f(Q_{ij}, a_{ij}(x), b_{ij}(x), c_{ij}(x)),$$

respectively, where $a_{ij}(x) = \tfrac{1}{4}[x_{ij} + x_{i-1,j} + x_{i,j-1} + x_{i-1,j-1}]$.

EXERCISES

E 1.5-1. Exhibit the numbers γ_i, α_{ij}, $\hat{\alpha}_{ij}$, β_{ij}, $\hat{\beta}_{ij}$, and M for which (12) reduces, respectively, to (8), (9), (10), and (11).

1.5. TWO-DIMENSIONAL VARIATIONAL PROBLEMS

E 1.5-2. Assume that $f: R^2 \to R^1$ is continuously differentiable. Compute the so-called gradient equations

$$\frac{\partial}{\partial x_{ij}} g(x) = 0, \quad i,j = 1,\ldots, m,$$

for each of the functions (8)–(10), and (17). Show that in the special case $f(p, q) = \frac{1}{2}(p^2 + q^2)$ all four of these systems are equivalent to the system

$$4x_{ij} - x_{i-1,j} - x_{i+1,j} - x_{i,j+1} - x_{i,j-1} = 0, \quad i,j = 1,\ldots, m.$$

E 1.5-3. Compute the gradient equations for (11) and show that for $f(p, q) = \frac{1}{2}(p^2 + q^2)$ these reduce to

$$4x_{ij} - x_{i-1,j-1} - x_{i-1,j+1} - x_{i+1,j-1} - x_{i+1,j+1} = 0, \quad i,j = 1,\ldots, m.$$

E 1.5-4. Consider the functional

$$g(x) = h^2 \sum_{i,j=1}^{m} f\big((2h)^{-1}[x_{i+1,j} - x_{i-1,j}], (2h)^{-1}[x_{i,j+1} - x_{i,j-1}]\big)$$

that results from approximating $u_s(P_{ij})$, $u_t(P_{ij})$ by central differences. Compute the gradient equations and show that for $f(p, q) = \frac{1}{2}(p^2 + q^2)$ the resulting system is not equivalent to the system given in **E 1.5-2**.

Chapter 2 / LINEAR ALGEBRA

2.1. A REVIEW OF BASIC MATRIX THEORY

We shall review in this section, without proof, a number of results from linear algebra which are assumed to be known. In the remaining sections of this chapter, additional, more specialized results will be collected.

As stated previously, R^n shall denote the real n-dimensional linear space of column vectors x with components $x_1, \ldots, x_n$, and C^n the corresponding space of complex column vectors. For $x \in R^n$, x^T denotes the transpose of x, while, for $x \in C^n$, x^H is the conjugate transpose; hence, x^T and x^H are row vectors.

A real $m \times n$ matrix $A = (a_{ij})$ defines a linear mapping from R^n to R^m, and we shall write $A \in L(R^n, R^m)$ to denote either the matrix or the linear operator, as context dictates. That is, we shall make no distinction, in general, between a linear operator and its concrete matrix representation in terms of the unit coordinate vectors $e^1 = (1, 0, \ldots, 0)^\mathrm{T}, \ldots, e^n = (0, \ldots, 0, 1)^\mathrm{T}$ of R^n. Similarly, the linear space of complex $m \times n$ matrices will be denoted by $L(C^n, C^m)$. In case $n = m$, we shall abbreviate $L(R^n, R^n)$ by $L(R^n)$ and, similarly, $L(C^n, C^n)$ by $L(C^n)$. Clearly, by assuming that R^n is imbedded in C^n in the natural way, we have, in terms of matrices, $L(R^n, R^m) \subset L(C^n, C^m)$, and the same relationship holds for linear operators.

If $A \in L(R^n, R^m)$ and $b \in R^m$, then the mapping $H: R^n \to R^m$ defined by $Hx = Ax + b$, $x \in R^n$, is an *affine* mapping of R^n into R^m.

A linear operator $A \in L(C^n)$ is *invertible* or *nonsingular* if A is one-to-

2.1. A REVIEW OF BASIC MATRIX THEORY

one; its inverse is denoted by A^{-1}. By A^T or A^H we mean the transpose or conjugate transpose of A, respectively.

For any $A \in L(C^n)$, a complex number λ is an *eigenvalue* of A if the equation

$$Ax = \lambda x \qquad (1)$$

has a nonzero solution x called an *eigenvector* of A corresponding to λ. Since a matrix B is singular if and only if the determinant, $\det B$, is zero, it follows that the eigenvalues of A are precisely the n roots (counting multiplicity) of the *characteristic polynomial*

$$\det(A - \lambda I) = 0, \qquad (2)$$

where I is the $n \times n$ identity matrix.

The determinant is a continuous function of the elements of the matrix, and the same is true for the eigenvalues of a matrix, but not, in general, for the eigenvectors (see **NR 2.1-2** and **E 2.1-7**).

If $p(t) = a_0 + a_1 t + \cdots + a_m t^m$ is any polynomial, a corresponding matrix polynomial $p(A) \in L(C^n)$ is defined by

$$p(A) = a_0 + a_1 A + \cdots + a_m A^m,$$

for any $A \in L(C^n)$. It follows directly from (1) that any eigenvector of A is an eigenvector of $p(A)$ and that the eigenvalues of $p(A)$ are $p(\lambda_i)$, $i = 1,\ldots,n$, where λ_i are the eigenvalues of A. Similarly, if A is invertible, then the eigenvalues of A^{-1} are λ_i^{-1}, $i = 1,\ldots,n$.

If A is real and symmetric ($A = A^T$), then all eigenvalues of A are real, and the inequality

$$\lambda_1 x^T x \leq x^T A x \leq \lambda_n x^T x, \qquad \forall x \in R^n, \qquad (3)$$

holds, where $\lambda_1 \leq \lambda_2 \leq \cdots \leq \lambda_n$ are the eigenvalues of A. More generally, if A is *Hermitian* ($A = A^H$), (3) holds for all $x \in C^n$ with x^T replaced by x^H.

If $A \in L(R^n)$ satisfies

$$x^T A x \geq 0, \qquad \forall x \in R^n, \qquad (4)$$

then A is *positive semidefinite*, and *positive definite* if strict inequality holds in (4) whenever $x \neq 0$. If A is positive (semi) definite and symmetric, then all its eigenvalues are positive (nonnegative). Note that, if $A \in L(R^n)$ is not symmetric, then A is positive (semi) definite if and only if the symmetric matrix $A + A^T$ is positive (semi) definite. Analogous considerations apply for complex matrices if (4) is replaced by $\operatorname{Re} x^H A x \geq 0$.

Two matrices A and B are *similar* if there is a nonsingular matrix P such that $P^{-1}AP = B$. Similar matrices necessarily have the same eigenvalues. A matrix U is *orthogonal* if $U^TU = I$. If A is real and symmetric, there is a real orthogonal matrix U such that

$$U^TAU = D, \tag{5}$$

where D is a diagonal matrix whose diagonal entries are the eigenvalues of A; that is, a real, symmetric matrix is *orthogonally similar* to a diagonal matrix. Moreover, the columns of U are necessarily eigenvectors of A, so that a real, symmetric matrix has n orthogonal eigenvectors.

More generally, if A is an arbitrary complex $n \times n$ matrix, there is a nonsingular matrix P such that

$$P^{-1}AP = J, \tag{6}$$

where J is the *Jordan canonical form* of A and has the following structure: J is a *block-diagonal* matrix

$$J = \begin{pmatrix} J_1 & & \bigcirc \\ & \ddots & \\ \bigcirc & & J_k \end{pmatrix}, \tag{7}$$

where each J_i is either one-dimensional or a matrix of the form

$$J_i = \begin{pmatrix} \lambda & 1 & & \bigcirc \\ & \lambda & \ddots & \\ & & \ddots & 1 \\ \bigcirc & & & \lambda \end{pmatrix},$$

and is called a *Jordan block*. Here, λ is an eigenvalue of A of multiplicity at least equal to the dimension of J_i. Note, however, that an eigenvalue may be associated with more than one Jordan block. (As an extreme example, consider the identity matrix, in which each Jordan block is one-dimensional.)

If all eigenvalues of A are distinct, then the Jordan form of A is a diagonal matrix and the columns of the similarity matrix P are n linearly independent eigenvectors of A. More generally, A has precisely as many linearly independent eigenvectors as there are Jordan blocks in its canonical form. In particular, if m_i is the dimension of J_i and p^i is the ith column of P, then $p^1, p^{m_1+1}, \ldots, p^{m_1+\cdots+m_{k-1}+1}$ are eigenvectors of A. The other columns of P are sometimes called *principal vectors* or *generalized eigenvectors*.

2.1. A REVIEW OF BASIC MATRIX THEORY

NOTES AND REMARKS

NR 2.1-1. The majority of the material in this section may be found in any of the numerous books on linear algebra. We mention, in particular, the excellent treatise of Gantmacher [1953], in which the Jordan canonical form is presented in great detail from both the algebraic (elementary divisor) and geometric (invariant subspace) points of view. We mention also the interesting book of Bellman [1960] for results on symmetric matrices, and also for its analytic point of view. Finally, most of this material, as well as some pertaining to the following sections, is reviewed in a form most suitable to numerical analysis by Faddeev and Faddeeva [1960], Householder [1964], and Wilkinson [1965].

NR 2.1-2. The only result of this section that is not usually available in the linear algebra textbooks is the fact that the roots of a polynomial are continuous functions of the coefficients, and, hence, that the eigenvalues of a matrix are continuous functions of the elements of the matrix. A proof of this fact may be found in Ostrowski [1966, Appendix K] together with error estimates which are, in themselves, of value (see also Wilkinson [1965]).

NR 2.1-3. The eigenvalue–eigenvector problem (1) may be written in the form of a system of $n + 1$ (in general, complex) nonlinear equations. Assume that $A \in L(C^n)$ and that the eigenvector condition $x \neq 0$ is normalized to $x^H x = 1$. Then, together with (1), this gives the system

$$\begin{pmatrix} Ax - \lambda x \\ x^H x - 1 \end{pmatrix} = 0$$

in the $n + 1$ unknowns x and λ. Although not a recommended approach to the numerical solution of matrix eigenvalue problems, the analogous phrasing of the problems in infinite-dimensional spaces may have merit. For recent work in this direction and further references, see Anselone and Rall [1968]. For a comprehensive study of computational methods for the matrix eigenvalue problem, see Wilkinson [1965].

EXERCISES

E 2.1-1. Compute $\det A$, A^{-1}, the characteristic polynomial, the eigenvalues and eigenvectors, and the Jordan canonical form for the matrix

$$A = \begin{pmatrix} 1 & 2 \\ 4 & 3 \end{pmatrix}.$$

E 2.1-2. Find the eigenvalues, eigenvectors, principal vectors, and Jordan form for the matrix

$$\begin{pmatrix} 1 & 0 & 0 \\ -1 & 0 & 1 \\ -1 & -1 & 2 \end{pmatrix}.$$

Comment on the uniqueness of the principal vectors.

E 2.1-3. Let $A \in L(R^n)$ be invertible. Show that $A^T A$ is symmetric positive definite.

E 2.1-4. Give an example of a 4×4, real, orthogonal matrix which has only imaginary eigenvalues.

E 2.1-5. Let $A \in L(R^n)$ be skew-symmetric (i.e., $A = -A^T$) and orthogonal. Describe the set of eigenvalues of A.

E 2.1-6. Let $A \in L(C^n)$ have eigenvalues $\lambda_1, \ldots, \lambda_n$ and assume that $\mu \neq -\lambda_i$, $i = 1, \ldots, n$. Show that, for any $k \geqslant 0$, the eigenvalues of $A^k(A + \mu I)^{-1}$ are $\lambda_i^k (\lambda_i + \mu)^{-1}$, $i = 1, \ldots, n$.

E 2.1-7. Let A be the 2×2 matrix

$$\begin{pmatrix} 1 + \epsilon \cos(2/\epsilon) & -\epsilon \sin(2/\epsilon) \\ -\epsilon \sin(2/\epsilon) & 1 - \epsilon \cos(2/\epsilon) \end{pmatrix}, \qquad \epsilon \neq 0.$$

Show that A has eigenvalues $1 \pm \epsilon$ and corresponding eigenvectors $(\sin(1/\epsilon), \cos(1/\epsilon))^T$, $(\sin(1/\epsilon), -\cos(1/\epsilon))^T$. Conclude that, as $\epsilon \to 0$, these eigenvectors do not tend to a limit. (J. W. Givens, unpublished.)

E 2.1-8. A matrix $A \in L(R^n)$ has a *square root* B if $B^2 = A$. Let A be symmetric, positive semidefinite and let D be the canonical form (5) of A where U is orthogonal. Show that $B = U D^{1/2} U^T$, with $D^{1/2} \equiv \mathrm{diag}(d_1^{1/2}, \ldots, d_n^{1/2})$, is a square root of A. Discuss the uniqueness of matrix square roots.

2.2. NORMS

A mapping $\|\cdot\|$ from R^n (or C^n) to R^1 which satisfies

(a) $\|x\| \geqslant 0$, $\forall x \in R^n$ (or C^n); $\quad \|x\| = 0$ only if $x = 0$;

(b) $\|\alpha x\| = |\alpha| \|x\|$, $\forall x \in R^n$ (or C^n), $\alpha \in R^1$ (or C^1); $\hfill (1)$

(c) $\|x + y\| \leqslant \|x\| + \|y\|$, $\forall x, y \in R^n$ (or C^n);

2.2. NORMS

is called a *norm*. Well-known examples of norms on either R^n or C^n are the l_p-norms:

$$\|x\|_p = \left(\sum_{i=1}^{n} |x_i|^p\right)^{1/p}, \quad 1 \leq p < \infty, \qquad (2)$$

and, the limiting case, the l_∞-norm:

$$\|x\|_\infty = \max_{1 \leq i \leq n} |x_i|. \qquad (3)$$

An *inner product* on R^n is a mapping $(\cdot, \cdot)$ from the product space $R^n \times R^n$ to R^1 which satisfies

(a) $(x, x) \geq 0, \quad \forall x \in R^n; \quad (x, x) = 0$ only if $x = 0;$
(b) $(x, y) = (y, x), \quad \forall x, y \in R^n;$ (4)
(c) $(x + y, z) = (x, z) + (y, z), \quad (\alpha x, y) = \alpha(x, y),$
$\quad \forall x, y, z \in R^n, \quad \alpha \in R^1.$

An inner product on R^n defines a norm by means of $\|x\| = (x, x)^{1/2}$; in particular, the l_2-norm, which we shall also call the *Euclidean* norm, derives from the inner product $(x, y) = x^T y$. For any inner product, the Cauchy–Schwarz inequality

$$|(x, y)| \leq \|x\| \|y\| \qquad (5)$$

holds and, in particular, for the l_2-norm, we have

$$\left|\sum_{i=1}^{n} x_i y_i\right| \leq \left(\sum_{i=1}^{n} |x_i|^2\right)^{1/2} \left(\sum_{i=1}^{n} |y_i|^2\right)^{1/2}. \qquad (6)$$

Similar considerations hold on C^n; in particular, (4) defines an inner product on C^n provided that (4b) is changed to $(x, y) = \overline{(y, x)}$.

As mentioned in the introduction, we will always assume that R^n is equipped with some norm. In line with this, it is important to note that all norms on R^n are equivalent in the sense of the following basic result.

2.2.1. Norm-Equivalence Theorem. Let $\|\cdot\|$ and $\|\cdot\|'$ be any two norms on R^n. Then there exist constants $c_2 \geq c_1 > 0$ such that

$$c_1 \|x\| \leq \|x\|' \leq c_2 \|x\|, \quad \forall x \in R^n. \qquad (7)$$

Proof. It suffices to show that (7) holds if $\|\cdot\|'$ is taken to be the l_2-norm; for, if the two relations

$$d_1 \|x\| \leq \|x\|_2 \leq d_2 \|x\|, \quad d_1' \|x\|' \leq \|x\|_2 \leq d_2' \|x\|'$$

hold with $d_2 \geqslant d_1 > 0$ and $d_2' \geqslant d_1' > 0$, then (7) holds with $c_1 = d_1/d_2'$ and $c_2 = d_2/d_1'$.

Now let e^i, $i = 1,\ldots, n$, again denote the coordinate vectors of R^n. Then, by (6),

$$\|x\| = \left\| \sum_{i=1}^n x_i e^i \right\| \leqslant \sum_{i=1}^n |x_i| \|e^i\| \leqslant \beta \|x\|_2, \quad \beta = \left(\sum_{i=1}^n \|e^i\|^2 \right)^{1/2}, \quad (8)$$

and, consequently, the left side of (7) holds with $c_1 = \beta^{-1}$. Moreover, (8) shows that

$$|\|x\| - \|y\|| \leqslant \|x - y\| \leqslant \beta \|x - y\|_2,$$

so that $\|\cdot\|$ is a continuous function with respect to the l_2-norm. Therefore, since the unit sphere $S = \{x \mid \|x\|_2 = 1\}$ is compact, $\|\cdot\|$ attains a (necessarily nonzero) minimum on S; that is, $\|x\| \geqslant \alpha > 0$ if $\|x\|_2 = 1$. But then $\|x\| \geqslant \alpha \|x\|_2$ for all $x \in R^n$, so that the right side of (7) holds with $c_2 = \alpha^{-1}$. ∎

We turn next to norms of linear operators. Given any two norms $\|\cdot\|$ and $\|\cdot\|'$ on R^n and R^m, respectively, and any $A \in L(R^n, R^m)$, the (operator) norm of A with respect to $\|\cdot\|$ and $\|\cdot\|'$ is defined by

$$\|A\| = \sup_{\|x\|=1} \|Ax\|'. \quad (9)$$

Such a matrix norm satisfies the properties

(a) $\|A\| \geqslant 0$, $\forall A \in L(R^n, R^m)$; $\|A\| = 0$ only if $A = 0$;

(b) $\|\alpha A\| = |\alpha| \|A\|$, $\forall A \in L(R^n, R^m)$, $\alpha \in R^1$; (10)

(c) $\|A + B\| \leqslant \|A\| + \|B\|$, $\forall A, B \in L(R^n, R^m)$;

therefore, $L(R^n, R^m)$ is itself a normed linear space. In the important special case that $R^n = R^m$ and the norms $\|\cdot\|$ and $\|\cdot\|'$ are the same, the multiplicative property $\|AB\| \leqslant \|A\| \|B\|$ also holds. The definition (9) as well as the properties (10) hold verbatim for C^n, C^m, and $A \in L(C^n, C^m)$.

In most concrete work in numerical analysis, the l_1-, l_2-, and l_∞-norms are the most useful. The following result describes precisely the corresponding matrix norms, which we denote by $\|A\|_1$, $\|A\|_2$, and $\|A\|_\infty$, respectively.

2.2.2. Let $A \in L(R^n, R^m)$ where both R^n and R^m are normed by their respective l_i-norms, $i = 1, 2, \infty$.

2.2. NORMS

Then

$$\|A\|_1 = \max_{1 \le j \le n} \sum_{i=1}^{m} |a_{ij}|, \tag{11}$$

$$\|A\|_\infty = \max_{1 \le i \le m} \sum_{j=1}^{n} |a_{ij}|, \tag{12}$$

and

$$\|A\|_2 = \sqrt{\lambda}, \tag{13}$$

where λ is the maximum eigenvalue of $A^T A$.

Proof. Consider first the l_1-norm. For any $x \in R^n$,

$$\|Ax\|_1 = \sum_{i=1}^{m} \left| \sum_{j=1}^{n} a_{ij} x_j \right| \le \sum_{i=1}^{m} \sum_{j=1}^{n} |a_{ij}| |x_j| \le \sum_{j=1}^{n} |x_j| \sum_{i=1}^{m} |a_{ij}|$$

$$\le \left(\max_{1 \le j \le n} \sum_{i=1}^{m} |a_{ij}| \right) \|x\|_1. \tag{14}$$

It then suffices to show that there is some $x \in R^n$ for which equality is attained in (14). Let k be an index for which the maximum in (11) is attained; then

$$\|Ae^k\|_1 = \sum_{i=1}^{m} |a_{ik}| = \max_{1 \le j \le n} \sum_{i=1}^{m} |a_{ij}|.$$

That is, the supremum in (9) is attained for the kth coordinate vector.

The proof for the l_∞-norm is similar. We note only that, in this case, the supremum is taken on for $x \in R^n$ defined by

$$x_i = \begin{cases} a_{ki}/|a_{ki}|, & a_{ki} \ne 0, \\ 1, & a_{ki} = 0, \end{cases} \quad i = 1, \dots, n,$$

where k is an index for which the maximum in (12) is attained.

Finally, for the l_2-norm, we have $\|Ax\|_2 = (x^T A^T A x)^{1/2}$, and the result follows from (2.1.3). ∎

We note that if $A \in L(R^n)$ is symmetric with eigenvalues $\lambda_1, \dots, \lambda_n$, then **2.2.2** shows that

$$\|A\|_2 = \max_{1 \le i \le n} |\lambda_i|.$$

Note also that the representations (11) and (12) hold verbatim on $L(C^n, C^m)$, while (13) holds if λ is the maximal eigenvalue of $A^H A$.

On the basis of (13), we have the following observation:

2.2.3. Let $P \in L(R^n)$ be orthogonal and $\alpha \neq 0$ an arbitrary scalar. Then $\|H\|_2 \|H^{-1}\|_2 = 1$, where $H = \alpha P$.

Proof. Clearly, $H^T H = \alpha^2 I$, while $(H^{-1})^T H^{-1} = \alpha^{-2} I$. The result then follows from **2.2.2**. ∎

For arbitrary nonsingular $A \in L(R^n)$, the number $\|A\| \|A^{-1}\|$ is called the *condition number* of A with respect to the particular norm used. Since $\|A\| \|A^{-1}\| \geq 1$, **2.2.3** shows that scalar multiples of orthogonal matrices have minimal condition numbers in the l_2-norm.

Although the norms of **2.2.2** are the most important for practical computation, we shall, from time to time, use other norms in the course of our analysis. For example, if $A \in L(R^n)$ is symmetric and positive definite, then an inner product is defined by $(x, y) = x^T A y$. The corresponding norm is sometimes useful in the context of minimization problems. This norm is, however, only a special case of the following general procedure for generating norms.

2.2.4. Let $\|\cdot\|$ be an arbitrary norm on R^n (or C^n) and P an arbitrary nonsingular, $n \times n$, real (or complex) matrix. Then the mapping defined by $\|x\|' = \|Px\|$, for all $x \in R^n$, is a norm on R^n (or C^n). Moreover, if $A \in L(R^n)$, then

$$\|A\|' = \|PAP^{-1}\|. \tag{15}$$

Proof. To show that $\|\cdot\|'$ is a norm requires only a simple calculation verifying the axioms (1). Then (15) results from

$$\|A\|' = \sup_{\|x\|'=1} \|Ax\|' = \sup_{\|Px\|=1} \|PAx\| = \sup_{\|y\|=1} \|PAP^{-1}y\| = \|PAP^{-1}\|. \ \blacksquare$$

By **2.2.1** arbitrary norms on $L(R^n)$ are, of course, equivalent. The following theorem relates the equivalence constants for vectornorms to those for their associated matrixnorms.

2.2.5. Let $\|\cdot\|$ and $\|\cdot\|'$ be arbitrary norms on R^n. Then, with c_1, c_2 from (7) and $d_1 = c_1/c_2$, $d_2 = c_2/c_1$,

$$d_1 \|A\| \leq \|A\|' \leq d_2 \|A\|, \quad \forall A \in L(R^n). \tag{16}$$

Proof. By means of (7), we obtain the inequality

$$\|A\|' = \sup_{x \neq 0} \{\|Ax\|'/\|x\|'\} \leq \sup_{x \neq 0} \{c_2 \|Ax\|/(c_1 \|x\|)\} = (c_2/c_1) \|A\|,$$

2.2. NORMS

so that the right half of (16) holds with $d_2 = c_2/c_1$. One finds, in a similar way, that $d_1 = c_1/c_2$. ∎

On the basis of **2.2.5**, we give a result which relates the norm of a matrix to its elements in the following way.

2.2.6. Let $\|\cdot\|$ be an arbitrary norm on R^n. Then there is a constant η_1 such that

$$\|A\| \leq \eta_1 \max_{1 \leq i,j \leq n} |a_{ij}|, \quad \forall A \in L(R^n). \tag{17}$$

Similarly, there is a constant η_2 such that

$$\|A\| \leq \eta_2 \sum_{i=1}^{n} \|a^i\|, \quad \forall A \in L(R^n), \tag{18}$$

where $a^1, \ldots, a^n$ are the columns of A.

Proof. By **2.2.5**, there is a constant $\hat{\eta}_1$ such that $\|A\| \leq \hat{\eta}_1 \|A\|_1$. Hence, (17) follows immediately from **2.2.2**. Moreover, by **2.2.1**,

$$\|A\| \leq \hat{\eta}_1 \sum_{i=1}^{n} \|a^i\|_1 \leq \eta_2 \sum_{i=1}^{n} \|a^i\|. \quad \blacksquare$$

We complete this section by relating the eigenvalues of a matrix to the value of a norm. For any complex $n \times n$ matrix A, the *spectral radius* of A is defined as the maximum of $|\lambda_1|, \ldots, |\lambda_n|$, where $\lambda_1, \ldots, \lambda_n$ are the eigenvalues of A. If $A \in L(C^n)$ and λ is any eigenvalue of A with corresponding eigenvector $u \neq 0$, then $\|Au\| = |\lambda| \|u\|$, so that $|\lambda| \leq \|A\|$. Therefore, the spectral radius of A, henceforth denoted by $\rho(A)$, satisfies $\rho(A) \leq \|A\|$. However, for a given norm, the spectral radius of A and the norm of A may be arbitrarily far apart; for example,

$$\left\| \begin{pmatrix} 0 & \alpha \\ 0 & 0 \end{pmatrix} \right\|_1 = |\alpha|, \tag{19}$$

while the spectral radius is zero. This large separation is due, in a sense, to a wrong choice of norm, as **2.2.8** shows. We first prove the following lemma, which will also be of independent interest.

2.2.7. Let $A \in L(C^n)$ have the Jordan form J. Then, for arbitrary $\epsilon > 0$, A is similar to a matrix $\hat{J}$ which is identical to J except that each off-diagonal 1 in (2.1.7) is replaced by ϵ.

Proof. Let P be a nonsingular matrix such that $P^{-1}AP = J$ and let D be the diagonal matrix diag$(1, \epsilon, ..., \epsilon^{n-1})$. Then it is easily verified that $D^{-1}JD = \hat{J}$. Hence, $\hat{J} = Q^{-1}AQ$, where $Q = PD$. ∎

2.2.8. Let $A \in L(C^n)$. Then, given any $\epsilon > 0$, there is a norm on C^n such that

$$\|A\| \leq \rho(A) + \epsilon. \tag{20}$$

Proof. Let $\hat{J} = Q^{-1}AQ$ be the modified Jordan form of **2.2.7**. Then **2.2.2** shows that

$$\|\hat{J}\|_1 \leq \rho(A) + \epsilon.$$

But, by **2.2.4**, $\|x\| = \|Q^{-1}x\|_1$ defines a norm on C^n, and (15) ensures that

$$\|A\| = \|Q^{-1}AQ\|_1 = \|\hat{J}\|_1.$$ ∎

As a corollary of **2.2.8**, we have the following important result.

2.2.9. Let $A \in L(C^n)$. Then $\lim_{k \to \infty} A^k = 0$ if and only if $\rho(A) < 1$.

Proof. If $\rho(A) < 1$, then, by **2.2.8**, there is a norm such that $\|A\| < 1$. Therefore, since $\|A^k\| \leq \|A\|^k$, it follows that $A^k \to 0$ as $k \to \infty$. For the converse, suppose that A has an eigenvalue λ with $|\lambda| \geq 1$ and the corresponding eigenvector $x \neq 0$. Then $A^k x = \lambda^k x$ for all k, so that $A^k x$ does not tend to zero. ∎

NOTES AND REMARKS

NR 2.2-1. All of the results of this section are standard. Excellent references for norms and their role in matrix numerical analysis are again Faddeev and Faddeeva [1960], Householder [1964], and Wilkinson [1965].

EXERCISES

E 2.2-1. Let $A \in L(C^n)$. Show that

$$\max_{1 \leq i,j \leq n} |a_{ij}| \leq \|A\|_2 \leq n \max |a_{ij}|, \quad 1 \leq i, j \leq n.$$

E 2.2-2. Let $A \in L(C^n)$. Show that there is a constant η, depending only on the norm, such that $\|A\| \leq \eta \max_i \|a^i\|$, where $a^1, ..., a^n$ are the columns of A.

E 2.2-3. A norm $\|\cdot\|$ is *uniformly convex* if, given $\epsilon > 0$, there is a $\delta > 0$ such that $\|x + y\| \leq 2(1 - \delta)$ whenever $\|x\| \leq 1$, $\|y\| \leq 1$, and $\|x + y\| \geq \epsilon$. A

2.3. INVERSES

norm is *strictly convex* if $\|x+y\| < 2$ whenever $\|x\| \leq 1$, $\|y\| \leq 1$, and $x \neq y$. Show that a norm on R^n is strictly convex if and only if it is uniformly convex.

2.3. INVERSES

We shall be concerned in various places with the problem of ascertaining that a given linear operator $A \in L(R^n)$ is invertible. In this section, we collect various sufficient conditions for this to be the case. A few additional results are also given in the following section.

2.3.1. Neumann Lemma. Let $B \in L(R^n)$ and assume that $\rho(B) < 1$. Then $(I - B)^{-1}$ exists and

$$(I - B)^{-1} = \lim_{k \to \infty} \sum_{i=0}^{k} B^i. \tag{1}$$

Proof. Since $\rho(B) < 1$, $I - B$ clearly has no zero eigenvalues and hence is invertible. To show that (1) holds, note that the identity $(I - B)(I + \cdots + B^{k-1}) = I - B^k$ yields

$$I + B + \cdots + B^{k-1} = (I - B)^{-1} - (I - B)^{-1} B^k.$$

By **2.2.9**, the right-hand side tends to $(I - B)^{-1}$. ∎

As an immediate corollary of **2.3.1**, we see that $I - B$ is invertible whenever $\|B\| < 1$, and from (1) that

$$\|(I - B)^{-1}\| \leq \sum_{i=0}^{\infty} \|B\|^i = 1/(1 - \|B\|). \tag{2}$$

This is a special case of the following, more general result.

2.3.2. Perturbation Lemma. Let $A, C \in L(R^n)$ and assume that A is invertible, with $\|A^{-1}\| \leq \alpha$. If $\|A - C\| \leq \beta$ and $\beta\alpha < 1$, then C is also invertible, and

$$\|C^{-1}\| \leq \alpha/(1 - \alpha\beta). \tag{3}$$

Proof. Since $\|I - A^{-1}C\| = \|A^{-1}(A - C)\| \leq \alpha\beta < 1$ and $A^{-1}C = I - (I - A^{-1}C)$, it follows from **2.3.1** that $A^{-1}C$ is invertible. Hence, C is invertible. Moreover, we conclude from (2) that

$$\|C^{-1}\| = \|[I - (I - A^{-1}C)]^{-1} A^{-1}\| \leq \alpha \sum_{i=0}^{\infty} (\alpha\beta)^i = \alpha/(1 - \alpha\beta). \blacksquare$$

In many of our later results, **2.3.2** will be applied to matrix-valued mappings $A: D \subset R^m \to L(R^n)$. We therefore give a simple corollary of **2.3.2** in this setting. Recall that we denote by $S(x, r)$ the open ball $\{y \in R^n \mid \|x - y\| < r\}$ and its closure by $\bar{S}(x, r)$.

2.3.3. Suppose that the mapping $A: D \subset R^m \to L(R^n)$ is continuous at a point $x^0 \in D$ for which $A(x^0)$ is invertible. Then there is a $\delta > 0$ and a $\gamma > 0$ so that $A(x)$ is invertible, and

$$\|A(x)^{-1}\| \leq \gamma, \qquad \forall x \in D \cap \bar{S}(x^0, \delta). \qquad (4)$$

Moreover, $A(x)^{-1}$ is continuous in x at x^0.

Proof. Set $\alpha = \|A(x^0)^{-1}\|$ and, for given $\beta < \alpha^{-1}$, choose δ so that $\|A(x^0) - A(x)\| \leq \beta$ whenever $x \in D \cap \bar{S}(x^0, \delta)$. Then **2.3.2** ensures that $A(x)$ is invertible and that (4) holds with $\gamma = \alpha/(1 - \beta\alpha)$. Therefore,

$$\|A(x^0)^{-1} - A(x)^{-1}\| = \|A(x^0)^{-1}[A(x) - A(x^0)]A(x)^{-1}\| \leq \alpha\gamma\|A(x^0) - A(x)\|,$$

and the continuity of A at x^0 ensures that of the inverse. ∎

We consider next a rather different situation which permits us to conclude the invertibility of a given linear operator. Recall that a permutation matrix has as its columns the coordinate vectors $e^1, \ldots, e^n$ in some arbitrary order.

2.3.4. Definition. An $n \times n$ real or complex matrix A is *reducible* if there is a permutation matrix P so that

$$PAP^T = \begin{pmatrix} B_{11} & B_{12} \\ 0 & B_{22} \end{pmatrix},$$

where B_{11} and B_{22} are square matrices; A is *irreducible* if it is not reducible. ∎

Clearly, any matrix all of whose elements are nonzero is irreducible. More generally, $A \in L(C^n)$ is reducible if and only if there is a nonempty subset of indices $J \subset \{1, \ldots, n\}$ such that

$$a_{kj} = 0, \qquad \forall k \in J, \ j \notin J. \qquad (5)$$

This is simply a restatement of **2.3.4** in terms of the components of A. A more useful equivalence is the following result.

2.3. INVERSES

2.3.5. The matrix $A \in L(C^n)$ is irreducible if and only if, for any two indices $1 \leqslant i, j \leqslant n$, there is a sequence of nonzero elements of A of the form
$$\{a_{i,i_1}, a_{i_1,i_2}, \ldots, a_{i_m j}\}. \tag{6}$$

Proof. If there is a sequence of nonzero elements of the form (6), we will say that there is a *chain* for i, j. Now, suppose that A is reducible, let J be such that (5) holds, and choose $i \in J$, $j \notin J$. If $a_{ik} \neq 0$, it follows that $k \in J$, so that it is impossible to construct a chain for i, j. Hence, the sufficiency is proved.

Conversely, assume that A is irreducible, and, for given i, set $J = \{k \mid \text{a chain exists for } i, k\}$. Clearly, J is not empty, since, otherwise, $a_{ik} = 0$, $k = 1, \ldots, n$, and this would contradict the irreducibility. Now, suppose that, for some j, there is no chain for i, j. Then J is not the entire set $\{1, \ldots, n\}$, and we claim that
$$a_{kl} = 0, \qquad \forall k \in J, \; l \notin J, \tag{7}$$

which would contradict the irreducibility of A. But (7) follows immediately from the fact that there is a chain for i, k, so that if $a_{kl} \neq 0$, we may add a_{kl} to obtain a chain for i, l. This implies that $l \in J$. ∎

As an example of the use of this result, we consider the matrices

$$A = \begin{pmatrix} 2 & -1 & & & O \\ -1 & \cdot & \cdot & & \\ & \cdot & \cdot & \cdot & \\ & & \cdot & \cdot & -1 \\ O & & & -1 & 2 \end{pmatrix}, \tag{8}$$

and

$$A = \begin{pmatrix} B & -I & & & O \\ -I & \cdot & \cdot & & \\ & \cdot & \cdot & \cdot & \\ & & \cdot & \cdot & -I \\ O & & & -I & B \end{pmatrix}, \quad B = \begin{pmatrix} 4 & -1 & & & O \\ -1 & \cdot & \cdot & & \\ & \cdot & \cdot & \cdot & \\ & & \cdot & \cdot & -1 \\ O & & & -1 & 4 \end{pmatrix}, \tag{9}$$

discussed in Sections 1.1 and 1.2. In the first case, if $1 \leqslant i < j \leqslant n$, then the elements $a_{i,i+1}, a_{i+1,i+2}, \ldots, a_{j-1,j}$ satisfy the conditions of **2.3.5**, while if $j < i$, we may take $a_{i,i-1}, a_{i-1,i-2}, \ldots, a_{j+1,j}$. For the matrix of (9) it is also possible (see **E 2.3-2**) to write down explicitly a chain of nonzero elements for any i, j, but the indexing becomes cumbersome. It is more transparent to return to the difference equation (1.2.5), with

$f \equiv 0$ and the corresponding grid points on the unit square. If we number these grid points sequentially from left to right, bottom to top, then the irreducibility of (9) follows immediately by constructing a chain whose indices correspond to the four possible paths illustrated in Fig. 2.1.

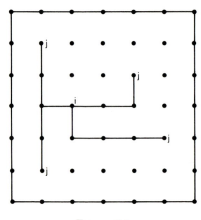

FIGURE 2.1

We summarize the above discussion in the following result.

2.3.6. The matrices (8) and (9) are irreducible. ∎

Irreducibility of a matrix implies nothing in general about its invertibility. We will be able to use the concept, however, in combination with the following condition.

2.3.7. Definition. An $n \times n$ real or complex matrix $A = (a_{ij})$ is *diagonally dominant* if

$$\sum_{j=1, j \neq i}^{n} |a_{ij}| \leqslant |a_{ii}|, \quad i = 1, \ldots, n, \tag{10}$$

and *strictly diagonally dominant* if strict inequality holds in (10) for all i; the matrix is *irreducibly diagonally dominant* if A is irreducible, diagonally dominant, and strict inequality holds in (10) for at least one i. ∎

The invertibility theorem we shall prove is the following:

2.3.8. Diagonal Dominance Theorem. Let $A \in L(C^n)$ be either strictly or irreducibly diagonally dominant. Then A is invertible.

2.3. INVERSES

Proof. Assume first that A is strictly diagonally dominant and suppose there is an $x \neq 0$ such that $Ax = 0$. Let

$$|x_m| = \max_{1 \leq j \leq n} |x_j|.$$

Then $|x_m| > 0$, and the inequality

$$|a_{mm}||x_m| = \left| \sum_{j \neq m} a_{mj} x_j \right| \leq |x_m| \sum_{j \neq m} |a_{mj}| \qquad (11)$$

contradicts the strict diagonal dominance. Similarly, if A is irreducibly diagonally dominant, again suppose there is an $x \neq 0$ so that $Ax = 0$, and let m be such that

$$|a_{mm}| > \sum_{j \neq m} |a_{mj}|. \qquad (12)$$

Next, define the set of indices

$$J = \{k \mid |x_k| \geq |x_i|, \; i = 1, \ldots, n, \; |x_k| > |x_j| \text{ for some } j\}.$$

Clearly, J is not empty, for this would imply that $|x_1| = \cdots = |x_n| \neq 0$, and hence the inequality (11) would contradict (12). Now, for any $k \in J$,

$$|a_{kk}| \leq \sum_{j \neq k} |a_{kj}| |x_j|/|x_k|,$$

and it follows that $a_{kj} = 0$ whenever $|x_k| > |x_j|$, or else the diagonal dominance is contradicted; that is,

$$a_{kj} = 0, \quad \forall k \in J, \; j \notin J.$$

But then A is reducible, and this is a contradiction. ∎

As a corollary of the last result, we have the following important isolation theorem for eigenvalues.

2.3.9. Gerschgorin Circle Theorem. Let $A \in L(C^n)$ and define the set of complex numbers

$$S = \bigcup_{i=1}^{n} \left\{ z \mid |a_{ii} - z| \leq \sum_{j \neq i} |a_{ij}| \right\}.$$

Then every eigenvalue of A lies in S.

Proof. Let λ be any eigenvalue and suppose that $\lambda \notin S$; that is,

$$|a_{ii} - \lambda| > \sum_{j \neq i} |a_{ij}|, \quad i = 1, \ldots, n.$$

But then $A - \lambda I$ is strictly diagonally dominant, and hence, by **2.3.8**, nonsingular. This is a contradiction. ∎

As a corollary of this result and of **2.3.8**, we obtain:

2.3.10. If $A \in L(R^n)$ is symmetric, irreducibly diagonally dominant, and has positive diagonal elements, then A is positive definite. In particular, the matrices (8) and (9) are positive definite.

Proof. Since the eigenvalues λ_i of A are real, **2.3.9** and the diagonal dominance ensure that $\lambda_i \geqslant 0$, $i = 1,\ldots, n$. But, by **2.3.8**, A is invertible, and hence $\lambda_i > 0$ for all i. The last statement is an immediate consequence of **2.3.6**. ∎

We complete this section with a commonly used formula for determining the inverse of matrices modified by rank-m matrices.

2.3.11. Sherman–Morrison–Woodbury Formula. Let $A \in L(R^n)$ be invertible and let $U, V \in L(R^m, R^n)$, $m \leqslant n$. Then $A + UV^T$ is invertible if and only if $I + V^T A^{-1} U$ is invertible, and then

$$(A + UV^T)^{-1} = A^{-1} - A^{-1}U(I + V^T A^{-1} U)^{-1} V^T A^{-1}. \tag{13}$$

The proof is immediate, since a direct computation shows that (13) is valid. In the important special case that $m = 1$, U and V may be taken as vectors $u, v \in R^n$, and (14) reduces to the *Sherman–Morrison formula*:

$$(A + uv^T)^{-1} = A^{-1} - [1/(1 + v^T A^{-1} u)] A^{-1} uv^T A^{-1}. \tag{14}$$

NOTES AND REMARKS

NR 2.3-1. Theorem **2.3.2**, and hence also **2.3.3**, remains valid in an arbitrary Banach space. Also, **2.3.1** extends, by a more sophisticated argument, to a complex Banach space (see, for example, Taylor [1958, pp. 164 and 260]).

NR 2.3-2. Definitions **2.3.4** and **2.3.7** and Theorems **2.3.8**, **2.3.9**, and **2.3.10** are classical. See, for example, Varga [1962] for a further discussion.

NR 2.3-3. The proof of **2.3.5** is taken from Henrici [1962], although the result had been previously stated by Varga. In fact, Varga [1962] discusses an essentially equivalent but more geometric approach to irreducibility by means of directed graphs.

NR 2.3-4. For further discussion of the formulas (13) and (14), see Householder [1964] and Zielke [1968].

EXERCISES

E 2.3-1. Let $A, B \in L(R^n)$ and assume that A is invertible. If there is a $C \in L(R^n)$ such that $\|A^{-1}(A - B)(I + CB)\| < 1$, then B is invertible. (Ostrowski [1967b]).

E 2.3-2. For given $1 \leq i, j \leq n$, exhibit explicitly a chain of nonzero elements for the matrix (9).

E 2.3-3. Let $A \in L(R^n)$ be symmetric, strictly diagonally dominant, and have positive diagonal elements. Show that A is positive definite.

E 2.3-4. Let $A \in L(R^n)$ be given by (8). Show that A has eigenvalues

$$\lambda_k = 2 - 2\cos[k\pi/(n+1)], \quad k = 1,\ldots, n,$$

and corresponding eigenvectors

$$(\sin[k\pi/(n+1)], \sin[2k\pi/(n+1)],\ldots, \sin[nk\pi/(n+1)])^T, \quad k = 1,\ldots, n.$$

E 2.3-5. Let $A, B \in L(R^n)$ be the tridiagonal matrices

$$A = \begin{pmatrix} a_1 & b_1 & & \bigcirc \\ c_1 & \cdot & \cdot & \\ & \cdot & \cdot & b_{n-1} \\ \bigcirc & & c_{n-1} & a_n \end{pmatrix}, \quad B = \begin{pmatrix} a_1 & \gamma_1 & & \bigcirc \\ \gamma_1 & \cdot & \cdot & \\ & \cdot & \cdot & \gamma_{n-1} \\ \bigcirc & & \gamma_{n-1} & a_n \end{pmatrix},$$

where $b_i c_i > 0$ and $\gamma_i = (b_i c_i)^{1/2}$, $i = 1,\ldots, n$. (a) Show that $B = DAD^{-1}$, where

$$D = \text{diag}[1, (b_1/c_1)^{1/2}, (b_1 b_2/c_1 c_2)^{1/2},\ldots, (b_1 \cdots b_{n-1}/c_1 \cdots c_{n-1})^{1/2}].$$

(b) If $a_i \geq |b_i| + |c_{i-1}|$, $i = 2,\ldots, n-1$, and $a_1 > |b_1|$, $a_n > |c_{n-1}|$, show that A is irreducibly diagonally dominant and that B is positive definite.

2.4. PARTIAL ORDERING AND NONNEGATIVE MATRICES

In many places in this book, it will be convenient to be able to compare vectors of R^n element by element. This we may do by means of the *natural* (or *component-wise*) *partial ordering* on R^n defined by

$$\text{For } x, y \in R^n, \quad x \leq y \quad \text{if and only if} \quad x_i \leq y_i, \quad i = 1,\ldots, n. \quad (1)$$

Two vectors $x, y \in R^n$ are said to be *comparable* under this ordering if $x \leq y$ or $y \leq x$. The following properties are immediately verified.

2.4.1. The ordering relation $\leq$ defined on R^n by (1) satisfies:
(a) $x \leq x$ for all $x \in R^n$;
(b) If $x \leq y$ and $y \leq x$, then $x = y$;
(c) If $x \leq y$ and $y \leq z$, then $x \leq z$;
(d) If $x \leq y$, then $\alpha x \leq \alpha y$ for all $\alpha \geq 0$;
(e) If $x \leq y$, then $x + z \leq y + z$ for all $z \in R^n$.

If $x \in R^n$ and $x \geq 0$, then x is *nonnegative*, and the set of nonnegative vectors in R^n is the *positive cone* of R^n. For many purposes, it is convenient to consider the nonnegative vector of absolute values defined by

$$|x| = (|x_1|, \dots, |x_n|)^T, \qquad x \in R^n. \tag{2}$$

It is again immediately verified that this absolute value satisfies the following normlike properties:

2.4.2. (a) For all $x \in R^n$, $|x| \geq 0$, and $|x| = 0$ only if $x = 0$; (b) for all $x \in R^n$ and $\alpha \in R^1$, $|\alpha x| = |\alpha| |x|$; (c) for all $x, y \in R^n$, $|x + y| \leq |x| + |y|$.

A norm on R^n is *monotonic* if, for any $x, y \in R^n$,

$$|x| \leq |y| \quad \text{implies} \quad \|x\| \leq \|y\|. \tag{3}$$

It is easy to see (**E 2.4-2**) that (3) is equivalent with

$$\| |x| \| = \|x\|, \qquad \forall x \in R^n, \tag{4}$$

and, in particular, that the l_p-norms ($1 \leq p \leq \infty$) satisfy (4).

By means of the partial ordering (1), we may define monotone operators on R^n.

2.4.3. Definition. A mapping $F: D \subset R^n \to R^m$ is *isotone* (*antitone*) on $D_0 \subset D$ if $Fx \leq Fy$ ($Fx \geq Fy$) whenever $x \leq y$, $x, y \in D_0$. ∎

We may introduce a partial ordering analogous to (1) on the space of $n \times m$ matrices by means of:

$$\text{For } A, B \in L(R^n, R^m), \quad A \leq B \text{ if and only if } a_{ij} \leq b_{ij}, \quad i = 1, \dots, m;$$
$$j = 1, \dots, n. \tag{5}$$

The corresponding absolute value is then defined by:

$$\text{For } A \in L(R^n, R^m), \quad |A| = (|a_{ij}|). \tag{6}$$

Clearly, the properties of **2.4.1** and **2.4.2** also hold in this case.

2.4. PARTIAL ORDERING AND NONNEGATIVE MATRICES

A matrix $A \in L(R^n, R^m)$ is *nonnegative* if $A \geq 0$. It is easy to see (**E 2.4-3**) that A is nonnegative if and only if A is isotone, which, in turn, is true if and only if $Ax \geq 0$ whenever $x \geq 0$.

By means of the partial ordering (5), we may define the following inverse-like property.

2.4.4. Definition. Let $A \in L(R^n)$. Then $B \in L(R^n)$ is a *left subinverse* (*left superinverse*) of A if

$$BA \leq I \qquad (BA \geq I), \tag{7}$$

and a *right subinverse* (*right superinverse*) if

$$AB \leq I \qquad (AB \geq I). \tag{8}$$

B is a *subinverse* (*superinverse*) if both (7) and (8) hold. ∎

Note that the null matrix is a subinverse of any matrix. Note also that if B is a subinverse (superinverse) of A, then A is a subinverse (superinverse) of B, while if A^{-1} exists, then A^{-1} is a subinverse and superinverse of A.

We shall be interested in conditions which ensure that a given matrix $A \in L(R^n)$ has a nontrivial nonnegative subinverse (superinverse) and, in particular, that A has a nonnegative inverse. We now collect some results on the latter question. The first is an analog of the Neumann lemma **2.3.1**.

2.4.5. Let $B \in L(R^n)$ and assume that $B \geq 0$. Then $(I - B)^{-1}$ exists and is nonnegative if and only if $\rho(B) < 1$.

Proof. If $\rho(B) < 1$, then **2.3.1** ensures that $(I - B)^{-1} = \sum_{i=0}^{\infty} B^i$, and, since each term of the sum is nonnegative, $(I - B)^{-1} \geq 0$. Conversely, assume that $(I - B)^{-1} \geq 0$ and let λ be any eigenvalue of B with corresponding eigenvector $x \neq 0$. Then $|\lambda| |x| \leq B |x|$, so that $(I - B)|x| \leq (1 - |\lambda|)|x|$. Consequently, $|x| \leq (1 - |\lambda|)(I - B)^{-1} |x|$, from which it follows, since $x \neq 0$ and $(I - B)^{-1} \geq 0$, that $|\lambda| < 1$. ∎

A lower (upper) triangular matrix L is *strictly lower* (*upper*) *triangular* if all diagonal elements are zero. As an immediate corollary of **2.4.5**, we may then state the following result.

2.4.6. If $D, L \in L(R^n)$ are nonnegative, D is diagonal and invertible, and L is strictly lower triangular, then $(D - L)^{-1} \geq 0$.

A particularly important class of matrices with nonnegative inverses is the following:

2.4.7. Definition. A matrix $A \in L(R^n)$ is an *M-matrix* if A is invertible, $A^{-1} \geq 0$, and $a_{ij} \leq 0$ for all $i, j = 1,\ldots, n$, $i \neq j$. A symmetric M-matrix is a *Stieltjes matrix*. ∎

The following gives a sometimes useful characterization of M-matrices:

2.4.8. Let $A \in L(R^n)$, with $a_{ij} \leq 0$, $i \neq j$. Then A is an M-matrix if and only if (a) the diagonal elements of A are positive and (b) the matrix $B = I - D^{-1}A$, where $D = \mathrm{diag}(a_{11},\ldots, a_{nn})$, satisfies $\rho(B) < 1$.

Proof. Suppose $\rho(B) < 1$. Then, because $B \geq 0$, **2.4.5** ensures that $(D^{-1}A)^{-1} = (I - B)^{-1} \geq 0$, so that A^{-1} exists, and, since $D \geq 0$, clearly, $A^{-1} \geq 0$. Conversely, if A is an M-matrix, then the diagonal elements are positive; for, if some $a_{ii} \leq 0$, then the ith column a^i is nonpositive, and, hence, $e^i = A^{-1}a^i \leq 0$, where e^i is the ith coordinate vector. This is a contradiction, and, hence, $D \geq 0$ and D is invertible. It follows that $B \geq 0$ and $(I - B)^{-1} = A^{-1}D \geq 0$; hence, **2.4.5** ensures that $\rho(B) < 1$. ∎

We next give some results concerning the derivation of M-matrices from given ones. These are consequences of the following general comparison theorem:

2.4.9. Let $B \in L(R^n)$ and $C \in L(C^n)$. If $|C| \leq B$, then $\rho(C) \leq \rho(B)$.

Proof. Set $\sigma = \rho(B)$, let $\epsilon > 0$ be arbitrary, and set $B_1 = (\sigma + \epsilon)^{-1}B$ and $C_1 = (\sigma + \epsilon)^{-1}C$. Clearly, $\rho(B_1) < 1$ and

$$|C_1|^k \leq B_1^k, \qquad k = 1, 2,\ldots .$$

Consequently, since **2.2.9** ensures that $\lim_{k \to \infty} B_1^k = 0$, it follows that $\lim_{k \to \infty} C_1^k = 0$. But then **2.2.9** also shows that $\rho(C_1) < 1$, and, therefore, that $\rho(C) < \sigma + \epsilon$. Hence, since ϵ was arbitrary, $\rho(C) \leq \sigma$. ∎

It is, in general, not true that the sum of M-matrices is again an M-matrix (**E 2.4-10**). The following result shows, however, that we may increase the diagonal elements arbitrarily, and also increase the off-diagonal elements as long as the correct sign pattern is retained.

2.4.10. Let $A_1 \in L(R^n)$ be an M-matrix with diagonal part D_1 and off-diagonal part $-B_1 = A_1 - D_1$. If $D_2 \in L(R^n)$ is any nonnegative diagonal matrix and $B_2 \in L(R^n)$ any nonnegative matrix with zero diagonal satisfying $B_2 \leq B_1$, then $A = D_1 + D_2 - (B_1 - B_2)$ is an M-matrix and $A^{-1} \leq A_1^{-1}$.

2.4. PARTIAL ORDERING AND NONNEGATIVE MATRICES

Proof. Set $D = D_1 + D_2$, $B = B_1 - B_2$, $H = D^{-1}B$, and $H_1 = D_1^{-1}B_1$. Then $D \geqslant D_1$ implies that $D^{-1} \leqslant D_1^{-1}$, and, clearly, $0 \leqslant B \leqslant B_1$. Hence, $0 \leqslant H \leqslant H_1$, so that, by **2.4.9**, $\rho(H) \leqslant \rho(H_1) < 1$, and then **2.4.8** ensures that A is an M-matrix. Since $A \geqslant A_1$, the inequality $A^{-1} \leqslant A_1^{-1}$ follows from multiplication by A^{-1} and A_1^{-1}. ∎

We state explicitly the following important special case of **2.4.10**.

2.4.11. Let $A \in L(R^n)$ be an M-matrix and let $D \in L(R^n)$ be any nonnegative diagonal matrix. Then $A + D$ is an M-matrix, and $(A + D)^{-1} \leqslant A^{-1}$.

As a corollary of **2.4.11**, we obtain, in turn, a result on Stieltjes matrices.

2.4.12. Let $A \in L(R^n)$ be a Stieltjes matrix. Then A is positive definite.

Proof. Suppose that A has an eigenvalue $\lambda \leqslant 0$. Then **2.4.11** shows that $A - \lambda I$ is an M-matrix, and this contradicts the singularity of $A - \lambda I$. ∎

Whereas **2.4.8** gives a characterization of M-matrices, it is useful to have more easily verified sufficient conditions. One such condition is a consequence of the following theorem, which is of interest by itself.

2.4.13. Let $B \in L(C^n)$ be irreducible. If

$$\sum_{j=1}^{n} |b_{ij}| \leqslant 1, \quad i = 1,\ldots, n, \tag{9}$$

and strict inequality holds for at least one i, then $\rho(B) < 1$.

Proof. Clearly, (9) implies that $\rho(B) \leqslant \|B\|_\infty \leqslant 1$. Now, suppose that $\rho(B) = 1$ and let λ be any eigenvalue of B with $|\lambda| = 1$. Then $\lambda I - B$ is singular but, by (9),

$$|\lambda - b_{ii}| \geqslant 1 - |b_{ii}| \geqslant \sum_{j \neq i} |b_{ij}|,$$

and strict inequality again holds for at least one i. Therefore, since $\lambda I - B$ is irreducible with B, **2.3.8** shows that $\lambda I - B$ is nonsingular. This is a contradiction, and we must have $\rho(B) < 1$. ∎

2.4.14. Let $A \in L(R^n)$ be strictly or irreducibly diagonally dominant and assume that $a_{ij} \leqslant 0$, $i \neq j$, and that $a_{ii} > 0$, $i = 1,\ldots, n$. Then A is an M-matrix.

Proof. Define $B \in L(R^n)$ by $B = I - D^{-1}A$ where, again, D is the diagonal part of A. Then, by **2.4.8**, it suffices to show that $\rho(B) < 1$. By the diagonal dominance of A, (9) holds for B, and, if A is strictly diagonally dominant, then strict inequality holds in (9) for all i. In this case $\|B\|_\infty < 1$. On the other hand, if A is irreducibly diagonally dominant, then the result follows immediately from **2.4.13**. ∎

We note that diagonal dominance of A is not a necessary condition that A be an M-matrix (**E 2.4-9**).

We complete this section with a result similar to **2.4.13**. These theorems on spectral radii will be of great importance in later chapters in the convergence study of iterative processes. We begin with the following terminology.

2.4.15. Definition. Let $A, B \in L(R^n)$. Then $A = B - C$ is a *regular splitting* of A if B is invertible, $B^{-1} \geqslant 0$, and $C \geqslant 0$; it is a *weak regular splitting* if the condition $C \geqslant 0$ is replaced by $B^{-1}C \geqslant 0$ and $CB^{-1} \geqslant 0$. ∎

Clearly, a regular splitting is a weak regular splitting, but the converse is not true (**E 2.4-11**).

Before proving the spectral radius result, we show that there is a close connection between weak regular splittings and nonnegative subinverses.

2.4.16. Assume that $B \in L(R^n)$ is invertible and $B^{-1} \geqslant 0$. Then B^{-1} is a subinverse of $A \in L(R^n)$ if and only if $A = B - C$ is a weak regular splitting.

Proof. Let $A = B - C$ be a weak regular splitting of A. Then $0 \leqslant B^{-1}C = B^{-1}(B - A) = I - B^{-1}A$, and thus $B^{-1}A \leqslant I$. Similarly, we obtain, from $CB^{-1} \geqslant 0$, that $AB^{-1} \leqslant I$, and, hence, B^{-1} is a subinverse of A. Conversely, if $B^{-1} \geqslant 0$ is a subinverse, then $0 \leqslant I - B^{-1}A = B^{-1}(B - A) = B^{-1}C$, and, in a similar fashion, it follows that $CB^{-1} \geqslant 0$. ∎

2.4.17. Let $A \in L(R^n)$ and suppose that $A = B - C$ is a weak regular splitting. Then $\rho(B^{-1}C) < 1$ if and only if A^{-1} exists and is nonnegative.

Proof. Set $H = B^{-1}C$. Then $H \geqslant 0$, and, by the relations

$$(I + H + \cdots + H^m)(I - H) = I - H^{m+1}, \quad B^{-1} = (I - H)A^{-1}, \quad (10)$$

we have, since $A^{-1} \geqslant 0$,

$$0 \leqslant (I + \cdots + H^m)B^{-1} = (I - H^{m+1})A^{-1} \leqslant A^{-1}, \quad \forall m \geqslant 0.$$

2.4. PARTIAL ORDERING AND NONNEGATIVE MATRICES

Clearly, B^{-1} must contain at least one positive element in each row, and it follows that $I + \cdots + H^m$ is bounded above for all m. Therefore, since $H \geqslant 0$, the sum converges, so that $\lim_{k \to \infty} H^k = 0$. Theorem **2.2.9** now shows that $\rho(H) < 1$. Conversely, if $\rho(H) < 1$, then **2.4.5** ensures that $(I - H)^{-1} \geqslant 0$. It follows that A^{-1} exists, and

$$A^{-1} = (I - H)^{-1} B^{-1} \geqslant 0. \quad \blacksquare$$

NOTES AND REMARKS

NR 2.4-1. Most of the results of this section may be found in Varga [1962] or Householder [1964], although the proofs given there use the deeper ideas of the Perron–Frobenius theory of nonnegative matrices. At the same time, by the use of this theory, several of the theorems presented here may be sharpened if irreducibility is also assumed. For example, if in **2.4.9** the matrix B is irreducible, then it can be shown that $\rho(C) = \rho(B)$ only if $|C| = B$. This, in turn, yields the frequently useful result that $\rho(B) < \rho(B + E)$ whenever $B \geqslant 0$ is irreducible and $E \geqslant 0$ is not zero. In a somewhat different vein, the irreducibility assumption in **2.4.14** allows the stronger conclusion that $A^{-1} > 0$, that is, all elements of A^{-1} are positive. Similarly, if B is irreducible in **2.4.5**, then $(I - B)^{-1} > 0$.

NR 2.4-2. The idea of a weak regular splitting as a generalization of the regular splitting of Varga, was introduced in Ortega and Rheinboldt [1967a] in order to deal with certain problems to be discussed in Chapter 13.

NR 2.4-3. Any relation $x \leqslant y$ defined between certain pairs of vectors $x, y \in R^n$ is a *partial ordering* on R^n if it satisfies conditions (a), (b), and (c) of **2.4.1**. The partial ordering is a *linear partial ordering* on R^n if, in addition, (d) and (e) of **2.4.1** hold. Thus, **2.4.1** shows that the natural ordering (1) is a linear partial ordering on R^n. Another such linear partial ordering is

$$x \leqslant y \quad \text{if and only if} \quad (Cx)_i \leqslant (Cy)_i, \quad i = 1, \ldots, n, \quad (11)$$

where $C \in L(R^n)$ is any nonsingular matrix (**E 2.4-4**). Several of the results of **2.4** can be extended to general linear partial orderings. (See, for example, Vandergraft [1968].)

EXERCISES

E 2.4-1. Verify the conclusions of **2.4.1** and **2.4.2**.

E 2.4-2. Show that (3) and (4) are equivalent and that any l_p-norm satisfies (3).

E 2.4-3. Let $A \in L(R^n)$. Show that A is a nonnegative matrix if and only if it is an isotone operator.

E 2.4-4. Let $C \in L(R^n)$ be nonsingular. Show that (11) of **NR 2.4-3** defines a linear partial ordering on R^n in the sense that **2.4.1** is satisfied.

E 2.4-5. Let $D \in L(R^n)$ be diagonal, nonsingular, and nonnegative. Show that $\|x\| \equiv \|Dx\|_\infty$ is a monotonic norm on R^n.

E 2.4-6. Let $A \in L(R^n)$. Show that A is invertible and $A^{-1} \geqslant 0$ if and only if there exist nonsingular, nonnegative matrices $P, Q \in L(R^n)$ such that $PAQ = I$. (Bramble and Hubbard [1964].)

E 2.4-7. Let $A \in L(R^n)$. Then $A^{-1} \geqslant 0$ if and only if there is a $B \in L(R^n)$ such that $C = A + B$ satisfies: (a) $C^{-1} \geqslant 0$; (b) $C^{-1}B \geqslant 0$; and (c) $\rho(C^{-1}B) < 1$. (Price [1968].)

E 2.4-8. Use **2.4.10** to conclude that if $A \in L(R^n)$ is an M-matrix and C is any matrix obtained from A by setting certain off-diagonal elements to zero, then C is an M-matrix and $C^{-1} \leqslant A^{-1}$.

E 2.4-9. Give an example of a 2×2 M-matrix which is not irreducible or diagonally dominant.

E 2.4-10. Give an example of two 2×2 M-matrices whose sum is not an M-matrix.

E 2.4-11. Let $C = B - A$, where

$$A = \begin{pmatrix} 2 & -1 & 0 \\ 0 & 1 & -1 \\ -1 & 0 & 1 \end{pmatrix}, \quad B = \begin{pmatrix} 2 & -2 & 2 \\ 0 & 2 & -2 \\ -1 & 0 & 1 \end{pmatrix}.$$

Show that A is an M-matrix and that $A = B - C$ is a weak regular splitting of A but not a regular splitting.

E 2.4-12. Let $A_k \in L(R^n)$, $k = 1, 2$, have elements $a_{ij}^{(k)}$, with $a_{ij}^{(k)} \leqslant 0$, $i \neq j$, $a_{ii}^{(k)} \geqslant 0$, and

$$a_{ii}^{(2)} a_{ij}^{(1)} \leqslant a_{ii}^{(1)} a_{ij}^{(2)}, \quad i, j = 1, \ldots, n; \quad i \neq j.$$

Assume that A_1 is irreducibly diagonally dominant. Show that

$$0 \leqslant (A_1 + A_2)^{-1} \leqslant A_1^{-1}.$$

Chapter 3 / ANALYSIS

3.1. DERIVATIVES AND OTHER BASIC CONCEPTS

We shall review in this chapter certain basic parts of n-dimensional calculus and, in particular, the theory of Gateaux and Frechet derivatives.

Recall that a real-valued function f of a single variable is differentiable at a point x if there is a real number $a = f'(x)$ such that

$$\lim_{t \to 0} (1/t)[f(x+t) - f(x) - at] = 0.$$

This definition extends in a natural way to n dimensions.

3.1.1. Definition. *A mapping $F: D \subset R^n \to R^m$ is Gateaux- (or G-) differentiable at an interior point x of D if there exists a linear operator $A \in L(R^n, R^m)$ such that, for any $h \in R^n$,*

$$\lim_{t \to 0} (1/t)\|F(x+th) - Fx - tAh\| = 0. \quad \blacksquare \qquad (1)$$

Note that, by **2.2.1**, the limit in (1) is independent of the particular norm on R^m; that is, if F is G-differentiable at x in some norm, then it is G-differentiable at x in any norm. Note also that **3.1.1** contains as a special case the definition of differentiability in one dimension. Finally, we stress that we define the derivative only at interior points of D, and, in subsequent results, the statement that F is G-differentiable on a set $D_0 \subset D$ implies that $D_0 \subset \text{int}(D)$.

As in the one-dimensional case, there is at most one linear operator A

for which (1) is satisfied. Indeed, if (1) holds for A_1 and A_2, then, for arbitrary $h \in R^n$ and suitably small $t > 0$,

$$\|(A_1 - A_2) h \| \leq t^{-1} \|F(x + th) - Fx - tA_1 h\| + t^{-1} \|F(x + th) - Fx - tA_2 h\|. \tag{2}$$

But since the right-hand side tends to zero as $t \to 0$, we must have $\|(A_1 - A_2) h\| = 0$, or $A_1 = A_2$, since h was arbitrary.

3.1.2. Definition. If $F: D \subset R^n \to R^m$ is G-differentiable at $x \in \text{int}(D)$, then the unique linear operator $A \in L(R^n, R^m)$ for which (1) holds is denoted by $F'(x)$ and called the *G-derivative* of F at x. ∎

If F is G-differentiable at each point of a subset $D_0 \subset D$, then, for each $x \in D_0$, $F'(x)$ is a linear operator; that is, F' is a mapping from D_0 into $L(R^n, R^m)$. In particular, F' is continuous at $x \in D_0$ if $\|F'(x + h) - F'(x)\| \to 0$ as $\|h\| \to 0$.

We consider next the concrete representation of $F'(x)$ in terms of the partial derivatives of the component functions $f_1, \ldots, f_m$ of F. If $A = (a_{ij})$, and h is chosen as the jth coordinate vector e^j, then clearly (1) implies that

$$\lim_{t \to 0} (1/t) | f_i(x + te^j) - f_i(x) - ta_{ij} | = 0,$$

which shows that the partial derivatives of each f_i exist at x and that

$$\partial_j f_i(x) \equiv \partial f_i(x)/\partial x_j = a_{ij}, \qquad i, j = 1, \ldots, n.$$

Hence, the matrix representation of $F'(x)$ is given by the *Jacobian matrix*:

$$F'(x) = \begin{pmatrix} \partial_1 f_1(x) & \cdots & \partial_n f_1(x) \\ \vdots & \ddots & \vdots \\ \partial_1 f_m(x) & \cdots & \partial_n f_m(x) \end{pmatrix}. \tag{3}$$

In the special case when $g: D \subset R^n \to R^1$ is a functional, $g'(x)$ is represented by the row vector $g'(x) = (\partial_1 g(x), \ldots, \partial_n g(x))$, and the column vector $g'(x)^\mathrm{T}$ is called the *gradient* of g at x.

It is of importance to note that the existence of the Jacobian matrix, that is, the existence of all partial derivatives, does not imply that F is G-differentiable (see **E 3.1-1**). More generally, not even the existence of the limit

$$\lim_{t \to 0} (1/t)[F(x + th) - Fx],$$

3.1. DERIVATIVES AND OTHER BASIC CONCEPTS

for *all* $h \in R^n$ implies that F has a G-derivative at x (see **E 3.1-5** and **NR 3.1-4**).

Many properties of differentiation in one variable carry over to the G-derivative. For example, if $F_1 : D_1 \subset R^n \to R^m$ and $F_2 : D_2 \subset R^n \to R^m$ both have a G-derivative at $x \in D_1 \cap D_2$ then, for any scalars α and β, $\alpha F_1 + \beta F_2$ also has a G-derivative at x, and

$$(\alpha F_1 + \beta F_2)'(x) = \alpha F_1'(x) + \beta F_2'(x) \tag{4}$$

(see **E 3.1-8**). On the other hand, many desirable properties do not carry over from one dimension. Indeed, the existence of a G-derivative at x does not imply that F is even continuous at x (see **E 3.1-4**). It can be proved, however, that F is continuous as y tends to x along straight lines.

3.1.3. Definition. A mapping $F: D \subset R^n \to R^m$ is *hemicontinuous* at $x \in D$ if, for any $h \in R^n$ and $\epsilon > 0$, there is a $\delta = \delta(\epsilon, h)$ so that whenever $|t| < \delta$ and $x + th \in D$, then $\|F(x + th) - Fx\| < \epsilon$. ∎

Now, if F is G-differentiable at x, then, for any fixed $h \in R^n$, the mapping $G(t) = F(x + th)$, defined for those t for which $x + th \in D$, is differentiable at 0, and

$$G'(0) = \lim_{t \to 0}(1/t)[G(t) - G(0)] = F'(x)h.$$

Consequently, G is continuous at 0, and we have proved:

3.1.4. If $F: D \subset R^n \to R^m$ is G-differentiable at $x \in D$, then F is hemicontinuous at x.

In order to overcome the lack of many desirable properties in the G-derivative, we consider next a stronger form of differentiation.

3.1.5. Definition. The mapping $F: D \subset R^n \to R^m$ is *Frechet-* (or *F-*) *differentiable* at $x \in \text{int}(D)$ if there is an $A \in L(R^n, R^m)$ such that

$$\lim_{h \to 0}(1/\|h\|)\|F(x + h) - Fx - Ah\| = 0. \tag{5}$$

The linear operator A is again denoted by $F'(x)$, and is called the *F-derivative* of F at x. ∎

Note that the limit (5) contains the special case (1); that is, F is G-differentiable at x whenever it is F-differentiable at x. It follows that any property of the G-derivative holds automatically for the F-derivative; in particular, the F-derivative is unique and its concrete representation is again the Jacobian matrix (3). We shall use $F'(x)$ to denote both the G- and F-derivatives, but explicit assumptions will always be made as to which derivative is being used. The exercise **E 3.1-6** gives an example of a mapping which has a G-derivative, but not an F-derivative.

The condition (5) is a uniformity condition which, when applied to a functional $f: D \subset R^n \to R^1$, ensures that f has a tangent plane at x in the usual geometric sense. It is this uniformity condition which allows most of the usual properties of derivatives in one dimension to be carried over to n dimensions. For example, in contrast to the rather weak result **3.1.4** for G-derivatives, we now have:

3.1.6. If $F: D \subset R^n \to R^m$ is F-differentiable at x, then F is continuous at x. More precisely, there is a $\delta > 0$ and a $c \geqslant 0$ such that $\bar{S}(x, \delta) \subset D$ and

$$\| F(x + h) - Fx \| \leqslant c \| h \| \quad \text{whenever} \quad \| h \| \leqslant \delta. \tag{6}$$

Proof. Since $x \in \text{int}(D)$, there is a $\delta_1 > 0$ so that $x + h \in D$ whenever $\| h \| < \delta_1$. Then, given $\epsilon > 0$, it follows by (5) that there is a $0 < \delta \leqslant \delta_1$ such that

$$\| F(x + h) - Fx - F'(x) h \| \leqslant \epsilon \| h \|$$

whenever $\| h \| \leqslant \delta$. Hence, with $c = \epsilon + \| F'(x) \|$, (6) holds by the triangle inequality. ∎

In the next section, we shall develop various analogs of the mean-value theorem and, by means of them, obtain several properties of both the F- and the G-derivatives; here, we consider one of the most important properties—the chain rule. As usual, for mappings $F: D_F \subset R^n \to R^m$ and $G: D_G \subset R^m \to R^p$, the *composite mapping* $H = G \cdot F$ is defined by $Hx = G(Fx)$ for all $x \in D_H = \{x \in D_F \mid Fx \in D_G\}$.

3.1.7. Chain Rule. If $F: D_F \subset R^n \to R^m$ has a G-derivative at x and $G: D_G \subset R^m \to R^p$ has an F-derivative at Fx, then the composite mapping $H = G \cdot F$ has a G-derivative at x, and

$$H'(x) = G'(Fx) F'(x). \tag{7}$$

If, in addition, $F'(x)$ is an F-derivative, then $H'(x)$ is an F-derivative.

Proof. Let $h \in R^n$ be fixed. By definition, we have $x \in \text{int}(D_F)$ and $Fx \in \text{int}(D_G)$; hence, since by **3.1.4** F is hemicontinuous at x, there is

3.1. DERIVATIVES AND OTHER BASIC CONCEPTS

a $\delta > 0$ such that $x + th \in D_F$ and $F(x + th) \in D_G$ for $|t| < \delta$. Therefore, for $0 < |t| < \delta$,

$$(1/|t|)\| H(x + th) - Hx - tG'(Fx) F'(x) h \|$$
$$\leqslant (1/|t|)\| G(F(x + th)) - G(Fx) - G'(Fx)[F(x + th) - Fx]\|$$
$$+ (1/|t|)\| G'(Fx)[F(x + th) - Fx - tF'(x) h]\|. \tag{8}$$

Now, clearly, the second term of (8) tends to zero as $t \to 0$, since F is G-differentiable. For any values of $0 < |t| < \delta$ such that $F(x + th) \neq Fx$, the first term may be both multiplied and divided by $\| F(x + th) - Fx \|$. But $\| F(x + th) - Fx \| \to 0$, since F is hemicontinuous at x while $(1/|t|) \| F(x + th) - Fx \|$ is bounded. Consequently, since G is F-differentiable at Fx, the first term of (8) tends to zero. Hence, the left-hand of (8) tends to zero which was to be shown. An analogous argument, using **3.1.6**, applies when $F'(x)$ is an F-derivative. ∎

Note that when G has only a G-derivative at x, then H need not be G-differentiable (see **E 3.1-7**).

We conclude this section with a discussion of a number of other important analytical concepts.

A mapping $F: D \subset R^n \to R^m$ is *one-to-one* on $U \subset D$ if $Fx \neq Fy$ whenever $x, y \in U$, $x \neq y$. Then the *restriction* $F_U : U \subset R^n \to R^m$ of F to U, defined by $F_U x = Fx$ for all $x \in U$, has an inverse F_U^{-1} on $F(U)$; that is, there is a mapping $F_U^{-1}: F(U) \to U$ for which $x = F_U^{-1}(F_U x)$ for all $x \in U$ and $y = F_U(F_U^{-1} y)$ for all $y \in F(U)$. In particular, if $U = D$, then F itself has an inverse F^{-1} defined on $F(D)$.

3.1.8. Definition. A mapping $F: D \subset R^n \to R^n$ is a *homeomorphism* of D onto $F(D)$ if F is one-to-one on D and F and F^{-1} are continuous on D and $F(D)$, respectively. ∎

In many of our considerations, we shall wish to single out different types of continuities.

3.1.9. Definition. A mapping $F: D \subset R^n \to R^m$ is *Hölder-continuous* on $D_0 \subset D$ if there exist constants $c \geqslant 0$ and $p \in (0, 1]$ so that, for all $x, y \in D$,

$$\| Fy - Fx \| \leqslant c \| y - x \|^p. \tag{9}$$

If $p = 1$, then F is *Lipschitz-continuous* on D_0. ∎

Note that if F is G-differentiable on D_0, then **3.1.9**, applied to the mapping $F'\colon D_0 \to L(R^n, R^m)$, states that the derivative is Hölder-continuous on D_0 if

$$\|F'(y) - F'(x)\| \leq c \|y - x\|^p, \qquad \forall x, y \in D_0.$$

We shall also occasionally apply **3.1.9** only at a point and say that F is Hölder-continuous at x if (9) holds for all y in a neighborhood of x. Note that, by **3.1.6**, an F-differentiable function at a point x is Lipschitz-continuous at that point.

If F is Hölder-continuous on D_0, then the quantity

$$\omega(t) = \sup\{\|Fx - Fy\| \mid x, y \in D_0, \ \|x - y\| \leq t\}, \tag{10}$$

is well defined and bounded by ct^p. More generally, whenever F is uniformly continuous on D_0, it follows immediately that the supremum in (10) is finite for all t in some interval $[0, a)$, $a > 0$. (Note, however, that if F is only continuous, this is not true; see **E 3.1-12**).

3.1.10. Definition. Assume that $F\colon D \subset R^n \to R^m$ is uniformly continuous on $D_0 \subset D$. Then the function ω defined by (10) for all $t \geq 0$ for which $\omega(t) < +\infty$, is the *modulus of continuity* of F on D_0. ∎

Clearly, ω is an isotone function, and $\omega(0) = 0$. It will be useful on occasion to know that ω is also defined and continuous on all of $[0, \infty)$; a convenient sufficient condition for this is given by the following condition on D_0.

3.1.11. Assume that $F\colon D \subset R^n \to R^m$ is uniformly continuous on a convex set $D_0 \subset D$. Then the modulus of continuity of F on D_0 is defined and uniformly continuous on $[0, \infty)$.

Proof. Let $t > 0$ be given and choose $\delta > 0$ so that $\|Fx - Fy\| \leq t$ whenever $\|x - y\| \leq \delta$. Now, choose m so that $\delta \leq t/m$. Then, for any $x, y \in D_0$ with $\|x - y\| \leq t$, the points $x^k = x + k\delta(y - x)$, $k = 0, \ldots, m-1$, $x^m = y$, all lie in D_0 and satisfy $\|x^k - x^{k-1}\| \leq \delta$, $k = 1, \ldots, m$. Hence,

$$\|Fx - Fy\| \leq \sum_{k=1}^{m} \|Fx^k - Fx^{k-1}\| \leq mt,$$

so that $\omega(t) < +\infty$ for all $t \in [0, \infty)$.

To prove the continuity, let t_1 and t_2 be any two points of $[0, \infty)$, not both zero, and, for arbitrary $x, y \in D_0$ with $\|x - y\| \leq t_1 + t_2$, set

3.1. DERIVATIVES AND OTHER BASIC CONCEPTS

$z = x + [t_1/(t_1 + t_2)](y - x)$. Then $z \in D_0$, $\|x - z\| \leq t_1$, and $\|y - z\| \leq t_2$, so that

$$\|Fx - Fy\| \leq \|Fx - Fz\| + \|Fz - Fy\| \leq \omega(t_1) + \omega(t_2).$$

Therefore,

$$\omega(t_1 + t_2) \leq \omega(t_1) + \omega(t_2), \qquad \forall t_1, t_2 \in [0, \infty).$$

It follows that

$$|\omega(t_2) - \omega(t_1)| \leq \omega(|t_2 - t_1|), \qquad \forall t_1, t_2 \in [0, \infty) \tag{11}$$

and, since the uniform continuity of F implies immediately that ω is continuous at 0, (11) proves the uniform continuity of ω. ∎

We note that the conclusions of **3.1.11** do not remain true, in general, if the convexity assumption for D_0 is removed (see **E 3.1-13**). However, a sufficient condition for ω to be at least defined on all of $[0, \infty)$ is that D_0 be compact (see **E 3.1-14**).

NOTES AND REMARKS

NR 3.1-1. Most of the material of this and the next two sections will be found in a number of advanced calculus books; see, for example, Apostol [1957].

NR 3.1-2. The definitions and results of this section go over almost verbatim to Banach spaces. That is, if $F: D \subset X \to Y$, where X and Y are Banach spaces, then the G- and F-derivatives are defined as in **3.1.1, 3.1.2**, and **3.1.5**, where A is a linear operator from X to Y. (Most authors assume that A is also bounded.) The results **3.1.4, 3.1.6**, and **3.1.7**, together with their proofs, are valid in this more general setting. For a discussion of the calculus in Banach spaces, we refer, for example, to Dieudonné [1960], Kantorovich and Akilov [1959], and Vainberg [1956].

NR 3.1-3. A more general approach to differentiation starts with the idea of a *differential*. If $F: D \subset R^n \to R^m$ and if for some $x \in \text{int}(D)$, and $h \in R^n$, the limit

$$\lim_{t \to 0}(1/t)[F(x + th) - Fx] = V(x, h)$$

exists, then F is said to have a *Gateaux-differential* at x in the direction h. If $V(x, h)$ exists for all $h \in R^n$ and is linear in h, that is, if $V(x, h) = A(x) h$, where $A(x) \in L(R^n, R^m)$, then, clearly, $A(x) = F'(x)$ is the G-derivative of F at x. However, the Gateaux differential may exist at x for all $h \in R^n$, and yet there is no G-derivative (see **E 3.1-5**).

If the G-differential exists at x for all h and if, in addition,

$$\lim_{h\to 0}(1/\|h\|)\|F(x+h) - Fx - V(x,h)\| = 0, \qquad (12)$$

then F has a *Frechet-differential* at x. The relation between the various concepts is shown schematically in Fig. 3.1, where "uniform in h" indicates the validity

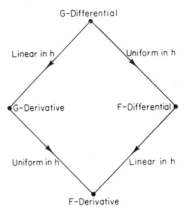

FIGURE 3.1

of (12). We note that if the G-differential $V(x, h)$ exists for all x in an open neighborhood of a point $x^0 \in \text{int}(D)$ and for all $h \in R^n$, then F has an F-derivative at x^0 provided that, for each fixed h, $V(x, h)$ is continuous in x at x^0. The proof of this result as well as a further discussion of differentials may be found, for example, in Vainberg [1956].

EXERCISES

E 3.1-1. Define $f: R^2 \to R^1$ by $f(x) = x_1$ if $x_2 = 0$, $f(x) = x_2$, if $x_1 = 0$, and $f(x) = 1$ otherwise. Show that the partial derivatives $\partial_1 f(0)$ and $\partial_2 f(0)$ exist, but that f does not have a G-derivative at 0.

E 3.1-2. Let $A \in L(R^n)$ and define $f: R^n \to R^1$ by $f(x) = x^T A x$. Show that f has an F-derivative at each $x \in R^n$ and compute $f'(x)$.

E 3.1-3. Define $f: R^2 \to R^1$ by $f(0) = 0$ and $f(x) = x_1 x_2^2/(x_1^2 + x_2^4)$ otherwise. Show that $\lim_{t\to 0}(1/t)[f(th) - f(0)]$ exists for all $h \in R^2$ but that f does not have a G-derivative at 0. Show, in addition, that f is not continuous at 0.

E 3.1-4. Define $f: R^2 \to R^1$ by $f(x) = 0$ if $x_1 = 0$ and

$$f(x) = 2x_2 \exp(-x_1^{-2})/(x_2^2 + \exp(-2x_1^{-2})), \qquad \text{if } x_1 \neq 0.$$

Show that f has a G-derivative at 0, but that f is not continuous at zero.

3.1. DERIVATIVES AND OTHER BASIC CONCEPTS

E 3.1-5. Define $f: R^2 \to R^1$ by
$$f(x) = \operatorname{sgn}(x_2) \min(|x_1|, |x_2|).$$
Show that, for any $h \in R^2$,
$$V(0, h) \equiv \lim_{t \to 0}(1/t)[f(th) - f(0)] = f(h),$$
but that f does not have a G-derivative at 0.

E 3.1-6. Define $f: R^2 \to R^1$ by $f(0) = 0$ if $x = 0$ and
$$f(x) = x_2(x_1^2 + x_2^2)^{3/2}/[(x_1^2 + x_2^2)^2 + x_2^2], \quad \text{if } x \neq 0.$$
Show that f has a G-derivative at 0, but not an F-derivative. Show, moreover, that the G-derivative is hemicontinuous at 0.

E 3.1-7. Define $f: R^2 \to R^1$ as in **E 3.1-6** and $G: R^2 \to R^2$ by $Gx = (x_1, x_2^2)^T$. Show that the composite mapping $f \cdot G$ does not have a G-derivative at 0.

E 3.1-8. Verify (4) for both G- and F-derivatives.

E 3.1-9. Complete the details of the proof of **3.1.7** when $F'(x)$ is an F-derivative.

E 3.1-10. Let $f(x) = \|x\|_p$ be the l_p-norm for some $p \in [1, \infty)$. Compute $\partial_i f(x)$ for any x with no component $x_j = 0$. Does f have an F- or G- derivative for such x? What more can be said if $p = 2$?

E 3.1-11. Suppose that $F: R^n \to R^n$ is G-differentiable in an open neighborhood of x^0 and F' is continuous at x^0. If $F'(x^0)$ is invertible, show that there is a $\delta > 0$ so that $F'(x)$ is invertible and $F'(x)^{-1}$ is bounded for all $x \in \bar{S}(x^0, \delta)$.

E 3.1-12. Consider the functions $F: D_0 \subset R^1 \to R^1$: (a) $D_0 = (0, 1)$, $Fx = x^{-1}$; (b) $D_0 = (0, \infty)$, $Fx = x^2$. Show that, in either case, the supremum in (10) is $+\infty$ for any $t > 0$.

E 3.1-13. Let $D_0 \subset R^1$ be the set $D_0 = \bigcup_{k=0}^{\infty}[2k, 2k+1]$ and define $F: D_0 \to R^1$ by $Fx = k^2$ if $x \in [2k, 2k+1]$. Show that F is uniformly continuous on D_0, but that the supremum in (10) is $+\infty$ for all $t > 1$.

E 3.1-14. Suppose that $F: D \subset R^n \to R^m$ is continuous on a compact set $D_0 \subset D$. Show that the modulus of continuity ω of F in D_0 is defined and bounded on $[0, \infty)$.

E 3.1-15. The tensor product of two matrices $A, B \in L(R^n)$ is defined as the $n^2 \times n^2$ matrix $A \times B = (a_{ij}B \mid i, j = 1, \ldots, n)$, where $A = (a_{ij})$. Consider two F-differentiable mappings $H, K: L(R^n) \to L(R^n)$ and set $F(X) = H(X) K(X)$ for all $X \in L(R^n)$. Show that
$$F'(X) = [H(X) \times I] K'(X) + [I \times K(X)^T] H'(X), \quad \forall X \in L(R^n).$$

3.2. MEAN-VALUE THEOREMS

The most frequently used results about derivatives will be mean-value theorems, and, in this section, we shall collect various results of this type, along with some applications. Throughout, we shall use the notation $[x, y]$, $x, y \in R^n$, to denote the *closed interval* $[x, y] = \{z \mid z = tx + (1 - t)y, 0 \leqslant t \leqslant 1\}$.

We begin by recalling the standard mean-value theorem for functions of a single variable.

3.2.1. If $\varphi: [a, b] \subset R^1 \to R^1$ is continuous on $[a, b]$ and differentiable on (a, b), then there is a $t \in (a, b)$ such that $\varphi(b) - \varphi(a) = \varphi'(t)(b - a)$.

As an immediate consequence of this one-dimensional result, we have the corresponding result for functionals.

3.2.2. Suppose that $f: D \subset R^n \to R^1$ is G-differentiable at each point of a convex set $D_0 \subset D$. Then, for any two points $x, y \in D_0$, there is a $t \in (0, 1)$ such that

$$f(y) - f(x) = f'(x + t(y - x))(y - x). \tag{1}$$

Proof. For given $x, y \in D_0$, it follows immediately that the function $\varphi(s) = f(x + s(y - x))$ is differentiable, and therefore continuous, on $[0, 1]$, and that

$$\varphi'(s) = f'(x + s(y - x))(y - x), \quad \forall s \in [0, 1].$$

Hence, by **3.2.1**,

$$f(y) - f(x) = \varphi(1) - \varphi(0) = f'(x + t(y - x))(y - x),$$

for some $t \in (0, 1)$. ∎

It is important to note that **3.2.2** does *not*, in general, hold for mappings $F: R^n \to R^m$, $m > 1$ (see **E 3.2-1**). There are, however, three alternatives, each of which is useful in certain circumstances. The first is an immediate consequence of **3.2.2** applied to each component functional of F. Thus, if $F: D \subset R^n \to R^m$ is G-differentiable on an open convex set $D_0 \subset D$ and $x, y \in D_0$, then

$$Fy - Fx = B(x, y)(y - x), \tag{2}$$

3.2. MEAN-VALUE THEOREMS

where $B(x, y) \in L(R^n, R^m)$ is constructed from the components $f_1, \ldots, f_m$ of F by 3.2.2. More precisely, there exist $t_1, \ldots, t_m \in (0, 1)$, such that

$$B(x, y) = \begin{pmatrix} f_1'(x + t_1(y - x)) \\ \vdots \\ f_m'(x + t_m(y - x)) \end{pmatrix}. \tag{3}$$

Note that, in general, the t_i will all be distinct and $B(x, y)$ will *not* be the G-derivative evaluated at an intermediate point.

The second alternative provides an upper bound for $Fy - Fx$ in terms of F'.

3.2.3. Assume that $F: D \subset R^n \to R^m$ is G-differentiable on a convex set $D_0 \subset D$. Then, for any $x, y \in D_0$,

$$\| Fy - Fx \| \leq \sup_{0 \leq t \leq 1} \| F'(x + t(y - x)) \| \, \| x - y \|. \tag{4}$$

Proof. Assume that $M = \sup_{0 \leq t \leq 1} \| F'(x + t(y - x)) \| < \infty$ and, for given $\epsilon > 0$, let Γ be the set of $t \in [0, 1]$ for which

$$\| F(x + t(y - x)) - Fx \| \leq Mt \| y - x \| + \epsilon t \| y - x \| \tag{5}$$

holds. Clearly, $0 \in \Gamma$, so that $\gamma = \sup_{t \in \Gamma} t$ is well defined, and, since 3.1.4 implies that $F(x + t(y - x))$ is continuous in t, we have

$$\| F(x + \gamma(y - x)) - Fx \| \leq M\gamma \| y - x \| + \epsilon \gamma \| y - x \|. \tag{6}$$

Since ϵ is arbitrary, clearly the result is proved if $\gamma = 1$.

Suppose that $\gamma < 1$. Then, since F' exists at $x + \gamma(y - x)$, there is a $\beta \in (\gamma, 1)$ such that

$$\| F(x + \beta(y - x)) - F(x + \gamma(y - x)) - F'(x + \gamma(y - x))(\beta - \gamma)(y - x) \|$$
$$\leq \epsilon(\beta - \gamma)\| y - x \|,$$

and hence

$$\| F(x + \beta(y - x)) - F(x + \gamma(y - x)) \|$$
$$\leq M(\beta - \gamma)\| y - x \| + \epsilon(\beta - \gamma)\| y - x \|.$$

But then, by (6), we have

$$\| F(x + \beta(y - x)) - Fx \| \leq (M\gamma + \epsilon\gamma)\| y - x \| + (M + \epsilon)(\beta - \gamma)\| y - x \|$$
$$= (M + \epsilon)\beta \| y - x \|;$$

that is, (5) holds for $1 > \beta > \gamma$, in contradiction to the definition of γ. ∎

Two immediate and useful corollaries of **3.2.3** are:

3.2.4. If $F: D \subset R^n \to R^m$ is G-differentiable on the convex set $D_0 \subset D$, and $\|F'(x)\| \leq M < +\infty$ for all $x \in D_0$, then F is Lipschitz-continuous on D_0.

3.2.5. If $F: D \subset R^n \to R^m$ is G-differentiable on the convex set $D_0 \subset D$, then, for any $x, y, z \in D_0$,

$$\|Fy - Fz - F'(x)(y - z)\|$$
$$\leq \sup_{0 \leq t \leq 1} \|F'(z + t(y - z)) - F'(x)\| \|y - z\| \qquad (7)$$

Proof. For fixed $x \in D_0$, define the mapping $Gw = Fw - F'(x)w$, $w \in D$. Clearly, the conditions of **3.2.3** are satisfied for G and, since $G'(w) = F'(w) - F'(x)$, (7) is precisely the relation

$$\|Gy - Gz\| \leq \sup_{0 \leq t \leq 1} \|G'(z + t(y - z))\| \|y - z\|. \quad \blacksquare$$

The third approach to mean-value theorems is based on the fundamental theorem of integral calculus:

3.2.6. If $f: [a, b] \subset R^1 \to R^1$ is continuous on $[a, b]$, and if f' exists and is Riemann-integrable on (a, b), then

$$\int_a^b f'(t)\, dt = f(b) - f(a).$$

For a mapping $G: [a, b] \subset R^1 \to R^m$, we define the integral of G in terms of its components $g_1, \ldots, g_m$ by

$$\int_a^b G(t)\, dt = \begin{pmatrix} \int_a^b g_1(t)\, dt \\ \vdots \\ \int_a^b g_m(t)\, dt \end{pmatrix} \qquad (8)$$

and say that G is Riemann-integrable on $[a, b]$ if each component is Riemann-integrable. Now, if $F: D \subset R^n \to R^m$ has a G-derivative at each point of an interval $[x, y] \subset D$, then, by **3.1.4**, each of the component functions $f_i(x + t(y - x))$ is continuous for $t \in [0, 1]$; therefore, if the derivatives $f_i'(x + t(y - x))(y - x)$ are Riemann-integrable in t on $[0, 1]$, then, by **3.2.6**,

$$f_i(y) - f_i(x) = \int_0^1 f_i'(x + t(y - x))(y - x)\, dt, \qquad i = 1, \ldots, n. \qquad (9)$$

3.2. MEAN-VALUE THEOREMS

Hence, using the definition (8), we may write (9) as

$$Fy - Fx = \int_0^1 F'(x + t(y - x))(y - x)\, dt. \tag{10}$$

A sufficient condition that f' is integrable on $[a, b]$ is that f' is continuous on $[a, b]$. Hence, a sufficient condition that (10) holds is that $F'(x + t(y - x))$ be continuous in t on $[0, 1]$. Since the definition 3.1.3 of hemicontinuity may, of course, be applied to the mapping F': $R^n \to L(R^n, R^m)$, we have:

3.2.7. If $F: D \subset R^n \to R^m$ has a G-derivative at each point of a convex set $D_0 \subset D$, and F' is hemicontinuous on D_0, then, for any $x, y \in D_0$, (10) holds.

We next examine some simple consequences of the mean-value theorems. The first is that continuity of F' at x implies that $F'(x)$ is an F-derivative.

3.2.8. If $F: D \subset R^n \to R^m$ has a G-derivative at each point of an open neighborhood of x, and F' is continuous at x, then F is F-differentiable at x.

Proof. For given $\epsilon > 0$, there is a $\delta > 0$ such that $\|F'(x + h) - F'(x)\| \leq \epsilon$ whenever $\|h\| \leq \delta$. Hence, by **3.2.5**,

$$\|F(x + h) - Fx - F'(x)h\| \leq \sup_{0 \leq t \leq 1} \|F'(x + th) - F'(x)\| \|h\| \leq \epsilon \|h\|. \quad \blacksquare$$

We note that F' is continuous at x if and only if all the partial derivatives $\partial_i f_j$ are continuous at x (**E 3.2-5**). However, the continuity of these partial derivatives is not a necessary condition for the existence of an F-derivative (**E 3.2-4**). We shall say that $F: D \subset R^n \to R^m$ is continuously differentiable on the open set $D_0 \subset D$ if F has a continuous G-derivative (and hence a continuous F-derivative) on D_0.

A useful condition on the F-derivative is given by the following concept, which is closely related to continuity of F'.

3.2.9. Definition. The F-derivative of $F: D \subset R^n \to R^m$ at $x^0 \in D$ is *strong* if, given any $\epsilon > 0$, there is a $\delta > 0$ so that $\bar{S}(x^0, \delta) \subset D$ and

$$\|Fy - Fx - F'(x^0)(y - x)\| \leq \epsilon \|y - x\|, \quad \forall x, y \in \bar{S}(x^0, \delta). \quad \blacksquare \tag{11}$$

3.2.10. Suppose that $F: D \subset R^n \to R^m$ has an F-derivative at each point of an open neighborhood of $x \in D$. Then, F' is strong at x if and only if F' is continuous at x.

Proof. Suppose first that F' is continuous at x and let $\epsilon > 0$ be given. Then, there is a $\delta > 0$ such that $\bar{S}(x, \delta) \subset D$ and $\|F'(x + h) - F'(x)\| \leq \epsilon$ if $\|h\| \leq \delta$. Hence, for $y, z \in \bar{S}(x, \delta)$ and $t \in [0, 1]$, we have

$$\|z + t(y - z) - x\| = \|t(y - x) + (1 - t)(z - x)\| \leq t\delta + (1 - t)\delta = \delta,$$

so that, by **3.2.5**,

$$\|Fy - Fz - F'(x)(y - z)\|$$
$$\leq \sup_{0 \leq t \leq 1} \|F'(z + t(y - z)) - F'(x)\| \|y - z\| \leq \epsilon \|y - z\|.$$

Conversely, if $F'(x)$ is strong, again let $\epsilon > 0$ be given, and choose $\delta > 0$ so that $F'(w)$ exists for all $w \in \bar{S}(x, \delta/2)$ and so that (11) holds. Then, for arbitrary $h \in R^n$ and $w \in \bar{S}(x, \delta/2)$, we may choose a $t > 0$, depending on w, so that $\|th\| \leq \delta/2$ and

$$\|F(w + th) - Fw - F'(w)(th)\| \leq \epsilon \|th\|.$$

Consequently, using (11) and noting that $\|w + th - x\| \leq \delta$, we have

$$\|[F'(w) - F'(x)](th)\| \leq \|F(w + th) - Fw - F'(x)(th)\|$$
$$+ \|F(w + th) - Fw - F'(w)(th)\| \leq 2\epsilon t \|h\|.$$

Therefore, $\|[F'(w) - F'(x)]h\| \leq 2\epsilon \|h\|$ and, since h was arbitrary, $\|F'(w) - F'(x)\| \leq 2\epsilon$. ∎

We note that **E 3.2-4** gives an example of a function F, in one dimension, which has a strong derivative at x even though F is not differentiable at all points of an open neighborhood of x.

For the next application, we first need the following lemma on integration:

3.2.11. If $G: [a, b] \subset R^1 \to R^m$ is continuous on $[a, b]$, then

$$\left\| \int_a^b G(t) \, dt \right\| \leq \int_a^b \|G(t)\| \, dt.$$

Proof. Since the norm is a continuous function, $\|G(\cdot)\|$ is Riemann-integrable, and, for arbitrary $\epsilon > 0$, there exists a partition $a < t_0 < \cdots < t_p < b$ such that

$$\left\| \int_a^b G(t) \, dt - \sum_{i=1}^p G(t_i)(t_i - t_{i-1}) \right\| \leq \epsilon,$$

3.2. MEAN-VALUE THEOREMS

and

$$\left| \int_a^b \|G(t)\| \, dt - \sum_{i=1}^p \|G(t_i)\|(t_i - t_{i-1}) \right| \leq \epsilon.$$

Hence,

$$\left\| \int_a^b G(t) \, dt \right\| \leq \left\| \sum_{i=1}^p G(t_i)(t_i - t_{i-1}) \right\| + \epsilon \leq \sum_{i=1}^p \|G(t_i)\|(t_i - t_{i-1}) + \epsilon$$

$$\leq \int_a^b \|G(t)\| \, dt + 2\epsilon,$$

and, since ϵ was arbitrary, the result follows. ∎

By means of **3.2.7** and **3.2.11**, we obtain the following estimate, which will be of continual value.

3.2.12. Let $F: D \subset R^n \to R^m$ be continuously differentiable on a convex set $D_0 \subset D$ and suppose that, for constants $\alpha \geq 0$ and $p \geq 0$, F' satisfies

$$\|F'(u) - F'(v)\| \leq \alpha \|u - v\|^p, \quad \forall u, v \in D_0. \tag{12}$$

Then, for any $x, y \in D_0$,

$$\|Fy - Fx - F'(x)(y - x)\| \leq [\alpha/(p + 1)] \|y - x\|^{p+1}. \tag{13}$$

Proof. It follows from **3.2.7** and **3.2.11** that

$$\|Fy - Fx - F'(x)(y - x)\| = \left\| \int_0^1 [F'(x + t(y - x)) - F'(x)](y - x) \, dt \right\|$$

$$\leq \int_0^1 \|F'(x + t(y - x)) - F'(x)\| \|y - x\| \, dt$$

$$\leq \alpha \|y - x\|^{p+1} \int_0^1 t^p \, dt. \quad \blacksquare$$

NOTES AND REMARKS

NR 3.2-1. The proofs of **3.2.1** and **3.2.6** may be found, for example, in Apostol [1957, p. 93 and p. 213].

NR 3.2-2. Theorem **3.2.2** requires, in essence, differentiability only in the direction $y - x$. Suppose that, in the terminology of **NR 3.1-3**, F has a G-differential $V(w, y - x)$ for all w in the interval $[x, y]$ and in the direction $y - x$. Then, again, the function $G(t) = F(x + t(y - x))$ is differentiable on $[0, 1]$, and there is some $t \in (0, 1)$ so that $Fy - Fx = V(x + t(y - x), y - x)$.

NR 3.2-3. The notion of a strong F-derivative has been used implicitly or explicitly by many authors, but **3.2.10** is apparently new.

NR 3.2-4. All of the results in this section hold if F is a mapping between Banach spaces. The only change necessary is the definition of the integral (10), which can no longer be defined in terms of the components of F. For a discussion of this integral and its properties, see, for example, Kantorovich and Akilov [1959], and Vainberg [1956].

EXERCISES

E 3.2-1. Let $F: R^2 \to R^2$ be defined by $f_1(x) = x_1^3$, $f_2(x) = x_2^2$. Set $x = 0$ and $y = (1, 1)^T$. Show that there is no $z \in [x, y]$ such that

$$Fy - Fx = F'(z)(y - x). \tag{14}$$

E 3.2-2. Let $F: R^n \to R^n$ be a diagonal mapping (Definition **1.1.1**) and assume that F is G-differentiable on R^n. Show that for any $x, y \in R^n$ there is a z such that (14) holds, but that z is not necessarily in $[x, y]$.

E 3.2-3. Assume that $F: D \subset R^n \to R^m$ has a G-derivative at each point of an open convex set $D_0 \subset D$ and that $F'(x) = 0$ for all $x \in D_0$. Show that F is constant on D_0.

E 3.2-4. Define $f: [-1, 1] \to R^1$ by $f(0) = 0$ and $f(x) = |x|/(n+1)$ if $1/(n+1) \leqslant |x| < 1/n$, $n = 1, 2, \ldots$. Show that $f'(0)$ exists and is strong.

E 3.2-5. Show that if $F: D \subset R^n \to R^m$ has a G-derivative at each point of an open neighborhood of x, then F' is continuous at x if and only if all partial derivatives $\partial_i f_j$ are continuous at x.

E 3.2-6. Let $F: D \subset R^n \to R^m$ and assume that F is continuously differentiable on a convex set $D_0 \subset D$. For any $x, y \in D_0$, show that

$$\|Fy - Fx - F'(x)(y - x)\| \leqslant \|y - x\| \omega(\|y - x\|),$$

where ω is the modulus of continuity of F' on $[x, y]$.

3.3. SECOND DERIVATIVES

If $F: D \subset R^n \to R^m$ has a G-derivative at each point of a set $D_0 \subset D$, then $F': D_0 \subset R^n \to L(R^n, R^m)$. Since the space $L(R^n, R^m)$ is again a normed linear space (of dimension $p = mn$), we may apply the definition of differentiation to the mapping F' in order to obtain the second derivative of F.

3.3. SECOND DERIVATIVES

3.3.1. Definition. Assume that $F: D \subset R^n \to R^m$ has a G-derivative at each point of an open set $D_0 \subset D$. If the mapping $F': D_0 \subset R^n \to L(R^n, R^m)$ has a G-derivative at $x \in D_0$, then $(F')'(x)$ is denoted by $F''(x)$ and called the *second G-derivative* of F at x. If F' has an F-derivative at x, then $F''(x)$ is called the *second F-derivative* at x. ∎

Derivatives of order higher than two can be defined in the analogous way by applying the basic definition of a derivative to $F^{(p-1)}$ in order to obtain $F^{(p)}$ (see **NR 3.3-2/3**).

It is important to note that the second F-derivative of F was defined in terms of the first G-derivative of F. However, by **3.1.6** applied to F' in place of F, it follows that if $F''(x)$ is an F-derivative, then F' is continuous at x; hence, by **3.2.8**, $F'(x)$ is itself an F-derivative. Thus, a necessary condition for F to have a second F-derivative at x is that F has a first F-derivative at x.

If the second G-derivative $F''(x)$ exists, then, by definition, $F''(x) \in L(R^n, L(R^n, R^m))$; that is, for each $h \in R^n$, $F''(x) h \in L(R^n, R^m)$ and, again by definition,

$$\lim_{t \to 0}(1/t)\| F'(x + th) - F'(x) - tF''(x) h \| = 0. \tag{1}$$

For convenience, we shall denote by $F''(x) hk$ the element $[F''(x) h] k$ in R^m. It is important to note that the mapping $B: R^n \times R^n \to R^m$ defined by $B(h, k) = [F''(x) h] k$ is linear separately in each vector variable h and k, and hence is *bilinear*. In other words, there is a natural interpretation of $F''(x)$ as a bilinear mapping from $R^n \times R^n$ into R^m.

In order to obtain the concrete representation of the second G-derivative $F''(x)$ in terms of the partial derivatives of the components $f_1, \ldots, f_m$ of F, consider first a functional $f: D \subset R^n \to R^1$. If $f''(x)$ exists, then the application of (1) to the coordinate vectors $e^1, \ldots, e^n$ yields

$$\lim_{t \to 0}(1/t)| f'(x + te^i) e^j - f'(x) e^j - tf''(x) e^i e^j | = 0,$$

so that

$$f''(x) e^i e^j = \lim_{t \to 0}(1/t)[\partial_j f(x + te^i) - \partial_j f(x)] = \partial_i \partial_j f(x).$$

Therefore, if $h = \sum_{i=1}^{n} h_i e^i$ and $k = \sum_{j=1}^{n} k_j e^j$, then

$$f''(x) hk = \sum_{i=1}^{n} \sum_{j=1}^{n} h_i k_j f''(x) e^i e^j = \sum_{i,j=1}^{n} h_i k_j \partial_i \partial_j f(x) = k^\mathrm{T} H_f(x) h,$$

where $H_f(x)$ is the $n \times n$ Hessian matrix:

$$H_f(x) = \begin{pmatrix} \partial_1 \partial_1 f(x) & \cdots & \partial_n \partial_1 f(x) \\ \vdots & & \vdots \\ \partial_1 \partial_n f(x) & \cdots & \partial_n \partial_n f(x) \end{pmatrix} \quad (2)$$

It is conceptually important to distinguish between the linear mappings $f''(x) \in L(R^n, L(R^n, R^1))$ and $H_f(x) \in L(R^n)$. Note that if $Fx = f'(x)^T$, then $F'(x) = H_f(x)$; that is, the Hessian matrix of f is the derivative of the gradient of f.

If $F: D \subset R^n \to R^m$, then by considering each component $f_1, \ldots, f_m$ separately, it follows immediately that the representation of $F''(x) hk \in R^m$ is

$$[F''(x) hk]^T = (k^T H_1(x) h, k^T H_2(x) h, \ldots, k^T H_m(x) h), \quad (3)$$

where $H_1(x), \ldots, H_m(x)$ are the Hessian matrices of $f_1, \ldots, f_m$ at x. If $F''(x)$ is an F-derivative, then it is also a G-derivative and, of course, has the same concrete representation (3).

The norm of $F''(x)$ on the space $L(R^n, L(R^n, R^m))$ is defined in the natural way; that is, for any $h \in R^n$,

$$\|F''(x) h\| = \sup\{\|F''(x) hk\| \mid \|k\| = 1, \quad k \in R^m\},$$

and thus

$$\|F''(x)\| = \sup_{\|h\|=1} \|F''(x) h\| = \sup_{\|h\|=1} \sup_{\|k\|=1} \|F''(x) hk\|.$$

Continuity of F'' is then defined, of course, in terms of the norm so that F'' is continuous at x if $\|F''(x) - F''(y)\| \to 0$ as $y \to x$. It is easy to verify that F'' is continuous at x if and only if all second partial derivatives of the components $f_1, \ldots, f_m$ are continuous at x (**E 3.3-1**).

As an immediate application of **3.2.8** applied to $(F')'$, we have the following result for the second derivative.

3.3.2. If $F: D \subset R^n \to R^m$ has a second G-derivative at all points of an open neighborhood of x in D, and F'' is continuous at x, then $F''(x)$ is an F-derivative.

An important property of the second derivative is symmetry.

3.3.3. Definition. If $F: D \subset R^n \to R^m$ has a second G-derivative at $x \in D$, then $F''(x)$ is *symmetric* if $F''(x) hk = F''(x) kh$ for all $h, k \in R^n$. ∎

It is easy to see that $F''(x)$ is symmetric if and only if each of the Hessian matrices $H_1(x), \ldots, H_n(x)$ is symmetric (see **E 3.3-2**). The basic, and initially surprising, result on symmetry of $F''(x)$ is the following.

3.3. SECOND DERIVATIVES

3.3.4. If $F: D \subset R^n \to R^m$ has a second F-derivative at x, then $F''(x)$ is symmetric.

Proof. For given $\epsilon > 0$, choose $\delta > 0$ so that $F'(y)$ exists and satisfies

$$\|F'(y) - F'(x) - F''(x)(y - x)\| \leqslant \epsilon \|x - y\|, \qquad (4)$$

whenever $\|x - y\| < \delta$. Now, let $h, k \in R^n$ lie in $S(0, \delta/2)$. Then, the mapping $G: [0, 1] \subset R^1 \to R^m$ defined by $G(t) = F(x + th + k) - F(x + th)$ is differentiable on $[0, 1]$ and

$$G'(t) = F'(x + th + k) h - F'(x + th) h.$$

Hence, for any $t \in [0, 1]$ we obtain, from (4),

$$\begin{aligned}\|G'(t) - F''(x) kh\| &\leqslant \|[F'(x + th + k) - F'(x) - F''(x)(th + k)] h\| \\ &\quad + \|[F'(x + th) - F'(x) - F''(x)(th)] h\| \\ &\leqslant \|h\|(\epsilon \|th + k\| + \epsilon \|th\|) \\ &\leqslant 2\epsilon \|h\|(\|h\| + \|k\|).\end{aligned}$$

Therefore,

$$\begin{aligned}\|G'(t) - G'(0)\| &\leqslant \|G'(t) - F''(x) kh\| + \|G'(0) - F''(x) kh\| \\ &\leqslant 4\epsilon \|h\|(\|h\| + \|k\|),\end{aligned}$$

and it follows by the mean-value theorem **3.2.5** that

$$\begin{aligned}\|G(1) - G(0) - F''(x) kh\| &\leqslant \|G(1) - G(0) - G'(0)\| + \|G'(0) - F''(x) kh\| \\ &\leqslant \sup_{0 \leqslant t \leqslant 1} \|G'(t) - G'(0)\| + 2\epsilon \|h\|(\|h\| + \|k\|) \\ &\leqslant 6\epsilon \|h\|(\|h\| + \|k\|). \qquad (5)\end{aligned}$$

Now, precisely the same argument, with h and k interchanged, may be applied to $\bar{G}(t) = F(x + h + tk) - F(x + tk)$, and we obtain

$$\|\bar{G}(1) - \bar{G}(0) - F''(x) hk\| \leqslant 6\epsilon \|k\|(\|h\| + \|k\|).$$

But $\bar{G}(1) - \bar{G}(0) = G(1) - G(0)$, which yields

$$\|F''(x) hk - F''(x) kh\| \leqslant 6\epsilon (\|h\| + \|k\|)^2. \qquad (6)$$

Thus, (6) holds for any h and k in $S(0, \delta/2)$. However, for arbitrary $h, k \in R^n$, we can choose $t > 0$ such that $\|th\| \leqslant \delta/2$ and $\|tk\| < \delta/2$, and then

$$t^2 \|F''(x) hk - F''(x) kh\| \leqslant 6\epsilon (\|th\|^2 + \|tk\|^2) \leqslant 6\epsilon t^2 (\|h\| + \|k\|)^2.$$

Therefore, (6) holds for arbitrary $h, k \in R^n$, and, since ϵ was arbitrary, it follows that $F''(x)\, hk - F''(x)\, kh = 0$. ∎

It is not true that a second G-derivative is necessarily symmetric (E. 3.3-3). Observe also that a corollary of **3.2.8** (applied to the mapping F') and **3.3.4** is that if all second partial derivatives of the components of F are continuous at x, then $F''(x)$ is symmetric.

We next apply the mean-value theorems of **3.2** to F''. It is important to note that the following three results are simply applications of the corresponding theorems of **3.2** to the mapping $F' \colon D_0 \subset R^n \to L(R^n, R^m)$. The first is an immediate corollary of **3.2.3**.

3.3.5. If $F \colon D \subset R^n \to R^m$ has a second G-derivative at each point of a convex set $D_0 \subset D$, then, for any $x, y \in D_0$,

$$\|F'(y) - F'(x)\| \leq \sup_{0 \leq t \leq 1} \|F''(x + t(y - x))\| \,\|y - x\|.$$

From **3.3.5** and **3.2.5**, we next obtain an estimate for $Fy - Fx - F'(x)(y - x)$.

3.3.6. If $F \colon D \subset R^n \to R^m$ has a second G-derivative at each point of a convex set $D_0 \subset D$, then, for any $x, y \in D_0$,

$$\|Fy - Fx - F'(x)(y - x)\| \leq \sup_{0 \leq t \leq 1} \|F''(x + t(y - x))\| \,\|y - x\|^2.$$

Proof. By **3.2.5** and **3.3.5**, we have

$$\|Fy - Fx - F'(x)(y - x)\| \leq \sup_{0 \leq t \leq 1} \|F'(x + t(y - x)) - F'(x)\| \,\|y - x\|$$

$$\leq \|y - x\| \sup_{0 \leq t \leq 1} \{ \sup_{0 \leq s \leq 1} \|F''(x + st(y - x))\| \,\|t(y - x)\| \}.$$

But

$$\sup_{0 \leq t \leq 1} \sup_{0 \leq s \leq 1} \|F''(x + st(y - x))\| = \sup_{0 \leq t \leq 1} \|F''(x + t(y - x))\|,$$

and the result follows. ∎

The definition **3.1.3** of hemicontinuity also pertains, of course, to F''; we may then apply **3.2.7** to F'.

3.3.7. If $F \colon D \subset R^n \to R^m$ has a hemicontinuous second G-derivative at each point of a convex set $D_0 \subset D$, then, for any $x, y \in D_0$,

$$F'(y) - F'(x) = \int_0^1 F''(x + t(y - x))(y - x)\, dt.$$

3.3. SECOND DERIVATIVES

We consider next some results based on the one-dimensional second-order Taylor formulas.

3.3.8. If $\varphi: [0, 1] \subset R^1 \to R^1$ is twice differentiable on $(0, 1)$ and φ and φ' are continuous on $[0, 1]$, then there is a $t \in (0, 1)$ so that

$$\varphi(1) - \varphi(0) - \varphi'(0) = \tfrac{1}{2}\varphi''(t). \qquad (7)$$

If, in addition, φ'' is Riemann-integrable on $(0, 1)$, then

$$\varphi(1) - \varphi(0) - \varphi'(0) = \int_0^1 (1-t)\varphi''(t)\,dt. \qquad (8)$$

Now note that **3.1.4** goes over immediately to second derivatives:

3.3.9. If $F: D \subset R^n \to R^m$ has a second G-derivative at $x \in D$, then F' is hemicontinuous at x.

Corresponding to **3.2.2**, we then obtain the following result involving the second derivative.

3.3.10. Let $f: D \subset R^n \to R^1$, and assume that f has a second G-derivative at each point of a convex set $D_0 \subset D$. Then, for any $x, y \in D_0$, there is a $t \in (0, 1)$ so that

$$f(y) - f(x) - f'(x)(y - x) = \tfrac{1}{2} f''(x + t(y - x))(y - x)(y - x).$$

Proof. For given $x, y \in D_0$, there is a $\delta > 0$ such that $x + t(y - x) \in D_0$ for all $t \in J = (-\delta, 1 + \delta)$. Hence, the mapping $\varphi: J \to R^1$ defined by $\varphi(t) = f(x + t(y - x))$ is differentiable on J, and $\varphi'(t) = f'(x + t(y - x))(y - x)$. Moreover, for any $t \in J$, $\varphi''(t)$ exists and

$$\varphi''(t) = \lim_{s \to 0}(1/s)[f'(x + (s + t)(y - x))(y - x) - f'(x + t(y - x))(y - x)]$$

$$= f''(x + t(y - x))(y - x)(y - x);$$

thus, in particular, φ' is continuous on $[0, 1]$. Hence, **3.3.8** applies, and by (7),

$$f(y) - f(x) - f'(x)(y - x) = \varphi(1) - \varphi(0) - \varphi'(0)$$
$$= \tfrac{1}{2} f''(x + t(y - x))(y - x)(y - x),$$

for some $t \in (0, 1)$. ∎

We next apply the integral representation (8). Note, first, that if $G: [0, 1] \to R^m$ has a second G-derivative on $[0, 1]$, then, by **3.1.4** and **3.3.9**, both G and G': $[0, 1] \to L(R^n, R^m)$ are continuous on $[0, 1]$, and, thus, if G'' is Riemann-integrable on $[0, 1]$, then we may apply **3.3.8** separately to each component g_i, $i = 1, \ldots, m$, of G to conclude that (8) holds for G itself; that is,

$$G(1) - G(0) - G'(0) = \begin{pmatrix} g_1(1) - g_1(0) - g_1'(0) \\ \vdots \\ g_m(1) - g_m(0) - g_m'(0) \end{pmatrix} = \begin{pmatrix} \int_0^1 (1-t) g_1''(t)\, dt \\ \vdots \\ \int_0^1 (1-t) g_m''(t)\, dt \end{pmatrix}$$

$$= \int_0^1 (1-t)\, G''(t)\, dt. \tag{9}$$

3.3.11. Suppose that $F: D \subset R^n \to R^m$ is twice G-differentiable on a convex set $D_0 \subset D$ and that F'' is hemicontinuous at each $x \in D_0$. Then, for any $x, y \in D_0$,

$$Fy - Fx - F'(x)(y-x) = \int_0^1 (1-t) F''(x + t(y-x))(y-x)(y-x)\, dt. \tag{10}$$

Proof. For given $x, y \in D_0$, define $G: J \subset R^1 \to R^m$ by $G(t) = F(x + t(y - x))$, $t \in J$, where $J = (-\delta, 1 + \delta)$ is an interval chosen as in the proof of **3.3.10**. Then, again as in **3.3.10**, it follows that $G'(t) = F'(x + t(y-x))(y-x)$ and $G''(t) = F''(x + t(y-x))(y-x)(y-x)$ for all $t \in J$. But the hemicontinuity of F'' on D_0 implies, in particular, that G'' is continuous on J and hence Riemann-integrable. Therefore, we may apply (9), which then gives (10). ∎

We complete this section with a result somewhat analogous to **3.3.6**.

3.3.12. Assume that $F: D \subset R^n \to R^m$ has a second G-derivative at $x \in D$. Then

$$\lim_{t \to 0} (1/t^2)[F(x + th) - Fx - F'(x)(th) - \tfrac{1}{2} F''(x)(th)(th)] = 0$$

for any $h \in R^n$. Moreover, if $F''(x)$ is an F-derivative, then

$$\lim_{h \to 0} (1/\|h\|^2)[F(x+h) - Fx - F'(x)h - \tfrac{1}{2} F''(x) hh] = 0. \tag{11}$$

Proof. For given $h \in R^n$, set

$$G(t) = F(x + th) - Fx - F'(x)(th) - \tfrac{1}{2} F''(x)(th)(th).$$

Since $F''(x)$ exists, G is well defined and differentiable for sufficiently small t, and

$$G'(t) = F'(x + th)\, h - F'(x)\, h - t F''(x)\, hh.$$

3.3. SECOND DERIVATIVES

It follows by the definition of the second G-derivative that, given $\epsilon > 0$, there is a $\delta > 0$ so that $\|G'(t)\| \leq \epsilon |t|$ whenever $|t| < \delta$, and consequently, by **3.2.3**,

$$\|G(t)\| = \|G(t) - G(0)\| \leq \sup_{0 \leq \theta \leq 1} \|G'(\theta t)\| |t| \leq \epsilon |t|^2, \qquad \forall |t| < \delta.$$

If $F''(x)$ is an F-derivative, then (11) results from a similar argument. In fact, if

$$R(h) = F(x + h) - Fx - F'(x)h - \tfrac{1}{2}F''(x)hh,$$

then R is well defined and G-differentiable in a neighborhood of x, and, given $\epsilon > 0$, there is a $\delta > 0$ so that

$$\|R'(h)\| = \|F'(x + h) - F'(x) - F''(x)h\| \leq \epsilon \|h\|, \qquad \forall h \in S(0, \delta).$$

Consequently, **3.2.3** ensures that whenever $\|h\| < \delta$

$$\|R(h)\| = \|R(h) - R(0)\| \leq \sup_{0 \leq t \leq 1} \|R'(th)\| \|h\| \leq \epsilon \|h\|^2. \quad \blacksquare$$

NOTES AND REMARKS

NR 3.3-1. The proof of **3.3.8** may be found, for example, in Apostol [1957, p. 96, p. 246]. The proof of **3.3.4** is taken from Dieudonné [1960].

NR 3.3-2. A mapping $M: R^{n_1} \times \cdots \times R^{n_p} \to R^m$ is called *multilinear* if, for any i and any fixed $x^j \in R^{n_j}$, $j = 1,\ldots, p$, $j \neq i$, $M(x^1,\ldots, x^{i-1},\cdot, x^{i+1},\ldots, x^p)$ is a linear operator from R^{n_i} to R^m. In the special case $p = 2$, M is bilinear. For an extensive discussion of multilinear operators, see, for example, Greub [1967]. If $F: R^n \to R^m$ has a pth G-derivative at x, then, as in the text for $p = 2$, there is a natural interpretation of $F^{(p)}(x)$ as a multilinear operator from $R^n \times \cdots \times R^n$ (p times) to R^m.

NR 3.3-3. Most of the results of this section have natural extensions to higher-order derivatives. For example, if $F^{(p)}(x)$ is an F-derivative, then it is symmetric in the sense that

$$F^{(p)}(x) h_1 \cdots h_p = F^{(p)}(x) h_{i_1} \cdots h_{i_p},$$

where $(1,\ldots, p) \to (i_1,\ldots, i_p)$ is any permutation of the indices $1,\ldots, p$. Also, the analogs of **3.3.6** and **3.3.11** hold: If $F: D \subset R^n \to R^m$ has a pth G-derivative at each point of a convex set $D_0 \subset D$, then for any $x, y \in D_0$,

$$\left\| Fy - Fx - \sum_{k=1}^{p-1} \frac{1}{k!} F^{(k)}(x)(y - x)^k \right\| \leq \frac{1}{p!} \sup_{0 \leq t \leq 1} \|F^{(p)}(x + t(y - x))\| \|y - x\|^p.$$

If, in addition, $F^{(p)}$ is hemicontinuous on $[x, y]$, then

$$Fy - Fx - \sum_{k=1}^{p-1} \frac{1}{k!} F^{(k)}(x)(y-x)^k = \int_0^1 \frac{(1-t)^{p-1}}{(p-1)!} F^{(p)}(x + t(y-x))(y-x)^p \, dt.$$

The proofs are easily carried out by an inductive argument using the methods of **3.3.6** and **3.3.11**. For a further discussion of higher derivatives, see, for example, Apostol [1957], Dieudonné [1960], and Kantorovich and Akilov [1959].

NR 3.3-4. All results of this section again hold if F is a mapping between Banach spaces, provided that the integrals are suitably defined (see **NR 3.2-4**).

EXERCISES

E 3.3-1. Let $F: D \subset R^n \to R^m$. Show that F'' is continuous at $x^0 \in D$ if and only if all second partial derivatives of the components $f_1, \ldots, f_m$ of F are continuous at x^0.

E 3.3-2. Let $F: D \subset R^n \to R^m$. Show that $F''(x^0)$ is symmetric if and only if each Hessian matrix $H_1(x^0), \ldots, H_m(x^0)$ is symmetric.

E 3.3-3. Define $f: R^2 \to R^1$ by $f(0) = 0$ and by

$$f(x) = \frac{x_1 x_2 (x_1^2 - x_2^2)}{(x_1^2 + x_2^2)}$$

for $x \neq 0$. Show that f has a G-derivative at 0 and that $\lim_{t \to 0}(1/t)[f'(th) - f'(0)]$ exists for all $h \in R^2$, but that f does not have a second G-derivative at 0. Show, moreover, that $\partial_1 \partial_2 f(0) \neq \partial_2 \partial_1 f(0)$.

3.4. CONVEX FUNCTIONALS

We shall collect in this section some of the basic properties of an extremely important class of functionals which, together with various generalizations, shall play a strong role throughout the book.

3.4.1. Definition. A functional $g: D \subset R^n \to R^1$ is *convex* on a convex set $D_0 \subset D$ if, for all $x, y \in D_0$ and $0 < \alpha < 1$

$$g(\alpha x + (1-\alpha)y) \leq \alpha g(x) + (1-\alpha)g(y). \tag{1}$$

The functional g is *strictly convex* on D_0 if strict inequality holds in (1)

3.4. CONVEX FUNCTIONALS

whenever $x \neq y$, and g is *uniformly convex* on D_0 if there is a constant $c > 0$ such that, for all $x, y \in D_0$ and $0 < \alpha < 1$

$$\alpha g(x) + (1 - \alpha) g(y) - g(\alpha x + [1 - \alpha] y) \geqslant c\alpha(1 - \alpha) \| x - y \|^2. \blacksquare \qquad (2)$$

It is clear that uniform convexity implies strict convexity which in turn, implies convexity. Note also that, by the equivalence-of-norms theorem **2.2.1**, if g is uniformly convex in one norm, it is uniformly convex in any norm.

The prototype convex functional is $g(x) = x^T A x$, where A is a symmetric, positive semidefinite matrix (see **E 3.4-1**). We shall see that a surprising number of the properties of this special class of functionals hold for convex functionals in general. We begin with a lemma which shows that (1) remains valid for a convex combination of an arbitrary number of points.

3.4.2. Assume that $g: D \subset R^n \to R^1$ is convex on a convex set $D_0 \subset D$, and let $x^0, \ldots, x^m$ be arbitrary points of D_0. Then, for any nonnegative numbers $\alpha_0, \ldots, \alpha_m$ with $\sum_{i=0}^m \alpha_i = 1$,

$$g\left(\sum_{i=0}^m \alpha_i x^i\right) \leqslant \sum_{i=0}^m \alpha_i g(x^i). \qquad (3)$$

Proof. The proof proceeds by induction on m. Clearly, (3) holds for $m = 1$ by the convexity of g. Now, assume that

$$g\left(\sum_{i=0}^{m-1} \beta_i x^i\right) \leqslant \sum_{i=0}^{m-1} \beta_i g(x^i), \qquad (4)$$

whenever $\sum_{i=0}^{m-1} \beta_i = 1$ and $\beta_i \geqslant 0$, $i = 0, \ldots, m - 1$, and write

$$\sum_{i=0}^m \alpha_i x^i = \gamma \sum_{i=0}^{m-1} (\alpha_i/\gamma) x^i + \alpha_m x^m,$$

where $\gamma = \sum_{i=0}^{m-1} \alpha_i$. We may assume that $\gamma > 0$, since otherwise the result is trivial; then, by the convexity of g and (4), it follows that

$$g\left(\sum_{i=0}^m \alpha_i x^i\right) \leqslant \gamma g\left(\sum_0^{m-1} (\alpha_i/\gamma) x^i\right) + \alpha_m g(x^m)$$

$$\leqslant \gamma \sum_{i=0}^{m-1} (\alpha_i/\gamma) g(x^i) + \alpha_m g(x^m) = \sum_{i=0}^m \alpha_i g(x^i). \blacksquare$$

By means of **3.4.2**, we next obtain an important continuity result.

3.4.3. Let $g: D \subset R^n \to R$ be convex on an open convex set $D_0 \subset D$. Then g is continuous on D_0.

Proof. Let x^0 be an arbitrary point in D_0. Since D_0 is open, we can find $n+1$ points $x^1, \ldots, x^{n+1} \in D_0$ such that the interior of the convex hull $C = \{x \mid x = \sum_{i=1}^{n+1} \alpha_i x^i, \alpha_i \geq 0, \sum_{i=1}^{n+1} \alpha_i = 1\}$ is not empty and $x^0 \in \text{int } C$. Now, let $\alpha = \max_{1 \leq i \leq n+1} g(x^i)$; then, by **3.4.2**,

$$g(x) = g\left(\sum_{i=1}^{n+1} \alpha_i x^i\right) \leq \sum_{i=1}^{n+1} \alpha_i g(x^i) \leq \alpha, \qquad \forall x \in C, \tag{5}$$

so that g is bounded above on C. Next, since $x^0 \in \text{int}(C)$, there is a $\delta > 0$ so that $\bar{S}(x^0, \delta) \subset C$; hence, for arbitrary $h \in \bar{S}(0, \delta)$ and $\lambda \in [0, 1]$, we have

$$x^0 = [1/(1+\lambda)](x^0 + \lambda h) + [\lambda/(1+\lambda)](x^0 - h),$$

and thus

$$g(x^0) \leq [1/(1+\lambda)] g(x^0 + \lambda h) + [\lambda/(1+\lambda)] g(x^0 - h),$$

or, by (5),

$$g(x^0 + \lambda h) - g(x^0) \geq \lambda[g(x^0) - g(x^0 - h)] \geq -\lambda[|g(x^0)| + \alpha].$$

Similarly, $g(x^0 + \lambda h) \leq \lambda g(x^0 + h) + (1 - \lambda) g(x^0)$, so that

$$g(x^0 + \lambda h) - g(x^0) \leq \lambda[g(x^0 + h) - g(x^0)] \leq \lambda [|g(x^0)| + \alpha],$$

and therefore

$$|g(x^0 + \lambda h) - g(x^0)| \leq \lambda[|g(x^0)| + \alpha].$$

Now, for given $\epsilon > 0$, choose $\delta' \leq \delta$ so that $\delta'[|g(x^0)| + \alpha] \leq \epsilon \delta$. Then, for any $k \in S(0, \delta')$, it follows, by setting $k = \lambda h$ with $\|h\| = \delta$, that $|g(x^0 + k) - g(x^0)| \leq \epsilon$. ∎

In general, a convex function whose domain is not open need not be continuous, as the example $g: (0, 1] \subset R^1 \to R^1$, $g(t) = 0$, $t \in (0, 1)$, $g(1) = 1$, shows.

The basic definition of a convex function is often unwieldy to use. Next, we consider certain differential inequalities which characterize convexity.

3.4.4. Suppose that $g: D \subset R^n \to R^1$ has a G-derivative on a convex set $D_0 \subset D$. Then g is convex on D_0 if and only if

$$g'(x)(y - x) \leq g(y) - g(x), \qquad \forall x, y \in D_0. \tag{6}$$

3.4. CONVEX FUNCTIONALS

Moreover, g is strictly convex on D_0 if and only if strict inequality holds in (6) whenever $x \neq y$, and g is uniformly convex on D_0 if and only if there exists a constant $c > 0$ such that

$$g(y) - g(x) \geqslant g'(x)(y - x) + c \| x - y \|^2, \qquad \forall x, y \in D_0. \tag{7}$$

Proof. Suppose, first, that (7) holds for some $c \geqslant 0$, and for given $x, y \in D_0$ and $0 \leqslant \alpha \leqslant 1$ set $z = \alpha x + (1 - \alpha) y$. Then $z \in D_0$, and, by (7),

$$g(x) - g(z) \geqslant g'(z)(x - z) + c \| x - z \|^2,$$
$$g(y) - g(z) \geqslant g'(z)(y - z) + c \| y - z \|^2,$$

so that multiplication by α and $(1 - \alpha)$, respectively, followed by addition gives

$$\alpha g(x) + (1 - \alpha) g(y) - g(z)$$
$$= \alpha[g(x) - g(z)] + (1 - \alpha)[g(y) - g(z)]$$
$$\geqslant g'(z)[\alpha(x - z) + (1 - \alpha)(y - z)] + c[\alpha \| x - z \|^2 + (1 - \alpha)\| y - z \|^2]$$
$$= c\alpha(1 - \alpha)\| x - y \|^2, \tag{8}$$

since the first term in brackets is zero, while $\| x - z \|^2 = (1 - \alpha)^2 \| x - y \|^2$ and $\| y - z \|^2 = \alpha^2 \| x - y \|^2$. Therefore, if $c > 0$, (8) shows that (7) implies uniform convexity, while, if $c = 0$, (8) shows that (6) implies that g is convex. Also, if $c = 0$, and strict inequality holds in (6) for $x \neq y$, then strict inequality holds in (8). Conversely, suppose that g is uniformly convex. Then, for $x, y \in D_0$ and sufficiently small $t > 0$, we have

$$g(y) - g(x) \geqslant (1/t)[g(x + t(y - x)) - g(x)] + (1 - t) c \| x - y \|^2, \tag{9}$$

so that (7) holds as $t \to 0$. But if g is convex, then (9) holds with $c = 0$, and (6) follows. Finally, if g is strictly convex, then (6) applied to x and $z = \tfrac{1}{2}(x + y)$ shows that

$$g'(x)(y - x) = 2g'(x)(z - x) \leqslant 2[g(z) - g(x)] < 2[\tfrac{1}{2}g(y) - \tfrac{1}{2}g(x)]. \blacksquare$$

The geometric interpretation of (6) is that a convex functional always lies above its tangent plane at any point.

As a consequence of **3.4.4**, we obtain the following relationships, which show that g' is a so-called monotone mapping (see Definition **5.4.2**, and **5.4.3**).

3.4.5. Suppose that $g: D \subset R^n \to R^1$ has a G-derivative on a convex set $D_0 \subset D$. Then g is convex on D_0 if and only if

$$[g'(y) - g'(x)](y - x) \geq 0, \qquad \forall x, y \in D_0, \tag{10}$$

and g is strictly convex on D_0 if and only if strict inequality holds in (10) whenever $x \neq y$. Finally, g is uniformly convex on D_0 if and only if

$$[g'(y) - g'(x)](y - x) \geq 2c \|y - x\|^2, \qquad \forall x, y \in D_0, \tag{11}$$

where $c > 0$ is the constant of (2).

Proof. If g is uniformly convex on D_0, then for any $x, y \in D_0$ we have, by **3.4.4**, that

$$\begin{aligned} g(y) - g(x) &\geq g'(x)(y - x) + c\|y - x\|^2, \\ g(x) - g(y) &\geq g'(y)(x - y) + c\|x - y\|^2, \end{aligned} \tag{12}$$

and if g is convex, then (12) is true with $c = 0$. Addition of these inequalities shows that (11) holds, and hence that, for $c = 0$, (10) is valid. Moreover, if g is strictly convex, then (12) holds with $c = 0$, but with strict inequality for $x \neq y$, and hence addition yields (10) with strict inequality for $x \neq y$.

To prove the converse, note that, for any fixed $x, y \in D_0$, the mean-value theorem **3.2.2** ensures that there exists a $t \in (0, 1)$ such that

$$g(y) - g(x) = g'(u)(y - x), \tag{13}$$

where $u = x + t(y - x)$. From (10), it follows that

$$[g'(u) - g'(x)](y - x) = (1/t)[g'(u) - g'(x)](u - x) \geq 0, \tag{14}$$

and hence

$$g(y) - g(x) = [g'(u) - g'(x)](y - x) + g'(x)(y - x) \geq g'(x)(y - x), \tag{15}$$

which, by **3.4.4**, shows that g is convex. If strict inequality holds in (10) for $x \neq y$, then the same will be true in (14), and thus g is strictly convex.

Finally, if (11) holds, set $t_k = k/(m + 1)$, $k = 0, 1, \ldots, m + 1$, with an arbitrary integer $m \geq 0$. Then, by the mean-value theorem **3.2.2**, there exist numbers s_k such that

$$\begin{aligned} g(x + t_{k+1}(y - x)) &- g(x + t_k(y - x)) \\ &= g'(x + s_k(y - x))(t_{k+1} - t_k)(y - x), \qquad t_k < s_k < t_{k+1}. \end{aligned}$$

3.4. CONVEX FUNCTIONALS

Hence,

$$g(y) - g(x) = \sum_{k=0}^{m} [g(x + t_{k+1}(y - x)) - g(x + t_k(y - x))]$$

$$= \sum_{k=0}^{m} [g'(x + s_k(y - x)) - g'(x)](t_{k+1} - t_k)(y - x) + g'(x)(y - x)$$

$$\geq 2c \| y - x \|^2 \sum_{k=0}^{m} (t_{k+1} - t_k) s_k + g'(x)(y - x).$$

But

$$\sum_{k=0}^{m} (t_{k+1} - t_k) s_k \geq \sum_{k=0}^{m} (t_{k+1} - t_k) t_k = \frac{1}{(m+1)^2} \sum_{k=0}^{m} k = \frac{1}{2} \frac{m}{m+1},$$

and, since m can be arbitrarily large, it follows that

$$g(y) - g(x) \geq c \| y - x \|^2 + g'(x)(y - x).$$

Therefore, **3.4.4** implies the uniform convexity of g. ∎

We complete this section with a characterization of convexity in terms of the second derivative. The second G-derivative of a functional g is called *positive definite* at x if $g''(x) hh > 0$ for all $h \in R^n$, $h \neq 0$, *positive semidefinite* at x if $g''(x) hh \geq 0$ for all $h \in R^n$, and *uniformly positive definite* on a set D_0 if there is a $c > 0$ so that

$$g''(x) hh \geq c \| h \|^2, \qquad \forall h \in R^n, \quad x \in D_0. \tag{16}$$

Note that these definitions do not require $g''(x)$ to be symmetric. Note also that $g''(x)$ is positive definite if and only if the Hessian matrix $H_g(x)$ is positive definite.

3.4.6. Assume that $g: D \subset R^n \to R^1$ has a second G-derivative at each point of a convex set $D_0 \subset D$. Then g is convex on D_0 if and only if $g''(x)$ is positive semidefinite for all $x \in D_0$. Moreover, g is strictly convex on D_0 if $g''(x)$ is positive definite for all $x \in D_0$, and g is uniformly convex on D_0 if and only if g'' is uniformly positive definite on D_0.

Proof. Let $x, y \in D_0$. Then, by the mean-value theorem **3.3.10**, there exists a point u in the interval $[x, y]$ so that

$$g(y) - g(x) - g'(x)(y - x) = \tfrac{1}{2} g''(u)(y - x)(y - x). \tag{17}$$

Hence, if g'' is positive semidefinite, positive definite, or uniformly positive definite on D_0, then **3.4.4** shows that g is convex, strictly convex, or uniformly convex on D_0, respectively. Conversely, if g is uniformly convex on D_0, then **3.4.5** shows that

$$g''(x)\, hh = \lim_{t \to 0}(1/t)[g'(x + th) - g'(x)]\, h \geqslant \lim_{t \to 0}(1/t^2)\, 2c \parallel th \parallel^2$$
$$= 2c \parallel h \parallel^2, \qquad \forall x \in D_0, \quad h \in R^n, \tag{18}$$

where c is the constant of (2); consequently, g'' is uniformly positive definite on D_0. Finally, if g is convex on D_0, then again **3.4.5** ensures that (18) holds with $c = 0$, so that $g''(x)$ is positive semidefinite. ∎

Observe that **3.4.6** does not contain the converse statement that strict convexity implies positive definiteness. Indeed, the function t^4 is strictly convex, but has a zero second derivative at $t = 0$. However, the statement of **3.4.6** can be strengthened somewhat to include this example (**E 3.4-4**).

NOTES AND REMARKS

NR 3.4-1. The concept of a convex functional dates back at least to Minkowski [1892]. There are several standard references on convex sets and functionals; see, for example, Eggleston [1958], Fenchel [1953], and parts of Berge [1959]. Note that the definitions of convexity and strict convexity extend immediately to an arbitrary linear space, while that of uniform convexity holds in a Banach space. For a discussion of more general definitions of uniform convexity, see **NR 4.3-3**.

NR 3.4-2. Various extensions of the notion of a convex functional will be introduced in later sections. In particular, in Chapter 4, we will deal with quasi- and pseudoconvex functionals, while in Chapter 13 the definition of convexity will be extended to mappings with range in R^m.

NR 3.4-3. Theorem **3.4.3** is only an indication of the regularity properties that a convex function must satisfy. For example, if $g: D \subset R^n \to R^1$ is convex on the open convex set D, then the F-derivative g' exists and is continuous almost everywhere. For the proof of this, and for additional results for higher derivatives, see, for example, Fenchel [1953]. For extensions to Banach spaces, see Apslund [1968].

NR 3.4-4. The basic differential inequality (6) is classical and probably due to Minkowski [1892]. The corresponding result for the second derivative—that is, g is convex if and only if $g''(x)$ is positive semidefinite—dates back to Rademacher [1922]. The results **3.4.4**, **3.4.5**, and **3.4.6** for strictly convex functionals are less well known, and, indeed, necessary conditions on g'' for g to be strictly convex do

3.4. CONVEX FUNCTIONALS

not seem to be known except in one dimension. (For the one-dimensional result, see Berge [1959] and **E 3.4-4**.) The results on uniform convexity were first stated by Poljak [1966], and also proved in more generality by Elkin [1968]. All of **3.4.4**, **3.4.5**, and **3.4.6** and their proofs extend to an arbitrary Banach space and were proved in that setting by Elkin [1968]. See also Kačurovskii [1960, 1962], and Minty [1964].

NR 3.4-5. A functional $g: D \subset R^n \to R^1$ is *midpoint convex* on a convex set $D_0 \subset D$ if

$$g(\tfrac{1}{2}[x+y]) \leq \tfrac{1}{2}g(x) + \tfrac{1}{2}g(y), \qquad \forall x, y \in D_0.$$

Surprisingly, midpoint convex functions are not necessarily convex unless g is also continuous (see **E 3.4-7**).

EXERCISES

E 3.4-1. Let $A \in L(R^n)$ be symmetric, and define $g: R^n \to R^1$ by $g(x) = x^T A x$. Show, directly from the definition **3.4.1**, that g is convex if and only if A is positive semidefinite and that g is both strictly and uniformly convex if and only if A is positive definite.

E 3.4-2. Show that $g: D \subset R^n \to R^1$ is convex on the convex set D if and only if, for any $x, y \in D$, the function $f: [0, 1] \to R^1$, $f(t) = g(tx + (1-t)y)$, is convex on $[0, 1]$.

E 3.4-3. Show that if $g_i : R^n \to R^1$ is convex and $\lambda_i \geq 0$, $i = 1, \ldots, M$, then $g = \sum_{i=1}^m \lambda_i g_i$ is convex. Show that if, in addition, $\lambda_k > 0$ and g_k is strictly or uniformly convex for some k, then g is strictly or uniformly convex, respectively.

E 3.4-4. Assume that $g: D \subset R^n \to R^1$ is twice-continuously differentiable on a convex set $D_0 \subset D$. Show that g is strictly convex on D_0 if $g''(x)$ is positive semidefinite on D_0, and is positive definite except on a set which contains no proper line segments. Conclude, in particular, that a twice-continuously differentiable function $f: R^1 \to R^1$ is strictly convex if and only if $f''(t) \geq 0$ and f'' does not vanish on any open interval.

E 3.4-5. Let $g: R^2 \to R^1$ be given by $g(x) = x_1^2 + x_2^4$. Show that g is strictly convex on R^n, but that $g''(x)$ is not positive definite on the ray $\{x \in R^n \mid x_2 = 0\}$ (Elkin [1968]).

E 3.4-6. Show that the function $g(t) = (1 + t^2)^{1/2}$, $t \in R^1$, is strictly, but not uniformly, convex on R^1.

E 3.4-7. Suppose that $g: D \subset R^n \to R^1$ is continuous on a convex set $D_0 \subset D$ and satisfies
$$\tfrac{1}{2}g(x) + \tfrac{1}{2}g(y) - g(\tfrac{1}{2}[x+y]) \geq \gamma \| x - y \|^2$$
for all $x, y \in D_0$. Show that g is convex on D_0 if $\gamma = 0$ and uniformly convex if $\gamma > 0$. (Hint: Show by induction that
$$m2^{-k}g(x) + (1 - m2^{-k})g(y) - g(m2^{-k}x + [1 - m2^{-k}]y)$$
$$\geq 4\gamma m 2^{-k}(1 - m2^{-k}) \| x - y \|^2$$
for any integers m and k with $0 < m \leq 2^k$.)

Part II / NONCONSTRUCTIVE EXISTENCE THEOREMS

In this part, we collect a sampling of various existence and uniqueness theorems which are proved by nonconstructive means. In contrast, Part V contains many other existence results which are proved by showing that a certain sequence generated by an iterative process converges and that its limit is a solution of the equation under consideration.

It is not the purpose of this part to give a comprehensive survey of existence theorems; that would be beyond the scope of this book. Nevertheless, most of the main approaches, at least in finite dimensions, are covered.

In Chapter 4, we deal with gradient operators, that is, with operators F which are the derivative of a real-valued functional g. Here, it is possible to study questions of existence and uniqueness of solutions of $Fx = 0$ by considering the related problem of finding minimizers of g. The main results of this chapter are for convex functionals, although the treatment is more general.

In Chapter 5, we consider two approaches to existence theorems when F is not a gradient operator. The first is the contraction mapping theorem (which is also studied in much more detail in Chapter 12); this, in turn, allows simple proofs of the inverse- and implicit-function theorems. The inverse-function theorem provides a classical sufficient condition that a mapping be a local homeomorphism and raises the question, which is studied by means of continuation techniques, of when a local homeomorphism is a global homeomorphism.

Finally, in Chapter 6, we develop the theory of the degree of a mapping —a powerful tool which allows, in particular, simple proofs of the classical Brouwer fixed point theorem as well as of more recent results on monotone mappings.

Chapter 4 / GRADIENT MAPPINGS AND MINIMIZATION

4.1. MINIMIZERS, CRITICAL POINTS, AND GRADIENT MAPPINGS

In this chapter, we consider various existence and uniqueness theorems relating to the minimization of functionals on R^n.

4.1.1. Definition. Let $g: D \subset R^n \to R^1$. A point $x^* \in D$ is a *local minimizer* of g if there is an open neighborhood S of x^* such that for all $x \in S \cap D$

$$g(x) \geq g(x^*); \tag{1}$$

x^* is a *proper* local minimizer of g if strict inequality holds in (1) for all $x \in S \cap D$, $x \neq x^*$. If (1) holds for all x in some subset D_0 of D which contains x^*, then x^* is a *global minimizer* of g on D_0. ∎

Local and global maximizers of g may be defined in an analogous way. However, the problem of finding maximizers of g is equivalent to finding minimizers of $-g$, and, accordingly, we can restrict our attention to minimizers.

The theory of local minimizers is closely related to the concept of critical points.

4.1.2. Definition. A point $x^* \in \text{int}(D)$ is a *critical point* of $g: D \subset R^n \to R^1$ if g has a G-derivative at x^* and $g'(x^*) = 0$ [or, equivalently, $g'(x^*)^T = 0$]. ∎

The relation between the two concepts is given by the following basic result.

4.1.3. Suppose that $x^* \in \text{int}(D)$ is a local minimizer of $g: D \subset R^n \to R^1$. If g has a G-derivative at x^*, then $g'(x^*) = 0$.

Proof. Since $x^* \in \text{int}(D)$ and x^* is a local minimizer, it follows that $g(x^* + th) - g(x^*) \geq 0$ for any $h \in R^n$ and sufficiently small t. Hence,

$$g'(x^*) h = \lim_{t \to 0+}(1/t)[g(x^* + th) - g(x^*)] \geq 0$$

and, since h was arbitrary, $g'(x^*) = 0$. ∎

It is, of course, not necessary that a local minimizer x^* be a critical point even if $x^* \in \text{int}(D)$. For example, $x^* = 0$ is a local minimizer of the function $g: R^1 \to R^1$, $g(x) = |x|$, but x^* is not a critical point, since $g'(0)$ does not exist. On the other hand, a critical point need not be a local minimizer, as the example $g: R^1 \to R^1$, $g(x) = x^3$ shows. The second derivative of g, if it exists, contains the necessary information to ensure that a critical point is also a minimizer.

4.1.4. Let $g: D \subset R^n \to R^1$ and suppose that the second G-derivative of g exists at $x^* \in \text{int}(D)$. If x^* is a critical point of g and $g''(x^*)$ is positive definite, then x^* is a proper local minimizer of g. Conversely, if x^* is a local minimizer, then $g''(x^*)$ is positive semidefinite.

Proof. Let x^* be a critical point and $g''(x^*)$ be positive definite. Then, for any fixed $h \in R^n$, $h \neq 0$, 3.3.12 shows that

$$\lim_{t \to 0}(1/t^2)[g(x^* + th) - g(x^*)] = \tfrac{1}{2} g''(x^*) hh. \tag{2}$$

Hence, $g(x^* + th) - g(x^*) > 0$ for sufficiently small t. Since h was arbitrary, this shows that x^* is a proper local minimizer. Conversely, let x^* be a local minimizer and suppose that $g''(x^*)$ is not positive semidefinite. Then there exists an h such that $g''(x^*) hh < 0$, and (2) shows that, for sufficiently small t, we have $g(x^* + th) - g(x^*) < 0$, which is a contradiction. ∎

It is, in general, not possible to interchange the roles of definite and semidefinite in **4.1.4**. For example, $g: R^1 \to R^1$, $g(x) = x^4$, has a proper local minimizer at 0, although $g''(0) = 0$; hence, $g''(x^*)$ need not be positive definite at a proper local minimizer. On the other hand, $g: R^1 \to R^1$, $g(x) = x^3$, shows that positive semidefiniteness of g'' at a critical point x^* is not sufficient to ensure that x^* is a local minimizer.

4.1. MINIMIZERS, CRITICAL POINTS, AND GRADIENT MAPPINGS

The problem of finding critical points of a functional is precisely the same as that of solving the system of equations $Fx = 0$ where F is the mapping $F: D \subset R^n \to R^n$ defined by $Fx = g'(x)^T$, $x \in D$. The last two results, in turn, show that the problem of determining critical points of a functional g is, in certain cases, equivalent to finding minimizers of g. In other words, if F is the derivative of a functional, then the problem of solving $Fx = 0$ can sometimes be replaced by that of minimizing the functional. As discussed briefly in Chapter 1, this is our primary motivation for considering the problem of finding minimizers of functionals.

The problem now arises: how general is the class of systems $Fx = 0$ which can be solved in this way?

4.1.5. Definition. A mapping $F: D \subset R^n \to R^n$ is a *gradient* (*potential*) *mapping* on a subset $D_0 \subset D$ if there exists a G-differentiable functional $g: D_0 \subset R^n \to R^1$ such that $Fx = g'(x)^T$ for all $x \in D_0$. ∎

The above question is, in essence, completely answered by the following result.

4.1.6. Symmetry Principle. Let $F: D \subset R^n \to R^n$ be continuously differentiable on an open convex set $D_0 \subset D$. Then F is a gradient mapping on D_0 if and only if $F'(x)$ is symmetric for all $x \in D_0$.

Proof. If F is a gradient mapping, then $F'(x) = H_g(x)$, where $H_g(x)$ is the Hessian matrix (3.3.2) of g, and the symmetry of $H_g(x)$ follows from **3.3.2** and **3.3.4**. Conversely, assume that $F'(x)$ is symmetric for all $x \in D_0$. For arbitrary but fixed $x^0 \in D_0$, we define a functional $g: D_0 \to R^1$ by

$$g(x) = \int_0^1 (x - x^0)^T F(x^0 + t(x - x^0))\, dt; \tag{3}$$

since D_0 is convex and F is continuous, g is clearly well defined on D_0. Now, for any two points $x, y \in D_0$, we have the identity

$$g(y) - g(x) = \int_0^1 (y - x)^T F(x^0 + t(y - x^0))\, dt$$
$$+ \int_0^1 (x - x^0)^T [F(x^0 + t(y - x^0)) - F(x^0 + t(x - x^0))]\, dt.$$

Denoting the second integral by J, we find, by the mean value theorem **3.2.7**, the change of variable $s \to s/t$, the symmetry of $F'(x)$, interchange of the order of integration, the change of variable $t \to (t - s)/(1 - s)$, and another application of 3.2.7, that

$$J = \int_0^1 \int_0^t (x - x^0)^T F'(x^0 + t(x - x^0) + s(y - x))(y - x)\, ds\, dt$$
$$= \int_0^1 \int_0^t (y - x)^T F'(x^0 + t(x - x^0) + s(y - x))(x - x^0)\, ds\, dt$$
$$= \int_0^1 \int_s^1 (y - x)^T F'(x^0 + t(x - x^0) + s(y - x))(x - x^0)\, dt\, ds$$
$$= \int_0^1 (y - x)^T [F(x^0 + (x - x^0) + s(y - x)) - F(x^0 + s(y - x^0))]\, ds.$$

Therefore, it follows that

$$g(y) - g(x) = \int_0^1 (y - x)^T F(x + s(y - x))\, ds.$$

Next, set $y = x + th$ for t sufficiently small; then

$$g'(x)\, h = \lim_{t \to 0} (1/t)[g(x + th) - g(x)] = \lim_{t \to 0} \int_0^1 h^T F(x + sth)\, ds = h^T F x,$$

and, since h is arbitrary, we have $Fx = g'(x)^T$. ∎

Note that (3) provides a means of constructing the functional g from F.

The symmetry principle just proved appears to place a severe limitation on the class of systems which can be solved by minimizing some nonlinear functional. However, there is always a simple way of converting the problem of solving $Fx = 0$ into a minimization problem, even if F is not a gradient operator.

Let $f\colon R^n \to R^1$ be a functional with the property that f has the unique global minimizer $x = 0$. For example, $f(x) = x^T A x$ with some symmetric, positive definite matrix A, or $f(x) = \|x\|$ with some arbitrary norm on R^n, are possible choices for such a functional. Now, for a given mapping $F\colon D \subset R^n \to R^n$, we define the functional

$$g\colon D \subset R^n \to R^1, \qquad g(x) = f(Fx), \quad x \in D. \tag{4}$$

If the system $Fx = 0$ has a solution $x^* \in D$, then, clearly, x^* is a global minimizer of g on D, and, hence, in order to find x^* it suffices to minimize g. In the case that $Fx = 0$ has no solution in D and x^* is a global minimizer of g on D, we call x^* an *f-minimal solution* of $Fx = 0$. When, in particular, $f(x) = x^T x$, then an f-minimal solution is called a *least-squares* solution of the problem.

4.1. MINIMIZERS, CRITICAL POINTS, AND GRADIENT MAPPINGS

Note that the concept of f-minimal solutions extends immediately to mappings F from R^m to R^n. In that case, g is a functional on a set D in R^m.

While it is possible to find solutions of $Fx = 0$ by determining minimizers of the functional g of (4), it is not always possible to find such solutions by determining critical points of g. The following result gives a sufficient condition for this problem in terms of the critical points of the functional f.

4.1.7. Suppose that $f: R^n \to R^1$ is F-differentiable on R^n and has a unique critical point at $x = 0$. Given $F: D \subset R^n \to R^n$, define $g: D \subset R^n \to R^1$, $g(x) = f(Fx)$, $x \in D$, and assume that at $x^* \in \text{int}(D)$ F has a nonsingular G-derivative. Then x^* is a critical point of g if and only if $Fx^* = 0$.

Proof. By the chain rule 3.1.7, g has a G-derivative at x^* and

$$g'(x^*) = f'(Fx^*) F'(x^*).$$

If $Fx^* = 0$, then $g'(x^*) = 0$, since 0 is a critical point of f. Conversely, if $g'(x^*) = 0$, then $f'(Fx^*) = 0$, since $F'(x^*)$ is nonsingular. But f has only the critical point 0, and therefore $Fx^* = 0$. ∎

NOTES AND REMARKS

NR 4.1-1. Definitions **4.1.1** and **4.1.2** and Theorems **4.1.3** and **4.1.4** are standard and may be found in advanced calculus books.

NR 4.1-2. Theorem **4.1.6** is essentially a result of Kerner [1933].

NR 4.1-3. The idea of solving the equation $Fx = 0$ by minimization of the functional $g(x) = (Fx)^T Fx$ dates back at least to Cauchy [1847].

EXERCISES

E 4.1-1. Let $F: R^2 \to R^2$ be defined by $Fx = (2x_1 x_2, x_1^2)^T$. Find a functional $g: R^2 \to R^1$ so that $Fx = g'(x)^T$ for all $x \in R^2$.

E 4.1-2. For $F: D \subset R^n \to R^n$, suppose that $x^* \in D$ is an isolated solution of $Fx = 0$. Assume that $f: R^n \to R^1$ is such that $f(0) < f(x)$ for all $x \in R^n$. Show that x^* is a proper local minimizer of $g(x) = f(Fx)$, $x \in D$.

E 4.1-3. Let $A \in L(R^n)$ be symmetric, positive definite and, for a given F-differentiable mapping $F: D \subset R^n \to R^n$, define $g: D \subset R^n \to R^1$ by

$g(x) = (Fx)^T A(Fx)$, $x \in D$. Assume that $F'(x^*)$ is nonsingular for some $x^* \in D$. Show that x^* is a critical point of g if and only if $Fx^* = 0$.

E 4.1-4. Assume that $g: D \subset R^n \to R^1$ has a G-derivative on an open, bounded set $D_0 \subset D$ and that g is continuous on $\bar{D}_0$. If there exists an $x^0 \in D_0$ such that $g(x^0) < g(x)$ for all x on the boundary of D_0, then g has a critical point in D_0. Apply this observation to show that if

$$[g'(x) - g'(y)](x - y) \geq c \| x - y \|_2^2, \qquad \forall x, y \in \bar{S}(x^0, r) \subset D,$$

and $\| g'(x^0) \| < \frac{1}{2} cr$, then g has a critical point in $S(x^0, r)$.

4.2. UNIQUENESS THEOREMS

Many of the questions concerning the existence and uniqueness of minimizers are closely connected with the concept of a level set.

4.2.1. Definition. If $g: D \subset R^n \to R^1$, then any nonempty set of the form $L(\gamma) = \{x \in D \mid g(x) \leq \gamma\}$, $\gamma \in R^1$, is a *level set* of g. ∎

As an immediate consequence of the fact that a continuous function on a compact set takes on its minimum, we have the following observation.

4.2.2. If $g: D \subset R^n \to R^1$ is continuous and has a compact level set, then there exists an $x^* \in D$ such that $g(x^*) \leq g(x)$ for all $x \in D$.

Note that the function $g: (0, 1) \to R^1$, $g(x) = x$, has bounded, but not closed, level sets and does not take on a minimum. On the other hand, $g(x) = e^x$, $x \in R^1$, shows that boundedness of some level set is, in general, necessary for g to have a global minimizer.

We shall return in the next section to a detailed discussion of the existence of minimizers. In the remainder of the present section, we consider the question of the uniqueness of local and global minimizers as well as of critical points, together with the related question of when a local minimizer is a global minimizer. These questions are closely associated with connectedness properties of the level sets, and we therefore introduce the following general class of functionals.

4.2.3. Definition. A functional $g: D \subset R^n \to R^1$ is *connected* on $D_0 \subset D$ if, given any $x, y \in D_0$, there exists a continuous function $p: [0, 1] \to D_0$ such that $p(0) = x$, $p(1) = y$, and

$$g(p(t)) \leq \max\{g(x), g(y)\}, \qquad \forall t \in (0, 1). \tag{1}$$

4.2. UNIQUENESS THEOREMS

The functional g is *strictly connected* if, whenever $x \neq y$, the function p may be chosen so that strict inequality holds in (1). ∎

We say that a set $S \subset R^n$ is *path-connected* if, for any $x, y \in S$, there is a continuous mapping $p \colon [0, 1] \to S$ such that $p(0) = x$ and $p(1) = y$. This leads to the following characterization of connected functionals in terms of level sets.

4.2.4. A functional $g \colon D \subset R^n \to R^1$ is connected on D if and only if every level set of g is path-connected.

Proof. Assume first that all level sets of g are path-connected and, for arbitrary $x, y \in D$, set $\gamma = \max(g(x), g(y))$. Then there exists a continuous $p \colon [0, 1] \to L(\gamma)$ so that $p(0) = x$, $p(1) = y$, and $g(p(t)) \leq \gamma$ for all $t \in (0, 1)$, since $p(t) \in L(\gamma)$. Conversely, suppose g is connected, and let x, y lie in an arbitrary level set $L(\gamma)$. Then there exists a continuous $p \colon [0, 1] \to D$ such that, by (1), $g(p(t)) \leq \max(g(x), g(y)) \leq \gamma$. Hence, $p(t) \in L(\gamma)$ for all $t \in (0, 1)$, and $L(\gamma)$ is path-connected. ∎

If $D \subset R^n$ is convex and $g \colon D \to R^1$ is a convex functional, then g is a connected functional; indeed, if $L(\gamma)$ is any level set of g and $x, y \in L(\gamma)$, then for all $t \in (0, 1)$

$$g(tx + (1-t)y) \leq tg(x) + (1-t)g(y) \leq t\gamma + (1-t)\gamma = \gamma,$$

so that $L(\gamma)$ is convex, and, hence, path-connected. More generally, any functional all of whose level sets are convex is connected. This gives rise to the following class of functionals.

4.2.5. Definition. A functional $g \colon D \subset R^n \to R^1$ is *quasiconvex* on a convex set $D_0 \subset D$ if, for any $x, y \in D_0$,

$$g(tx + (1-t)y) \leq \max\{g(x), g(y)\}, \qquad \forall t \in (0, 1). \tag{2}$$

The functional g is *strictly quasiconvex* if strict inequality holds in (2) when $x \neq y$. ∎

It follows immediately that any convex functional is also quasiconvex, but the converse is not true; for example, the function $\ln t$ is quasiconvex on $(0, \infty)$, but not convex. It is also easy to see, by a slight modification of the proof of **4.2.4**, that a functional $g \colon D \subset R^n \to R^1$ is quasiconvex on D if and only if all level sets of g are convex (see **E 4.2-4**). Finally, note that, in contrast to **3.4.3** stating the continuity of convex functionals on open sets, quasiconvex, and hence connected, functionals need not be continuous, as the example $f(x) = x$, $x > 0$, $f(x) = x - 1$, $x \leqslant 0$, shows.

We turn now to the question of the uniqueness of minimizers.

4.2.6. Let $g: D \subset R^n \to R^1$ be connected on D. Then g has at most one proper local minimizer x^*, and, moreover, $g(x^*) < g(x)$ for all $x \in D$, $x \neq x^*$. If g is strictly connected, then g has at most one local minimizer x^* and $g(x^*) < g(x)$ for all $x \in D$, $x \neq x^*$.

Proof. Suppose that x^* is a proper local minimizer and that there is a $y \neq x^*$ such that $g(y) \leqslant g(x^*) = \gamma$. Since g is connected, there is a continuous map $p: [0, 1] \to L(\gamma)$ such that $p(0) = x^*$ and $p(1) = y$. The continuity of p now ensures that for any open neighborhood S of x^* there is a t such that $p(t) \in S \cap D$, $p(t) \neq x^*$, and $g(p(t)) \leqslant \gamma$. This contradicts that x^* is a proper local minimizer. It follows that $g(x^*) < g(x)$ for all $x \in D$, which also implies that there can be at most one proper local minimizer. Now suppose that g is strictly connected and that $x^* \neq y^*$ are two local minimizers. Assume, for definiteness, that $g(x^*) \leqslant g(y^*) = \gamma$. Then, since g is strictly connected, there is a continuous $p: [0, 1] \to L(\gamma)$ such that $g(p(t)) < \gamma$ for all $t \in (0, 1)$, and $p(0) = x^*$, $p(1) = y^*$. Hence, for any open neighborhood S of y^*, there exist $t \in (0, 1)$ such that $p(t) \in S \cap D$. But then $g(p(t)) < g(y^*)$, and this contradicts that y^* is a local minimizer. Hence, there is at most one local minimizer, and since any global minimizer is also a local minimizer, there is at most one global minimizer. ∎

Clearly, any strictly quasiconvex functional, and hence any strictly convex functional, is strictly connected. Therefore, we have the following immediate corollary of **4.2.6**.

4.2.7. If $g: D \subset R^n \to R^1$ is strictly quasiconvex (in particular, strictly convex) on a convex set $D_0 \subset D$, then g has at most one local minimizer in D_0, and any local minimizer in D_0 is a global minimizer on D_0.

We complete this section with two simple results on uniqueness of critical points of convex functionals.

4.2.8. Suppose that $g: D \subset R^n \to R^1$ is convex and G-differentiable on an open convex set $D_0 \subset D$. Then $x^* \in D_0$ is a critical point of g if and only if x^* is a global minimizer on D_0. Moreover, if g is strictly convex on D_0, there is at most one critical point in D_0.

4.2. UNIQUENESS THEOREMS

Proof. If $x^* \in D_0$ is a critical point, then, by **3.4.4**,

$$g(x) \geq g(x^*) + g'(x^*)(x - x^*) = g(x^*), \qquad \forall x \in D_0.$$

Conversely, if $x^* \in D_0$ is a global minimizer, then x^* is a local minimizer, and, since D_0 is open, **4.1.3** ensures that $g'(x^*) = 0$. Finally, if g is strictly convex on D_0, then the inequality $[g'(x) - g'(y)](x - y) > 0$ of **3.4.5** precludes the possibility of distinct critical points. ∎

As an immediate corollary of **4.2.8** and **3.4.6**, we obtain the following useful result.

4.2.9. If $g: D \subset R^n \to R^1$ has a positive definite second G-derivative at each point of an open convex set $D_0 \subset D$, then g has at most one critical point (or local or global minimizer) in D_0.

NOTES AND REMARKS

NR 4.2-1. The concept of a connected functional appears to be new, but is a natural extension of the idea of a quasiconvex functional, which, in turn, dates back at least to Fenchel [1953].

NR 4.2-2. Various definitions have been put forth for strictly quasiconvex functionals. Ponstein [1967] (see also Mangasarian [1965], and Hanson [1964]) uses the relation

$$g(x) < g(y), \quad x, y \in D_0, \quad \text{implies} \quad g(tx + (1-t)y) < g(y), \quad \forall t \in (0, t), \quad (3)$$

while Poljak [1966] requires that

$$g(\tfrac{1}{2}[x + y]) < \max\{g(x), g(y)\}, \qquad x \neq y, \quad x, y \in D_0. \tag{4}$$

We have followed Elkin [1968], who advocated **4.2.5** for the following reasons. It is easy to see (**E 4.2-7**) that (3) does not imply the uniqueness of local minimizers, and we prefer to associate the terminology "strict" with uniqueness of minimizers. Moreover, as shown by Elkin, whenever g is continuous on D_0, then (4) also implies that g is strictly quasiconvex in the sense of **4.2.5**. We point out, however, that Ponstein in fact introduced the relation

$$g(tx + (1-t)y) < \max\{g(x), g(y)\}, \qquad \forall t \in (0, 1), \quad x \neq y,$$

but under the terminology "x-convexity."

NR 4.2-3. As shown by Ponstein [1967], (see **E 4.2-6**), any local minimizer of a functional $g: D \subset R^n \to R^1$ which satisfies (3) on an open convex set is also a global minimizer of g on D_0. Interestingly, the converse holds in the sense that if any local minimizer of g in D_0 is a global minimizer, then (3) holds. This has been proved independently by Martos [1967] and Elkin [1968].

NR 4.2-4. Intermediate between quasiconvex and convex functionals is the class of pseudoconvex functionals, which may be defined as follows: A functional $g\colon D \subset R^n \to R^1$ is *pseudoconvex* on a convex set $D_0 \subset D$ if whenever $g(x) > g(y)$, $x, y \in D_0$, there exist $\alpha > 0$ and $\tau \in (0, 1]$, both depending, in general, on x and y, such that

$$g((1-t)x + ty) \leqslant g(x) - t\alpha, \qquad \forall t \in [0, \tau]. \tag{5}$$

In addition, g is *strictly pseudoconvex* if (5) holds whenever $g(x) \geqslant g(y)$, $x \neq y$.

Pseudoconvexity was introduced by Mangasarian [1965] for continuously differentiable functionals $g\colon R^n \to R^1$ by means of the definition

$$g(x) > g(y) \quad \text{implies} \quad g'(x)(x - y) > 0. \tag{6}$$

It is easy to see (**E 4.2-12**) that (5) implies (6) whenever g is G-differentiable. It is also true that a convex functional is pseudoconvex, a pseudoconvex functional is quasiconvex, and any critical point of a pseudoconvex functional is a local minimizer. These and other properties of pseudoconvex functionals have been studied by Mangasarian [1965], Ponstein [1967], and Elkin [1968], and, together with other results, are given in **E 4.2-13**–**E 4.2-18**.

The concept of pseudoconvexity is also closely related to that of supportable convexity introduced by Nashed [1967].

EXERCISES

E 4.2-1. Let $g\colon R^n \to R^1$ be strictly convex. Show that any level set of g is strictly convex.

E 4.2-2. Let $g\colon D \subset R^n \to R^1$ be connected on D. Modify the proof of **4.2.6** to show that any local minimizer of g is also a global minimizer with respect to D.

E 4.2-3. Prove **4.2.6** directly for convex functionals.

E 4.2-4. Show that $g\colon D \subset R^n \to R^1$, where D is convex, is quasiconvex if and only if all level sets of g are convex.

E 4.2-5. Show that (a) $g\colon R^n \to R^1$ is quasiconvex if and only if $g(x) \leqslant g(y)$ implies $g(\alpha x + (1-\alpha)y) \leqslant g(y)$, for all $\alpha \in (0, 1)$. (b) If g is G-differentiable, then g is quasiconvex if and only if $g(x) \leqslant g(y)$ implies $g'(y)(y - x) \geqslant 0$. (Ponstein [1967]).

4.2. UNIQUENESS THEOREMS

E 4.2-6. Suppose that $g: R^n \to R^1$ has the property that

$$g(x) < g(y) \quad \text{implies} \quad g(tx + (1-t)y) < g(y), \quad \forall t \in (0, 1).$$

Show that any local minimizer of g is a global minimizer and also give an example to show that local minimizers need not be unique (Ponstein [1967]; Elkin [1968]).

E 4.2-7. Show that a strictly quasiconvex functional $g: R^n \to R^1$ satisfies the condition of **E 4.2-6**, but not conversely. Show that the conclusion of **E 4.2-6** does not hold, in general, for quasiconvex functionals.

E 4.2-8. Let $g: R^n \to R^1$ be continuous, and satisfy the condition of **E 4.2-6**. Show that g is quasiconvex (Elkin [1968]).

E 4.2-9. Let $g: R^n \to R^1$ be convex. Show that the set of local minimizers of g is convex (or empty). Show that this does not necessarily hold for quasiconvex functionals. Show, finally, that if g is convex and G-differentiable, then the set of critical points of g is convex.

E 4.2-10. Let $g: R^n \to R^1$ be continuous and strictly quasiconvex. Show that any level set of g has a nonempty interior or consists of precisely one point.

E 4.2-11. A level set $L(\gamma)$ of $g: R^n \to R^1$ with boundary $\dot{L}(\gamma)$ is called *proper* if either $E(\gamma) \equiv \{x \mid g(x) = \gamma\} \subset \dot{L}(\gamma)$ or $E(\gamma) = L(\gamma)$, and *strictly proper* if $E(\gamma) = L(\gamma)$. Assume that g is continuous. Then g satisfies the condition of **E 4.2-6** if and only if $L(\gamma)$ is convex and proper for all γ, and g is strictly quasiconvex if and only if $L(\gamma)$ is convex and strictly proper for all γ (Elkin [1968]).

E 4.2-12. Suppose that $g: R^n \to R^1$ is G-differentiable on D. Show that g is pseudoconvex in the sense of **NR 4.2-4** if and only if

$$g(x) > g(y), \quad x, y \in D \quad \text{implies} \quad g'(x)(x-y) > 0$$

and strictly pseudoconvex if and only if

$$g(x) \geq g(y), \quad x \neq y, \quad \text{implies} \quad g'(x)(x-y) > 0.$$

E 4.2-13. Show that $g: R^1 \to R^1$, $g(x) = x^2(1 + x^2)^{-1}$, is strictly pseudoconvex.

E 4.2-14. If $g: R^n \to R^1$ is (strictly) convex on the convex set D, show that g is (strictly) pseudoconvex on D.

E 4.2-15. If $g: R^n \to R^1$ is continuous and pseudoconvex on a convex set D, then g satisfies the condition of **E 4.2-6** on D. If g is continuous and strictly pseudoconvex on D, then g is strictly quasiconvex on D. (Elkin [1968]).

E 4.2-16. Show that the discontinuous function $g: R^1 \to R^1$,

$$g(x) = x, \quad x \geq 0, \qquad g(x) = x - 1, \quad x < 0,$$

is strictly pseudoconvex on R^1.

E 4.2-17. Suppose that $g: R^n \to R^1$ is G-differentiable and (strictly) pseudoconvex on an open convex set D. Show that any critical point of g in D is a (unique) global minimizer of g on D (Mangasarian [1965]).

E 4.2-18. Assume that $g: R^n \to R^1$ is continuous and pseudoconvex on an open convex set D. Show that any local minimizer of g in D is a global minimizer of g on D (Ponstein [1967]).

4.3. EXISTENCE THEOREMS

In **4.2.2**, we observed that whenever g has a compact level set, then g has a global minimizer. In this section, we shall examine conditions on g which ensure that g has a compact level set. Note first that if D is closed and g is continuous on D, then every level set is closed, and, in this case, the question reduces to the boundedness of level sets. In general, it is difficult to give useful conditions for g which ensure that *some* level set is bounded. This is so because of the essential equivalence of this question with the existence problem for minimizers.

4.3.1. Assume that $g: D \subset R^n \to R^1$ is continuous on the closed set D. Then g has a bounded level set if and only if the set of global minimizers of g is nonempty and bounded.

Proof. If g has a bounded level set $L(\gamma)$, then, because of the continuity of g and the closedness of D, $L(\gamma)$ is compact and, by **4.2.2**, the set of global minimizers is nonempty and bounded. For the converse, note that if x^* is a global minimizer, then the level set $L(g(x^*))$ is precisely the set of global minimizers, and, hence, bounded. ∎

We consider next a simple necessary and sufficient condition for *all* level sets to be bounded. Clearly, if D itself is bounded, then all level sets are bounded, so we shall assume that D is unbounded.

4.3.2. Let $g: D \subset R^n \to R^1$, where D is unbounded. Then all level sets of g are bounded if and only if $\lim_{k \to \infty} g(x^k) = +\infty$ whenever $\{x^k\} \subset D$ and $\lim_{k \to \infty} \| x^k \| = \infty$.

Proof. Assume first that all level sets of g are bounded. Then the existence of a sequence $\{x^k\} \subset D$ with $\lim_{k \to \infty} \| x^k \| = \infty$ for which $g(x^k) \leq \gamma < +\infty$ for all k would constitute a contradiction, since $\{x^k\} \subset L(\gamma)$. Conversely, suppose that $\lim_{k \to \infty} g(x^k) = +\infty$ whenever $\{x^k\} \subset D$ and $\lim_{k \to \infty} \| x^k \| = \infty$. Then the existence of an unbounded level set $L(\gamma)$ implies the existence of a sequence $\{x^k\} \subset L(\gamma)$ with $\lim_{k \to \infty} \| x^k \| = \infty$, and hence $g(x^k) \leq \gamma$ is again contradictory. ∎

4.3. EXISTENCE THEOREMS

We note that an analogous infinity condition on mappings $F: R^n \to R^n$ will also play an important role in the next two chapters (see, for example, **5.3.6** and **5.3.8**).

As an immediate corollary of **4.3.2**, **4.2.2**, and **4.2.6**, we have the following result.

4.3.3. If $g: D \subset R^n \to R^1$ is continuous on a closed set $D_0 \subset D$ and $\lim_{k \to \infty} g(x^k) = +\infty$ whenever $\lim_{k \to \infty} \|x^k\| = \infty$ for $\{x^k\} \subset D_0$, then g has a global minimizer $x^* \in D_0$. If, in addition, g is strictly connected on D_0, then x^* is also a unique local minimizer, and $g(x^*) < g(x)$ for all $x \in D_0$, $x \neq x^*$.

By **3.4.3**, a convex functional on an open set is continuous. Hence, another useful corollary derives from **4.3.3** and **4.2.7**.

4.3.4. Suppose that $g: R^n \to R^1$ is strictly convex (or continuous and strictly quasiconvex) and that $\lim_{k \to \infty} g(x^k) = +\infty$ whenever $\lim_{k \to \infty} \|x^k\| = \infty$. Then g has a unique minimizer x^*.

Similar results using the differential inequalities of **3.4** are given in E **4.3-7**.

Next we consider a specific, but still very general, condition which ensures that g tends to infinity when $\|x\|$ does.

4.3.5. Definition. A functional $g: D \subset R^n \to R^1$ is *uniformly connected* on $D_0 \subset D$ if an isotone function $d: [0, \infty) \to [0, \infty)$ with $d(t) > 0$ for $t > 0$ can be found such that, for any $x, y \in D_0$, there is a continuous mapping $p: [0, 1] \subset R^1 \to D_0$ which satisfies $p(0) = x$, $p(1) = y$, and, for all $t \in (0, 1)$,

$$g(p(t)) \leq \max\{g(x), g(y)\}$$
$$- \max\{\|x - p(t)\| d(\|y - p(t)\|), \|y - p(t)\| d(\|x - p(t)\|)\}. \quad \blacksquare \quad (1)$$

We note that functions d with the properties of **4.3.5** will play a strong role in the convergence analysis of minimization methods in Chapter 14.

4.3.6. Assume that $g: D \subset R^n \to R^1$ is continuous and uniformly connected on a closed set $D_0 \subset D$. Then g has a unique local minimizer $x^* \in D_0$ and $g(x^*) < g(x)$ for all $x \in D_0$, $x \neq x^*$.

Proof. By **4.3.3**, it suffices to show that $\lim_{k \to \infty} g(x^k) = +\infty$ whenever $\lim_{k \to \infty} \|x^k\| = \infty$ for $\{x^k\} \subset D_0$. For arbitrary $y^0 \in D_0$, the closedness of D_0 implies the compactness of $S_0 = \bar{S}(y^0, 1) \cap D_0$, and the continuity of g ensures that there is an α such that $g(x) \geq \alpha$ for all $x \in S_0$.

Now let $\{x^k\} \subset D_0$ with $\lim_{k\to\infty} \|x^k\| = \infty$, and assume that $\|x^k - y^0\| > 1$ for all $k \geq k_0$. Next let $p_k: [0, 1] \to D_0$, $k = k_0, \ldots$, be continuous mappings such that $p_k(0) = x^k$, $p_k(1) = y^0$, and such that (1) holds for each p_k with $x = x^k$ and $y = y^0$. By the continuity of the p_k, there exist $t_k \in (0, 1)$ such that $\|p_k(t_k) - y^0\| = 1$, and therefore, $g(p_k(t_k)) \geq \alpha$, $k = k_0, \ldots$. Hence, we have by (1) that

$$\max\{g(x^k), g(y^0)\} \geq g(p_k(t_k)) + \|x^k - p_k(t_k)\| d(\|p_k(t_k) - y^0\|)$$
$$\geq \alpha + \{\|x^k - y^0\| - \|y^0 - p_k(t_k)\|\} d(1).$$

But $d(1) > 0$, $\lim_{k\to\infty} \|x^k - y^0\| = +\infty$, and $\|y^0 - p_k(t_k)\| = 1$, so that $\lim_{k\to\infty} g(x^k) = +\infty$. ∎

As an important corollary of **4.3.6**, we have the following result for uniformly convex functionals; similar results are given in E 4.3-8.

4.3.7. Assume that $g: R^n \to R^1$ is uniformly convex on R^n. Then g has a unique local and global minimizer, Moreover, $\lim_{k\to\infty} g(x^k) = +\infty$ whenever $\lim_{k\to\infty} \|x^k\| = +\infty$, and every level set of g is compact.

Proof. There exists a constant $c > 0$ such that

$$g(p(t)) \leq \max\{g(x), g(y)\} - c(1-t) t \|x - y\|^2, \quad \forall x, y \in R^n,$$

where $p(t) = tx + (1-t)y$, $t \in [0, 1]$. But, with $d(t) = ct$ for $t \in [0, \infty)$,

$$\|x - p(t)\| d(\|y - p(t)\|) = \|y - p(t)\| d(\|x - p(t)\|) = ct(1-t)\|x - y\|^2,$$

so that g is uniformly connected. Since **3.4.3** ensures that g is continuous on R^n, **4.3.6** applies. Finally, we showed in the proof of **4.3.6** that g has the infinity property, and the last statement is then a consequence of **4.3.2**. ∎

When g is differentiable, **4.3.7** provides the basis for showing not only that g has a unique critical point, but that g' is even a homeomorphism of the whole space onto itself. We first give the following lemma on inverse mappings.

4.3.8. Let $F: D \subset R^n \to R^n$. If, for some $\gamma > 0$,

$$\|Fx - Fy\| \geq \gamma \|x - y\|, \quad \forall x, y \in D, \qquad (2)$$

then F^{-1} exists on $F(D)$ and

$$\|F^{-1}u - F^{-1}v\| \leq \gamma^{-1} \|u - v\|, \qquad \forall u, v \in F(D).$$

4.3. EXISTENCE THEOREMS

Proof. Clearly, (2) implies that F is one-to-one, so that F^{-1} exists on $F(D)$. Now, for any $u, v \in F(D)$, there exists $x, y \in D$ such that $Fx = u$ and $Fy = v$. Hence,

$$\| F^{-1}u - F^{-1}v \| = \| x - y \| \leq \gamma^{-1} \| Fx - Fy \|$$
$$= \gamma^{-1} \| u - v \|, \quad \forall u, v \in F(D). \quad \blacksquare$$

4.3.9. Assume that $g: R^n \to R^1$ is uniformly convex and continuously differentiable on R^n. Then the mapping $F: R^n \to R^n$, defined by $Fx = g'(x)^T$, $x \in R^n$, is a homeomorphism from R^n onto R^n.

Proof. Let b be an arbitrary element in R^n and define the functional $g_b: R^n \to R^1$ by $g_b(x) = g(x) - b^T x$. Then, since

$$g_b(tx + (1-t)y) - [tg_b(x) + (1-t)g_b(y)]$$
$$= g(tx + (1-t)y) - [tg(x) + (1-t)g(y)],$$

it follows that g_b is also uniformly convex, and, by **4.3.7**, has a unique local minimizer x^*. Therefore, by **4.1.3** and **4.2.8**, x^* is a unique critical point of g_b, and, since $g_b'(x) = g'(x) - b^T$, x^* is the unique solution of $Fx = b$; that is, F is one-to-one and onto R^n. Finally, to show that F is a homeomorphism, **3.4.5** implies the existence of a constant $\gamma > 0$ such that

$$\| g'(x) - g'(y) \| \| x - y \| \geq [g'(x) - g'(y)](x - y) \geq \gamma \| x - y \|^2$$

for all $x, y \in R^n$. Hence, $\| Fx - Fy \| \geq \gamma \| x - y \|$ for all $x, y \in R^n$, and **4.3.8** shows that F^{-1} is continuous. $\blacksquare$

As an easy consequence of this result together with the symmetry principle **4.1.6**, we have the following important corollary.

4.3.10. Let $F: R^n \to R^n$ be continuously differentiable on R^n and assume that $F'(x)$ is symmetric for each $x \in R^n$. If there is a constant $c > 0$ such that

$$h^T F'(x) h \geq c h^T h, \quad \forall x, h \in R^n, \tag{3}$$

then F is a homeomorphism from R^n onto itself.

Proof. The symmetry principle **4.1.6** ensures that there is a functional $g: R^n \to R^1$ such that $Fx = g'(x)^T$ for all $x \in R^n$. Condition (3) implies, by **3.4.6**, that g is uniformly convex on R^n, and hence the result follows from **4.3.9**. $\blacksquare$

Note that either **4.3.9** or **4.3.10** implies, of course, that for any $y \in R^n$ the equation $Fx = y$ has a unique solution which varies continuously with y.

NOTES AND REMARKS

NR 4.3-1. The results **4.3.4** and **4.3.10**, although "well known" and used implicitly as lemmas in various analyses, do not seem to be precisely stated in the literature. Theorem **4.3.7** is, essentially, a result of Poljak [1966].

NR 4.3-2. The notion of a uniformly connected functional seems new, but is a natural extension of uniformity concepts discussed by Poljak [1966] and Elkin [1968] for quasiconvex functionals. More specifically, Poljak calls a functional $g: R^n \to R^1$ *uniformly quasiconvex* if

$$g(\tfrac{1}{2}[x+y]) \leq \max\{g(x), g(y)\} - d(\|x-y\|), \quad \forall x, y \in R^n, \tag{4}$$

where $d: [0, \infty) \to [0, \infty)$ is such that $d(t) > 0$ for $t > 0$, whereas Elkin [1968] uses the definition

$$g(tx + (1-t)y) \leq \max\{g(x), g(y)\}$$
$$- \|x-y\| \min\{(1-t)\,d((1-t)\|x-y\|),\, t\,d(t\|x-y\|)\},$$
$$\forall t \in (0, 1),\ x, y \in R^n. \tag{5}$$

Note that our definition **4.3.5** reduces, for quasiconvex functionals, to

$$g(tx + (1-t)y) \leq \max\{g(x), g(y)\}$$
$$- \|x-y\| \max\{(1-t)\,d(t\|x-y\|),\, t\,d((1-t)\|x-y\|)\},$$

which differs somewhat from both (4) and (5). Neither Elkin nor Poljak used their definitions to obtain existence theorems, however, and **4.3.6** appears to be new even for uniformly quasiconvex functionals. A related concept of uniform pseudoconvexity has also been introduced by Elkin. (See **NR 14.6-2**.)

NR 4.3-3. The discussion of **NR 4.3-2** suggests the possibility of a more general definition of uniform convexity. Such a definition arises naturally from **4.3.5** in the form

$$g(tx + (1-t)y) \leq tg(x) + (1-t)g(y)$$
$$- \|x-y\| \max\{td((1-t)\|x-y\|),\, (1-t)\,d(t\|x-y\|)\} \tag{6}$$

and has been considered by Elkin [1968]. Note that if d is the function $d(t) = ct$ for some $c > 0$, then (6) reduces to (3.4.2). Note also that if $g: R^n \to R^1$ satisfies (6) for $x, y \in R^n$, and d satisfies the conditions of **4.3.5**, then **4.3.6** ensures that g

4.3. EXISTENCE THEOREMS

has a unique local and global minimizer. There are certain difficulties present in the weakening of (3.4.2) to (6), however. Assume that $g: R^n \to R^1$ is twice-continuously differentiable and that the function d of (6) is integrable. Then, proceeding as in **3.4.6**, it is easy to see that (6) implies that

$$g''(x) hh \geqslant \|h\| \hat{d}(\|h\|), \qquad \forall x, h \in R^n$$

where $\hat{d}: [0, \infty) \to [0, \infty)$ is again an isotone positive function. But $g''(x)$, as a bilinear operator, can only have quadratic lower bounds in $\|h\|$, so that $\hat{d}$ must be of the form $\hat{d}(t) = ct$ for some constant $c > 0$. Hence, as shown by **3.4.6**, g must be uniformly convex in the sense of **3.4.1**. This indicates that definitions of the form (6) are probably not meaningful, and casts a certain shadow on the very weak properties required of d in **4.3.5** as well.

EXERCISES

E 4.3-1. Let $A \in L(R^n)$ be symmetric. Show that the functional $g(x) = x^T A x$ has a bounded level set if and only if A is positive definite.

E 4.3-2. Give examples of functions $g: R^1 \to R^1$ (a) which have a local minimizer, but no bounded level set; (b) for which every level set is bounded, but there is no local minimizer.

E 4.3-3. Show that $g: R^1 \to R^1$, $g(x) = x + e^x$, is strictly convex, but that $g(x) \to -\infty$ as $x \to \infty$.

E 4.3-4. If $g: R^n \to R^1$ is convex and has a proper local minimizer x^*, or if g is strictly convex and has a local minimizer x^*, then $g(x) \to +\infty$ as $\|x\| \to \infty$.

E 4.3-5. Let $g: R^n \to R^1$ be uniformly convex and twice G-differentiable on R^n. Show directly, using **3.3.10** and **3.4.6**, that $g(x) \to +\infty$ as $\|x\| \to +\infty$.

E 4.3-6. Suppose that $g: R^n \to R^1$ is convex and has a bounded level set. Show that $g(x) \to +\infty$ as $\|x\| \to +\infty$. Conclude from this that if any level set of a convex functional is bounded, then every level set is bounded.

E 4.3-7. Assume that $g: R^n \to R^1$ satisfies $g(x) \to +\infty$ as $\|x\| \to \infty$ and either (a) g is G-differentiable on R^n and $[g'(x) - g'(y)](x - y) > 0$ for all $x \neq y$; or (b) g is twice G-differentiable on R^n and $g''(x)$ is positive definite for all $x \in R^n$. Show that g has a unique critical point which is also a unique local and global minimizer.

E 4.3-8. Assume that $g: R^n \to R^1$ satisfies either

(a) g is continuously differentiable on R^n and, for some $c > 0$,

$$[g'(x) - g'(y)](x - y) \geqslant c \|x - y\|^2, \qquad \forall x, y \in R^n,$$

or

(b) g is twice-continuously differentiable and, for some $c > 0$,
$$g''(x) hh \geq c \|h\|^2, \quad \forall x, h \in R^n.$$
Show that $(g')^T$ is a homeomorphism of R^n onto itself.

E 4.3-9. Let $g: R^n \to R^1$. Show that $g(x) - x^T b \to +\infty$ as $\|x\| \to \infty$ for every $b \in R^n$ if and only if $g(x)/\|x\| \to +\infty$, $\|x\| \to \infty$.

E 4.3-10. Let $g: R^n \to R^1$ be G-differentiable on R^n and suppose that one of the following conditions holds for some ball $S = S(0, \delta)$:
  (a) $g(x) \geq c_1 x^T x$, $c_1 > \frac{1}{2}$, $\forall x \in S$,
  (b) $g(x) \leq c_2 x^T x$, $c_2 < \frac{1}{2}$, $\forall x \in S$.

Show that the equation $x = g'(x)^T$ has a solution (Goldstein [1967]).

4.4. APPLICATIONS

In this section, the previous general results are applied to more specialized situations and to some of the sample problems of Chapter 1. We begin with a corollary of **4.3.9** and **4.3.10**. Recall that diagonal and isotone mappings are defined in **1.1.1** and **2.4.3**, respectively.

4.4.1. Let $A \in L(R^n)$ be symmetric, positive definite and suppose that either:

(a) $\phi: R^n \to R^n$ is continuously differentiable on R^n and $\phi'(x)$ is symmetric, positive semidefinite for all x, or

(b) $\phi: R^n \to R^n$ is continuous, diagonal, and isotone on R^n.

Then the mapping $F: R^n \to R^n$ defined by $Fx = Ax + \phi x$ is a homeomorphism of R^n onto R^n.

Proof. If (a) holds, then clearly F is continuously differentiable, and the inequality (4.3.3) is valid with c the minimal eigenvalue of A. Hence, **4.3.10** applies. If (b) holds, define

$$g(x) = \tfrac{1}{2} x^T A x + \sum_{i=1}^{n} \int_0^{x_i} \varphi_i(t)\, dt.$$

Clearly, g is continuously differentiable on R^n, and $g'(x) = (Fx)^T$. The isotonicity of ϕ then implies that

$$[g'(x) - g'(y)](x - y) \geq (x - y)^T A(x - y),$$

4.4. APPLICATIONS

so that, by **3.4.5**, g is uniformly convex. The result now follows from **4.3.9**. ∎

We note that the conditions on ϕ in **4.4.1** may be relaxed somewhat (see **E 4.4-1**).

We consider next the application of **4.4.1** to discrete analogs of boundary value problems of the form

$$u'' = f(t, u), \quad t \in [0, 1], \quad u(0) = \alpha, \quad u(1) = \beta. \tag{1}$$

In particular, the next result treats the discretization

$$x_{j+1} - 2x_j + x_{j-1} = h^2 f(jh, x_j), \quad j = 1, \ldots, n,$$
$$x_0 = \alpha, \quad x_{n+1} = \beta, \quad h = 1/(n+1), \tag{2}$$

of (1) discussed in detail in **1.1**.

4.4.2. Let $f: [0, 1] \times R^1 \to R^1$ and assume that, for each $t \in [0, 1]$, $f(t, \cdot)$ is continuous and isotone. Then, for any $n \geq 1$ and any α and β, the system of equations (2) has a unique solution which is a continuous function of α and β.

Proof. As discussed in **1.1**, (2) may be written in the form $Ax + \phi(x) = b$, where $A \in L(R^n)$, $b \in R^n$, and $\phi: R^n \to R^n$ are given by

$$A = \begin{pmatrix} 2 & -1 & & \bigcirc \\ -1 & 2 & \cdot & \\ & \cdot & \cdot & \cdot \\ & & \cdot & -1 \\ \bigcirc & & -1 & 2 \end{pmatrix}, \quad \phi x = h^2 \begin{pmatrix} f(h, x_1) \\ \vdots \\ \vdots \\ f(nh, x_n) \end{pmatrix}, \quad b = \begin{pmatrix} \alpha \\ 0 \\ \vdots \\ 0 \\ \beta \end{pmatrix} \tag{3}$$

Clearly, the conditions on f imply that ϕ is continuous and isotone on R^n. Moreover, we have shown in **2.3.10** that A is positive definite. Hence, **4.4.1** ensures that $A + \phi$ is a homeomorphism, so that, in particular, for any $b \in R^n$, $Ax + \phi x = b$ has a unique solution which is a continuous function of b. ∎

Note that the condition on f in **4.4.2** is, of course, satisfied if f has a continuous partial derivative $\partial_2 f$ for which

$$\partial_2 f(t, s) \geq 0, \quad \forall t \in [0, 1], \quad s \in (-\infty, \infty). \tag{4}$$

Indeed, this condition may be weakened somewhat to allow that $\partial_2 f(t, s)$ be negative (see **E 4.4-2**).

The result **4.4.2** is primarily meant to be illustrative and extends almost immediately to more general situations. For example, under the same conditions on f, the system

$$\sum_{j=0}^{n+1} a_{ij}x_j + c_i f(t_i, x_i) = 0, \quad i = 1,\ldots, n, \tag{5}$$

has a unique solution if x_0 and x_{n+1} are known, $t_1,\ldots, t_n$ are arbitrary points of $[0, 1]$, $c_1,\ldots, c_n$ are arbitrary nonnegative constants, and the $n \times n$ matrix $(a_{ij})_1^n$ is symmetric, positive definite. Such systems might arise, for example, from the boundary value problem (1) if, again, central differences are used to approximate u'' but the grid points $t_1,\ldots, t_n$ are not equally spaced, or, more generally, from discretization of problems of the form $Lu = f(t, u)$, where L is a linear differential operator other than d^2/dt^2.

In complete analogy to **4.4.2**, we obtain next the corresponding result for the discretization of the two-dimensional problem $\Delta u = f(s, t, u)$ discussed in **1.2**. Again, this result is meant to be only illustrative, and immediate extensions are possible to more general domains, discretizations, and differential operators provided that the conditions of **4.4.1** can be satisfied.

4.4.3. Let $f: [0, 1] \times [0, 1] \times R^1 \to R^1$ and assume that, for each $s, t \in [0, 1]$, $f(s, t, \cdot)$ is continuous and isotone. Then the system of equations

$$4x_{ij} - x_{i+1,j} - x_{i-1,j} - x_{i,j+1} - x_{i,j-1} + h^2 f(ih, jh, x_{ij}) = 0, \quad i,j = 1,\ldots, m, \tag{6}$$

where $h = (m + 1)^{-1}$, has a unique solution for any given values of $x_{0,j}$, $x_{m+1,j}$, $x_{j,0}$, and $x_{j,m+1}$, $j = 0,\ldots, m + 1$. Moreover, the solution is a continuous function of these boundary values.

Proof. As discussed in **1.2**, the system (6) may be written as $Ax + \phi x = b$, where $A \in L(R^n)$, $n = m^2$, is given by (1.2.7, 8), $\phi: R^n \to R^n$ is defined by (1.2.10), and $b \in R^n$ is the vector containing the boundary values. We showed in **2.3.10** that A is positive definite. The proof now proceeds precisely as in **4.2.2**. ∎

We turn next to variational problems. Recall that in **1.5** we showed that various discretizations of the problem: minimize $\iint f(u_s, u_t)\, ds\, dt$, give rise to a functional $g: R^n \to R^1$ of the form

$$g(x) = \sum_{i=1}^{M} \gamma_i f\left(\sum_{j=1}^{n} \alpha_{ij}x_j + \sum_{j=1}^{N} \hat{\alpha}_{ij}\eta_j, \sum_{j=1}^{n} \beta_{ij}x_j + \sum_{j=1}^{N} \hat{\beta}_{ij}\eta_j\right). \tag{7}$$

4.4. APPLICATIONS

We shall now be concerned with giving sufficient conditions for f and the coefficients α_{ij}, β_{ij} in order that g have a unique minimizer.

In order to write (7) in a more compact way, it is convenient to introduce the following notation. Define the matrices $H_i \in L(R^n, R^2)$,

$$H_i = \begin{pmatrix} \alpha_{i1} & \cdots & \alpha_{in} \\ \beta_{i1} & \cdots & \beta_{in} \end{pmatrix}, \quad i = 1,\ldots, M, \tag{8}$$

and the vectors $b^i \in R^2$,

$$b^i = \left(\sum_{j=1}^{N} \hat{\alpha}_{ij}\eta_j, \sum_{j=1}^{N} \hat{\beta}_{ij}\eta_j \right)^T, \quad i = 1,\ldots, M. \tag{9}$$

Then, clearly, (7) may be written in the form

$$g(x) = \sum_{i=1}^{M} \gamma_i f(H_i x + b^i). \tag{10}$$

A further condensation is possible if we also introduce the functional

$$h: (R^2)^M \to R^1, \quad h(y) = \sum_{i=1}^{M} \gamma_i f(y^i), \quad \forall y = (y^1,\ldots, y^M) \in (R^2)^M, \tag{11}$$

and the linear mapping $H \in L(R^n, (R^2)^M)$ defined by the block matrix

$$H = \begin{pmatrix} H_1 \\ \vdots \\ H_M \end{pmatrix}. \tag{12}$$

Therefore, we may write (10), and hence (7), in the form

$$g(x) = h(Hx + b), \quad x \in R^n, \tag{13}$$

where $b \in (R^2)^M$ is defined by

$$b^T = ((b^1)^T,\ldots, (b^M)^T). \tag{14}$$

One advantage of rewriting (7) as (13) is that g is now seen to be a composition of an affine operator and a functional, which permits general results on such compositions to be brought to bear. We consider next one such general result in which h and H need not be the form (11) and (12).

4.4.4. Let $h: R^m \to R^1$ be strictly convex and satisfy $\lim_{\|x\|\to\infty} h(x) = +\infty$. Assume that $H \in L(R^n, R^m)$ has rank n. Then, for any $b \in R^m$, the functional $g: R^n \to R^1$, $g(x) \equiv h(Hx + b)$, has a unique minimizer.

Proof. We show first that g is strictly convex. Let $x, y \in R^n$, $x \neq y$, and $\alpha \in (0, 1)$. Since rank $H = n$, it follows that $Hx \neq Hy$, and, therefore, by the strict convexity of h, that

$$\begin{aligned} g(\alpha x + [1-\alpha]y) &= h(H(\alpha x + [1-\alpha]y) + b) \\ &= h(\alpha[Hx + b] + [1-\alpha][Hy + b]) \\ &< \alpha h(Hx + b) + (1-\alpha) h(Hy + b); \end{aligned}$$

hence, g is strictly convex. Again, since H is one-to-one, we find that $\lim_{\|x\|\to\infty} \| Hx + b \| = \infty$, and therefore that $\lim_{\|x\|\to\infty} g(x) = +\infty$. The result is now a consequence of **4.3.4**. ∎

We note that various modifications in the hypotheses of **4.4.4** are possible; some of these are given in **E 4.4-6–E 4.4-8**.

We return now to the functional (7).

4.4.5. Suppose that $f: R^2 \to R^1$ is strictly convex and that $\lim_{\|y\|\to\infty} f(y) = +\infty$. Assume that $\gamma_i > 0$, $i = 1,\ldots, M$, and that the matrix H of (12) and (8) has rank n. Then the functional (7) has a unique minimizer.

Proof. For the functional h of (11) we have

$$\alpha h(x) + (1-\alpha) h(y) - h(\alpha x + [1-\alpha]y)$$
$$= \sum_{i=1}^{M} \gamma_i \{\alpha f(x^i) + (1-\alpha) f(y^i) - f(\alpha x^i + [1-\alpha] y^i)\}. \tag{15}$$

Now, if $x \neq y$, then $x^i \neq y^i$ for some i, and, therefore, since $\gamma_i > 0$, (15) shows that h is strictly convex. Similarly, if $\|x\| \to \infty$, then $\|x^i\| \to \infty$ for some i, and, consequently, $\gamma_i f(x^i) \to +\infty$. But all of the terms of the sum (11) either tend to $+\infty$ or remain bounded, depending on whether $\|x^j\| \to \infty$ or not. Hence, $\lim_{\|x\|\to\infty} h(x) = +\infty$, and the result follows from **4.4.4**. ∎

We illustrate **4.4.5** by means of the sample problems discussed in **1.5**:

$$\text{minimize} \iint (1 + u_s^2 + u_t^2)^{1/2} \, ds \, dt; \tag{16}$$

$$\text{minimize} \iint [u_s^2 + u_t^2 - (c-d) \ln(c + u_s^2 + u_t^2)] \, ds \, dt, \quad c > d > 0. \tag{17}$$

In the first case, we have $f(x) = (1 + x_1^2 + x_2^2)^{1/2}$ and, clearly, the Hessian matrix $H_f(x)$ of f is

$$H_f(x) = (1 + x_1^2 + x_2^2)^{-3/2} \begin{pmatrix} 1 + x_2^2 & -x_1 x_2 \\ -x_1 x_2 & 1 + x_1^2 \end{pmatrix}.$$

4.4. APPLICATIONS

Therefore $f''(x)$ is positive definite for all $x \in R^n$ and, by **3.4.6**, f is strictly convex. Clearly, $f(x) \to +\infty$ as $\|x\| \to \infty$. Similarly, for (17) we have $f(x) = x_1^2 + x_2^2 - (c - d)\ln(c + x_1^2 + x_2^2)$, and it is easy to see (**E 4.4-10**) that f is strictly convex and $f(x) \to +\infty$ as $x_1^2 + x_2^2 \to \infty$. Therefore, we can say: *The discrete analog (7) of either of the problems (16) or (17) has a unique minimizer provided only that the discretization matrix H of (8), (12) has rank n.*

In order to exhibit a class of suitable discretizations with the rank n property, we restrict our attention to two-dimensional problems on the unit square. In the following, Ω_{ij} denotes a mesh square with vertices P_{ij}, $P_{i-1,j-1}$, $P_{i,j-1}$, $P_{i-1,j}$ as used in Section **1.5**.

4.4.6. Let $[0, 1] \times [0, 1] = \bigcup_{i,j=1}^{m} \Omega_{ij}$, and let $Q_{ij} \in \Omega_{ij}$. Then any approximation of the form

$$u_s(Q_{ij}) \doteq \alpha_1 u(P_{ij}) + \alpha_2 u(P_{i-1,j}) + \alpha_3 u(P_{i,j-1}) + \alpha_4 u(P_{i-1,j-1})$$

with at least one nonzero α_i leads to a discretization matrix H of rank n.

Proof. We need to show that H contains n linearly independent rows. Let us suppose first that $\alpha_1 \neq 0$. Then, clearly, the n equations

$$\alpha_1 u(P_{ij}) + \alpha_2 u(P_{i-1,j}) + \alpha_3 u(P_{i,j-1}) + \alpha_4 u(P_{i-1,j-1}) = 0, \quad i,j = 1,\ldots,m,$$

yield a lower triangular coefficient matrix of rank n. Hence, the other equations are redundant. If $\alpha_1 = 0$, but $\alpha_2 \neq 0$, then the set of indices $i = 2,\ldots, m+1, j = 1,\ldots, m$ gives the lower triangular matrix. Similarly, if $\alpha_1 = \alpha_2 = 0$, and $\alpha_3 \neq 0$, then we take $i = 1,\ldots, m, j = 2,\ldots, m+1$, while if $\alpha_1 = \alpha_2 = \alpha_3 = 0$ and $\alpha_4 \neq 0$, we take $i,j = 2,\ldots, m+1$. ∎

Note that **4.4.6** covers most of the discretizations discussed in **1.5**. Note also that only a very weak condition on the approximation of u_s (or, alternatively, on the approximation of u_t) is required.

NOTES AND REMARKS

NR 4.4-1. The results of this section from **4.4.4** on are due, in this generality, to Stepleman [1969]. However, the existence of minimizers for the discrete Plateau problem as well as for uniformly convex f was shown by Schechter [1962] for the backward difference approximation.

NR 4.4-2. The results of this section extend immediately to more general problems

$$Ju = \iint f(s, t, u(s, t), u_s(s, t), u_t(s, t))\, ds\, dt,$$

provided that $f(s, t, r, p, q)$ is again strictly convex in r, p, q for each s, t; see **E 4.4-12**. However, Stepleman [1969] has shown that the following is also true: Suppose that $f(s, t, r, p, q)$ is convex in r, p, q for each fixed s, t, and strictly convex in p and q. Assume further that $f(s, t, r, p, q) \to +\infty$ as $p^2 + q^2 \to \infty$ and that the matrix H of (8), (12) has rank n. Then the functional

$$g(x) \equiv \sum_{i=1}^{M} \gamma_i f\left(s_i, t_i, \sum_{j=1}^{n} \xi_{ij} x_j + \sum_{j=1}^{N} \hat{\xi}_{ij} \eta_j, \sum_{j=1}^{n} \alpha_{ij} x_j \right.$$

$$\left. + \sum_{j=1}^{N} \hat{\alpha}_{ij} \eta_j, \sum_{j=1}^{n} \beta_{ij} x_j + \sum_{j=1}^{N} \hat{\beta}_{ij} \eta_j \right), \tag{18}$$

with $\gamma_i > 0$, $i = 1, \ldots, M$, has a unique minimizer. (Note that this holds regardless of the ξ_{ij} and $\hat{\xi}_{ij}$.)

More generally, Stepleman has also given results when f is not convex in r, as well as a treatment of the "nonlinear" discretization (1.5.17).

EXERCISES

E 4.4-1. Conclude that **4.4.1** remains valid provided that there is a constant $c > -\lambda$, where λ is the minimal eigenvalue of A, such that either (a) ϕ is continuously differentiable and $\phi'(x) - cI$ is symmetric, positive semidefinite for all $x \in R^n$, or (b) ϕ is continuous and diagonal and ϕ-cI is isotone.

E 4.4-2. Let $f: [0, 1] \times R^1 \to R^1$ have a continuous partial derivative $\partial_2 f$ which satisfies $\partial_2 f(t, s) \geq \eta > -\pi^2$ for all $t \in [0, 1]$ and $s \in R^1$. Use **E 4.4-1** and **E 2.3-4** to conclude that the system (2) has a unique solution for sufficiently small $h = (n + 1)^{-1}$. Apply this result to the pendulum problem (1.1.1) with $|c| < \pi^2$.

E 4.4-3. Consider the boundary value problem

$$u'' + a(t) u' = f(t, u), \quad u(0) = \alpha, \quad u(1) = \beta,$$

where a is a continuous function on $[0, 1]$ and f satisfies the hypotheses of **4.4.2**. Set $a_i = a(ih)$, $i = 1, \ldots, n$, and show, by applying **4.4.1** and **E 2.3-5**, that, for all $h \leq h_0 < (\max |a_i|)^{-1}$, the system of equations

$$h^{-2}[x_{i+1} - 2x_i + x_{i-1}] + a_i(2h)^{-1}(x_{i+1} - x_{i-1}) = f(ih, x_i), \quad i = 1, \ldots, n,$$

has a unique solution.

E 4.4-4. Let $B \in L(R^n)$ be symmetric, negative definite and suppose that $\phi: R^n \to R^n$ is continuously differentiable and that $\phi'(x)$ is symmetric, positive semidefinite for all x. For any $b \in R^n$, show that the equation $x = B\phi x + b$ has a unique solution.

4.4. APPLICATIONS

E 4.4-5. Let $f: R^n \to R^1$ be G-differentiable on R^n and let $B \in L(R^n)$ be symmetric. Show that the equation $x = Bf'(x)^T + b$ has a solution if any one of the following conditions hold:

(a) B is positive definite and
$$2f(x) \leq \alpha x^T x + \beta (x^T x)^\delta + \gamma,$$
where β and γ are arbitrary constants, $|\delta| < 1$, $\alpha < \lambda^{-1}$, and λ is the (algebraically) largest eigenvalue of B (α need not be positive).

(b) B is nonsingular and
$$f(x) \geq \lambda_0^{-1} x^T x + \beta (x^T x)^\delta + \gamma,$$
where β, γ, and δ are as in (a), and λ_0 is the smallest positive eigenvalue of B.

(c) f is twice F-differentiable and
$$f''(x) hh \leq \alpha h^T h, \qquad \forall x, h \in R^n,$$
where α is such that $I - \alpha B$ is positive definite.

(d) f is twice F-differentiable, B is nonsingular, and
$$f''(x) hh \geq 2\lambda_0^{-1} h^T h,$$
where λ_0 is as in (b).

E 4.4-6. Let $f: R^m \to R^1$ be uniformly convex and assume that $H \in L(R^n, R^m)$ has rank n. For any $b \in R^m$, show that $g(x) \equiv f(Hx + b)$ is uniformly convex (Stepleman [1969]).

E 4.4-7. Let $f: R^m \to R^1$ be uniformly convex and let $H \in L(R^n, R^m)$ be one-to-one. Show that, for any fixed $b \in R^n$, the functional $f(Hx + b)$ has a unique minimizer.

E 4.4-8. Let $H \in L(R^n, R^m)$, assume that $f: R^m \to R^1$ is twice F-differentiable on R^m, and, for arbitrary $b \in R^m$, define $g(x) = f(Hx + b)$. Show that
$$g''(x) hh = f''(Hx + b)(Hh)(Hh), \qquad \forall x, h \in R^n,$$
and, hence, conclude that, if H has rank n and $f''(x)$ is positive definite for all $x \in R^n$, then $g''(x)$ is positive definite for all $x \in R^n$.

E 4.4-9. Let $f: R^2 \to R^1$ be uniformly convex and $\gamma_i > 0$, $i = 1,..., M$. Show that the functional (11) is uniformly convex.

E 4.4-10. Let $f: R^2 \to R^1$ be defined by $f(p, q) = \psi(p^2 + q^2)$, where $\psi: R^1 \to R^1$ satisfies $0 < m \leq \psi'(t) \leq M < \infty$, $m \leq \psi'(t) + 2t\psi''(t) \leq M$.

Show that f is uniformly convex, and, in particular, that if $c > d > 0$, then
$$f(p, q) = p^2 + q^2 - (c - d) \ln(c + p^2 + q^2)$$
is uniformly convex.

E 4.4-11. Consider the central difference approximations
$$u_s(P_{ij}) = (2h)^{-1} [u(P_{i+1,j}) - u(P_{i-1,j})], \quad u_t(P_{ij}) = (2h)^{-1} [u(P_{i,j+1}) - u(P_{i,j-1})].$$
Let $\Omega = [0, 1] \times [0, 1]$. Show that the discretization matrix H of (8), (12) is one-to-one if and only if the number m^2 of interior grid points is even.

E 4.4-12. Consider the functional $g: R^n \to R^1$ defined by equation (18) of NR 4.4-2. Assume that $f: [0, 1]^2 \times R^3 \to R^1$ and that, for each $s, t \in [0, 1]$, $f(s, t, r, p, q)$ is strictly convex in r, p, q and tends to $+\infty$ as $r^2 + p^2 + q^2 \to \infty$. State and prove the analogs of **4.4.5** and **4.4.6**, and thus give sufficient conditions that g has a unique minimizer.

E 4.4-13. Assume that $A \in L(R^n)$ is symmetric, positive definite, and that $\phi: R^n \to R^n$ is continuously differentiable, diagonal, and isotone on R^n. Show that the unique solution x^* of $Ax + \phi x = 0$ satisfies
$$\| x^* \|_2 \leq \| A^{-1} \|_2 \| \phi(0) \|_2 .$$

Chapter 5 / CONTRACTIONS AND THE CONTINUATION PROPERTY

5.1. CONTRACTIONS

If A is a linear operator from R^n to R^n, then **2.3.1** shows that the equation $x = Ax$ has a unique solution if $\|A\| < 1$. This extends in various ways to nonlinear operators and, in particular, to one of the most basic results of nonlinear analysis—the contraction mapping theorem **5.1.3**. We begin by introducing the following concept.

5.1.1. Definition. A mapping $G: D \subset R^n \to R^n$ is *nonexpansive* on a set $D_0 \subset D$ if

$$\|Gx - Gy\| \leq \|x - y\|, \quad \forall x, y \in D_0, \tag{1}$$

and *strictly nonexpansive* on D_0 if strict inequality holds in (1) whenever $x \neq y$. ∎

Note that any nonexpansive mapping on D_0 is Lipschitz-continuous on D_0. Observe also that a linear operator $A \in L(R^n)$ is nonexpansive (strictly nonexpansive) if and only if $\|A\| \leq 1$ ($\|A\| < 1$).

In the following, we shall consider equations of the form $x - Gx = 0$. Any solution x^* of this equation, that is, any point x^* in the domain of G for which $x^* = Gx^*$, is a *fixed point* of G.

If $G: D \subset R^n \to R^n$ is strictly nonexpansive on D_0 and $x^*, y^* \in D_0$ are any two distinct fixed points, then

$$\|x^* - y^*\| = \|Gx^* - Gy^*\| < \|x^* - y^*\|$$

is a contradiction, which implies that $x^* = y^*$; that is, a strictly nonexpansive mapping can have at most one fixed point. However, strict nonexpansivity is not sufficient to guarantee the existence of a fixed point, as the following one-dimensional example shows:

$$Gx = \begin{cases} x + \exp(-x/2), & x \geq 0. \\ \exp(x/2), & x \leq 0. \end{cases}$$

This leads to the following strengthening of Definition **5.1.1**.

5.1.2. Definition. A mapping $G: D \subset R^n \to R^n$ is *contractive* on a set $D_0 \subset D$ if there is an $\alpha < 1$ such that $\| Gx - Gy \| \leq \alpha \| x - y \|$ for all $x, y \in D_0$. ∎

If G is contractive on D_0, we shall also call G a contraction mapping, or simply a contraction, on D_0. Clearly, a linear operator $A \in L(R^n)$ is contractive on all of R^n if and only if $\| A \| < 1$. Note, however, that the contractive property is norm dependent, so that a mapping may be contractive in one norm on R^n but not in another (**E 5.1-1**).

Clearly, a contractive mapping G is strictly nonexpansive, and, thus, in particular, G is Lipschitz-continuous and has at most one fixed point. The existence of a fixed point is given by the following basic result.

5.1.3. Contraction-Mapping Theorem. Suppose that $G: D \subset R^n \to R^n$ is contractive on a closed set $D_0 \subset D$ and that $GD_0 \subset D_0$. Then G has a unique fixed point in D_0.

Proof. Let x^0 be an arbitrary point in D_0, and form the sequence $x^k = Gx^{k-1}$, $k = 1, 2, \ldots$. Since $GD_0 \subset D_0$, $\{x^k\}$ is well defined and lies in D_0, and

$$\| x^{k+1} - x^k \| = \| Gx^k - Gx^{k-1} \| \leq \alpha \| x^k - x^{k-1} \|$$

so that

$$\| x^{k+p} - x^k \| \leq \sum_{i=1}^{p} \| x^{k+i} - x^{k+i-1} \| \leq (\alpha^{p-1} + \cdots + 1) \| x^{k+1} - x^k \|$$

$$\leq [\alpha^k/(1-\alpha)] \| x^1 - x^0 \|.$$

Hence, $\{x^k\}$ is a Cauchy sequence and has a limit x^* in D_0. Moreover, by the continuity of G, it follows that $\lim_{k\to\infty} Gx^k = Gx^*$, so that x^* is a fixed point. ∎

The contraction-mapping theorem will play a major role in many parts of this book. In particular, Chapter 12 is devoted to a thorough discussion

5.1. CONTRACTIONS

of this theorem along with its many variants and generalizations. In the present chapter, we shall use **5.1.3** primarily as a tool for obtaining other existence theorems.

As a first application of **5.1.3**, we prove the following fixed-point theorem for nonexpansive mappings.

5.1.4. Assume that $G: D \subset R^n \to R^n$ is nonexpansive on a closed convex set $D_0 \subset D$ and that $GD_0 \subset D_0$. Then G has a fixed point in D_0 if and only if the sequence $x^{k+1} = Gx^k$, $k = 0, 1, \ldots$, is bounded for at least one $x^0 \in D_0$.

Proof. We denote by G^k the kth power of G defined recursively by $G^0 = I$, $G^k x = G^{k-1}(Gx)$, $k = 1, 2, \ldots$, $x \in D_0$. If G has a fixed point $x^* \in D_0$, then the sequence $G^k x^* = x^*$, $k = 0, 1, \ldots$, is certainly bounded. For the converse, we show first that D_0 contains a compact, convex set $\bar{C}$ such that $G\bar{C} \subset \bar{C}$. Choose r so that

$$\| G^k x^0 - x^0 \| \leq r, \qquad k = 1, 2, \ldots,$$

and set

$$Q_k = \{ x \in D_0 \mid \| x - G^j x^0 \| \leq r, \; j = k, k+1, \ldots \}.$$

Then $Q_k \subset Q_{k+1}$ and $x^0 \in Q_k$ for all k. Moreover, if $x \in Q_k$, then $\| x - G^k x^0 \| \leq r$ and

$$\| x - x^0 \| \leq \| x - G^k x^0 \| + \| G^k x^0 - x^0 \| \leq 2r,$$

so that

$$C = \bigcup_{k=0}^{\infty} Q_k \subset \{ x \in R^n \mid \| x - x^0 \| \leq 2r \}.$$

Clearly the closure $\bar{C}$ is compact, and, since each Q_k is convex, $\bar{C}$ is also convex. Furthermore, if $x \in C$, then $x \in Q_k$ for some k, and $Gx \in Q_{k+1}$, since

$$\| Gx - G^{j+1} x^0 \| \leq \| x - G^j x^0 \| \leq r, \qquad j = k, k+1, \ldots.$$

It follows that if $y \in \bar{C}$ and $\{y^i\} \subset C$ converges to y, then $\{Gy^i\} \subset C$, so that $Gy \in \bar{C}$ and, thus, $G\bar{C} \subset \bar{C}$.

Now, let $\alpha \in (0, 1)$ and $z \in \bar{C}$ be fixed and define

$$G_\alpha x = \alpha Gx + (1 - \alpha) z, \qquad \forall x \in \bar{C}.$$

Since $\bar{C}$ is convex, G maps $\bar{C}$ into itself, and

$$\| G_\alpha x - G_\alpha y \| = \alpha \| Gx - Gy \| \leq \alpha \| x - y \|, \qquad \forall x, y \in \bar{C},$$

shows that G_α is a contraction on $\bar{C}$. Hence, by **5.1.3**, G_α has a unique fixed point x^α in $\bar{C}$. From

$$x^\alpha - \alpha G x^\alpha = G_\alpha x^\alpha - \alpha G x^\alpha = (1-\alpha)z,$$

it follows that

$$\lim_{\alpha \to 1}[(1/\alpha)\, x^\alpha - G x^\alpha)] = \lim_{\alpha \to 1}[(1-\alpha)/\alpha]\, z = 0. \tag{2}$$

Let $\{\alpha_k\} \subset (0, 1)$ be any sequence such that $\lim_{k \to \infty} \alpha_k = 1$, and set $x^k = x^{\alpha_k}$. Since $\bar{C}$ is compact, there exists a convergent subsequence $\{x^{k_i}\}$ with limit point $x^* \in \bar{C}$. Then, also $\lim_{i \to \infty}(1/\alpha_{k_i})\, x^{k_i} = x^*$ and, by the continuity of G, (2) implies that $x^* = Gx^*$. ∎

Note that an important special case of **5.1.4** is that if a nonexpansive mapping G maps a compact convex set D_0 into itself, then G has a fixed point in D_0. This is also a special case of the Brouwer fixed-point theorem **6.3.2** to be proved in the next chapter. We also note that the convexity assumption in **5.1.4** cannot, in general, be removed, as the simple one-dimensional example $Gx = -x$, $D_0 = \{-1, 1\}$ shows.

As another application of **5.1.3**, we prove the following result, which may be considered a nonlinear analog of the Neumann lemma **2.3.1**.

5.1.5. Let $F = I - G$, where $G: R^n \to R^n$ and I is the identity on R^n. If G is contractive on R^n, then F is a homeomorphism from R^n onto R^n.

Proof. It follows immediately from **5.1.3** that for any $y \in R^n$ the mapping G_y defined by $G_y x = Gx + y$ has a unique fixed point in R^n; that is, the equation $Fx = y$ has a unique solution for any $y \in R^n$, so that F is one-to-one and onto R^n. Now, clearly, F is continuous, and it follows from

$$\|Fx - Fy\| = \|x - y - (Gx - Gy)\| \geq (1 - \alpha)\|x - y\| \tag{3}$$

and **4.3.8** that F^{-1} is continuous. ∎

Next, we generalize **5.1.5** in two ways. First, the identity operator is replaced by an arbitrary nonsingular linear operator, and, secondly, the contractive condition is not assumed to hold on the whole space.

5.1.6. Suppose that $A \in L(R^n)$ is nonsingular and that $G: D \subset R^n \to R^n$ is a mapping such that in a closed ball $S_0 = \bar{S}(x^0, \delta) \subset D$

$$\|Gx - Gy\| \leq \alpha \|x - y\|, \quad \forall x, y \in S_0, \tag{4}$$

5.1. CONTRACTIONS

where
$$0 < \alpha < \beta^{-1}, \qquad \beta = \|A^{-1}\|. \tag{5}$$

Then the mapping $F: S_0 \to R^n$, defined by $Fx = Ax - Gx$, $x \in S_0$, is a homeomorphism between S_0 and $F(S_0)$. Moreover, for any $y \in S_1 = S(Fx^0, \sigma)$, where $\sigma = (\beta^{-1} - \alpha)\delta$, the equation $Fx = y$ has a unique solution in S_0. Hence, in particular, $S_1 \subset F(S_0)$.

Proof. For fixed $y \in S_1$, define the mapping $H: S_0 \to R^n$ by $Hx = A^{-1}[Gx + y] = x - A^{-1}[Fx - y]$. Clearly, $Fx = y$ has a unique solution in S_0 if and only if H has a unique fixed point. But, for any $x, y \in S_0$

$$\|Hx - Hz\| = \|A^{-1}(Gx - Gz)\| \leq \beta\alpha \|x - z\|,$$

so that, by (5), H is contractive on S_0. Moreover, for any $x \in S_0$,

$$\|Hx - x^0\| \leq \|Hx - Hx^0\| + \|Hx^0 - x^0\| \leq \beta\alpha \|x - x^0\| + \beta \|Fx - y\|$$
$$\leq \beta\alpha\delta + \beta\sigma = \delta$$

by definition of σ. Hence, H maps S_0 into S_0, and by **5.1.3**, H has a unique fixed point x in S_0. Finally, to show that F is a homeomorphism, note that, for any $x, y \in S_0$,

$$\|x - y\| = \|A^{-1}(Gx - Gy) + A^{-1}(Fx - Fy)\| \leq \alpha\beta \|x - y\| + \beta \|Fx - Fy\|,$$

so that

$$\|Fx - Fy\| \geq (\beta^{-1} - \alpha)\|x - y\|. \tag{6}$$

Therefore, by **4.3.8**, F^{-1} is continuous while, clearly, F itself is also continuous. ∎

If $A = I$ in **5.1.6**, then the result may be considered a local version of **5.1.5**. The important point to note is that when G is contractive on only a subset of R^n, then G need not have fixed points (**E 5.1-3**); however, **5.1.6** ensures that the equation $x - Gx = y$ has a unique solution provided y is sufficiently close to $x^0 - Gx^0$. On the other hand, if (4) holds for all $x, y \in R^n$, we have the following extension of **2.3.1** and **5.1.5**.

5.1.7. If $A \in L(R^n)$ is nonsingular and $G: R^n \to R^n$ satisfies $\|Gx - Gy\| \leq \alpha \|x - y\|$ for all $x, y \in R^n$, where $\alpha < \|A^{-1}\|^{-1}$, then $A - G$ is a homeomorphism from R^n onto R^n.

The content of this result is that the linear homeomorphism A may be perturbed by a nonlinear mapping G provided G is "suitably small." We next give another interpretation of **5.1.6** in a different context. Let $F: D \subset R^n \to R^n$ and suppose that the equation

$$Fx = y \tag{7}$$

has a solution x^*. Then, three questions arise immediately. First, is x^* an *isolated solution*, that is, is there a neighborhood of x^* which contains no other solutions of (7)? This question is, of course, answered affirmatively if F is one-to-one in a neighborhood of x^*. Second, if y is changed by a small amount, will (7) still have a solution? This is indeed the case if $F(D)$ contains an open neighborhood of y. Finally, does this solution of the perturbed equation vary continuously with y? This, as well as the previous questions, will be true if F is a "local homeomorphism." Recall from Section 3.1 that the restriction F_U of F to $U \subset D$ is defined by $F_U x = Fx$, $x \in U$.

5.1.8. Definition. The mapping $F: D \subset R^n \to R^n$ is a *local homeomorphism* at $x \in \operatorname{int}(D)$ if there exist open neighborhoods U and V of x and Fx, respectively, such that the restriction of F to U is a homeomorphism between U and V.

Now note that a corollary of **5.1.6** in this terminology is that F is a local homeomorphism at x^0. Indeed, we have shown that F is a homeomorphism between S_0 and $F(S_0)$ and that $F(S_0)$ contains the open ball $V = S(Fx^0, \sigma)$. Hence, if we set $U = F^{-1}(V)$, then U is open and the restriction F_U is a homeomorphism between U and V. We may therefore state:

5.1.9. Let $F: D \subset R^n \to R^n$. If, for some $x^0 \in D$, there is a nonsingular $A \in L(R^n)$ and a $\delta > 0$ so that

$$\| Fx - Fy - A(x - y) \| \leqslant \alpha \| x - y \|, \quad \forall x, y \in \bar{S}(x^0, \delta) \subset D,$$

where $\alpha < \| A^{-1} \|^{-1}$, then F is a local homeomorphism at x_0.

NOTES AND REMARKS

NR 5.1-1. All of the results of this section, as well as their proofs, remain valid for arbitrary Banach spaces. Indeed, the contraction-mapping theorem holds on an arbitrary complete metric space; that is, if X is a complete metric space with metric d and $G: X \to X$ such that $d(Gx, Gx) \leqslant \alpha d(x, y)$ for some $\alpha < 1$ and all $x, y \in X$, then G has a unique fixed point. The proof is word for word the same. For complete normed linear spaces, the theorem was first formulated and proved by Banach [1922] in his famous dissertation.

NR 5.1-2. The proof of **5.1.3** shows that the iterates $x^{k+1} = Gx^k$, $k = 0, 1,\ldots$, actually converge to the fixed point x^*. This convergence aspect of the contraction theorem, as well as related error estimates, will be discussed in detail in Chapter 12.

NR 5.1-3. Theorem **5.1.4** was given in Browder and Petryshyn [1966], for uniformly convex Banach spaces (see **E 2.2-3**), as an extension of earlier work of Browder [1965b], Kirk [1965], and Belluce and Kirk [1966]. Belluce and Kirk [1969] have more recently introduced an additional condition for G which allows results of this type for more general domains.

EXERCISES

E 5.1-1. Give an example of a linear operator $B \in L(R^2)$ and of two norms on R^2 such that B is contractive in one norm, but not in the other.

E 5.1-2. Assume that $F: D \subset R^n \to R^n$ has a G-derivative which satisfies $\|F'(x)\| \leq \alpha < 1$ for all x in a convex set $D_0 \subset D$. Use **3.2.3** to show that F is contractive on D_0.

E 5.1-3. Define $f: [0, 1] \subset R^1 \to R^1$ by $f(x) = \tfrac{1}{2}x + 2$, $x \in [0, 1]$. Show that f is contractive on $[0, 1]$, but has no fixed point.

E 5.1-4. Suppose that $F: D \subset R^n \to R^n$ has a continuous F-derivative in an open neighborhood S of x^0 and that $\rho(F'(x^0)) < 1$. Show that there is another open neighborhood S_1 of x^0 and a norm on R^n so that F is contractive in S_1.

E 5.1-5. Assume that $F: R^n \to R^n$ is G-differentiable on an open set D and, for some $\gamma > 0$, satisfies $\|Fx - Fy\| \geq \gamma \|x - y\|$, $x, y \in D$. Show that, for any $x \in D$, $F'(x)$ is invertible and $\|F'(x)^{-1}\| \leq \gamma^{-1}$.

5.2. THE INVERSE AND IMPLICIT FUNCTION THEOREMS

One answer to the question of when a mapping is a local homeomorphism at x can be phrased in terms of the derivative. Recall that a strong derivative is defined in **3.2.9**.

5.2.1. Inverse Function Theorem. Suppose that $F: D \subset R^n \to R^n$ has a strong F-derivative at $x^0 \in \text{int}(D)$ (or that F has an F-derivative in a neighborhood of x^0 which is continuous at x^0) and that $F'(x^0)$ is nonsingular. Then F is a local homeomorphism at x^0. Moreover, if F_U is the restriction of F to any open neighborhood U of x^0 on which F is one-to-one, then F_U^{-1} has a strong F-derivative at Fx^0 and

$$(F_U^{-1})'(Fx^0) = [F'(x^0)]^{-1}. \tag{1}$$

In addition, if F' exists and is continuous in an open neighborhood U of x^0 then $(F_U^{-1})'$ exists and is continuous in an open neighborhood of Fx^0.

Proof. Set $A = F'(x^0)$ and let α satisfy $0 < \alpha < \beta^{-1}$, $\beta = \|A^{-1}\|$. Then, since $F'(x^0)$ is strong, there is a $\delta > 0$ so that $S_0 = \bar{S}(x^0, \delta) \subset D$ and

$$\|Fx - Fy - A(x - y)\| \leq \alpha \|x - y\|, \qquad \forall x, y \in S_0.$$

Therefore, by **5.1.9**, F is a local homeomorphism at x^0. [If F' exists in a neighborhood of x^0 and is continuous at x^0, then **3.2.10** ensures that $F'(x^0)$ is strong, so that the result still holds.] Now, let $\epsilon > 0$ be given, and let U be any open neighborhood of x^0 on which F is one-to-one. Then there is a $\delta' > 0$ such that $S' = S(x^0, \delta') \subset U$,

$$\|Fx - Fy - A(x - y)\| \leq \epsilon \|x - y\| \qquad \text{for} \quad x, y \in S',$$

and $V = FS'$ is an open neighborhood of Fx^0. Hence, for any $u, v \in V$, there exist $x, y \in S'$ such that $Fx = u$, $Fy = v$, and, therefore,

$$\|F_U^{-1}u - F_U^{-1}v - F'(x^0)^{-1}(u - v)\| \leq \|A^{-1}\| \|A(x - y) - [Fx - Fy]\|$$
$$\leq \epsilon\beta\|x - y\| \leq \epsilon[\beta/(\beta^{-1} - \alpha)]\|u - v\|,$$

where the last inequality follows from (5.1.6). Hence, (1) holds and $(F_U^{-1})'(Fx^0)$ is strong.

Finally, if F' exists and is continuous in a neighborhood of x^0 then it follows from the perturbation lemma **2.3.3** that there is an open ball $S_1 = S(x^0, \delta_1)$ such that $F'(x)$ is nonsingular for $x \in S_1$ and $F'(x)^{-1}$ is continuous in x in S_1. Also, **3.2.10** ensures that $F'(x)$ is strong for each $x \in S_1$. Hence, the first part of the theorem applies, and

$$(F_U^{-1})'(Fx) = F'(x)^{-1}. \quad \blacksquare$$

We next consider the more general situation in which F is a function of two vector variables and the equation $F(x, y^0) = 0$ is known to have a solution x^0 for some given y^0. Then, as before, we ask whether this equation also has solutions when y is close to y^0, and, if so, what is the behavior of the corresponding x as a function of y? In order to discuss this question in suitable generality, we first introduce the notion of a partial derivative with respect to a subspace.

5.2. THE INVERSE AND IMPLICIT FUNCTION THEOREMS

5.2.2. Definition. Let R^n be the product space $R^{n_1} \times \cdots \times R^{n_p}$, where $n_1 + \cdots + n_p = n$, and denote the elements of R^n by $x = (x^1,...,x^p)$ with $x^i \in R^{n_i}$, $i = 1,...,p$. Let $F: D \subset R^n \to R^m$ and, for given $x = (x^1,..., x^p) \in D$, set

$$D_i = \{y \in R^{n_i} \mid (x^1,..., x^{i-1}, y, x^{i+1},..., x^p) \in D\},$$

and define $F_i: D_i \to R^m$ by $F_i y = F(x^1,..., y,..., x^p)$, $y \in D_i$. Then F has a *partial F-derivative* $\partial_i F(x) \equiv F_i'(x^i)$ at x with respect to R^{n_i} if $x^i \in \text{int } D_i$ and F_i has an F-derivative at x^i. In addition, $\partial_i F$ is *strong* at x if, given $\epsilon > 0$, there is a $\delta > 0$ so that

$$\| F(y^1,..., y^{i-1}, y^i + h^i, y^{i+1},..., y^p) - F(y^1,..., y^{i-1}, y^i + k^i, y^{i+1},..., y^p)$$
$$- \partial_i F(x)(h^i - k^i) \| \leq \epsilon \| h^i - k^i \| \tag{2}$$

whenever $\| x - y \| \leq \delta$, $\| h^i \| \leq \delta$, and $\| k^i \| \leq \delta$. ∎

Note that if $n_1 = \cdots = n_p = 1$ and $m = 1$, then **5.2.2** reduces to the usual definition of partial derivatives. As in that case, $\partial_i F$ is defined on a subset of R^n, not on R^{n_i}, and continuity and similar statements refer to R^n. On the other hand, $\partial_i F$ inherits all the properties of the F-derivative from F_i and, in particular, the mean-value theorems of **3.2** apply immediately to $\partial_i F$. Note, finally, that if $F_i'(x^i)$ is strong, it does not follow that $\partial_i F(x)$ is strong, since (2) allows that $y \neq x$.

Continuity of the partial F-derivative at a point implies strongness at that point. We prove this lemma only for a product of two spaces; the general case as well as other properties of partial derivatives are given in **E 5.2-2**–**E 5.2-4**.

5.2.3. Let $F: D \subset R^p \times R^q \to R^m$. If $\partial_i F$, $i = 1$ or 2, exists in an open neighborhood of $(x^0, y^0) \subset D$ and is continuous at (x^0, y^0), then $\partial_i F(x^0, y^0)$ is strong.

Proof. Assume that $i = 1$ (the proof for $i = 2$ is analogous). For given $\epsilon > 0$ choose $\delta_1, \delta_2 > 0$ so that $\partial_1 F(x, y)$ exists for all $x \in S_1 = S(x^0, \delta_1)$ and $y \in S_2 = S(y^0, \delta_2)$, and so that

$$\| \partial_1 F(x, y) - \partial_1 F(x^0, y^0) \| \leq \epsilon, \quad \forall x \in S_1, \ y \in S_2.$$

Now, for given $y \in S_2$, define $G: S_1 \to R^m$ by $Gx = F(x, y) - \partial_1 F(x^0, y^0)x$; then

$$\| G'(x) \| = \| \partial_1 F(x, y) - \partial_1 F(x^0, y^0) \| \leq \epsilon, \quad \forall x \in S_1.$$

Consequently, for any $x, z \in S_1$, it follows from the mean-value theorem **3.2.5** that

$$\|F(x, y) - F(z, y) - \partial_1 F(x^0, y^0)(x - z)\|$$
$$= \|Gx - Gz\| \leq \sup_{0 \leq t \leq 1} \|G'(x + t(z - x))\| \|x - z\| \leq \epsilon \|x - z\|. \quad \blacksquare$$

On the basis of **5.2.3**, we may also phrase the implicit function theorem either in terms of a strong or a continuous partial derivative.

5.2.4. Implicit Function Theorem. Suppose that $F: D \subset R^n \times R^p \to R^n$ is continuous on an open neighborhood $D_0 \subset D$ of a point (x^0, y^0) for which $F(x^0, y^0) = 0$. Assume that $\partial_1 F$ exists and is strong at (x^0, y^0) [or that $\partial_1 F$ exists in a neighborhood of (x^0, y^0) and is continuous at (x^0, y^0)], and that $\partial_1 F(x^0, y^0)$ is nonsingular. Then there exist open neighborhoods $S_1 \subset R^n$ and $S_2 \subset R^p$ of x^0 and y^0, respectively, such that, for any $y \in \bar{S}_2$, the equation $F(x, y) = 0$ has a unique solution $x = Hy \in \bar{S}_1$ and the mapping $H: S_2 \to R^n$ is continuous. Moreover, if $\partial_2 F$ exists at (x^0, y^0), then H is F-differentiable at y^0 and

$$H'(y_0) = -[\partial_1 F(x^0, y^0)]^{-1} \partial_2 F(x^0, y^0). \tag{3}$$

Proof. Set $A = \partial_1 F(x^0, y^0)$, $\beta = \|A^{-1}\|$, and let $0 < \alpha < \beta^{-1}$. Then, since $\partial_1 F$ is strong at (x^0, y^0), we may choose $\delta_1, \delta_2 > 0$ so that

$$\|F(x, y) - F(z, y) - A(x - z)\| \leq \alpha \|x - z\| \tag{4}$$

for all $x, z \in \bar{S}_1 = \bar{S}(x^0, \delta_1)$ and $y \in \bar{S}_2 = \bar{S}(y^0, \delta_2)$, and that $\bar{S}_1 \times \bar{S}_2 \subset D_0$. Now, for fixed $y \in \bar{S}_2$, define the mapping $G_y: \bar{S}_1 \to R^n$ by

$$G_y x = Ax - F(x, y) - F(x^0, y), \quad \forall x \in \bar{S}_1.$$

Then, by (4),

$$\|G_y x - G_y z\| \leq \alpha \|x - z\|, \quad \forall x, z \in \bar{S}_1,$$

and, by the continuity of F at (x^0, y^0), we may also assume that δ_2 has been chosen sufficiently small that

$$\|F(x^0, y)\| = \|F(x^0, y) - F(x^0, y^0)\| < \sigma \equiv (\beta^{-1} - \alpha) \delta_1.$$

Then **5.1.6** ensures that the equation $Ax - G_y x = F(x^0, y)$ has a unique solution in $\bar{S}_1$; that is, for any $y \in \bar{S}_2$, the equation $F(x, y) = 0$ has a unique solution in $\bar{S}_1$. We denote this solution by Hy and show next that the mapping $H: S_2 \to R^n$ is continuous.

5.2. THE INVERSE AND IMPLICIT FUNCTION THEOREMS

Let $y, z \in S_2$. Then, by the definition of H, $F(Hy, y) = F(Hz, z) = 0$, and it follows from (4) that

$$\|Hy - Hz\| \leq \|A^{-1}[F(Hy, y) - F(Hz, y) - A(Hy - Hz)]\|$$
$$+ \|A^{-1}[F(Hz, y) - F(Hz, z)]\|$$
$$\leq \beta\alpha \|Hy - Hz\| + \beta \|F(Hz, y) - F(Hz, z)\|.$$

Hence, since $\beta\alpha < 1$, we have

$$\|Hy - Hz\| \leq [\beta/(1 - \beta\alpha)] \|F(Hz, y) - F(Hz, z)\|, \tag{5}$$

and the continuity of F ensures that of H.

Assume, finally, that $\partial_2 F(x^0, y^0)$ exists. Then, for given $\epsilon > 0$, we may choose $\delta > 0$, so that (5) becomes

$$\|Hy - Hy^0\| \leq \gamma \|y - y^0\|, \quad \forall y \in S(y^0, \delta),$$
$$\gamma = [\beta/(1 - \alpha\beta)][\|\partial_2 F(x^0, y^0)\| + \epsilon]. \tag{6}$$

Consequently,

$$\|Hy - Hy^0 + [\partial_1 F(x^0, y^0)]^{-1} \partial_2 F(x^0, y^0)(y - y^0)\|$$
$$\leq \beta \|\partial_1 F(x^0, y^0)(Hy - Hy^0) + \partial_2 F(x^0, y^0)(y - y^0)\|$$
$$\leq \beta \|F(Hy, y) - F(Hy^0, y) - \partial_1 F(x^0, y^0)(Hy - Hy^0)\|$$
$$+ \beta \|F(Hy^0, y) - F(x^0, y^0) - \partial_2 F(x^0, y^0)(y - y^0)\|$$
$$\leq \beta\epsilon \|Hy - Hy^0\| + \beta\epsilon \|y - y^0\| \leq (\beta\gamma + \beta) \epsilon \|y - y^0\|,$$

which shows that H is F-differentiable at y^0 and that (3) holds. ∎

The inverse function theorem is based on the fundamental assumption that the derivative $F'(x^0)$ is nonsingular. For this to be meaningful, we had to assume that F maps R^n into itself, so that $F'(x)$ is an $n \times n$ square matrix. This raises the question of what happens when F maps R^n into R^m, where $m \neq n$, that is, when we have a system $Fx = y$ with fewer or more equations than unknowns.

For the case $m < n$, the implicit function theorem provides an answer. In fact, if in that case the $m \times n$ matrix $F'(x^0)$ has rank m, then we can split $R^n = R^m \times R^{n-m}$ and set $G(u, v) = Fx - Fx^0$ with $x = (u, v)$, $u \in R^m$, $v \in R^{n-m}$, such that $\partial_1 G(u^0, v^0)$ is a nonsingular $m \times m$ submatrix of $F'(x^0)$. Then, under the assumptions of the implicit function theorem, we have for all v in a neighborhood V of v^0 a solution $u = Hv$ of $G(u, v) = 0$. In other words, $Fx = Fx^0$ has in this case infinitely many solutions, namely, those given by $x = (Hv, v), v \in V$.

This behavior is entirely analogous to that of a linear system $Ax = 0$, which, in the case of a rectangular $m \times n$ matrix A with $m < n$, always has infinitely many solutions. If we continue this analogy, then we should expect that for systems $Fx = y$ with more equations than unknowns there exist solutions only for certain right-hand sides y. As we shall see, this is indeed the case.

For the formulation of the next theorem, recall that a set $S \subset R^n$ is said to have *measure zero* if, given any $\epsilon > 0$, there is a countable number of hyperrectangles Q_j with volume q_j such that $S \subset \bigcup_{j=1}^{\infty} Q_j$ and $\sum_{j=1}^{\infty} q_j \leqslant \epsilon$. Here, by a hyperrectangle we mean any set of the form

$$Q = \left\{ x \in R^n \,\middle|\, x = x^0 + \sum_{j=1}^n \alpha_j h^j, \; \alpha_j \in [0, 1] \right\},$$

where the vectors $h^1, \ldots, h^n$ are mutually orthogonal and nonzero. If $\|h^1\| = \cdots = \|h^n\|$, we shall say that Q is a *cube*. The volume q of a hyperrectangle Q is, of course, defined by $q = \prod_{i=1}^n \|h^i\|$.

5.2.5. Sard Theorem. Let $F: D \subset R^n \to R^n$ be continuously differentiable on the open set D. Let S be a compact subset of D and set $C = \{x \in S \mid F'(x) \text{ is singular}\}$. Then $F(C)$ has measure zero.

Proof. Since S is compact and D open, we may cover S with finitely many cubes $Q_j \subset D, j = 1, \ldots, p$. Therefore, it clearly suffices to assume that S is a cube Q. Assume that Q has sides of length γ, and divide Q into m^n cubes P_j of side γ/m. Now, let P be any of these smaller cubes and suppose there is a $u \in P$ for which $F'(u)$ is singular. For given $\epsilon > 0$, we may assume that m has been chosen so large that

$$\|Fx - Fu - F'(u)(x - u)\|_\infty \leqslant \epsilon \|x - u\|_\infty \leqslant (\epsilon/m)\gamma, \quad \forall x \in P. \tag{7}$$

Now, define the affine mapping $B: R^n \to R^n$ by $Bx = Fu + F'(u)(x - u)$; then, with $\beta = \sup_{x \in Q} \|F'(x)\|$,

$$\|Bx - Fu\|_\infty \leqslant \beta \|x - u\|_\infty \leqslant \gamma\beta/m, \quad \forall x \in P. \tag{8}$$

Since $F'(u)$ is singular, $B(P)$ lies in a hyperplane of dimension at most $n - 1$ and, by (7), $F(P)$ is contained in an $(\epsilon\gamma/m)$-neighborhood of $B(P)$; in particular, $F(P)$ is contained in a hyperrectangle of volume at most equal to

$$[2(\beta + \epsilon)\gamma/m]^{n-1} (2\gamma\epsilon/m) = (2\gamma/m)^n (\beta + \epsilon)^{n-1} \epsilon.$$

Therefore, $F(C)$ is contained in hyperrectangles with total volume at most equal to $(2\gamma)^n (\beta + \epsilon)^{n-1}\epsilon$ and, since ϵ was arbitrary, this shows that $F(C)$ has measure zero. ∎

If $A \in L(R^n)$ is singular, then the range of A is at most $(n - 1)$-dimensional, and, hence, a set of measure zero. As a direct corollary of **5.2.5**, we have the natural extension of this result to nonlinear mappings.

5.2. THE INVERSE AND IMPLICIT FUNCTION THEOREMS

5.2.6. Assume that $F: R^n \to R^n$ is continuously differentiable and $F'(x)$ is singular for all $x \in R^n$. Then $F(R^n)$ has measure zero.

Proof. Let $\{Q_j\}$ be a countable collection of cubes such that $\bigcup_{j=1}^{\infty} Q_j = R^n$. Then **5.2.6** shows that $F(Q_j)$ has measure zero for all j, and the result follows. ∎

This result has immediate application to the problem of overdetermined systems mentioned previously. Suppose that $F: R^n \to R^m$, $m > n$, is continuously differentiable, and consider the natural extension of F to R^m defined by

$$\hat{F}(x, u) = Fx, \quad \forall x \in R^n, \ u \in R^{m-n}.$$

Then $\hat{F}'(x, u) = (F'(x), 0)$, where 0 is an $m \times (m - n)$ zero matrix, and, hence, $\hat{F}'(x, u)$ is singular for all $(x, u) \in R^m$. Thus, **5.2.6** shows that the set of all y for which there is a solution to $Fx = y$ is a set of measure zero in R^m. This shows, in particular, that even if $Fx = y$ has a solution, arbitrarily small changes in y will give a system which has no solution. Thus, the problem of solving m equations in $n < m$ unknowns is numerically an ill-posed problem.

NOTES AND REMARKS

NR 5.2-1. The inverse and implicit function theorems for continuously differentiable functions may be found in a number of advanced calculus books; see, for example, Apostol [1957]. Under only the assumption that $F'(x^0)$ is strong, **5.2.1** is a special case of a more general result of Leach [1961], while the use of strong partial derivatives in **5.2.4** appears to be new. Both **5.2.1** and **5.2.4**, together with their proofs, remain valid in arbitrary Banach spaces provided the nonsingularity assumption for the derivatives is replaced by that of the existence of bounded inverses. Results of this type for Banach spaces were given by Hildebrandt and Graves [1927].

NR 5.2-2. Sard's Theorem as stated in **5.2.5** is a special case of the more general result: Let $F: D \subset R^n \to R^n$ be continuously differentiable on the open set D. Then, for any (Lebesque) measurable set $D_0 \subset D$, $F(D_0)$ is measurable and

$$\text{meas}\{F(D_0)\} \leqslant \int_{D_0} |\det F'(x)| \, dx.$$

For a proof, see, for example, Schwartz [1964] or Sard [1942].

EXERCISES

E 5.2-1. Define $f: R^1 \to R^1$ by $f(0) = 0$, $f(x) = \frac{1}{2}x + x^2 \sin(1/x)$, $x \neq 0$. Show that $f'(0) \neq 0$, but that f is not a local homeomorphism at zero.

E 5.2-2. Assume that $F: R^{n_1} \times \cdots \times R^{n_p} \to R^m$, $n = n_1 + \cdots + n_p$, has partial F-derivatives $\partial_i F(x)$ at $x = (x^1,...,x^p)$, and that all, except possibly one, of the $\partial_i F(x)$ are strong. Show that F has an F-derivative at x and that

$$F'(x) h = \sum_{i=1}^{p} \partial_i F(x) h^i, \qquad h = (h^1,..., h^p).$$

E 5.2-3. Assume that, in **E 5.2-2**, all $\partial_i F(x)$ are strong. Show that $F'(x)$ itself is strong.

E 5.2-4. Assume that $F: R^{n_1} \times \cdots \times R^{n_p} \to R^m$ has a partial F-derivative $\partial_i F$ at each point of an open neighborhood of $x \in R^n$. Show that if $\partial_i F$ is continuous at x, then $\partial_i F(x)$ is strong. Conclude, therefore, that if all $\partial_i F$, $i = 1,...,p$, are continuous at x, then the conclusion of **E 5.2-2** holds.

E 5.2-5. Assume in Theorem **5.2.4** that $\partial_1 F$ and $\partial_2 F$ exist and are continuous on an open neighborhood of (x^0, y^0). Show that H is continuously differentiable on an open neighborhood S of y^0 and that

$$H'(y) = -[\partial_1 F(Hy, y)]^{-1} \partial_2 F(Hy, y), \qquad \forall y \in S.$$

E 5.2-6. Assume in Theorem **5.2.4** that $\partial_2 F(x^0, y^0)$ is also strong. Show that $H'(y^0)$ is strong.

E 5.2-7. Under the assumptions of **5.2.1**, show that if F is twice-continuously differentiable in a neighborhood of x^0, then F_U^{-1} is twice-continuously differentiable in a neighborhood of Fx^0.

5.3. THE CONTINUATION PROPERTY

Suppose that $F: R^n \to R^n$ is a local homeomorphism at each point x in R^n. We shall be concerned in this section with additional conditions on F which ensure that it is a homeomorphism of the whole space onto itself.

We note, first, that additional conditions are indeed necessary. For example, the function e^x in one variable has a nonzero derivative at each point of R^1, and hence, by **5.2.1**, is a local homeomorphism; but the range of e^x is not R^1. Similarly, **E 5.3-1** and **E 5.3-2** give examples of continuously differentiable mappings $F: R^2 \to R^2$ such that $F'(x)$ is nonsingular for all x, but F is not one-to-one. Note, however, that in one

5.3. THE CONTINUATION PROPERTY

dimension the mean-value theorem shows immediately that F is one-to-one if $F'(x) \neq 0$ for all x; this argument fails to hold in higher dimensions (see, however, E 5.3-4 for a simple sufficiency condition).

We begin with a basic lemma which allows the major deductions of this section from the following simple, but powerful, condition on F.

5.3.1. Definition. The mapping $F: D \subset R^n \to R^n$ has the *continuation property* for a given continuous function $q: [0, 1] \subset R^1 \to R^n$ if the existence of a continuous function $p: [0, a) \to D$, $a \in (0, 1]$ such that $Fp(t) = q(t)$ for all $t \in [0, a)$ implies that $\lim_{t \to a^-} p(t) = p(a)$ exists with $p(a) \in D$, and $Fp(a) = q(a)$. ∎

5.3.2. Suppose that $F: D \subset R^n \to R^n$ is a local homeomorphism at each point of the open set D. If F has the continuation property for a continuous function $q: [0, 1] \to R^n$ such that $Fx^0 = q(0)$ for some $x^0 \in D$, then there exists a unique continuous function $p: [0, 1] \to D$ which satisfies $p(0) = x^0$ and $Fp(t) = q(t)$ for all $t \in [0, 1]$.

Proof. Let U and V be open neighborhoods of x^0 and Fx^0, respectively, such that the restriction F_U of F to U is a homeomorphism of U onto V. Then there is a $t_1 \in (0, 1]$ so that $q(t) \in V$ for $t \in [0, t_1)$, and, hence, we can define a continuous function $p: [0, t_1) \to U \subset D$ by $p(t) = F_U^{-1}q(t)$, $t \in [0, t_1)$. Moreover, by the continuation property, $p(t_1) = \lim_{t \to t_1^-} p(t)$ exists and $Fp(t_1) = q(t_1)$. If $t_1 < 1$, then we can repeat the process and continue p successively to points $t_2 < t_3 < \cdots \leq 1$. Now, let $\hat{t} \leq 1$ be the maximum value of t for which p may be continued in this way, that is, $\hat{t} = \sup t_i$. Then, either $\hat{t} = t_N$ for some N, in which case $Fp(t) = q(t)$ for all $t \in [0, \hat{t}]$, or else $Fp(t) = q(t)$ for all $t \in [0, \hat{t})$, so that,

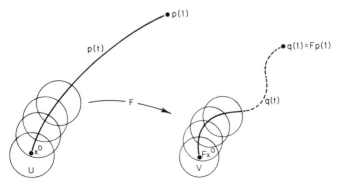

FIGURE 5.1

by the continuation property, again $Fp(\hat{t}) = q(\hat{t})$. Since $p(\hat{t}) \in D$ and D is open, we may therefore apply the same process to continue p beyond $\hat{t}$. But this would contradict the maximality of $\hat{t}$, and hence $\hat{t}$ must equal 1.

To show the uniqueness of p, suppose that $r: [0, 1] \subset R^1 \to D$ is another continuous mapping such that $r(0) = x^0 = p(0)$ and $Fr(t) = q(t)$, $t \in [0, 1]$. Then the set $J_0 = \{t \in [0, 1] \mid p(s) = r(s) \text{ for all } s \in [0, t]\}$ is not empty, and, hence, $\bar{t} = \sup\{t \mid t \in J_0\}$ is well defined. Moreover, by the continuity of r and p, we have $\bar{t} \in J_0$. Now, if $\bar{t} < 1$, there exist points $t_k \in (\bar{t}, 1]$ with $\lim_{k \to \infty} t_k = \bar{t}$ such that $p(t_k) \neq r(t_k)$. But $\lim_{k \to \infty} p(t_k) = p(\bar{t}) = r(\bar{t}) = \lim_{k \to \infty} r(t_k)$, which contradicts that F is a local homeomorphism at $p(\bar{t})$. ∎

By considering the continuation property only for linear mappings q, we obtain the following important corollary of **5.3.2**,

5.3.3. Let $F: D \subset R^n \to R^n$ be a local homeomorphism at each point of the open set D. If F has the continuation property for all linear functions $q(t) = (1 - t)y^0 + ty^1$, $t \in [0, 1]$, where y^0, $y^1 \in R^n$ are arbitrary, then $FD = R^n$.

Proof. Let $x^0 \in D$ and $y \in R^n$ be arbitrary. Then, since F has the continuation property for $q(t) = (1 - t)Fx^0 + ty$, $t \in [0, 1]$, **5.3.2** ensures the existence of a mapping $p: [0, 1] \to D$ such that $Fp(t) = q(t)$ for all $t \in [0, 1]$. In particular, $Fp(1) = y$, and, hence, $FD = R^n$. ∎

We turn next to the question of when F is one-to-one. Surprisingly, it is the continuation property which plays the crucial role here also. We first prove the following lemma, which extends **5.3.2** to functions of two variables.

5.3.4. Suppose $F: D \subset R^n \to R^n$ is a local homeomorphism at each point of the open set D in R^n, and let $q: [0, 1] \times [0, 1] \subset R^2 \to R^n$ and $r: [0, 1] \subset R^1 \to D$ be continuous functions such that $Fr(s) = q(s, 0)$ for all $s \in [0, 1]$. If, for each fixed $s \in [0, 1]$, F has the continuation property for $q_s(t) = q(s, t)$, $t \in [0, 1]$, then there exists a unique continuous mapping $p: [0, 1] \times [0, 1] \to D$ such that $p(s, 0) = r(s)$ and $Fp(s, t) = q(s, t)$ for $s, t \in [0, 1]$. Moreover, if $q(s, 1) = q(0, t) = q(1, t) = y$ for all $s, t \in [0, 1]$, then $r(0) = r(1)$.

Proof. Set $J = [0, 1]$. For each $s \in J$, **5.3.2** guarantees the existence of a unique continuous mapping $p_s: J \to D$ such that $p_s(0) = r(s)$ and $Fp_s(t) = q(s, t)$ for all $t \in J$. Hence, with $p(s, t) = p_s(t)$ for all $s, t \in J$, it remains only to show that p is continuous. Suppose that p is discontinuous

5.3. THE CONTINUATION PROPERTY

at $(s_0, t_1) \in J \times J$. Let t_0 be the infimum of all $t \in J$ such that p is discontinuous at (s_0, t), and let U and V be open neighborhoods of $p(s_0, t_0)$ and $q(s_0, t_0)$, respectively, for which the restriction F_U is a homeomorphism of U onto V. Now, clearly, $t_0 \neq 0$, for the continuity of r and the uniqueness of each p_s ensures that $p(s, t) = F_U^{-1} q(s, t)$ for all $q(s, t) \in V$. Hence, p is continuous in those s and t for which $p(s, t) \in U$. Assume, then, that $t_0 > 0$, and let $J(s_0)$ and $J(t_0)$ be neighborhoods of s_0 and t_0 in J such that $q(s, t) \in V$ whenever $(s, t) \in J(s_0) \times J(t_0)$; such intervals exist by the continuity of q. Since p_{s_0} is continuous, we may choose $t' < t_0$ such that $t' \in J(t_0)$ and $p(s_0, t') \in U$. Moreover, since p is continuous in both variables at (s_0, t'), there is an interval $J'(s_0) \subset J(s_0)$ such that $p(s, t') \in U$ for all $s \in J'(s_0)$. But then $p_s(t) = p(s, t) = F_U^{-1} q(s, t)$ for all $(s, t) \in J'(s_0) \times J(t_0)$, so that p is continuous in a neighborhood of (s_0, t_0). This contradicts the construction of t_0.

For the proof of the last statement, note, first, that $Fp(0, t) = q(0, t) = y$ and the continuity of p imply that $p(0, t) = r(0)$ for all $t \in [0, 1]$; otherwise, there would exist points arbitrarily close to $r(0)$ which F maps into y, and this contradicts the local homeomorphism property of F. Similarly, $q(1, t) = y$ shows that $p(1, t) = r(1)$ for all $t \in [0, 1]$. But then, $p(s, 1)$ is a continuous function of s such that $Fp(s, 1) = q(s, 1) = y$ for all $s \in [0, 1]$ and $p(0, 1) = r(0)$, $p(1, 1) = r(1)$. Hence if $r(0) \neq r(1)$, then there exist points $p(s, 1)$ arbitrarily close to $r(0)$ which are mapped into y, and this would again contradict the local homeomorphism property of F. Therefore, $r(0) = r(1)$. ∎

A continuous mapping $q: [0, 1] \to R^n$ is a *path* in R^n with the end points $q(0)$ and $q(1)$. A continuous mapping $q: [0, 1] \times [0, 1] \to R^n$ may then be considered a continuous deformation, or *homotopy*, of the path $q(0, \cdot)$ into the path $q(1, \cdot)$. The content of **5.3.2** is that, given a path in R^n with one end point in the range of F, there is a path in D which F maps onto the given path. Similarly, the content of **5.3.4** is that a homotopy in D exists which maps onto a given homotopy in R^n.

We may now prove the main result of this section. Recall that a set D is path-connected if, for any two points $x, y \in D$, there is a continuous mapping $p: [0, 1] \to D$ such that $p(0) = x$ and $p(1) = y$.

5.3.5. Let $D \subset R^n$ be open and path-connected, and assume that $F: D \to R^n$ is a local homeomorphism at each point of D. Then F is a homeomorphism from D onto R^n if and only if F has the continuation property for all continuous functions $q:[0, 1] \subset R^1 \to R^n$.

Proof. Assume that F is a homeomorphism of D onto R^n and let $p: [0, 1) \to D$ and $q: [0, 1] \to R^n$ be continuous mappings such that

$Fp(t) = q(t)$ for $t \in [0, 1)$. Set $p(1) = F^{-1}q(1)$. Then $p(1) \in D$, and the continuity of F^{-1} ensures that $\lim_{t \to 1-} p(t) = \lim_{t \to 1-} F^{-1}q(t) = p(1)$. Hence, F has the continuation property for any continuous mapping $q: [0, 1] \to R^n$.

Conversely, assume that F has the continuation property for all continuous $q: [0, 1] \subset R^1 \to R^n$. Then it follows from 5.3.3 that $FD = R^n$ and, since the continuity of F and F^{-1} is a consequence of the fact that F is a local homeomorphism on D, it remains only to show that F is one-to-one. Suppose that $Fx^0 = Fx^1 = y$ for some $x^0, x^1 \in D$. Let $r: [0, 1] \subset R^1 \to D$ be a continuous mapping such that $r(0) = x^0$, $r(1) = x^1$, and define $q: [0, 1] \times [0, 1] \subset R^2 \to R^n$ by $q(s, t) = ty + (1 - t)Fr(s)$. Clearly, q is continuous and satisfies $q(0, t) = q(1, t) = q(s, 1) = y$ for all $s, t \in [0, 1]$. Hence, it follows from the last part of 5.3.4 that $x^0 = r(0) = r(1) = x^1$. ∎

The continuation property is an operational condition that may be difficult to verify in concrete situations. We therefore next consider other conditions which imply the continuation property.

5.3.6. Definition. The mapping $F: D \subset R^n \to R^n$ is *norm-coercive* on an open set $D_0 \subset D$ if, for any $\gamma > 0$, there is a closed, bounded set $D_\gamma \subset D_0$ such that $\|Fx\| > \gamma$ for all $x \in D_0 \sim D_\gamma$. ∎

Note that if $D_0 = R^n$, then F is norm-coercive if and only if $\lim_{\|x\| \to \infty} \|Fx\| = \infty$.

Before showing that the continuation property may be replaced in 5.3.5 by norm-coerciveness, we first prove the following lemma.

5.3.7. Let $F: D \subset R^n \to R^n$ be a local homeomorphism at each point of D and let $p: [0, a) \subset [0, 1] \to D$ be a continuous function. If $\lim_{t \to a-} Fp(t) = y$ exists, and if there is a sequence $\{t_k\} \subset [0, a)$ with $\lim_{k \to \infty} t_k = a$ such that $\lim_{k \to \infty} p(t_k) = x \in D$, then $\lim_{t \to a-} p(t) = x$.

Proof. From the continuity of F, we have $Fx = y$. Now, let U and V be open neighborhoods of x and y, respectively, such that the restriction F_U of F to U is a homeomorphism from U onto V. Clearly, there is a $t' < a$ so that $p(t_k) \in U$ for $t_k \in (t', a)$ and $Fp(t) \in V$ for $t \in (t', a)$. Therefore, the function $\hat{p}(t) = F_U^{-1}Fp(t)$, $t \in (t', a)$, satisfies $\hat{p}(t_k) = p(t_k)$ for all $t_k \in (t', a)$, and it follows, by precisely the same argument as in the uniqueness proof of 5.3.2, that $\hat{p}(t) = p(t)$ for all $t \in (t', a)$. By the continuity of F_U^{-1}, we therefore have $\lim_{t \to a-} p(t) = \lim_{t \to a-} F_U^{-1}(Fp(t)) = x$. ∎

5.3.8. Norm-Coerciveness Theorem. Let D be open and path-connected, and assume that $F: D \subset R^n \to R^n$ is a local homeomorphism at

5.3. THE CONTINUATION PROPERTY

each point of D. Then F is a homeomorphism of D onto R^n if and only if F is norm-coercive on D.

Proof. Suppose, first, that F is a homeomorphism from D onto R^n, let $\gamma > 0$ be given, and set $S = \bar{S}(0, \gamma)$. Then, since F^{-1} is continuous, $D_\gamma = F^{-1}S$ is closed and bounded, and, if $x \in D \sim D_\gamma$, then $Fx \notin S$; that is, $\|Fx\| > \gamma$. Conversely, let F be norm-coercive on D; then, by **5.3.5**, it suffices to show that F has the continuation property for any continuous mapping $q\colon [0, 1] \to R^n$. Let q be given, and suppose that for some continuous $p\colon [0, a) \subset [0, 1] \to D$ we have $Fp(t) = q(t)$ for all $t \in [0, a)$. Set $\gamma = \max\{\|q(t)\| \mid t \in [0, a]\}$; then the norm-coerciveness of F implies that there is a compact set $D_\gamma \subset D$ such that $\|Fx\| > \gamma$ for all $x \in D \sim D_\gamma$. This implies that $p(t) \in D_\gamma$ for all $t \in [0, a)$, and hence it follows from the compactness of D_γ that there is a sequence $\{t_k\} \subset [0, a)$ such that $\lim_{k\to\infty} t_k = a$ and $\lim_{k\to\infty} p(t_k) = \hat{x} \in D$. Therefore, **5.3.7** ensures that $\lim_{t \to a-} p(t) = \hat{x}$, so that, by the continuity of F, $F\hat{x} = q(a)$. ∎

We note that in **5.3.8**, as well as in **5.3.5**, the set D is necessarily "simply" path-connected (see **E 5.3-9**).

As an immediate corollary of **5.2.1** and **5.3.8**, we have the following very useful result.

5.3.9. Suppose that $F\colon R^n \to R^n$ is continuously differentiable on all of R^n, and that $F'(x)$ is nonsingular for all $x \in R^n$. Then F is a homeomorphism from R^n onto R^n if and only if $\lim_{\|x\|\to\infty} \|Fx\| = \infty$.

We conclude this section with two results of the same type in which the coerciveness condition is replaced by a bound for $F'(x)^{-1}$.

5.3.10. Hadamard Theorem. Assume that $F\colon R^n \to R^n$ is continuously differentiable on R^n and that $\|F'(x)^{-1}\| \leq \gamma < +\infty$ for all $x \in R^n$. Then F is a homeomorphism of R^n onto R^n.

Proof. Observe that **5.2.1** implies that F is a local homeomorphism at each $x \in R^n$. We show next that F has the continuation property for any linear function $q(t) = (1 - t)y^0 + ty^1$, $t \in [0, 1]$, $y^0, y^1 \in R^n$. Suppose that for some continuous $p\colon [0, a) \subset [0, 1] \to R^n$ we have $Fp(t) = q(t)$ for $t \in [0, a)$. For any fixed $t \in [0, a)$, let U and V be open neighborhoods of $p(t)$ and $q(t)$, respectively, such that the restriction F_U of F to U is a homeomorphism of U onto V. Then, by **5.2.1**, F_U^{-1} is continuously differentiable in a neighborhood of $q(t)$ and

$$(F_U^{-1})'(Fx) = F'(x)^{-1}, \qquad \forall x \in U. \tag{1}$$

Thus, by the chain rule **3.1.7**,

$$p'(t) = F'(p(t))^{-1} q'(t),$$

which shows that p is continuously differentiable on $[0, a)$. Now, let $\{t_k\} \subset [0, a)$ be any monotone increasing sequence such that $\lim_{k \to \infty} t_k = a$. Then, for any $k < j$, we have

$$\| p(t_j) - p(t_k) \| = \left\| \int_{t_k}^{t_j} p'(t) \, dt \right\| = \left\| \int_{t_k}^{t_j} F'(p(t))^{-1} q'(t) \, dt \right\|$$
$$\leq \gamma \| y^1 - y^0 \| \, | t_k - t_j |. \tag{2}$$

Therefore, $\{p(t_k)\}$ is a Cauchy sequence and, hence, convergent. If $\lim_{k \to \infty} p(t_k) = x$, then it follows from **5.3.7** that $\lim_{t \to a-} p(t) = x$, and, by continuity of F, that $Fx = q(a)$. Hence, F has the continuation property for all linear functions in R^n, and **5.3.3** implies that $FR^n = R^n$. In order to show that F is one-to-one, suppose that $Fx^0 = Fx^1 = y$, and set

$$q(s, t) = ty + (1 - t) F((1 - s) x^0 + sx^1), \qquad s, t \in [0, 1].$$

Clearly, for each fixed $s \in [0, 1]$, $q_s = q(s, \cdot)$ is linear, so that F has the continuation property for each q_s. Moreover, $q(s, 1) = q(0, t) = q(1, t) = y$ for all $s, t \in [0, 1]$, so that, by the last part of **5.3.4**, we conclude that $x^0 = x^1$. ∎

The following theorem can be regarded as a local version of the Hadamard theorem.

5.3.11. Let $F \colon D \subset R^n \to R^n$ be continuously differentiable on D and suppose that there is an open ball $S = S(x^0, r) \subset D$ so that $\| F'(x)^{-1} \| \leq \gamma$ for $x \in S$ and $r > \gamma \| Fx^0 \|$. Then $Fx = 0$ has a solution in S.

Proof. The proof is analogous to that of **5.3.10**. Clearly, **5.2.1** implies again that F is a local homeomorphism at each $x \in S$. We show that F has the continuation property for $q(t) = (1 - t) y^0$, $t \in [0, 1]$, where $y^0 = Fx^0$. Suppose, therefore, that, for some continuous $p \colon [0, a) \subset [0, 1] \to S$ we have $Fp(t) = q(t)$, $t \in [0, a)$, and $p(0) = x^0$. As in the proof of **5.3.10**, it follows that p is continuously differentiable on $[0, a)$ and that (1) holds. Hence, if again $\{t_k\} \subset [0, a)$ is any monotone increasing sequence which converges to a, (2) shows that $\{p(t_k)\}$ is a Cauchy sequence, and therefore $\lim_{k \to \infty} p(t_k) = x$ exists. From

$$\| p(t_k) - x^0 \| = \int_0^{t_k} \| p'(t) \| \, dt \leq a\gamma \| Fx^0 \| < ar, \qquad \forall k \geq 0,$$

5.3. THE CONTINUATION PROPERTY

we find that $x \in S(x^0, ar) \subset S$; thus, **5.3.7** implies that $\lim_{t \to a^-} p(t) = x$, and, by continuity of F, that $Fx = q(a)$. This proves that F has the continuation property for q, or, in other words, that p can be continued to $t = 1$ and that $p(1) \in S$ and $Fp(1) = q(1) = 0$. ∎

NOTES AND REMARKS

NR 5.3-1. The presentation of this section is taken from Rheinboldt [1969a], where the theory is developed in much greater generality. In particular, the results and definitions **5.3.1**–**5.3.7** hold for arbitrary Banach spaces. However, the continuation principle itself dates back at least to the last century, and is, of course, a standard tool in the theory of ordinary differential equations. In one form or another, it has been used by several authors to obtain existence theorems for operator equations; see, for example, Ehrmann [1963], and, for a review of earlier work, Ficken [1951].

NR 5.3-2. Theorem **5.3.10** was first given by Hadamard [1906] and extended to Hilbert spaces by Levy [1920]. The result holds as stated in arbitrary Banach spaces (see Rheinboldt [1969a] or, for a direct proof along similar lines, Schwartz [1964]). On the other hand, the norm-coerciveness theorem **5.3.8** requires the following modification to hold in a Banach space: If $F: X \to X$ is of the form $F = I - G$, where G is continuous and compact (that is, G maps closed, bounded sets into compact sets) and if F is a norm-coercive local homeomorphism, then F is a homeomorphism from X onto itself. This result appears to be due to Cacciopoli [1932]; see also Rheinboldt [1969a].

NR 5.3-3. The proof of **5.3.10** is easily modified to obtain the following more general result of Meyer [1968]: The conclusion of **5.3.10** remains valid if the uniform boundedness of $F'(x)^{-1}$ is replaced by $\|F'(x)^{-1}\| \leq \alpha \|x\| + \beta$, for all $x \in R^n$.

For the proof, proceed as in **5.3.10** to obtain (5.3.2) with γ now given by $\gamma = \alpha\mu + \beta$, where $\mu = \sup_{0 \leq t < a} \|p(s)\|$. Hence, the proof of **5.3.10** will remain valid provided that $\mu < +\infty$. In order to show this, note that, as in (2) and with $\eta = \|y^1 - y^0\|$,

$$\|p(t) - p(0)\| \leq \eta \int_0^t [\alpha \|p(s)\| + \beta]\, ds \leq c_1 + c_2 \int_0^t \|p(s) - p(0)\|\, ds,$$

where $c_1 = \eta[\alpha \|p(0)\| + \beta]$, and $c_2 = \eta\alpha$. Then it follows from the Gronwall inequality (see, for example, Bellman [1953, p. 35]) that

$$\|p(t) - p(0)\| \leq c_1 \exp\left(\int_0^t c_2\, dt\right) \leq c_1 \exp c_2, \quad \forall t \in [0, a) \subset [0, 1].$$

NR 5.3-4. It would be interesting to prove directly that the hypotheses of **5.3.10** imply that F has the continuation property for all continuous functions (note that this is, of course, true as a consequence of the theorem, since F is a homeomorphism). Such a proof has so far not been found, but we note that a trivial modification of the proof of **5.3.10** permits showing directly that the continuation property holds for all continuously differentiable functions.

EXERCISES

E 5.3-1. Define $F: R^2 \to R^2$ by $f_1(x) = (\exp x_1) \sin x_2$, $f_2(x) = (\exp x_1) \cos x_2$. Show that $F'(x)$ is nonsingular for all $x \in R^2$, but that F is not one-to-one.

E 5.3-2. Define $F: R^2 \to R^2$ by

$$f_1(x) = (\exp 2x_1) - x_2^2 + 3, \qquad f_2(x) = 4x_2(\exp 2x_1) - x_2^3.$$

Show that the leading principal minor determinants of $F'(x)$ are positive for all $x \in R^2$, but that F is not one-to-one (Gale and Nikaido [1965]).

E 5.3-3. Define $F: R^2 \to R^2$ by $f_1(x) = \arctan x_1$, $f_2(x) = (1 + x_1^2) x_2$. Show that $\det F'(x) = 1$ for all $x \in R^2$, but that F is not onto.

E 5.3-4. Assume that $F: D \subset R^n \to R^n$ is continuously differentiable on the open convex set D and that, for any n points $x^1, ..., x^n \in D$, the matrix $(f_1'(x^1)^T, ..., f_n'(x^n)^T)$ is invertible. Show that F is one-to-one.

E 5.3-5. Define $f: (0, \infty) \subset R^1 \to R^1$ by $f(x) = 1 - (1/x)$. Show that f does not have the continuation property for all linear functions.

E 5.3-6. Define $p: [0, 1) \to R^1$ by $p(t) = \sin[1/(1 - t)]$. Show that there is a sequence $\{t_k\} \subset [0, 1)$, $t_k \to 1$, such that $\lim_{k \to \infty} p(t_k) = 0$, while $\lim_{t \to 1^-} p(t)$ does not exist.

E 5.3-7. Let $F: D \subset R^n \to R^n$, where $D = D_1 \cup D_2$ and D_1, D_2 are open, path-connected, disjoint sets. Suppose that F is a local homeomorphism at each point of D and has the continuation property for all continuous mappings $q: [0, 1] \to R^n$. Show that F maps each D_i homeomorphically onto R^n. Formulate and prove the corresponding result when D is an arbitrary union of open, disjoint, path-connected sets (Rheinboldt [1969a]).

E 5.3-8. Let $F: D \subset R^n \to R^n$ be a local homeomorphism at each point of the open set D. Suppose that FD is path-connected and that F has the continuation property for all continuous mappings $q: [0, 1] \to R^n$. Show that for any two points $y^0, y^1 \in FD$ the sets $\Gamma_i = \{x \in D \mid Fx = y^i\}$, $i = 0, 1$, have the same cardinality (Rheinboldt [1969a]).

5.4. MONOTONE OPERATORS AND OTHER APPLICATIONS

E 5.3-9. Let $F: D \subset R^n \to R^n$ be a local homeomorphism at each point of the open path-connected set D, and assume that F has the continuation property for all continuous functions $q: [0, 1] \subset R^1 \to R^n$. Show that D is simply path-connected; that is, that when $p_1, p_2: [0, 1] \to D$ are any two continuous functions, with $p_1(0) = p_2(0)$, $p_1(1) = p_2(1)$, then there exists a homotopy $q: [0, 1] \times [0, 1] \to D$ such that $q(0, t) = p_1(t)$, $q(1, t) = p_2(t)$ for all $t \in [0, 1]$.

5.4. MONOTONE OPERATORS AND OTHER APPLICATIONS

We now consider some specializations of the results of the previous section. The following theorem may be considered as a natural complement of **4.4.1**, where A was assumed to be symmetric, positive definite. Recall from **2.4.7** that A is an M-matrix if $a_{ij} \leq 0$, $i \neq j$, and $A^{-1} \geq 0$.

5.4.1. Let $A \in L(R^n)$ be an M-matrix and assume that $\phi: R^n \to R^n$ is continuously differentiable and that $\phi'(x)$ is nonnegative and diagonal for all $x \in R^n$. Then the mapping $F: R^n \to R^n$ defined by $Fx \equiv Ax + \phi x$ is a homeomorphism of R^n onto itself.

Proof. Clearly, F continuously differentiable and $F'(x) = A + \phi'(x)$. It follows from **2.4.11** that

$$0 \leq F'(x)^{-1} \leq A^{-1}, \quad \forall x \in R^n,$$

and, consequently, in any norm, $\| F'(x)^{-1} \|$ is uniformly bounded in x. The Hadamard theorem **5.3.10** now applies. ∎

Theorem **5.4.1** has immediate application to boundary value problems of the form considered in **1.1**, **1.2**, and **4.4** whenever the discretization of the linear differential operator leads to an M-matrix, rather than to a symmetric matrix.

Another modification of **4.4.1** results if we assume that A is only positive definite, but no longer symmetric. This will be a consequence of the following theory.

5.4.2. Definition. A mapping $F: D \subset R^n \to R^n$ is *monotone* on $D_0 \subset D$ if

$$(Fx - Fy)^T (x - y) \geq 0, \quad \forall x, y \in D_0; \tag{1}$$

F is *strictly monotone* on D_0 if strict inequality holds in (1) whenever $x \neq y$, and *uniformly monotone* if there is a $\gamma > 0$ so that

$$(Fx - Fy)^T (x - y) \geq \gamma (x - y)^T (x - y), \quad \forall x, y \in D_0. \quad \blacksquare \tag{2}$$

If $F: R^1 \to R^1$, then, clearly, F is monotone on $D_0 \subset R^1$ if and only if F is isotone on D_0. A matrix operator $F = A \in L(R^n)$ is monotone on R^n if and only if A is positive semidefinite (but not necessarily symmetric) Some properties of positive definite matrices are given in **E 5.4-1,2**.

Now, suppose that F is a gradient mapping (see **4.1.5**), so that there is a G-differentiable functional $g: R^n \to R^1$ for which $g'(x)^T = Fx$. Then the differential inequalities **3.4.5** show that F is monotone, strictly monotone, or uniformly monotone if and only if g is convex, strictly convex, or uniformly convex, respectively. For nongradient operators F, monotonicity may therefore be regarded as a natural generalization of convexity. We shall see that this is more than a superficial generalization by showing that essentially the same existence theorems hold as in **4.3** for gradient operators. Before proceeding to the first of these theorems, we observe that the analog of **3.4.6** holds for monotone operators.

5.4.3. Let $F: D \subset R^n \to R^n$ be continuously differentiable on an open convex set $D_0 \subset D$. Then

(a) F is monotone on D_0 if and only if $F'(x)$ is positive semidefinite for all $x \in D_0$.

(b) If $F'(x)$ is positive definite for all $x \in D_0$, then F is strictly monotone on D_0.

(c) F is uniformly monotone on D_0 if and only if there is a $\gamma > 0$ so that

$$h^T F'(x) h \geq \gamma h^T h, \quad \forall x \in D_0, \quad h \in R^n. \tag{3}$$

Proof. We prove (a) and (c) together. Suppose that (2) holds. Then, from the definition of F', it follows that, for any $x \in D_0$ and $h \in R^n$,

$$h^T F'(x) h = h^T \lim_{t \to 0} t^{-1}[F(x + th) - Fx] \geq \lim_{t \to 0} t^{-2} \gamma \| th \|_2^2 = \gamma h^T h. \tag{4}$$

Consequently, if F is uniformly monotone, then (3) holds, while if F is monotone, then $\gamma = 0$, and $F'(x)$ is positive semidefinite. Conversely, if (3) holds, then the mean-value theorem **3.2.7** shows that

$$(x - y)^T (Fx - Fy) = \int_0^1 (x - y)^T F'(y + t(x - y))(x - y) \, dt$$

$$\geq \gamma (x - y)^T (x - y), \tag{5}$$

so that F is monotone or uniformly monotone depending on whether $\gamma = 0$ or $\gamma > 0$. Finally, if $F'(u)$ is positive definite for all $u \in D_0$,

5.4. MONOTONE OPERATORS AND OTHER APPLICATIONS

and $x \neq y$, then the integrand in (5) is positive for all $t \in [0, 1]$ and F is strictly monotone. ∎

Note that a strictly monotone mapping F on a set D_0 is necessarily one-to-one on D_0. In particular, **5.4.3**(b) yields the following uniqueness result.

5.4.4. If $F: D \subset R^n \to R^n$ is continuously differentiable on an open, convex set $D_0 \subset D$, and $F'(x)$ is positive definite for all $x \in D_0$, then F is one-to-one on D_0.

The conditions of **5.4.4** do not suffice for the existence of solutions, as the example $Fx = e^x$ in one dimension shows. We may guarantee existence, however, by strengthening the monotonicity assumption.

5.4.5. If $F: R^n \to R^n$ is continuously differentiable and uniformly monotone on R^n, then F is a homeomorphism of R^n onto R^n.

Proof. If A is any linear operator in $L(R^n)$ such that $h^T A h \geqslant \gamma h^T h$ for all $h \in R^n$ and some $\gamma > 0$, then A is invertible and by the Cauchy-Schwarz inequality, $\| Ah \|_2 \geqslant \gamma \| h \|_2$. Consequently, from **4.3.8**, $\| A^{-1} \|_2 \leqslant \gamma^{-1}$. Therefore, it follows from **5.4.3**(c) that

$$\| F'(x)^{-1} \|_2 \leqslant 1/\gamma, \qquad \forall x \in R^n,$$

and the Hadamard theorem **5.3.10** applies. ∎

Note that **5.4.5** contains, as a special case, the corresponding result **4.3.10** for gradient operators. However, the tools required to prove **5.4.5**—the Hadamard theorem and the continuation theory of **5.3**—are somewhat deeper than those employed in Chapter 4. On the other hand, by using still deeper results, namely, the degree theory of the next chapter, we will be able to show that **5.4.5** remains valid if the requirement of continuous differentiability is weakened to continuity (see **6.4.4**).

As an immediate corollary of **5.4.5**, we conclude that **5.4.1** holds if A is positive definite rather than an M-matrix. Slightly more generally, we have:

5.4.6. Let $A \in L(R^n)$ be positive definite and assume that $\phi: R^n \to R^n$ is continuously differentiable and that $\phi'(x)$ is positive semidefinite for all $x \in R^n$. Then $A + \phi$ is a homeomorphism of R^n onto itself.

Proof. Since $h^T A h > 0$ whenever $h \neq 0$, it follows, by continuity, that $h^T A h \geqslant \gamma > 0$ when $h^T h = 1$, and, consequently, that $h^T A h \geqslant \gamma h^T h$

for all $h \in R^n$. Now set $Fx \equiv Ax + \phi x$. Then, clearly, F is also continuously differentiable, and

$$h^T F'(x) h = h^T[A + \phi'(x)] h \geq h^T A h \geq \gamma h^T h, \qquad \forall x, h \in R^n.$$

Therefore, the result follows directly from **5.4.3** and **5.4.5**. ∎

We end this section with a result on the solution set of $Fx = b$ when F is monotone. This is a natural generalization of the fact (see **E 4.2-9**) that the set of critical points of a convex functional is convex.

5.4.7. Let $F: D \subset R^n \to R^n$ be continuous and monotone on the open, convex set $D_0 \subset D$. Then, for any $b \in R^n$, the solution set $\Gamma = \{x \in D_0 \mid Fx = b\}$ is convex if it is not empty.

Proof. For given $b \in R^n$, define the sets

$$Q = \{x \in D_0 \mid (y - x)^T (Fx - b) \geq 0, \quad \forall y \in R^n\},$$

and

$$Q^* = \{x \in D_0 \mid (y - x)^T (Fy - b) \geq 0, \quad \forall y \in D_0\}.$$

It is clear that $Q = \Gamma$ and, since D_0 is convex, that Q^* is also convex. Hence, it suffices to show that $Q = Q^*$. If $x \in Q$, then, in particular,

$$(y - x)^T (Fx - b) \geq 0, \qquad \forall y \in D_0,$$

and, by the monotonicity of F,

$$(y - x)^T (Fy - Fx) \geq 0, \qquad \forall y \in D_0,$$

so that, by addition, it follows that $x \in Q^*$. Conversely, let $x \in Q^*$ and, for any $y \in R^n$, set $x^t = ty + (1 - t)x$, $t \in [0, 1]$. Since D_0 is open, there is a $\delta \in (0, 1]$ such that $x^t \in D_0$ for all $t \in [0, \delta)$, and, thus,

$$0 \leq (x^t - x)^T (Fx^t - b) = t(y - x)^T (Fx^t - b),$$

so that

$$(y - x)^T (Fx^t - b) \geq 0, \qquad \forall t \in (0, \delta). \tag{6}$$

Then the continuity of F implies that (6) holds in the limit as $t \to 0$, and, hence, that $x \in Q$. ∎

As an immediate consequence of this result, we also obtain the corresponding statement for fixed points of a nonexpansion.

5.4. MONOTONE OPERATORS AND OTHER APPLICATIONS

5.4.8. Assume that $G: D \subset R^n \to R^n$ is nonexpansive in the l_2-norm on the open, convex set D. Then the set of fixed points of G is convex if it is not empty.

Proof. Define $F: D \subset R^n \to R^n$ by $Fx = x - Gx$, $x \in D$. Then the nonexpansivity of G and the Cauchy–Schwarz inequality show that

$$(x-y)^T(Fx - Fy) = \|x-y\|_2^2 - (x-y)^T(Gx - Gy)$$
$$\geq \|x-y\|_2^2 - \|x-y\|_2 \|Gx - Gy\|_2 \geq 0, \quad \forall x, y \in D.$$

Hence, F is monotone on D and the result follows immediately from **5.4.7**. ∎

NOTES AND REMARKS

NR 5.4-1. Theorem **5.4.1** is valid under the assumption that ϕ is only continuous provided it is still isotone and diagonal. This has been shown by Caspar [1969] using **5.3.8** as well as the domain-invariance theorem (see **NR 6.2-1**). In Chapter 13, however, we shall obtain this extension using much simpler arguments (see **13.5.6** and also **E 13.1-2**).

NR 5.4-2. See **NR 6.4-1,2** for further discussion of monotone mappings.

NR 5.4-3. For additional existence and uniqueness theorems for discrete analogs of two-point boundary value problems, see Lees [1966] and Lees and Schultz [1966].

NR 5.4-4. A matrix $A \in L(R^n)$ is called a *P-matrix* if all principal minor determinants are positive. Since any positive definite matrix is a *P*-matrix (see **E 5.4-2**), the following result of Gale and Nikaido [1965] represents an interesting extension of the uniqueness theorem **5.4.4**: If $F: R^n \to R^n$ is F-differentiable and $F'(x)$ is a P-matrix for all $x \in R^n$, then F is one-to-one.

NR 5.4-5. Theorem **5.4.7** was proved by Minty [1962]. The corresponding result **5.4.8** holds more generally in a strictly convex Banach space (see, for example, Opial [1967a]).

EXERCISES

E 5.4-1. Suppose that $A \in L(R^n)$ is positive (semi) definite. Prove that all eigenvalues of A have positive (nonnegative) real part. By means of the example

$$A = \begin{pmatrix} 200 & -100 \\ -1 & 2 \end{pmatrix},$$

show that the converse is, in general, not true, but prove that the converse holds if A is normal (i.e., $AA^T = A^TA$). (Hint: A normal matrix has n orthonormal eigenvectors in C^n.)

E 5.4-2. Let $A \in L(R^n)$ be positive definite. Prove that all principal minor determinants are positive. Give a 2×2 example which shows that the converse is not true.

E 5.4-3. Show that **5.4.5** may also be proved by the norm-coerciveness theorem 5.3.9.

E 5.4-4. Let $\phi\colon R^n \to R^n$ be monotone and $B \in L(R^n)$ be positive definite. Show that $F = I + B\phi$ is one-to-one. Similarly, if ϕ is strictly monotone and B is positive semidefinite, show that F is one-to-one.

E 5.4-5. Assume that $F_1, F_2 : R^n \to R^n$ are both continuously differentiable and monotone, and that F_1 is uniformly monotone. Then $F = F_1 + F_2$ is a homeomorphism of R^n onto R^n. Conclude, in particular, that if $A \in L(R^n)$ is positive definite, then $F = A + F_2$ is a homeomorphism.

E 5.4-6. Let $A \in L(R^n)$ be skew-symmetric ($A = -A^T$) and $\phi\colon R^n \to R^n$ continuously differentiable and monotone. If $c > 0$, then $F = A + cI + \phi$ is a homeomorphism of R^n onto itself.

E 5.4-7. Let $\phi\colon R^n \to R^n$ be continuously differentiable and monotone, and assume that $B \in L(R^n)$ is positive definite. Then $I + B\phi$ is one-to-one and onto R^n. (Hint: Consider the mapping $Fx \equiv B^Tx + B\phi B^Tx$.)

Chapter 6 / THE DEGREE OF A MAPPING

6.1. ANALYTIC DEFINITION OF THE DEGREE

When $F: D \subset R^n \to R^n$ is a continuous mapping and $y \in R^n$ is a given vector, it is frequently of considerable interest to know in advance the number of solutions of the system $Fx = y$ in some specified set $C \subset R^n$. It might be tempting to devise methods for calculating this solution count, or at least for finding suitable bounds for it. But such an attempt will be thwarted by the fact that the number of solutions of $Fx = y$ does not, in general, vary continuously with either F or y. We shall see that this undesirable property can be avoided if we first count the number of solutions $x \in C$ for which F preserves the "orientation" in a neighborhood of x, and then subtract from the result the count of those solutions in C for which the orientation is reversed. The number obtained in this way is the so-called degree of F at y with respect to C.

Clearly, this does not constitute a precise definition of the degree; in order to provide such a definition, it would be necessary to clarify the meaning of such concepts as "preservation" or "reversal" of "orientation," and to specify the sets C which may be considered. While the degree no longer constitutes a precise count of all solutions of the system $Fx = y$ in C, it nevertheless plays a considerable role in the existence theory of such systems. The above vague definition already indicates that when the degree of F at y with respect to some set C is known to be nonzero, then there must be at least one solution of $Fx = y$ in C. Moreover, the continuous dependence of the degree on F and y allows us to transfer this knowledge of the existence of a solution to neighboring systems.

In its original form, the degree of a mapping is a concept of combinatorial topology (see **NR 6.1-2**), but there are also purely analytic definitions for it. We shall present here one such analytic approach to degree theory. For this we require, particularly in the proof of one lemma, some analytical tools not needed elsewhere in this book. In order not to interrupt the flow of the presentation, we have relegated a discussion of these analytical tools, as well as the proof of this particular lemma, to an appendix of this chapter.

The degree concept will first be developed for mappings $F: D \subset R^n \to R^n$ which are continuously differentiable on D. For the sake of convenience, we assume D always to be an open set. The degree will be defined with respect to open, bounded sets C for which $\bar{C} \subset D$. The openness assumption for C is rather natural, since we are concerned with the behavior of F in full neighborhoods of the solution of $Fx = y$ in C; the boundedness will assure that the degree remains finite.

Under these assumptions about F and C, we now associate with any $x \in C$ the local coordinate system $x + \epsilon e^j$, $j = 1,\ldots, n$, spanning a neighborhood of x. For sufficiently small $\epsilon > 0$, the image vectors $F(x + \epsilon e^j)$, $j = 1,\ldots, n$, are, by definition, approximately equal to $Fx + \epsilon F'(x) e^j$, $j = 1,\ldots, n$, and, hence, if $F'(x)$ is nonsingular, will span a neighborhood of $y = Fx$. This shows that both the local coordinate system at x and its image have the same orientation if $\det F'(x)$ has a positive sign and that their orientation will be opposite to each other if $\operatorname{sgn} \det F'(x) = -1$.

Assume now that, for a given $y \in R^n$, there are no solutions of $Fx = y$ on the boundary $\dot{C}$ of C and that, for every solution x in C, $F'(x)$ is nonsingular. Then **5.2.1** shows that F maps some neighborhood $U(x)$ of each solution $x \in C$ homeomorphically onto a certain neighborhood of y. This means that $U(x)$ does not contain another solution, or, in other words, that the solutions of $Fx = y$ have no accumulation point in C or in $\bar{C}$. Since $\bar{C}$ is compact, there are, therefore, only finitely many such solutions, say $x^1,\ldots, x^m$. In line with our general discussion above, this suggests that, in this case, we should define the degree of F at y with respect to C by the sum

$$\sum_{j=1}^{m} \operatorname{sgn} \det F'(x^j). \tag{1}$$

The question now is how this tentative definition of the degree could be extended to the case when there are solutions of $Fx = y$ in C for which $F'(x)$ is singular, and, then, more generally, to the case when F is only continuous.

If, at first, we retain the differentiability of F, but permit the derivative

6.1. ANALYTIC DEFINITION OF THE DEGREE

to become singular at some solution, then it may be expected that the sum (1) will have to be replaced by an integral. The appropriate form of such an integral is not evident, and it appears to be simplest to present here, without further motivation, a certain integral and to prove that it indeed reduces to the sum (1) if $F'(x)$ is nonsingular at all solutions of $Fx = y$ in C.

For given $\alpha > 0$, let W_α be the set of all real functions $\varphi \colon [0, \infty) \subset R^1 \to R^1$ which are continuous on $[0, \infty)$ and for which there exists a $\delta \in (0, \alpha)$ such that $\varphi(t) = 0$ whenever $t \notin [\delta, \alpha]$. We call every $\varphi \in W_\alpha$ a *weight function of index* α. Clearly, if $\varphi \in W_\alpha$, then $g \colon R^n \to R^1$, $g(x) = \varphi(\|x\|_2)$, is a continuous function of *compact support* on R^n, that is, $g(x) = 0$ for all x outside of some compact set S, called the support of g. Hence, the (Riemann) integral $\int_{R^n} \varphi(\|x\|_2)\,dx$ and the set

$$W_\alpha^1 = \left\{ \varphi \in W_\alpha \,\Big|\, \int_{R^n} \varphi(\|x\|_2)\,dx = 1 \right\}$$

are well-defined.

6.1.1. Definition. Let $F \colon D \subset R^n \to R^n$ be continuously differentiable on the open set D, C an open, bounded set with $\bar{C} \subset D$, and $y \notin F(\dot{C})$ a given point. For any weight function φ of index $\alpha < \dot{\gamma} = \min\{\|Fx - y\|_2 \mid x \in \dot{C}\}$, define the mapping

$$\phi \colon R^n \to R^n, \quad \phi(x) = \begin{cases} \varphi(\|Fx - y\|_2) \det F'(x), & x \in C \\ 0, & \text{otherwise.} \end{cases} \quad (2a)$$

Then the integral

$$d_\varphi(F, C, y) = \int_{R^n} \phi(x)\,dx \quad (2b)$$

is called the *degree integral* of F on C with respect to y and the weight function φ. ∎

Note that, since F is continuous on D and $\dot{C}$ is compact, there is an open neighborhood D_0 of $\dot{C}$ such that $\|Fx - y\|_2 \geq \tfrac{1}{2}(\dot{\gamma} + \alpha) > \alpha$ for all $x \in D_0$. Hence, by definition of φ and ϕ, ϕ is zero whenever $x \in D_0$ or $x \notin C$. Moreover, ϕ is clearly continuous on C, and, hence, on all of R^n; that is, ϕ is a continuous function with compact support, and therefore the integral in (2b) is well-defined.

The next result shows that the degree integral indeed reduces to the sum (1) when the derivative is nonsingular at all solutions.

6.1.2. Let $F: D \subset R^n \to R^n$ be continuously differentiable on the open set D, and C an open, bounded set with $\bar{C} \subset D$. Suppose further that, for given $y \notin F(\dot{C})$, $F'(x)$ is nonsingular for all $x \in \Gamma = \{x \in C \mid Fx = y\}$. Then Γ consists of at most finitely many points, and there exists an $\hat{\alpha}$ with $0 < \hat{\alpha} \leq \dot{\gamma} = \min\{\|Fx - y\|_2 \mid x \in \dot{C}\}$ such that, for any $\varphi \in W_\alpha^1$ with $\alpha \in (0, \hat{\alpha})$,

$$d_\varphi(F, C, y) = \begin{cases} \sum_{j=1}^m \operatorname{sgn} \det F'(x^j), & \text{if } \Gamma = \{x^1, ..., x^m\}, \\ 0, & \text{if } \Gamma \text{ is empty.} \end{cases} \qquad (3)$$

Proof. We have already seen that under the assumptions of this theorem, there can be at most finitely many solutions $x^1, ..., x^m$ of $Fx = y$ in C. If Γ is empty, that is, if $y \notin F(\bar{C})$, then, by the compactness of $\bar{C}$, we have $\dot{\gamma} = \min\{\|Fx - y\|_2 \mid x \in \bar{C}\} > 0$, and, hence, for $\hat{\alpha} = \dot{\gamma}$ and $\alpha \in (0, \hat{\alpha})$, $\varphi(\|Fx - y\|_2) = 0$ for $x \in \bar{C}$, and $\varphi \in W_\alpha^1$. Thus, $\phi(x) = 0$ for $x \in R^n$, and (3) holds. If $m > 0$, then, by the inverse function theorem **5.2.1**, there exist open neighborhoods $U(x^j) \subset C$ and $V_j(y)$ of x^j and y, respectively, $j = 1, ..., m$, such that the restriction F_j of F to $U(x^j)$ is a homeomorphism from $U(x^j)$ onto $V_j(y)$. We may assume that $U(x^j)$ is so small that $\operatorname{sgn} \det F'(x)$ is constant for $x \in U(x^j)$.

Since there are only finitely many $V_j(y)$, there exists an $\hat{\alpha} \in (0, \dot{\gamma})$ such that $K = \bar{S}(y, \hat{\alpha}) \subset V_j(y)$ for $j = 1, ..., m$. Let $U_j = F_j^{-1}(K)$ and $\alpha \in (0, \hat{\alpha})$. Then, evidently, $U_j \subset C$, and, for any $\varphi \in W_\alpha^1$ we have $\varphi(\|Fx - y\|_2) = 0$ whenever $x \notin \bigcup_{j=1}^m U_j$. Hence, it follows from the change-of-variable theorem (see **6.5.1**) applied to each F_j that

$$d_\varphi(F, C, y) = \sum_{j=1}^m \int_{U_j} \varphi(\|Fx - y\|_2) \det F'(x)\, dx$$

$$= \sum_{j=1}^m \int_{F_j^{-1}(K)} \varphi(\|F_j x - y\|_2) \det F_j'(x)\, dx$$

$$= \sum_{j=1}^m \operatorname{sgn} \det F'(x^j) \int_K \varphi(\|x\|_2)\, dx = \sum_{j=1}^m \operatorname{sgn} \det F'(x^j),$$

since $\int_K \varphi(\|x\|_2)\, dx = \int_{R^n} \varphi(\|x\|_2)\, dx = 1$. ∎

This theorem suggests that we should formally define the degree of F at y with respect to C by the degree integral $d_\varphi(F, C, y)$. But, for that, we still have to specify the choice of the weight function φ.

Theorem **6.1.2** already shows that, under the conditions of that theo-

6.1. ANALYTIC DEFINITION OF THE DEGREE

rem, the degree integral is independent of φ for all $\varphi \in W_\alpha^1$ provided that α is sufficiently small. It turns out that this is actually true in general for all $\alpha \in (0, \dot\gamma)$. To prove this fact, we require the following lemma.

6.1.3. Let $F: D \subset R^n \to R^n$ be continuously differentiable on the open set D, and C an open, bounded set with $\bar C \subset D$. Suppose further that $y \in R^n$ satisfies $\dot\gamma = \min\{\|Fx - y\|_2 \mid x \in \dot C\} > 0$. Then, for $\alpha \in (0, \dot\gamma)$ and any $\varphi \in W_\alpha$, the integral

$$\eta(\varphi) = \int_0^\infty t^{n-1}\varphi(t)\, dt \tag{4}$$

is well defined, and $\eta(\varphi) = 0$ implies that $d_\varphi(F, C, y) = 0$.

This is the lemma mentioned at the outset which requires the use of some—actually well-known—analytical tools not used elsewhere in the book. We have therefore included its proof in the appendix to this chapter.

Based on Lemma **6.1.3**, we can now prove that the degree integral is independent of the choice of the weight function.

6.1.4. Let $F: D \subset R^n \to R^n$ be continuously differentiable on the open set D, and C an open, bounded set with $\bar C \subset D$. If $y \notin F(\dot C)$, then

$$d_{\varphi_1}(F, C, y) = d_{\varphi_2}(F, C, y) \tag{5}$$

for any $\varphi_1, \varphi_2 \in W_\alpha^1$ with $0 < \alpha < \dot\gamma = \min\{\|Fx - y\|_2 \mid x \in \dot C\}$.

Proof. If **6.1.3** is applied to the identity mapping $I: R^n \to R^n$ and the set $C^0 = S(y, 2\alpha)$, then $\eta(\varphi) = 0$ for some $\varphi \in W_\alpha$ implies that

$$d_\varphi(I, C^0, y) = \int_{R^n} \varphi(\|x\|_2)\, dx = 0.$$

Now, let $\varphi_1, \varphi_2 \in W_\alpha^1$. Then, clearly, $\varphi = \eta(\varphi_1)\varphi_2 - \eta(\varphi_2)\varphi_1 \in W_\alpha$, and

$$\eta(\varphi) = \int_0^\infty s^{n-1}\varphi_2(s) \int_0^\infty t^{n-1}\varphi_1(t)\, dt\, ds - \int_0^\infty s^{n-1}\varphi_1(s) \int_0^\infty t^{n-1}\varphi_2(t)\, dt\, ds = 0.$$

Therefore, by the above observation,

$$0 = \int_{R^n} \varphi(\|x\|_2)\, dx = \eta(\varphi_1) \int_{R^n} \varphi_1(\|x\|_2)\, dx - \eta(\varphi_2) \int_{R^n} \varphi_2(\|x\|_2)\, dx$$
$$= \eta(\varphi_1) - \eta(\varphi_2) = \eta(\varphi_1 - \varphi_2).$$

Hence, again by **6.1.3**, we find that

$$d_{\varphi_1}(F, C, y) - d_{\varphi_2}(F, C, y) = d_{\varphi_1-\varphi_2}(F, C, y) = 0. \quad \blacksquare$$

This result shows that the following definition is now meaningful.

6.1.5. Definition. Let $F: D \subset R^n \to R^n$ be continuously differentiable on the open set D, and C any open, bounded set with $\bar{C} \subset D$. Then the degree of F at any point $y \notin F(\dot{C})$ with respect to C is defined by

$$\deg(F, C, y) = d_\varphi(F, C, y), \tag{6}$$

where $\varphi \in W_\alpha^1$ is any weight function of index α such that $0 < \alpha < \dot{\gamma} = \min\{\|Fx - y\|_2 \mid x \in \dot{C}\}$. $\quad \blacksquare$

With this definition, we can now rephrase **6.1.2** as follows: Under the assumptions of Theorem **6.1.2**, if $\{x \in C \mid Fx = y\} = \{x^1, ..., x^m\}$, then

$$\deg(F, C, y) = \sum_{j=1}^m \operatorname{sgn} \det F'(x^j). \tag{7}$$

This shows that, at least in this particular case, the degree introduced in **6.1.5** is integer-valued. For the general case, we shall prove this in the next section.

At the outset, we mentioned as one of the fundamental properties of the degree its continuous dependence on F. This is the content of the next result.

6.1.6. Let $F, G: D \subset R^n \to R^n$ be two continuously differentiable maps on the open set D, and C an open, bounded set with $\bar{C} \subset D$. Suppose, further, that $y \in R^n$ satisfies $\dot{\gamma} = \min\{\|Fx - y\|_2 \mid x \in \dot{C}\} > 0$. If $\alpha \in (0, \dot{\gamma})$ and

$$\sup\{\|Fx - Gx\|_2 \mid x \in \bar{C}\} < \tfrac{1}{7}\alpha, \tag{8}$$

then

$$\deg(F, C, y) = \deg(G, C, y). \tag{9}$$

Proof. Set $\alpha_0 = \tfrac{1}{7}\alpha$ and let $\mu: [0, \infty) \to [0, 1]$ be continuously differentiable on $[0, \infty)$ and satisfy $\mu(t) = 1$ for $t \in [0, 2\alpha_0]$ as well as $\mu(t) = 0$ for $t \geq 3\alpha_0$. Then, since $\|\cdot\|_2$ is continuously differentiable except at 0, the mapping

$$H: D \subset R^n \to R^n, \quad Hx = [1 - \mu(\|Fx - y\|_2)] Fx + \mu(\|Fx - y\|_2) Gx,$$

6.1. ANALYTIC DEFINITION OF THE DEGREE

is continuously differentiable on D. Moreover, since

$$\| Hx - Fx \|_2 \leqslant \mu(\| Fx - y \|_2) \| Gx - Fx \|_2 < \alpha_0, \quad \forall x \in \bar{C}, \quad (10)$$

we have

$$\| Hx - y \|_2 \geqslant \| Fx - y \|_2 - \| Hx - Fx \|_2 > 6\alpha_0 > 0, \quad \forall x \in \dot{C},$$

so that $\deg(H, C, y)$ is well defined for any $\varphi \in W_{6\alpha_0}$.

Now choose $\varphi_1 \in W^1_{5\alpha_0}$ such that $\varphi_1(t) = 0$ for $t \in [0, 4\alpha_0]$. Then, since (10) implies that

$$\| Hx - y \|_2 \leqslant \| Fx - y \|_2 + \| Fx - Hx \|_2 < \| Fx - y \|_2 + \alpha_0, \quad \forall x \in \bar{C},$$

we have

$$\varphi_1(\| Fx - y \|_2) = \varphi_1(\| Hx - y \|_2) = 0 \quad \text{if} \quad \| Fx - y \|_2 < 3\alpha_0.$$

But, if $\| Fx - y \|_2 \geqslant 3\alpha_0$, it follows from the definition of H and μ that $Hx = Fx$. Consequently,

$$\varphi_1(\| Hx - y \|_2) \det H'(x) = \varphi_1(\| Fx - y \|_2) \det F'(x), \quad \forall x \in \bar{C},$$

so that $\deg(H, C, y) = \deg(F, C, y)$.

Similarly, let $\varphi_2 \in W^1_{\alpha_0}$. Then

$$\| Gx - y \|_2 \geqslant \| Fx - y \|_2 - \| Fx - Gx \|_2 > \| Fx - y \|_2 - \alpha_0 \quad (11)$$

shows that $\| Gx - y \|_2 > 6\alpha_0$ if $x \in \dot{C}$, and, consequently, $d_\varphi(G, C, y)$ is well-defined for φ_2. Moreover, if $\| Fx - y \|_2 > 2\alpha_0$, then (11) implies that $\| Gx - y \|_2 > \alpha_0$, while, by (10), $\| Hx - y \|_2 > \alpha_0$; hence, $\varphi_2(\| Gx - y \|_2) = \varphi_2(\| Hx - y \|_2) = 0$. But, if $\| Fx - y \|_2 \leqslant 2\alpha_0$, then $Gx = Hx$, so that

$$\varphi_2(\| Gx - y \|_2) \det G'(x) = \varphi_2(\| Hx - y \|_2) \det H'(x), \quad \forall x \in \bar{C}.$$

Therefore, $\deg(G, C, y) = \deg(H, C, y) = \deg(F, C, y)$. ∎

This result permits us to extend the definition of the degree to continuous mappings by using the standard Weierstrass approximation theorem (see **6.5.4** for a precise statement). For any continuous mapping $G: \bar{C} \subset R^n \to R^n$, define

$$\| G \|_C = \sup_{x \in \bar{C}} \| Gx \|_2, \quad (12)$$

where C is an open set and $\bar{C}$ is compact. Then, for any continuous $F: \bar{C} \subset R^n \to R^n$, there exists a sequence $\{F_k\}$ of continuously differentiable maps $F_k: R^n \to R^n$, $k = 0, 1,...$, such that $\lim_{k\to\infty} \| F_k - F \|_C = 0$. If $y \notin F(\dot{C})$, and, more specifically, $0 < \alpha < \min\{\| Fx - y \|_2 \mid x \in \dot{C}\}$, then also $\min\{\| F_k x - y \|_2 \mid x \in \dot{C}\} > \alpha > 0$ for $k \geq k_0$, and, thus, $\deg(F_k, C, y)$ is well defined for all large k. Now, certainly, $\| F_k - F_j \|_C < \frac{1}{7}\alpha$ for $k, j \geq k_1 \geq k_0$, and, thus by **6.1.6**, $\deg(F_k, C, y) = $ const for $k \geq k_1$. This implies that $\lim_{k\to\infty} \deg(F_k, C, y)$ trivially exists. Moreover, this limit does not depend on the choice of the sequence $\{F_k\}$. In fact, let $D \supset \bar{C}$ be any open set and $G_k: D \subset R^n \to R^n$, $k = 0, 1,...$, any sequence of mappings which are continuously differentiable on D and for which $\lim_{k\to\infty} \| G_k - F \|_C = 0$. Then, clearly, $\| F_k - G_j \|_C < \frac{1}{7}\alpha$ and, hence, $\deg(F_k, C, y) = \deg(G_j, C, y)$, for all sufficiently large k, j.

This warrants the following definition.

6.1.7. Definition. Let $F: \bar{C} \subset R^n \to R^n$ be continuous and C an open, bounded set. Then, for any $y \notin F(\dot{C})$, the degree of F at y with respect to C is defined by

$$\deg(F, C, y) = \lim_{k\to\infty} \deg(F_k, C, y), \qquad (13)$$

where $F_k: D \subset R^n \to R^n$ is any sequence of maps which are continuously differentiable on an open set $D \supset \bar{C}$ and for which $\lim_{k\to\infty} \|F_k - F\|_C = 0$. ∎

As shown before, the limit in (13) is always reached after finitely many steps and is independent of the choice of the sequence $\{F_k\}$.

NOTES AND REMARKS

NR 6.1-1. The concept of a local degree, that is, of the degree with respect to a neighborhood of an isolated solution of the system $Fx = y$, goes back to Kronecker [1869], who introduced his "index" or "characteristic" of such a solution in terms of an integral now called the Kronecker integral. A detailed discussion of Kronecker's index and of some of its applications was given by Hadamard [1910]; see also Alexandroff and Hopf [1935].

NR 6.1-2. The extension of this local degree to a degree in the large is due to Brouwer [1912], who used it as the basis for many of his famous results. This "global" degree is by nature a concept of combinatorial topology. Briefly, for every simplex of a polyhedral complex, an orientation can be readily defined. Accordingly, when a simplical map F of such a complex K_1 into a second one, K_2, is given, the degree of F for a certain simplex σ in K_2 is set equal to the count of

6.1. ANALYTIC DEFINITION OF THE DEGREE

the number of times a simplex of K_1 is mapped into σ preserving the orientation minus the count of those times the orientation is reversed. When F is continuous on a compact set $\bar{C}$ in R^n, the simplicial-mapping theorem allows F to be approximated by simplicial mappings, and their degree is inherited by F.

For a thorough definition of the degree along this line, or, more precisely, in terms of singular homology theory, see, for example, Alexandroff and Hopf [1935] or Cronin [1964].

NR 6.1-3. Since Brouwer's basic paper in 1912, many efforts have been made toward defining the degree of a mapping by strictly analytic methods not involving the concepts of combinatorial topology. Nagumo [1951] based his approach essentially on the sum (1) and on the result of Theorem **6.2.10**. Then Heinz [1959] developed an integral approach related to that of Kronecker, and it is this approach of Heinz that has been followed here. We refer also to Schwartz [1964] for a presentation of degree theory which merges the approaches of Nagumo and Heinz.

NR 6.1-4. All of the results of this section, indeed, of the whole chapter, remain valid under any inner product on R^n.

EXERCISES

E 6.1-1. Suppose the conditions of **6.1.2** hold. Let $f_t : R^n \to R^1$, $t \in (0, \infty)$, be a family of continuous mappings of compact support such that (a) for any $t \in (0, \infty)$, the support of f_t is the ball $\bar{S}(y, t)$, and (b) $\int_{R^n} f_t(Fx)\,dx = 1$. Show that there exists a $t_0 > 0$ such that

$$\deg(F, C, y) = \int_C f_t(Fx) \det F'(x)\,dx, \qquad \forall t \in (0, t_0)$$

(see Schwartz [1964, p. 79]).

E 6.1-2. Let p be any real polynomial for which all real roots are simple. Show that, for any interval $[a, b] \subset R^1$ such that $p(a)\,p(b) \neq 0$,

$$\deg(p, (a, b), 0) = \begin{cases} +1 & \text{if } \operatorname{sgn} p(a) = -1, \ \operatorname{sgn} p(b) = +1 \\ 0 & \text{if } \operatorname{sgn} p(a) = \operatorname{sgn} p(b) \\ -1 & \text{if } \operatorname{sgn} p(a) = +1, \ \operatorname{sgn} p(b) = -1. \end{cases}$$

E 6.1-3. Consider the mapping $F: R^2 \to R^2$ given by $f_1(x) = x_1^3 - 3x_1 x_2^2$, $f_2(x) = -x_2^3 + 3x_1^2 x_2$. Show that $\deg(F, S(0, 2), e^1) = 3$, where $e^1 = (1, 0)^T$.

E 6.1-4. Let $A \in L(R^n)$. Show that

$$\deg(A, S(0, 1), 0) = \begin{cases} +1 & \text{if } \det A > 0 \\ -1 & \text{if } \det A < 0 \\ \text{not defined} & \text{if } \det A = 0. \end{cases}$$

6.2. PROPERTIES OF THE DEGREE

In this section, we prove some of the simpler properties of the degree of a continuous function as defined in **6.1.7**. As a first result, Theorem **6.1.6** is extended to continuous functions.

6.2.1. Let C be open and bounded, and $F: \bar{C} \subset R^n \to R^n$ a continuous mapping. If $y \in R^n$ is such that $\min\{\|Fx - y\|_2 \mid x \in \dot{C}\} > \alpha > 0$, then $\deg(F, C, y) = \deg(G, C, y)$ for any continuous map $G: \bar{C} \subset R^n \to R^n$ for which $\|G - F\|_C < \frac{1}{7}\alpha$.

Proof. Choose sequences F_k, $G_k: D \subset R^n \to R^n$, $k = 0, 1,\ldots$, of continuously differentiable mappings on an open set $D \supset \bar{C}$ such that $\lim_{k \to \infty} \|F_k - F\|_C = \lim_{k \to \infty} \|G_k - G\|_C = 0$. Then there is a k_0 so that

$$\|G_j x - F_k x\|_2 \leq \|G_j x - Gx\|_2 + \|Gx - Fx\|_2 + \|Fx - F_k x\|_2 < \tfrac{1}{7}\alpha$$

for $k, j \geq k_0$ and all $x \in \bar{C}$. Moreover, k_0 can be chosen sufficiently large that $\min\{\|F_k x - y\|_2 \mid x \in \dot{C}\} > \alpha$ as well as $\min\{\|G_k x - y\|_2 \mid x \in \dot{C}\} > \alpha$ for $k \geq k_0$. Then **6.1.6** states that

$$\deg(F_k, C, y) = \deg(G_j, C, y), \qquad j, k \geq k_0$$

and the result is a direct consequence of Definition **6.1.7**. ∎

From this, the following result, which plays a central role in degree theory, follows almost immediately.

6.2.2. Homotopy Invariance Theorem. Let C be open and bounded, and $H: \bar{C} \times [0, 1] \subset R^{n+1} \to R^n$ a continuous map from $\bar{C} \times [0, 1]$ into R^n. Suppose, further, that $y \in R^n$ satisfies $H(x, t) \neq y$ for all $(x, t) \in \dot{C} \times [0, 1]$. Then $\deg(H(\cdot, t), C, y)$ is constant for $t \in [0, 1]$.

Proof. Since $\dot{C} \times [0, 1]$ is compact, we have

$$\min\{\|H(x, t) - y\|_2 \mid (x, t) \in \dot{C} \times [0, 1]\} > \alpha > 0,$$

for some α, and, from the uniform continuity of H on $\bar{C} \times [0, 1]$, it follows that there exists a $\delta > 0$ such that $\sup_{x \in \bar{C}} \|H(x, s) - H(x, t)\|_2 < \tfrac{1}{7}\alpha$ for all $s, t \in [0, 1]$ such that $|s - t| < \delta$. Hence, by **6.2.1**,

$$\deg(H(\cdot, s), C, y) = \deg(H(\cdot, t), C, y)$$

whenever $|s - t| < \delta$, and $s, t \in [0, 1]$. Since $[0, 1]$ can be covered by a finite number of intervals of length δ, the result follows. ∎

6.2. PROPERTIES OF THE DEGREE

As an interesting application of this result, we find that the degree depends only the on values of F on the boundary of C.

6.2.3. Boundary-Value Theorem. Let C be open and bounded, and $F: \bar{C} \subset R^n \to R^n$ continuous. Suppose, further, that $G: \bar{C} \subset R^n \to R^n$ is any continuous map such that $Fx = Gx$ for $x \in \dot{C}$. Then for any $y \notin F(\dot{C})$ we have $\deg(F, C, y) = \deg(G, C, y)$.

Proof. Consider the homotopy

$$H: \bar{C} \times [0, 1] \subset R^{n+1} \to R^n, \qquad H(x, t) = tFx + (1-t)\, Gx. \tag{1}$$

Then $H(x, t) = Fx \neq y$ for $(x, t) \in \dot{C} \times [0, 1]$, and, hence, the result follows directly from **6.2.2**. ∎

Theorem **6.2.3** is actually a special case of the following more general result whose proof proceeds precisely as that of **6.2.3** by considering the homotopy (1).

6.2.4. Poincaré–Bohl Theorem. Let $F, G: \bar{C} \subset R^n \to R^n$ be two continuous maps and C an open, bounded set. If $y \in R^n$ is any point such that

$$y \notin \{u \in R^n \mid u = tFx + (1-t)\, Gx, \ x \in \dot{C}, \ t \in [0, 1]\}, \tag{2}$$

then $\deg(G, C, y) = \deg(F, C, y)$.

So far, we have concerned ourselves only with the influence of the change of F upon the degree. We turn now to its variation under changes of y. For this, we need the following lemma, which shows that the degree is invariant under translations. In the sequel, $F - z$ denotes the mapping $Fx - z$, $x \in D$.

6.2.5. Let $F: \bar{C} \subset R^n \to R^n$ be continuous and C an open, bounded set. If $y \notin F(\dot{C})$ and $z \in R^n$ is any point, then $\deg(F - z, C, y - z) = \deg(F, C, y)$.

Proof. Observe, first, that when F_k is continuously differentiable on an open set $D \supset \bar{C}$, then clearly with $G_k = F_k - z$

$$\varphi(\| F_k x - y \|_2) \det F_k'(x) = \varphi(\| G_k x - (y - z)\|_2) \det G_k'(x),$$

and, hence, $\deg(G_k, C, y - z) = \deg(F_k, C, y)$. Evidently, this property carries over to continuous F if we consider sequences of continuously differentiable functions on D which converge to F uniformly on $\bar{C}$. ∎

Using this lemma, we find that the degree does not vary under rather extensive changes of y.

6.2.6. Let $F: \bar{C} \subset R^n \to R^n$ be continuous and C an open, bounded set. Suppose $y^0, y^1 \in R^n$ are any two points which can be connected by a continuous path $p: [0, 1] \subset R^1 \to R^n$ avoiding $F(\dot{C})$, that is, for which $p(0) = y^0, p(1) = y^1$, and $p(t) \notin F(\dot{C})$ for $t \in [0, 1]$. Then $\deg(F, C, y^0) = \deg(F, C, y^1)$.

Proof. Consider the homotopy
$$H: \bar{C} \times [0, 1] \subset R^{n+1} \to R^n, \qquad H(x, t) = Fx - p(t).$$
Then, by assumption, $H(x, t) \neq 0$ for $(x, t) \in \dot{C} \times [0, 1]$, and, hence, by **6.2.2** and **6.2.5**,
$$\deg(F, C, y^0) = \deg(H(\cdot, 0), C, 0) = \deg(H(\cdot, 1), C, 0) = \deg(F, C, y^1). \quad\blacksquare$$

We now consider the influence of the set C on the degree. As a first result, we see that—not surprisingly—the degree shares with the integral the following additivity property.

6.2.7. Let $F: \bar{C} \subset R^n \to R^n$ be continuous and C, as well as $C_1, \ldots, C_m \subset C$, open, bounded sets such that $C_i \cap C_j = \emptyset$, $i \neq j$, and $\bigcup_{j=1}^{m} \bar{C}_j = \bar{C}$. Then, for any $y \notin \bigcup_{j=1}^{m} F(\dot{C}_j)$, $\deg(F, C, y) = \sum_{j=1}^{m} \deg(F, C_j, y)$.

Proof. For continuously differentiable F_k on some open set $D \subset \bar{C}$, the result is an immediate consequence of the additivity of the integral (6.1.2b), and, clearly, the statement remains valid as we go to the limit in the definition **6.1.7** of the degree for continuous F. $\quad\blacksquare$

The following theorem shows that C can be reduced by removing any closed set which does not intersect $\{x \in C \mid Fx = y\}$ without changing the degree.

6.2.8. Excision Theorem. Let $F: \bar{C} \subset R^n \to R^n$ be continuous, and C an open, bounded set. Moreover, let $y \notin F(\dot{C})$. Then, for any closed set $Q \subset \bar{C}$ for which $y \notin F(Q)$, we have $\deg(F, C, y) = \deg(F, C \sim Q, y)$. In particular, if $Q = \bar{C}$, then $\deg(F, C, y) = 0$.

Proof. Suppose F_k is continuously differentiable on some open set $D \supset \bar{C}$ such that
$$\dot{\gamma} = \min\{\|F_k x - y\|_2 \mid x \in \dot{C}\} > 0, \qquad \eta = \min\{\|F_k x - y\|_2 \mid x \in Q\} > 0.$$

6.2. PROPERTIES OF THE DEGREE

By assumption, α exists such that $\min(\dot{\gamma}, \eta) > \alpha > 0$. If $\varphi \in W_\alpha^1$, then $\varphi(\|F_k x - y\|_2) = 0$ for $x \in Q$, and, thus, $d_\varphi(F_k, C, y) = d_\varphi(F_k, C \sim Q, y)$. Therefore, $\deg(F_k, C, y) = \deg(F_k, C \sim Q, y)$. In particular, if $Q = \bar{C}$, then $d_\varphi(F_k, C, y) = 0$, so that $\deg(F_k, C, y) = 0$. The result for continuous F now follows readily by selecting a sequence $\{F_k\}$ of continuously differentiable maps converging uniformly to F on $\bar{C}$. ∎

It remains to prove that the degree is integer-valued. Recall that we have shown this already under the assumptions of Theorem **6.1.2**. Let us now rephrase that result in a slightly more concise way by introducing for continuously differentiable $F: D \subset R^n \to R^n$ and any subset $Q \subset D$ the set $\mathscr{C}(Q) = \{x \in Q \mid F'(x) \text{ singular}\}$. Then **6.1.2** and **6.1.5** state:

6.2.9. Let $F: D \subset R^n \to R^n$ be continuously differentiable on the open set D, and C an open, bounded set such that $\bar{C} \subset D$. If $y \notin F(\dot{C}) \cup F(\mathscr{C}(\bar{C}))$, either $\Gamma = \{x \in C \mid Fx = y\}$ is empty and $\deg(F, C, y) = 0$, or Γ consists of finitely many points $x^1, \ldots, x^m$ and

$$\deg(F, C, y) = \sum_{j=1}^{m} \operatorname{sgn} \det F'(x^j). \tag{3}$$

Now recall that, by Sard's theorem **5.2.5**, the set $F(\mathscr{C}(\bar{C}))$ always has measure zero in R^n. In other words, if $y \in F(\mathscr{C}(\bar{C}))$, then, in any neighborhood of y, there are infinitely many points not contained in $F(\mathscr{C}(\bar{C}))$. This leads us to the following theorem.

6.2.10. Let $F: D \subset R^n \to R^n$ be continuously differentiable on the open set D, and C an open, bounded set with $\bar{C} \subset D$. If $y \notin F(\dot{C})$, then there exists a sequence $y^k \notin F(\dot{C}) \cup F(\mathscr{C}(C))$, $k = 1, 2, \ldots$, such that $\lim_{k \to \infty} y^k = y$, and, for any such sequence $\{y^k\}$, there is a k_0 such that

$$\deg(F, C, y) = \deg(F, C, y^k), \quad \forall k \geq k_0. \tag{4}$$

Proof. As already mentioned, $F(\mathscr{C}(\bar{C}))$ has measure zero by **5.2.5**, and this ensures the existence of a sequence $\{y^k\}$ as specified in the theorem. Let $\{y^k\}$ be any such sequence. By assumption, there exists a ball $S(y, \epsilon)$ such that $S(y, \epsilon) \cap F(\dot{C})$ is empty. Then $y^k \in S(y, \epsilon)$ for $k \geq k_0$, and the paths $p_k(t) = (1-t) y^k + ty$, $0 \leq t \leq 1$, are contained in $S(y, \epsilon)$, and, hence, do not intersect $F(\dot{C})$. Therefore, **6.2.6** states that (4) holds. ∎

As a direct corollary, we obtain now the announced result:

6.2.11. Let $F: \bar{C} \subset R^n \to R^n$ be continuous and C an open, bounded set. Then, for any $y \notin F(\dot{C})$, $\deg(F, C, y)$ is integer-valued.

Proof. If F_k is continuously differentiable on an open set $D \subset \bar{C}$, then, by **6.2.9** and **6.2.10**, the degree is integer-valued. But then, by Definition **6.1.7**, the same will hold for continuous F. ∎

Note that the same argument also shows that the degree is norm-independent, since the sum on the right of (3) has this property.

NOTES AND REMARKS

NR 6.2-1. We have given in **6.2** only the most fundamental properties of the degree. In particular, we did not include the product theorem due to Leray [1950], which gives a formula for the degree of a composite function $G \circ F$ in terms of G and F. For a proof along our lines, see, for example, Heinz [1959] or Schwartz [1964]. As Schwartz shows, this product theorem permits an analytical proof of the generalized Jordan theorem in R^n which—as a corollary—leads, for example, to a proof of the famous *domain-invariance theorem*: "A continuous, one-to-one mapping from an open set in R^n into R^n maps open sets into open sets."

NR 6.2-2. The extension of the degree concept to infinite-dimensional spaces presents serious problems due to the fact that bounded, closed sets are no longer compact. However, if certain compactness assumptions are made for the mapping, an extension is possible, and was first given by Leray and Schauder [1934] for mappings on Banach spaces. This Leray–Schauder degree theory has numerous applications throughout analysis, and, in particular, to differential equations (see Cronin [1964]) and integral equations (see Krasnoselskii [1956]). It has also been extended to more general topological linear spaces; for references, see, for example, Cronin [1964].

EXERCISES

E 6.2-1. Use **6.2.3** to show that the conclusion of **E 6.1-2** holds for any real polynomial.

E 6.2-2. Let $\bar{S} = \bar{S}(0, 1) \subset R^n$ be the unit ball under the Euclidean norm and $\dot{S}$ its boundary. Show that there exists no continuous mapping $F: S \to \dot{S}$ such that $Fx = x$ for all $x \in \dot{S}$ (See Schwartz [1964, p. 93]).

E 6.2-3. Let $F, G: \bar{C} \subset R^n \to R^n$ be continuous and C an open, bounded set. We denote the restrictions of F and G to $\dot{C}$ by $\dot{F}$ and $\dot{G}$. Suppose there exists a homotopy $H: \dot{C} \times [0, 1] \to R^n$ such that $H(x, 0) = \dot{F}x$ and $H(x, 1) = \dot{G}x$ or all $x \in \dot{C}$. If $y \in R^n$ is such that $H(x, t) \neq y$, $\forall x \in \dot{C}$, $t \in [0, 1]$, then $\deg(F, C, y) = \deg(G, C, y)$. (See Schwartz [1964, p. 93]).

6.3. BASIC EXISTENCE THEOREMS

We now use the general results of the previous two sections to prove existence theorems for equations of the form $Fx = y$ or $x = Gx$. The first result has already been discussed in Section **6.1** as one of the motivations for the entire degree theory and is an immediate consequence of **6.2.8**.

6.3.1. Kronecker Theorem. Let $F: \bar{C} \subset R^n \to R^n$ be continuous and C an open, bounded set. If $y \notin F(\dot{C})$ and if $\deg(F, C, y) \neq 0$, then the equation
$$Fx = y \tag{1}$$
has a solution in C.

Proof. Suppose that (1) has no solutions in C. Then $y \notin F(\bar{C})$ and **6.2.8** gives the contradiction that $\deg(F, C, y) = 0$. ∎

A direct application of **6.3.1** is rarely possible, since the computation of the degree is a nontrivial problem. Rather, **6.3.1** is a useful tool for the proof of other existence theorems, as is illustrated by the next result, which is one of the most famous theorems in analysis.

6.3.2 Brouwer Fixed-Point Theorem. Let $G: \bar{C} \subset R^n \to R^n$ be continuous on the compact, convex set $\bar{C}$, and suppose that $G\bar{C} \subset \bar{C}$. Then G has a fixed point in $\bar{C}$.

Proof. We first prove the result under the assumption that $\bar{C}$ is the ball $\bar{S}(0, r) = \{x \mid \|x\|_2 \leqslant r\}$. Define the homotopy
$$H: \bar{C} \times [0, 1] \subset R^{n+1} \to R^n, \quad H(x, t) = x - tGx, \quad t \in [0, 1], \quad x \in \bar{C},$$
and observe that, because $\|Gx\|_2 \leqslant r$ for all $x \in \bar{C}$,
$$\|H(x, t)\|_2 \geqslant \|x\|_2 - t\|Gx\|_2 \geqslant r(1 - t), \quad \forall t \in [0, 1], \quad x \in \dot{C}. \tag{2}$$
Now, it follows directly from the definition **6.1.5** that
$$\deg(H(\cdot, 0), C, 0) = \deg(I, C, 0) = 1.$$
Consequently, if G has no fixed point in $\bar{C}$, then $H(x, 1) \neq 0$, so that, by (2), $H(x, t) \neq 0$ for all $t \in [0, 1]$ and $x \in \dot{C}$; thus, by **6.2.2**,
$$\deg(I - G, C, 0) = \deg(H(\cdot, 1), C, 0) = \deg(H(\cdot, 0), C, 0) = 1.$$
Hence, **6.3.1** provides a contradiction.

We next consider an arbitrary compact convex set $\bar{C}$. Since $\bar{C}$ is bounded, we may choose an r sufficiently large so that $\bar{C} \subset \bar{C}_0 = \bar{S}(0, r)$. For fixed $y \in \bar{C}_0$, define the functional $g_y \colon \bar{C} \to R^1$ by $g_y(x) = \|x - y\|_2^2$. Clearly, since g_y is continuous and $\bar{C}$ is compact, g_y has a minimizer $\hat{G}y$, that is

$$\|\hat{G}y - y\|_2 = \min\{\|x - y\|_2 \mid x \in \bar{C}\}.$$

Moreover, by **4.2.7**, $\hat{G}y$ is unique, since $g_y''(x) = 2I$, so that by **3.4.6** g_y is strictly convex. Hence, the mapping $\hat{G} \colon \bar{C}_0 \to \bar{C}$ is well-defined, and, clearly, $\hat{G}y = y$ for all $y \in \bar{C}$. We show next that $\hat{G}$ is continuous. Let $\{x^k\}$ be any convergent sequence in $\bar{C}_0$ with limit x. Then it suffices to show that every convergent subsequence of $\{\hat{G}x^k\}$ has the limit $\hat{G}x$. The compactness of $\bar{C}$ ensures that there is a convergent subsequence $\{\hat{G}x^{k_i}\}$ with limit $z \in \bar{C}$, and it suffices to show that $\|x - z\|_2 = \|x - \hat{G}x\|_2$ since, by the uniqueness of minimizers of g_y, this implies that $\hat{G}x = z$. Assume the contrary and choose $\epsilon > 0$ so that $3\epsilon + \|\hat{G}x - x\|_2 < \|z - x\|_2$. Then, for sufficiently large k_i,

$$\|x - z\|_2 \leq \|x - x^{k_i}\|_2 + \|x^{k_i} - \hat{G}x^{k_i}\|_2 + \|\hat{G}x^{k_i} - z\|_2$$
$$\leq 2\epsilon + \|x^{k_i} - \hat{G}x\|_2$$
$$\leq 2\epsilon + \|x^{k_i} - x\|_2 + \|x - \hat{G}x\|_2 < \|x - z\|_2,$$

where we have used the fact that $\|x^{k_i} - \hat{G}x^{k_i}\|_2 \leq \|x^{k_i} - y\|_2$ for all $y \in \bar{C}$ and, in particular, for $y = \hat{G}x$. This is a contradiction, and, hence, $\hat{G}$ is continuous on $\bar{C}_0$. Consequently, the composite mapping $G \circ \hat{G} \colon \bar{C}_0 \to \bar{C} \subset \bar{C}_0$ is continuous, and, therefore, by the first part of the proof, has a fixed point $x^* \in \bar{C}_0$. But, since $\hat{G}\bar{C}_0 \subset \bar{C}$, it follows that $x^* = G(\hat{G}x^*) \in \bar{C}$, and, because $\hat{G}x = x$ for all $x \in \bar{C}$, that $x^* = Gx^*$. ∎

It is important to note that no uniqueness can be asserted in **6.3.2**; indeed, every point of $\bar{C}$ is a fixed point of the identity mapping.

We next prove another important consequence of the Kronecker theorem **6.3.1** which, in turn, will be an important tool for obtaining further existence results.

6.3.3. Leray–Schauder Theorem. Let C be an open, bounded set in R^n containing the origin and $G \colon C \subset R^n \to R^n$ a continuous mapping. If $Gx \neq \lambda x$ whenever $\lambda > 1$ and $x \in \dot{C}$, then G has a fixed point in $\bar{C}$.

Proof. Consider again the homotopy H defined by $H(x, t) = x - tGx$, $x \in \bar{C}$, $t \in [0, 1]$. Then, provided that $x \neq Gx$,

$$H(x, t) = t(t^{-1}x - Gx) \neq 0, \quad \forall t \in (0, 1], \quad x \in \dot{C},$$

while $0 \in C$ implies that $H(x, 0) \neq 0$ for $x \in \dot{C}$. Since $\deg(I, C, 0) = 1$, it therefore follows from **6.2.2** that

$$1 = \deg(H(\cdot, 0), C, 0) = \deg(H(\cdot, 1), C, 0) = \deg(I - G, C, 0)$$

and hence the result is a consequence of **6.3.1**. ∎

If $f: [a, b] \subset R^1 \to R^1$ is continuous and $f(a) \leqslant 0$ while $f(b) \geqslant 0$, then $f(x) = 0$ has a solution in $[a, b]$. As a first application of **6.3.3**, we prove a natural extension of this result to n dimensions. Note that the condition $f(a) \leqslant 0, f(b) \geqslant 0$ may be written as $(x - x^0)f(x) \geqslant 0$, $x = a$ and $x = b$, where x^0 is an arbitrary point in (a, b).

6.3.4. Let C be an open, bounded set in R^n and assume that $F: \bar{C} \subset R^n \to R^n$ is continuous and satisfies $(x - x^0)^T Fx \geqslant 0$ for some $x^0 \in C$ and all $x \in \dot{C}$. Then $Fx = 0$ has a solution in $\bar{C}$.

Proof. Set $C_0 = \{x \mid x + x^0 \in C\}$ and define $G: \bar{C}_0 \to R^n$ by $Gx = x - F(x + x^0)$. Clearly, G is continuous. Now, let $x \in \dot{C}_0$; then $x + x^0 \in \dot{C}$, and, therefore, for any $\lambda > 1$,

$$x^T(\lambda x - Gx) = x^T[(\lambda - 1) x + F(x + x^0)] \geqslant (\lambda - 1) x^T x > 0,$$

since $x \neq 0$. It follows from **6.3.3** that G has a fixed point $x^* \in \bar{C}_0$ so that $F(x^* + x^0) = 0$. ∎

NOTES AND REMARKS

NR 6.3-1. Theorem **6.3.1** was first given by Kronecker [1869] in the context of **NR 6.1-1**.

NR 6.3-2. Theorem **6.3.2** was proved by Brouwer [1912] using his topologically based degree theory as described in **NR 6.1-2**. The proof given here for a ball is that of Heinz [1959]. A direct proof, not using degree theory explicitly, but some closely-related ideas, is contained in Dunford and Schwartz [1958]. The theorem extends to an arbitrary Banach space provided that G is also compact (that is, that G maps closed, bounded sets into compact sets); this is the famous Schauder fixed-point theorem (Schauder [1930]).

NR 6.3-3. Theorem **6.3.3** is due to Leray and Schauder [1934] and also extends to Banach spaces for mappings $F = I - G$ with a continuous and compact operator G.

NR 6.3-4. A simple proof of **6.3.3** in the case that C is a ball has been given by S. Karlin (see Lees and Schultz [1966]) by using the corresponding Brouwer theorem for balls: Suppose that $C = S(0, r)$ and $Gx \neq \lambda x$ when $\lambda > 1$ and $x \in \dot{C}$. If G has no fixed point in $\bar{C}$, then the mapping $\hat{G}x = r(Gx - x)/\|Gx - x\|$ is well-defined and continuous on $\bar{C}$, and, clearly, $\|\hat{G}x\| = r$ whenever $x \in \bar{C}$. Hence, $\hat{G}$ has a fixed point $x^* \in \bar{C}$ and $\|x^*\| = \|\hat{G}x^*\| = r$. But x^* satisfies

$$Gx^* = [1 + (1/r) \|Gx^* - x^*\|] x^*,$$

which is a contradiction.

NR 6.3-5. Theorem **6.3.4** was first given by Minty [1963] in a Hilbert space X as follows (see Dolph and Minty [1964] for the proof): If $F: X \to X$ is continuous and monotone, and satisfies $(x, Fx) \geq 0$ for all $x \notin \bar{S}(0, r)$ for some $r > 0$, then $Fx = 0$ has a solution in this ball. Note that in R^n the monotonicity is not needed, as **6.3.4** shows.

NR 6.3-6. An interesting combination of the contraction theorem and the Schauder theorem (or in R^n the Brouwer theorem) is given by Krasnoselskii [1955]: Let C be a closed, bounded convex set in a Banach space X and $G_1, G_2 : C \to X$ such that (a) $G_1 x + G_2 y \in C$ for all $x, y \in C$, (b) G_1 is contractive on C, (c) G_2 is continuous and compact. Then $G_1 + G_2$ has a fixed point in C.

A generalization of this result, due to J. H. Bramble (private communication), is the following: Suppose $H: C \times C \to C$ satisfies the two conditions: (a) There is an $\alpha < 1$ so that, for each fixed $y \in C$, $H(\cdot, y)$ is a contraction with contraction constant α; (b) there is a continuous compact mapping $H_1 : C \to X_1$, where X_1 is another Banach space, such that $\|H(x, y) - H(x, z)\| \leq \|H_1 x - H_1 z\|$ for all $x, y, z \in C_1$. Then there exists an $x^* \in C$ such that $x^* = H(x^*, x^*)$.

EXERCISES

E 6.3-1. Let $G: R^n \to R^n$ be continuous and satisfy $\|Gx\| \leq \alpha \|x\| + \beta$, $\forall x \in R^n$, where $0 \leq \alpha < 1, \beta > 0$. Then G has a fixed point.

E 6.3-2. Let C be an open, bounded set in R^n, and suppose that $F_1, F_2 : \bar{C} \to R^n$ are continuous. If $\|F_1 x - F_2 x\|_2 < \|F_2 x\|_2$, $\forall x \in \dot{C}$, and $\deg(F_2, C, 0) \neq 0$, then $F_1 x = 0$ has a solution in $\bar{C}$. (Altman [1957a]).

E 6.3-3. Let the mappings $A: R^n \to L(R^n)$, $B: R^n \to L(R^m, R^n)$, and $F: R^n \to R^n$ be continuous and assume that

$$\|A(x)^{-1}\| \leq c_1 < \infty, \quad \|A(x)^{-1} B(x)\| \leq c_2 < \infty, \quad \|Fx\| \leq c_3 < \infty, \quad \forall x \in R^n.$$

Then, for any $b \in R^n$, the equation $A(x) x = B(x) b + Fx$ has a solution. (Stepleman [1969]).

E 6.3-4. Let $C \subset R^n$ be an open, bounded set containing the origin and assume that $F: \bar{C} \to R^n$ is continuous. If $Fx \neq 0$ for all $x \in \bar{C}$, then there exist

6.4. MONOTONE AND COERCIVE MAPPINGS

$y^1, y^2 \in \dot{C}$ and $\lambda_1 < 0, \lambda_2 > 0$, so that $Fy^i = \lambda_i y^i$, $i = 1, 2$. (Hint: Consider the homotopies $H_{\pm}(x, t) = tFy \pm (1 - t)y$ and use **6.2.4**).

E 6.3-5. Extend **6.3.3** as follows: Let $C \subset R^n$ be an open, bounded set and assume that $G: \bar{C} \to R^n$ is continuous. Suppose there is an $x^0 \in C$ such that $Gx \neq \lambda x + (1 - \lambda)x^0$ whenever $x \in \dot{C}$ and $\lambda > 1$. Then G has a fixed point in $\bar{C}$. (Yamamuro [1963]).

E 6.3-6. (*Alternative Form of Leray–Schauder Theorem*). Let $C \subset R^n$ be an open, bounded set containing the origin and $G: \bar{C} \to R^n$ a continuous mapping. If $Gx \neq \lambda x$ whenever $\lambda < 1$ and $x \in \dot{C}$, then G has a fixed point in $\bar{C}$. [Hint: Consider the homotopy $H(x, t) = (2t - 1)x - tGx$].

E 6.3-7. Let $C \subset R^n$ be an open, bounded set containing the origin and assume that $G: \bar{C} \to R^n$ is continuous. If $x^T G x \leq x^T x$ for all $x \in \dot{C}$, then G has a fixed point in $\bar{C}$. (Shinbrot [1964]).

E 6.3-8. Let $C \subset R^n$ be a convex, open set containing the origin, and assume that $G: \bar{C} \to R^n$ is continuous. If $G\dot{C} \subset \bar{C}$, then G has a fixed point in $\bar{C}$. (Rothe [1937]).

E 6.3-9. Let $A \in L(R^n)$ be nonsingular and assume that $\phi: R^n \to R^n$ is continuous. Suppose that for some $r > 0$ every solution of $Ax - t\phi x = 0$ lies in $\bar{S}(0, r)$ for all $t \in [0, 1]$. Show, by the Leray–Schauder theorem, that $Ax = \phi x$ has a solution in $\bar{S}(0, r)$. (Bers [1953]).

6.4. MONOTONE AND COERCIVE MAPPINGS

Recall from **5.3.8** that if $F: R^n \to R^n$ is norm-coercive and a local homeomorphism at each point of R^n, then F is a homeomorphism. By means of the degree theory established in this chapter, we are now able to give several related results which replace the explicit assumption that F is a local homeomorphism by a stronger coerciveness condition, as well as by monotonicity properties. This will allow, in particular, a strengthening of the results of **5.4** on monotone mappings.

6.4.1. Definition. A mapping $F: D \subset R^n \to R^n$ is *weakly coercive* on the open set $D_0 \subset D$ if there exists a point $z \in D_0$ with the property that, for any $\gamma > 0$, there is an open, bounded set $D_\gamma \subset D_0$ containing z for which $\bar{D}_\gamma \subset D_0$ and

$$(x - z)^T Fx > \gamma \| x - z \|_2, \qquad \forall x \in D_0 \sim \bar{D}_\gamma. \tag{1}$$

If $D = D_0 = R^n$, and $z = 0$, F is said to be *coercive*. ∎

From the Cauchy–Schwarz inequality

$$(x - z)^T Fx \leq \| x - z \|_2 \| Fx \|_2,$$

it follows immediately that any weakly coercive mapping is also norm-coercive in the sense of **5.3.6**. Moreover, if $D = D_0 = R^n$, then, clearly, F is weakly coercive if and only if

$$\lim_{\|x\|_2 \to \infty} [(x - z)^T Fx / \| x - z \|_2] = \infty \qquad (2)$$

for some $z \in R^n$, while F is coercive if and only if

$$\lim_{\|x\|_2 \to \infty} (x^T Fx / \| x \|_2) = +\infty. \qquad (3)$$

Of course, a coercive mapping on R^n is, by definition, weakly coercive (with $z = 0$), but the converse is not true (see **E 6.4-1**).

If $f : (a, b) \subset R^1 \to R^1$ is weakly coercive, then it is easy to see that

$$\lim_{x \to a+} f(x) = -\infty, \qquad \lim_{x \to b-} f(x) = +\infty.$$

Hence, if f is also continuous, then it maps (a, b) onto R^1. This extends to n dimensions in the following way.

6.4.2. Coerciveness Theorem. Let $F : D \subset R^n \to R^n$ be continuous and weakly coercive on the open set D. Then $FD = R^n$, and for any $y \in R^n$ the solution set $\Gamma = \{x \in D \mid Fx = y\}$ is bounded.

Proof. For arbitrary $y \in R^n$, let $\gamma = \| y \|_2$ and choose $\bar{D}_\gamma \subset D$ according to **6.4.1**. Then, by the Cauchy–Schwarz inequality,

$$(x - z)^T (Fx - y) > \gamma \| x - z \|_2 - \| y \|_2 \| x - z \|_2 > 0, \qquad \forall x \in D \sim \bar{D}_\gamma,$$

so that, by the continuity of F,

$$(x - z)^T (Fx - y) \geq 0, \qquad \forall x \in \dot{D}_\gamma.$$

It now follows from **6.3.4** that $Fx - y = 0$ has a solution in $\bar{D}_\gamma$.

For the proof of the final statement, let

$$\gamma > \eta = \max_{\|u\|_2 = 1} u^T y;$$

then for any $x^* \in \Gamma$ we must have $x^* \in \bar{D}_\gamma$, since otherwise, by (1),

$$\eta \geq (x^* - z)^T y / \| x^* - z \|_2 \geq \gamma > \eta,$$

which is a contradiction. The statement now follows from the boundedness of $\bar{D}_\gamma$. ∎

6.4. MONOTONE AND COERCIVE MAPPINGS

Theorem **6.4.2** guarantees only existence, and not uniqueness, of solutions. If $D = R^n$, one sufficient condition for F also to be one-to-one is that F be a local homeomorphism at each point of R^n; this follows immediately from **5.3.8**, since weak coerciveness implies norm-coerciveness. Another sufficient condition is given in the following immediate corollary of **6.4.2**. Recall that monotone mappings are defined in **5.4.2**.

6.4.3. If $F: R^n \to R^n$ is continuous and weakly coercive, then $FR^n = R^n$. If, in addition, F is strictly monotone, then F is also one-to-one.

As a final corollary of **6.4.2**, we give the following strengthening of **5.4.5**, where continuous differentiability was assumed.

6.4.4. Uniform Monotonicity Theorem. If $F: R^n \to R^n$ is continuous and uniformly monotone, then F is a homeomorphism of R^n onto itself.

Proof. By the uniform monotonicity, there is a $c > 0$ so that

$$x^\mathrm{T}(Fx - y) \geqslant c \| x \|_2^2, \quad \forall x \in R^n,$$

where $y = F(0)$. Hence,

$$\frac{x^\mathrm{T} Fx}{\| x \|_2} = \frac{x^\mathrm{T}(Fx - y)}{\| x \|_2} + \frac{x^\mathrm{T} y}{\| x \|_2} \geqslant c \| x \|_2 - \| y \|_2$$

shows, by (3), that F is coercive. Therefore, it follows from **6.4.3** that F is one-to-one and onto. Finally, the uniform monotonicity implies that

$$\| Fx - Fy \|_2 \geqslant c \| x - y \|_2, \quad \forall x, y \in R^n,$$

and, hence, by **4.3.8**, F^{-1} is (Lipschitz) continuous. ∎

NOTES AND REMARKS

NR 6.4-1. The concept of a monotone mapping was apparently first introduced by Kačurovskii [1960], and, in the same year, the first fixed-point theorem for monotone mappings in Hilbert space satisfying a uniform Lipschitz condition was given by Vainberg [1960]. This continued earlier work of various Russian authors on related fixed-point results for gradient operators on Banach spaces (see, e.g., Vainberg [1956]). Independently, Zarantonello [1960] had proved a theorem which is practically equivalent to the result of Vainberg. A major step was then taken by Minty [1962], who removed the uniform Lipschitz condition in the Vainberg–Zarantonello theorem and proved that when G is a continuous, monotone mapping on a Hilbert space, then $I - G$ is a homeomorphism of the space onto itself. (Note, in this connection, that Definition **5.4.2** extends immediately to an arbitrary innerproduct on R^n, and, more generally, to a real Hilbert space.) Browder [1963a] then pointed out that Minty's result could be rephrased in terms of uniformly monotone mappings; on R^n this result reduces to our Theorem **6.4.4**. However, Minty's original proof proceeded along entirely different lines than those used in the text. Subsequent to these two papers by Minty and Browder, a considerable literature has now developed in which these results have been extended to more general spaces and mappings and to weaker continuity requirements. For summaries of this work, see the surveys of Browder [1965a], de Figueiredo [1967], and Opial [1967a].

NR 6.4-2. The definition of a coercive mapping extends immediately to Hilbert spaces and has been used by several authors. However, the idea of a weakly coercive mapping defined on only a subset of R^n seems to be new, and **6.4.2** extends a theorem, apparently first proved by Browder [1963a], that a continuous coercive mapping on R^n is onto. By means of the result mentioned in **NR 6.3-5**, **6.4.3** extends to a Hilbert space provided that F is assumed to be monotone. For $z = 0$, this represents a special case of more general results of Browder [1963a] and Minty [1963].

EXERCISES

E 6.4-1. Define $F: R^2 \to R^2$ by $f_1(x) = (x_1 - 1) - x_2[(x_1 - 1)^2 + x_2^2]^{1/2}$, $f_2(x) = (x_1 - 1)[(x_1 - 1)^2 + x_2^2]^{1/2} + x_2$. Show that F is weakly coercive (with $z = e^1$) but not coercive. (Opial [1967a]).

E 6.4-2. Let $A \in L(R^n)$. Show that (2) holds for each $z \in R^n$ if and only if A is positive definite.

E 6.4-3. If $F: R^n \to R^n$ is monotone and weakly coercive, then F is coercive. (Opial [1967a)].

E 6.4-4. Give an example of a continuously differentiable mapping $F: R^2 \to R^2$ such that $\| F'(x)^{-1} \| \leqslant \gamma$ for all $x \in R^2$ but $| x^T F x |/\| x \| \nrightarrow \infty$ as $\| x \| \to \infty$.

E 6.4-5. Let $F: R^n \to R^n$. Show that $x^T(Fx - y) \to +\infty$ as $\| x \| \to +\infty$ for all $y \in R^n$ if and only if F is coercive.

E 6.4-6. If $g: R^n \to R^1$ is uniformly convex and G-differentiable on R^n, then $(g')^T$ is coercive.

E 6.4-7. Let $g: R^n \to R^1$ be continuously differentiable. Then $(g')^T$ is coercive if and only if $g(x)/\| x \| \to +\infty$ as $\| x \| \to \infty$. Hence, conclude, by **E 4.3-5**, that $(g')^T$ is coercive if and only if $g(x) - b^T x \to +\infty$ as $\| x \| \to \infty$ for any $b \in R^n$.

6.5. APPENDIX. ADDITIONAL ANALYTIC RESULTS

E 6.4-8. Suppose that $g\colon R^n \to R^1$ is continuously differentiable and strictly convex. Then $g(x) - b^T x$ has a unique minimizer for any $b \in R^n$ if and only if $(g')^T$ is coercive.

E 6.4-9. Let $F\colon D \subset R^n \to R^n$, where D is open, and consider any subset $D_0 \subset D$. Suppose there is a $z \in D_0$ so that, for any $\gamma > 0$, there is a subset $D_\gamma \subset D_0$ such that (1) holds. Show that D_0 is open.

E 6.4-10. Extend the results of **E 5.4-5**–**E 5.4-7** to continuous mappings.

E 6.4-11. Assume that in **5.4.6** ϕ is assumed to be only continuous and monotone. Show that $A + \phi$ is a homeomorphism.

E 6.4-12. Suppose that $F\colon R^n \to R^n$ is continuous and there exists a function $\varphi\colon [0, \infty) \to [0, \infty)$ such that $\lim_{t \to +\infty} \varphi(t) = +\infty$, $\varphi(t) > 0$ for $t > 0$, and

$$(x - y)^T (Fx - Fy) \geq \|x - y\| \varphi(\|x - y\|), \quad \forall x, y \in R^n.$$

Show that F is one-to-one and onto R^n.

E 6.4-13. $F\colon R^n \to R^n$ is *asymptotically monotone on rays* if there is a t_0 so that, for any x with $x^T x = 1$, the function $\alpha_x(t) = x^T F(tx)$ is isotone on $[t_0, +\infty)$. Show that if F is continuous, asymptotically monotone on rays, and $c > 0$, then $(cI + F) R^n = R^n$. [Hint: For arbitrary $y \in R^n$, define $Gy = c^{-1}(y - Fx)$, set $r_0 = \sup\{x^T G(t_0 x) \mid x^T x = 1\}$, $r = \max(r_0, t_0)$, and apply **6.3.3** on the ball $\bar{S}(0, r)$] (Shinbrot [1964]).

6.5. APPENDIX. ADDITIONAL ANALYTIC RESULTS

In this appendix, we collect some remarks about the additional analytical tools used in this chapter, and we present a proof of Lemma **6.1.3** together with a summary of the results needed in that proof.

Besides the other usual properties of Riemann integrals, we used in the proof of **6.1.2** the following well-known change-of-variable theorem (see, for instance, Apostol [1957, p. 271]).

6.5.1. Suppose that $F\colon D \subset R^n \to R^n$ is continuously differentiable and one-to-one on an open, bounded set $C \subset D$ and that $F'(x)$ is nonsingular for all $x \in C$. Let $f\colon F(C) \subset R^n \to R^1$ be continuous and K a Jordan-measurable, compact subset of C. Then

$$\int_K f(x)\, dx = \int_{F_C^{-1}(K)} f(Fx) \,|\det F'(x)|\, dx,$$

where F_C is the restriction of F to C.

Actually, we used this theorem only for the case when K is a ball under the Euclidean norm on R^n. Such balls are certainly Jordan-measurable.

In preparation for the proof of Lemma **6.1.3**, we need the following special case of the well-known divergence theorem.

6.5.2. Let $F: R^n \to R^n$ be a continuously differentiable function of compact support. Then

$$\int_{R^n} \operatorname{div} Fx \, dx = \int_{R^n} \sum_{i=1}^{n} \partial_i f_i(x) \, dx = 0.$$

The proof is in this case a simple computation using the well-known theorem on repeated integration (see again Apostol [1957]). In fact, if $Q = \{x \in R^n \mid -\alpha \leqslant x_i \leqslant \alpha\}$ contains the support of F, then

$$\int_{R^n} \operatorname{div} Fx \, dx = \sum_{i=1}^{n} \int_{Q} \partial_i f_i(x) \, dx = \sum_{i=1}^{n} \int_{-\alpha}^{+\alpha} \cdots \int_{-\alpha}^{+\alpha} \partial_i f_i(x) \, dx_1 \ldots dx_n$$

and $\int_{-\alpha}^{+\alpha} \partial_i f_i(x) \, dx_i = f_i(\alpha) - f_i(-\alpha) = 0$.

Another result needed in the proof of **6.1.3** is the following lemma about Jacobian determinants.

6.5.3. Let $F: D \subset R^n \to R^n$ be twice-continuously differentiable in the open set D, and let $a_{ij}(x)$ denote the cofactor of the (i,j)th element of $F'(x)$, $x \in D$. Then

$$\sum_{j=1}^{n} \partial_j a_{ij}(x) = 0, \quad i = 1,\ldots, n, \quad \forall x \in D. \tag{1}$$

The proof is given in Muir [1933]. It uses the following standard result from determinant theory: If $(\alpha_{ij}) \in L(R^n)$, $(\beta_{ij}) = (\alpha_{ij})^{-1}$, and the cofactors of the (i,j)th element of (α_{ij}) are denoted by a_{ij}, then $\mu = \det(\alpha_{ij})$ satisfies

$$\left.\begin{array}{l} \delta_{ij}\mu = \sum_{k=1}^{n} \alpha_{ik} a_{jk}, \\[4pt] \partial \mu / \partial \alpha_{ij} = a_{ij}, \\[4pt] \mu \beta_{ij} = a_{ji}, \end{array}\right\} \quad i,j = 1,\ldots, n. \quad \begin{array}{r}(2)\\(3)\\(4)\end{array}$$

Equation (2) is the expansion of $\det A$ in terms of the elements of the ith row, (3) follows from (2) by differentiation, and (4) is just the usual formula for $(\alpha_{ij})^{-1}$ in terms of μ and the cofactors.

6.5. APPENDIX. ADDITIONAL ANALYTIC RESULTS

For the proof of **6.5.3**, suppose first that, for given $x^0 \in D$, $F'(x^0)$ is nonsingular. Then, by the inverse function theorem **5.2.1**, the restriction F_U of F to some open neighborhood U of x^0 maps U onto a neighborhood of $y^0 = Fx^0$, and $G = F_U^{-1}$ satisfies $F'(x)\, G'(y) = I$, where $y = Fx$, $x \in U$. By **E 5.2-7**, G is even twice-continuously differentiable, since this is assumed for F. For abbreviation, set $F'(x) = (\alpha_{ij})$, $G'(y) = (\beta_{ij})$, $\mu = \det F'(x)$, $\eta = \det G'(y)$, and denote the cofactors of the (i,j)th element of $G'(y)$ by b_{ij}. Then, using (3) and (4), we find that

$$\frac{\partial \eta}{\partial y_k} = \sum_{i,j=1}^n \frac{\partial \eta}{\partial \beta_{ij}} \frac{\partial \beta_{ij}}{\partial y_k} = \sum_{i,j=1}^n b_{ij} \frac{\partial \beta_{ij}}{\partial y_k} = \eta \sum_{i,j=1}^n \alpha_{ji} \frac{\partial \beta_{ij}}{\partial y_k}.$$

But

$$\frac{\partial \beta_{ik}}{\partial x_i} = \frac{\partial^2 g_i}{\partial x_i \partial y_k} = \sum_{j=1}^n \frac{\partial}{\partial y_k} \frac{\partial g_i}{\partial y_j} \frac{\partial y_j}{\partial x_i} = \sum_{j=1}^n \alpha_{ji} \frac{\partial \beta_{ij}}{\partial y_k},$$

and, hence,

$$\frac{\partial \eta}{\partial y_k} = \eta \sum_{i=1}^n \frac{\partial \beta_{ik}}{\partial x_i}.$$

Since $\mu\eta = 1$, it follows that $\mu\, \partial\eta/\partial y_k + \eta\, \partial\mu/\partial y_k = 0$, and therefore that

$$0 = \frac{\partial \mu}{\partial y_k} + \mu \sum_{i=1}^n \frac{\partial \beta_{ik}}{\partial x_i} = \sum_{i=1}^n \left[\frac{\partial \mu}{\partial x_i} \beta_{ik} + \mu \frac{\partial \beta_{ik}}{\partial x_i} \right]$$

$$= \sum_{i=1}^n \frac{\partial}{\partial x_i}(\mu \beta_{ik}) = \sum_{i=1}^n \frac{\partial}{\partial x_i} a_{ki}, \qquad k=1,\dots,n,$$

as stated.

Suppose, now, that $F'(x^0)$ is singular; then we consider, instead of F, the mapping $F_\epsilon x = Fx + \epsilon x$. Clearly, for small ϵ, $F_\epsilon'(x) = F'(x) + \epsilon I$ is nonsingular, and we have, by the first part of the proof, $\sum_{i=1}^n (\partial/\partial x_i)\, a_{ji}^\epsilon = 0$, where a_{ji}^ϵ is the cofactor of the (i,j)th element in $F_\epsilon'(x)$. By continuity, then, also,

$$\sum_{i=1}^n \frac{\partial}{\partial x_i} a_{ji} = \lim_{\epsilon \to 0} \sum_{i=1}^n \frac{\partial}{\partial x_i} a_{ji}^\epsilon = 0,$$

which is (1).

In Section **6.1**, we used the well-known Weierstrass approximation theorem:

6.5.4. Let $F: D \subset R^n \to R^n$ be continuous on the compact set $\bar{C} \subset D$. Then there exists for any $\epsilon > 0$ a continuously differentiable function $G: R^n \to R^n$ such that $\|F - G\|_C < \epsilon$.

Actually, the components of G can be chosen as polynomials on R^n. Proofs can be found in many texts; see, for example, Dieudonné [1960].

Considerably less known is the fact that, when F is continuously differentiable, one can choose G such that the derivative of G also approximates that of F on $\bar{C}$. For the proof of Lemma **6.1.3**, we need this result in the following form.

6.5.5. Let $F: D \subset R^n \to R^m$ be continuously differentiable on the open set D, and $\bar{C}$ a compact subset of D. Then, for any $\epsilon > 0$, there exists a twice-continuously differentiable mapping $G: R^n \to R^m$ such that

$$\max(\|Gx - Fx\|, \|G'(x) - F'(x)\|) < \epsilon, \qquad \forall x \in \bar{C}. \tag{5}$$

We shall give here a proof of this result using Bernstein polynomials. In one dimension, this proof is well known (see Lorentz [1953]); it is relatively easily extended to n dimensions using known one-dimensional estimates.

If $f: Q_n: R^n \to R^1$ is continuous on the unit cube

$$Q_n = \{x \in R^n \mid 0 \leqslant x_i \leqslant 1, \quad i = 1,\ldots, n\},$$

then the mth Bernstein polynomial on Q_n is defined by

$$B_m(f, x) = \sum_{j_1,\ldots,j_n=0}^{m} f(j_1/m,\ldots, j_n/m)\, p_{mj_1}(x_1) \cdots p_{mj_n}(x_n),$$

where

$$p_{mj}(t) = \binom{m}{j} t^j (1 - t)^{m-j}, \qquad t \in [0, 1]. \tag{6}$$

We first prove the following special case of **6.5.5**.

6.5.6. Let $f: Q_n \to R^1$ be continuous on Q_n and continuously differentiable on some closed ball $\bar{S}(u, r_0) \subset Q_n$. Then

$$\lim_{m \to \infty} B_m(f, x) = f(x), \qquad \lim_{m \to \infty} B_m'(f, x) = f'(x), \tag{7}$$

uniformly on any ball $\bar{S}(u, r)$ with $r < r_0$. ∎

6.5. APPENDIX. ADDITIONAL ANALYTIC RESULTS

Proof. Denote by $P_{m\mathbf{j}}(x)$ the product $p_{mj_1}(x_1) \cdots p_{mj_n}(x_n)$, and by $\sum_{\mathbf{j}=0}^{m}$ the sum $\sum_{j_1,\ldots,j_n=0}^{m}$. Clearly, by the binomial theorem, $\sum_{j=0}^{m} p_{mj}(t) = 1$, and, hence,

$$\sum_{\mathbf{j}=0}^{m} P_{m\mathbf{j}}(x) = \prod_{k=1}^{n} \left\{ \sum_{j=0}^{m} p_{mj}(x_k) \right\} = 1.$$

For given $\delta \in (0, 1)$ and $k \leqslant n$, set

$$J_k(x) = \{(j_1,\ldots,j_k) \mid 0 \leqslant j_i \leqslant m, \ |j_i - mx_i| < m\delta, \ i = 1,\ldots, k\},$$

and

$$K_k(x) = \{(j_1,\ldots,j_k) \mid 0 \leqslant j_i \leqslant m, \ (j_1,\ldots,j_k) \notin J_k(x)\}.$$

It is known (see, e.g., Lorentz [1953, p. 15]) that when $n = 1$, there is a constant c so that

$$\sum_{j \in K_1(t)} p_{mj}(t) \leqslant c/(m^2 \delta^4), \quad \forall t \in [0, 1], \tag{8}$$

and, by induction on n, we extend this inequality to n dimensions, with a right-hand side of $cn/m^2\delta^4$. In fact, using (8) and the induction hypothesis, it follows that

$$\sum_{\mathbf{j} \in K_n(x)} P_{m\mathbf{j}}(x) \leqslant \sum_{j_1,\ldots,j_{n-1}=0}^{m} p_{mj_1}(x_1) \cdots p_{mj_{n-1}}(x_{n-1}) \sum_{j_n \in K_1(x_n)} p_{mj_n}(x_n)$$

$$+ \sum_{(j_1,\ldots,j_{n-1}) \in K_{n-1}(x)} p_{mj_1}(x_1) \cdots p_{mj_{n-1}}(x_{n-1}) \sum_{j_n=0}^{m} p_{mj_n}(x_n)$$

$$\leqslant \frac{c}{m^2\delta^4} + \frac{c(n-1)}{m^2\delta^4} = \frac{cn}{m^2\delta^4}, \quad \forall x \in Q_n. \tag{9}$$

Now, for given $\epsilon > 0$, choose $\delta \in (0, 1)$ so that $|f(x) - f(y)| \leqslant \epsilon$ whenever $\|x - y\| \leqslant \delta$. Then

$$|B_m(f, x) - f(x)| \leqslant \sum_{\mathbf{j}=0}^{m} |f(\mathbf{j}/m) - f(x)| P_{m\mathbf{j}}(x)$$

$$\leqslant \epsilon \sum_{\mathbf{j} \in J_n(x)} P_{m\mathbf{j}}(x) + 2M \sum_{\mathbf{j} \in K_n(x)} P_{m\mathbf{j}}(x)$$

$$\leqslant \epsilon + (2Mcn/m^2\delta^4), \quad \forall x \in Q_n,$$

where $M = \max\{|f(x)| \mid x \in Q_n\}$. Hence, $B_m(f, x)$ converges uniformly to $f(x)$ for all $x \in Q_n$.

The important point is now the uniform convergence of the derivative. We first note that B_m is linear in f; that is, for any $f, g: Q_n \to R^1$, we have

$B_m(f+g, x) = B_m(f, x) + B_m(g, x)$. In particular, for an affine functional $\alpha + a^T x$, we obtain

$$B_m(\alpha + a^T x, x) = \alpha + \sum_{k=1}^{n} a_k \sum_{j=0}^{m} (j/m) p_{mj}(x_k),$$

and, therefore, from the easily proved relation

$$\sum_{j=0}^{m} j p_{mj}(t) = mt, \qquad \forall t \in [0, 1],$$

it follows that

$$B'_m(\alpha + a^T x, x) = a^T. \tag{10}$$

Now, for given fixed $x \in \bar{S}(u, r)$, define

$$g(y) = f(y) - f(x) - f'(x)(y - x), \qquad \forall y \in Q_n.$$

Then, from (10), it follows that

$$B'_m(f, x) - f'(x) = B'_m(g, x),$$

and it suffices to show that $\lim_{m \to \infty} \| B'_m(g, x) \| = 0$ uniformly for $x \in \bar{S}(u, r)$.

For given $\epsilon > 0$, the uniform continuity of f' on $\bar{S}(u, r)$ implies, by the mean-value theorem **3.2.5**, that we may choose $\delta \in (0, r - r_0)$ such that

$$|g(y)| = |f(y) - f(x) - f'(x)(y - x)| \leq \epsilon \| y - x \|, \qquad \forall y \in S(x, \delta),$$

holds uniformly for $x \in \bar{S}(u, r)$. Moreover, since $\bar{S}(u, r) \subset \text{int}(Q_n)$, we may choose γ so that $0 < \gamma \leq x_i(1 - x_i)$, $i = 1, \ldots, n$, for all $x \in \bar{S}(u, r)$. Now, note that

$$p'_{mj}(t) = [1/t(1 - t)](j - mt) p_{mj}(t), \qquad 0 < t < 1.$$

Therefore,

$$| \partial_i B_m(g, x)| = \left| \sum_{j=0}^{m} g(j/m)[(j_i - mx_i)/x_i(1 - x_i)] P_{mj}(x) \right|$$

$$\leq (\epsilon/m\gamma) \sum_{j \in J_n(x)} \left(\sum_{k=1}^{n} |j_k - mx_k| \right) |j_i - mx_i| P_{mj}(x)$$

$$+ (M/\gamma) \sum_{j \in K_n(x)} |j_i - mx_i| P_{mj}(x), \qquad \forall x \in \bar{S}(u, r), \tag{11}$$

where $M = \max\{g(y) \mid y \in Q_n, \ x \in \bar{S}(u, r)\}$. Since $|j_i - mx_i| \leq 2m$, (9) shows that the second sum is bounded by $2cn/m\delta^4$. In order to bound the first sum, we apply the Cauchy–Schwarz inequality to the one-dimensional estimate (see Lorentz [1953, p. 5])

$$\sum_{j=0}^{m} (j - mt)^2 p_{mj}(t) \leq \tfrac{1}{4}m, \qquad \forall t \in [0, 1],$$

to obtain

$$\sum_{j=0}^{m} |j - mt| \, p_{mj}(t) \leq \tfrac{1}{2}m^{1/2}.$$

Consequently, the first sum in (11) is bounded by

$$\sum_{k=1}^{n} \sum_{j=0}^{m} |j_k - mx_k| \, |j_i - mx_i| \, P_{mj}(x)$$

$$= \sum_{k=1}^{n} \left\{ \sum_{j=0}^{m} |j - mx_k| \, p_{mj}(x_k) \right\} \left\{ \sum_{j=0}^{m} |j - mx_i| \, p_{mj}(x_i) \right\} \leq \tfrac{1}{4}nm.$$

Altogether, then, we obtain

$$|\partial_i B_m(g, x)| \leq (n/4\gamma)\,\epsilon + (2cnM/\gamma\delta^4)(1/m),$$

and the proof of **6.5.6** is complete. ∎

This result can be easily extended to prove **6.5.5** by using the following wellknown Tietze–Urysohn extension theorem, whose proof may be found, for example, in Dieudonné [1960]:

6.5.7. Let $f: \bar{C} \subset R^n \to R^1$ be continuous on the compact set $\bar{C}$; then there exists a continuous mapping g defined on all of R^n such that $f(x) = g(x)$ for $x \in \bar{C}$. ∎

The proof of **6.5.5** is now fairly evident. There exists an open, bounded set C_1 such that $\bar{C} \subset C_1 \subset \bar{C}_1 \subset D$. Since $\bar{C}_1$ is compact, it is contained in some cube, and it is no restriction to assume that $\bar{C}_1 \subset \operatorname{int}(Q_n)$, since this can always be ensured after a simple affine transformation. Clearly, $r_0 > 0$ exists such that $\bar{S}(x, r_0) \subset \bar{C}_1$ for all $x \in \bar{C}$, and, by the compactness, we can find points $x^1, \ldots, x^k \in \bar{C}$ such that $\bar{C} \subset \bigcup_{i=1}^{k} \bar{S}(x^i, r_0/2) \subset \bar{C}_1 \subset D$.

Now, by **6.5.7**, we may extend each component f_j of F to a continuous mapping $g_j: R^n \to R^1$ such that $g_j(x) = f_j(x)$ for all $x \in \bar{C}_1$. Hence, **6.5.6** applies to each f_j, and we have

$$\|B'_m(f_j, x) - f_j'(x)\| < \epsilon, \qquad \forall x \in \bar{S}(x^i, \tfrac{1}{2}r_0), \quad j = 1,\ldots, n,$$

for all $m \geqslant m_i(\epsilon)$. Since there are only finitely many balls $\bar{S}(x^i, \tfrac{1}{2}r_0)$, the result is proved.

We now turn to the proof of Lemma **6.1.3**, that is, of the following result:

6.5.8. Let $F: D \subset R^n \to R^n$ be continuously differentiable on the open set D, and C an open, bounded set with $\bar{C} \subset D$. Suppose, further, that $y \in R^n$ satisfies $\dot{\gamma} = \min\{\|Fx - y\|_2 \mid x \in \dot{C}\} > 0$. Then, for $\alpha \in (0, \dot{\gamma})$ and $\varphi \in W_\alpha$, $\eta(\varphi) = \int_0^\infty t^{n-1}\varphi(t)\,dt = 0$ implies $d_\varphi(F, C, y) = 0$.

Proof. By **6.5.5**, there exists for any $\epsilon > 0$ a twice-continuously differentiable mapping $G: R^n \to R^n$ such that (5) holds uniformly for $x \in \bar{C}$. For sufficiently small ϵ, we evidently also have $\min\{\|Gx - y\|_2 \mid x \in \dot{C}\} > \alpha$ which shows that $d_\varphi(G, C, y)$ is well defined. Moreover, given $\epsilon_1 > 0$, the continuity of φ and the continuous dependence of the determinant on its elements implies that $|d_\varphi(F, C, y) - d_\varphi(G, C, y)| < \epsilon_1$ whenever ϵ is small enough. Thus, if we can prove that $\eta(\varphi) = 0$ indeed implies that $d_\varphi(G, C, y) = 0$ for the twice-continuously differentiable G, then $|d_\varphi(F, C, y)| < \epsilon_1$, and, hence—because ϵ_1 was arbitrary—$d_\varphi(F, C, y) = 0$. It is therefore no restriction of the generality to assume that F itself is already twice-continuously differentiable on D.

Given $\varphi \in W_\alpha$ such that $\eta(\varphi) = 0$, set $\psi(t) = t^{-n} \int_0^t s^{n-1}\varphi(s)\,ds$, for $0 < t < +\infty$, and $\psi(0) = 0$. It is evident that ψ is continuously differentiable on $[0, \infty)$ and that

$$t\psi'(t) + n\psi(t) = \varphi(t), \qquad t \in [0, \infty). \tag{12}$$

Consider the mappings

$$H:\ R^n \to R^n, \qquad Hx = (h_1(x),\ldots, h_n(x))^T = \psi(\|x\|_2)x$$

$$G:\ D \subset R^n \to R^n, \qquad Gx = (g_1(x),\ldots, g_n(x))^T = H(Fx).$$

Since ψ is zero near $t = 0$ and the Euclidean norm is continuously differentiable for $x \neq 0$, it follows that H and, hence, also G are both continuously differentiable on D. Moreover, by (12), it follows that

$$\operatorname{div} H(x) = \|x\|_2\, \psi'(\|x\|_2) + n\psi(\|x\|_2) = \varphi(\|x\|_2). \tag{13}$$

6.5. APPENDIX. ADDITIONAL ANALYTIC RESULTS

Again, let $a_{ij}(x)$, $x \in D$, denote the cofactor of the (i,j)th element of $F'(x)$; then (2) becomes

$$\delta_{kj} \det F'(x) = \sum_{i=1}^{n} a_{ji}(x) \, \partial_i f_k(x), \quad j = 1,\ldots, n, \quad x \in D, \quad (14)$$

and, moreover, **6.5.3** holds. Hence, we obtain, for $x \in D$, using (13), (14), and **6.5.3**, that

$$\sum_{i=1}^{n} \partial_i \sum_{j=1}^{n} a_{ji}(x) g_j(x) = \sum_{j=1}^{n} \left\{ \sum_{i=1}^{n} \partial_i a_{ji}(x) \right\} g_j(x) + \sum_{i,j=1}^{n} a_{ji}(x) \, \partial_i g_j(x)$$

$$= \sum_{i,j=1}^{n} a_{ji}(x) \sum_{k=1}^{n} \partial_k h_j(Fx) \, \partial_i f_k(x)$$

$$= \sum_{j=1}^{n} \sum_{k=1}^{n} \left\{ \sum_{i=1}^{n} a_{ji} \, \partial_i f_k(x) \right\} \partial_k h_j(Fx) = \sum_{j=1}^{n} \partial_j h_j(Fx) \det F'(x)$$

$$= (\operatorname{div} H)(Fx) \det F'(x) = \varphi(\|Fx\|_2) \det F'(x).$$

This shows that for

$$P: R^n \to R^n, \quad Px = (p_1(x),\ldots, p_n(x))^{\mathrm{T}}, \quad p_i(x) = \begin{cases} \sum_{i=1}^{n} a_{ji}(x) g_j(x), & x \in C \\ 0 & \text{otherwise}, \end{cases}$$

we have

$$\operatorname{div} P(x) = \phi(x), \quad \forall x \in R^n,$$

where ϕ is given by (6.1.2a). Then **6.5.2** implies that $d_\varphi(F, C, y) = 0$. ∎

Part III / ITERATIVE METHODS

In this part, we shall survey the most common iterative methods for solving nonlinear systems of equations. The treatment will be largely descriptive, and the analysis of these methods is postponed until Parts IV and V. In this part, only as much analysis is included as is needed to describe the methods effectively.

In Chapter 7, we begin by discussing processes which may be viewed as generalizations of methods for the solution of one equation in one unknown, such as the n-dimensional counterparts of Newton's method, the secant method, Steffensen's method, and their variations. Then we consider generalizations of iterative methods for linear systems of equations, with particular emphasis on successive overrelaxation methods. Finally, a number of other processes are discussed which are either motivated by the form of the iterations previously discussed, or which are attempts to provide a technique for overcoming the problem of finding a suitable initial approximation.

In many applications, the system of equations to be solved arises in the attempt to find a minimizer or a critical point of a related nonlinear functional. Alternatively, one can always convert the problem of solving $Fx = 0$ into the problem of minimizing $f(Fx)$, where f is, for example, a norm. Accordingly, in Chapter 8, we describe various minimization methods, including the gradient and conjugate-gradient iterations.

Chapter 7 / GENERAL ITERATIVE METHODS

7.1. NEWTON'S METHOD AND SOME OF ITS VARIATIONS

If f is a real-valued function of a single variable with a root x^*, the *parallel-chord method* consists in replacing f at some approximation x^0 of x^* by a linear function

$$l(x) = \alpha(x - x^0) + f(x^0)$$

with a suitable slope $\alpha \neq 0$, and then in taking the root x^1 of l as a new approximation to x^*. Repeating this procedure with fixed α, we obtain the iteration

$$x^{k+1} = x^k - \alpha^{-1} f(x^k), \quad k = 0, 1, \ldots. \quad (1)$$

The geometric situation is shown in Fig. 7.1.

We can extend (1) immediately to an n-dimensional function $F: D \subset R^n \to R^n$ by replacing α with a constant, nonsingular matrix A. Accordingly, the n-dimensional *parallel-chord method* is defined by

$$x^{k+1} = x^k - A^{-1} F x^k, \quad k = 0, 1, \ldots. \quad (2)$$

Here, (2) is equivalent to replacing F at x^k by the affine function

$$L_k x = A(x - x^k) + F x^k,$$

and setting x^{k+1} equal to the (unique) solution of $L_k x = 0$. Geometrically, this means that x^{k+1} is the intersection of the n hyperplanes

$$\sum_{j=1}^{n} a_{ij}(x_j - x_j^k) + f_i(x^k) = 0, \quad i = 1,\ldots, n,$$

in R^{n+1} with the hyperplane $x = 0$.

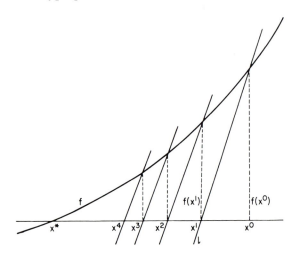

FIGURE 7.1

The crucial point in applying the iteration (2) is, of course, the proper choice of A. A particularly simple possibility is $A = \alpha I$ with some scalar $\alpha \neq 0$, so that, in essence, the one-dimensional iteration (1) is applied to each component f_i of F separately. A more sophisticated choice is motivated by the fact that in one dimension the slope $f'(x^0)$ of the tangent of f at x^0 is a reasonable selection for α. If we take $A = F'(x^0)$, where $F'(x)$ denotes the G-derivative at x, then (2) becomes the *simplified Newton method*

$$x^{k+1} = x^k - F'(x^0)^{-1} F x^k, \quad k = 0, 1, \ldots . \tag{3}$$

On the other hand, in many problems, a special form of F may lead to a natural choice for A. For example, if

$$Fx = Ax - Gx, \tag{4}$$

where G is nonlinear and A some nonsingular matrix, a natural iteration, often called the *Picard iteration*, is

$$x^{k+1} = A^{-1} G x^k, \quad k = 0, 1, \ldots . \tag{5}$$

7.1 NEWTON'S METHOD AND SOME OF ITS VARIATIONS

Since $A^{-1}Gx = x - A^{-1}Fx$, (5) can evidently be rewritten in the form (2).

There are many other possible choices for the matrix A in the parallel-chord method (2), and we shall examine some of them in other sections of this part. An underlying requirement for any A, however, is that the iteration (2) be at least locally convergent. This means that when x^0 is sufficiently close to a solution x^* of $Fx = 0$, then we should be assured that $\lim_{k \to \infty} x^k = x^*$. In Chapter 10, we will show that, when $F'(x^*)$ exists, a sufficient (and essentially necessary) condition for such local convergence is that

$$\sigma = \rho(I - A^{-1}F'(x^*)) < 1, \tag{6}$$

where ρ denotes the spectral radius of the matrix; moreover, we will also show that the smaller σ, the more rapid will be the convergence. Since x^* is unknown, it is, in general, very difficult to choose A in advance so that (6) holds; but an ideal choice would clearly be $A = F'(x^*)$. This leads us to consider iterations where A is allowed to vary from step to step, that is

$$x^{k+1} = x^k - A_k^{-1}Fx^k, \quad k = 0, 1,\dots,$$

and to choose the matrices A_k in such a way that $\lim_{k \to \infty} A_k = F'(x^*)$. The prototype of such methods is the famous Newton method which will play a central role throughout the book.

Recall that if f is again a real-valued function of one variable, then the iteration

$$x^{k+1} = x^k - f'(x^k)^{-1}f(x^k), \quad k = 0, 1,\dots, \tag{7}$$

is known as *Newton's method* and may be represented geometrically as in Fig. **7.2**.

Formally, this method can be immediately extended to n-dimensional mappings $F: D \subset R^n \to R^n$ by replacing $f(x^k)$ in (7) with the G-derivative $F'(x^k)$ of F at x^k. This gives the n-dimensional *Newton method*:

$$x^{k+1} = x^k - F'(x^k)^{-1}Fx^k, \quad k = 0, 1,\dots. \tag{8}$$

In analogy with Fig. **7.2**, the step from x^k to x^{k+1} in (8) can be viewed geometrically as follows: Each component f_i of F is approximated by the affine function

$$f_i'(x^k)(x - x^k) + f_i(x^k), \tag{9}$$

which describes the tangent hyperplane of f_i at x^k, and then x^{k+1} is taken as the intersection of the n hyperplanes (9) in R^{n+1} with the hyperplane $x = 0$.

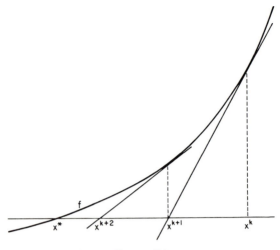

FIGURE 7.2

When $n = 1$, then (8) reduces to (7), and, in that sense, (8) is a reasonable generalization of the one-dimensional Newton method to n dimensions. Note, however, that there are arbitrarily many other n-dimensional methods which for $n = 1$ reduce to (7); for example, consider the iteration

$$x^{k+1} = x^k - F'(x^k)^{-1} Fx^k + (n-1) Gx^k, \qquad k = 0, 1, ..., \qquad (10)$$

where $G \colon R^n \to R^n$ is some "arbitrary" mapping. The importance of the Newton iteration (8) rests on the fact that, under certain natural conditions on F, an estimate of the form

$$\| x^{k+1} - x^* \| \leqslant c \, \| x^k - x^* \|^2 \qquad (11)$$

holds, provided the x^k are sufficiently close to a solution x^*. This shows that the $(k+1)$th error is proportional to the square of the kth error, so that the convergence is very rapid once the errors are small. This so-called property of "quadratic convergence" is, in general, not shared by iterations of the form (10); together with the simplicity and elegance of (8), it makes Newton's method a focal point in the study of iterative methods for nonlinear equations.

A precise proof of the estimate (11) will be given in **10.2.2**, but an understanding of this estimate can already be obtained from the following derivation of Newton's method. If F has an F-derivative at x^k, then

$$0 = Fx^* = Fx^k + F'(x^k)(x^* - x^k) + R(x^* - x^k), \qquad (12)$$

7.1. NEWTON'S METHOD AND SOME OF ITS VARIATIONS

where $\lim_{h \to 0} R(h)/\| h \| = 0$. Therefore, if x^k is close to the solution x^*, it is natural to drop the remainder term $R(x^* - x^k)$ and to approximate the difference $x^* - x^k$ by the solution h of the linear system

$$F'(x^k) h = -Fx^k. \tag{13}$$

In other words, $x^{k+1} = x^k + h = x^k - F'(x^k)^{-1} Fx^k$ is taken as a new approximation to x^*. If the second derivative of F is bounded in a neighborhood of x^*, then Theorem **3.3.6** shows that

$$\| R(x^k - x^*) \| \leqslant \alpha \| x^k - x^* \|^2,$$

and the additional assumption that $F'(x^*)$ is nonsingular leads to the estimate (11).

A related derivation of Newton's method can be obtained by means of the inverse function theorem **5.2.1**. In fact, if F is continuously differentiable and $F'(x)$ is nonsingular in a neighborhood U of x^*, then, by **5.2.1**, the restriction F_U of F to U has a differentiable inverse $G = F_U^{-1}$, and $G'(Fx) = F'(x)^{-1}$. Hence, if we expand G about Fx^k, we obtain

$$\begin{aligned} x^* &= G(0) = G(Fx^k) - G'(Fx^k) Fx^k + \hat{R}(Fx^k) \\ &= x^k - F'(x^k)^{-1} Fx^k + \hat{R}(Fx^k), \end{aligned} \tag{14}$$

which again yields the Newton approximation when the remainder term is dropped.

Although Newton's method is theoretically attractive, it may be difficult to use in practice. In fact, each step requires the solution of the linear system (13) [the inverse of $F'(x^k)$ is rarely computed explicitly] and—especially for problems arising from partial differential equations in which the dimension of the system may be several thousand—this may be a difficult task. Moreover, at each step, not only the n components of Fx^k but also the n^2 entries of $F'(x^k)$ are needed, and, unless the partial derivatives $\partial_j f_i(x^k)$ have a simple functional form, it may be desirable to avoid their computation altogether. In **7.4**, we will describe various modifications of Newton's method which mitigate the first difficulty. On the other hand, most of the remaining methods of this chapter avoid the explicit computation of derivatives.

The most direct approach to avoiding the computation of $F'(x)$ is simply to approximate the partial derivatives $\partial_j f_i(x)$ by difference quotients. Two typical and frequently used difference approximations are

$$\partial_j f_i(x) \doteq (1/h_{ij}) \left[f_i \left(x + \sum_{k=1}^{j} h_{ik} e^k \right) - f_i \left(x + \sum_{k=1}^{j-1} h_{ik} e^k \right) \right], \tag{15}$$

and

$$\partial_j f_i(x) \doteq (1/h_{ij})[f_i(x + h_{ij}e^j) - f_i(x)], \tag{16}$$

where the h_{ij} are given discretization parameters and e^j is the jth coordinate vector.

More generally, let $h \in R^p$ represent a parameter vector and let $\varDelta_{ij}(x, h)$ denote any difference approximation of $\partial_j f_i(x)$ with the property that whenever $\partial_j f_i(x)$ exists, then

$$\lim_{h \to 0} \varDelta_{ij}(x, h) = \partial_j f_i(x), \qquad i, j = 1,..., n. \tag{17}$$

Then, with the difference matrix

$$J(x, h) = (\varDelta_{ij}(x, h)), \tag{18}$$

the process

$$x^{k+1} = x^k - J(x^k, h^k)^{-1} Fx^k, \qquad k = 0, 1,..., \tag{19}$$

is called a *discretized Newton iteration*. Note that the parameter vectors $h^k \in R^p$ are permitted to vary with the iteration index.

The simplest choice of the parameters is $h^k \equiv h$; that is, the differencing parameters are held constant throughout the iteration. In this case, we shall see in Chapter 11 that the iterates will have, in general, only a linear rate of convergence, and that in order to approximate the rapid convergence of Newton's method it is necessary that $\lim_{k \to \infty} h^k = 0$. This can be achieved in many ways; for example, we may set $h^k = \gamma_k h$ with some fixed h and a sequence $\{\gamma_k\}$ such that $\lim_{k \to \infty} \gamma_k = 0$. Particularly interesting methods arise, however, if the h^k are chosen as functions of the iterates, and we shall examine some important classes of these methods in the next section.

There are numerous other modifications of Newton's method; indeed, essentially every method to be discussed in this chapter is a modification of Newton's method in some sense. In the remainder of this section, we discuss some particularly simple and natural variations.

One requirement which may be imposed upon any iterative method is that it be *norm-reducing* in the sense that

$$\|Fx^{k+1}\| \leq \|Fx^k\|, \qquad k = 0, 1,..., \tag{20}$$

holds in some norm. The Newton method does not necessarily satisfy this requirement, even in one dimension, and one simple modification of (8) is

$$x^{k+1} = x^k - \omega_k F'(x^k)^{-1} Fx^k, \qquad k = 0, 1,..., \tag{21}$$

7.1. NEWTON'S METHOD AND SOME OF ITS VARIATIONS

where the factors ω_k are chosen to ensure that (20) holds. Sufficient conditions that such ω_k can indeed be so chosen are given in **E 7.1-1** and in the next chapter. Another, somewhat similar modification is

$$x^{k+1} = x^k - [F'(x^k) + \lambda_k I]^{-1} F x^k, \quad k = 0, 1, \ldots . \quad (22)$$

Here, the parameters λ_k are selected to ensure that $F'(x^k) + \lambda_k I$ is nonsingular, if $F'(x^k)$ itself is singular, as well as to ensure, under certain conditions, that (20) holds.

The last modification of Newton's method that we shall discuss here is to reevaluate $F'(x)$ only occasionally; then the iteration becomes

$$x^{k+1} = x^k - F'(x^{p(k)})^{-1} F x^k, \quad k = 0, 1, \ldots, \quad (23)$$

where $p(k)$ is some integer less than or equal to k. The limiting cases of (23) are, of course, $p(k) \equiv k$, which gives Newton's method, and $p(k) \equiv 0$, which gives the simplified Newton method (3). Now, suppose that F' is reevaluated every m steps. Then, if we renumber the iterates so that now x^k denotes the km-th iterate of (23), the iteration is equivalent to

$$x^{k+1} = x^{k,m}, \quad x^{k,i} = x^{k,i-1} - F'(x^k)^{-1} F x^{k,i-1}, \quad i = 1, \ldots, m, \quad x^{k,0} = x^k, \quad (24)$$

which, in the special case $m = 2$, may be written as

$$x^{k+1} = x^k - F'(x^k)^{-1} \{ F x^k + F(x^k - F'(x^k)^{-1} F x^k) \}, \quad k = 0, 1, \ldots . \quad (25)$$

The iteration (24) may be considered as the composition of one Newton step with $m - 1$ simplified Newton steps, and, as we shall see in **10.2.4**, represents a simple way of generating higher-order methods.

NOTES AND REMARKS

NR 7.1-1. For a thorough discussion of iterative methods in one dimension, see Ostrowski [1966] and Traub [1964].

NR 7.1-2. The Newton and discretized Newton iterations have been studied by many authors, and, in other sections, references will be given to a number of them. In particular, the rate of convergence of Newton's method is studied in **10.2**, and other convergence theorems are given in Chapters 12–14. The method is also discussed from a different viewpoint in **8.1**. Local convergence results for certain discretized Newton methods are given in **11.2** and **11.3**.

NR 7.1-3. The Newton method extends immediately to mappings $F: D \subset X \to Y$, where X and Y are arbitrary Banach spaces. Here, F' is to be interpreted as the Frechet derivative and $F'(x)$ is assumed to have a bounded inverse defined on all of Y. More generally, the method has been considered on topological linear spaces (Hirasawa [1954]). Newton's method has been studied extensively in Banach spaces by many authors, most notably L. Kantorovich and his colleagues (see Kantorovich and Akilov [1959]), L. Collatz and his students (see Collatz [1964]), and R. Bellman and his associates under the name *quasilinearization* (see Bellman and Kalaba [1965]).

NR 7.1-4. Crockett and Chernoff [1955] were apparently the first to consider the iteration (21) in a formal way; see also Gleyzal [1959]. The iteration (22) dates back at least to Levenberg [1944]. Convergence results for these two iterations are given in **10.2**, **11.2**, and **14.4**.

NR 7.1-5. Traub [1964] has studied the iteration (24) and shown that it possesses the property of "cubic convergence" for $m = 2$ and even higher-order convergence for $m > 2$. We give these results in **10.2**. Šamanski [1967a] has considered the same procedure with $F'(x)$ replaced by the approximation (16) with $h_{ij} = h, i, j = 1, ..., n$.

NR 7.1-6. Several authors (see, e.g., Altman [1961a]; Jankó [1962b]; Lika [1965]; Mertvecova [1953]; Nečepurenko [1954]; Šafiev [1964]; and Traub [1964]) have considered extensions to n dimensions, and also infinite dimensions, of other one-dimensional methods which possess a high rate of convergence. Such a typical process is the *method of tangent hyperbolas*, which may be formulated as

$$x^{k+1} = x^k - \{I - \tfrac{1}{2}F'(x^k)^{-1}F''(x^k)F'(x^k)^{-1}Fx^k\}^{-1}F'(x^k)^{-1}Fx^k, \quad (26)$$

and can be shown to exhibit cubic convergence. However, methods of this type, which require second- and higher-order derivatives, are rather cumbersome from the computational viewpoint. Note that, while computation of $F'(x)$ involves only the n^2 first partial derivatives $\partial_j f_i$, computation of F'' requires the n^3 second derivatives $\partial_j \partial_k f_i$, in general, an exorbitant amount of work. Indeed, much recent research has been devoted to finding methods needing *fewer* derivative computations than Newton's method. One, not very satisfactory approach to reducing the derivative requirements in methods such as (26) is to consider difference analogs similar to the secant analogs of Newton's method; for work along these lines, see Ul'm [1963b, 1965b].

EXERCISES

E 7.1-1. Assume that $F: R^n \to R^n$ is G-differentiable at x and that $F'(x)$ is invertible. Set $y = F'(x)^{-1}Fx$ and show that there exist $\lambda > 0$ so that

$$\|F(x - \lambda y)\| \leq \|Fx\|. \quad (27)$$

7.2. SECANT METHODS

Show, moreover, that if
$$\|F'(x - ty) - F'(x)\| \leqslant Kt \|y\|, \quad \forall t \in [0, 1],$$
then (27) holds for all $\lambda \in [0, \|Fx\| / (K\|y\|^2)]$.

E 7.1-2. For $A, B \in L(R^n)$, the eigenvalue problem $Ax = \lambda Bx$, $x^T x = 1$, is equivalent with the equation $Fx = 0$, where

$$F: R^n \times R^1 \to R^{n+1}, \quad F\binom{x}{\lambda} = \binom{Ax - \lambda Bx}{x^T x - 1}.$$

Write down Newton's method for this mapping F.

7.2. SECANT METHODS

The discretized Newton methods discussed in the last section constitute n-dimensional generalizations of the one-dimensional discretized Newton methods:

$$x^{k+1} = x^k - \left[\frac{f(x^k + h^k) - f(x^k)}{h^k}\right]^{-1} f(x^k), \quad k = 0, 1, \dots. \tag{1}$$

Two important special cases of (1) are the *regula falsi* iteration

$$x^{k+1} = x^k - \left[\frac{f(\bar{x}) - f(x^k)}{\bar{x} - x^k}\right]^{-1} f(x^k), \quad k = 0, 1, \dots, \tag{2}$$

in which $h^k = \bar{x} - x^k$ for some fixed $\bar{x}$, and the *secant method*

$$x^{k+1} = x^k - \left[\frac{f(x^{k-1}) - f(x^k)}{x^{k-1} - x^k}\right]^{-1} f(x^k), \quad k = 0, 1, \dots, \tag{3}$$

where $h^k = x^{k-1} - x^k$.

It is possible to use similar choices of h^k in the n-dimensional discretized Newton methods. However, in order to discuss the resulting methods in suitable generality, it is desirable to begin with a somewhat different approach to the one-dimensional methods (1). As drawn in Fig. 7.3, the next iterate x^{k+1} of (1) is the solution of the linearized equation

$$l(x) = \left[\frac{f(x^k + h^k) - f(x^k)}{h^k}\right](x - x^k) + f(x^k) = 0.$$

The important point now is that l can be viewed in two different ways: either it is regarded as an approximation of the tangent line

$l_T(x) = f'(x^k)(x - x^k) + f(x^k)$, or as a linear interpolation of f between the points x^k and $x^k + h^k$. In generalizing (1) to n dimensions, we arrive at different methods depending on which view we take. In the case of the discretized Newton methods, the first interpretation was used, and we replaced the derivative $F'(x^k)$ by a matrix $J(x^k, h^k)$ of difference quotients approximating it.

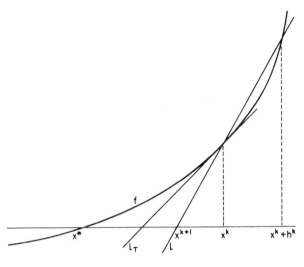

FIGURE 7.3

In order to extend the second point of view to n dimensions, we replace each "component surface" f_i, $i = 1,..., n$, in R^{n+1} by a hyperplane which interpolates f_i at $n + 1$ given points $x^{k,j}$, $j = 0,..., n$, in a neighborhood of x^k. That is, a vector a^i and a scalar α_i are to be found such that the affine mapping $L_i x = \alpha_i + x^T a^i$ satisfies

$$L_i x^{k,j} = f_i(x^{k,j}), \quad j = 0, 1,..., n.$$

The next iterate x^{k+1} is then obtained as the intersection of these n hyperplanes in R^{n+1} with the hyperplane $x = 0$; that is, x^{k+1} is the solution of the linear system $L_i x = 0$, $i = 1,..., n$. This describes the *general secant method* in n dimensions. Depending on the choice of the interpolation points $x^{k,j}$, $j = 0,..., n$, there are numerous possible different specific methods, but, before giving any of these, we develop some results on n-dimensional linear interpolation in order to see how the next iterate can actually be calculated.

7.2. SECANT METHODS

7.2.1. Definition. Any $n+1$ points $x^0,\ldots, x^n$ in R^n are in *general position* if the vectors $x^0 - x^j$, $j = 1,\ldots, n$ are linearly independent. ∎

This definition appears to depend upon the order of enumeration of the x^j, but this is not the case. Indeed, we have the following equivalent conditions for points to be in general position.

7.2.2. Let $x^0,\ldots, x^n$ be any $n+1$ points in R^n. Then the following statements are equivalent:

(a) $x^0,\ldots, x^n$ are in general position.

(b) For any j, $0 \leq j \leq n$, the vectors $x^j - x^i$, $i = 0,\ldots, n$, $i \neq j$, are linearly independent.

(c) The $(n+1) \times (n+1)$ matrix (e, X^T), where $e^T = (1,\ldots, 1)$ and $X = (x^0,\ldots, x^n)$, is nonsingular.

(d) For any $y \in R^n$, there exist scalars $\alpha_0,\ldots, \alpha_n$ with $\sum_{i=0}^n \alpha_i = 1$ such that $y = \sum_{i=0}^n \alpha_i x^i$.

Proof. By the matrix identity

$$\begin{pmatrix} 1 & 0 & \cdots & 0 & 0 & \cdots & 0 \\ x^j & x^0 - x^j & \cdots & x^{j-1} - x^j & x^{j+1} - x^j & \cdots & x^n - x^j \end{pmatrix}$$

$$= \begin{pmatrix} 1 & 1 & \cdots & 1 & 1 & \cdots & 1 \\ x^j & x^0 & \cdots & x^{j-1} & x^{j+1} & \cdots & x^n \end{pmatrix} \begin{pmatrix} 1 & -1 & \cdots & & & & -1 \\ 0 & 1 & 0 & \cdots & & & 0 \\ \vdots & & \ddots & & & & \vdots \\ & & & & & & 0 \\ 0 & & \cdots & & 0 & & 1 \end{pmatrix}$$

we have

$$\det(x^0 - x^j,\ldots, x^{j-1} - x^j, x^{j+1} - x^j,\ldots, x^n - x^j)$$

$$= \det\begin{pmatrix} 1 & 1 & \cdots & 1 & 1 & \cdots & 1 \\ x^j & x^0 & \cdots & x^{j-1} & x^{j+1} & \cdots & x^n \end{pmatrix} = (-1)^j \det(e, X^T),$$

for any $j = 0,\ldots, n$, which shows the equivalence of (a), (b) and (c).

Finally, (d) is equivalent to the statement that the linear system

$$\begin{pmatrix} e^T \\ X \end{pmatrix} \begin{pmatrix} \alpha_0 \\ \vdots \\ \alpha_n \end{pmatrix} = \begin{pmatrix} 1 \\ y \end{pmatrix} \qquad (4)$$

has a solution for any y, so that, clearly, (c) implies (d). Conversely, by solving (4) for y successively set equal to $0, e^1,\ldots, e^n$, we see that (e, X^T) is nonsingular. ∎

The geometrical interpretation of general position is that the points $x^0,\ldots, x^n$ do not lie in an affine subspace of dimension less than n. Thus, for $n = 2$, the points x^0, x^1, x^2 are in general position if they are not colinear, that is, if they do not lie on a line in R^2. Note, however, that the vectors $x^0,\ldots, x^n$ may span R^n even if they are not in general position (see **E 7.2-6**).

The following result now gives a complete answer to the linear interpolation problem in R^n.

7.2.3. Let $x^0,\ldots, x^n$ and $y^0,\ldots, y^n$ be given points in R^n. Then there exists a unique affine function $Lx = a + Ax$, where $a \in R^n$ and $A \in L(R^n)$, such that $Lx^j = y^j$, $j = 0,\ldots, n$, if and only if $x^0, x^1,\ldots, x^n$ are in general position. Moreover, A is nonsingular if and only if $y^0,\ldots, y^n$ are in general position.

Proof. The conditions $Lx^j = y^j$, $j = 0, 1,\ldots, n$, can be written in matrix form as

$$(e, X^T) \begin{pmatrix} a^T \\ A^T \end{pmatrix} = (y^0, \cdots, y^n)^T, \tag{5}$$

where again $e^T = (1, 1,\ldots, 1)$ and $X = (x^0,\ldots, x^n)$. Hence, the first part is an immediate consequence of **7.2.2**. Now, $Lx^j = y^j$, $j = 0,\ldots, n$, implies that

$$A(x^j - x^0) = y^j - y^0, \quad j = 1,\ldots, n, \tag{6}$$

and, since $x^j - x^0, j = 1,\ldots, n$, are linearly independent, it follows that A is nonsingular if and only if the vectors $y^j - y^0$, $j = 1,\ldots, n$ are linearly independent, and, hence, if and only if $y^0,\ldots, y^n$ are in general position. ∎

In line with these results, one step of any general secant method can now be phrased as follows.

7.2.4. Definition. Let $F: D \subset R^n \to R^n$ and assume that the two sets of points $x^0,\ldots, x^n \in D$ and $Fx^0,\ldots, Fx^n$ are in general position. Then the point

$$x^s = -A^{-1}a, \tag{7}$$

where a and A satisfy

$$a + Ax^j = Fx^j, \quad j = 0,\ldots, n, \tag{8}$$

is a *basic secant approximation* with respect to $x^0,\ldots, x^n$. ∎

7.2. SECANT METHODS

Note that **7.2.3** ensures that x^s is well defined. Note also that in one dimension the conditions reduce to $x^0 \neq x^1$ and $f(x^0) \neq f(x^1)$, which are just the conditions which ensure that the unique secant line intersects the x-axis.

The computation of a basic secant approximation may be carried out by finding a and A to satisfy (8)—which, in turn, requires solving the linear system (5) with $y^j = Fx^j$—and then by solving $a + Ax = 0$. However, it turns out that there is no need to compute the interpolating function $a + Ax$ explicitly. We consider next two alternative formulations, both of which show that x^s can be obtained by solving only *one* linear system.

7.2.5. Wolfe Secant Formulation. Let $x^0, \ldots, x^n$, as well as $Fx^0, \ldots, Fx^n$, be in general position. Then the basic secant approximation satisfies

$$x^s = Xz = \sum_{j=0}^{n} z_j x^j, \qquad (9)$$

where $z = (z_0, \ldots, z_n)^T$ is the unique solution of the $(n+1) \times (n+1)$ linear system

$$\begin{pmatrix} 1 & \cdots & 1 \\ Fx^0 & & Fx^n \end{pmatrix} z = (1, 0, \ldots, 0)^T. \qquad (10)$$

Proof. Since the Fx^j are in general position, **7.2.2**(b) ensures that (10) has a unique solution which satisfies $\sum_{i=0}^{n} z_i = 1$ and $\sum_{i=0}^{n} z_i Fx^i = 0$. Hence, by (8),

$$0 = \sum_{j=0}^{n} z_j Fx^j = \sum_{j=0}^{n} z_j (a + Ax^j) = a + A\left(\sum_{j=0}^{n} z_j x^j\right),$$

and, since x^s is the unique solution of $Ax + a = 0$, (9) holds. ∎

Note that (9) and (10) uniquely determine a vector x^s provided only that $Fx^0, \ldots, Fx^n$ are in general position; therefore, the Wolfe formulation can be carried out even if $x^0, \ldots, x^n$ are not in general position. In that case, however, x^s will lie in the lower-dimensional affine subspace $\{x \mid x = \sum_{i=0}^{n} c_i x^i, \sum_{i=0}^{n} c_i = 1\}$ and no linear interpolator $a + Ax$ can exist such that $a + Ax^j = Fx^j, j = 0, \ldots, n$ (see **E 7.2-7**).

For the next formulation, it is convenient to introduce the operator

$$J: D_K \subset R^n \times L(R^n) \to L(R^n)$$

defined by

$$J(x, H) = (F(x + He^1) - Fx, \ldots, F(x + He^n) - Fx) H^{-1} \qquad (11)$$

where, if D is the domain of definition of F,

$$D_K = \{(x, H) \mid x + He^i \in D, \quad i = 1,\ldots, n; \quad H \text{ nonsingular}\}.$$

7.2.6. Newton Formulation. Assume that $x^0,\ldots, x^n$ and $Fx^0,\ldots, Fx^n$ are in general position, and set

$$H = (x^1 - x^0,\ldots, x^n - x^0). \tag{12}$$

Then $J(x^0, H)$ is nonsingular and the basic secant approximation x^s is given by

$$x^s = x^0 - J(x^0, H)^{-1} Fx^0. \tag{13}$$

Proof. Since $Fx^i = F(x^0 + He^i)$, it follows from (6) with $y^j = Fx^j$ that

$$AH = (F(x^0 + He^1) - Fx^0,\ldots, F(x^0 + He^n) - Fx^0)$$

so that, since H is nonsingular, $A = J(x^0, H)$. Thus **7.2.3** ensures that $J(x^0, H)$ is nonsingular, and, from $x^s = -A^{-1}a$ and $a = Fx^0 - Ax^0$, we obtain

$$x^s = -A^{-1}(-Ax^0 + Fx^0) = x^0 - J(x^0, H)^{-1} Fx^0. \quad\blacksquare$$

Note that if we set $\Gamma = (Fx^1 - Fx^0,\ldots, Fx^n - Fx^0)$, then (13) may be written in the form

$$x^s = x^0 - H\Gamma^{-1} Fx^0. \tag{14}$$

Therefore, as with the Wolfe formulation, the Newton formulation may be carried out provided only that $Fx^0,\ldots, Fx^n$ are in general position. Again, however, x^s will then lie in the affine subspace

$$\left\{x \mid x = \sum_{i=0}^{n} c_i x^i, \sum_{i=0}^{n} c_i = 1\right\}$$

(see **E 7.2-7**).

Observe also that a basic secant approximation, by either the Wolfe or Newton formulation, requires indeed only the solution of one linear system of equations—namely, (10) in the first case, and $\Gamma x = Fx^0$ in the second—followed by the calculation of a linear combination of the vectors $x^0,\ldots, x^n$ by means of (9) or (14), respectively.

It is of interest to note that the following representation of x^s is also valid:

$$x^s = x^0 - [(Fx^1 - Fx^0, Fx^2 - Fx^1,\ldots, Fx^n - Fx^{n-1})(x^1 - x^0,\ldots, x^n - x^{n-1})^{-1}]^{-1} Fx^0. \tag{15}$$

7.2. SECANT METHODS

This is an immediate consequence of the following lemma, which will be of use in Chapter 11, since (15) follows from (16) if $x = x^0$ and H is defined by (12).

7.2.7. Let $J(x, H)$ be defined by (11) with $H = (h^1, \ldots, h^n)$. Then

$$J(x, H) = (F(x + h^1) - Fx, F(x + h^2) - F(x + h^1), \ldots, F(x + h^n) - F(x + h^{n-1}))\hat{H}^{-1} \quad (16)$$

where

$$\hat{H} = (h^1, h^2 - h^1, \ldots, h^n - h^{n-1}).$$

Proof. We note that, for any $n + 1$ vectors $r^0, r^1, \ldots, r^n \in R^n$, we have

$$(r^1 - r^0, r^2 - r^0, \ldots, r^n - r^0) P = (r^1 - r^0, r^2 - r^1, \ldots, r^n - r^{n-1}),$$

where $P \in L(R^n)$ is defined by

$$P = \begin{bmatrix} 1 & -1 & & & \\ & 1 & -1 & & \\ & & \ddots & \ddots & \\ & & & & -1 \\ & & & & 1 \end{bmatrix}.$$

Clearly, P is nonsingular, and (16) then follows from

$$J(x, H) = (F(x + h^1) - Fx, \ldots, F(x + h^n) - Fx) P(HP)^{-1}. \quad \blacksquare$$

The Newton formulation allows the general secant method described at the beginning of this section to be expressed in the compact form

$$\begin{cases} x^{k+1} = x^k - J(x^k, H_k)^{-1} Fx^k, & k = 0, 1, \ldots, \\ H_k = (x^{k,1} - x^k, \ldots, x^{k,n} - x^k), \end{cases} \quad (17)$$

where we have set $x^{k,0} = x^k$.

We consider next several possible ways of choosing the auxiliary points $x^{k,1}, \ldots, x^{k,n}$. A first simple choice is given by

$$x^{k,j} = x^k + (x_j^{k-1} - x_j^k) e^j, \quad j = 1, \ldots, n. \quad (18)$$

In this case, H_k is the diagonal matrix

$$H_k = \operatorname{diag}(x_1^{k-1} - x_1^k, \ldots, x_n^{k-1} - x_n^k),$$

and, if we set $h_j^k = x_j^{k-1} - x_j^k$, $j = 1,\ldots, n$, then

$$J(x^k, H_k) = \big((1/h_1^k)[F(x^k + h_1^k e^1) - Fx^k],\ldots, (1/h_n^k)[F(x^k + h_n^k e^n) - Fx^k]\big).$$

Substituting this into the Newton formulation (17), we see that the resulting method is exactly the same as the discretized Newton method (7.1.19) using the difference approximation (7.1.16) with $h_{ij} = h_i$, $i = 1,\ldots, n$. For future use, it is convenient to redefine J in this case as a mapping

$$\begin{cases} J: D_J \times D_h \subset R^n \times R^n \to L(R^n) \\ D_J \times D_h = \{(x, h) \in R^n \times R^n \mid x + h_i e^i \in D, \; h_i \neq 0, \; i = 1,\ldots, n\} \\ J(x, h) = \big(h_1^{-1}[F(x + h_1 e^1) - Fx],\ldots, h_n^{-1}[F(x + h_n e^n) - Fx]\big), \end{cases} \quad (19)$$

and to write the method as

$$x^{k+1} = x^k - J(x^k, x^{k-1} - x^k)^{-1} Fx^k, \qquad k = 0, 1,\ldots. \quad (20)$$

If we choose instead of (18) the points

$$x^{k,j} = x^k + \sum_{i=1}^j (x_i^{k-1} - x_i^k) e^i, \qquad j = 1,\ldots, n, \quad (21)$$

then a simple computation shows that the iteration (17) is exactly the same as the discretized Newton method using the difference approximation (7.1.15) with $h_{ij}^k = x_j^{k-1} - x_j^k$. In this case, we may define the iteration by (20), where J is now given by

$$J(x, h) = \Big(h_1^{-1}[F(x + h_1 e^1) - Fx],\ldots, h_n^{-1}\Big[F\Big(x + \sum_{j=1}^n h_j e^j\Big) - F\Big(x + \sum_{j=1}^{n-1} h_j e^j\Big)\Big]\Big). \quad (22)$$

More generally, we can consider the choice of auxiliary points

$$x^{k,j} = x^k + P_{j,k}(x^{k-1} - x^k), \qquad j = 1,\ldots, n, \quad (23)$$

where $P_{j,k} \in L(R^n)$ are given linear operators. Clearly, (18) is the special case of (23) in which $P_{j,k} = (0,\ldots, 0, e^j, 0,\ldots, 0)$, $j = 1,\ldots, n$, $k = 0, 1,\ldots$, while (23) reduces to (21) if $P_{j,k} = (e^1,\ldots, e^j, 0,\ldots, 0)$, $j = 1,\ldots, n$, $k = 0, 1,\ldots$.

The choice (23) of the auxiliary points depends only on x^k and x^{k-1}. In general, if the auxiliary points $x^{k,j}$ depend on precisely p of the previous iterates $x^k,\ldots, x^0$, we say that the iteration (17) is *a p-point secant method*, while if the $x^{k,j}$ depend upon $x^k,\ldots, x^{k-p+1}$, it is called a *sequential p-point secant method*.

7.2. SECANT METHODS

The iterations (17) with $x^{k,i}$ given by (18) or (21) are examples of sequential two-point methods, while the iteration

$$x^{k+1} = x^k - J(x^k, H_k)^{-1} F x^k, \quad H_n = (x^{k-1} - x^k, \ldots, x^{k-n} - x^k), \quad k = 0, 1, \ldots, \tag{24}$$

is a sequential $(n + 1)$-point method. As an example of a nonsequential $(n + 1)$-point method, the auxiliary points may be chosen from the set of previous iterates by the criterion that $x^{k,1}, \ldots, x^{k,n}$ are those n vectors among $x^0, x^1, \ldots, x^{k-1}$ for which the $\| F x^j \|$ are smallest.

General $(p + 1)$- point methods may be generated in a variety of ways. For example, in analogy with (23), we may choose

$$x^{k,j} = x^k + \sum_{i=1}^{p} P_{i,j,k}(x^{k-i} - x^k), \quad j = 1, \ldots, n, \tag{25}$$

where again the $P_{i,j,k}$ are given linear operators.

In general, the secant method requires $n + 1$ evaluations of the function F at each stage—namely at the points $x^k, x^{k,1}, \ldots, x^{k,n}$. In particular, this is true for the two-point method defined by (18). This amount of computation is comparable to that of Newton's method if the evaluation of $f_i(x)$ takes about as much work as that of $\partial_j f_i(x)$.

In certain cases, however, the particular choice of the $x^{k,i}$ permits fewer function evaluations. For example, if (21) is used, then $x^{k,n} = x^{k-1}$, and, since $F x^{k-1}$ is available from the previous stage, only n new function evaluations are required. The most spectacular saving, however, is available through the $(n + 1)$-point method (24). Here, since $F x^{k-1}, \ldots, F x^{k-n}$ are already available (except at the first stage, when $F x^0, \ldots, F x^n$ must all be calculated), only one new function evaluation is required—namely, $F x^k$. Moreover, another possible computational savings is available in the solution of the linear system of (24). To see this, recall from **7.2.7** that (24) can be written in the alternative form

$$x^{k+1} = x^k - H_k \Gamma_k^{-1} F x^k, \tag{26}$$

where now

$$\begin{aligned} H_k &= (x^k - x^{k-1}, x^{k-1} - x^{k-2}, \ldots, x^{k-n+1} - x^{k-n}), \\ \Gamma_k &= (F x^k - F x^{k-1}, \ldots, F x^{k-n+1} - F x^{k-n}). \end{aligned} \tag{27}$$

Then we find:

7.2.8. Assume that the matrices Γ_p and Γ_{p+1} defined by (27) for

198 7. GENERAL ITERATIVE METHODS

$k = p, p+1$, are both nonsingular, and denote the rows of Γ_p^{-1} by $v^1,\ldots, v^n$. Then

$$\Gamma_{p+1}^{-1} = B - \frac{B(q^p - q^{p-n})v^n}{1 + v^n(q^p - q^{p-n})}, \tag{28}$$

where $q^i = Fx^{i+1} - Fx^i$, and B is the matrix with the rows $v^n, v^1,\ldots, v^{n-1}$.

Proof. Let P be a permutation matrix such that

$$\Gamma_p P = (q^{p-n}, q^{p-1},\ldots, q^{p-n+1});$$

then

$$\Gamma_{p+1} = (q^p,\ldots, q^{p-n+1}) = \Gamma_p P + (q^p - q^{p-n})(e^1)^{\mathrm{T}}.$$

As a permutation matrix, P is nonsingular, and, hence, the same is true for $\Gamma_p P$. It then follows, by the Sherman–Morrison formula (2.3.14), that

$$\Gamma_{p+1}^{-1} = P^{-1}\Gamma_p^{-1} - (1/\alpha)\, P^{-1}\Gamma_p^{-1}(q^p - q^{p-n})(e^1)^{\mathrm{T}}\, P^{-1}\Gamma_p^{-1}, \tag{29}$$

where $\alpha = 1 + (e^1)^{\mathrm{T}} P^{-1}\Gamma_p^{-1}(q^p - q^{p-n}) \neq 0$, since, by hypothesis, Γ_{p+1}^{-1} exists. But the effect of multiplication on the left by P^{-1} is to place the last row in the position of the first row and move all other rows down one place. ∎

Although the $(n+1)$-point sequential secant method requires the least amount of computation per step, it will be shown in Chapter 11 that the method is prone to unstable behavior and that no satisfactory convergence results can be given. In contrast, the two-point methods defined by (18) and (21) will be shown to retain the essential properties of Newton's method and, in particular, satisfactory local convergence theorems will be obtained for them in Section **11.2.**

We end this section by describing a closely related class of iterative processes known as Steffensen methods.

Consider again the basic one-dimensional secant method (1). If we set $h^k = f(x^k)$, we obtain Steffensen's method in one dimension:

$$x^{k+1} = x^k - \frac{f(x^k)}{f(x^k + f(x^k)) - f(x^k)}f(x^k), \qquad k = 0, 1,\ldots. \tag{30}$$

This iteration is of interest, since—under suitable conditions—it exhibits the same quadratic convergence as Newton's method while not requiring any derivatives of f. (See Chapter 11).

The concepts already developed for the secant method now permit immediate and natural extensions of (30) to n dimensions. In corre-

7.2. SECANT METHODS

spondence with the two-point secant methods defined by the choice of points (23), we can define the analogous Steffensen method by (17) with the choice of auxiliary points

$$x^{k,j} = x^k + P_{j,k} F x^k, \qquad j = 1,\ldots, n.$$

For example, if $P_{j,k} = (0,\ldots, 0, e^j, 0,\ldots,0)$, we obtain the particular Steffensen method

$$x^{k+1} = x^k - J(x^k, Fx^k)^{-1} F x^k, \tag{31}$$

where J is defined by (19); this is the direct analog of the two-point secant method (19)–(20). Similarly, corresponding to the secant method (20), (22), the choice $P_{j,k} = (e^1,\ldots, e^j, 0,\ldots, 0)$ gives a Steffensen method of the form (31) in which J is defined by (22).

More generally, corresponding to (25), we can choose

$$x^{k,j} = x^k + \sum_{i=1}^{p} P_{i,j,k} F x^{k-i+1}, \qquad j = 1,\ldots, n. \tag{32}$$

As a special case of (32), let

$$x^{k,j} = x^k + F x^{k-j+1}, \qquad j = 1,\ldots, n,$$

which leads to the method

$$x^{k+1} = x^k - J(x^k, H_k)^{-1} F x^k, \qquad H_k = (Fx^k,\ldots, Fx^{k-n+1}), \tag{33}$$

corresponding to the $(n+1)$-point secant method (24). Here, of course, $J(x, H)$ is defined by (11).

Note that in the Steffensen method (31), (19) it is necessary to evaluate F at the points $x^k + f_j(x^k) e^j, j = 1,\ldots, n$, as well as at x^k, so that precisely the same number of function evaluations are needed as for the corresponding secant method defined by (19)–(20). On the other hand, in the case of (33), it is necessary to obtain $F(x^k + Fx^{k-j}), j = 1,\ldots, n$, so that, again, $n + 1$ function evaluations are required. Thus, the advantage of (24), in which only one new evaluation of F is needed at each stage, does not carry over to (33).

Another form of Steffensen's method arises in connection with the fixed-point equation $x = Gx$. Here, the auxiliary points $x^{k,i}$ may be taken as the iterates $x^{k,i} = G^i x^k, i = 1,\ldots, n$, generated by the operator G. This then leads to the iteration

$$x^{k+1} = x^k - J(x^k, H_k)^{-1} [x^k - Gx^k], \qquad H_k = (Gx^k - x^k,\ldots, G^n x^k - x^k) \tag{34}$$

by setting $Fx = x - Gx$ in (17). Note that the evaluation of $J(x^k, H_k)$ involves the computation of the n vectors

$$F(x^k + H_k e^i) - Fx^k = F(G^i x^k) - Fx^k = G^i x^k - G^{i+1} x^k - x^k + Gx^k,$$
$$i = 1,..., n,$$

so that $n + 1$ evaluations of G are required. The iteration (34) is, of course, not restricted to equations in fixed-point form, since the conversion $Gx = x - Fx$ may always be made. Note also that, in contrast to the multistep method (33), (34) is a one-step method.

NOTES AND REMARKS

NR 7.2-1. For a thorough discussion of the one-dimensional secant method and of related higher-order methods, such as that of Muller, the reader is referred to Ostrowski [1966] and Traub [1964].

NR 7.2-2. The idea of replacing F by a linear interpolating function in order to extend the secant method to higher dimensions dates back to Gauss in the two-dimensional case (see Ostrowski [1966, Appendix D]). Its modern revival and generalization to n dimensions is apparently due to H. Heinrich in unpublished lectures (circa 1955), and its first rigorous analysis due to Bittner [1959]. However, Wolfe [1959] independently suggested the $(n + 1)$-point method, described in the text, in which the vector with largest function value is dropped. More recent works on the $(n + 1)$-point or related methods include Tornheim [1964], Anderson [1965] (see **NR 7.2-10**), and Barnes [1965] (see **NR 7.3-2**).

NR 7.2-3. The Steffensen iteration (34) was first considered by Ludwig [1952], and, more recently, by Henrici [1964], but from an entirely different point of view. One way of deriving the Steffensen iteration in one dimension is by means of the Aitken δ^2-process, which may be extended to n dimensions as follows: Given $n + 2$ points $y^0,..., y^{n+1}$, define the matrices

$$H = (y^1 - y^0, y^2 - y^1,..., y^n - y^{n-1})$$

and

$$S = (y^2 - 2y^1 + y^0,..., y^{n+1} - 2y^n + y^{n-1}),$$

and then introduce an "extrapolated" vector $\hat{x}$ by $\hat{x} = y^0 - HS^{-1}(h^1 - y^0)$. For the fixed-point equation $x = Gx$, set $y^0 = x^k$, $y^i = G^i x^k$, $i = 1,..., n + 1$, and $x^{k+1} = \hat{x}$. Although slightly different in form, it is easy to see, by **7.2.7**, that the x^{k+1} thus produced is precisely that of (34).

7.2. SECANT METHODS

NR 7.2-4. The two-point secant and related Steffensen methods have been considered by a variety of authors. Korganoff [1961] handles the particular method in which the auxiliary points $x^{k,i}$ are given by (18), while Maergoiz [1967] and Wegge [1966] treat the corresponding Steffensen method (31), (19). Other authors have considered the method (22), (31) in the context of divided differences (**NR 7.2-6**).

NR 7.2-5. There is a fundamental difficulty in extending the interpolation approach to infinite dimensions, since, presumably, we would require that the linear interpolator L agree with F at infinitely many points. One possibility is to assume that $\{x^i\}$ span the space in some sense and that $Lx^i = Fx^i$, $i = 0, 1, \ldots$. Then the basic secant approximation is defined as the solution of $Lx = 0$. This type of extension to infinite dimensions does not seem very promising, and has not been explored in the literature.

However, extensions of the two-point secant methods to infinite dimensions have been made by two related approaches. For one such approach, see **NR 7.2-6**; the other might be called "pointwise extension," and has been treated, for example, by Collatz [1964]. Consider the two-point boundary problem

$$u''(t) = f(u(t)), \quad u(0) = u(1) = 0, \quad t \in [0, 1].$$

"Newton's method" applied directly to this equation gives the sequence of linear boundary value problems

$$u''_{k+1} = f(u_k) + (u_{k+1} - u_k) f'(u_k), \quad u_{k+1}(0) = u_{k+1}(1) = 0.$$

Now let h_k be a function on $[0, 1]$ and define the corresponding discrete Newton method by

$$u''_{k+1} = f(u_k) + (u_{k+1} - u_k) \frac{f(u_k + h_k) - f(u_k)}{h_k}.$$

Hence, for $h_k = u_{k-1} - u_k$, we have the two-point secant method:

$$u''_{k+1} = f(u_k) + (u_{k+1} - u_k) \frac{f(u_k) - f(u_{k-1})}{u_k - u_{k-1}}.$$

Related Steffensen-type procedures, treated as extrapolation formulas, have been given by Bellman, Kagiwada, and Kalaba [1965], and Noble [1964, p. 274].

NR 7.2-6. A more axiomatic approach to the extension of the secant method was taken by Schmidt [1961, 1963a] by means of the concept of a divided-difference operator. Briefly, Schmidt defines a *first divided difference* of F on a Banach space X as a mapping $J: D \times D \subset X \times X \to L(X)$ which satisfies

$$J(x, h) h = F(x + h) - Fx, \quad \forall x, x + h \in D, \quad (35)$$

and

$$\| J(x, h) - J(x + h, k) \| \leq a \| h + k \| + b \| h \| + b \| k \|, \quad \forall x, x + h, x + k \in D, \tag{36}$$

where a and b are certain constants. Second and higher-order divided differences may be defined analogously. One may now consider the two-point secant method (20).

It is easy to verify (**E 7.2-9**) that the operator J of (22) satisfies the algebraic condition (35), while the J of (19) does not. In Chapter 11, we shall analyze both choices (19) and (22) by means of the notion of a "consistent approximation." This eliminates the requirement (35), while, in essence, retaining (36) (see **NR 11.2-2** for further discussion).

Schmidt has shown that certain natural derivative approximations, such as that of the previous note, yield divided-difference operators for various differential and integral equations. Consider, for example, the Banach space $X = C[0, 1]$ and the mapping $G: X \to X$ given by

$$(Gx)(s) = \int_0^1 f(s, t, x(t))\, dt, \quad s \in [0, 1].$$

Now, set $F = I - G$ and introduce $J: X \times X \to L(X)$ by

$$[J(x, h)\, w](s) = w(s) - \int_0^1 \left[\frac{f(s, t, x(t) + h(t)) - f(s, t, x(t))}{h(t)} \right] w(t)\, dt; \tag{37}$$

that is, the divided-difference operator on X is defined in the natural way by means of divided differences of f with respect to its third variable. It is easy to see that (35) holds, and, under appropriate smoothness conditions for f as well as restrictions on D, Schmidt shows that (36) is also valid.

It is of interest to note that the corresponding device for the discrete integral equation leads back to the operator J of (22). More specifically, consider the mapping $F: R^n \to R^n$ defined by

$$f_i(x) = x_i - \sum_{j=1}^n \gamma_j f(s_i, s_j, x_j), \quad i = 1, \ldots, n,$$

as discussed in **1.3**. Now, define $J: R^n \times R^n \to L(R^n)$, in analogy to (37), by

$$[J(x, h)\, w]_i = w_i - \sum_{j=1}^n (\gamma_j/h_j)[f(s_i, s_j, x_j + h_j) - f(s_i, s_j, x_j)]\, w_j,$$
$$i = 1, \ldots, n. \tag{38}$$

If we note that

$$f_i\left(x + \sum_{k=1}^j h_k e^k\right) = y_{i,j} - \sum_{k=1}^j \gamma_k f(s_i, s_k, x_k + h_k) - \sum_{k=j+1}^n \gamma_k f(s_i, s_k, x_k),$$

7.2. SECANT METHODS

where $y_{i,j} = x_i + h_i$ if $i \leqslant j$ and x_i if $i > j$, then it follows immediately that the operator J of (38) is precisely that of (22).

NR 7.2-7. Various authors have applied and extended the idea of the divided-difference operators of **NR 7.2-6**. Chen [1964] used (35) and (36) to consider the corresponding Steffensen method (31), as did Ul'm [1964c]. However, Ul'm also required that the divided difference be symmetric in the sense that

$$J(x, y - x) = J(y, x - y).$$

This symmetry condition, also considered by Sergeev [1961], is very strong and is not satisfied by either one of the operators (19) or (22), except under special conditions. Koppel' [1966] considered extensions of the results of Chen and Ul'm. For additional discussion, see **NR 11.2-2**.

Other definitions of divided differences have also been advanced; for example, $J(x, h) = \int_0^1 F'(x + th) \, dt$. For a summary of other definitions and relations between them, see Ul'm [1967a].

NR 7.2-8. In all of the concrete examples of secant methods in the text, the auxiliary points have been chosen as linear functions of $x^k, \ldots, x^{k-n}$. However, Robinson [1966] has given an example of a two-point method in which the auxiliary points are nonlinear functions of x^k and x^{k-1}. Set $q_i = x_i^k - x_i^{k-1}$,

$$P = \begin{bmatrix} q_1 q_2 & q_1 q_3 & \cdots & q_1 q_n & -q_1 \\ -q_1^2 & q_2 q_3 & & q_2 q_n & -q_2 \\ 0 & -q_1^2 - q_2^2 & & \vdots & \vdots \\ \vdots & 0 & \ddots & & \\ \vdots & \vdots & & q_{n-1} q_n & -q_{n-1} \\ 0 & 0 & & -q_1^2 - \cdots - q_n^2 & -q_n \end{bmatrix},$$

and $H_k = \| x^k - x^{k-1} \| PD$, where D is a diagonal matrix chosen so that the columns of PD have Euclidean length one. This amounts to a choice of the auxiliary points as

$$x^{k,i} = x^k + \| x^k - x^{k-1} \| (p^i / \| p^i \|_2), \quad i = 1, \ldots, n,$$

where $p^1, \ldots, p^n$ are the columns of P. It is easy to verify that PD is an orthogonal matrix, which, as will be seen in Chapter 11 (see, in particular, **E 11.3-4**), is desirable.

In case some column of P is zero, the above prescription cannot be carried out. It is clear that the first column is zero if and only if $q^1 = 0$, and, more generally, the ith column is zero if and only if $q^j = 0, j = 1, \ldots, i$, in which case all previous columns are also zero. One simple modification of the procedure is then to set the first i columns of P equal to the coordinate vectors $e^1, \ldots, e^i$, and the resulting matrix PD is again orthogonal.

Note, finally, that one step of this process requires evaluation of Fx^k and $Fx^{k,i}$, $i = 1,..., n-1$; since $x^{k,n} = x^k - (x^k - x^{k-1}) = x^{k-1}$, no evaluation is required for $i = n$. Hence, the same number of function evaluations is needed as for the two-point method (20), (22), and there appear to be no advantages that would indicate preference of this method over (20), (22).

NR 7.2-9. For an important class of equations, the two-point methods (20), (19), and (20), (22) require only *two* evaluations of F at each step, rather than n or $n + 1$. Let F be of the form

$$f_i(x) = \sum_{j=1}^{n} f_{ij}(x_j) \qquad i = 1,..., n;$$

that is, each component of F is the sum of n functions each of a single variable. It was noted in Chapter 1 that such equations typically result from discretization of integral and differential equations. Now, suppose that x and y differ only in one component—say, the kth. Then

$$f_i(x) - f_i(y) = f_{ik}(x_k) - f_{ik}(y_k),$$

and, in particular, the (i, j)th component of the operator J of (19) is simply

$$[J(x, h)]_{ij} = (1/h_j)[f_{ij}(x_j + h_j) - f_{ij}(x_j)].$$

Hence, the evaluation of $J(x, h)$ requires each f_{ij} to be computed twice, which is equivalent, in general, to two evaluations of F itself. Note, however, that the analogous considerations also hold for Newton's method, since $F'(x) = (f'_{ij}(x_j))$.

NR 7.2-10. One of the disadvantages of the $(n + 1)$-point sequential secant method, besides its possible poor behavior, is the necessity of retaining $n + 1$ points. Recently, Anderson [1965] has proposed a modification in which only $m + 1 < n + 1$ points are retained. Set

$$H_k = (x^{k-1} - x^k,..., x^{k-m} - x^k), \qquad \Gamma_k = (Fx^{k-1} - Fx^k,..., Fx^{k-m} - Fx^k),$$

and suppose that rank $\Gamma_k = m$. Then x^{k+1} is defined by

$$x^{k+1} = x^k - H_k y^k, \qquad y^k = (\Gamma_k^T \Gamma_k)^{-1} \Gamma_k^T Fx^k. \tag{39}$$

If $m = n$, and Γ_k and H_k are nonsingular, then this reduces to the $(n + 1)$-point sequential method (24). More generally, if rank $\Gamma_k < m$, then y^k may be defined as a least-squares solution of $\Gamma_k y^k = Fx^k$.

Now note that if x^{k+1} is defined by (39), then

$$x^{k+1} = x^k - \sum_{j=1}^{m} y_j^k (x^{k-j} - x^k),$$

so that x^{k+1} is a linear combination of $x^k,..., x^{k-m}$. Hence, all iterates will be

7.2. SECANT METHODS

linear combinations of the starting vectors $x^0,\ldots, x^{-m+1}$ and there is no hope of convergence unless x^* lies in the subspace spanned by these initial vectors. To circumvent this difficulty, Anderson proposes the addition of certain linear combinations of the Fx^i, as, for example,

$$x^{k+1} = x^k - H_k y^k + \beta[Fx^k - \Gamma_k y^k], \quad k = 0, 1,\ldots, \qquad (40)$$

where again y^k is given by (39) and $\beta \neq 0$. If $m = n$, then (40) also reduces to the $(n+1)$-point secant method, since $Fx^k = \Gamma_k y^k$.

One could also consider (40) with $m > n$ in order to obtain a possibly more stable behavior of the iterates.

None of the methods of this note have, as yet, been completely analyzed, although Voigt [1969] has given results which indicate that (40) will not exhibit a superlinear rate of convergence except under the most stringent conditions.

NR 7.2-11. The one-dimensional *regula falsi* iteration (2) may also be extended to n dimensions in a variety of ways. Corresponding to (20), with J given by either (19) or (22), one may consider the iteration

$$x^{k+1} = x^k - J(x^k, \bar{x} - x^k)^{-1} Fx^k, \quad k = 0, 1,\ldots, \qquad (41)$$

where $\bar{x}$ is held fixed. Similarly, for the general secant method (17), various *regula falsi*-type methods are obtained by holding fixed one or more of the auxiliary points $x^{k,i}$.

EXERCISES

E 7.2-1. Let $F: R^n \to R^n$ be the affine mapping $Fx = Ax - b$, where $A \in L(R^n)$ and $b \in R^n$. Show that if J is defined by (11) or (16), then $J(x, H) = A$, for all $x \in R^n$.

E 7.2-2. Let $F: R^n \to R^n$ and assume that $x^0,\ldots, x^n$ are in general position. Show that the linear interpolator may be written in Lagrangian form $Lx = \sum_{i=0}^n L_i(x) Fx^i$, where each L_i is an affine functional of the form

$$L_i x = \beta_i + b_i^T x \quad \text{and} \quad L_i x^j = \delta_{ij}, \quad i,j = 0,\ldots, n.$$

E 7.2-3. Let $f: R^n \to R^1$, and $g_i : R^n \to R^1$, $i = 0,\ldots, n$, be given functionals and $x^0,\ldots, x^n$ given points in R^n. Show that there exists a unique functional g of the form $g(x) = \sum_{i=0}^n \alpha_i g_i(x)$ such that $g(x^i) = f(x^i)$, $i = 0,\ldots, n$, if and only if the $(n+1) \times (n+1)$ matrix $(g_i(x^j))$ is nonsingular.

E 7.2-4. Define $F: R^2 \to R^2$ by $f_1(x) = x_1 + x_2^2$, $f_2(x) = x_1^2 - x_2$. Show that the points $x^0 = (0, 2)^T$, $x^1 = (0, 1)^T$, $x^2 = (0, -1)^T$ are not in general position, but that the points Fx^i, $i = 0, 1, 2$ are in general position.

E 7.2-5. Define $F: R^2 \to R^2$ by $f_1(x) = x_1^2$, $f_2(x) = x_1^2 - \frac{1}{2}x_2$. Show that the points $x^0 = (0, -1)^T$, $x^1 = (-1, 2)^T$, $x^2 = (1, 2)^T$ are in general position, but that Fx^i, $i = 0, 1, 2$ are not in general position.

E 7.2-6. Let $x^0 = (1, 0)^T$, $x^1 = (1, 1)^T$, $x^2 = (1, -1)^T$. Show that x^0, x^1, x^2 span R^2 but are not in general position.

E 7.2-7. Let $F: R^n \to R^n$. Assume that the points $x^0,\ldots, x^n$ are not in general position, but that $Fx^0,\ldots, Fx^n$ are. Show that x^s, as defined by (13), lies in the affine subspace generated by $\sum_{i=0}^{n} z_i x^i$, $\sum_{i=0}^{n} z_i = 1$. Show also that no linear interpolator $Lx = a + Ax$, $A \in L(R^n)$, $a \in R^n$, can exist such that $Lx^i = Fx^i$, $i = 0,\ldots, n$.

E 7.2-8. Formulate all of the methods of this section for the fixed-point equation $x = Gx$.

E 7.2-9. Show that the operator J of (22) satisfies the relation (35) while, in general, the operator J of (19) does not.

7.3. MODIFICATION METHODS

Consider again the $(n + 1)$-point sequential secant method

$$x^{k+1} = x^k - A_k^{-1} Fx^k, \qquad k = 0, 1,\ldots, \tag{1}$$

where

$$A_k = \Gamma_k H_k^{-1}, \quad H_k = (p^{k-1},\ldots, p^{k-n}), \quad \Gamma_k = (q^{k-1},\ldots, q^{k-n}), \quad k = 0, 1,\ldots, \tag{2}$$

$$p^i = x^{i+1} - x^i, \qquad q^i = Fx^{i+1} - Fx^i, \tag{3}$$

and the initial points $x^0, x^{-1},\ldots, x^{-n}$ are given. In **7.2.8**, we showed that Γ_{k+1}^{-1} may be obtained in a simple way from Γ_k^{-1} by adding a rank one matrix to a matrix formed by permuting the rows of Γ_k^{-1}. We observe next that A_{k+1} and A_k also differ by a matrix of rank one.

7.3.1. Let the two sets of points $x^k,\ldots, x^{k-n}$ and $x^{k+1},\ldots, x^{k-n+1}$ be in general position and define A_k and A_{k+1} by (2) and (3). If $A_k p^k = -Fx^k$, that is, if (1) holds, then

$$A_{k+1} = A_k - Fx^{k+1}(v^k)^T, \tag{4}$$

where $(v^k)^T$ is the first row of H_{k+1}^{-1}.

Proof. From the relations $A_{k+i} H_{k+i} = \Gamma_{k+i}$, $i = 0, 1$, we obtain directly that

$$A_k p^j = q^j, \qquad j = k - 1,\ldots, k - n,$$

7.3. MODIFICATION METHODS

and
$$A_{k+1}p^j = q^j, \quad j = k,\ldots, k - n + 1,$$

so that
$$(A_{k+1} - A_k)p^j = 0, \quad j = k - 1,\ldots, k - n + 1, \tag{5}$$

while
$$(A_{k+1} - A_k)p^k = q^k - A_k p^k = q^k + Fx^k = Fx^{k+1}. \tag{6}$$

Now, by **7.2.2**, it follows that $p^k,\ldots, p^{k-n+1}$ are linearly independent, and, hence, the relations (5) and (6) completely specify $A_{k+1} - A_k$. Indeed, (5) shows that $A_{k+1} - A_k$ has at most rank one, and thus has the form $A_{k+1} - A_k = u^k(v^k)^T$, with certain $u^k, v^k \in R^n$. But, then, (5) also requires that v^k be orthogonal to $p^{k-1},\ldots, p^{k-n+1}$, and therefore $(v^k)^T$ must be proportional to the first row of H_{k+1}^{-1}. If we take $(v^k)^T$ precisely equal to this first row, then $(v^k)^T p^k = 1$, and, hence, (6) implies that $u^k = Fx^{k+1}$. ∎

This result is not meant to be a computational tool for modifying A_k —the direct modification of Γ_k^{-1} according to **7.2.8** is more efficient. Rather, **7.3.1** suggests the formulation of other iterative methods of the form (1) by requiring that A_{k+1} is to be obtained from A_k by addition of a matrix of rank one. Since any matrix of rank one may be written in the form uv^T, where $u, v \in R^n$, this means that

$$A_{k+1} = A_k + u^k(v^k)^T, \quad u^k, v^k \in R^n, \quad k = 0, 1,\ldots. \tag{7}$$

One of the advantages of allowing the A_k to change according to (7) is that A_{k+1}^{-1} is easily obtained in terms of A_k^{-1}; indeed, if

$$1 + (v^k)^T A_k^{-1} u^k \neq 0, \tag{8}$$

then the Sherman–Morrison formula (2.3.14) shows that

$$A_{k+1}^{-1} = A_k^{-1} - \frac{A_k^{-1} u^k (v^k)^T A_k^{-1}}{1 + (v^k)^T A_k^{-1} u^k}. \tag{9}$$

In the sequel, we shall require that A_0 and all A_k generated by (7) be nonsingular. Since **2.3.11** shows that, when A_k^{-1} exists, (8) is also a necessary condition for the existence of A_{k+1}^{-1}, we see that (8) provides a first constraint on u^k and v^k. This condition is rather weak, however, and may be satisfied by scaling u^k or v^k once their directions are given. Hence, with no further requirements on u^k and v^k, a wide variety of iterative methods is possible.

One way to specify completely the choice of u^k and v^k is to impose the conditions (6) and (5) of the secant method; in terms of u^k and v^k, these take the form

$$u^k(v^k)^T p^k = Fx^{k+1}, \quad k = 0, 1,..., \tag{10}$$

and

$$(v^k)^T p^j = 0, \quad j = k-1,..., k-n+1, \quad k = n-1, n,..., \tag{11}$$

where again

$$p^j = x^{j+1} - x^j = -A_j^{-1} Fx^j, \quad j = 0, 1,.... \tag{12}$$

Since (11) holds only for $k \geq n-1$, we also impose the conditions

$$(v^k)^T p^j = 0, \quad j = 0,..., k-1, \quad k < n-1. \tag{13}$$

This latter condition gives a certain latitude in choosing $v^0,..., v^{n-2}$; in particular, v^k is only required to be orthogonal to the k vectors $p^0,..., p^{k-1}$. However, whenever p^j, $j = k-1,..., k-n$, $k \geq n$, are linearly independent, then A_k is completely determined.

7.3.2. Assume that the matrices $A_0,..., A_m$ are nonsingular for some $m \geq n-1$ and satisfy (7), (10), (11), and (13), where the vectors $p^0,..., p^m$ satisfy (12) and $p^m,..., p^{m-n+1}$ are linearly independent. Then

$$A_{m+1} = (q^m,..., q^{m-n+1})(p^m,..., p^{m-n+1})^{-1}, \tag{14}$$

where $q^i = Fx^{i+1} - Fx^i$.

Proof. Using (7), (10), and (12), it follows that

$$A_{j+1} p^j = A_j p^j + u^j(v^j)^T p^j = -Fx^j + Fx^{j+1} = q^j, \quad j = 0,..., m,$$

so that, by (11),

$$A_{m+1} p^j = A_m p^j + u^m(v^m)^T p^j = A_m p^j = \cdots = A_{j+1} p^j = q^j,$$
$$j = m-1,..., m-n+1.$$

Hence,

$$A_{m+1}(p^m,..., p^{m-n+1}) = (q^m,..., q^{m-n+1}).$$

and, since $p^m,..., p^{m+n-1}$ are linearly independent, (14) holds. ∎

This result shows that the algorithm specified by (1), (7), (10), (11), and (13) must reduce to the $(n+1)$-point sequential secant method after, at most, n steps provided that $p^0,..., p^{n-1}$ are linearly independent. Thus, in order to obtain different methods, it is necessary to relinquish

7.3. MODIFICATION METHODS

some of the conditions (10) and (11). One possibility is to require that v^k be orthogonal to only l previous p^j, or, in other words, to replace (11) by

$$(v^k)^T p^j = 0, \quad j = k-1, \ldots, k-l, \quad l < n-1. \tag{15}$$

As an extreme case, assume that $l = 0$ so that only (10), together with the nonsingularity requirement (8), is specified; then we may write (10) as

$$(v^k)^T p^k \neq 0, \quad u^k = Fx^{k+1}/(v^k)^T p^k, \quad k = 0, 1, \ldots, \tag{16}$$

so that, since $p^k + A_k^{-1} F x^{k+1} = A_k^{-1} q^k$, (8) becomes

$$(v^k)^T A_k^{-1} q^k \neq 0, \quad k = 0, 1, \ldots. \tag{17}$$

If both (16) and (17) hold, then (9) gives

$$A_{k+1}^{-1} = A_k^{-1} - \frac{A_k^{-1} F x^{k+1} (v^k)^T A_k^{-1}}{(v^k)^T A_k^{-1} q^k}, \quad k = 0, 1, \ldots. \tag{18}$$

Note that (16) and (17) specify that v^k is not orthogonal to either p^k or $A_k^{-1} q^k$, and, in particular, that neither p^k nor $A_k^{-1} q^k$ is zero. If $p^k = 0$, then $A_k^{-1} F x^k = 0$, and, thus x^k is already a solution of $Fx = 0$, so that we may ignore this case. Now, assume that $A_k^{-1} q^k \neq 0$. Then the choice

$$v^k = \alpha_k p^k - A_k^{-1} q^k, \quad k = 0, 1, \ldots, \tag{19}$$

where $\alpha_k \neq 0$ if $(p^k)^T A_k^{-1} q^k = 0$ and

$$\alpha_k \neq (p^k)^T A_k^{-1} q^k / (p^k)^T p^k, \quad \alpha_k \neq (A_k^{-1} q^k)^T (A_k^{-1} q^k)/(p^k)^T A_k^{-1} q^k$$

otherwise, ensures that both (16) and (17) are satisfied. Hence, (19), together with the specification

$$u^k = Fx^{k+1}/(v^k)^T p^k, \quad k = 0, 1, \ldots, \tag{20}$$

guarantees that the sequence $\{A_k\}$, defined by (7), remains nonsingular.

We could, of course, consider many other conditions that might be imposed upon u^k and v^k. Instead, we shall continue to require that (6) holds, but that the difference $A_{k+1} - A_k$ now be of rank m. Since any $n \times n$ matrix of rank m may be written in the form UV^T, where U and V are $n \times m$ matrices of rank m, this means that

$$A_{k+1} = A_k + U_k V_k^T, \quad U_k V_k \in L(R^m, R^n), \quad \text{rank } U_k = \text{rank } V_k = m, \tag{21}$$

and

$$U_k V_k^T p^k = F x^{k+1}, \quad k = 0, 1, \ldots. \tag{22}$$

Geometrically, (21) means, of course, that $A_{k+1}x = A_k x$ for all x in an $(n - m)$-dimensional subspace of R^n, namely, the null space of $U_k V_k^T$; hence, the requirement (21) allows the operator A_k to change at each step only on a subspace of dimension m.

If A_{k+1} is given by (21), then the Sherman–Morrison–Woodbury formula (2.3.13) shows that

$$A_{k+1}^{-1} = A_k^{-1} - A_k^{-1} U_k (I + V_k^T A_k^{-1} U_k)^{-1} V_k^T A_k^{-1}, \quad k = 0, 1, \ldots, \quad (23)$$

provided that

$$(I + V_k^T A_k^{-1} U_k)^{-1}, \quad k = 0, 1, \ldots, \quad \text{exist.} \quad (24)$$

Indeed, (24) is also a necessary condition that A_{k+1}^{-1} exists if A_k^{-1} exists, so that we shall also require that (24) holds. Clearly, (8) and (9) are the special case of (24) and (23) for $m = 1$.

The modification formula (23) is not nearly as convenient as (9) even when $m = 2$. However, since the matrices

$$Y_k = -A_k^{-1} U_k (I + V_k^T A_k^{-1} U_k)^{-1} \qquad W_k = (A_k^{-1})^T V_k$$

are clearly $n \times m$ matrices of rank m, (23) shows that

$$B_{k+1} = B_k + Y_k W_k^T, \quad k = 0, 1, \ldots, \quad (25)$$

where we have set

$$B_k = A_k^{-1}, \quad k = 0, 1, \ldots. \quad (26)$$

Thus, if the difference $A_{k+1} - A_k$ is of rank m, then the difference $B_{k+1} - B_k$ is also of rank m. Since it is really the B_k that are of interest, it is then convenient to work directly with them, rather than with the A_k, and to specify that B_k is to change by (25), where Y_k, W_k are $n \times m$ matrices of rank m. Of course, we will also require that all B_k remain nonsingular, so that the condition (24) becomes

$$(I + W_k^T A_k Y_k)^{-1}, \quad k = 0, 1, \ldots, \quad \text{exist.} \quad (27)$$

Finally, because of $p^k = -A_k^{-1} F x^k$, the condition (22) assumes the form

$$A_{k+1} p^k = A_k p^k + F x^{k+1} = q^k,$$

which may be written in terms of the B_k as

$$B_{k+1} q^k = p^k, \quad k = 0, 1, \ldots, \quad (28)$$

7.3. MODIFICATION METHODS

or, equivalently, since $p^k = -B_k F x^k = -B_k[F x^{k+1} - q^k]$, as

$$Y_k W_k^T q^k = -B_k F x^{k+1}. \tag{29}$$

We remark that working in terms of the B_k is only a convenience and the modifications $Y_k W_k^T$ may be converted, by the formula (2.3.13), to the corresponding modifications $U_k V_k^T$ for the A_k.

In the remainder of this section, we shall deal only with the case $m = 2$; hence,

$$Y_k = (y^{1,k}, y^{2,k}), \quad W_k = (w^{1,k}, w^{2,k}), \quad w^{i,k}, y^{i,k} \in R^n, \quad i = 1, 2, \tag{30}$$

and the condition (29) becomes

$$\gamma_{1,k} y^{1,k} + \gamma_{2,k} y^{2,k} = -B_k F x^{k+1}, \quad k = 0, 1, \ldots, \tag{31}$$

where

$$\gamma_{i,k} = (w^{i,k})^T q^k, \quad i = 1, 2.$$

Now, $-B_k F x^{k+1} = p^k - B_k q^k$, and one simple way to satisfy (31) is to take

$$\gamma_{i,k} \neq 0, \quad i = 1, 2; \quad y^{1,k} = (1/\gamma_{1,k}) p^k; \quad y^{2,k} = -(1/\gamma_{2,k}) B_k q^k; \tag{32}$$

so that

$$B_{k+1} = B_k + \frac{p^k (w^{1,k})^T}{(w^{1,k})^T q^k} - \frac{B_k q^k (w^{2,k})^T}{(w^{2,k})^T q^k}. \tag{33}$$

The nonsingularity requirement (27) must still be satisfied, and a simple calculation shows that (27) holds if and only if

$$(w^{2,k})^T F x^k \neq 0, \quad k = 0, 1, \ldots. \tag{34}$$

Hence, the prescription (33) generates a sequence of nonsingular matrices B_k whose differences satisfy $(B_{k+1} - B_k) q^k = -B_k F x^{k+1}$, provided that the vectors $w^{i,k}$ satisfy

$$(w^{i,k})^T q^k \neq 0, \quad i = 1, 2; \quad (w^{2,k})^T F x^k \neq 0; \quad k = 0, 1, \ldots. \tag{35}$$

We note that it is not necessary for $w^{1,k}$ and $w^{2,k}$ to be linearly independent. In fact, if $w^{1,k} = w^{2,k} = w^k$, then (33) and (35) reduce to

$$B_{k+1} = B_k - \frac{B_k F x^{k+1} (w^k)^T}{(w^k)^T q^k}, \quad (w^k)^T q^k \neq 0, \quad (w^k)^T F x^k \neq 0,$$
$$k = 0, 1, \ldots, \tag{36}$$

since $p^k - B_k q^k = -B_k F x^{k+1}$. Moreover, if we set $(w^k)^T = (v^k)^T B_k$, then (36) becomes

$$B_{k+1} = B_k - \frac{B_k F x^{k+1} (v^k)^T B_k}{(v^k)^T B_k q^k}, \quad (v^k)^T B_k q^k \neq 0, \quad (v^k)^T p^k \neq 0,$$
$$k = 0, 1, \ldots, \quad (37)$$

since $p^k = -A_k^{-1} F x^k$, and this is simply the method described by (16)–(18).

Finally, we consider another special choice of $w^{1,k}$ and $w^{2,k}$ in (33) —namely,

$$w^{1,k} = p^k, \quad w^{2,k} = B_k^T q^k, \quad k = 0, 1, \ldots, \quad (38)$$

which yields

$$B_{k+1} = B_k + \frac{p^k (p^k)^T}{(p^k)^T q^k} - \frac{B_k q^k (q^k)^T B_k}{(q^k)^T B_k q^k}, \quad k = 0, 1, \ldots. \quad (39)$$

For the solution of gradient equations, (39) is the specification of the *Davidon–Fletcher–Powell* method, which will be discussed in more detail in **8.2**. In particular, in conjunction with (39), we require that

$$(p^k)^T q^k \neq 0, \quad (q^k)^T B_k q^k \neq 0, \quad (q^k)^T B_k F x^k \neq 0, \quad k = 0, 1, \ldots, \quad (40)$$

and it will be shown in **8.2.4** that (40) may be satisfied under certain natural conditions on F.

NOTES AND REMARKS

NR 7.3-1. Methods of the type discussed in this section have been proposed recently by several authors under the title "quasi-Newton methods." The first of these methods to be studied, other than the $(n+1)$-point sequential secant method, was the Davidon–Fletcher–Powell method (Davidon [1959], Fletcher and Powell [1963]), in which the change in A_k^{-1} is prescribed by (39). This method will be discussed in more detail in **8.2**.

NR 7.3-2. The secant-type method defined by (1), (7), (10), (11), and (13) was proposed by Barnes [1965]. However, as **7.3.2** (due to Barnes) shows, this method reduces to the $(n+1)$-point sequential secant method after, at most, n steps, provided the vectors $x^{j+1} - x^j$, $j = 0, \ldots, n-1$ remain linearly independent, so that the only difference is that the initial matrix A_0 may be picked arbitrarily.

NR 7.3-3. The special case of the method defined by (16), in which v^k is taken equal to p^k, was given by Broyden [1965]. However, this choice of v^k may not

7.3. MODIFICATION METHODS

satisfy the nonsingularity criterion (17), which now reduces to $(p^k)^T A_k^{-1} q^k \neq 0$, $k = 0, 1, \ldots$; the same comments apply, of course, to the method (36) if one chooses there $w^k = q^k$.

NR 7.3-4. The general prescription (33) for a change of B_k by a matrix of rank two was proposed independently by Broyden [1967] and Zeleznik [1968] as an attempt to generalize the Davidon–Fletcher–Powell method.

NR 7.3-5. Rather than the particular choice (38), which leads to the formula (39), one could consider taking $w^{1,k}$ and $w^{2,k}$ as linear combinations of p^k and $B_k^T q^k$; that is,

$$w^{i,k} = \alpha_{i,k} p^k + \beta_{i,k} B_k^T q^k, \quad i = 1, 2, \quad k = 0, 1, \ldots .$$

Concrete methods based on this general choice have not yet been considered.

NR 7.3-6. All of the methods of this section may, of course, be modified to $x^{k+1} = x^k - \omega_k A_k^{-1} F x^k$, where the parameter ω_k is chosen to ensure that $\|F x^{k+1}\| \leq \|F x^k\|$; this is the form in which they are generally discussed in the literature.

NR 7.3-7. Another motivation for the condition

$$A_{k+1}(x^{k+1} - x^k) = F x^{k+1} - F x^k \tag{41}$$

arises from the divided-difference operator requirement (7.2.35) discussed in **NR 7.2-6.** In this case, $Fy - Fx = J(x, y - x)(y - x)$ for all x, y in a certain domain, so that, with $A_{k+1} = J(x^k, x^k - x^{k+1})$, the condition (41) is automatically satisfied.

NR 7.3-8. Note that the orthogonality requirements $(A_{k+1} - A_k) p^j = 0$, $j = k - 1, \ldots, k - l$, together with the condition $A_{k+1} p^k = q^k$, imply that $A_{k+1} p^i = q^i$, $i = k, \ldots, k - l$, which is perhaps a more natural assumption to use from the outset.

NR 7.3-9. A related method based on the iteration

$$x^{k+1} = x^k - F'(x^k)^+ F x^k, \quad k = 0, 1, \ldots,$$

is discussed by Fletcher [1968]. Here, A^+ denotes a generalized inverse (see **NR 8.5-4**) of A, and $F: R^n \to R^m$. Fletcher proposes an iteration of the form

$$x^{k+1} = x^k - A_k^+ F x^k, \quad k = 0, 1, \ldots,$$

where A_k^+ is to be modified by a formula similar to (33).

EXERCISES

E 7.3-1. Assume that A_k and A_{k+1} are defined by (2) and (3), and that H_k, H_{k+1}, Γ_k, and Γ_{k+1} are all nonsingular. Show that

$$A_{k+1}^{-1} = A_k^{-1} + [A_k^{-1} F x^{k+1} (v^k)^\mathrm{T} A_k^{-1} / (1 + (v^k)^\mathrm{T} p^k)],$$

where $(v^k)^\mathrm{T}$ is the first row of H_{k+1}^{-1} and $p^k = x^{k+1} - x^k$.

E 7.3-2. Carry out the calculation that leads to (34); that is, show that, for $y^{1,k}$ and $y^{2,k}$ given by (32), the matrix

$$I + \begin{pmatrix} (w^{1,k})^\mathrm{T} A_k y^{1,k} & (w^{1,k})^\mathrm{T} A_k y^{2,k} \\ (w^{2,k})^\mathrm{T} A_k y^{1,k} & (w^{2,k})^\mathrm{T} A_k y^{2,k} \end{pmatrix}$$

is nonsingular if and only if $(w^{2,k})^\mathrm{T} F x^k \neq 0$.

7.4. GENERALIZED LINEAR METHODS

The Newton and secant methods discussed in the previous sections represent generalizations of one-dimensional iteration methods. In this section, we will consider another class of processes which arise from iterative methods used for linear systems of equations. Intuitively, we may think of a *generalized linear method* as one which, upon application to a linear system $Ax = b$, reduces to a feasible iteration for the solution of that system. Newton's method, for example, would not fall into this category, since the iteration becomes $x^{k+1} = x^k - A^{-1}(Ax^k - b) = A^{-1}b$, which cannot be considered a feasible iteration for the linear system. Another characteristic of these generalized linear methods, as we shall see in Chapter 10, is that they all exhibit only a linear rate of convergence except under special circumstances.

We shall begin by examining an extension of one of the basic iterative methods for a linear system of equations $Ax = b$. If we assume that the diagonal elements a_{ii} of A are all nonzero, then the *Gauss–Seidel* iteration (sometimes called the method of successive displacements) for $Ax = b$ proceeds as follows: Suppose that the kth iterate $x^k = (x_1{}^k, \ldots, x_n{}^k)^\mathrm{T}$ and the first $i - 1$ components $x_1^{k+1}, \ldots, x_{i-1}^{k+1}$ of the $(k + 1)$th iterate x^{k+1} have been determined. Then, to obtain the next component x_i^{k+1}, the linear equation in the single variable x_i,

$$\sum_{j=1}^{i-1} a_{ij} x_j^{k+1} + a_{ii} x_i + \sum_{j=i+1}^{n} a_{ij} x_j{}^k = b_i, \qquad (1)$$

is solved for x_i and the solution taken as x_i^{k+1}.

7.4. GENERALIZED LINEAR METHODS

In order to write the Gauss–Seidel iteration in compact form, we first write the matrix A as

$$A = D - L - U, \qquad (2)$$

where D, L, U are diagonal, strictly lower triangular, and strictly upper triangular matrices, respectively; that is, L and U both have zero diagonal elements. The assumption that the diagonal elements of A are nonzero then ensures that $(D - L)^{-1}$ exists, and it is easy to verify that the Gauss–Seidel iteration may be written as

$$x^{k+1} = (D - L)^{-1}[Ux^k + b] = x^k - (D - L)^{-1}(Ax^k - b), \quad k = 0, 1, \ldots. \qquad (3)$$

In many problems for which the Gauss–Seidel iteration is potentially useful, it is desirable to make an important modification. Rather than taking x_i^{k+1} as the solution x_i of (1), we set

$$x_i^{k+1} = x_i^k + \omega(x_i - x_i^k) \qquad (4)$$

where ω is a *relaxation parameter*. If we substitute the solution x_i of (1) into (4), we obtain, after some rearranging,

$$a_{ii}x_i^{k+1} + \omega \sum_{j=1}^{i-1} a_{ij}x_j^{k+1} = (1 - \omega) a_{ii}x_i^k - \omega \sum_{j=i+1}^{n} a_{ij}x_j^k + \omega b_i,$$

and, by means of (2), we may then write the iteration in the form

$$\begin{aligned} x^{k+1} &= (D - \omega L)^{-1}[(1 - \omega) D + \omega U] x^k + \omega(D - \omega L)^{-1}b \\ &= x^k - \omega(D - \omega L)^{-1}(Ax^k - b), \quad k = 0, 1, \ldots. \end{aligned} \qquad (5)$$

Clearly, (5) reduces to the Gauss–Seidel iteration (2) when $\omega = 1$. For all values of ω, we shall call (5) the *SOR* or *successive overrelaxation* method, although in the literature this nomenclature is sometimes reserved for the case $\omega > 1$.

One way to utilize the *SOR* iteration in connection with nonlinear equations is as a means of approximating solutions of the linear systems which must be solved to carry out Newton's method. In this case, we would have a composite *Newton–SOR* iteration, with Newton's method as the *primary* iteration and SOR as the *secondary* iteration.

In order to give a concrete representation of this iteration, we first express the iterate x^{k+1} of (5) in terms of x^0. If we set

$$B = \omega^{-1}(D - \omega L), \qquad C = \omega^{-1}[(1 - \omega) D + \omega U], \qquad H = B^{-1}C,$$

then (5) can be written as

$$x^{k+1} = Hx^k + B^{-1}b = H^{k+1}x^0 + (H^k + H^{k-1} + \cdots + I)B^{-1}b$$
$$= x^0 + (H^{k+1} - I)x^0 + (H^k + \cdots + I)B^{-1}(Ax^0 - Ax^0 + b). \quad (6)$$

Then, since $B^{-1}A = B^{-1}(B - C) = I - H$ and

$$(I + \cdots + H^k)(I - H) = I - H^{k+1},$$

it follows that

$$x^{k+1} = x^0 - \omega(H^k + \cdots + I)(D - \omega L)^{-1}(Ax^0 - b). \quad (7)$$

We now continue with the description of the Newton–SOR iteration for the mapping $F: D \subset R^n \to R^n$. Assume that x^k has been determined. Then the next Newton iterate would be the solution of the linear system

$$F'(x^k)x = F'(x^k)x^k - Fx^k, \quad (8)$$

and we wish to approximate this Newton iterate by means of the SOR iteration. Therefore, the Jacobian matrix is decomposed as

$$F'(x^k) = D_k - L_k - U_k \quad (9)$$

where, again, D_k, L_k, and U_k are diagonal, strictly lower triangular, and strictly upper triangular matrices, respectively. We next assume that D_k is nonsingular and define

$$H_k = [D_k - \omega_k L_k]^{-1}[(1 - \omega_k)D_k + \omega_k U_k], \quad (10)$$

for some relaxation parameter ω_k. Now apply the SOR iteration to (8) and denote the SOR iterates by $x^{k,m}$. Then, by (7), noting that now $b = F'(x^k)x^k - Fx^k$ and $A = F'(x^k)$, we have, for $m = 1, 2, \ldots$,

$$x^{k,m} = x^{k,0} - \omega_k[H_k^{m-1} + \cdots + I][D_k - \omega_k L_k]^{-1}[F'(x^k)(x^{k,0} - x^k) + Fx^k]. \quad (11)$$

In general, the most reasonable choice of the starting approximation $x^{k,0}$ for the SOR iteration will be x^k. Hence, if we set $x^{k,0} = x^k$ and $x^{k,m_k} = x^{k+1}$, where m_k is the total number of SOR iterations at the kth stage, then (11) reduces to

$$x^{k+1} = x^k - \omega_k(H_k^{m_k-1} + \cdots + I)(D_k - \omega_k L_k)^{-1}Fx^k, \quad (12)$$

which we shall call a *general Newton–SOR iteration*.

7.4. GENERALIZED LINEAR METHODS

If the SOR iteration (11) is terminated by some convergence criterion, as, for example, $\| x^{k,m} - x^{k,m-1} \| \leq \epsilon_k$, then the number m_k of secondary iterations will not be known in advance. On the other hand, we may also choose the m_k before the iteration begins. For the simplest such choice, $m_k = 1$, $k = 0, 1,...$, (12) reduces to the *one-step Newton–SOR iteration*:

$$x^{k+1} = x^k - \omega_k (D_k - \omega_k L_k)^{-1} F x^k, \qquad k = 0, 1,..., \qquad (13)$$

which is (5) when $Fx = Ax - b$. More generally, we can set $m_k = m$, $k = 0, 1,...$, to obtain the *m-step Newton–SOR* method, or choose any sequence m_k in advance, as, for example, $m_k = k + 1$, $k = 0, 1,....$

The Newton–SOR iterations are only a special case of the general principle of combining nonlinear iterative methods with linear iterations in order to form composite methods. There are, of course, numerous other iterative methods for linear equations, and, at least in principle, the majority of these methods may be combined with Newton's method in a manner analogous to the use of SOR. On the other hand, Newton's method itself may be replaced as the primary iteration by, for example, any of the secant methods of the previous sections. Rather than describe very many of these possible methods, we shall develop a general generating principle and then illustrate it by a few additional composite methods of this type.

One of the basic principles used in the generation and analysis of iterative methods for linear equations is that of *splitting*. That is, if $Ax = b$ is the linear system, then A is decomposed, or split, into the sum

$$A = B - C \qquad (14)$$

of two matrices, where B is nonsingular and the linear system $Bx = d$ is "easy" to solve. Then an iterative method is defined by

$$x^{k+1} = B^{-1} C x^k + B^{-1} b = x^k - B^{-1}(Ax^k - b), \qquad k = 0, 1,.... \qquad (15)$$

For example, the SOR method (5) is defined by the splitting

$$B = \omega^{-1}(D - \omega L), \qquad C = \omega^{-1}[(1 - \omega) D + \omega U], \qquad (16)$$

where, again, $A = D - L - U$ is the decomposition of A into diagonal, strictly lower triangular, and strictly upper triangular parts. Similarly, the *Jacobi iteration* (sometimes called the method of simultaneous displacements),

$$x^{k+1} = D^{-1}(L + U)x^k + D^{-1}b, \qquad (17)$$

is defined by the splitting $B = D$, $C = L + U$.

Another important class of linear iterative methods, the alternating-direction methods, may also be formulated in this way. For example, the method of Peaceman and Rachford has the form

$$x^{k+\frac{1}{2}} = (H + \alpha I)^{-1}(\alpha I - V)x^k + (H + \alpha I)^{-1}b,$$
$$x^{k+1} = (V + \alpha I)^{-1}(\alpha I - H)x^{k+\frac{1}{2}} + (V + \alpha I)^{-1}b, \qquad k = 0, 1, \ldots. \qquad (18)$$

Here $A = H + V$, α is a parameter, and it is assumed that the linear systems $(H + \alpha I)x = d$ and $(V + \alpha I)x = d$ are "easy" to solve; in many cases of interest, H and V are either tridiagonal or equivalent to tridiagonal matrices, and meet this assumption. The iteration (18) may also be written in the form (15) with

$$B = (2\alpha)^{-1}(H + \alpha I)(V + \alpha I), \qquad C = (2\alpha)^{-1}(H - \alpha I)(V - \alpha I)$$

if $\alpha \neq 0$. In fact,

$$B - C = (2\alpha)^{-1}(HV + \alpha H + \alpha V + \alpha^2 I) - (2\alpha)^{-1}(HV - \alpha H - \alpha V + \alpha^2 I)$$
$$= H + V = A, \qquad (19)$$

and, since $(H + \alpha I)^{-1}$ and $H - \alpha I$ commute, we have

$$x^{k+1} = (V + \alpha I)^{-1}(H - \alpha I)(H + \alpha I)^{-1}(V - \alpha I)x^k$$
$$\quad + (V + \alpha I)^{-1}[(\alpha I - H)(H + \alpha I)^{-1} + I]b$$
$$= B^{-1}Cx^k + (V + \alpha I)^{-1}[\alpha I - H + \alpha I + H](H + \alpha I)^{-1}b$$
$$= x^k - 2\alpha[V + \alpha I]^{-1}[H + \alpha I](Ax^k - b), \qquad k = 0, 1, \ldots. \qquad (20)$$

Now, for any linear iterative method of the form (15), and any nonlinear iterative method which can be written as

$$x^{k+1} = x^k - A_k^{-1}Fx^k, \qquad (21)$$

we may consider a composite nonlinear–linear method, completely analogous to the Newton–SOR method (12), by subjecting the matrix A_k to the splitting of the linear method:

$$A_k = B_k - C_k. \qquad (22)$$

The derivation of the Newton–SOR method carries over, with obvious changes, to this more general case, and, provided that the matrices B_k are nonsingular, we obtain the explicit representation

$$x^{k+1} = x^k - [I + \cdots + H_k^{m_k-1}]B_k^{-1}Fx^k \quad H_k = B_k^{-1}C_k, \quad k = 0, 1, \ldots, \qquad (23)$$

7.4. GENERALIZED LINEAR METHODS

which generalizes (12). This represents the composite iteration in which, at the kth stage starting from x^k, m_k steps of the secondary linear iteration are taken in order to approximate a solution of the linear system $A_k x = A_k x^k - F x^k$ defined by the primary iteration. The Newton–SOR iteration is the special case of (23) in which $A_k = F'(x^k)$ and $B_k = \omega_k^{-1}(D_k - \omega_k L_k)$. As other examples of (23), we mention the *m-step Newton–Jacobi* iteration

$$x^{k+1} = x^k - [I + \cdots + H_k^{m-1}] D_k^{-1} F x^k, \qquad k = 0, 1, \ldots, \qquad (24)$$

where $F'(x^k) = D_k - C_k$ is the decomposition of $F'(x^k)$ into diagonal and nondiagonal parts and $H_k = D_k^{-1} C_k$, and the *one-step Newton–Peaceman–Rachford* method

$$x^{k+1} = x^k - 2\alpha [V_k + \alpha I]^{-1} [H_k + \alpha I]^{-1} F x^k, \qquad k = 0, 1, \ldots, \qquad (25)$$

where, now $F'(x^k) = H_k + V_k$. This latter iteration results immediately from (20) by noting that $Ax - b \equiv F'(x^k) x - F'(x^k) x^k + F x^k = 0$ is the linear system to be solved, and, thus, $Ax^k - b = F x^k$. General Newton–Peaceman–Rachford iterations, in which starting from x^k m_k steps of the Peaceman–Rachford method are taken, may also be written in the form (23) by means of (20) (see **E 7.4-6**).

The primary iteration (21) may represent not only the Newton method, but also, for example, any of the secant or Steffensen iterations discussed in **7.2**. Hence, as special cases of (23), we may consider Secant–SOR methods, Steffensen–Peaceman–Rachford methods, etc. The formulation of a few of these possible methods is left to **E 7.4-4**.

We have, up to now, considered the linear iterative methods only in their traditional role of solving linear systems. However, it is also possible to give a direct extension of these methods to nonlinear equations.

If we interpret the Gauss–Seidel iteration (1) in terms of obtaining x_i^{k+1} as the solution of the ith equation of the system with the other $n - 1$ variables held fixed, then we may immediately consider the same prescription for nonlinear equations. That is, if $F: D \subset R^n \to R^n$ has components $f_1, \ldots, f_n$, then the basic step of the *nonlinear Gauss–Seidel* iteration is to solve, in analogy to (1), the ith equation

$$f_i(x_1^{k+1}, \ldots, x_{i-1}^{k+1}, x_i, x_{i+1}^k, \ldots, x_n^k) = 0, \qquad (26)$$

for x_i, and to set $x_i^{k+1} = x_i$. Thus, in order to obtain x^{k+1} from x^k, we solve successively the n one-dimensional nonlinear equations (26) for $i = 1, \ldots, n$. More generally, we may set $x_i^{k+1} = x_i^k + \omega_k(x_i - x_i^k)$ in order to obtain a *nonlinear SOR method*.

In a completely analogous fashion, the Jacobi iteration (17) can be extended. Written in component form, (17) becomes

$$x_i^{k+1} = (1/a_{ii}) \sum_{j=1, j \neq i}^{n} a_{ij} x_j^k + b_i, \quad i = 1,\ldots, n, \quad k = 0, 1,\ldots,$$

which is equivalent to solving the ith equation for x_i with all other x_j held at the values x_j^k. Hence, the kth stage of the *nonlinear Jacobi* iteration may be defined by solving the equations

$$f_i(x_1^k,\ldots, x_{i-1}^k, x_i, x_{i+1}^k, \ldots, x_n^k) = 0, \quad i = 1,\ldots, n, \qquad (27)$$

for x_i and setting $x_i^{k+1} = x_i$, $i = 1,\ldots, n$.

Note that the above prescriptions have meaning, of course, only if the equations (26) or (27) have unique solutions in some specific domain under consideration; in later chapters, we shall consider various conditions on F, analogous to the requirement that $a_{ii} \neq 0$, $i = 1,\ldots, n$, in the linear case, which ensure that this is true. However, even when these equations have unique solutions, there will not, in general, be an explicit representation or even a finite algorithm for obtaining them, and a one-dimensional iterative method must be applied. If we use, for example, the one-dimensional Newton method to approximate a solution of (26), then the Newton method plays the role of a secondary iteration, while the SOR method is now the primary iteration. Hence, the roles of the Newton and SOR iterations have been reversed from those in the previous discussion.

We can, in principle, write explicit formulas analogous to (12) for a general SOR–Newton iteration in which $m_{k,i}$ Newton steps are applied to (26) to yield an approximate solution $\hat{x}_i$ and then x_i^{k+1} is set equal to $x_i^k + \omega(\hat{x}_i - x_i^k)$. However, these formulas became rather cumbersome (see **E 7.4-2**), and we restrict our attention to the *one-step SOR–Newton* (or Gauss–Seidel–Newton if $\omega = 1$) iteration in which $m_{k,i} = 1$, $i = 1,\ldots, n$, $k = 0, 1,\ldots$, and the Newton step is always taken from x_i^k. In this case, it is easily verified (**E 7.4-3**) that the explicit form of the iteration is

$$x_i^{k+1} = x_i^k - \omega f_i(x^{k,i})/\partial_i f_i(x^{k,i}), \quad i = 1,\ldots, n, \quad k = 0, 1,\ldots, \qquad (28)$$

where we have set

$$x^{k,i} = (x_1^{k+1},\ldots, x_{i-1}^{k+1}, x_i^k,\ldots, x_n^k)^T. \qquad (29)$$

Similarly, the one-step Jacobi–Newton iteration is defined by

$$x_i^{k+1} = x_i^k - f_i(x^k)/\partial_i f_i(x^k), \quad k = 0, 1,\ldots, \quad i = 1,\ldots, n, \qquad (30)$$

7.4. GENERALIZED LINEAR METHODS

which (E 7.4-5) is identical to the one-step Newton–Jacobi iteration (24). Note that (28) and (30) reduce to the SOR and Jacobi iterations, respectively, when $Fx = Ax - b$ (E 7.4-2).

In the iterations (28) and (30), we may, of course, replace Newton's method by other one-dimensional iterations. For example, corresponding to (28), we may formulate the *one-step SOR–secant* method

$$x_i^{k+1} = x_i^k - \omega \left[\frac{f_i(x^{k,i}) - f_i(x^{k,i} + [x_i^{k-1} - x_i^k] e^i)}{x_i^k - x_i^{k-1}} \right]^{-1} f_i(x^{k,i}), \quad (31)$$

or the *one-step SOR–Steffensen* iteration

$$x_i^{k+1} = x_i^k - \omega \left[\frac{f_i(x^{k,i}) - f_i(x^{k,i} - f_i(x^{k,i}) e^i)}{f_i(x^{k,i})} \right]^{-1} f(x^{k,i}), \quad (32)$$

where $x^{k,i}$ is again defined by (29). Analogous Jacobi–secant and Jacobi–Steffensen methods are immediately formulated by replacing $x^{k,i}$ in (31) and (32) by x^k. In the context of Section 7.4, these composite Jacobi iterations may be written in the form

$$x^{k+1} = x^k - J(x^k, x^k - x^{k-1})^{-1} F x^k, \quad k = 0, 1, \ldots, \quad (33)$$

and

$$x^{k+1} = x^k - J(x^k, F x^k)^{-1} F x^k, \quad k = 0, 1, \ldots, \quad (34)$$

where $J(x, h)$ is the diagonal $n \times n$ matrix

$$J(x, h) = \text{diag}\{[f_i(x) - f_i(x - h_i e^i)]/h_i\}.$$

We conclude this section with analogous extensions of the Peaceman–Rachford iteration (18). Consider now a decomposition of the form $F = F_H + F_V$; then, corresponding to (18), the *nonlinear* Peaceman–Rachford method is defined by

$$\begin{aligned} \alpha x^{k+\frac{1}{2}} + F_H x^{k+\frac{1}{2}} &= \alpha x^k - F_V x^k, \\ \alpha x^{k+1} + F_V x^{k+1} &= \alpha x^{k+\frac{1}{2}} - F_H x^{k+\frac{1}{2}}, \end{aligned} \quad k = 0, 1, \ldots. \quad (35)$$

If $Fx = Ax - b$ and $A = H + V$ is a decomposition of the matrix A, then the splitting $F_H x = Hx - \frac{1}{2}b$ and $F_V x = Vx - \frac{1}{2}b$ causes (35) to reduce to (18). In general, however, both F_H and F_V will be nonlinear, and one step of the nonlinear Peaceman–Rachford method requires the solution of the two nonlinear systems (35) for $x^{k+\frac{1}{2}}$ and x^{k+1}, respectively. If we attempt to solve these nonlinear equations by Newton's method, we may consider a general Peaceman–Rachford–Newton method in which m_k Newton steps are taken in the first equation, starting from x^k, to obtain an approximation $\hat{x}^{k+\frac{1}{2}}$ to $x^{k+\frac{1}{2}}$, and then n_k steps are taken on the second equation, starting from $\hat{x}^{k+\frac{1}{2}}$, to obtain the next iterate x^{k+1} of the composite method. If $m_k \equiv n_k \equiv 1$, then the resulting *one-step Peaceman–Rachford–Newton* iteration takes the explicit form

$$x^{k+\frac{1}{2}} = x^k - [\alpha I + F_H'(x^k)]^{-1} F x^k,$$
$$x^{k+1} = x^{k+\frac{1}{2}} - [\alpha I + F_V'(x^{k+\frac{1}{2}})]^{-1} F x^{k+\frac{1}{2}}, \qquad k = 0, 1, \ldots. \qquad (36)$$

Clearly, various other composite methods utilizing, for example, the secant and Steffensen iterations are possible.

NOTES AND REMARKS

NR 7.4-1. The literature on iterative methods for linear equations is vast; for a fairly complete summary and for a detailed treatment of the linear methods of this section, see Forsythe and Wasow [1960], Varga [1962], and Wachspress [1966].

NR 7.4-2. The use of the SOR iteration as a secondary iteration has been standard for a number of years. Various analyses and computational results are reported by Bellman and Kalaba [1965, p. 118], Greenspan [1965a], Greenspan and Yohe [1963], Greenspan and Parter [1965], and Ortega and Rheinboldt [1967a], among others.

NR 7.4-3. The nonlinear Gauss–Seidel method was first discussed rigorously by Bers [1953] for discrete analogs of mildly nonlinear elliptic boundary value problems of the form $\Delta u = f(x, y, u, u_x, u_y)$, and, more recently, by Schechter [1962, 1968], Ortega and Rockoff [1966], Ortega and Rheinboldt [1970a], and several other authors.

NR 7.4-4. The SOR–Newton and Jacobi–Newton iterations were proposed by Lieberstein [1959, 1960], and were subsequently studied by a number of authors including Schechter [1962, 1968], Bryan [1964], and Ortega and Rockoff [1966]. Computational results are given by Greenspan [1965a], and Greenspan and Yohe [1963].

NR 7.4-5. The use of the Peaceman–Rachford method as a secondary iteration was discussed by Douglas [1961, 1962] for discrete analogs $Ax = \phi x$ of a mildly nonlinear boundary value problem for $\Delta u = f(u)$. Here, the outer iteration was of the Picard type: $(A + \gamma I) x^{k+1} = \phi x^k + \gamma x^k$. A similar situation was analyzed by Gunn [1964a, 1965]. More recently, Kellogg [1969] has considered a nonlinear Peaceman–Rachford iteration similar to (35) (see **12.1.7**).

7.4. GENERALIZED LINEAR METHODS

NR 7.4-6. The one-step Jacobi–secant and Jacobi–Steffensen iterations (33) and (34) have been considered by Wegge [1966], but from a rather different point of view. Indeed, Wegge considers these to be natural generalizations of the corresponding one-dimensional iterations to n dimensions. A similar interpretation would lead to the conclusion that the Jacobi–Newton iteration is a natural extension of the Newton method. Note that these interpretations are contrary to the thesis of this section that such methods are natural extensions of the linear method, while the nonlinear method plays only a subsidiary role.

NR 7.4-7. Each iteration of the SOR–Newton process requires only the evaluation of each f_i as well as of the n partial derivatives $\partial_i f_i$. By contrast, the one-step Newton–SOR process requires the evaluation of $D_k - \omega_k L_k$, that is, of $n(n-1)/2$ partial derivatives, as well as the solution of a triangular system of equations. The m-step Newton–SOR process, with $m > 1$, requires the evaluation of all partial derivatives and the solution of m triangular systems. Note that the one-step Newton–SOR iteration avoids the evaluation of U_k because of the choice of x^k as the starting approximation for the secondary iteration; in general, to obtain the ith secondary iterate $x^{k,i}$, we solve, for $\omega = 1$, the the triangular system

$$[D(x^k) - L(x^k)][x^{k,i} - x^k] = U(x^k)[x^{k,i-1} - x^k] - Fx^k.$$

In Section **10.3**, it will be shown that the asymptotic rate of convergence of the one-step SOR–Newton and one-step Newton–SOR iterations are identical, while that of the m-step Newton–SOR iteration is m times the rate of convergence of the one-step process. Based on this asymptotic rate of convergence and the above operation counts, the one-step SOR–Newton iteration would seem to be the most efficient of the SOR methods discussed in this section, if the derivatives are not too difficult to evaluate. If derivatives are difficult to evaluate, then a three- or four-step Newton–SOR process might be more efficient.

In addition to the rate of convergence results, various other convergence theorems for these methods are given in Chapters 12–14.

NR 7.4-8. We have considered only the cyclic Gauss–Seidel process in which the equations $f_i = 0$ are solved in their natural order. Schechter [1962] considers "free-steering" methods in which the equations are solved in an essentially arbitrary order (see Section **14.6**). Another procedure, classical for linear equations and sometimes called the *Seidel method*, is to choose the equation which has the largest function value. Then the kth stage is

(a) Choose j so that $|f_j(x^k)| \geq |f_i(x^k)|$, $i = 1,\ldots, n$.

(b) Solve $f_j(x_1^k,\ldots, x_{j-1}^k, x_j, x_{j+1}^k,\ldots, x_n^k) = 0$ for x_j

and set $x^{k+1} = x^k + \omega(x_j - x_j^k) e^j$.

Note that here, in contrast with our description of the cyclic SOR method, it is convenient to consider the solution of a single one-dimensional equation as a complete iteration step.

NR7.4-9. One way of summarizing the "Gauss–Seidel principle" is that new information is used as soon as it is available. In the Newton–SOR processes, we have not fully utilized this principle, since F' is evaluated at x^k. Consider the one-step Newton–SOR iteration (13) written in the implicit component form:

$$x_i^{k+1} = x_i^k - \frac{\omega}{\partial_i f_i(x^k)} \left[\sum_{j=1}^{i-1} \partial_j f_i(x^k)(x_j^{k+1} - x_j^k) + f_i(x^k) \right], \quad i = 1,\ldots, n.$$

Here, f_i and $\partial_j f_i$, $j = 1,\ldots, i-1$, are not needed until the computation of x_i^{k+1}, and could be evaluated at $x^{k,i} = (x_1^{k+1},\ldots, x_{i-1}^{k+1}, x_i^k,\ldots, x_n^k)^T$. This leads to a modified iteration with the kth stage defined by

$$x_i^{k+1} = x_i^k - \frac{\omega}{\partial_i f_i(x^{k,i})} \sum_{j=1}^{i-1} [\partial_j f_i(x^{k,i})(x_j^{k+1} - x_j^k) + f_i(x^{k,i})], \quad i = 1,\ldots, n.$$

There seems to be no natural m-step version of this method.

Another manifestation of the Gauss–Seidel principle involves the iteration $x^{k+1} = Gx^k$, $k = 0, 1,\ldots$. Here, a Gauss–Seidel-type modification of this procedure would be the process

$$x_i^{k+1} = g_i(x_1^{k+1},\ldots, x_{i-1}^{k+1}, x_i^k,\ldots, x_n^k), \quad i = 1,\ldots, n, \quad k = 0, 1,\ldots,$$

in which the new components x_j^{k+1} are used as they are available.

NR 7.4-10. In applying the SOR methods to a linear problem $Ax = b$, it is advantageous, in certain situations, to use *block* (line) methods. In this case, A is partitioned in the form

$$A = \begin{pmatrix} A_{11} & \cdots & A_{1m} \\ \vdots & & \vdots \\ A_{m1} & \cdots & A_{mm} \end{pmatrix},$$

where A_{ij} is a $q_i \times q_j$ matrix and $n = q_1 + \cdots + q_m$. The block SOR method is then defined by (3), where now

$$D = \begin{pmatrix} A_{11} & & O \\ & \ddots & \\ O & & A_{mm} \end{pmatrix}, \quad L = -\begin{pmatrix} 0 & & & O \\ A_{21} & & & \\ \vdots & \ddots & & \\ A_{m1} & \cdots & A_{m,m-1} & 0 \end{pmatrix},$$

$$U = -\begin{pmatrix} 0 & A_{12} & \cdots & A_{1m} \\ & & \ddots & \vdots \\ & & & A_{m-1,m} \\ O & & & 0 \end{pmatrix}. \tag{38}$$

7.4. GENERALIZED LINEAR METHODS

For a further discussion of block methods for linear equations, see, for example, Varga [1962].

The block SOR method extends in a natural way to nonlinear equations. First, note that if $F'(x^k) = D_k - L_k - U_k$ is partitioned in block form, as in (38), then (12), (10) gives a general block Newton–SOR iteration. On the other hand, the nonlinear SOR method can be extended to block form by partitioning x as $x = (x^1, \ldots, x^m)$, with $x^i \in R^{q_i}$, and by grouping, correspondingly, the components f_i of F into mappings $F_i : R^n \to R^{q_i}$, $i = 1, \ldots, m$. Then

$$F_i((x^1)^{k+1}, \ldots, (x^i)^{k+1}, (x^{i+1})^k, \ldots, (x^m)^k) = 0, \quad i = 1, \ldots, m$$

describes a nonlinear block Gauss–Seidel process in which a complete iteration requires the solution of m nonlinear system of dimension q_i, $i = 1, \ldots, m$; a nonlinear block SOR iteration is formulated analogously. Similarly, a one-step block Gauss–Seidel–Newton method would take the form

$$(x^i)^{k+1} = (x^i)^k - [\partial_i F_i(y^{k,i})]^{-1} F_i y^{k,i}, \quad k = 0, 1, \ldots, \quad i = 1, \ldots, m, \quad (39)$$

where $y^{k,i} = ((x^1)^{k+1}, \ldots, (x^{i-1})^{k+1}, (x^i)^k, \ldots, (x^m)^k)$, and $\partial_i F_i(x)$ denotes the Jacobian matrix of F_i with respect to x^i.

These block methods may be advantageous for systems of the form $Ax + \phi x = b$ which arise from elliptic boundary value problems. For example, in the model problem (1.2.5), A is of the form (37) with $A_{ij} = 0$ for $|j - i| > 1$, each A_{ii} is itself tridiagonal, and $\phi'(x)$ is diagonal. Hence, (39) requires only the solution of tridiagonal linear systems.

NR 7.4-11. It is important to realize that, although there is, in general, no finite algorithm to solve the one-dimensional equations in the nonlinear SOR method, it may be, in practice, just as easy to obtain an approximate solution of these equations with a prescribed accuracy, as to obtain an approximate solution of the linear systems arising in Newton's method. At the same time, it is fruitful to interpret the nonlinear SOR method as a guide to explicit and feasible composite methods such as the one-step SOR–Newton iteration. Similar remarks apply, of course, to the nonlinear Jacobi and Peaceman–Rachford iterations.

NR 7.4-12. The splitting principle used in describing the composite nonlinear–linear iterations may also be applied directly to the nonlinear operator F. Indeed, most of the iterative methods, including those of the previous sections, may be viewed in this way. For example, in analogy to the decomposition $A = B - C$ for linear operators, we may consider the splitting $F = P - R$, and the corresponding iteration $Px^{k+1} = Rx^k$, $k = 0, 1, \ldots$. Here, P is not necessarily linear, although it is assumed that, for each k, the equation $Px = Rx^k$ has a solution x^{k+1} which is unique in some domain under consideration. The splitting

$F = I - G$ which leads to the successive approximation scheme $x^{k+1} = Gx^k$, $k = 0, 1,...$, is the simplest example of this, while the splitting $F = A - (A - F)$, for some nonsingular linear operator A, leads to the general parallel-chord method (7.1.2).

More generally, we may consider the sequence of splittings

$$F = P_k - R_k, \qquad (40)$$

and the corresponding iteration

$$P_k x^{k+1} = R_k x^k, \qquad k = 0, 1,.... \qquad (41)$$

The most important special case of (40) is when P_k is a nonsingular linear operator A_k so that (41) becomes $x^{k+1} = x^k - A_k^{-1} Fx^k$.

On the other hand, if we define the components $P_{k,i}$ of P_k by

$$P_{k,i}(x) = f_i(x_1,..., x_i, x_{i+1}^k,..., x_n^k), \qquad i = 1,..., n,$$

and set $R_k x = P_k x - Fx$, then (41) describes the nonlinear Gauss–Seidel iteration (26).

NR 7.4-13. A natural extension of the nonlinear Gauss–Seidel iteration is the following method:

(a) Choose an index $i \in \{1,..., n\}$ and a vector $p^k \in R^n$;
(b) Solve $f_i(x^k - \alpha p^k) = 0$ for $\alpha = \alpha_k$; (42)
(c) Set $x^{k+1} = x^k - \alpha_k p^k$.

For $p^k = e^{k(\text{mod} n)+1}$ this reduces to the nonlinear Gauss–Seidel method (with a different numbering of the iterates), but it is also related to the method of functional averaging (see Luchka [1963]) developed for integral and differential equations. For a nonlinear mapping $G: R^n \to R^n$, the natural interpretation of the method of functional averaging for the fixed-point equation $x = Gx$ requires x^{k+1} to be obtained from

$$x^{k+1} = G(x^k + P_k(x^{k+1} - x^k)), \qquad k = 0, 1,..., \qquad (43)$$

where $\{P_k\}$ is a sequence of projection operators. If P_k is chosen as the projection onto the $k' = k(\text{mod} n) + 1$ coordinate direction, that is, $P_k x = x_{k'} e^{k'}$, then (43) reduces to one step of the Gauss–Seidel process. More generally, if P_k is a projection onto p^k, then (43) is equivalent to (42) for the fixed-point equation $x = Gx$.

On the other hand, (42) leads naturally to composite methods analogous to the SOR–Newton iterations. In fact, if one Newton step is taken toward a solution of (42), we obtain the explicit iteration

$$x^{k+1} = x^k - [f_i(x^k)/f_i'(x^k) p^k] p^k.$$

7.4. GENERALIZED LINEAR METHODS

Similarly, we may consider a Jacobi-type iteration based on (42) in which x^k is not modified until after a complete sweep through the equations; that is, we solve

$$f_i(x^k - \alpha_i p^{k,i}) = 0, \qquad i = 1,\ldots, n, \qquad (44)$$

and set $x^{k+1} = x^k - \sum_{i=1}^n \alpha_i p^{k,i}$. Again, we may take only one Newton step toward the solutions of (44), and this gives the iteration

$$x^{k+1} = x^k - \sum_{i=1}^n [f_i(x^k)/f_i'(x^k) p^{k,i}] p^{k,i}, \qquad k = 0, 1,\ldots. \qquad (45)$$

A special case of (45) is a method of Hart and Motzkin [1956] in which $p^{k,i} = [f_i'(x^k)]^T$, so that (45) may be written as

$$x^{k+1} = x^k - F'(x^k)^T D_k^{-1} F x^k,$$

where

$$D_k = \text{diag}(\|f_1'(x^k)\|_2^2, \ldots, \|f_n'(x^k)\|_2^2).$$

This is related to the Gauss–Newton iteration to be discussed in the next chapter.

NR 7.4-14. It is possible to formulate the nonlinear Peaceman–Rachford iteration in a somewhat more general way. For the linear equation $Ax = b$ and the decomposition $A = H + V$, the iteration (18) may be written as

$$H_1 x^{k+\frac{1}{2}} = -V_1 x^k, \qquad V_2 x^{k+1} = -H_2 x^{k+\frac{1}{2}},$$

where

$$H_1 x = (H + \alpha I) x - \tfrac{1}{2}b, \qquad V_1 x = (V - \alpha I) x - \tfrac{1}{2}b,$$
$$H_2 x = (H - \alpha I) x - \tfrac{1}{2}b, \qquad V_2 x = (V + \alpha I) x - \tfrac{1}{2}b.$$

If $Fx = Ax - b$, then, clearly, $Fx = (H_1 + V_1) x = (H_2 + V_2) x$.

When $F: R^n \to R^n$ is nonlinear, we may consider a corresponding double splitting of F:

$$F = F_{H,1} + F_{V,1} = F_{H,2} + F_{V,2},$$

where $F_{H,i}$ and $F_{V,i}$, $i = 1, 2$, are, in general, all nonlinear, and the iteration is

$$F_{H,1} x^{k+\frac{1}{2}} = -F_{V,1} x^k, \qquad F_{V,2} x^{k+1} = -F_{H,2} x^{k+\frac{1}{2}}, \qquad k = 0, 1,\ldots. \qquad (46)$$

Clearly, (46) reduces to (35) if

$$F_{H,1} = F_H x + \alpha x, \quad F_{V,1} = F_V x - \alpha x, \quad F_{H,2} = F_H x - \alpha x, \quad F_{V,2} = F_V x + \alpha x.$$

NR 7.4-15. All of the methods of this section have been generalizations of methods for solving linear systems of equations. However, the central method for linear systems, Gaussian elimination, has not been considered. A straightforward—but, in general, impractical—generalization of this important method proceeds as follows. Give the system $f_i(x_1, \ldots, x_n) = 0$, $i = 1, \ldots, n$, solve the

the first equation, say, for x_1 in terms of $x_2, \ldots, x_n$ and substitute the resulting function $x_1 = h_1(x_2, \ldots, x_n)$ into the remaining equations to obtain $n - 1$ new functions of $x_2, \ldots, x_n$:

$$f_i^2(x_2, \ldots, x_n) \equiv f_i(h_1(x_2, \ldots, x_n), x_2, \ldots, x_n), \qquad i = 2, \ldots, n. \tag{47}$$

Now, repeat the process by solving $f_2^2 = 0$ for x_2 to obtain $x_2 = h_2(x_3, \ldots, x_n)$ and substitute $h_2(x_3, \ldots, x_n)$ into f_j^2, $j = 3, \ldots, n$, to obtain $n - 2$ new functions f_j^3, $j = 3, \ldots, n$, of $x_3, \ldots, x_n$. Continuing in this fashion, we obtain, finally, a single equation $f_n{}^n(x_n) = 0$ whose solution $x_n{}^*$ is the nth component of the solution x^*. The remaining components are then obtained by "back substitution" into the functions h_i; that is, $x_i{}^* = h_i(x_{i+1}^*, \ldots, x_n{}^*)$, $i = n - 1, \ldots, 1$.

This procedure reduces to Gaussian elimination (without interchanges) when the f_i are affine. In this case, or when n is small and the f_i are suitably simple, the mappings h_i are obtainable explicitly. In general, however, it will be necessary to proceed by an iterative procedure, first to solve the equation $f_n{}^n(x_n) = 0$, and then to obtain the remaining $x_i{}^*$. But, as we show next, even the approximate solution of $f_n{}^n(x_n) = 0$ presents almost insurmountable difficulties, since just the *evaluation* of $f_n{}^n(x_n{}^0)$ at some trial point $x_n{}^0$ requires the solution of the $(n-1) \times (n-1)$ system of equations $f_i(x_1, \ldots, x_{n-1}, x_n{}^0) = 0$, $i = 1, \ldots, n - 1$. For simplicity, we shall show that only for $n = 3$.

Because $f_3{}^3(x_3{}^0) = f_3{}^2(h_2(x_3{}^0), x_3{}^0)$, it is necessary to obtain $h_2(x_3{}^0)$, and, for this, we must solve the equation

$$f_2^2(x_2, x_3{}^0) = 0 \tag{48}$$

for x_2. Since

$$f_2^2(x_2, x_3{}^0) = f_2(h_1(x_2, x_3{}^0), x_2, x_3{}^0),$$

it follows that it is necessary to obtain $h_1(x_2, x_3{}^0)$, which requires solving

$$f_1(x_1, x_2, x_3{}^0) = 0 \tag{49}$$

for x_1 in terms of x_2 and $x_3{}^0$. But solving (48) and (49) is equivalent to solving the 2×2 system

$$f_1(x_1, x_2, x_3{}^0) = 0, \qquad f_2(x_1, x_2, x_3{}^0) = 0.$$

The simplest approach to overcome the above difficulties is to linearize the system $Fx = 0$ by expansion about an initial approximation x^0 and then to apply Gaussian elimination to the linearized system $F'(x^0)(x - x^0) + Fx^0 = 0$; this is simply Newton's method.

Alternatively, Brown [1966] (see also Brown and Conte [1967] and Brown [1969]) suggests that the individual components $f_k{}^k$ be linearized as the Gaussian elimination proceeds. Again, let x^0 be an initial approximation, and set

$$h_1(x_2, \ldots, x_n) = x_1{}^0 - [1/\partial_1 f_1(x^0)][f_1(x^0) + \sum_{i=2}^{n} \partial_i f_1(x^0)(x_i - x_i{}^0)].$$

7.4. GENERALIZED LINEAR METHODS

Hence, h_1 is simply the result of setting $f_1(x^0) + f_1'(x^0)(x - x^0) = 0$ and solving for x_1. Now, again define $n - 1$ new functions f_i^2, $i = 2,..., n$, by (47), and repeat the above process with f_2^2 in order to get $h_3(x_3 ,..., x_n)$. Continuing in this way, we obtain the functions $h_i(x_{i+1} ,..., x_n)$, $i = 1,..., n - 1$, and a final single linear equation in x_n whose solution is now, because of the linearization, only an approximation x_n^1 to x_n^*. The remaining components of x^1 are then found from $x_i^1 = h_i(x_{i+1}^1 ,..., x_n^1)$, $i = n - 1,..., 1$. The whole process is now repeated to obtain the next approximation x^2, and so on. Brown has given convergence proofs for this procedure and shown that the rate of convergence is quadratic under the expected conditions.

The above process requires the partial derivatives of the reduced functions f_i^i, and a simple calculation shows that these may be obtained by recursion relations in terms of the partial derivatives of the original f_i (although not evaluated at x^0). Indeed, it is easy to see that, in general, all n^2 partial derivatives $\partial_j f_i$ need to be evaluated, so that, in this respect, there is no gain over Newton's method. However, Brown [1966] has also shown that, by approximating the derivatives of f_i^i by certain difference quotients, the evaluation of the n^2 partial derivatives $\partial_j f_i$ can be replaced by only $\tfrac{1}{2}n^2$ additional evaluations of the f_i. This difference method shows promise, but has not, as yet, been rigorously analyzed.

EXERCISES

E 7.4-1. Show that the one-step Newton–SOR iteration reduces to Newton's method when $F: R^1 \to R^1$ and $\omega = 1$.

E 7.4-2. Formulate a general SOR–Newton iteration in which $m_{k,i}$ Newton steps are taken to obtain an approximate solution $\hat{x}_i^k$ to (26) and x_i^{k+1} is then set equal to $x_i^k + \omega(\hat{x}_i^k - x_i^k)$. Show that this reduces to the SOR iteration if F is affine.

E 7.4-3. If $m_{k,i} = 1$, for all i, k in **E 7.4-2**, verify that the iteration takes the explicit form (28).

E 7.4-4. With a secant or Steffensen method as primary iteration and the SOR or Peaceman-Rachford method as secondary one, write down various combined methods corresponding to the Newton-SOR iteration.

E 7.4-5. (a) Show that the one-step Newton–Jacobi and one-step Jacobi–Newton processes are identical, but that the m-step processes, in general, differ. (b) Show that the one-step SOR–Newton and one-step Newton–SOR processes are, in general, different, but are identical for systems of the form $Ax = \phi x$, where $A \in L(R^n)$ and $\phi'(x)$ is diagonal.

E 7.4-6. Formulate a general m_k-step Newton–Peaceman–Rachford method corresponding to (25).

E 7.4-7. Formulate all of the methods of this section in terms of the fixed-point equation $x = Gx$.

7.5. CONTINUATION METHODS

As will be seen in later chapters, most of the methods discussed so far will, in general, converge to a solution x^* of $Fx = 0$ only if the initial approximations are sufficiently close to x^*. The continuation methods to be discussed in this section may be considered as an attempt to widen the domain of convergence of a given method, or, alternatively, as a procedure to obtain sufficiently close starting points.

In many practical situations, the problem may depend in a natural way on a parameter t, and, when this parameter is set equal to a specific value, say, 1, then the mapping F results, while for $t = 0$ the resulting system $F_0 x = 0$ has a known solution x^0. More precisely, instead of a single mapping F, we have an entire family $H: D \times [0, 1] \subset R^{n+1} \to R^n$ such that

$$H(x, 0) = F_0 x, \quad H(x, 1) = Fx, \quad \forall x \in D, \tag{1}$$

where a solution x^0 of $H(x, 0) = 0$ is known, and the equation $H(x, 1) = 0$ is to be solved.

Even if F does not depend naturally on a suitable parameter t, we can always define a family H satisfying (1) in various ways. For example, we might set

$$H(x, t) = tFx + (1 - t)F_0 x, \quad x \in D, \quad t \in [0, 1], \tag{2}$$

with a given mapping F_0 for which a solution x^0 of $F_0 x = 0$ is known, or

$$H(x, t) = Fx + (t - 1)Fx^0, \quad x \in D, \quad t \in [0, 1], \tag{3}$$

where x^0 is fixed. Note that (3) is obtained from (2) for $F_0 x = Fx - Fx^0$. No matter how H was obtained, let us consider the equation

$$H(x, t) = 0, \quad t \in [0, 1], \tag{4}$$

and suppose that (4) has for each $t \in [0, 1]$ a solution $x = x(t)$ which depends continuously on t. In other words, assume that there exists a continuous mapping $x: [0, 1] \to D$ such that

$$H(x(t), t) = 0, \quad \forall t \in [0, 1]. \tag{5}$$

7.5. CONTINUATION METHODS

Then x describes a space curve in R^n with one end point at some given point x^0 and with the other end point at a solution $x^* = x(1)$ of $Fx = H(x, 1) = 0$.

As an example of how it may be ascertained that such a continuous solution curve exists, we have the following simple consequence of the norm-coerciveness theorem **5.3.9** or the Hadamard theorem **5.3.10**. Recall, from Definition **5.3.6**, that a mapping $F: R^n \to R^n$ is norm-coercive if $\lim_{\|x\|\to\infty} \|Fx\| = \infty$.

7.5.1. Let $F: R^n \to R^n$ be continuously differentiable on R^n and assume that $F'(x)$ is nonsingular for all $x \in R^n$. Assume, also, that either F is norm-coercive or that $\|F'(x)^{-1}\| \leq \beta$ for all $x \in R^n$. Then, for any fixed $x^0 \in R^n$ there exists a unique mapping $x: [0, 1] \to R^n$ such that (5) holds for H defined by (3). Moreover, x is continuously differentiable and

$$x'(t) = -F'(x(t))^{-1}Fx^0, \quad \forall t \in [0, 1], \quad x(0) = x^0. \tag{6}$$

Proof. Clearly, the equation (4) is, in this case, equivalent with

$$Fx = (1 - t)Fx^0, \quad t \in [0, 1]. \tag{7}$$

But either **5.3.9** or **5.3.10**, respectively, shows that F is a homeomorphism from R^n onto R^n, and, therefore, for each $t \in [0, 1]$, (7) has a unique solution

$$x(t) = F^{-1}([1 - t]Fx^0), \quad t \in [0, 1].$$

Now, the inverse function theorem **5.2.1** implies that F^{-1} is also continuously differentiable on R^n. Thus, by the chain rule **3.1.7**, the same holds for $x = x(t)$ on $[0, 1]$, and, moreover, (6) is valid. ∎

Henceforth, we will assume that $H: D \times [0, 1] \subset R^{n+1} \to R^n$ is a given homotopy, that is, a continuous mapping, such that there is a continuous $x: [0, 1] \to D$ for which (5) holds with some known x^0. As a first approach to obtaining $x = x(1)$, we partition the interval $[0, 1]$ by

$$0 = t_0 < t_1 < t_2 \cdots < t_N = 1, \tag{8}$$

and consider solving the problems

$$H(x, t_i) = 0, \quad i = 1, \ldots, N, \tag{9}$$

by some iterative method which uses the solution x^{i-1} of the $(i-1)$th problem as a starting approximation to solve the ith problem. If $t_{i+1} - t_i$ is sufficiently small, then, hopefully, x^{i-1} will be a sufficiently good approximation to x^i so that convergence will occur.

Now, clearly, only finitely many steps of the iterative method can be taken toward the solution of the ith problem (9). If, for example, Newton's method is used for solving (9), and if $m_i \geq 1$ steps are taken, then we have the explicit representation

$$x^{i,k+1} = x^{i,k} - \partial_1 H(x^{i,k}, t_i)^{-1} H(x^{i,k}, t_i), \qquad k = 0, 1,\ldots, m_i - 1,$$
$$x^{1,0} = x^0, \qquad x^{i+1,0} = x^{i,m_i}, \qquad i = 1,\ldots, N - 1. \tag{10}$$

where ∂_1 denotes the partial derivative with respect to x (see 5.2.2). In Section **10.4**, we shall show, under suitable conditions on H, that a partition (8) of $[0, 1]$ and integers $m_1,\ldots, m_N$ can be found such that the entire sequence $x^{i,k}$ of (10) is well-defined, and, moreover, that the Newton iterates

$$x^{k+1} = x^k - \partial_1 H(x^k, 1)^{-1} H(x^k, 1), \qquad k = k_0,\ldots, \quad x^{k_0} = x^{N,m_N}, \tag{11}$$

converge to $x(1)$.

As an example of (10) and (11), consider the homotopy (3) and suppose that only one step of the Newton iteration is taken toward the solution of each problem (9). Then we have the process

$$x^{k+1} = x^k - F'(x^k)^{-1} [Fx^k + (t_k - 1) Fx^0], \qquad k = 0, 1,\ldots, N - 1,$$
$$x^{k+1} = x^k - F'(x^k)^{-1} Fx^k, \qquad k = N, N + 1,\ldots. \tag{12}$$

Note that, in this discussion, the use of Newton's method is only illustrative, and the same principles may be applied to other iterative processes; see **E 7.5-1**.

We consider next a somewhat different approach to the solution of (4). Assume that the mapping $x\colon [0, 1] \to D$ satisfying (5) is continuously differentiable on $[0, 1]$ and that H has continuous partial derivatives with respect to x and t. If we define

$$\phi(t) = H(x(t), t), \qquad \forall t \in [0, 1], \tag{13}$$

then it follows from **E 5.2-5** and the chain rule **3.1.7** that ϕ is continuously differentiable on $[0, 1]$ and that

$$\phi'(t) = \partial_1 H(x(t), t) x'(t) + \partial_2 H(x(t), t), \qquad \forall t \in [0, 1].$$

Since $x = x(t)$ was assumed to satisfy (5), we have $\phi'(t) = 0$ for all t, and, hence, x satisfies the differential equation

$$\partial_1 H(x(t), t) x'(t) = -\partial_2 H(x(t), t), \qquad \forall t \in [0, 1]. \tag{14}$$

7.5. CONTINUATION METHODS

Conversely, if $x\colon [0, 1] \to R^n$ is a continuously differentiable solution of the differential equation (14), and satisfies the initial condition $H(x(0), 0) = 0$, then, with ϕ again defined by (13), the mean-value theorem **3.2.3** shows that

$$\| H(x(t), t) \| = \| \phi(t) - \phi(0) \| \leqslant \sup_{0 \leqslant s \leqslant t} \| \phi'(s) \| = 0$$

so that $H(x(t), t) = 0$ for $t \in [0, 1]$. Therefore, a solution of the differential equation (14) under the initial condition $H(x(0), 0) = 0$ gives a solution of the functional equation (4).

In the sequel, we shall assume that $\partial_1 H$ is nonsingular for all x, t under consideration and write (14) in the form

$$x'(t) = -\partial_1 H(x, t)^{-1} \partial_2 H(x, t), \quad \forall t \in [0, 1], \quad H(x(0), 0) = 0. \tag{15}$$

Note that when H is defined by (3), the differential equation (15) is the same as (6), and, under the conditions of **7.5.1**, has a unique solution for any fixed x^0.

We now consider approximating a solution of (14) by numerical integration. Recall that one of the simplest integration schemes is the method of Euler, which, for the differential equation

$$x' = f(x, t), \quad x(0) = x^0, \quad t \in [0, 1],$$

and a partition (8), takes the form

$$x^{k+1} = x^k + (t_{k+1} - t_k) f(x^k, t_k), \quad k = 0, 1, \ldots, N - 1. \tag{16}$$

For Eq. (15), this method becomes

$$x^{k+1} = x^k - (t_{k+1} - t_k) \partial_1 H(x^k, t_k)^{-1} \partial_2 H(x^k, t_k), \quad k = 0, 1, \ldots, N - 1, \tag{17}$$

and in the case of the homotopy (3), and thus Eq. (6), (17) assumes the simple form

$$x^{k+1} = x^k - h_k F'(x^k)^{-1} F x^0, \quad k = 0, 1, \ldots, N - 1, \quad h_k = t_{k+1} - t_k. \tag{18}$$

Under the conditions of **7.5.1**, there is a continuous solution curve $x = x(t)$ of (6), and, if the h_k are sufficiently small, the x^k defined by (18) should approximate this curve. In other words, x^N should, hopefully, be sufficiently close to $x^* = x(1)$ so that, starting from x^N, an iterative process, such as Newton's method, will converge to x^* (Fig. 7.4). In Section **10.4**, we shall give sufficient conditions that this is indeed true.

Note that the process (18) bears a certain resemblance to Newton's method. To analyze the relationship between the two processes, we

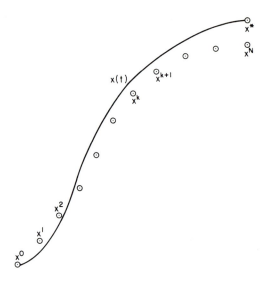

FIGURE 7.4

introduce the change of variable $t = 1 - e^{-\tau}$. Then τ varies from 0 to $+\infty$ as t varies from 0 to 1, and, hence, (3) assumes the form

$$H(x, \tau) = Fx - e^{-\tau}Fx^0, \quad \tau \in [0, \infty).$$

The differential equation corresponding to (16) is therefore

$$x' = -F'(x)^{-1} e^{-\tau}Fx^0 = -F'(x)^{-1}Fx, \quad \tau \in [0, \infty), \quad x(0) = x^0. \tag{19}$$

If we again integrate (19) by Euler's method, this time using the step-size $h_k = 1$, we obtain

$$x^{k+1} = x^k - F'(x^k)^{-1} Fx^k, \quad k = 0, 1, ..., \tag{20}$$

which is Newton's method.

Note that the use of Euler's method for the numerical integration of (14) is again only illustrative, and more sophisticated methods could also be employed (see **NR 7.5-3** and **E 7.5-2**).

NOTES AND REMARKS

NR 7.5-1. The use of continuation methods—or imbedding methods, as they are also called—for the study of existence problems for operator equations goes back to the last century; see **NR 5.3-1** and Ficken [1951] for a historical summary.

7.5. CONTINUATION METHODS

The earliest application of the continuation techniques for the numerical solution of equations appears to have been made by Lahaye [1934, 1935] for a single equation, using Newton's method to move along the solution curve. Later, Lahaye [1948] also considered systems of equations. The basic idea has been rediscovered several times since; see, for example, Freudenstein and Roth [1963], Deist and Sefor [1967], and, in particular, Sidlovskaya [1958], who discussed the use of Newton's method for a stepwise approximation to the solution curve in Banach spaces.

NR 7.5-2. Independently of the work of Lahaye, Davidenko [1953a, b] introduced the differential equation (15) as a means of solving $H(x, t) = 0$ for $t \in [0, 1]$ and has applied this idea to a wide variety of problems, including integral equations, matrix inversion, determinant evaluation, and matrix eigenvalue problems, as well as systems of equations; see, for example, Davidenko [1965a, b]. Davidenko's approach is now frequently called the method of differentiation with respect to a parameter. There also appears to be earlier and incompletely referenced work of the Russians Fok and Kirija along these lines. Recent applications of the method to two-point boundary value problems have been given by Roberts and Shipman [1967b] and Bosarge [1968].

NR 7.5-3. We have discussed in the text only the simplest numerical method —namely, Euler's method—for numerically integrating the differential equation (15). Bittner [1967], Kizner [1964], Kleinmichel [1968], and Bosarge [1968] have investigated the application of more sophisticated methods, such as the Runge–Kutta methods, in this context.

NR 7.5-4. Gavurin [1958] has used the differential equation (15) as a means of obtaining existence theorems for solutions of $H(x, t) = 0$ in Banach spaces. Additional results along this line have been given by Jakovlev [1964c, 1965], and Meyer [1968]. Gavurin considers the differential equation (15) to be a "continuous iterative process." Such continuous processes have also been studied by Rosenbloom [1956] and Poljak [1964b].

NR 7.5-5. Davis [1966] has discussed the problems arising when $\partial_1 H(x(t), t)$ is singular at some point on the solution curve and considers the use of coordinate transformations for overcoming them. Meyer [1969] attacks the same problem by modifying the homotopy H.

EXERCISES

E 7.5-1. Formulate iterations of the form (10)–(11) and (12) using the Newton–SOR iteration of **7.2** in place of the Newton iteration.

E 7.5-2. Consider the initial value problem $u' = g(u)$, $u(0) = 0$. A well-known Runge–Kutta formula is given by

$$u_{j+1} = u_j + \tfrac{1}{6}h[K_{1j} + 2K_{2j} + 2K_{3j} + K_{4j}]$$

$K_{1j} = g(u_j)$, $K_{2j} = g(u_j + \tfrac{1}{2}K_{1j})$, $K_{3j} = g(u_j + \tfrac{1}{2}K_{2j})$, $K_{4j} = g(u_j + K_{3j})$.

Apply this formula to (6) and (19).

7.6. GENERAL DISCUSSION OF ITERATIVE METHODS

In the preceding sections, we have discussed several concrete iterative processes, and additional ones will be the subject of the next chapter. In subsequent parts, however, it will be useful, on occasion, to refer to iterative processes in a generic way. Therefore, in this section, we give a general definition and classification of such processes.

Intuitively, an iterative process is a rule by which the new iterate x^{k+1} is obtained from the previous iterates. We make this precise in the following way:

7.6.1. Definition. A family of operators $\{G_k\}$,

$$G_k : D_k \subset (R^n)^{k+p} = \underbrace{R^n \times \cdots \times R^n}_{k+p \text{ times}} \to R^n, \quad k = 0, 1, \ldots, \quad (1)$$

defines an *iterative process* $\mathscr{I} = (\{G_k\}, D^*, p)$ with p *initial points* and with *domain* $D^* \subset D_0$, if D^* is not empty and if for any point $(x^0, \ldots, x^{-p+1}) \in D^*$, the sequence $\{x^k\}$ generated by

$$x^{k+1} = G_k(x^k, \ldots, x^{-p+1}), \quad k = 0, 1, \ldots, \quad (2)$$

exists, that is, if $(x^k, \ldots, x^{-p+1}) \in D_k$ for all $k \geq 0$. A point x^* such that $\lim_{k \to \infty} x^k = x^*$ is called a *limit* of the process and the set of all sequences $\{x^k\}$ which are generated by $\mathscr{I}$ and converge to x^* is denoted by $C(\mathscr{I}, x^*)$. ∎

Note that by this definition, an iterative process is not defined unless there is some nonempty set D^* such that, for $(x^0, \ldots, x^{-p+1}) \in D^*$, the entire sequence $\{x^k\}$ may be generated. Of course, if G_k is defined on all of $(R^n)^{k+p}$ for all $k \geq 0$, then this condition is automatically satisfied. But this need not always be the case. Consider, for example, the Newton method (7.1.8) and assume that an x^0 in the domain D of F has been found so that $F'(x^0)$ exists and is nonsingular. This will then guarantee the existence of x^1, but x^1 need not be in D, nor, if it is, does $F'(x^1)^{-1}$ have to

7.6. GENERAL DISCUSSION OF ITERATIVE METHODS

exist. Hence, the process may stop after the first step. In general, it is an extremely difficult problem to determine precisely the domain D^* of an iterative process. However, an important consideration in subsequent chapters will be the specification of certain subsets of D^*, that is, the specification of certain sets of points $(x^0,..., x^{-p+1})$ for which one can guarantee that the whole sequence can be generated.

Iterations of the extremely general form (2) are rarely encountered in practice, and it is useful to introduce a classification of the most important types of processes.

7.6.2. Definition. An iterative process $\mathscr{I} = (\{G_k\}, D^*, p)$ is an *m-step method* if $p = m$ and if the mappings G_k are of the form

$$G_k : D_k \subset (R^n)^m \to R^n, \quad k = 0, 1,....$$

An *m*-step process is *sequential* if the iterates are generated by

$$x^{k+1} = G_k(x^k,..., x^{k-m+1}), \quad k = 0, 1,...,$$

and a sequential *m*-step process is *stationary with iteration function* G if $G_k \equiv G$, $D_k \equiv D$, $k = 0, 1,....$ ∎

The conceptually simplest, and in many ways most important, iterations are the one-step stationary processes which are described by $x^{k+1} = Gx^k$, $k = 0, 1,....$ These include, for example, the Newton iteration (7.1.8), the one-step Newton–SOR iteration (7.4.13) for fixed ω, and the Steffensen iteration (7.2.34). On the other hand, the Newton–SOR iteration (7.4.12) with fixed ω but $m_k \equiv k+1$, $k = 0, 1,...,$ is a nonstationary one-step process with the sequence $\{G_k\}$ defined by

$$G_k x = x - \omega[I + \cdots + H(x)^k][D(x) - \omega L(x)]^{-1} Fx, \quad k = 0, 1,....$$

Here, $F'(x) = D(x) - L(x) - U(x)$ is the decomposition into diagonal, strictly lower triangular, and strictly upper triangular parts, and

$$H(x) = [D(x) - \omega L(x)]^{-1} [(1 - \omega) D(x) + \omega U(x)].$$

In this case, an infinite sequence of mappings G_k is necessary to describe the process.

Any process that is not one-step will be called *multistep*. Examples of multistep methods are the various two-point and $(n + 1)$-point secant methods discussed in **7.2**, where examples were given of both sequential and nonsequential methods of this type.

Although **7.6.1** specifies the meaning of a general iterative process, it is sometimes convenient to rewrite (2) in the implicit form

$$H_k(x^{k+1},\ldots, x^{-p+1}) = 0, \qquad k = 0, 1,\ldots. \tag{3}$$

For example, in some ways, a more natural form of the Newton iteration is

$$F'(x^k)\, x^{k+1} = F'(x^k)\, x^k - Fx^k, \qquad k = 0, 1,\ldots, \tag{4}$$

since, in practice, one would rarely compute $F'(x^k)^{-1}$ explicitly but, instead, would solve the linear system (4). More importantly, processes such as the nonlinear SOR method (7.4.26) occur naturally in the form (3). In this case, a nonlinear problem must be solved in order to obtain the next iterate x^{k+1}, and the iterative process is not completely defined until a subsidiary algorithm for obtaining x^{k+1} is specified; for example, in connection with the nonlinear SOR prescription, we may specify that the one-dimensional problems are to be solved by Newton's method. In general, there are then two main problems in connection with (3):

(a) Given $x^{-p+1},\ldots, x^k$, does there exist an x^{k+1} which satisfies (3)?

(b) If x^{k+1} does exist, does the subsidiary algorithm produce a sequence $x^{k,i}$, $i = 0, 1,\ldots$, which converges to x^{k+1}?

For the purpose of analyzing the first question, it is often convenient to consider (3) as a nonlinear vector difference equation. Hence, the existence of all x^k is equivalent to the existence of a solution $\{x^k\}$ of the difference equation, while the questions of convergence and rate of convergence of the x^k may be considered as questions about the asymptotic behavior of the solution of the difference equation. One advantage of this point of view is that the methods and results of the theory of difference equations may be brought to bear on the problems of iterative processes. Another advantage is that, in some sense, the necessity for giving a complete specification of an iterative method is bypassed. In other words, even if the answers to questions (a) and (b) are affirmative, x^{k+1} will, in general, only be obtained as the limit of an infinite sequence. Therefore, although the requirement (3) together with the specification of the subsidiary algorithm implicitly define the mappings G_k of (2), these G_k cannot, in general, be evaluated. Hence, a true iterative procedure will be obtained only by requiring that the subsidiary algorithm is to take finitely many steps. For example, we may specify, as we did in **7.4** to obtain the one-step Newton–SOR iteration, that the subsidiary algorithm is to take only a fixed number of steps and that the resulting approximation to the solution of (3) is to be taken as the next iterate

7.6. GENERAL DISCUSSION OF ITERATIVE METHODS

x^{k+1} of the process. Alternatively, we may specify that the subsidiary iteration is to proceed until its ith iterate, $x^{k,i}$, satisfies a convergence condition, such as, for example,

$$\| H_k(x^{k,i}, x^k, \ldots, x^{-p+1}) \| \leqslant \epsilon_k .$$

In either case, the analysis of the existence and behavior of a solution $\{x^k\}$ to the difference equation (3) may be considered as an analysis of an "ideal," but usually unattainable, iterative process. But this analysis will, hopefully, yield insight into concrete iterative processes which, in some sense, are approximations to the ideal.

Although these considerations are prompted by our inability to compute a solution of a nonlinear problem in finitely many steps, they also hold for a realistic analysis of processes (for example, Newton's method) in which only linear equations need to be solved to obtain the iterates. For, except in the most trivial cases, the iterates $\{x^k\}$ produced by an implementation of Newton's method on a computer will satisfy the defining relation (4) only approximately, due to roundoff error. Hence, an analysis of the difference equation (4) in the real-number field again represents only an idealization.

NOTES AND REMARKS

NR 7.6-1. The definition **7.6.1** is closely related to a definition of a *numerical process* given by Prager and Vitasek [1963].

NR 7.6-2. It is not necessary for the mapping H_k of (3) to be linear in x^{k+1} in order that x^{k+1} may be represented explicitly. Consider, for example, the one-step SOR-Newton process (7.4.28), which may be represented by the difference equations $H(x^{k+1}, x^k) = 0$, $k = 0, 1, \ldots$, with the components h_i of H defined by

$$h_i(x^{k+1}, x^k) = \partial_i f_i(x^{k,i})(x_i^{k+1} - x_i^k) + f_i(x^{k,i}), \qquad i = 1, \ldots, n,$$

and $x^{k,i} = (x_1^{k+1}, \ldots, x_{i-1}^{k+1}, x_i^k, \ldots, x_n^k)$. Here, H will, in general, be nonlinear in x^{k+1}, but, clearly, an explicit representation of x^{k+1} in terms of x^k may be easily obtained from (7.4.28), at least in principle (see also **10.3**).

Chapter 8 / MINIMIZATION METHODS

8.1. PARABOLOID METHODS

As we saw in Chapters 1 and 4, the problem of solving a system of nonlinear equations may be replaced by an equivalent problem of minimizing a functional $g: D \subset R^n \to R^1$. In this chapter, we describe the most commonly used iterative methods for performing such minimizations. As mentioned in the introduction to Part III, the convergence analysis of most of these methods will be the topic of Chapter 14, although, in some cases, we shall discuss here the existence of the next iterate if this provides insight into the method under discussion.

Consider a quadratic functional

$$g: R^n \to R^1, \qquad g(x) = c - b^T x + \tfrac{1}{2} x^T A x. \tag{1}$$

If $A \in L(R^n)$ is a symmetric, positive definite matrix, then g has a unique global minimizer x^* which is the solution of the linear system

$$g'(x)^T = Ax - b = 0. \tag{2}$$

In this section, we shall discuss various iterative methods for minimizing a nonquadratic functional $g: R^n \to R^1$ by replacing g at the kth stage of the iteration by a quadratic functional g_k which approximates g in a neighborhood of the kth iterate x^k. The minimizer of g_k is then taken to be the next iterate x^{k+1}.

One of the simplest ways of obtaining such a quadratic functional g_k is by means of the Taylor expansion of g at x^k. Suppose that

8.1. PARABOLOID METHODS

$g: D \subset R^n \to R^1$ is twice F-differentiable at $x^k \in D$, and let $H_g(x)$ denote the Hessian matrix (3.3.2) of g. Then we have, by **3.3.12**, that, for every $x \in D$,

$$g(x) = g(x^k) + g'(x^k)(x - x^k) + \tfrac{1}{2}(x - x^k)^T H_g(x^k)(x - x^k) + R(x - x^k),$$

where $\lim_{h \to 0} R(h)/\| h \|^2 = 0$. Hence, for x sufficiently close to x^k, the quadratic functional

$$g_k(x) = g(x^k) + g'(x^k)(x - x^k) + \tfrac{1}{2}(x - x^k)^T H_g(x^k)(x - x^k) \quad (3)$$

represents an approximation of g near x^k.

By **3.3.4**, we know that the second derivative $g''(x^k)$, and, hence, also the Hessian matrix $H_g(x^k)$, is symmetric; if it is also positive definite, then the unique global minimizer of g_k is the solution of the linear system

$$H_g(x^k)(x - x^k) = g'(x^k)^T,$$

and we take

$$x^{k+1} = x^k - H_g(x^k)^{-1} g'(x^k)^T \quad (4)$$

as the next iterate. Note that this is simply Newton's method for the equation $Fx = g'(x)^T = 0$, and therefore a Newton step for this system is equivalent to obtaining the minimum of the osculating paraboloid g_k at x^k. If $H_g(x^k)$ is not positive definite but nonsingular, then a Newton step (4) corresponds to obtaining the unique critical point of the osculating quadratic functional g_k.

In case $H_g(x^k)$ is singular, or indefinite, it is sometimes useful to consider, in place of (3), the approximating quadratic

$$g_k(x) = g(x^k) + g'(x^k)(x - x^k) + \tfrac{1}{2}(x - x^k)^T [H_g(x^k) + \lambda_k I](x - x^k). \quad (5)$$

If the scalar λ_k is chosen so that $H_g(x^k) + \lambda_k I$ is positive definite, then g_k has a unique global minimizer x^{k+1}. With $Fx = g'(x)^T$, this corresponds to the modified Newton iteration

$$x^{k+1} = x^k - [F'(x^k) + \lambda_k I]^{-1} Fx^k, \quad k = 0, 1, \ldots, \quad (6)$$

already mentioned in **7.1**.

More generally, we may consider approximating quadratics of the form

$$g_k(x) = g(x^k) + b_k^T(x - x^k) + \tfrac{1}{2}(x - x^k)^T A_k(x - x^k), \quad (7)$$

where A_k is symmetric. In particular, in analogy with the discrete Newton and secant methods of **7.1** and **7.2**, we may obtain the quadratic

(7) by interpolation, and, in general, this requires $n + \tfrac{1}{2}n(n+1)$ interpolating points $x^{k,i}$ in order to compute the unknown coefficients of b_k and A_k from the system of linear equations

$$g_k(x^{k,i}) = g(x^{k,i}), \quad i = 1,\ldots, 1 + n + \tfrac{1}{2}n(n+1). \tag{8}$$

The system (8) is, in general, difficult to solve, and we restrict our attention here to a special choice of interpolation points.

Consider the $1 + n + \tfrac{1}{2}n(n+1)$ points

$$x^k, \quad x^k + h_i e^i, \quad i = 1,\ldots, n, \quad x^k + h_i e^i + h_j e^j, \quad i = 1,\ldots, n, \quad j = i,\ldots, n, \tag{9}$$

and the corresponding differences

$$\Delta_j g(x^k) = g(x^k + h_j e^j) - g(x^k),$$

$$\Delta_i \Delta_j g(x^k) = g(x^k + h_i e^i + h_j e^j) - g(x^k + h_i e^i) - g(x^k + h_j e^j) + g(x^k).$$

Then it is easily verified (**E 8.1-1**) that the interpolating quadratic

$$g_k(x) = g(x^k) + \sum_{i=1}^n (1/h_i)[\Delta_i g(x^k) - \tfrac{1}{2}\Delta_i^2 g(x^k)](x_i - x_i^k)$$

$$+ \tfrac{1}{2} \sum_{i,j=1}^n (h_i h_j)^{-1} \Delta_i \Delta_j g(x^k)(x_i - x_i^k)(x_j - x_j^k) \tag{10}$$

agrees with g at the points (9). Using this interpolating quadratic g_k, we then obtain the iterative process $x^{k+1} = x^k - A_k^{-1} b_k$, where

$$A_k = ((h_i h_j)^{-1} \Delta_i \Delta_j g(x^k)),$$

and

$$b_k^T = \left(h_1^{-1}[\Delta_1 g(x^k) - \tfrac{1}{2}\Delta_1^2 g(x^k)], \ldots, h_n^{-1}[\Delta_n g(x^k) - \tfrac{1}{2}\Delta_n^2 g(x^k)]\right).$$

NOTES AND REMARKS

NR 8.1-1. The idea of adding the scalar matrix $\lambda_k I$ to $H_g(x^k)$ in (5) dates back to Levenberg [1944] in a slightly different setting (see **NR 8.5-2**), and has subsequently been rediscovered several times (see, for example, Goldfeld, Quandt, and Trotter [1966]). A choice of λ_k which ensures that $H_g(x^k) + \lambda_k I$ is nonsingular is to take $\lambda_k > \| H_g(x^k) \|$, although this may yield a much larger λ_k than is desirable.

NR 8.1-2. The interpolating quadratic functional (10) has been considered by Schmidt and Trinkaus [1966].

NR 8.1-3. A program utilizing the approximation (3) has been given by Späth [1967].

EXERCISE

E 8.1-1. For given $g: R^n \to R^1$, let g_k be the quadratic functional of (10), where $h_1, \ldots, h_n$ are any nonzero numbers. Show that g and g_k assume the same values at the $1 + n + \frac{1}{2}n(n+1)$ points (9).

8.2. DESCENT METHODS

A very broad and important class of minimization algorithms are those for which the iterates decrease the function value at each stage; that is, for which

$$g(x^{k+1}) \leqslant g(x^k), \qquad k = 0, 1, \ldots. \tag{1}$$

Such algorithms shall be called *descent methods* for g, and the analysis of Chapter 14 will be restricted exclusively to this class.

As simple examples show, the methods of the preceding section do not necessarily satisfy (1) (see **E 8.2-1**). However, it is usually possible to modify them, in a way to be described, so that they become descent methods.

Consider, for example, the Newton iteration (8.1.4) for the system $Fx = g'(x)^T = 0$, but with an added parameter α_k:

$$x^{k+1} = x^k - \alpha_k F'(x^k)^{-1} Fx^k, \qquad k = 0, 1, \ldots. \tag{2}$$

If $Fx^k \neq 0$, then the parameter α_k in (2) may be chosen so that (1) holds. This is a consequence of the following lemma.

8.2.1. Suppose that $g: D \subset R^n \to R^1$ is G-differentiable at $x \in \text{int}(D)$ and that, for some $p \in R^n$, $g'(x)p > 0$. Then there is a $\delta > 0$ so that

$$g(x - \alpha p) < g(x), \qquad \forall \alpha \in (0, \delta).$$

Proof. Since

$$\lim_{\alpha \to 0} (1/\alpha)[g(x - \alpha p) - g(x)] + g'(x)p = 0,$$

and $g'(x)p > 0$, it follows that we may choose a $\delta > 0$ so that $x - \alpha p \in D$ for all $\alpha \in (0, \delta)$ and $(1/\alpha)[g(x - \alpha p) - g(x)] + g'(x)p < g'(x)p$. ∎

For Newton's method, set $p^k = F'(x^k)^{-1} Fx^k$. Then the condition

$$g'(x^k) p^k > 0 \qquad (3)$$

is satisfied if, for example, the Hessian matrix $F'(x^k) = H_g(x^k)$ is positive definite; for, then, $F'(x^k)^{-1}$ is also positive definite and

$$g'(x^k) p^k = (Fx^k)^{\mathrm{T}} F'(x^k)^{-1} Fx^k > 0, \qquad (4)$$

provided that $Fx^k \neq 0$. In this case, **8.2.1** ensures that $\alpha_k > 0$ may be chosen so that x^{k+1}, as given by (2), satisfies (1). But if α_k is taken to be equal to unity, then (1) may fail, and the Newton iterate is said to "overshoot." The modification (2), with sufficiently small α_k, is sometimes called the *damped Newton method*.

The iteration (2) is of the general form

$$x^{k+1} = x^k - \alpha_k p^k, \qquad k = 0, 1, \ldots, \qquad (5)$$

and, clearly, *any* iterative method may be written in this way with suitable vectors p^k and scalars α_k. In the setting of this chapter, it is convenient to think of the vector $-p^k$ as defining a *direction* along which a new iterate x^{k+1} will be chosen, and the scalar α_k as determining the *steplength* from x^k to x^{k+1}. If $\| p^k \| = 1$, then α_k is simply the distance from x^k to x^{k+1}.

The Newton iteration, as well as the other methods of the previous section, gives means of determining the direction vectors $-p^k$. However, perhaps the conceptually simplest minimization methods are those of *univariate relaxation*, in which the p^k are chosen from among the coordinate vectors $e^1, \ldots, e^n$, and, hence, the iterates $\{x^k\}$ change in only one component at each stage. The classical choice of a particular coordinate vector at the kth stage is that for which the local decrease is maximal; that is, the index i of the vector e^i is chosen so that

$$| g'(x^k) e^i | = \max_{1 \leqslant j \leqslant n} | g'(x^k) e^j |.$$

If the sign of $\pm e^i$ is selected so that $g'(x^k)(\pm e^i) > 0$, then clearly (3) holds, and hence by **8.2.1**, the steplength α_k may be taken so that (1) is valid. Another common principle for the choice of the coordinate vectors is that of *cyclic* selection, that is,

$$p^k = \pm e^{k(\bmod n)+1}, \qquad k = 0, 1, \ldots. \qquad (6)$$

Here, g is decreased cyclically in each variable separately. Note that, in this case, it may happen that $g'(x^k) p^k = 0$, even though $g'(x^k) \neq 0$.

8.2. DESCENT METHODS

However, unless a critical point has been reached, there will clearly be an index $k + m$, $m \leq n - 1$, so that $g'(x^{k+m}) p^{k+m} > 0$.

Note that the choice of the coordinate vectors $e^1, \ldots, e^n$ as the basic direction vectors is only illustrative. We can, instead, use any set $q^1, \ldots, q^m$ of nonzero vectors which span R^n, and select the p^k from among these q^j by any of the principles discussed above.

Another natural choice of p^k, which ensures that (3) holds whenever $g'(x^k)^T \neq 0$, is $p^k = g'(x^k)^T$; that is, the direction of p^k is that of the gradient vector of g, and such methods are accordingly called *gradient methods*.

A closely related class of methods is obtained if $-p^k$ is chosen as the direction of maximal local decrease of g. If g is differentiable at x^k, then this direction is that for which $-g'(x^k) p / \| p \|$ takes on its minimum as a function of $p \neq 0$. Since $g'(x^k) p$ is a continuous function of p and $\{p \mid \| p \| = 1\}$ is compact, it follows that such a "best" direction p^k always exists, although it need not be unique. Note that, because $g'(x^k) \in L(R^n, R^1)$, it is true from (2.2.9) that

$$\| g'(x^k) \| = \sup_{\|h\|=1} | g'(x^k) h |,$$

and thus p^k is a best direction if

$$\| g'(x^k) \| = g'(x^k) p^k / \| p^k \|.$$

We call any such p^k a direction of *steepest descent*, and any method of the form (5) which uses only such directions, a *method of steepest descent*. Clearly, (3) holds for these directions unless $g'(x^k) = 0$.

The direction of steepest descent depends on the particular norm being used. For elliptic norms, we give the following simple result.

8.2.2. If $C \in L(R^n)$ is symmetric, positive definite and $g : D \subset R^n \to R^1$ is G-differentiable at x, then the direction of steepest descent of g at x in the norm $\| x \| = (x^T C x)^{1/2}$, is given by

$$-C^{-1} g'(x)^T. \tag{7}$$

Proof. Recall that, since C is positive definite, there is a real symmetric matrix B, denoted by $C^{1/2}$, so that $B^2 = C$ (see **E 2.1-8**). Then, for arbitrary $p \neq 0$, it follows from the Cauchy–Schwarz inequality that

$$[g'(x) p]^2 = [g'(x) C^{-1} C^{\frac{1}{2}} C^{\frac{1}{2}} p]^2 \leq [g'(x) C^{-1} C^{\frac{1}{2}}][g'(x) C^{-1} C^{\frac{1}{2}}]^T (C^{\frac{1}{2}} p)^T C^{\frac{1}{2}} p$$
$$= \| C^{-1} g'(x)^T \|^2 \| p \|^2,$$

and, clearly, equality holds if $-p$ is given by (7). ∎

Note that when C is the identity, then $p = g'(x)^T$; that is, the direction of steepest descent in the l_2-norm is the negative of the gradient vector. As a consequence of **8.2.2**,

$$x^{k+1} = x^k - \alpha_k C^{-1} g'(x^k)^T, \quad k = 0, 1, \ldots,$$

is a steepest descent method in the norm $\|x\| = (x^T C x)^{1/2}$. This suggests considering methods of the form

$$x^{k+1} = x^k - \alpha_k C_k^{-1} g'(x^k)^T, \quad k = 0, 1, \ldots, \tag{8}$$

where the matrix C_k may change at each step, but is always symmetric and positive definite. One interpretation of (8) is that, at the kth stage, the new iterate x^{k+1} is chosen along the direction of steepest descent in the norm defined by $\|x\|_k = (x^T C_k x)^{1/2}$.

A special case of (8) is the damped Newton iteration (2), for which $C_k = H_g(x^k)$, provided that the Hessian matrix $H_g(x^k)$ is positive definite for each x^k. Another example is the *Davidon–Fletcher–Powell method*, introduced in **7.3**, in which

$$x^{k+1} = x^k - \alpha_k B_k g'(x^k)^T, \tag{9}$$

and the matrices B_k are defined recursively by

$$B_{k+1} = B_k + \frac{r^k (r^k)^T}{(r^k)^T q^k} - \frac{(B_k q^k)(B_k q^k)^T}{(q^k)^T B_k q^k}, \quad k = 0, 1, \ldots. \tag{10}$$

Here, B_0 is an arbitrary, symmetric, positive definite matrix, usually taken to be the identity, and

$$r^k = x^{k+1} - x^k, \quad q^k = g'(x^{k+1})^T - g'(x^k)^T, \quad k = 0, 1, \ldots. \tag{11}$$

In order to conclude that (9) is of the form (8), and, indeed, that the B_k are even well defined, it is necessary to know that all B_k are positive definite. A sufficient condition for this is given by **8.2.4**, which is based on the following lemma.

8.2.3. Let $B \in L(R^n)$ be symmetric, positive definite and let $r, q \in R^n$ satisfy $r^T q > 0$. Then

$$\hat{B} = B + \frac{r r^T}{r^T q} - \frac{(Bq)(Bq)^T}{q^T B q}$$

is symmetric, positive definite.

8.2. DESCENT METHODS

Proof. The symmetry is obvious. For the positive definiteness, let $x \neq 0$ be arbitrary and set $y = B^{1/2}x$, $z = B^{1/2}q$. Then

$$x^{\mathrm{T}}\hat{B}x = x^{\mathrm{T}}Bx + \frac{(r^{\mathrm{T}}x)^2}{r^{\mathrm{T}}q} - \frac{(x^{\mathrm{T}}Bq)^2}{q^{\mathrm{T}}Bq}$$

$$= \frac{(y^{\mathrm{T}}y)(z^{\mathrm{T}}z) - (y^{\mathrm{T}}z)^2}{z^{\mathrm{T}}z} + \frac{(r^{\mathrm{T}}x)^2}{r^{\mathrm{T}}q} \geq 0,$$

by the Cauchy–Schwarz inequality and the fact that $r^{\mathrm{T}}q > 0$. But $(z^{\mathrm{T}}z)(y^{\mathrm{T}}y) - (y^{\mathrm{T}}z)^2 > 0$ unless $y = \beta z$, that is, unless $B^{1/2}x = \beta B^{1/2}q$, or $x = \beta q$. In this case $\beta \neq 0$ (since $x \neq 0$), and, therefore, $(r^{\mathrm{T}}x)^2 = \beta^2(r^{\mathrm{T}}q)^2 > 0$, so that again $x^{\mathrm{T}}\hat{B}x > 0$. ∎

8.2.4. Let $g: R^n \to R^1$ be G-differentiable and satisfy

$$[g'(x) - g'(y)](x - y) > 0, \quad \forall x, y \in R^n, \quad x \neq y. \tag{12}$$

Then, for any x^0, there is a sequence of $\alpha_k > 0$ so that the iterates (9)–(11) are well-defined for all k [unless $g'(x^k)^{\mathrm{T}} = 0$ for some k, in which case the process stops], and $g(x^{k+1}) < g(x^k)$.

Proof. We proceed by induction. Assume that $g'(x^k)^{\mathrm{T}} \neq 0$ and that B_k is positive definite. Then $g'(x^k) B_k g'(x^k)^{\mathrm{T}} > 0$, so that **8.2.1** ensures the existence of an $\alpha_k > 0$ such that $g(x^{k+1}) < g(x^k)$. Clearly, $x^{k+1} \neq x^k$, and, hence, (12) ensures that $r_k^{\mathrm{T}} q_k > 0$. Therefore, **8.2.3** shows that B_{k+1} is positive definite. ∎

If g is a quadratic functional $\frac{1}{2}x^{\mathrm{T}}Ax - b^{\mathrm{T}}x + c$, where $A \in L(R^n)$ is symmetric, positive definite, and α_k is chosen to minimize g on the ray $\{x \mid x = x^k - \alpha B_k g'(x^k)^{\mathrm{T}}; \alpha \in (-\infty, \infty)\}$, then the Davidon–Fletcher–Powell iterates will converge to $A^{-1}b$ in at most n steps. A proof of this result is given in Appendix 1 (Section **8.6**) of this chapter.

NOTES AND REMARKS

NR 8.2-1. Many of the basic ideas in minimization methods date back to a paper by Cauchy [1847] in which the gradient and steepest-descent methods, as well as some of the steplength algorithms of the next section, were first discussed.

NR 8.2-2. The damped Newton method was described by Crockett and Chernoff [1955], but had probably been employed in actual computations prior to that time.

NR 8.2-3. The interpretation of (8) as a steepest-descent method in a sequence of norms has been exploited, for the purposes of analysis, by Nashed [1964, 1965]. On the other hand, our analysis of these methods in Chapter 14 will be based on the observation that if C_k^{-1} is positive definite, then $p^k = C_k^{-1} g'(x^k)^T$ is not orthogonal to the gradient direction $g'(x^k)^T$.

NR 8.2-4. Still another interpretation of algorithms of the form

$$x^{k+1} = x^k - \alpha_k C^{-1} g'(x^k)^T,$$

with symmetric, positive definite C, is as gradient methods in some other basis (see Crockett and Chernoff [1955]). In fact, let $C^{-1} = PP^T$, and define the change of variable $y = P^{-1}x$ and the new functional $\hat{g}(y) = g(Py)$, $y \in R^n$. Then $\hat{g}'(y) = g'(Py) P$, and the iteration becomes

$$y^{k+1} = y^k - \alpha_k \hat{g}'(y^k)^T.$$

Transformations of the form $y = P^{-1}x$, where P is a diagonal matrix, are widely used in practice to scale the variables.

NR 8.2-5. The basic idea of the Davidon–Fletcher–Powell method was proposed by Davidon [1959] and, in the form presented here, by Fletcher and Powell [1963]; Lemma **8.2.3** is taken from the latter paper. For a derivation of this method by another approach, see Section **7.3**. A related method has also been suggested by Davidon [1967]. It again has the form of a modification method in the sense of **7.3** and is given by

$$x^{k+1} = x^k - \omega_k B_k g'(x^k)^T, \qquad k = 0, 1, \ldots,$$

$$B_{k+1} = B_k + \frac{\lambda_k [B_k g'(\bar{x}^{k+1})^T]^T [B_k g'(\bar{x}^{k+1})^T]}{g'(\bar{x}^{k+1}) B_k g'(\bar{x}^{k+1})^T},$$

where $\bar{x}^{k+1} = x^k - B_k g'(x)^T$, $\omega_k = 1$ if $g(\bar{x}^{k+1}) < g(x^k)$, while $\omega_k = 0$ otherwise, and the factors λ_k are to be chosen in a suitable manner. No numerical experience has been presented for this method, and, in fact, Davidon states that his proposed choise of λ_k may cause the iteration to cycle.

NR 8.2-6. A version of the Davidon–Fletcher–Powell method which does not require the derivative g' has been discussed by Stewart [1967].

EXERCISES

E 8.2-1. Define $g: R^1 \to R^1$ by $g(x) = (1 + x^2)^{1/2}$. Show that the Newton iterates for the equation $g'(x) = 0$ have the property:
(a) If $|x^0| < 1$, then $g(x^{k+1}) < g(x^k)$, $k = 0, 1, \ldots$, and $\lim_{k \to \infty} x^k = 0$.
(b) If $|x^0| > 1$, then $g(x^{k+1}) > g(x^k)$, $k = 0, 1, \ldots$, and $\lim_{k \to \infty} |x^k| = +\infty$.

8.3. STEPLENGTH ALGORITHMS

E 8.2-2. Let $F: R^n \to R^n$ be continuously differentiable and have a nonsingular derivative $F'(x)$ for all $x \in R^n$. Use **8.2.1** to show that there exist α_k, $k = 0, 1,...$, so that the damped Newton iterates (2) satisfy

$$\| Fx^{k+1} \|_2 \leqslant \| Fx^k \|_2, \quad k = 0, 1,\ldots.$$

(Note that this complements **E 7.1-1**.) [Hint: Apply **8.2.1** to $g(x) = (Fx)^T Fx$.]

E 8.2-3. Show that the Davidon–Fletcher–Powell algorithm is not invariant under scaling of the functional. That is, if $\{x^k\}$ are the iterates for the functional $g: R^n \to R^1$ and $\{\hat{x}^k\}$ those for $\hat{g} = cg$, with $c \neq 1$, then $x^k \neq \hat{x}^k$. (Bard [1968])

8.3. STEPLENGTH ALGORITHMS

We consider now in more detail various ways of choosing the steplength α_k in the general iteration

$$x^{k+1} = x^k - \alpha_k p^k, \quad k = 0, 1,\ldots, \qquad (1)$$

assuming that the direction p^k is given.

(a). Minimization Principles. We saw in **8.2.1** that if $g'(x^k) p^k > 0$, then it is always possible to choose α_k so that $g(x^{k+1}) < g(x^k)$. The maximal possible decrease occurs for given p^k when α_k is chosen so as to minimize g along the ray $\{x \mid x = x^k - \alpha p^k, \alpha \in R^1\}$, that is, when α_k satisfies the *minimization principle*

$$g(x^k - \alpha_k p^k) = \min\{g(x^k - \alpha p^k) \mid x^k - \alpha p^k \in D\}. \qquad (2)$$

For many purposes, the set of α over which the minimization in (2) is taken is too large, and we next consider other possibilities. Let L_k denote the level set $\{x \in D \mid g(x) \leqslant g(x^k)\}$ and let L_k^0 be the path-connected component of L_k which contains x^k. Then there are two natural modifications of (2):

$$g(x^k - \alpha_k p^k) = \min\{g(x^k - \alpha p^k) \mid x^k - \alpha p^k \in L_k^0\}, \qquad (3)$$

$$g(x^k - \alpha_k p^k) = \min\{g(x^k - \alpha p^k) \mid [x^k, x^k - \alpha p^k] \subset L_k^0\}, \qquad (4)$$

where, as usual, $[x, y]$ denotes the line segment

$$\{z \mid z = tx + (1 - t) y, t \in [0, 1]\}.$$

The difference between (2), (3), and (4) is illustrated in Fig. **8.1**, which shows that the minimization of (4) is over the line segment Λ_1, while those of (3) and (2) are over $\Lambda_1 \cup \Lambda_3$ and $\Lambda_1 \cup \Lambda_2 \cup \Lambda_3$, respectively.

If a situation as depicted in Fig. 8.1 should prevail, then either (2) or (3) may be difficult to implement numerically, since it is necessary to "climb" out of the set $L_k{}^0$. Moreover, in some problems, it may be required to find the minimum of g only over the component $L_k{}^0$, so that (2) would constitute a wrong algorithm.

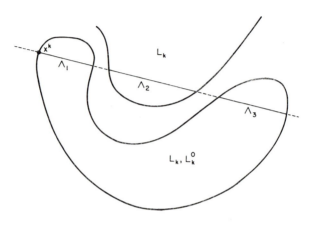

FIGURE 8.1

We note that if g is a connected functional (see **4.2.3**), then all level sets of g are path-connected, so that (2) and (3) become identical. Similarly, if g is quasiconvex (see **4.2.5**), then, by **E 4.2-4**, all level sets of g are convex, and all three possibilitites (2), (3), and (4) are identical.

(b). The Curry and Altman Principles If, for any one of the minimization choices (2)–(4), $x^k - \alpha_k p^k$ is an interior point of D, then α_k is a root of the derivative of the function

$$\varphi(\alpha) = g(x^k - \alpha p^k);$$

that is, α_k is a solution of the one-dimensional equation

$$g'(x^k - \alpha p^k) p^k = 0, \tag{5}$$

where we assume now that g is continuously differentiable on D. If, for example, g is strictly convex on D, then (5) has at most one solution and α_k is a solution if and only if (2) holds. More generally, in connection with the restricted minimization choice (4), if g is strictly convex on the line segment Λ_1 of Fig. 8.1, then again α_k is the only solution of (5) in the set $\{\alpha \mid x^k - \alpha p^k \in \Lambda_1\}$.

8.3. STEPLENGTH ALGORITHMS

For more general functionals, however, (5) may have solutions which are not associated with minimizers of φ, as shown, for example, in Fig. 8.2. Nevertheless, it is still possible to define the steplength α_k to be any solution of (5). Indeed, for the choice (8.2.6) of direction vectors, this is simply the nonlinear Gauss–Seidel iteration (7.4.26) applied to $Fx = g'(x)^T$.

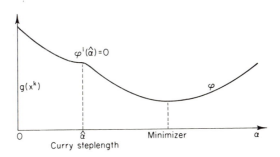

FIGURE 8.2

On the other hand, it is desirable to specify completely which root of (5) is to be taken. If we assume that the sign of p^k is taken so that $g'(x^k)p^k > 0$, then the *Curry principle* states that α_k is to be chosen as the smallest positive root of (5). In this case, it is clear from the mean-value theorem **3.2.7**, since $g'(x^k - \alpha p^k)p^k > 0$ for all $\alpha \in [0, \alpha_k)$, that

$$g(x^k - \alpha_k p^k) = g(x^k) - \alpha_k \int_0^1 g'(x^k - s\alpha_k p^k) p^k \, ds < g(x^k),$$

so that the Curry principle leads to a descent method.

Closely related to the Curry principle is the *Altman principle*, in which, for a fixed $\mu \in [0, 1)$, α_k is chosen as the smallest positive root of

$$[g'(x^k - \alpha p^k) - \mu g'(x^k)] p^k = 0. \tag{6}$$

If $\mu = 0$, then (6) reduces to (5); if $\mu > 0$, then the interpretation of the Altman principle is that α_k is the first positive α for which $g'(x^k - \alpha p^k) p^k$ is equal to the fraction μ of its original value $g'(x^k) p^k$. It is easy to see that, again, this choice of α_k leads to a descent method under nominal conditions on g (see **E 8.3-2**).

The Altman principle would seem to offer little to recommend it as a practical procedure. However, we shall see in Chapter 14 that it provides a useful theoretical tool for the analysis of other steplength algorithms.

(c). Approximate Minimization and Root-finding. The minimization principles as well as the Curry–Altman principles require the exact solution of one-dimensional problems, and, in practice, we must usually settle for only approximate solutions of these problems. In Appendix 2 of this chapter (8.7), various one-dimensional search algorithms are described for solving the minimization problem (4); here, we consider other techniques.

In order to approximate the solution of the one-dimensional minimization problem, we can, of course, proceed by the methods of 8.1 and take α_k as the minimizer of the osculating parabola

$$\psi(\alpha) = g(x^k) - \alpha g'(x^k) p^k + \tfrac{1}{2}\alpha^2 g''(x^k) p^k p^k. \tag{7}$$

If $g''(x^k) p^k p^k > 0$, then α_k is well-defined and given explicitly by

$$\alpha_k = g'(x^k) p^k / g''(x^k) p^k p^k. \tag{8}$$

In case $g''(x^k)$ and $g'(x^k)$ are difficult to evaluate, it may be desirable to use, instead of the osculating polynomial (7), a quadratic interpolation polynomial $\tilde{\psi}$ which agrees with $g(x^k - \alpha p^k)$ at three points a_1, a_2, a_3 (usually, $a_1 = 0$), and to take α_k as the minimizer of $\tilde{\psi}$. The precise formula for this steplength is given in **E 8.3-3**.

The minimization of $g(x^k - \alpha p^k)$ may be carried out iteratively by successive minimization of osculating (or interpolating) parabolas. On the other hand, we may start from the Curry principle and attempt to solve Eq. (5) by a one-dimensional iterative method such as that of Newton. If, in this case, α_k is taken to be the result of precisely one Newton step starting from $\alpha = 0$, then, clearly, α_k is again given by (8). It follows that when the directions p^k are those of (8.2.6), we obtain again the one-step SOR–Newton method (7.4.28) (with $\omega = 1$) expressed now as

$$x_j^{k+1} = x_j^k - \begin{cases} 0, & j \neq i, \\ \partial_i g(x^k)/\partial_i^2 g(x^k), & j = i, \end{cases} \quad j = 1,\ldots,n; \quad i = k(\bmod n) + 1. \tag{9}$$

Similarly, if p^k is the gradient direction, we obtain with the steplength choice (8), the (one-step) *gradient–Newton method*

$$x^{k+1} = x^k - [g'(x^k) g'(x^k)^\mathrm{T}/g''(x^k) g'(x^k)^\mathrm{T} g'(x^k)^\mathrm{T}] g'(x^k)^\mathrm{T}, \quad k = 0, 1,\ldots. \tag{10}$$

Clearly, various methods of this type may be constructed by using other one-dimensional root-finding methods. For example, the choice

$$\alpha_k = \bar{\alpha} g'(x^k) p^k / [g'(x^k - \bar{\alpha} p^k) - g'(x^k)] p^k$$

8.3. STEPLENGTH ALGORITHMS

results from applying one step of the secant method using a trial point $\bar{\alpha}$.

The algorithm (8) for α_k does not necessarily lead to a descent method, even when $g''(x^k)p^k p^k > 0$, as illustrated in Fig. 8.3. However, if

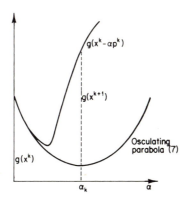

FIGURE 8.3

$g'(x^k)p^k > 0$, it is always possible to choose, on the basis of 8.2.1, a *damping* (or *underrelaxation*) *factor* ω_k so that

$$g(x^k - \omega_k \alpha_k p^k) < g(x^k).$$

Clearly, the introduction of a factor ω_k is not restricted to the particular steplength choice (8), and may be used as well with, for example, the minimization or the Curry–Altman principles. In that case, since these are already descent methods, the reason for introducing ω_k is usually to enhance the rate of convergence, and we may wish to choose $\omega_k > 1$.

(d). The Majorization Principle. As just pointed out, the steplength algorithms considered in **(c)** do not necessarily lead to descent methods without the introduction of an underrelaxation factor; the difficulty, as shown in Fig. 8.3, is that the function ψ whose minimizer is taken as α_k may lie below $g(x^k - \alpha p^k)$. This suggests the following *majorization principle*. If there is a function ψ: $[0, \hat{\alpha}] \to R^1$ such that

$$g(x^k - \alpha p^k) \leq \psi(\alpha) < g(x^k), \quad \forall \alpha \in (0, \hat{\alpha}), \quad (11)$$

then any steplength

$$\alpha_k \in (0, \hat{\alpha}) \quad (12)$$

decreases g. One way of obtaining such a function ψ, at least in principle, is given by the following lemma.

8.3.1. Suppose that $g: D \subset R^n \to R^1$ is continuously differentiable on $D_0 \subset D$ and that the modulus of continuity,

$$\omega(t) \equiv \sup\{\|g'(x) - g'(y)\| \mid \|x - y\| \leqslant t, \; x, y \in D_0\},$$

of $g': D_0 \subset R^n \to L(R^n, R^1)$, is well defined and continuous on $[0, \infty)$. Then, for any $x \in D_0$ and $p \neq 0$,

$$g(x - \alpha p) \leqslant g(x) - \alpha g'(x) p + \alpha \|p\| \int_0^1 \omega(t\alpha \|p\|) \, dt, \tag{13}$$

for all α such that $[x, x - \alpha p] \subset D_0$. In particular, if g' satisfies, for some $\lambda \in (0, 1]$,

$$\|g'(x) - g'(y)\| \leqslant \gamma \|x - y\|^\lambda, \qquad \forall x, y \in D_0, \tag{14}$$

then

$$g(x - \alpha p) \leqslant g(x) - \alpha g'(x) p + [\gamma/(1 + \lambda)](\|p\| \alpha)^{1+\lambda}. \tag{15}$$

Proof. For any α such that $x - \alpha p \in D_0$, it follows, by the definition of $\|g'(x) - g'(y)\|$, that

$$|[g'(x) - g'(x - \alpha p)] p | \leqslant \|g'(x) - g'(x - \alpha p)\| \|p\| \leqslant \|p\| \omega(\alpha \|p\|).$$

Hence, if $[x, x - \alpha p] \subset D_0$, the mean-value theorem **3.2.7** gives

$$g(x - \alpha p) = g(x) - \alpha g'(x) p + \alpha \int_0^1 [g'(x) - g'(x - t\alpha p)] p \, dt$$

$$\leqslant g(x) - \alpha g'(x) p + \alpha \|p\| \int_0^1 \omega(\alpha t \|p\|) \, dt.$$

Finally, if (14) holds, then $\omega(t) \leqslant \gamma t^\lambda$, and (15) follows directly from (13). ∎

Suppose, now, that the assumptions of **8.3.1** hold and that $x^k \in D_0$. Further, assume that $g'(x^k) p^k > 0$ and set

$$\hat{\alpha} = \sup\{\alpha > 0 \mid [x^k, x^k - \alpha p^k] \subset D_0, \; \eta(\alpha) < g'(x^k) p^k\},$$

where

$$\eta(\alpha) = \|p^k\| \int_0^1 \omega(\alpha t) \|p^k\| \, dt.$$

8.3. STEPLENGTH ALGORITHMS

Since $\eta(0) = 0$ and η is isotone, and since $g'(x^k)p^k > 0$, $\hat{\alpha}$ is well-defined, and, clearly, $\hat{\alpha} > 0$. Hence, if $[x^k, x^k - \hat{\alpha}p^k] \subset D_0$, 8.3.1 ensures that (11) holds with

$$\psi(\alpha) = g(x^k) - \alpha g'(x^k)p^k + \alpha \eta(\alpha).$$

An admissible steplength choice is then specified by (12), and, in particular, α_k may be chosen as that value of $\alpha \in (0, \hat{\alpha})$ which minimizes ψ, that is,

$$\psi(\alpha_k) = \min\{\psi(\alpha) \mid \alpha \in (0, \hat{\alpha})\}. \tag{16}$$

The situation is depicted in Fig. 8.4; note that the interval $(0, \hat{\alpha})$ need not contain either the first zero of $g'(x^k - \alpha p^k)p^k$ nor the minimizer of $\varphi(\alpha) = g(x^k - \alpha p^k)$.

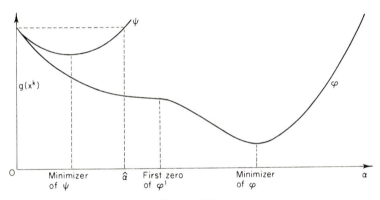

FIGURE 8.4

If g' is Hölder-continuous on D_0, that is, if (14) holds, then, by 8.3.1, ψ may be taken as

$$\psi(\alpha) = g(x^k) - \alpha g'(x^k)p^k + [\beta/(1+\lambda)](\| p^k \| \alpha)^{1+\lambda}.$$

Here, the steplength α_k for which (16) holds is given explicitly by

$$\alpha_k = (1/\| p^k \|)[g'(x^k)p^k/(\beta \| p^k \|)]^{1/\lambda}. \tag{17}$$

(e). Steplength Sets and the Goldstein Principle. The majorization principle gives an admissible interval (12) for α_k which ensures that the descent condition (1) holds. We consider next a somewhat different criterion for such admissible steplength sets.

Suppose, again, that g is G-differentiable at $x^k \in \text{int}(D)$ and that

$g'(x^k) p^k > 0$. Then the tangent line to the curve $g(x^k - \alpha p^k)$ at $\alpha = 0$ is given by

$$\tau(\alpha) = g(x^k) - \alpha g'(x^k) p^k.$$

Thus, for arbitrary μ_1, μ_2 satisfying $0 < \mu_1 \leqslant \mu_2 < 1$, the lines

$$\sigma_i(\alpha) = g(x^k) - \mu_i \alpha g'(x^k) p^k, \qquad i = 1, 2,$$

lie above the tangent line; more specifically,

$$\tau(\alpha) < \sigma_2(\alpha) < \sigma_1(\alpha), \qquad \forall \alpha > 0.$$

Consider now the set

$$J_k = \{\alpha > 0 \mid [x^k, x^k - \alpha p^k] \subset D, \quad \sigma_2(\alpha) \leqslant g(x^k - \alpha p^k) \leqslant \sigma_1(\alpha)\}.$$

Clearly, for any $\alpha \in J_k$, we have

$$0 < \mu_1 \alpha g'(x^k) p^k \leqslant g(x^k) - g(x^k - \alpha p^k) \leqslant \mu_2 \alpha g'(x^k) p^k,$$

which shows, in particular, that for any choice of $\alpha_k \in J_k$, g decreases, and we have a descent method. This is the *Goldstein principle*. Here, J_k is not necessarily an interval, nor is there necessarily any overlap with the interval (12) of the majorization principle; the situation is illustrated in Fig. 8.5.

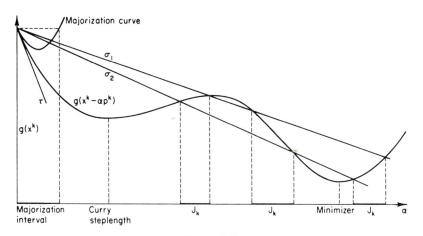

FIGURE 8.5

8.3. STEPLENGTH ALGORITHMS

Note that the smaller $\mu_2 - \mu_1$, the smaller is J_k, and, in the limiting case $\mu_1 = \mu_2 = \mu$, we have $\sigma_1(\alpha) = \sigma_2(\alpha)$, so that α_k must satisfy

$$g(x^k - \alpha_k p^k) + \mu \alpha_k g'(x^k) p^k = g(x^k).$$

The Goldstein principle was proposed originally for use in conjunction with another iteration (for example, Newton's method) in the following way. Suppose that, at the kth iterate x^k, a direction p^k such that $g'(x^k)p^k > 0$ and a tentative steplength α_k are given. Moreover, let $0 < \mu_1 \leqslant \mu_2 < 1$ be fixed constants. If

$$g(x^k) - g(x^k - \alpha_k p^k) \geqslant \alpha_k \mu_1 g'(x^k) p^k, \tag{18}$$

then we accept the point

$$x^{k+1} = x^k - \alpha_k p^k$$

as the next iterate. If, on the other hand, (18) does not hold, then we choose a parameter ω_k such that

$$0 < \mu_1 \omega_k \alpha_k g'(x^k) p^k \leqslant g(x^k) - g(x^k - \omega_k \alpha_k p^k) \leqslant \mu_2 \omega_k \alpha_k g'(x^k) p^k \tag{19}$$

and set

$$x^{k+1} = x^k - \omega_k \alpha_k p^k.$$

The following lemma shows that it is always possible to select an ω_k for which (19) is satisfied.

8.3.2. Let $g: D \subset R^n \to R^1$ be continuously differentiable on D and suppose that x^k, $\alpha_k > 0$, and p^k satisfy $g'(x^k) p^k > 0$, $[x^k, x^k - \alpha_k p^k] \subset D$, and

$$g(x^k) - g(x^k - \alpha_k p^k) < \mu_1 \alpha_k g'(x^k) p^k \tag{20}$$

for some $\mu_1 \in (0, 1)$. Then, for any $\mu_2 \in [\mu_1, 1)$, there is an $\omega_k \in (0, 1)$ so that (19) holds.

Proof. Define $\psi: [0, 1] \to R^1$ by $\psi(0) = 1$ and

$$\psi(\omega) = [g(x^k) - g(x^k - \omega \alpha_k p^k)]/[\omega \alpha_k g'(x^k) p^k], \quad \omega \in (0, 1].$$

By L'Hospital's rule, we have $\lim_{\alpha \to 0} \psi(\alpha) = 1$, so that ψ is continuous on $[0, 1]$. Therefore, since $\psi(1) < \mu_1$, ψ takes on all values between μ_1 and 1. In particular, there is an $\omega_k \in (0, 1)$ so that $\mu_1 \leqslant \psi(\omega_k) \leqslant \mu_2$ which is (19). ∎

The condition (20) indicates that α_k was chosen too large. A natural procedure is to decrease α_k geometrically and the simplest possibility is to use a constant factor $1/q$ with $q > 1$. This means that ω_k is taken as the largest number in the sequence $\{q^{-j}\}$ such that

$$g(x^k) - g(x^k - \omega_k \alpha_k p^k) \geqslant \mu_1 \omega_k \alpha_k g'(x^k) p^k.$$

We shall analyze this in **14.2**.

NOTES AND REMARKS

NR 8.3-1. The minimization principles of **(a)** may produce very slow convergence in certain cases. For example, if $g(x) = \frac{1}{2} x^T A x - b^T x + c$ is a quadratic functional with symmetric, positive definite $A \in L(R^n)$, and γ is the ratio of the smallest to the largest eigenvalue of A, then the gradient method (1), with $p^k = Ax^k - b$ and α_k chosen by (2), converges for small γ at a rate roughly proportional to $(1 - \gamma)^{1/2}$. As is easily illustrated in two dimensions, this slow rate of convergence corresponds to an oscillation of the iterates back and forth across the "narrow valley" of g.

NR 8.3-2. Gleyzal [1959] has suggested choosing α_k in the damped Newton method (8.2.1) by the minimization principle (2). This suggestion was also implicitly contained in Crockett and Chernoff [1955].

NR 8.3-3. The Curry and Altman principles were used by Curry [1944] and Altman [1966a], respectively.

NR 8.3-4. The iteration (10) was first advocated by Cauchy [1847], and, more recently, it has been discussed by several authors, including Altman [1957f], Kivistik [1960a], and Krasnoselskii and Rutickii [1961].

NR 8.3-5. In the case that $g: R^n \to R^1$ satisfies $g(x) \geqslant 0$ for all $x \in R^n$ (see, for example, Section **8.5**), Cauchy [1847] and, later, Booth [1949] have suggested determining the steplength α_k by a Newton step for the equation $g(x^k - \alpha p^k) = 0$; that is,

$$\alpha_k = g(x^k)/g'(x^k) p^k.$$

With the choice $p^k = g'(x^k)^T$, the iteration (1) then becomes

$$x^{k+1} = x^k - [g(x^k)/\|g'(x^k)\|_2^2] g'(x^k)^T, \quad k = 0, 1, \ldots,$$

which is a special gradient method studied by Altman [1961d].

NR 8.3-6. The prescription (17) for α_k was first studied by Goldstein [1962] in the special case $p^k = g'(x^k)^T$; for this choice of p^k and Lipschitz-continuous

8.3. STEPLENGTH ALGORITHMS

g', (17) reduces to $\alpha_k = 1/\beta$ in the l_2-norm. It has also been used by Ostrowski [1966] for essentially arbitrary directions p^k. In both cases, β was assumed to be a bound for $g''(x)$ rather than a Lipschitz constant for g'.

EXERCISES

E 8.3-1. Let $g(x) = x^T A x - b^T x + c$ be a quadratic functional, where A is symmetric, positive definite. If, for arbitary $p \neq 0$ and $x \in R^n$, $g(x - \alpha p) = \min\{g(x - \beta p) \mid \beta \in (-\infty, \infty)\}$, show that $\alpha = p^T(Ax - b)/p^T Ap$.

E 8.3-2. Assume that $g: D \subset R^n \to R^1$ is continuously differentiable on D and that, for $x^k \in D$ and p^k such that $g'(x^k) p^k > 0$, the steplength α_k is chosen by the Altman principle (and exists) for some $\mu \in [0, 1)$. Show that $g(x^k - \alpha_k p^k) < g(x^k)$.

E 8.3-3. Show that the quadratic function $\psi = \psi(\alpha)$ which agrees with $g(x - \alpha p)$ at the distinct points a_1, a_2, a_3 is given by

$$\psi(\alpha) = \frac{(\alpha - a_2)(\alpha - a_3)}{(a_1 - a_2)(a_1 - a_3)} g_1 + \frac{(\alpha - a_1)(\alpha - a_3)}{(a_2 - a_1)(a_2 - a_3)} g_2 + \frac{(\alpha - a_1)(\alpha - a_2)}{(a_3 - a_1)(a_3 - a_2)} g_3,$$

where $g_i = g(x - a_i p)$, $i = 1, 2, 3$. Show that ψ has a minimizer $\bar{\alpha}$ if and only if

$$\frac{2g_1}{(a_1 - a_2)(a_1 - a_3)} + \frac{2g_2}{(a_2 - a_1)(a_2 - a_3)} + \frac{2g_3}{(a_3 - a_1)(a_3 - a_2)} > 0.$$

and that, then,

$$\bar{\alpha} = \frac{1}{2} \frac{(a_2{}^2 - a_3{}^2) g_1 + (a_3{}^2 - a_1{}^2) g_2 + (a_1{}^2 - a_2{}^2) g_3}{(a_2 - a_3) g_1 + (a_3 - a_1) g_2 + (a_1 - a_2) g_3}.$$

Write down ψ and $\bar{\alpha}$ for the special cases $a_1 = 0$, $a_3 = 2a_2$, and $a_2 = 0$, $a_1 = -a_3$.

E 8.3-4. Assume that $g: D \subset R^n \to R^1$ is continuously differentiable and that $\|g'(x) - g'(y)\| \leq \gamma \|x - y\|$ for all $x, y \in D_0$, where D_0 is a convex subset of D. Let $A \in L(R^n)$ be symmetric. Use **8.3.1** to show that, for any $x \in D_0$ such that $g'(x)^T \neq 0$, there is a λ so that $A + \lambda I$ is nonsingular and

$$g(x - (A + \lambda I)^{-1} g'(x)^T) < g(x).$$

(Marquardt [1963])

E 8.3-5. Suppose that $g: D \subset R^n \to R^1$ is uniformly convex on the convex set $D_0 \subset D$, and that the F-derivative g' exists and is Lipschitz-continuous with constant K on D_0. Let $\epsilon \in [0, 1)$. Show that there exist constants $0 < \mu_1 < \mu_2 < 1$ such that, whenever $g'(x)p > 0$ for $x \in D_0$ and $\|p\| = 1$, and

$$[(1-\epsilon)/K]g'(x)p \leqslant \gamma \leqslant [(1+\epsilon)/K]g'(x)p,$$

then

$$\mu_1 \gamma g'(x)p \leqslant g(x) - g(x - \alpha p) \leqslant \mu_2 \gamma g'(x)p.$$

8.4. CONJUGATE-DIRECTION METHODS

The conjugate-direction methods originally developed for linear systems of equations are, explicitly or implicitly, the basis for a number of general minimization methods. We first review the basic properties of these methods for the minimization of a quadratic functional

$$g(x) = c - b^\mathrm{T}x + \tfrac{1}{2}x^\mathrm{T}Ax, \tag{1}$$

where $A \in L(R^n)$ is symmetric, positive definite.

8.4.1. Definition. Let $A \in L(R^n)$ be symmetric, positive definite. Then any two nonzero vectors $p, q \in R^n$ are *conjugate* (with respect to A) if $p^\mathrm{T}Aq = 0$. A set of n nonzero vectors $p^1,..., p^n$ is a *conjugate basis* if p^i and p^j are conjugate for all $i \neq j$. ∎

Since A is positive definite, we may define an inner product by $(x, y)_A = x^\mathrm{T}Ay$. Hence, p and q are conjugate with respect to A if and only if they are *A-orthogonal*, that is, orthogonal in this inner product. Therefore, $p^1,..., p^n$ form a conjugate basis if and only if it is an orthogonal set in this inner product.

Any conjugate-direction method for the functional (1) has the form

$$x^{k+1} = x^k - \alpha_k p^k, \qquad k = 0, 1,..., \tag{2}$$

where the p^k are A-orthogonal vectors in R^n, and α_k is determined by the minimization principle (see **E 8.3.1**):

$$\alpha_k = (Ax^k - b)^\mathrm{T} p^k / (Ap^k)^\mathrm{T} p^k. \tag{3}$$

The fundamental result for these methods when g is quadratic is that the iterates $\{x^k\}$ converge to the unique minimizer of g in at most n steps.

8.4.2. Let g be given by (1) with symmetric, positive definite $A \in L(R^n)$, and suppose that $p^0,..., p^{n-1}$ is a conjugate basis with respect to A. Then the vectors x^k of (2)–(3) satisfy $x^m = A^{-1}b$ for some $m \leqslant n$.

8.4. CONJUGATE-DIRECTION METHODS

Proof. From (2) and (3), it follows that

$$(Ax^{k+1} - b)^T p^j = (Ax^k - b)^T p^j - [(Ax^k - b)^T p^k/(Ap^k)^T p^k](Ap^k)^T p^j$$

for any $0 \leqslant j \leqslant n - 1$. Hence, the A-orthogonality of the p^j implies that

$$(Ax^{k+1} - b)^T p^j = \begin{cases} (Ax^k - b)^T p^j, & j \neq k, \\ 0, & j = k. \end{cases} \quad (4)$$

Therefore,

$$(Ax^n - b)^T p^j = (Ax^{n-1} - b)^T p^j = \cdots = (Ax^{j+1} - b)^T p^j = 0$$

for $j = 0, \ldots, n - 1$, and, since the p^j are linearly independent, we have that $Ax^n = b$. Of course, it may happen that $Ax^m = b$ for some $m < n$. In this case $\alpha_m = 0$, and it follows that $x^m = x^{m+1} = \cdots = x^n$. ∎

The essential problem in carrying out a conjugate-direction algorithm is to obtain the conjugate directions $p^0, \ldots, p^{n-1}$. A direct approach is to compute $p^0, \ldots, p^{n-1}$ by Gram–Schmidt orthogonalization in the inner product $(x, y)_A$. This, however, is a very inefficient procedure. A much more elegant scheme is the *conjugate-gradient* procedure, in which the p^j are generated simultaneously with the x^j according to the prescription

$$x^{k+1} = x^k - \alpha_k p^k, \quad \alpha_k = (Ax^k - b)^T p^k/(Ap^k)^T p^k, \quad (5)$$

$$p^0 = Ax^0 - b, \quad p^{k+1} = Ax^{k+1} - b - \beta_k p^k, \quad \beta_k = (Ax^{k+1} - b)^T Ap^k/(Ap^k)^T p^k. \quad (6)$$

Note that, again, α_k is the steplength obtained by the minimization principle, while β_k is chosen so that $(p^k)^T A p^{k+1} = 0$.

The basic result for the conjugate gradient method (5)–(6) is the following. The proof is given in Appendix 1 of this chapter (Section 8.6).

8.4.3. Given the quadratic functional (1) with symmetric, positive definite matrix A, the conjugate-gradient iterates (5)–(6) satisfy $x^m = A^{-1}b$, for some $m \leqslant n$. Moreover,

$$(Ap^i)^T p^j = 0, \quad i \neq j, \quad 0 \leqslant i, j \leqslant m. \quad (7)$$

Note that the relation (7) shows that the directions p^i generated by (6) are indeed conjugate.

We consider, now, various generalizations of the conjugate-gradient algorithm (5)–(6) to nonquadratic functionals. For the first method, assume that $g: D \subset R^n \to R^1$ is twice F-differentiable. Then the *Daniel algorithm* is given by

$$\begin{cases} x^{k+1} = x^k - \alpha_k p^k, & g(x^{k+1}) = \min\{g(x^k - \alpha p^k) \mid x^k - \alpha p^k \in D\}, \\ p^0 = g'(x^0)^T, & \\ p^{k+1} = g'(x^{k+1})^T - \beta_k p^k, & \beta_k = g''(x^{k+1}) g'(x^k)^T p^k / g''(x^{k+1}) p^k p^k. \end{cases} \quad (8)$$

If $g(x) = c - b^T x + \tfrac{1}{2} x^T A x$, then $g'(x)^T = Ax - b$ and $g''(x) hk = k^T Ah$, so that (8) clearly reduces to (5). Note that, in connection with (8), a natural assumption is that $g''(x)$ is positive definite for all x; this ensures that the denominators of the β_k do not vanish.

Another generalization of the conjugate-gradient method results from an equivalent form (see **8.6**) of the specification (6) for β_k, namely,

$$\beta_k = -(Ax^{k+1} - b)^T (Ax^{k+1} - b)/(Ax^k - b)^T (Ax^k - b). \quad (9)$$

In line with this, the *Fletcher–Reeves algorithm* has the form

$$x^{k+1} = x^k - \alpha_k p^k, \quad g(x^{k+1}) = \min\{g(x^k - \alpha p^k) \mid x^k - \alpha p^k \in D\}, \quad (10)$$

$$p^0 = g'(x^0)^T, \quad p^{k+1} = g'(x^{k+1})^T - \beta_k p^k,$$

$$\beta_k = -g'(x^{k+1}) g'(x^{k+1})^T / g'(x^k) g'(x^k)^T. \quad (11)$$

Note that (11) has the advantage over (8) that only g', and not g'', is needed to form the next p^k. Note, also, that p^{k+1} can always be formed if $g'(x^k)^T \neq 0$ [if $g'(x^k)^T = 0$ we have found a critical point of g, and the algorithm terminates].

We end this section with a description of a process which does not reduce to a conjugate–direction method when g is quadratic, but which nevertheless bears a certain resemblance to the previous algorithms. This is the *method of Rosenbrock*.

Assume, for simplicity, that $g: R^n \to R^1$ is defined on all of R^n. The process begins with n orthogonal vectors $q^{0,0},\ldots, q^{0,n-1}$ and an initial approximation x^0. At the end of the kth stage, these $n+1$ vectors will have been transformed into x^k and n orthogonal vectors $q^{k,0},\ldots, q^{k,n-1}$, and the $(k+1)$th step of the algorithm consists of two parts:

I. Compute

$$y^{k,0} = x^k, \quad y^{k,j+1} = y^{k,j} - \alpha_{kj} q^{k,j}, \quad j = 0,\ldots, n-1, \quad (12)$$

where the α_{kj} are determined by the minimization principle

$$g(y^{k,j+1}) = \min\{g(y^{k,j} - \alpha q^{k,j}) \mid \alpha \in (-\infty, \infty)\}. \quad (13)$$

8.4. CONJUGATE-DIRECTION METHODS

II. Set $x^{k+1} = y^{k,n}$ and determine a new set of orthogonal vectors $q^{k+1,0},..., q^{k+1,n-1}$ by (Gram–Schmidt) orthogonalization of the set of vectors

$$x^{k+1} - y^{k,j}, \quad j = 0,..., n-1. \tag{14}$$

In case the vectors (14) are not linearly independent, the orthogonalization is done for the set

$$x^{k+1} - y^{k,0},..., x^{k+1} - y^{k,n-i-1}, e^1,..., e^i, \tag{15}$$

where i is the smallest integer for which these vectors are linearly independent.

We note that the usual choice of the initial orthogonal vectors is $q^{0,j} = e^{j+1}$, $j = 0,..., n-1$, so that $y^{1,j}$, $j = 1,..., n$ are just the vectors given by the nonlinear Gauss–Seidel procedure. The intent of the orthogonalization of (15) is to obtain a set of directions $q^{k+1,j}$ which, when g is the quadratic functional $\frac{1}{2}x^\mathrm{T}Ax - b^\mathrm{T}x + c$, approximate the n eigenvectors of A. Since these eigenvectors are a conjugate basis for A (see **E 8.4-4**), Rosenbrock's method may be thought of as an "asymptotic conjugate-direction method."

NOTES AND REMARKS

NR 8.4-1. The conjugate-direction methods, and, in particular, the conjugate gradient method, were first developed for linear systems of equations by Hestenes and Stiefel in a series of individual and joint papers; see, for example, Hestenes and Stiefel [1952].

NR 8.4-2. The initially surprising result **8.4.2** is actually geometrically obvious. It is clear that for a "diagonal" quadratic functional

$$g(x) = \sum_{i=1}^{n} (\lambda_i x_i^2 - b_i x_i) + c,$$

successive minimization along the coordinate directions $e^1,..., e^n$ will produce the minimizer in, at most, n steps. Now, for the general quadratic functional $g(x) = \frac{1}{2}x^\mathrm{T}Ax - b^\mathrm{T}x + c$, the change of variable $x = Py$, where $P = (p^0,..., p^{n-1})$, yields the new functional

$$f(y) \equiv g(Py) = \tfrac{1}{2} y^\mathrm{T} P^\mathrm{T} A P y - b^\mathrm{T} P y + c.$$

But the definition of a conjugate basis implies that $P^\mathrm{T}AP$ is a diagonal matrix, so that $f(y)$ is in "diagonal form." Therefore, since

$$g(x - \alpha p^j) = g(Py - \alpha P e^j) = f(y - \alpha e^j),$$

it follows that minimization of g along the direction p^j is equivalent to minimization of f along a coordinate direction.

NR 8.4-3. In dealing with the diagonal functional of **NR 8.4-2**, the successive steps give the minimum of g in the expanding sequence of affine subspaces $V(x^0; e^1,..., e^k)$, $k = 1,..., n$, where, in general,

$$V(z; y^1,..., y^k) = \left\{ x \in R^n \mid x = z + \sum_{i=1}^{k} \alpha_i y^i, \ \forall \alpha_1,..., \alpha_k \in R^1 \right\}.$$

Correspondingly, for general quadratic g, we obtain minima successively in the subspaces $V(x^0; p^1,..., p^k)$. (See **E 8.4-2**.)

NR 8.4-4. The Davidon–Fletcher–Powell method (8.2.9)–(8.2.11) may be considered to be a conjugate-direction method in the sense that the directions $x^{k+1} - x^k$, $k = 0,..., n-1$, are conjugate when g is a quadratic functional; this is shown in **8.6**. A somewhat similar situation occurs in the *Partan* (parallel tangents) *method* of Shah, Buehler, and Kempthorne [1964], and in a related method of Powell [1962]. The description of the Partan algorithm for a differentiable functional $g: R^n \to R^1$ is as follows:

(a) For given x^0, choose $p^1 \in R^n$ so that $g'(x^0) p^1 \neq 0$ and obtain x^1 so that $g(x^1) = \min\{g(x^0 - \alpha p^1) \mid \alpha \in R^1\}$.

(b) Choose $\hat{p}^1 \in R^n$ so that $g'(x^0) \hat{p}^1 = 0$ and $g'(x^1) \hat{p}^1 \neq 0$, and compute $\hat{x}^1$ such that $g(\hat{x}^1) = \min\{g(x^0 - \alpha \hat{p}^1) \mid \alpha \in R^1\}$. Then set $\tilde{p}^1 = x^0 - \hat{x}^1$ and obtain x^2 to satisfy $g(x^2) = \min\{g(\hat{x}^1 - \alpha \tilde{p}^1) \mid \alpha \in R^1\}$.

(c) At the kth stage, $k \leq n - 1$, select $\hat{p}^k$ so that

$$g'(x^i) \hat{p}^k = 0, \quad i = 0,..., k-1, \quad g'(x^k) \hat{p}^k \neq 0.$$

Set $\hat{x}^k = x^k - \hat{\alpha}_k \hat{p}^k$, $\tilde{p}^k = x^{k-1} - \hat{x}^k$, and $x^{k+1} = \hat{x}^k - \alpha_k \tilde{p}^k$, where $\hat{\alpha}_k$ and α_k are again chosen by the minimization principle.

If g is a positive-definite quadratic functional with unique minimizer x^*, it may be shown that $x^m = x^*$ for some $m \leq n$ and that $p^k = x^{k+1} - x^k$, $k = 0, 1,..., m-1$, are conjugate with respect to A. If g is not quadratic, then the algorithm will not, in general, terminate in finitely many steps, and the process may by continued beyond step (c) by restarting after $k = n - 1$ or by:

(d) if $k \geq n$, choose $\hat{p}^k$ such that

$$g'(x^i) \hat{p}^k = 0, \quad i = k - n + 1,..., k-1, \quad g'(x^k) \hat{p}^k \neq 0$$

and proceed as in (c).

For an elementary geometrical discussion of Partan, see Wilde [1964].

8.4. CONJUGATE-DIRECTION METHODS

NR 8.4-5. The algorithm (8) was introduced by Daniel [1967a], and that of (10)–(11) by Fletcher and Reeves [1964]. Still another choice for β_k is that for which $g'(x^{k+2}) p^k = 0$. This was proposed by Takahashi [1965].

NR 8.4-6. The Rosenbrock method was described in Rosenbrock [1960] in a slightly different form than presented in the text. In particular, it was Swann [1964] who suggested choosing the $\alpha_{k,j}$ by the minimization principle (13), while Elkin [1968] advocated the orthogonalization of (15) if the set (14) is linearly dependent. A major expenditure of time in the Rosenbrock method is required by the orthogonalization, and, for this case, Powell [1968] has recently suggested a more efficient orthogonalization procedure than the usual Gram–Schmidt process. A method related to the Rosenbrock algorithm has also been given by Baer [1962a].

NR 8.4-7. Powell [1964] presented a conjugate-direction type of algorithm with the advantage that no derivatives of the functional g are required. It was claimed that when g was quadratic and positive definite, the method would produce convergence in finitely many steps, but Zangwill [1967] produced a counterexample to this claim and also analyzed a revised form of the algorithm based upon a suggestion in Powell's paper.

In order to describe the Zangwill process, assume, for simplicity, that $g\colon R^n \to R^1$ is defined on all of R^n. As in the Rosenbrock method, the iteration begins with an initial point x^0 and n linearly independent unit vectors $q^{0,0},\ldots, q^{0,n-1}$; in addition, an index m_0 is set equal to one. At the end of the $(k-1)$th stage ($k \geqslant 1$), an index m_k will have been obtained and the $n+1$ vectors transformed into x^k and n linearly independent unit vectors $q^{k,0},\ldots, q^{k,n-1}$. The next stage of the algorithm then consists of two parts:

I. Let m_{k+1} be the smallest index following m_k (mod n) $+ 1$ such that

$$y^{k,0} = x^k - \beta_k e^{m_{k+1}} \neq x^k,$$

where β_k is determined by the minimization principle:

$$g(y^{k,0}) = \min\{g(x^k - \beta e^{m_{k+1}}) \mid \beta \in (-\infty, +\infty)\}.$$

Then compute $y^{k,j}$, $j = 1,\ldots, n$ by (12) and (13).

II. Set

$$p^k = (y^{k,n} - x^k)/\| y^{k,n} - x^k \|,$$

and

$$x^{k+1} = x^k - \alpha_k p^k,$$

where

$$g(x^{k+1}) = \min\{g(x^k - \alpha p^k) \mid \alpha \in (-\infty, +\infty)\}.$$

and update the $q^{k,j}$ as follows:

$$q^{k+1,j} = q^{k,j+1}, \quad j = 0,..., n-2, \quad q^{k+1,n-1} = p^k.$$

Zangwill shows that when g is a quadratic functional with symmetric, positive definite matrix A, then the vectors

$$q^{k,n-k-1}, q^{k,n-k},..., q^{k,n-1}, \quad k = 0, 1,...,$$

are conjugate under A, and, hence, the process stops at the solution after at most n steps. Zangwill also proves a convergence theorem for the method when g is strictly convex and continuously differentiable, and when all points $y^{k,0},..., y^{k,n}$, x^k remain in a compact set.

Another modification of the Powell method is discussed in Kowalik and Osborne [1968].

NR 8.4-8. Box [1966] and Leon [1966] have made numerical comparisons of several of the algorithms discussed here, including, in particular, the Davidon–Fletcher–Powell, Rosenbrock, steepest-descent, and Partan algorithms.

EXERCISES

E 8.4-1. Let $g: R^n \to R^1$ be the quadratic functional $g(x) = \frac{1}{2}x^T Ax - b^T x + c$, where $A \in L(R^n)$ is symmetric, positive definite. For arbitrary $x^0 \in R^n$ and linearly independent vectors $p^0,..., p^{k-1}$, let $V_k = V(x^0; p^0,..., p^{k-1})$ be the affine subspace defined in **NR. 8.4-3**. Show that the unique minimizer of the restriction of g to V_k is given by

$$x^0 + P_k(P_k^T A P_k)^{-1} P_k^T(b - Ax^0),$$

where $P_k \in L(R^k, R^n)$ is the matrix with columns $p^0,..., p^{k-1}$.

E 8.4-2. Assume that the conditions and notation of **E 8.4-1** hold, and let $p^0,..., p^{n-1}$ be a conjugate basis with respect to A. Show that the conjugate-direction iterates (2)–(3) satisfy

$$g(x^k) = \min\{g(x) \mid x \in V_k\}, \quad k = 1,..., n.$$

E 8.4-3. Let $g(x) = \frac{1}{2}x^T Ax - b^T x + c$ be a quadratic functional, where $A \in L(R^n)$ is symmetric, positive definite. Let V be any linear subspace of R^n, and for any $z^1, z^2 \in R^n$ define y^1, y^2 by

$$g(y^i) = \min_{x \in V} g(z^i + x), \quad i = 1, 2.$$

Show that $y^2 - y^1$ is A-orthogonal to any $w \in V$.

E 8.4-4. Let $A \in L(R^n)$ be symmetric, positive definite. Show that any n linearly independent nonzero eigenvectors of A form a conjugate basis for A. More generally, show that $p^1,\ldots, p^n$ is a conjugate basis with respect to A if and only if $P = QUD$, where $P = (p^1,\ldots, p^n)$, Q is a matrix of eigenvectors of A scaled so that $Q^T A Q = I$, U is an arbitrary orthogonal matrix, and D is an arbitrary nonsingular diagonal matrix.

8.5. THE GAUSS-NEWTON AND RELATED METHODS

For given mappings $F: D \subset R^n \to R^m$ and $f: R^m \to R^1$, define $g: D \subset R^n \to R^1$ by

$$g(x) = f(Fx), \quad \forall x \in D. \tag{1}$$

As discussed in **4.1**, if f is the quadratic functional $\frac{1}{2} x^T x$, then a minimizer of

$$g(x) = \frac{1}{2}(Fx)^T Fx \tag{2}$$

is a least-squares solution of the equation $Fx = 0$. Alternatively, if $m = n$, and $x = 0$ is a unique global minimizer of f, then the functional g of (1) has a global minimizer x^* if and only if $Fx^* = 0$. Hence, procedures for minimizing (1) are procedures for solving $Fx = 0$.

Any of the methods described in the previous sections may, of course, be applied to (1), and we shall consider here only some additional iterations which utilize the special form of (1).

Assume that F is F-differentiable on D and that x^k is an approximation to a minimizer of (1). Then a natural approach to the computation of the next approximation x^{k+1} is to minimize the functional

$$g_k(x) = f(Fx^k + F'(x^k)(x - x^k)), \tag{3}$$

obtained by linearizing F about x^k. If f is twice F-differentiable, then we may approximate a minimizer of (3) by one step of the Newton iteration (8.1.4) starting from x^k. Since

$$g_k'(x) = f'(Fx^k + F'(x^k)(x - x^k)) F'(x^k),$$

and the Hessian matrix is

$$H_{g_k}(x) = F'(x^k)^T H_f(Fx^k + F'(x^k)(x - x^k)) F'(x^k),$$

this gives the algorithm

$$\begin{aligned} x^{k+1} &= x^k - H_{g_k}(x^k)^{-1} g_k'(x^k)^T \\ &= x^k - [F'(x^k)^T H_f(Fx^k) F'(x^k)]^{-1} F'(x^k)^T f'(Fx^k)^T, \end{aligned} \tag{4}$$

where we assume, of course, that the indicated inverse exists. In the special case that $f(x) = \tfrac{1}{2} x^\mathrm{T} x$, (4) reduces to the *Gauss–Newton method*

$$x^{k+1} = x^k - [F'(x^k)^\mathrm{T} F'(x^k)]^{-1} F'(x^k)^\mathrm{T} F x^k, \qquad k = 0, 1, \ldots, \tag{5}$$

and, therefore, we shall call (4) a *generalized Gauss–Newton method*. Note that the iterate x^{k+1} of (5) is simply the unique global minimizer of the quadratic functional

$$[F x^k + F'(x^k)(x - x^k)]^\mathrm{T} [F x^k + F'(x^k)(x - x^k)],$$

and, hence, (5) is a paraboloid method for (2) in the sense of **8.1**.

In place of (5), we may also consider the modified Gauss–Newton iteration with parameters ω_k and λ_k:

$$x^{k+1} = x^k - \omega_k [F'(x^k)^\mathrm{T} F'(x^k) + \lambda_k I]^{-1} F'(x^k)^\mathrm{T} F x^k. \tag{6}$$

Since $F'(x^k)^\mathrm{T} F'(x^k)$ is symmetric, positive semidefinite, the inverse in (6) will always exist provided only that $\lambda_k > 0$. As in **8.2**, the parameter ω_k may be chosen to ensure that $g(x^{k+1}) \leqslant g(x^k)$. Alternatively, this decrease may also be achieved by a suitable selection of λ_k (see **E 8.3-4**).

If $F'(x)$ is difficult to evaluate, we may, of course, as in **7.1**, approximate the derivatives by difference quotients in order to obtain discrete Gauss–Newton methods. Similarly, as in **7.2**, we also may use secant-type approximations to $F'(x)$. At the kth stage, let $x^{k,j}$, $j = 0, 1, \ldots, n$, be $n+1$ points in general position (see **7.2.1**) and consider the interpolating affine function $L_k : R^n \to R^m$ defined by

$$L_k x = a^k + A_k x, \tag{7}$$

where $a^k \in R^m$ and $A_k \in L(R^n, R^m)$ are chosen so that

$$L_k x^{k,j} = F x^{k,j}, \qquad j = 0, 1, \ldots, n, \tag{8}$$

(see **E 8.5-1**). Next, approximate the quadratic functional (2) by

$$g_k(x) \equiv \tfrac{1}{2} (L_k x)^\mathrm{T} L_k x, \tag{9}$$

and take x^{k+1} as a minimizer of g_k. If $m \geqslant n$ and if $F x^{k,j}$, $j = 0, 1, \ldots, n$, are in general position in R^m, then it follows, as in **7.4.3**, that A_k has rank n, and, hence, that $A_k^\mathrm{T} A_k$ is nonsingular (see **E 8.5-1**). Thus, (9) has a unique global minimizer given by

$$x^{k+1} = -(A_k^\mathrm{T} A_k)^{-1} A_k^\mathrm{T} a^k. \tag{10}$$

8.5. THE GAUSS-NEWTON AND RELATED METHODS

In order to put (9) into the form of the Gauss–Newton method (5), we proceed as in **7.2** and introduce the matrices

$$H_k = (x^{k,1} - x^{k,0},\ldots, x^{k,n} - x^{k,0}), \qquad \Gamma_k = (Fx^{k,1} - Fx^{k,0},\ldots, Fx^{k,n} - Fx^{k,0}).$$

As in **7.2.6**, it follows from (7) and (8) that

$$A_k H_k = \Gamma_k, \qquad a^k = -A_k x^k + Fx^k,$$

where we have set $x^{k,0} = x^k$. Consequently, (10) may be written as

$$\begin{aligned}x^{k+1} &= x^k - [(\Gamma_k H_k^{-1})^T \Gamma_k H_k^{-1}]^{-1} (\Gamma_k H_k^{-1})^T Fx^k \\ &= x^k - H_k(\Gamma_k^T \Gamma_k)^{-1} \Gamma_k^T Fx^k,\end{aligned} \qquad (11)$$

which is the basic form of the *Gauss–secant method*.

Any of the choices of points indicated in **7.2** can be used here as well, and, as in the case of the secant method, certain choices can reduce the computational work considerably. For example, if we select some index j_0 and set

$$x^{k+1,j} = x^{k,j}, \quad j = 0,\ldots, n, \quad j \neq j_0, \qquad x^{k+1,j_0} = x^{k+1},$$

then $\Gamma_{k+1}^T \Gamma_{k+1}$ differs from $\Gamma_k^T \Gamma_k$ only in the j_0th row and column; hence, $(\Gamma_{k+1}^T \Gamma_{k+1})^{-1}$ may be obtained from $(\Gamma_k^T \Gamma_k)^{-1}$ by applying the Sherman–Morrison formula (2.3.14) twice. As with the corresponding secant method, however, this procedure may be unstable and should be used with caution.

NOTES AND REMARKS

NR 8.5-1. The Gauss–Newton method (5), which is also sometimes called the Gauss method or method of differential corrections, has been discussed by several authors; see, for example, Hartley [1961], who considers (6) with $\lambda_k = 0$ and ω_k chosen by the minimization principle. Note that (5) is *not* Newton's method applied to the gradient equation of $g(x) = \frac{1}{2}(Fx)^T Fx$. In fact, if $F: R^n \to R^m$ is twice F-differentiable, then

$$g'(x)^T = F'(x)^T Fx, \qquad H_g(x) = F''(x) Fx + F'(x)^T F'(x),$$

and, hence, Newton's method for $g'(x)^T = 0$ has the form

$$x^{k+1} = x^k - [F''(x^k) Fx^k + F'(x^k)^T F'(x^k)]^{-1} F'(x^k)^T Fx^k. \qquad (12)$$

Note, also, that if $m = n$, and $F'(x^k)$ is nonsingular, then (5) reduces to Newton's method for $Fx = 0$, whereas (12) does not.

NR 8.5-2. The introduction of the parameter λ_k into the method (6) is due to Levenberg [1944]. Marquardt [1963] has also considered the use of this parameter to ensure that $g(x^{k+1}) \leqslant g(x^k)$ (see **E 8.3-4**).

NR 8.5-3. There are various steplength algorithms which may be used in conjunction with minimizing $g(x) = f(Fx)$ and which utilize this special form. The exact minimization principle, which requires that $f(F(x^k - \alpha p^k))$ be minimized, may be modified in a way analogous to the Gauss–Newton method by minimizing $f(Fx^k - \alpha F'(x^k) p^k)$ instead. In the case $f(x) = x^\mathrm{T} x$, α_k is then given explicitly by

$$\alpha_k = (Fx^k)^\mathrm{T} F'(x^k) p^k / [F'(x^k) p^k]^\mathrm{T} F'(x^k) p^k.$$

A different approach is to choose α_k to minimize $f(x^k - \alpha p^k - x^*)$, where x^* is the solution of $Fx = 0$. Since x^* is not known, this cannot be carried out exactly, but, if x^k is close to x^*, then $x^k - x^* \doteq F'(x^k)^{-1} Fx^k$ and we can determine α_k to minimize $f(F'(x^k)^{-1} Fx^k - \alpha p^k)$. Again, in the case $f(x) = x^\mathrm{T} x$, α_k is given by

$$\alpha_k = (p^k)^\mathrm{T} F'(x^k)^{-1} Fx^k / (p^k)^\mathrm{T} p^k;$$

with the choice $p^k = g'(x^k)^\mathrm{T} = F'(x^k)^\mathrm{T} Fx^k$, this gives the algorithm

$$x^{k+1} = x^k - [(Fx^k)^\mathrm{T} Fx^k / (Fx^k)^\mathrm{T} F'(x^k) F'(x^k)^\mathrm{T} Fx^k] F'(x^k)^\mathrm{T} Fx^k,$$

which has been treated by Fridman [1961].

NR 8.5-4. If $F: R^n \to R^m$, $m > n$, and $F'(x^k)$ has rank less than n, then $F'(x^k)^\mathrm{T} F'(x^k)$ is singular and the next step of the Gauss–Newton iteration (5) is not defined. In this case, Ben-Israel [1965] has suggested replacing (5) by

$$x^{k+1} = x^k - F'(x^k)^+ Fx^k, \tag{13}$$

where, for arbitrary $A \in L(R^n, R^m)$, A^+ denotes a generalized inverse of A, defined as the unique linear operator in $L(R^m, R^n)$ which satisfies

$$AA^+A = A, \qquad A^+AA^+ = A^+, \qquad (A^+A)^\mathrm{T} = A^+A, \qquad (AA^+)^\mathrm{T} = AA^+,$$

(see, for example, Penrose [1955]). In particular, if $A^\mathrm{T} A$ is nonsingular, then $A^+ = (A^\mathrm{T} A)^{-1} A^\mathrm{T}$, so that whenever $F'(x^k)$ has rank n, (13) reduces to (5). (See also **NR 7.3-9**.)

NR 8.5-5. Modification methods in the sense of Section **7.3** are also easily constructed in the setting of this section. Consider, for example, the iteration

$$x^{k+1} = x^k - (B_k^\mathrm{T} B_k)^{-1} B_k^\mathrm{T} Fx^k, \qquad k = 0, 1, \ldots,$$

where, again, $F: R^n \to R^m$ and $B_k \in L(R^n, R^m)$. If B_{k+1} is obtained by replacing one column of B_k by a vector b^k, then $(B_{k+1}^\mathrm{T} B_{k+1})^{-1}$ is easily calculated from

8.6. APPENDIX 1. THE DAVIDON-FLETCHER-POWELL METHOD

$(B_k^T B_k)^{-1}$ by applying the Sherman–Morrison formula (2.3.14) twice. One possible choice of b^k is $F'(x^k)(x^{k+1} - x^k)$, as was also used in **7.3**. A similar choice, involving difference quotients in place of derivatives, has been suggested by Powell [1965].

NR 8.5-6. Computational aspects of the methods of this section have been discussed by Powell [1965] and Maddison [1966], among others.

EXERCISES

E 8.5-1. Let $x^0,..., x^n \in R^n$ be in general position and let $y^0,..., y^n \in R^m$, where $m \geqslant n$. Modify the proof of **7.2.3** to show that there exist $a \in R^m$ and $A \in L(R^n, R^m)$ such that $a + Ax^i = y^i$, $i = 0,..., n$. Show, moreover, that rank$(A) = n$ if and only if $y^0,..., y^n$ are in general position.

8.6. APPENDIX 1.
CONVERGENCE OF THE CONJUGATE GRADIENT AND THE DAVIDON–FLETCHER–POWELL ALGORITHMS FOR QUADRATIC FUNCTIONALS

In this appendix, we show that the Davidon–Fletcher–Powell method discussed in **8.2** and the conjugate gradient method of **8.4** converge in finitely many steps for quadratic functionals. The proof of the following result for the conjugate gradient method is adapted from that given in Hestenes [1956].

8.6.1. Given the quadratic functional $g(x) = \frac{1}{2}x^T A x - b^T x + c$, where $A \in L(R^n)$ is symmetric, positive definite, the conjugate gradient iterates (8.4.5)–(8.4.6) satisfy $x^m = A^{-1}b$ for some $m \leqslant n$. Moreover,

$$(Ap^i)^T p^j = 0, \quad i \neq j, \quad 0 \leqslant i, \ j \leqslant m. \tag{1}$$

Proof. We note first that α_k and β_k, and hence x^{k+1} and p^{k+1}, are well-defined provided that $p^k \neq 0$. If $p^0 = Ax^0 - b = 0$, then x^0 is the solution and (1) holds vacuously. Assume, then, that $p^k \neq 0$, $k = 0,..., m-1$, for some $m \leqslant n$ and set $r^k = Ax^k - b$, $k = 0,..., m$. It follows that $r^k \neq 0$, and hence that $\alpha_k \neq 0$, for $k = 0,..., m-1$; for, if some $r^j = 0$, then $\beta_{j-1} = 0$, which leads to the contradiction $p^j = r^j = 0$.

Now note that $x^{j+1} = x^j - \alpha_j p^j$, $j = 0,..., m-1$, is equivalent to

$$r^{j+1} = r^j - \alpha_j A p^j, \quad j = 0,..., m-1, \tag{2}$$

and, therefore, by the definition of α_j,

$$(r^{j+1})^\mathrm{T} p^j = (r^j)^\mathrm{T} p^j - [(r^j)^\mathrm{T} p^j/(Ap^j)^\mathrm{T} p^j](Ap^j)^\mathrm{T} p^j = 0, \quad j = 0,\ldots, m-1. \quad (3)$$

Note also that, by the definition of β_j,

$$(Ap^{j+1})^\mathrm{T} p^j = (Ar^{j+1})^\mathrm{T} p^j - \beta_j (Ap^j)^\mathrm{T} p^j = 0, \quad j = 0,\ldots, m-1. \quad (4)$$

We now introduce the induction hypothesis

$$\left.\begin{array}{l}(Ap^k)^\mathrm{T} p^j = 0,\\ (r^k)^\mathrm{T} r^j = 0,\end{array}\right\} \quad j = 0,\ldots, k-1, \quad \begin{array}{l}(5)\\ (6)\end{array}$$

for some $k \leqslant m$. Clearly, (4) and (3) show that (5) and (6), respectively, hold for $k = 1$. It then follows from (5) and (6) as well as (2) that

$$(r^{k+1})^\mathrm{T} r^j = (r^k - \alpha_k Ap^k)^\mathrm{T} r^j = (r^k)^\mathrm{T} r^j - \alpha_k (Ap^k)^\mathrm{T} (p^j + \beta_{j-1} p^{j-1}) = 0,$$

$$j = 0,\ldots, k-1, \quad (7)$$

where we have used the convention that $\beta_{-1} = 0$. Moreover, by (3)–(5),

$$(r^{k+1})^\mathrm{T} r^k = (r^{k+1})^\mathrm{T} (p^k + \beta_{k-1} p^{k-1}) = \beta_{k-1} (r^{k+1})^\mathrm{T} p^{k-1}$$
$$= \beta_{k-1} (r^k - \alpha_k Ap^k)^\mathrm{T} p^{k-1} = 0.$$

Hence, (6) holds for $k+1$.

In order to show that (5) holds for $k+1$, we observe that, since $\alpha_j \neq 0$,

$$(Ap^{k+1})^\mathrm{T} p^j = (r^{k+1} - \beta_k p^k)^\mathrm{T} Ap^j = (r^{k+1})^\mathrm{T} Ap^j$$
$$= (r^{k+1})^\mathrm{T} [(1/\alpha_j)(r^j - r^{j+1})] = 0, \quad j = 0,\ldots, k-1,$$

while (4) shows that $(Ap^{k+1})^\mathrm{T} p^k = 0$. The induction is now complete and (1) is valid.

To complete the proof, assume that $m < n$ and $p^m = 0$. It then follows from (3) that

$$0 = (p^m)^\mathrm{T} p^m = (r^m)^\mathrm{T} r^m - 2\beta_{m-1}(r^m)^\mathrm{T} p^{m-1} + \beta_{m-1}^2 (p^{m-1})^\mathrm{T} p^{m-1} \geqslant (r^m)^\mathrm{T} r^m.$$

Hence, $x^m = A^{-1}b$. On the other hand, if $m = n$, then $p^0,\ldots, p^{n-1}$ form a conjugate basis for A and the result follows from **8.4.2**. ∎

As a consequence of (5) and (6), we may now establish the validity of the alternative form (8.4.9) for the computation of β_k.

8.6. APPENDIX 1. THE DAVIDON-FLETCHER-POWELL METHOD

8.6.2. The computation of β_k given by (8.4.6) is equivalent to

$$\beta_k = -(Ax^{k+1} - b)^T (Ax^{k+1} - b)/(Ax^k - b)^T (Ax^k - b).$$

Proof. By (3), we obtain

$$(r^{k+1})^T p^{k+1} = (r^{k+1})^T (r^{k+1} - \beta_k p^k) = (r^{k+1})^T r^{k+1},$$

and, by (5) and (6),

$$(p^{k+1})^T r^{k+1} = (p^{k+1})^T (r^k - \alpha_k A p^k) = (p^{k+1})^T r^k$$
$$= (r^{k+1} - \beta_k p^k)^T r^k = -\beta_k (p^k)^T r^k.$$

Consequently,

$$\beta_k = -(p^{k+1})^T r^{k+1}/(p^k)^T r^k = -(r^{k+1})^T r^k/(r^k)^T r^k \quad \blacksquare$$

We finish this appendix by showing that the Davidon–Fletcher–Powell iterates converge in at most n steps when $g: R^n \to R^1$ is the quadratic functional

$$g(x) = \tfrac{1}{2} x^T A x - b^T x + c. \tag{8}$$

In this case, the iteration (8.2.9)–(8.2.11) takes the form

$$x^{k+1} = x^k - \alpha_k H_k (Ax^k - b), \tag{9}$$

$$H_{k+1} = H_k + \frac{r^k (r^k)^T}{(r^k)^T q^k} - \frac{(H_k q^k)(H_k q^k)^T}{(q^k)^T H_k q^k} \tag{10}$$

where

$$r^k = x^{k+1} - x^k, \qquad q^k = A r^k. \tag{11}$$

We assume, moreover, that α_k is determined by the minimization principle, that is,

$$\alpha_k = (Ax^k - b)^T H_k (Ax^k - b)/(Ax^k - b)^T H_k A H_k (Ax^k - b). \tag{12}$$

The following result was proved by Fletcher and Powell [1963].

8.6.3. Assume that A, $H_0 \in L(R^n)$ are symmetric, positive definite. Then, for any $x^0 \in R^n$, the iterates (9)–(12) are well defined and $x^m = A^{-1} b$ for some $m \leq n$. Moreover, if $x^k \neq A^{-1} b$ for $k = 0, ..., n - 1$, then $H_n = A^{-1}$.

Proof. Assume that H_k is well defined and symmetric, positive definite, and that $x^k \neq A^{-1} b$. Then $\alpha_k > 0$, so that $(r^k)^T q^k > 0$. Hence 8.2.3 shows that H_{k+1} is well defined and symmetric, positive definite. Therefore, by induction, all x^k are well defined provided that $x^{k-1} \neq A^{-1} b$, and we may assume that $x^k \neq A^{-1}b$, $k = 0,\ldots, n - 1$, or else the result is proved.

We note first that α_k is the unique minimizer of

$$\varphi(\alpha) = g(x^k - \alpha H_k(Ax^k - b)),$$

and, therefore, the root of $\varphi'(\alpha) = 0$. Hence,

$$(Ax^{k+1} - b)^T r^k = 0, \qquad k = 0,\ldots, n - 1. \tag{13}$$

Moreover, from (10) and (11), we obtain

$$H_{k+1} A r^k = H_k q^k + r^k - H_k q^k = r^k, \qquad k = 0, 1,\ldots, n - 1, \tag{14}$$

so that, in particular, $H_1 A r^0 = r^0$, and, by (13),

$$(r^1)^T A r^0 = -\alpha_1 [H_1(Ax^1 - b)]^T A r^0 = -\alpha_1 (Ax^1 - b)^T r^0 = 0.$$

We will now prove by induction that

$$H_k A r^j = r^j, \qquad j = 0, 1,\ldots, k - 1, \tag{15}$$

and

$$(r^i)^T A r^j = 0, \qquad i \neq j, \quad 0 \leqslant i,j \leqslant k, \tag{16}$$

for $k = 1,\ldots, n$. For $k = 1$ we have shown that (15) and (16) are valid; assume they hold for $k < n$. Then, for any $0 \leqslant j \leqslant k - 1$, we have, by (10), that

$$H_{k+1} A r^j = H_k A r^j + \frac{(r^k)^T A r^j}{(r^k)^T q^k} r^k - \frac{(H_k q^k)(q^k)^T H_k A r^j}{(q^k)^T H_k q^k}$$

$$= r^j - \frac{(H_k q^k)(q^k)^T r^j}{(q^k)^T H_k q^k} = r^j - \frac{(H_k q^k)(r^k)^T A r^j}{(q^k)^T H_k q^k} = r^j,$$

which, by (14), also holds for $j = k$. Hence, (15) remains valid with k replaced by $k + 1$. Next, note that

$$Ax^{k+1} - b = Ax^{j+1} - b + A(r^{j+1} + \cdots + r^k), \qquad j < k,$$

and, hence, by (15) and (13),

$$(r^j)^T (Ax^{k+1} - b) = (r^j)^T (Ax^{j+1} - b) = 0, \qquad \forall j \leqslant k. \tag{17}$$

Thus,

$$(r^{k+1})^T A r^j = -\alpha_{k+1}(Ax^{k+1} - b)^T H_{k+1} A r^j = -\alpha_{k+1}(Ax^{k+1} - b)^T r^j = 0$$

for $j = 1,\ldots, k$. This completes the induction, and (15) and (16) hold for all $k \leqslant n$.

By assumption, $x^k \neq A^{-1}b$, for $k = 0,..., n - 1$, and, therefore, $r^k \neq 0$, $k = 0,..., n - 1$. Hence, it follows from **8.4.2** that $x^n = A^{-1}b$. Finally, (15) shows that $H_n A$ has n linearly independent eigenvectors $r^0,..., r^{n-1}$ with eigenvalue unity, so that $H_n A = I$. ∎

8.7. APPENDIX 2. SEARCH METHODS FOR ONE-DIMENSIONAL MINIMIZATION

In **(a)** of **8.3** we introduced the exact minimization principles for determining the steplength. In this appendix, we discuss some methods for the approximate minimization of one-dimensional functions and, in particular, of the following special class of functions.

8.7.1. Definition. A function $\varphi: [a, b] \subset R^1 \to R^1$ is *strictly unimodal* on $[a, b]$ if there exists a $t^* \in [a, b]$ such that $\varphi(t^*) = \min\{\varphi(t) \mid t \in [a, b]\}$, and if, for any $a \leqslant t_1 < t_2 \leqslant b$,

$$\begin{cases} t_2 \leqslant t^* & \text{implies that} \quad \varphi(t_1) > \varphi(t_2), \\ t^* \leqslant t_1 & \text{implies that} \quad \varphi(t_2) > \varphi(t_1). \end{cases} \quad (1)$$ ∎

Clearly, any strictly unimodal function on $[a, b]$ has a unique global minimizer on $[a, b]$.

Strict unimodality is equivalent to the concept of strict quasiconvexity (see **4.2.5**), as the following result of Elkin [1968] shows.

8.7.2. Suppose that $\varphi: [a, b] \subset R^1 \to R^1$ and there is a $t^* \in [a, b]$ for which $\varphi(t^*) = \min\{\varphi(t) \mid t \in [a, b]\}$. Then φ is strictly unimodal on $[a, b]$ if and only if φ is strictly quasiconvex.

Proof. Suppose that φ is strictly quasiconvex and $a \leqslant t_1 < t_2 \leqslant b$. Then:

$t_1 < t_2 \leqslant t^*$ implies that $\varphi(t_2) < \max\{\varphi(t_1), \varphi(t^*)\} = \varphi(t_1)$,

$t^* \leqslant t_1 < t_2$ implies that $\varphi(t_1) < \max\{\varphi(t_2), \varphi(t^*)\} = \varphi(t_2)$,

which shows that φ is strictly unimodal. Conversely, suppose that φ is strictly unimodal on $[a, b]$, and set $t_\alpha = (1 - \alpha) t_1 + \alpha t_2$, $0 < \alpha < 1$. Then

$t^* \geqslant t_\alpha > t_1$ implies that $\varphi(t_\alpha) < \varphi(t_1) \leqslant \max\{\varphi(t_1), \varphi(t_2)\}$,

$t^* \leqslant t_\alpha < t_2$ implies that $\varphi(t_\alpha) < \varphi(t_2) \leqslant \max\{\varphi(t_1), \varphi(t_2)\}$,

which shows that φ is strictly quasiconvex. ∎

Now assume that φ is strictly unimodal on $[a, b]$ with a minimizer t^*. Then, if t_1, t_2 are any two points such that $a \leqslant t_1 < t_2 \leqslant b$, we have:

$$\begin{cases} \varphi(t_1) > \varphi(t_2) & \text{implies that} \quad t^* \in (t_1, b), \\ \varphi(t_1) = \varphi(t_2) & \text{implies that} \quad t^* \in (t_1, t_2), \\ \varphi(t_1) < \varphi(t_2) & \text{implies that} \quad t^* \in (a, t_2). \end{cases} \quad (2)$$

Therefore, by evaluating φ at t_1 and t_2 and comparing function values, (2) allows us to reduce the size of the interval known to contain t^*.

This idea can easily be used to construct sequences which converge to t^*. The simplest approach is to start at the midpoint $t_0 = \frac{1}{2}(a + b)$ and, if φ is, say, decreasing for $t > t_0$, we test φ at $t_0 + jh_0$, $j = 1, 2,...$, until we find a point t_1 from where φ begins to increase again (or until we reach b). Then we repeat this procedure starting at t_1 and using a smaller steplength h_1. A precise algorithm of this type, due to Berman [1966], is given in the flow chart of Fig. **8.6**.

The total number of evaluations of φ needed for executing the algorithm up to some index k depends on the location of t^*. If, for example, $t^* = b$, then we clearly need the maximal number of evaluations at each step, say q, and, thus, after k steps, we have used kq evaluations. This number will decrease the closer t^* is to t_0, and Berman [1966] has shown that, for example, if $q = 4$, the "expected" number of evalutations is three and that, asymptotically, for $q \to \infty$, this expected number tends to $\frac{1}{2}q$.

Another important minimization process is the *Fibonaccian search* method. The basic idea of this method is to perform a sequence of two-point searches to reduce the uncertainty interval, and to place the search points t_1^k, t_2^k, $k = 0, 1,...$, in such a way that, if, for example, $[a, t_2^0]$ is the first reduced uncertainty interval, then t_1^0 is used as one of the next search points; that is, we set $t_2^1 = t_1^0$, and so on. The problem is to select the sequence $\{t_1^k, t_2^k\}$ in such a way that the decrease in the length of the uncertainty interval is maximal.

Kiefer [1953] showed that the optimal choice of the t_1^k, t_2^k is as follows: Let

$$\tau_{k+1} = \tau_k + \tau_{k-1}, \quad \tau_0 = \tau_1 = 1, \quad k = 1, 2,..., \quad (3)$$

be the Fibonacci sequence and $M > 0$ the maximal number of two-point searches which are to be performed; then we use the points

$$\left.\begin{array}{l} t_1^{k+1} = (\tau_{M-1-k}/\tau_{M+1-k})(b^k - a^k) + a^k, \\ t_2^{k+1} = (\tau_{M-k}/\tau_{M+1-k})(b^k - a^k) + a^k, \end{array}\right\} \quad k = 0, 1,..., M - 2. \quad (4)$$

8.7. APPENDIX 2. ONE-DIMENSIONAL MINIMIZATION

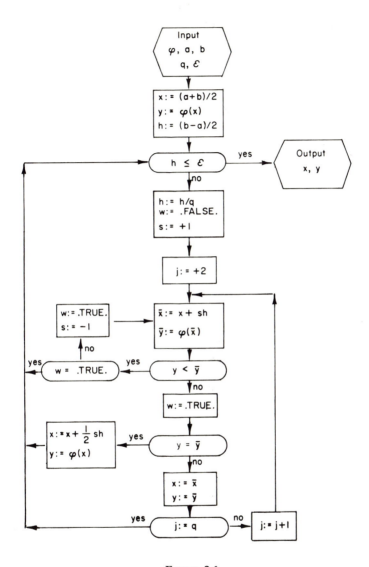

FIGURE 8.6

Here $a^0 = a$, $b^0 = b$, and

$$a^{k+1} = a^k, \quad b^{k+1} = t_2^{k+1}, \quad \text{if } \varphi(t_1^{k+1}) < \varphi(t_2^{k+1}),$$
$$a^{k+1} = t_1^{k+1}, \quad b^{k+1} = b^k, \quad \text{if } \varphi(t_1^{k+1}) \geqslant \varphi(t_2^{k+1}),$$
(5)

for $k = 0, 1,..., M-2$. For $k = M-1$, (4) has to be modified, and if $\epsilon > 0$ is the maximal roundoff error in the evaluation of φ, set

$$t_1^M = \tfrac{1}{2}(a^{M-1} + b^{M-1}) - \epsilon,$$
$$t_2^M = \tfrac{1}{2}(a^{M-1} + b^{M-1}) + \epsilon. \tag{6}$$

It is easy to prove that, for $k = 1, 2,..., M-2$,

$$\begin{aligned} t_1^{k+1} = a^k + (t_2^k - t_1^k), \quad t_2^{k+1} = t_1^k, & \quad \text{if} \quad \varphi(t_1^k) < \varphi(t_2^k), \\ t_1^{k+1} = t_2^k, \quad t_2^{k+1} = b^k - (t_2^k - t_1^k), & \quad \text{if} \quad \varphi(t_1^k) \geq \varphi(t_2^k). \end{aligned} \tag{7}$$

This shows that the process indeed requires only one function evaluation per step, except for $k = 0$.

Finally, it is easily verified that

$$b^{k+1} - a^{k+1} = \begin{cases} (\tau_{M-k}/\tau_{M+1-k})(b^k - a^k), & \text{for} \quad k \leq M-2, \\ \tfrac{1}{2}(b^{M-1} - a^{M-1}) + \epsilon, & \text{for} \quad k = M-1, \end{cases}$$

which implies that

$$b^M - a^M = (2\tau_{M+1})^{-1}(b - a) + \epsilon.$$

This permits us to determine M once an upper bound for the desired length is known.

We note that Overholt [1965] showed that the computation of the Fibonacci sequence by means of (7) leads to numerical instabilities. Hence, the original formulas (4) should be used for computation.

It is interesting to compare the efficiency of the Berman algorithm with the Fibonaccian search method. If, in the first case, we use $q = 4$, then, as stated earlier, the "expected" number of evaluations per step is three, and, hence, per evaluation, the uncertainty interval decreases by a factor $4^{-1/3} \doteq 0.63$. In comparison, the Fibonaccian search method has a reduction factor $2/(1 + \sqrt{5}) \doteq 0.62$. Of course, in the case of the Berman method, the factor 0.63 represents only an average; it can be considerably lower but also as high as $4^{-1/4} \doteq 0.87$.

Part IV / LOCAL CONVERGENCE

As mentioned in the Introduction, two of the most important problems in the study of iterative processes are: (a) When will the iterates converge? and (b) How fast is this convergence? In the following three chapters, we will discuss these problems from a "local" or "asymptotic" viewpoint.

When a sequence $\{x^k\}$ converges to x^*, it is sometimes relatively easy to draw conclusions about the asymptotic behavior of the error vectors $x^k - x^*$, that is, about their behavior as k tends to infinity, while, usually, little can be said about them when k is small. Similarly, it is frequently much simpler to assert convergence for an iterative process if the initial approximations are assumed to be close to the desired limit x^* than if they are allowed to vary in some larger domain. These two observations are the basis for the discussions in this part. The resulting local theorems are of considerable importance not only because they provide a first, and perhaps only, convergence statement, but, more importantly, because they characterize the theoretical behavior of certain iterative processes in the neighborhood of a solution. In the following part, we shall study the convergence of iterative processes from a "nonlocal" viewpoint.

The organization of this part is as follows. In Chapter 9, we introduce two different measures—analogous to the root and ratio tests for the convergence of series—of the asymptotic rate of convergence of sequences and iterative processes. These measures provide a precise means of comparing the asymptotic rate of convergence of different iterative processes, and hence a means of determining which of two processes is "faster."

In Chapters 10 and 11, the rate of convergence, as well as local convergence, of iterative processes is discussed, first from a general viewpont, and then for several important special processes. In particular, in Chapter 10, we consider one-step stationary processes of the form $x^{k+1} = Gx^k$, $k = 0, 1,...$, while, in Chapter 11, the more difficult analysis of nonstationary and multistep methods is presented.

Most of the results on rate of convergence treated in this book appear in this part. However, there are also several results requiring special techniques distributed throughout Part V. In particular, Chapter 14 contains most of the results on rates of convergence of minimization methods.

Chapter 9 / RATES OF CONVERGENCE—GENERAL

9.1. THE QUOTIENT CONVERGENCE FACTORS

Let $\mathscr{I}$ denote a general iterative process, as introduced in Definition 7.6.1, and let x^* be one of its limit points. For the purposes of this chapter, it will suffice to consider such a process simply as a collection of sequences in R^n and to disregard how these sequences are generated. Our aim is to attach to $\mathscr{I}$ precise indicators of the asymptotic rate of convergence of the process at x^*. In particular, the definition of the indicators discussed in this section is motivated by the fact that estimates of the form

$$\| x^{k+1} - x^* \| \leqslant \gamma \| x^k - x^* \|^p, \quad \forall k \geqslant k_0, \tag{1}$$

often arise naturally in the study of certain iterative processes, as will be seen, for example, in Chapter 10. We first define these indicators for an arbitrary convergent sequence which need not have been generated by an iterative process.

9.1.1. Definition. Let $\{x^k\} \subset R^n$ be any convergent sequence with limit x^*. Then the quantities

$$Q_p\{x^k\} = \begin{cases} 0, & \text{if } x^k = x^*, \text{ for all but finitely many } k, \\ \limsup_{k \to \infty} \dfrac{\| x^{k+1} - x^* \|}{\| x^k - x^* \|^p}, & \text{if } x^k \neq x^*, \text{ for all but finitely many } k, \\ +\infty, & \text{otherwise,} \end{cases}$$

defined for all $p \in [1, \infty)$, are the *quotient convergence factors*, or *Q-factors*, for short, of $\{x^k\}$ with respect to the norm $\|\cdot\|$ on R^n. ∎

Note that if $Q_p = Q_p\{x^k\} < +\infty$ for some $p \in [1, \infty)$, then, for any $\epsilon > 0$, there exists a k_0 such that (1) holds with $\gamma = Q_p + \epsilon$.

When considering not just one sequence but an iterative process $\mathscr{I}$, it is desirable that the rate-of-convergence indicator measure the worst possible asymptotic rate of convergence of any sequence of $\mathscr{I}$ with the the same limit point.

9.1.2. Definition. Let $C(\mathscr{I}, x^*)$ denote the set of all sequences with limit x^* generated by an iterative process $\mathscr{I}$. Then

$$Q_p(\mathscr{I}, x^*) = \sup\{Q_p\{x^k\} \mid \{x^k\} \in C(\mathscr{I}, x^*)\}, \quad 1 \leqslant p < +\infty,$$

are the *Q-factors* of $\mathscr{I}$ at x^* with respect to the norm in which the $Q_p\{x^k\}$ are computed. ∎

It is important to note that, in general, the Q-factors of a process depend upon the limit point x^* and that, for different limit points of $\mathscr{I}$, they may be different, (see **E 9.1-5**).

For the following discussion, note that our present interpretation of iterative processes permits us to associate with any single sequence $\{x^k\} \subset R^n$ converging to x^* an artificial process $\mathscr{I}$ for which $C(\mathscr{I}, x^*)$ consists of $\{x^k\}$ alone. As a consequence, all definitions and results of this chapter phrased in terms of iterative processes hold verbatim for sequences, and it is therefore unnecessary to include corresponding statements for single sequences.

For a given iterative process $\mathscr{I}$ and limit point x^*, we consider first the behavior of $Q_p(\mathscr{I}, x^*)$ as a function of p. The following basic result shows that Q_p is an isotone function of p which takes on only the values 0 and ∞ except at possibly one point.

9.1.3. Let $Q_p(\mathscr{I}, x^*)$, $p \in [1, \infty)$, denote the Q-factors of an iterative process at x^* in some fixed norm on R^n. Then exactly one of the following conditions holds:

(a) $Q_p(\mathscr{I}, x^*) = 0, \forall p \in [1, \infty)$;
(b) $Q_p(\mathscr{I}, x^*) = \infty, \forall p \in [1, \infty)$;
(c) There exists a $p_0 \in [1, \infty)$ such that $Q_p(\mathscr{I}, x^*) = 0, \forall p \in [1, p_0)$, and $Q_p(\mathscr{I}, x^*) = \infty, \forall p \in (p_0, \infty)$.

Proof. Let $\{x^k\} \in C(\mathscr{I}, x^*)$ be any sequence generated by $\mathscr{I}$ which

9.1. THE QUOTIENT CONVERGENCE FACTORS

converges to x^*, and set $\epsilon_k = \| x^k - x^* \|$, $k = 0, 1,\ldots$ If $\epsilon_k = 0$ for all but finitely many k, then $Q_p\{x^k\} = 0$, $p \in [1, \infty)$. Assume, therefore, that $\epsilon_k > 0$ for $k \geqslant k_0$, and suppose, further, that $Q_p\{x^k\} < +\infty$ for some $p \in (1, \infty)$. Then we have, for any $q \in [1, p)$,

$$Q_q\{x^k\} = \lim_{k \to \infty} \sup(\epsilon_{k+1} \epsilon_k^{p-q}/\epsilon_k^p) \leqslant Q_p\{x^k\} \lim_{k \to \infty} \sup \epsilon_k^{p-q} = 0,$$

which, in turn, implies that $Q_p\{x^k\} = +\infty$ whenever $Q_q\{x^k\} > 0$ and $p > q \geqslant 1$; in fact, if $Q_p < +\infty$, then we just saw that $Q_q = 0$. Hence, one and only one of the properties (a), (b), (c) holds for any sequence $\{x^k\} \in C(\mathscr{I}, x^*)$.

Now assume that neither (a) nor (b) holds, and set

$$p_0 = \inf\{p \in [1, \infty) \mid Q_p(\mathscr{I}, x^*) = +\infty\}.$$

Suppose there is a $p > p_0$ such that $Q_p(\mathscr{I}, x^*) < +\infty$, that is, $Q_p\{x^k\} < +\infty$ for every sequence $\{x^k\}$ generated by $\mathscr{I}$ which converges to x^*. Then, by the definition of p_0, there exists a $p' \in [p_0, p)$ such that $Q_{p'}(\mathscr{I}, x^*) = +\infty$, and, hence, we have $Q_{p'}\{x^k\} > 0$ for some $\{x^k\} \in C(\mathscr{I}, x^*)$. This implies that $Q_p\{x^k\} = +\infty$ which is a contradiction, and therefore it follows that $Q_p(\mathscr{I}, x^*) = +\infty$ for all $p \in (p_0, \infty)$. On the other hand, if $p_0 > 1$ and $Q_p(\mathscr{I}, x^*) \neq 0$ for some $p \in [1, p_0)$, then it follows, by a similar argument, that $Q_{p'}(\mathscr{I}, x^*) = +\infty$ for $p' \in (p, p_0)$, and this would contradict the definition of p_0. Hence, we have $Q_p(\mathscr{I}, x^*) = 0$ for all $p \in [1, p_0)$. ∎

The primary motivation for introducing the Q-factors of an iterative process is to have a precise means of comparing the rate of convergence of different iterations. We utilize the Q-factors in this connection as follows

9.1.4. Definition. Let $\mathscr{I}_1$ and $\mathscr{I}_2$ denote two iterative processes with the same limit point x^*, and let $Q_p(\mathscr{I}_1, x^*)$ and $Q_p(\mathscr{I}_2, x^*)$ be the corresponding Q-factors computed in the same norm on R^n. Then $\mathscr{I}_1$ is Q-*faster* than $\mathscr{I}_2$ at x^* if there is a $p \in [1, \infty)$ such that $Q_p(\mathscr{I}_1, x^*) < Q_p(\mathscr{I}_2, x^*)$. ∎

Note that by **9.1.3** the concept of "Q-faster" is well-defined, that is, it is not possible to find $p, p' \in [1, \infty)$ such that $Q_p(\mathscr{I}_1, x^*) < Q_p(\mathscr{I}_2, x^*)$ but $Q_{p'}(\mathscr{I}_1, x^*) > Q_{p'}(\mathscr{I}_2, x^*)$. This follows from the fact that if, say, $p' > p$ and $Q_p(\mathscr{I}_2, x^*) > Q_p(\mathscr{I}_1, x^*)$, then $Q_p(\mathscr{I}_2, x^*) > 0$, and, hence, $Q_{p'}(\mathscr{I}_2, x^*) = +\infty$. Note, furthermore, that **9.1.3** also shows the transitivity of the "Q-faster" relation; that is, if $\mathscr{I}_1$ is Q-faster than $\mathscr{I}_2$ and $\mathscr{I}_2$ is Q-faster than $\mathscr{I}_3$, then $\mathscr{I}_1$ is Q-faster than $\mathscr{I}_3$.

The concept of "Q-faster" depends upon the norm on R^n, and it is possible that in one norm a process $\mathscr{I}_1$ is Q-faster than another process $\mathscr{I}_2$, while in another norm $\mathscr{I}_2$ is Q-faster than $\mathscr{I}_1$. This simply reflects the fact (see **E 9.1-2**) that when $\{x^k\}$ is an arbitrary convergent sequence such that $0 < Q_p\{x^k\} < +\infty$, then the magnitude of $Q_p\{x^k\}$ is dependent upon the norm. However, there is a very important norm-independent measure that we now introduce.

9.1.5. Definition. Let $Q_p(\mathscr{I}, x^*)$ be the Q-factors of the iterative process $\mathscr{I}$ at x^* in some norm on R^n. Then

$$O_Q(\mathscr{I}, x^*) = \begin{cases} +\infty, & \text{if } Q_p(\mathscr{I}, x^*) = 0, \quad \forall p \in [1, \infty), \\ \inf\{p \in [1, \infty) \mid Q_p(\mathscr{I}, x^*) = +\infty\}, & \text{otherwise,} \end{cases}$$

is the *Q-order* of $\mathscr{I}$ at x^*. ∎

9.1.6. Let $Q_p(\mathscr{I}, x^*)$ be the Q-factors of the iterative process $\mathscr{I}$ at x^*. Then the three relations $Q_p(\mathscr{I}, x^*) = 0$, $0 < Q_p(\mathscr{I}, x^*) < +\infty$, and $Q_p(\mathscr{I}, x^*) = +\infty$ are independent of the norm on R^n. Hence, the Q-order of $\mathscr{I}$ at x^* is also independent of the norm.

Proof. Let $\{x^k\} \subset R^n$ be any sequence that converges to x^*, and denote by $Q_p\{x^k\}$ and $Q'_p\{x^k\}$ the Q-factors of $\{x^k\}$ in the norms $\|\cdot\|$ and $\|\cdot\|'$. Suppose that $x^k \neq x^*$, for all $k \geq k_0$; otherwise, we would have, by definition, either $Q_p\{x^k\} = 0$ or $Q_p\{x^k\} = +\infty$, in every norm. By **2.2.1**, there exist constants $d \geq c > 0$ such that

$$d\|x\| \geq \|x\|' \geq c\|x\|, \quad \forall x \in R^n; \tag{2}$$

therefore,

$$Q_p\{x^k\} = \limsup_{k \to \infty} \frac{\|x^{k+1} - x^*\|}{\|x^k - x^*\|^p} \leq \limsup_{k \to \infty} \frac{d^p}{c} \frac{\|x^{k+1} - x^*\|'}{\|x^k - x^*\|'^p} = \frac{d^p}{c} Q'_p\{x^k\}. \tag{3}$$

Hence, $Q_p\{x^k\} = 0$ whenever $Q'_p\{x^k\} = 0$, and $Q'_p\{x^k\} = +\infty$ whenever $Q_p\{x^k\} = +\infty$. This shows that the relations $Q_p\{x^k\} = 0, +\infty$ are both norm-independent, and, consequently, $0 < Q_p\{x^k\} < +\infty$ must also be norm-independent.

Now, suppose that $Q_p(\mathscr{I}, x^*) = 0$ in the norm $\|\cdot\|$. Then, for any sequence $\{x^k\}$ generated by $\mathscr{I}$ and converging to x^*, we must have $Q_p\{x^k\} = 0$. But we have just seen that this remains valid in any norm, and therefore $Q_p(\mathscr{I}, x^*) = 0$ in any norm. On the other hand, suppose that $Q_p(\mathscr{I}, x^*) = +\infty$ in the norm $\|\cdot\|$. Then (3) implies an immediate

9.1. THE QUOTIENT CONVERGENCE FACTORS

contradiction to the proposition that $Q_p'(\mathscr{I}, x^*) < \infty$ in another norm $\|\cdot\|'$. It follows then that the set $\{p \in [1, \infty) \mid Q_p(\mathscr{I}, x^*) = +\infty\}$, and, hence, also the Q-order $O_Q(J, x^*)$, is independent of the norm. ∎

As direct consequences of **9.1.4** and **9.1.6**, we obtain the following important observations.

9.1.7. Let $\mathscr{I}_1$ and $\mathscr{I}_2$ be iterative processes with limit x^*. If

$$O_Q(\mathscr{I}_1, x^*) > O_Q(\mathscr{I}_2, x^*),$$

then $\mathscr{I}_1$ is Q-faster than $\mathscr{I}_2$ at x^* in every norm.

9.1.8. Let $\mathscr{I}$ be an iterative process with limit x^*. If $Q_p(\mathscr{I}, x^*) < +\infty$ for some $p \in [1, \infty)$, then $O_Q(\mathscr{I}, x^*) \geq p$. If $Q_q(\mathscr{I}, x^*) > 0$ for some $q \in [1, \infty)$, then $O_Q(\mathscr{I}, x^*) \leq q$. Hence, if $0 < Q_p(\mathscr{I}, x^*) < +\infty$ for some $p \in [0, \infty)$, then $O_Q(\mathscr{I}, x^*) = p$.

We may now interpret our results so far as follows. The comparison of two iterative processes $\mathscr{I}_1$ and $\mathscr{I}_2$ with the same limit point x^* consists of two stages. First, we compare the Q-orders $O_Q(\mathscr{I}_1, x^*)$ and $O_Q(\mathscr{I}_2, x^*)$; if they are different, the process with the larger Q-order is Q-faster than the other one in every norm. If $O_Q(\mathscr{I}_1, x^*) = O_Q(\mathscr{I}_2, x^*) = p$, then we compare the two Q-factors. If, say, $Q_p(\mathscr{I}_1, x^*) = 0 < Q_p(\mathscr{I}_2, x^*)$, or, if $Q_p(\mathscr{I}_1, x^*) < Q_p(\mathscr{I}_2, x^*) = +\infty$, then $\mathscr{I}_1$ is Q-faster than $\mathscr{I}_2$ in every norm. However, if $0 < Q_p(\mathscr{I}_1, x^*) < Q_p(\mathscr{I}_2, x^*) < +\infty$ in some norm, then $\mathscr{I}_1$ is Q-faster than $\mathscr{I}_2$ in that norm, but there may exist other norms in which the relation is reversed.

Whenever we have $Q_p(\mathscr{I}, x^*) < +\infty$ for some iterative process $\mathscr{I}$ with limit x^*, an estimate of the form (1) holds, with $\gamma = Q_p + \epsilon$ and any $\epsilon > 0$, for every sequence $\{x^k\}$ generated by $\mathscr{I}$ that converges to x^*; note that the index k_0 will, in general, depend on $\{x^k\}$.

Iterative processes with Q-orders 1, 2, or 3 play an especially important role in the theory, and we introduce some additional terminology that will be useful on occasion. Whenever $Q_1(\mathscr{I}, x^*) = 0$, we say that the process has Q-*superlinear* convergence at x^*, while, if $0 < Q_1(\mathscr{I}, x^*) < 1$ in some norm, the convergence is called Q-*linear*. Note that any process of Q-order greater than one is Q-superlinear, and that, by **9.1.6**, the concept of Q-superlinear convergence is norm-independent. In contrast, the Q-linear convergence of a process $\mathscr{I}$ depends on the norm. But we can say that, for a Q-linearly convergent process $\mathscr{I}$ there is always a $\gamma \in (0, 1)$ and a norm such that (1) holds for any sequence generated by

$\mathscr{I}$ and converging to x^*; that is, there is a norm in which the error vectors ultimately decrease at each step by a factor $\gamma < 1$. Any process $\mathscr{I}$ for which $Q_1(\mathscr{I}, x^*) \geq 1$ in some norm is called Q-*sublinear* in that norm.

Similarly, any process $\mathscr{I}$ of Q-order two for which

$$0 < Q_2(\mathscr{I}, x^*) < +\infty$$

is called Q-*quadratic* at x^*, while, for $Q_2(\mathscr{I}, x^*) = 0$ or $Q_2(\mathscr{I}, x^*) = +\infty$, we refer to Q-superquadratic or Q-subquadratic convergence, respectively. Finally, an analogous terminology is sometimes applied to Q-order three convergence with "quadratic" replaced by "cubic."

NOTES AND REMARKS

NR 9.1-1. The convergence factors for sequences and iterative processes discussed in this chapter depend only on the real numbers $\epsilon_k = \| x^k - x^* \|$, and not on the vectors x^k themselves. Hence, these concepts are independent of the dimension of the space and apply equally well in infinite dimensional normed linear spaces. Except for the norm-invariance theorems **9.1.6** and **9.2.2**, which are, of course, intrinsically finite-dimensional, all results of this chapter remain unchanged in infinite dimensions.

NR 9.1-2. The Q-order has previously been used by Schmidt [1963a], as well as by Feldstein and Firestone [1967]. Traub [1964] and many other authors define, instead of the Q_p, the asymptotic error constants

$$C_p = \lim_{k \to \infty}(\| x^{k+1} - x^* \|/\| x^k - x^* \|^p) \tag{4}$$

under the assumption that $x^k \neq x^*$ for all k. Then the sequence $\{x^k\}$ is said to have order p if $0 < C_p < +\infty$. Simple examples, however, show that the limit in (4) need not exist even though Q_p is finite; hence, this definition is rather restrictive. But note that, by definition, $Q_p = 0$ implies that $C_p = 0$, while the existence of C_p ensures that $C_p = Q_p$. Moreover, it follows that when $0 < C_p < \infty$ for some $p \in [1, \infty)$, then the Q-, R-, and C-orders are identical (see **E 9.3-4**).

EXERCISES

E 9.1-1. Let $\{x^k\}, \{y^k\} \subset R^n$ be sequences which converge to some x^*, and let $\| \cdot \|$ be any norm on R^n. Define the "scaled" norm $\| \cdot \|' = c \| \cdot \|$, where $c > 0$, and denote by Q_p and Q_p' the Q-factors with respect to $\| \cdot \|$ and $\| \cdot \|'$. Prove the following statements:

9.2. THE ROOT-CONVERGENCE FACTORS

(a) If $0 < Q_p\{x^k\} < +\infty$ for $p > 1$, then $Q'_p\{x^k\} = Q_p\{x^k\}$ if and only if $c = 1$.

(b) There is a p such that $Q_p\{x^k\} < Q_p\{y^k\}$ if and only if $Q'_p\{x^k\} < Q'_p\{y^k\}$, that is, the Q-faster relation is invariant under scaling of the norm.

E 9.1-2. In R^2, let $e^1 = (1, 0)^T$ and $u = (1, 1)^T$, and define the sequences

$$x^k = \begin{cases} (1/2)^k e^1 & \text{for } k \text{ odd,} \\ (1/2)^{-1/2}(1/2)^k u, & \text{for } k \text{ even,} \end{cases} \quad y^k = (5/8)^k u, \quad k = 0, 1, \ldots.$$

Show that $Q_1\{x^k\} < Q_1\{y^k\}$ under the norm $\|x\|_2 = (x_1^2 + x_2^2)^{1/2}$, while $Q_1\{y^k\} < Q_1\{x^k\}$ under the norm $\|x\|_\infty = \max(|x_1|, |x_2|)$.

E 9.1-3. Compute the Q-factors and the corresponding orders of the following sequences in R^1:

(a) $x^k = 2^{-p^k}$;

(b) $x^k = 4^{-p^k}$;

(c) $x^k = 2 \cdot 2^{-p^k}$;

(d) $x^k = k^{-p^k}$;

(e) $x^k = 2^{-p^k}$ for k even, and $x^k = 3^{-p^{k-1}}$ for k odd;

(f) $x^k = k^{-p^k}$ for k even, and $x^k = \alpha(k-1)^{-p^{k-1}}$ for k odd, with $0 < \alpha < 1$;

(g) $x^k = 2^{-2^{2^k}}$;

(h) $x^k = (3k)^{-k}$ for k even, and $x^k = (2k)^{-k}$ for k odd;

(i) $x^k = c^{pk} 2^{-p^k}$, with $c > 0$ and $p > 1$;

(j) $x^k = c^{-k^2}$, with $c > 1$.

E 9.1-4. Let $\{x^k\} \subset R^n$ be convergent, and, for some $k' \geqslant 1$, define the sequence $y^k = x^{k+k'}$, $k = 0, 1, \ldots$. Show that, in any norm, $Q_p\{x^k\} = Q_p\{y^k\}$, $\forall p \in [1, \infty)$.

E 9.1-5. Consider the Newton iteration $\mathscr{I}$ for the one-dimensional problem $f(x) = x^2(1-x)$. Show that this process has the limit points 0 and 1, and that $Q_1(\mathscr{I}, 0) = \frac{1}{2}$, while $O_O(\mathscr{I}, 1) = 2$.

9.2. THE ROOT-CONVERGENCE FACTORS

In this section, we consider another measure of the rate of convergence of iterative processes; it is obtained by taking appropriate roots of successive errors rather than quotients. Recall that the definition of the

Q-factors was motivated by the estimate (9.1.1), which, for $p = 1$ and $k_0 = 0$, reduces to

$$\| x^k - x^* \| \leqslant \gamma \| x^{k-1} - x^* \| \leqslant \cdots \leqslant \gamma^k \| x^0 - x^* \|. \tag{1}$$

If $\gamma < 1$, then (1) shows that the norms of the error vectors $x^k - x^*$ are decreasing as rapidly as a geometric progression with ratio γ. Hence, in analogy with the root test for the convergence of series, we are led to consider geometric averages of the $\| x^k - x^* \|$.

9.2.1. Definition. Let $\{x^k\} \subset R^n$ be any sequence that converges to x^*. Then the numbers

$$R_p\{x^k\} = \begin{cases} \limsup_{k \to \infty} \| x^k - x^* \|^{1/k}, & \text{if } p = 1, \\ \limsup_{k \to \infty} \| x^k - x^* \|^{1/p^k}, & \text{if } p > 1, \end{cases}$$

are the *root-convergence factors*, or *R-factors*, for short, of the sequence. If $\mathscr{I}$ is an iterative process with limit point x^*, and $C(\mathscr{I}, x^*)$ is the set of all sequences generated by $\mathscr{I}$ which converge to x^*, then

$$R_p(\mathscr{I}, x^*) = \sup\{R_p\{x^k\} \mid \{x^k\} \in C(\mathscr{I}, x^*)\}, \quad 1 \leqslant p < \infty,$$

are the R-factors of $\mathscr{I}$ at x^*. ∎

Note that when $\{x^k\}$ converges to x^*, there is always a $k_0 \geqslant 0$ such that

$$0 \leqslant \| x^k - x^* \| < 1, \quad k \geqslant k_0,$$

and, hence, we have $0 \leqslant R_p\{x^k\} \leqslant 1$ for all $p \geqslant 1$.

In contrast to the Q-factors of a sequence, the R-factors are always independent of the norm on R^n.

9.2.2. Let $\{x^k\} \subset R^n$ be any sequence that converges to x^*. Then $R_p\{x^k\}$ is independent of the norm on R^n for any $p \in [1, \infty)$.

Proof. Let $\| \cdot \|$ and $\| \cdot \|'$ be any two norms on R^n and let $\{\gamma_k\}$ be any sequence of positive real numbers with $\lim_{k \to \infty} \gamma_k = 0$. Then (9.1.2) again holds with certain constants $d \geqslant c > 0$ and, since $\lim_{k \to \infty} a^{\gamma_k} = 1$ for any real $a > 0$, we have

$$\limsup_{k \to \infty} \| x^k - x^* \|^{\gamma_k} \leqslant \lim_{k \to \infty} (1/c)^{\gamma_k} \limsup_{k \to \infty} \| x^k - x^* \|'^{\gamma_k}$$

$$\leqslant \lim_{k \to \infty} (d/c)^{\gamma_k} \limsup_{k \to \infty} \| x^k - x^* \|^{\gamma_k} = \limsup_{k \to \infty} \| x^k - x^* \|^{\gamma_k}. \ \blacksquare$$

9.2. THE ROOT-CONVERGENCE FACTORS

As an immediate corollary of **9.2.2**, it follows, of course, that $R_p(\mathscr{I}, x^*)$ is also always independent of the norm.

The following result shows that, as functions of p, the R-factors behave in a manner analogous to the Q-factors.

9.2.3. Let $\mathscr{I}$ be an iterative process with limit x^*. Then exactly one of the following conditions holds:

(a) $R_p(\mathscr{I}, x^*) = 0, \forall p \in [1, \infty)$.

(b) $R_p(\mathscr{I}, x^*) = 1, \forall p \in [1, \infty)$.

(c) There is a $p_0 \in [1, \infty)$ such that $R_p(\mathscr{I}, x^*) = 0, \forall p \in [1, p_0)$, and $R_p(\mathscr{I}, x^*) = 1, \forall p \in (p_0, \infty)$.

Proof. Let $\{x^k\}$ be an arbitrary sequence that converges to x^*, and set $\gamma_{1k} = 1/k, \gamma_{pk} = 1/p^k$ for $p > 1, k = 1, 2,...$. Then $\lim_{k\to\infty}(\gamma_{qk}/\gamma_{pk}) = \infty$ whenever $1 \leqslant q < p$. Now, suppose that $R_p\{x^k\} < 1$ for some $p \in (1, \infty)$, and choose $\epsilon > 0$ so that $R_p\{x^k\} + \epsilon = \alpha < 1$. Next, set $\epsilon_k = \| x^k - x^* \|$ and choose $k_0 \geqslant 0$ so that

$$\epsilon_k^{\gamma_{pk}} \leqslant \alpha, \quad \forall k \geqslant k_0.$$

Then, we find that, for any $q \in [1, p)$,

$$R_q\{x^k\} = \limsup_{k\to\infty}(\epsilon_k^{\gamma_{pk}})^{\gamma_{qk}/\gamma_{pk}} \leqslant \lim \alpha^{\gamma_{qk}/\gamma_{pk}} = 0;$$

that is, $R_q\{x^k\} = 0$ whenever $q < p$ and $R_p\{x^k\} < 1$. This, in turn, also shows that $R_q\{x^k\} = 1$ whenever $q > p$ and $R_p\{x^k\} > 0$. Hence, one and only one of the properties (a), (b), or (c) holds for any convergent sequence. Now assume that neither (a) nor (b) holds; then $p_0 = \inf\{p \in [1, \infty) \mid R_p(\mathscr{I}, x^*) = 1\}$ is well-defined. Suppose there is a $p > p_0$ such that $R_p(\mathscr{I}, x^*) < 1$. Then $R_p\{x^k\} < 1$ for all $\{x^k\} \in C(\mathscr{I}, x^*)$, while, by definition of p_0, there exists a $p' \in [p_0, p)$ such that $R_{p'}(\mathscr{I}, x^*) = 1$. In particular, therefore, $R_{p'}\{x^k\} > 0$ for some sequence in $C(\mathscr{I}, x^*)$, which, by the first part of the proof, implies that $R_p\{x^k\} = 1$ for this sequence; this is a contradiction. Thus, $R_p(\mathscr{I}, x^*) = 1$ for $p > p_0$, and, similarly, it follows that $R_p(\mathscr{I}, x^*) = 0$ for $p < p_0$. ∎

We may now proceed along the same lines as in Section **9.1**.

9.2.4. Definition. Let $\mathscr{I}_1$ and $\mathscr{I}_2$ be two iterative processes with limit x^*. Then $\mathscr{I}_1$ is *R-faster* than $\mathscr{I}_2$ at x^* if there exists a $p \in [1, \infty)$ such that $R_p(\mathscr{I}_1, x^*) < R_p(\mathscr{I}_2, x^*)$. ∎

Note that **9.2.3** shows that the concept of R-faster is well defined, while **9.2.2** ensures that when $\mathscr{I}_1$ is R-faster than $\mathscr{I}_2$ in some norm, the same relation holds in every norm on R^n.

9.2.5. Definition. Let $\mathscr{I}$ be an iterative process with limit point x^*. Then the quantity

$$O_R(\mathscr{I}, x^*) = \begin{cases} \infty, & \text{if } R_p(\mathscr{I}, x^*) = 0, \ \forall p \in [1, \infty), \\ \inf\{p \in [1, \infty) \mid R_p(\mathscr{I}, x^*) = 1\}, & \text{otherwise,} \end{cases}$$

is called the *R-order* of $\mathscr{I}$ at x^*. ∎

Again, **9.2.2** and **9.2.3** show that the R-order is well defined and norm-independent. Moreover, the following results are immediate consequences of **9.2.3**.

9.2.6. Let $\mathscr{I}_1$ and $\mathscr{I}_2$ be iterative processes with the same limit point x^*. If $O_R(\mathscr{I}_1, x^*) > O_R(\mathscr{I}_2, x^*)$, then $\mathscr{I}_1$ is R-faster than $\mathscr{I}_2$ at x^*.

9.2.7. Let $\mathscr{I}$ be an iterative process with limit x^*. If $R_p(\mathscr{I}, x^*) < 1$ for some $p \in [1, \infty)$, then $O_R(\mathscr{I}, x^*) \geq p$. If $R_q(\mathscr{I}, x^*) > 0$ for some $q \in [1, \infty)$, then $O_R(\mathscr{I}, x^*) \leq q$. Hence, if $0 < R_p(\mathscr{I}, x^*) < 1$ for some $p \in [1, \infty)$, then $O_R(\mathscr{I}, x^*) = p$.

This shows that, as in the case of the Q-factors, the comparison of two iterative processes $\mathscr{I}_1, \mathscr{I}_2$ in terms of the R-factors consists again of two stages. First, we compare the R-orders $O_R(\mathscr{I}_1, x^*)$ and $O_R(\mathscr{I}_2, x^*)$; if they are different, the process with the larger R-order is R-faster than the other one. If $O_R(\mathscr{I}_1, x^*) = O_R(\mathscr{I}_2, x^*) = p_0$, then we compare the R-factors for p_0. If, say, $R_{p_0}(\mathscr{I}_1, x^*) < R_{p_0}(\mathscr{I}_2, x^*)$, then $\mathscr{I}_1$ is R-faster than $\mathscr{I}_2$.

We note that whenever $R_p = R_p(\mathscr{I}, x^*) < 1$, then, for any $\epsilon > 0$ with $R_p + \epsilon < 1$, there is a $k_0 \geq 0$, depending on each sequence $\{x^k\} \in C(\mathscr{I}, x^*)$, such that either

$$\| x^k - x^* \| \leq (R_p + \epsilon)^{p^k}, \quad \forall k \geq k_0, \quad \text{if } p > 1,$$

or

$$\| x^k - x^* \| \leq (R_1 + \epsilon)^k, \quad \forall k \geq k_0, \quad \text{if } p = 1.$$

In the latter case, the convergence of any sequence generated by $\mathscr{I}$ and converging to x^* is ultimately as rapid as a geometric progression with ratio $R_1 + \epsilon < 1$.

If $0 < R_1(\mathscr{I}, x^*) < 1$, we shall say that the convergence of $\mathscr{I}$ at x^*

9.2. THE ROOT-CONVERGENCE FACTORS

is *R-linear*, while for $R_1 = 1$ or $R_1 = 0$ we call the convergence *R-sublinear* or *R-superlinear*, respectively. Similarly, if

$$0 < R_2(\mathscr{I}, x^*) < 1,$$

we say that the convergence is *R-quadratic* at x^*. Note that these concepts are all norm-independent.

We conclude this section with a result about *R*-orders which will be of value in connection with multistep iterative processes. For this we need the following lemma.

9.2.8. For any integer $m \geq 1$, the polynomial $p_m(t) = t^{m+1} - t^m - 1$ has a unique positive root τ_m. Moreover, $\tau_m \in (1, 2)$, $\tau_m > \tau_{m+1}$, and $\lim_{m \to \infty} \tau_m = 1$.

Proof. Since $p_m(1) = -1$ and $p_m(2) = 2^m - 1 > 0$, there is a root τ_m in $(1, 2)$, and, by Descartes' rule of signs, there can be no other positive root. To show that τ_m is monotone decreasing with m, note that $p_{m+1}(\tau_m) = \tau_m - 1 > 0$, so that $\tau_{m+1} \in (1, \tau_m)$. Finally, suppose that $\lim_{m \to \infty} \tau_m = \tau > 1$. Then, for m sufficiently large,

$$p_m(\tau) = \tau^m(\tau - 1) - 1 > 0,$$

and, hence, p_m has a root in $(1, \tau)$, which is a contradiction. ∎

9.2.9. Let $\mathscr{I}$ be an iterative process and $C(\mathscr{I}, x^*)$ the set of sequences generated by $\mathscr{I}$ which converge to the limit x^*. Furthermore, let $\gamma_0, \gamma_1, ..., \gamma_m$ be nonnegative constants. If, for any $\{x^k\} \in C(\mathscr{I}, x^*)$, there is a $k_0 \geq m$ such that

$$\| x^{k+1} - x^* \| \leq \| x^k - x^* \| \sum_{j=0}^{m} \gamma_j \| x^{k-j} - x^* \|, \quad \forall k \geq k_0, \qquad (2)$$

then $O_R(\mathscr{I}, x^*) \geq \tau$, where τ is the unique positive root of

$$t^{m+1} - t^m - 1 = 0.$$

If, in addition, there exists a $\beta > 0$ and some sequence $\{x^k\} \in C(\mathscr{I}, x^*)$ such that, for a $k_0 \geq m$,

$$\| x^{k+1} - x^* \| \geq \beta \| x^k - x^* \| \| x^{k-m} - x^* \| > 0, \quad \forall k \geq k_0, \qquad (3)$$

then $O_R(\mathscr{I}, x^*) = \tau$.

Proof. We may assume that $\gamma = \sum_{j=0}^{m} \gamma_j > 0$, or else the first result is trivial and (3) could not hold. Let $\{x^k\} \in C(\mathscr{I}, x^*)$ and set $\epsilon_k = \| x^k - x^* \|$,

$\eta_k = \gamma \epsilon_k$, and $\delta_j = \gamma_j / \gamma$. Then $\sum_{j=0}^{m} \delta_j = 1$, and the sequence $\{\eta_k\}$ satisfies

$$\eta_{k+1} \leqslant \eta_k \sum_{j=1}^{m} \delta_j \eta_{k-j}, \qquad \forall k \geqslant k_0 \geqslant m. \tag{4}$$

Since $\{\epsilon_k\}$ converges to zero, we have $\eta_k \leqslant \eta < 1$ for $k \geqslant k'$ ($\geqslant k_0$). Hence, it follows, from (4), that

$$\eta_{k'+m+1} \leqslant \eta \sum_{j=0}^{m} \delta_j \eta = \eta^2,$$

$$\eta_{k'+m+2} \leqslant \delta_0 \eta^4 + \eta^2 \sum_{j=1}^{m} \delta_j \eta \leqslant \eta^3,$$

and, by induction,

$$\eta_{k'+i} \leqslant \eta^{\mu_i}, \qquad i = 0, 1, \ldots, \tag{5}$$

where

$$\mu_{i+1} = \mu_i + \mu_{i-m}, \qquad i = m, m+1, \ldots, \quad \mu_0 = \mu_1 = \cdots = \mu_m = 1. \tag{6}$$

In fact, if (5) holds up to some $i \geqslant m$, then

$$\eta_{k'+i+1} \leqslant \eta^{\mu_i} \sum_{j=0}^{m} \delta_j \eta^{\mu_{i-j}} \leqslant \eta^{\mu_i + \mu_{i-m}},$$

since the μ_i are monotonically increasing.

We show next that the μ_i satisfy

$$\mu_i \geqslant \alpha \tau^i, \qquad i = 0, 1, \ldots, \tag{7}$$

where $\alpha = \tau^{-m}$. In fact, since, by **9.2.8**, $\tau > 1$, (7) holds for $i = 0, 1, \ldots, m$, and, if it is valid up to some $i \geqslant m$, then

$$\mu_{i+1} = \mu_i + \mu_{i-m} \geqslant \alpha \tau^i + \alpha \tau^{i-m} = \alpha \tau^{i+1}(\tau^{-1} + \tau^{-m-1}) = \alpha \tau^{i+1},$$

which completes the induction. Therefore, (5) and (7) together imply that

$$\epsilon_{k'+i} \leqslant \gamma^{-1} \eta_{k'+i} \leqslant \gamma^{-1} \eta^{\alpha \tau^i}, \qquad \forall i \geqslant 0,$$

so that

$$R_\tau \{x^k\} = \limsup_{i \to \infty} \epsilon_{k'+i}^{1/\tau^{k'+i}} \leqslant \eta^{\alpha/\tau^{k'}} < 1.$$

Hence, for any $\epsilon > 0$, we have $R_{\tau-\epsilon}\{x^k\} = 0$ for all $\{x^k\} \in C(\mathcal{I}, x^*)$. Thus, $R_{\tau-\epsilon}(\mathcal{I}, x^*) = 0$, and, by **9.2.7**, since ϵ is arbitrary, $O_R(\mathcal{I}, x^*) \geqslant \tau$.

9.2. THE ROOT-CONVERGENCE FACTORS

Now, suppose that (3) holds for some $\{x^k\} \in C(\mathscr{I}, x^*)$; then we have, for $\eta_k = \beta\epsilon_k = \beta \| x^k - x^* \|$,

$$\eta_{k+1} \geq \eta_k \eta_{k-m}, \qquad \forall k \geq k_0. \tag{8}$$

Let $k' \geq k_0$ be such that $\eta_k \leq \hat\eta < 1$, for all $k \geq k'$, and set $\eta = \min(\eta_{k'}, \ldots, \eta_{k'+m})$. Then, in a similar fashion as before, an induction shows that

$$\eta_{k'+i} \geq \eta^{\mu_i}, \qquad i = 0, 1, \ldots, \tag{9}$$

where $\{\mu_i\}$ is defined by (6). In fact, (9) clearly holds for $i = 0, \ldots, m$, and, if it is valid for $i \geq m$, then

$$\eta_{k'+i+1} \geq \eta_{k'+i}\eta_{k'+i-m} \geq \eta^{\mu_i + \mu_{i-m}}.$$

Moreover, we have

$$\mu_i \leq \tau^i, \qquad i = 0, 1, \ldots, \tag{10}$$

since this is certainly true for $i = 0, \ldots, m$ because of $\tau > 1$. Thus, if (10) holds for some $i \geq m$, then

$$\mu_{i+1} = \mu_i + \mu_{i-m} \leq \tau^i + \tau^{i-m} = \tau^{i+1}(\tau^{-1} + \tau^{-m-1}) = \tau^{i+1}.$$

Therefore, altogether,

$$\eta_{k'+i} \geq \eta^{\mu_i} \geq \eta^{\tau^i}, \qquad \forall i \geq 0,$$

or

$$R_\tau\{x^k\} \geq \eta^{1/\tau^{k'}} > 0,$$

so that $R_\tau(\mathscr{I}, x^*) > 0$, and, by **9.2.7**, $O_R(\mathscr{I}, x^*) \geq \tau$. ∎

We note that **9.2.9** does not hold for the Q-order, and, in particular, (2) does not imply that $O_Q(\mathscr{I}, x^*) \geq \tau$ (see **E 9.2-8**).

NOTES AND REMARKS

NR 9.2-1. The R_1-factor has been used implicitly in much work concerning iterative processes for linear systems of equations; see, for example, Varga [1962]. For nonlinear systems, it was used explicitly by Ortega and Rockoff [1966].

NR 9.2-2. Wall [1956] (see also Tornheim [1964]) introduced the quantity

$$r = \lim_{k\to\infty}(\ln \| x^{k+1} - x^* \|/\ln \| x^k - x^* \|) \tag{11}$$

as the order of $\{x^k\}$ provided the limit exists. It is easy to see that r is then equal to the R-order of $\{x^k\}$ (see **E 9.2-5**).

NR 9.2-3. Ostrowski [1966] considered convergence factors more general than Q_1 by introducing the quantities

$$\alpha_r = \limsup_{k\to\infty}(\|x^{k+r} - x^*\|/\|x^k - x^*\|), \qquad x^k \neq x^* \quad \text{for all} \quad k, \quad (12)$$

where $r \geq 1$ is some integer. The sequence $\{x^k\}$ is then called *weakly linearly convergent* to x^* if there exists an r such that $\alpha_r < 1$. In contrast to Q-linearity, this property is norm-invariant; more precisely, (see **E 9.2-6**) if $\alpha_r < 1$ and $\|\cdot\|'$ is another norm on R^n, then there exists an integer m such that

$$\limsup_{k\to\infty}(\|x^{k+mr} - x^*\|'/\|x^k - x^*\|') < 1.$$

A weakly linearly convergent sequence need not be Q-linearly convergent, but it is always R-linearly convergent (see **E 9.2-7**). The converse is not true, as is seen by the example

$$x^k = \begin{cases} 3^{-k}, & k = j(j+1), \quad j = 1, 2, ..., \\ 2^{-k}, & \text{otherwise.} \end{cases}$$

Here, $R_1\{x^k\} = \tfrac{1}{2}$, while $\alpha_r = \infty$ for any $r \geq 1$.

NR 9.2-4. Theorem **9.2.9** is, in essence, a result about the asymptotic behavior of solutions of the difference inequality

$$\epsilon_{k+1} \leq \epsilon_k \sum_{j=0}^{m} \gamma_j \epsilon_{k-j}, \qquad k = m, m+1, ...$$

For further results on related difference inequalities and difference equations, we refer to Ostrowski [1966], Tornheim [1964], and Traub [1964]. In particular, the crucial step in the proof of **9.2.9**—namely that $\mu_i \geq \alpha \tau^i$—is a special case of a more general result of Ostrowski [1966; Theorem 12.3] while Lemma **9.2.8** is a special case of more general results of Ostrowski [1966] and Traub [1964] on roots of the equation $t^{m+1} = a_m t^m + a_{m-1} t^{m-1} + \cdots + a_0$, where $a_j \geq 0$, $j = 0, 1, ..., m$.

EXERCISES

E 9.2-1. Compute the R-convergence factors for the sequences of **E 9.1-3**.

E 9.2-2. Let $\{x^k\}$ and $\{y^k\}$ be defined as in **E 9.1-4**. Show that
(a) $R_1\{x^k\} = R_1\{y^k\}$,
(b) $R_p\{y^k\} = [R_p\{x^k\}]^{p^{k'}}$, for $p > 1$.

9.3. RELATIONS BETWEEN THE R AND Q CONVERGENCE FACTORS

E 9.2-3. For the process $\mathscr{I}$ of **E 9.1-5**, show that $R_1(\mathscr{I}, 0) = \frac{1}{2}$, while $R_2(\mathscr{I}, 1) = 2$.

E 9.2-4. Let $\{x^k\} \subset R^n$ be a convergent sequence with limit x^* and suppose that

$$\| x^k - x^* \| = \epsilon_k = \gamma^{r_1 r_2 \cdots r_k}, \quad k = 0, 1, \ldots,$$

where $0 < \gamma < 1$ and $\lim_{k \to \infty} r_k = r \geqslant 1$. Show that the R-order of $\{x^k\}$ is equal to r.

E 9.2-5. Let $\{x^k\} \subset R^n$ be a convergent sequence with limit x^* such that $x^k \neq x^*$ for all $k \geqslant 0$ and

$$r = \lim_{k \to \infty}(\ln \| x^{k+1} - x^* \|/\ln \| x^k - x^* \|)$$

exists. Show that the R-order of $\{x^k\}$ is equal to r.

E 9.2-6. Let $\{x^k\} \subset R^n$ be a convergent sequence with limit x^*, and suppose that

$$\limsup_{k \to \infty}(\| x^{k+r} - x^* \|/\| x^k - x^* \|) = \alpha_r < 1$$

for some integer $r \geqslant 1$. If $\| \cdot \|'$ is another norm on R^n such that

$$c \| x \| \leqslant \| x \|' \leqslant d \| x \|, \quad \forall x \in R^n, \quad 0 < c \leqslant d,$$

show that

$$\limsup_{k \to \infty}(\| x^{k+mr} - x^* \|/\| x^k - x^* \|) \leqslant (d/c) \alpha^m, \quad m = 1, 2, \ldots,$$

for some $\alpha < 1$.

E 9.2-7. Let $\alpha_r < 1$ be defined as in **E 9.2-6**. Show that $R_1\{x^k\} < 1$.

E 9.2-8. Define the sequence $\{x^k\} \subset R^1$ by $x^{k+1} = x^k x^{k-1}$, k odd, $x^{k+1} = (x^k)^2 x^{k-1}$, k even, where $| x^0 |$, $| x^1 | < 1$. Show that (2) is satisfied with $k_0 = m = 1$, $\gamma_0 = 0$, $\gamma_1 = 1$, and, hence, $O_R\{x^k\} \geqslant (1 + \sqrt{5})/2$, but that $O_Q \leqslant \frac{3}{2}$ (Voigt [1969]).

9.3. RELATIONS BETWEEN THE R AND Q CONVERGENCE FACTORS

We turn now to the important question of the relation between $Q_p(\mathscr{I}, x^*)$ and $R_p(\mathscr{I}, x^*)$. When $\{x^k\} \subset R^n$ converges to x^* and $0 < Q_p\{x^k\} < +\infty$ as well as $0 < R_p\{x^k\} < 1$ for some $p > 1$, then we can always achieve either one of the relations $R_p\{x^k\} < Q_p\{x^k\}$ or $R_p\{x^k\} > Q_p\{x^k\}$ by selecting a suitable norm (see **E 9.3-1**). It is an important fact, however, that this cannot be done if $p = 1$.

9.3.1. Let $\{x^k\} \subset R^n$ be a sequence convergent to x^*. Then

$$R_1\{x^k\} \leqslant Q_1\{x^k\}$$

in every norm. Consequently, if $\mathscr{I}$ is an iterative process with limit point x^*, then also $R_1(\mathscr{I}, x^*) \leqslant Q_1(\mathscr{I}, x^*)$ in every norm.

Proof. Assume that $Q_1\{x^k\} < \infty$, and set $\epsilon_k = \| x^k - x^* \|$. Then, for any $\epsilon > 0$ and $\gamma = Q_1\{x^k\} + \epsilon$, there is a $k_0 \geqslant 0$ such that

$$\epsilon_k \leqslant \gamma \epsilon_{k-1} \leqslant \cdots \leqslant \gamma^{k-k_0} \epsilon_{k_0}, \qquad \forall k \geqslant k_0.$$

Therefore,

$$R_1\{x^k\} \leqslant \gamma \lim_{k\to\infty} \sup [\epsilon_{k_0}/\gamma^{k_0}]^{1/k} = \gamma,$$

and, since ϵ was arbitrary, it follows that $R_1\{x^k\} \leqslant Q_1\{x^k\}$. But then, if $C(\mathscr{I}, x^*)$ again denotes all sequences generated by $\mathscr{I}$ and converging to x^*, we find that

$$R_1(\mathscr{I}, x^*) = \sup\{R_1\{x^k\} \mid \{x^k\} \in C(\mathscr{I}, x^*)\}$$
$$\leqslant \sup\{Q_1\{x^k\} \mid \{x^k\} \in C(\mathscr{I}, x^*)\} = Q_1(\mathscr{I}, x^*). \quad \blacksquare$$

As an immediate consequence of **9.3.1**, we see that $R_1\{x^k\}$ is a lower bound for any possible constant γ in the estimate

$$\| x^{k+1} - x^* \| \leqslant \gamma \| x^k - x^* \|, \qquad \forall k \geqslant k_0.$$

As mentioned before, for $p > 1$ the relation between R_p and Q_p will, in general, depend on the norm. Nevertheless, we can always compare the R- and Q-orders.

9.3.2. Let $\mathscr{I}$ be an iterative process with limit x^*. Then

$$O_Q(\mathscr{I}, x^*) \leqslant O_R(\mathscr{I}, x^*).$$

Proof. Suppose that $\{x^k\} \subset R^n$ converges to x^*. We show first that $Q_p\{x^k\} < \infty$ for $p > 1$ implies that $R_p\{x^k\} < 1$. Set $\epsilon_k = \| x^k - x^* \|$, $k = 0, 1, \ldots$, and, for given $\epsilon > 0$, let $\gamma = Q_p\{x^k\} + \epsilon$. Then there is a k_0 such that

$$\epsilon_{k+1} \leqslant \gamma \epsilon_k^p \leqslant \cdots \leqslant \gamma^{1+p+\cdots+p^{k-k_0}} \epsilon_{k_0}^{p^{k-k_0+1}}, \qquad \forall k \geqslant k_0,$$

and, hence,

$$\epsilon_{k+1}^{1/p^{k+1}} \leqslant \gamma^{\sum_{j=k_0+1}^{k+1}(1/p^j)} \epsilon_{k_0}^{1/p^{k_0}} \leqslant (\gamma' \epsilon_{k_0})^{1/p^{k_0}},$$

9.3. RELATIONS BETWEEN THE R AND Q CONVERGENCE FACTORS

where $\gamma' = \max(1, \gamma^{1/(p-1)})$. But, since $\lim_{k \to \infty} \epsilon_k = 0$, we may assume that k_0 has been chosen so that $\gamma' \epsilon_{k_0} < 1$; then

$$R_p\{x^k\} \leq (\gamma' \epsilon_{k_0})^{1/p^{k_0}} < 1.$$

Now, suppose that $q = O_Q(\mathscr{I}, x^*) > O_R(\mathscr{I}, x^*) = r$. Then it follows immediately from **9.1.3**, **9.2.3**, and the definition of the orders, that $Q_p(\mathscr{I}, x^*) = 0$ and $R_p(\mathscr{I}, x^*) = 1$ for all $p \in (r, q)$. Hence, for any sequence $\{x^k\}$ generated by $\mathscr{I}$ and converging to x^*, we have $Q_{p'}\{x^k\} = 0$ for $p' = (r + q)/2$, and, by the first part of the proof, $R_{p'}\{x^k\} < 1$. Then it follows from **9.2.3** that $R_p\{x^k\} = 0$ for any $p \in (r, p')$ and all $\{x^k\}$ generated by $\mathscr{I}$; that is, $R_p(\mathscr{I}, x^*) = 0$, which is a contradiction. ∎

In **E 9.3-3**, we give an example which shows that it is possible to have $O_Q(\mathscr{I}, x^*) < O_R(\mathscr{I}, x^*)$. However, in many important instances, the orders will be identical. The following result, similar to **9.2.9**, gives one simple sufficient condition for this.

9.3.3. Let $\mathscr{I}$ be an iterative process and $C(\mathscr{I}, x^*)$ the set of sequences generated by $\mathscr{I}$ which converge to the limit x^*. Suppose there exists a $p \in [1, \infty)$ and a constant c_2 such that, for any $\{x^k\} \in C(\mathscr{I}, x^*)$,

$$\| x^{k+1} - x^* \| \leq c_2 \| x^k - x^* \|^p, \quad \forall k \geq k_0 = k_0(\{x^k\}). \quad (1)$$

Then $O_R(\mathscr{I}, x^*) \geq O_Q(\mathscr{I}, x^*) \geq p$. On the other hand, if there is a constant $c_1 > 0$ and some sequence $\{x^k\} \in C(\mathscr{I}, x^*)$ such that

$$\| x^{k+1} - x^* \| \geq c_1 \| x^k - x^* \|^p > 0, \quad \forall k \geq k_0 = k_0(\{x^k\}), \quad (2)$$

then $O_Q(\mathscr{I}, x^*) \leq O_R(\mathscr{I}, x^*) \leq p$. Hence, if (1) and (2) both hold, then $O_Q(\mathscr{I}, x^*) = O_R(\mathscr{I}, x^*) = p$.

Proof. If (1) holds, then $Q_p\{x^k\} \leq c_2$ for all $\{x^k\} \in C(\mathscr{I}, x^*)$, so that $Q_p(\mathscr{I}, x^*) \leq c_2 < +\infty$, and, by **9.1.8** and **9.3.2**,

$$O_R(\mathscr{I}, x^*) \geq O_Q(\mathscr{I}, x^*) \geq p.$$

Now, suppose that (2) holds for some sequence $\{x^k\} \in C(\mathscr{I}, x^*)$. Then $\epsilon_k = \| x^k - x^* \| > 0$ for $k \geq k_0$ and

$$\epsilon_{k+1} \geq c_1^{1+p+\cdots+p^{k-k_0}} \epsilon_{k_0}^{p^{k-k_0+1}}, \quad \forall k \geq k_0.$$

Therefore, for $p = 1$,

$$R_1\{x^k\} \geq \lim_{k \to \infty}(c_1^{k-k_0+1} \epsilon_{k_0})^{1/k} = c_1 > 0,$$

while, for $p > 1$,

$$\epsilon_{k+1}^{1/p^{k+1}} \geq c_1^{\sum_{j=k_0}^{k}(1/p^{j+1})} \epsilon_{k_0}^{1/p^{k_0}} \geq \min(1, c_1^{1/(p-1)}) \epsilon_{k_0}^{1/p^{k_0}} > 0,$$

so that, again, $R_p\{x^k\} > 0$. Hence, $R_p(\mathscr{I}, x^*) > 0$, and, by **9.2.7** and **9.3.2**, $O_Q(\mathscr{I}, x^*) \leq O_R(\mathscr{I}, x^*) \leq p$. ∎

EXERCISES

E 9.3-1. Assume that the conditions of **E 9.1-1** hold. Show that if $0 < R_p\{x^k\} < 1$ and $0 < Q_p\{x^k\} < +\infty$ for some $p > 1$, then either one of the relations $Q_p'\{x^k\} < R_p'\{x^k\}$ or $Q_p'\{x^k\} > R_p'\{x^k\}$ holds for suitable choices of the scaling factor c.

E 9.3-2. Let $\{x^k\} \subset R^1$ be defined by $x^k = 1/k$. Show that $R_1\{x^k\} = Q_1\{x^k\} = 1$. Hence, $Q_p\{x^k\} < +\infty$ does not imply that $R_p\{x^k\} < 1$ when $p = 1$.

E 9.3-3. In R^1, consider the sequence

$$x^k = \begin{cases} (\alpha/2)^{p^k} & \text{for } k \text{ even,} \\ (\alpha^{q/p})^{p^k} & \text{for } k \text{ odd,} \end{cases} \quad k = 0, 1, \ldots,$$

where $0 < \alpha < 1$ and $1 < q < p$. Show that $O_R(\{x^k\}) = p$ and $O_Q(\{x^k\}) \leq q$.

E 9.3-4. Suppose that for the sequence $\{x^k\} \subset R^n$ there is some $p \in [1, \infty)$ such that the limit

$$C_p\{x^k\} = \lim_{k \to \infty}(\| x^{k+1} - x^* \|/\| x^k - x^* \|^p)$$

exists and $0 < C_p < \infty$. Show that both the Q- and the R- orders of $\{x^k\}$ are equal to p.

E 9.3-5. Suppose that the sequence $\{x^k\} \subset R^n$ satisfies

$$\| x^{k+1} - x^* \| = \alpha_k \| x^k - x^* \|, \quad k = 0, 1, \ldots,$$

where $\lim_{k \to \infty} \alpha_k = \alpha \in (0, 1)$. Show that $\lim_{k \to \infty} x^k = x^*$ and that

$$Q_1\{x^k\} = R_1\{x^k\} = \alpha.$$

Chapter 10 / ONE-STEP STATIONARY METHODS

10.1. BASIC RESULTS

In this chapter, we consider local-convergence and rate-of-convergence results for one-step stationary iterations of the form

$$x^{k+1} = Gx^k, \quad k = 0, 1,..., \tag{1}$$

where $G : D \subset R^n \to R^n$. This includes, of course, the Newton method, the m-step Newton–SOR and m-step SOR–Newton methods, and some of the minimization methods of Chapter 8. Results for particular methods of the form (1) will be given in Sections **10.2** and **10.3**.

By local convergence of the process (1) to a point x^*, we mean that the iterates of (1) will converge to x^* whenever x^0 is sufficiently close to x^*. This is made precise by the following concept.

10.1.1. Definition. Let $G: D \subset R^n \to R^n$. Then x^* is a *point of attraction* of the iteration (1) if there is an open neighborhood S of x^* such that $S \subset D$ and, for any $x^0 \in S$, the iterates $\{x^k\}$ defined by (1) all lie in D and converge to x^*. ∎

The determination of points of attraction and estimation of the rate of convergence will, in this chapter, usually depend upon showing that the conditions of the following simple lemma are satisfied. In all the remaining results, $\mathscr{I}$ shall denote the iteration (1), while $R_p(\mathscr{I}, x^*)$, $Q_p(\mathscr{I}, x^*)$ and $O_R(\mathscr{I}, x^*)$, $O_Q(\mathscr{I}, x^*)$ are, respectively, the R and Q

convergence factors and orders of $\mathscr{I}$ at x^*, as defined in the previous chapter.

10.1.2. Let $G\colon D \subset R^n \to R^n$, and suppose that there is a ball $S = S(x^*, \delta) \subset D$ and a constant $\alpha < 1$ such that

$$\| Gx - x^* \| \leqslant \alpha \| x - x^* \|, \quad \forall x \in S. \tag{2}$$

Then, for any $x^0 \in S$, the iterates given by (1) remain in S and converge to x^*. Hence, x^* is a point of attraction of the iteration (1), and, moreover, $R_1(\mathscr{I}, x^*) \leqslant Q_1(\mathscr{I}, x^*) \leqslant \alpha$.

Proof. Whenever $x^0 \in S$, then

$$\| x^1 - x^* \| = \| Gx^0 - x^* \| \leqslant \alpha \| x^0 - x^* \|.$$

Therefore, $x^1 \in S$ and a simple induction shows that all remaining iterates x^k lie in S and satisfy $\| x^k - x^* \| \leqslant \alpha^k \| x^0 - x^* \|$. Thus, $\lim_{k \to \infty} x^k = x^*$, and x^* is a point of attraction of (1). Now, let $\{x^k\}$ be any sequence generated by (1) such that $\lim_{k \to \infty} x^k = x^*$. Then $x^k \in S$ for $k \geqslant k_0$, and, hence, $\| x^{k+1} - x^* \| \leqslant \alpha \| x^k - x^* \|$ for $k \geqslant k_0$. From this, it follows immediately that $Q_1(\mathscr{I}, x^*) \leqslant \alpha$, and **9.3.1** shows that $R_1(\mathscr{I}, x^*) \leqslant Q_1(\mathscr{I}, x^*)$. ∎

One way of ensuring that the condition (2) is satisfied is to assume that $G'(x^*)$ exists and has sufficiently small eigenvalues.

10.1.3. Ostrowski Theorem. Suppose that $G\colon D \subset R^n \to R^n$ has a fixed point $x^* \in \text{int}(D)$ and is F-differentiable at x^*. If the spectral radius of $G'(x^*)$ satisfies $\rho(G'(x^*)) = \sigma < 1$, then x^* is a point of attraction of the iteration (1).

Proof. For arbitrary $\epsilon > 0$, **2.2.8** ensures the existence of a norm on R^n such that

$$\| G'(x^*) \| \leqslant \sigma + \epsilon. \tag{3}$$

Moreover, the F-differentiability of G at x^* implies that there is a $\delta > 0$ so that $S = S(x^*, \delta) \subset D$ and

$$\| Gx - Gx^* - G'(x^*)(x - x^*) \| \leqslant \epsilon \| x - x^* \|, \quad \forall x \in S. \tag{4}$$

Therefore,

$$\| Gx - x^* \| \leqslant \| Gx - Gx^* - G'(x^*)(x - x^*) \| + \| G'(x^*) \| \| x - x^* \|$$
$$\leqslant (\sigma + 2\epsilon) \| x - x^* \|, \quad \forall x \in S. \tag{5}$$

10.1. BASIC RESULTS

Since $\sigma < 1$, we may assume that ϵ was chosen so that $\sigma + 2\epsilon < 1$, and the result follows from **10.1.2**. ∎

Note that, under the conditions of **10.1.3**, **10.1.2** yields an upper bound for $R_1(\mathscr{I}, x^*)$ and $Q_1(\mathscr{I}, x^*)$. More importantly, the conditions of **10.1.3** already suffice for a precise determination of $R_1(\mathscr{I}, x^*)$.

10.1.4. Linear Convergence Theorem. Suppose that the conditions of **10.1.3** hold. Then $R_1(\mathscr{I}, x^*) = \rho(G'(x^*))$. Moreover, if $\rho(G'(x^*)) > 0$, then $O_R(\mathscr{I}, x^*) = O_Q(\mathscr{I}, x^*) = 1$.

Proof. By the estimate (5), it follows from **10.1.2** that

$$R_1(\mathscr{I}, x^*) \leqslant \sigma + 2\epsilon.$$

But, for arbitrary $\epsilon > 0$, there is a norm on R^n for which (3), and hence (5), holds. Therefore, since, by **9.2.2**, $R_1(\mathscr{I}, x^*)$ is independent of the norm, it follows that $R_1(\mathscr{I}, x^*) \leqslant \sigma$. If $\sigma = 0$, this completes the proof.

If $\sigma > 0$, let $\lambda_1, \ldots, \lambda_n$ denote the eigenvalues of $G'(x^*)$, ordered so that $|\lambda_i| = \sigma$, $i = 1, \ldots, m$, $|\lambda_i| < \sigma$, $i = m+1, \ldots, n$, and set $\alpha = \max\{|\lambda_i| \mid i = m+1, \ldots, n\}$ or $\alpha = 0$ if $m = n$. Clearly, $\alpha < \sigma$, and we may select $\epsilon > 0$ so that $\alpha + 3\epsilon < \sigma - 3\epsilon$, $\sigma + 2\epsilon < 1$. Now choose a basis $u^1, \ldots, u^n$ for the complex space C^n such that, in this basis, $G'(x^*)$ has the representation

$$G'(x^*) = \begin{pmatrix} \lambda_1 & \epsilon_1 & & \bigcirc \\ & \ddots & \ddots & \\ & & \ddots & \epsilon_{n-1} \\ \bigcirc & & & \lambda_n \end{pmatrix}, \tag{6}$$

where the ϵ_i are either 0 or ϵ, and, in particular, $\epsilon_m = 0$; this is simply the modified Jordan form of **2.2.7**. Then we may express any $y \in R^n$ in terms of $u^1, \ldots, u^n$ by $y = \sum_{i=1}^n \hat{y}_i u^i$, where, in general, the coefficients $\hat{y}_i$ are complex, and define a norm by $\|y\| = \sum_{i=1}^n |\hat{y}_i|$. In this norm, it follows, by direct computation, that $\|G'(x^*)\| \leqslant \sigma + \epsilon$. Hence, if $\delta > 0$ is chosen so that

$$\|Gx - Gx^* - G'(x^*)(x - x^*)\| \leqslant \epsilon \|x - x^*\|, \quad \forall x \in S = S(x^*, \delta) \subset D, \tag{7}$$

then **10.1.2** shows that for any $x^0 \in S$, $x^k = Gx^{k-1} \in S$ for all k, and $\lim_{k \to \infty} x^k = x^*$. Now, set

$$y^k = x^k - x^*, \quad k = 0, 1, \ldots, \quad \gamma_k = \sum_{i=1}^m |\hat{y}_i^k|, \quad \beta_k = \sum_{i=m+1}^n |\hat{y}_i^k|,$$

where, again, $\hat{y}_1^k,\ldots,\hat{y}_n^k$ are the coefficients of y^k in the basis $u^1,\ldots,u^n$, and choose $x^0 \in S$ so that $\beta_0 < \gamma_0$. We will show by induction that

$$\beta_p < \gamma_p, \qquad p = 0, 1,\ldots. \tag{8}$$

Assume that (8) holds for $p = 0, 1,\ldots, k$. Then

$$y^{k+1} = G'(x^*)y^k + R(x^k),$$

where $R(x) = Gx - Gx^* - G'(x^*)(x - x^*)$, and it follows from the representation (6) of $G'(x^*)$ that

$$\beta_{k+1} = \sum_{i=m+1}^{n} |\hat{y}_i^{k+1}| = \sum_{i=m+1}^{n} |\lambda_i \hat{y}_i^k + \epsilon_i \hat{y}_{i+1}^k + r_i(x^k)| \leq (\alpha + \epsilon)\beta_k + \|R(x^k)\|.$$

Therefore, since $\|y^k\| = \gamma_k + \beta_k$, it follows from (7) and the induction hypothesis that

$$\beta_{k+1} \leq (\alpha + \epsilon)\beta_k + \epsilon(\gamma_k + \beta_k) \leq (\alpha + 3\epsilon)\gamma_k.$$

Similarly, recalling that $\epsilon_m = 0$,

$$\gamma_{k+1} = \sum_{i=1}^{m} |\lambda_i \hat{y}_i^k + \epsilon_i \hat{y}_{i+1}^k + r_i(x^k)| \geq \sigma\gamma_k - \epsilon\gamma_k - \epsilon(\gamma_k + \beta_k)$$

$$\geq (\sigma - 3\epsilon)\gamma_k. \tag{9}$$

Hence, by the choice of ϵ,

$$\beta_{k+1} \leq [(\alpha + 3\epsilon)/(\sigma - 3\epsilon)]\gamma_{k+1} < \gamma_{k+1},$$

so that (8) holds for $p = k + 1$, and, by induction, for all p. It follows that (9) holds for all k, and, therefore, that

$$\|x^k - x^*\| = \gamma_k + \beta_k \geq \gamma_k \geq (\sigma - 3\epsilon)\gamma_{k-1} \geq \cdots \geq (\sigma - 3\epsilon)^k \gamma_0.$$

Hence, $R_1\{x^k\} \geq \sigma - 3\epsilon$, and, since ϵ was arbitrary, $R_1\{x^k\} \geq \sigma$, so that $R_1(\mathscr{I}, x^*) \geq \sigma$; thus, $R_1(\mathscr{I}, x^*) = \sigma$. Finally, if $\sigma \neq 0$, then, by 9.2.7, $O_R(\mathscr{I}, x^*) = 1$. Moreover, 9.3.1 ensures that $Q_1(\mathscr{I}, x^*) > 0$, and the proof of **10.1.3** exhibited a norm in which $Q_1(\mathscr{I}, x^*) < 1$. Hence $O_Q(\mathscr{I}, x^*) = 1$. ∎

The size of the open neighborhood about x^* for which the starting points allow convergence depends, in general, on the nonlinearity of G in a neighborhood of x^*. In the extreme case of an affine operator, we may conclude convergence for any starting point.

10.1. BASIC RESULTS

10.1.5. For $B \in L(R^n)$, $b \in R^n$, define $G: R^n \to R^n$ by $Gx = Bx + b$, $x \in R^n$. If $\rho(B) < 1$, then G has a unique fixed point x^*, the iterates $x^{k+1} = Gx^k$, $k = 0, 1$, converge to x^* for any $x^0 \in R^n$, and

$$R_1(\mathscr{I}, x^*) = \rho(B).$$

Proof. The Neumann lemma **2.3.1** ensures the existence of a unique fixed point x^*. Hence, for any x^0, we have

$$x^{k+1} - x^* = B(x^k - x^*) = \cdots = B^k(x^0 - x^*), \quad k = 0, 1, \ldots,$$

and **2.2.9** shows that $\lim_{k\to\infty}(x^k - x^*) = 0$. The rate-of-convergence statement is an immediate corollary of **10.1.4**. ∎

Note that **2.2.9** also shows that $\rho(B) < 1$ is a necessary, as well as sufficient, condition for the convergence of the iterates $x^{k+1} = Bx^k + b$, $k = 0, 1, \ldots$, for *any* $x^0 \in R^n$. If G is not affine, however, x^* may be a point of attraction of the iteration (1) even though $\rho(G'(x^*)) = 1$. (See **E 10.1-2**.)

If $\rho(G'(x^*)) = 0$, then **10.1.4** shows that the rate of convergence is R-superlinear. However, it is not necessarily true that the convergence is also Q-superlinear (see **E10.1-6**). The following simple lemma gives sufficient conditions for Q-superlinear convergence as well as for higher-order convergence.

10.1.6. Let $G: D \subset R^n \to R^n$ have a fixed point $x^* \in \text{int}(D)$. Suppose that G is F-differentiable at x^* and that $G'(x^*) = 0$. Then x^* is a point of attraction of the process (1) and $R_1(\mathscr{I}, x^*) = Q_1(\mathscr{I}, x^*) = 0$. Moreover, if in some ball $S = S(x^*, r) \subset D$ the estimate

$$\|Gx - Gx^*\| \leq \alpha \|x - x^*\|^p, \quad \forall x \in S, \tag{10}$$

holds for some $p > 1$, then $O_R(\mathscr{I}, x^*) \geq O_Q(\mathscr{I}, x^*) \geq p$. If, in addition, the estimate

$$\|Gx - Gx^*\| \geq \beta \|x - x^*\|^p, \quad \forall x \in S, \tag{11}$$

holds for some $\beta > 0$, then $O_R(\mathscr{I}, x^*) = O_Q(\mathscr{I}, x^*) = p$.

Proof. **10.1.3** ensures that x^* is a point of attraction. Now, let $\{x^k\}$ be any sequence generated by G such that $\lim_{k\to\infty} x^k = x^*$. Since $G'(x^*) = 0$, it follows that $\lim_{x\to x^*}(\|Gx - Gx^*\|/\|x - x^*\|) = 0$, and, hence,

$$\lim_{k\to\infty}(\|x^{k+1} - x^*\|/\|x^k - x^*\|) = \lim_{k\to\infty}(\|Gx^k - Gx^*\|/\|x^k - x^*\|) = 0,$$

so that $Q_1(\mathscr{I}, x^*) = 0$. Suppose, next, that (10) holds, and again let

$\{x^k\}$ be any sequence generated by G such that $\lim_{k \to \infty} x^k = x^*$. Then there is a k_0 so that $x^k \in S$, for all $k \geq k_0$, and, hence,

$$\| x^{k+1} - x^* \| = \| Gx^k - Gx^* \| \leq \alpha \| x^k - x^* \|^p, \qquad \forall k \geq k_0.$$

Similarly, if (11) holds, then $\| x^{k+1} - x^* \| \geq \beta \| x^k - x^* \|$, for all $k \geq k_0$, and the result follows directly from **9.3.3**. ∎

By means of **3.2.12**, it is easy to show (see **E 10.1-9**) that the estimate (10) holds provided that G is F-differentiable in a ball $S(x^*, r)$ about x^*, that $G'(x^*) = 0$, and that G' satisfies an estimate of the form

$$\| G'(x) \| \leq c \| x - x^* \|^{p-1}, \qquad \forall x \in S(x^*, r),$$

with constants $c \geq 0$ and $p > 1$. However, we shall see in the analysis of Newton's method in the next section that it is possible to obtain (10) without G being differentiable in a neighborhood of x^*.

We conclude this section with a specific and important application of **10.1.6**.

10.1.7. For $G: D \subset R^n \to R^n$, assume that $x^* \in \text{int}(D)$ is a fixed point of G and that, on some open ball $S = S(x^*, \delta) \subset D$, G is continuously differentiable. If $G'(x^*) = 0$ and G is twice F-differentiable at x^*, then x^* is a point of attraction of the iteration $\mathscr{I}$ defined by (1), and $O_R(\mathscr{I}, x^*) \geq O_Q(\mathscr{I}, x^*) \geq 2$. If, in addition,

$$G''(x^*) hh \neq 0, \qquad \forall h \in R^n, \ h \neq 0, \tag{12}$$

then $O_R(\mathscr{I}, x^*) = O_Q(\mathscr{I}, x^*) = 2$.

Proof. Set

$$H(x) = G'(x) - G'(x^*) - G''(x^*)(x - x^*), \qquad \forall x \in S.$$

Clearly, H is continuous on S, and, by definition of $G''(x^*)$, for any $\epsilon > 0$ there is a $\delta' \in (0, \delta)$ such that

$$\| H(x) \| \leq \epsilon \| x - x^* \|, \qquad \forall x \in S' = S(x^*, \delta'). \tag{13}$$

It then follows from the mean-value theorem **3.2.7**, together with $G'(x^*) = 0$, that, for any $x \in S'$,

$$\| Gx - x^* \| = \| Gx - Gx^* \| = \left\| \int_0^1 [G'(x^* + t(x - x^*)) - G'(x^*)](x - x^*) \, dt \right\|$$
$$= \left\| \int_0^1 [tG''(x^*)(x - x^*) + H(x^* + t(x - x^*))](x - x^*) \, dt \right\|$$
$$\leq \tfrac{1}{2} \| G''(x^*) \| \, \| x - x^* \|^2 + \tfrac{1}{2}\epsilon \| x - x^* \|^2 = \alpha \| x - x^* \|^2, \tag{14}$$

10.1. BASIC RESULTS

where $\alpha = \frac{1}{2}(\| G''(x^*) \| + \epsilon)$. The first part of the theorem is now an immediate consequence of **10.1.6**.

Suppose, next, that (12) also holds. Then, since $\| G''(x^*)hh \|$ is a continuous function of h which is nonzero on the compact set

$$\{h \in R^n \mid \| h \| = 1\},$$

there is a constant $c > 0$ such that

$$\| G''(x^*) hh \| \geq c \| h \|^2, \qquad \forall h \in R^n.$$

We may assume that the constant ϵ in (13) satisfies $\epsilon \leq c/2$, and, then, estimating the second line of (14) downward, we obtain

$$\| Gx - x^* \| = \left\| G''(x^*)(x - x^*)(x - x^*) \int_0^1 t \, dt \right\|$$

$$- \left\| \int_0^1 H(x^* + t(x - x^*))(x - x^*) \, dt \right\|$$

$$\geq \tfrac{1}{2} c \| x - x^* \|^2 - \tfrac{1}{4} c \| x - x^* \|^2 = \tfrac{1}{4} c \| x - x^* \|^2, \qquad \forall x \in S'.$$

Therefore, the result again follows from **10.1.6**. ∎

We note that $O_0(\mathscr{I}, x^*)$ may be greater than two if (12) is removed, or even weakened in a certain natural way (see **NR 10.1-8** and **E 10.1-12**).

NOTES AND REMARKS

NR 10.1-1. The concept of a point of attraction and the sufficiency of the condition $\rho(G'(x^*)) < 1$ for a point of attraction were announced by Ostrowski [1957], and a proof of Theorem **10.1.3** was given in Ostrowski [1960] under the condition that G' is continuous in a neighborhood of x^*. This differentiability requirement was reduced to that of **10.1.3** in Ostrowski [1966], Ortega and Rockoff [1966], and Kitchen [1966].

However, this result had, in essence, already been obtained by Perron [1929] in the setting of perturbed linear difference equations of the form

$$z^{k+1} = Az^k + \phi(z^k), \qquad k = 0, 1, \dots. \tag{15}$$

Here, $A \in L(R^n)$, and $\phi \colon S \subset R^n \to R^n$ satisfies

$$\lim_{\|x\| \to 0} \phi(x)/\| x \| = 0 \tag{16}$$

on some open neighborhood S of 0. Perron showed that if $\rho(A) < 1$ and

$\|z^0\|$ is sufficiently small, then $\lim_{k\to\infty} z^k = 0$. Under the conditions of **10.1.3**, the iteration $x^{k+1} = Gx^k$ may be written in the form (16) with $A = G'(x^*)$, $z^k = x^k - x^*$, and

$$\phi(x) = G'(x^*)(x - x^*) + Gx - Gx^*.$$

The F-differentiability of G at x^* thus ensures that (16) holds.

NR 10.1-2. The concept of a point of attraction extends verbatim to a Banach space. Theorem **10.1.3** also extends to an arbitrary Banach space, but a somewhat different proof is required (see Kitchen [1966]).

NR 10.1-3. The result **10.1.2** extends verbatim to a Banach space. Moreover, **E 10.1-1** gives various assumptions on a nonlinear function $\varphi: R^1 \to R^1$ which ensure that x^* is a point of attraction if the more general estimate

$$\| Gx - x^* \| \leqslant \varphi(\| x - x^* \|), \quad \forall x \in S(x^*, \delta),$$

holds.

NR 10.1-4. The first part of **10.1.4** was given by Ortega and Rockoff [1966] under the assumption that

$$\| Gx - Gx - G'(x^*)(x - x^*) \| \leqslant c \| x - x^* \|^{1+\epsilon}, \quad \forall x \in S(x^*, r), \quad (17)$$

for some $\epsilon > 0$. This condition holds if, for example, G' satisfies a Hölder condition $\| G'(x) - G'(x^*) \| \leqslant \alpha \| x - x^* \|^\epsilon$ at x^*. Under the condition (17), it was shown, moreover, that there exists a constant β such that

$$\| x^k - x^* \| \leqslant \beta k^p \sigma^k, \quad \forall k \geqslant k_0, \quad (18)$$

for any sequence $\{x^k\}$ satisfying $x^{k+1} = Gx^k$, $k = 0, 1, ...$, and $\lim_{k\to\infty} x^k = x^*$; here, $\sigma = \rho(G'(x^*))$, k_0 depends on the particular sequence $\{x^k\}$, and $p+1$ is the dimension of the largest block in the Jordan form of $G'(x^*)$ which is associated with σ. The estimate (18) is the best (up to the constant β) which may be obtained for a linear iterative process, and, moreover, it does not hold under the weaker differentiability requirements of **10.1.4** (see **E 10.1-5**). The proof of **10.1.4** as given in the text has been adapted from a result of Coffman [1964] on the asymptotic behavior of solutions of perturbed linear difference equations (see also Panov [1964]), and was given by Ortega and Rheinboldt [1970a].

NR 10.1-5. Theorem **10.1.4** gives the rate of convergence only in the R-measure. However, the proof of **10.1.3** already shows that, for any $\epsilon > 0$, we may choose a norm such that $Q_1(\mathscr{I}, x^*) \leqslant \sigma + \epsilon$, where, again, $\sigma = \rho(G'(x^*))$. The following additional assumption on $G'(x^*)$ suffices to give equality with σ: Suppose that G satisfies the conditions of **10.1.4** and that $G'(x^*)$ is similar to a block diagonal matrix with blocks H_1 and H_2, where H_1 is diagonal and $\rho(H_1) > \rho(H_2)$. Then there exists a norm on R^n for which $Q_1(\mathscr{I}, x^*) = \sigma$.

10.1. BASIC RESULTS

The proof requires only the fact that there is a norm such that $\lVert G'(x^*) \rVert = \sigma$ (see Householder [1964]) together with the observation that

$$\sigma = R_1(\mathscr{I}, x^*) \leqslant Q_1(\mathscr{I}, x^*).$$

The condition on $G'(x^*)$ is equivalent to saying that any Jordan block corresponding to an eigenvalue of modulus σ is diagonal.

NR 10.1-6. Theorem **10.1.5** is a standard result for the iteration $x^{k+1} = Bx^k + b$, $k = 0, 1,...$; see, for example, Forsythe and Wasow [1960] and Varga [1962] for further discussions of linear iterations.

NR 10.1-7. Theorem **10.1.6** and its proof extend verbatim to Banach spaces.

NR 10.1-8. The first part of **10.1.7** is essentially a standard result, although the assumption is always made in the literature that G is twice-continuously differentiable in a neighborhood of x^* and only the rate of convergence in the Q-measure is given (see, for example, Durand [1960]; Kuntzmann [1959]; Korganoff [1961], and Traub [1964]). A sufficient condition to obtain precisely second-order convergence has not previously been given, although Traub [1964] *defines* the order to be two if $G''(x^*) \neq 0$. It is easy to see, however, that $G''(x^*) \neq 0$ is not sufficient to ensure precisely second-order convergence under our definitions [see **E 10.1-11** for an example in which $G''(x^*) \neq 0$, but both orders of convergence are infinite]. Indeed, even the stronger condition that $G''(x^*)$ [as a linear operator from R^n to $L(R^n)$] is one-to-one is not sufficient (see **E 10.1-12**).

NR 10.1-9. The result **10.1.6** is actually somewhat stronger than stated; indeed, the proof already shows that $0 < Q_p(\mathscr{I}, x^*) < \infty$. A similar remark holds for **10.1.7**.

NR 10.1-10. Theorem **10.1.7** and the method of proof extend to higher-order iterations. More precisely, if G is p-times continuously differentiable on D, $x^* = Gx^*$, $G^{(i)}(x^*) = 0$, $i = 1,..., p$, and $G^{(p+1)}(x^*)$ exists as an F-derivative and satisfies $G^{(p+1)}(x^*) h^{p+1} \neq 0$, for all $h \in R^n$, $h \neq 0$, then

$$O_R(\mathscr{I}, x^*) = O_Q(\mathscr{I}, x^*) = p + 1.$$

This is the n-dimensional analog of a famous theorem of Schröder [1870] in one dimension. Voigt [1969] has shown that one cannot replace the condition $G'(x^*) = 0$ by $\rho(G'(x^*)) = 0$ in this result and retain even Q-superlinear convergence, even if G is infinitely differentiable at x^* and $G^{(p)}(x^*) = 0$ for all $p \geqslant 2$.

NR 10.1-11. Theorem **10.1.7** and its proof extend to Banach spaces provided the condition $G''(x^*) hh \neq 0$ is replaced by $\lVert G''(x^*) hh \rVert \geqslant c \lVert h \rVert^2$, $c > 0$.

EXERCISES

E 10.1-1. Let $G: D \subset R^n \to R^n$ and $\varphi: [0, \infty) \subset R^1 \to R^1$. Suppose that, for some $x^* \in \text{int}(D)$, $\|Gx - x^*\| \leqslant \varphi(\|x - x^*\|)$, for all $x \in S(x^*, \delta) \subset D$. Show that x^* is a point of attraction for the iteration $x^{k+1} = Gx^k$, $k = 0, 1, \ldots$, if any one of the following conditions on φ holds:
  (a) For some $s_0 \in (0, \delta)$, any sequence $\{\sigma_k\}$ such that $0 \leqslant \sigma_{k+1} \leqslant \varphi(\sigma_k)$, $k = 0, 1, \ldots$, and $\sigma_0 \leqslant s_0$, satisfies $\sigma_k \leqslant s_0$, $k = 1, 2, \ldots$, and $\lim_{k \to \infty} \sigma_k = 0$.
  (b) For some $s_0 \in (0, \delta)$, $\varphi(s) < s$ for all $s \in (0, s_0)$.
  (c) φ is isotone on some interval $[0, \hat{s}]$, and, for some $s_0 \in (0, \min(\hat{s}, \delta))$, $\varphi(s_0) \leqslant s_0$ and $\lim_{k \to \infty} s_k = 0$, where $s_{k+1} = \varphi(s_k)$, $k = 0, 1, \ldots$.

E 10.1-2. Let $G: R^1 \to R^1$ be defined by $Gx = x - x^3$. Show that $x^* = 0$ is a point of attraction of $x^{k+1} = Gx^k$, $k = 0, 1, \ldots$, although $G'(x^*) = 1$. Show also that $R_1(\mathscr{I}, x^*) = Q_1(\mathscr{I}, x^*) = 1$. On the other hand, if $Gx = x + x^3$, show that 0 is not a point of attraction.

E 10.1-3. Define $G: R^n \to R^n$ by $Gx = Hx$, where $H \in L(R^n)$ is nilpotent (that is, $H^m = 0$ for some m). Show that $Q_1(\mathscr{I}, 0) = 0$, although $G'(x) = H$ need not be zero.

E 10.1-4. Define $G: R^1 \to R^1$ by $Gx = -x/\ln|x|$, $x \neq 0$, $G(0) = 0$. Show that $Q_1(\mathscr{I}, 0) = R_1(\mathscr{I}, 0) = 0$, but that $Q_p(\mathscr{I}, 0) = +\infty$ for any $p > 1$. (This shows that Q-superlinear convergence does not necessarily imply convergence of any higher order.)

E 10.1-5. Define $G: R^1 \to R^1$ by $Gx = \lambda x - (x/\ln|x|)$, $x \neq 0$, and $G(0) = 0$, for some $\lambda \in (0, 1)$. Show that $x^* = 0$ is a point of attraction, but that an estimate of the form $|x^k| \leqslant \beta \lambda^k$, $k \geqslant k_0$, is impossible if $x^0 \neq 0$.

E 10.1-6. Define $G: R^2 \to R^2$ by $g_1(x) = x_1^2 - x_2$, $g_2(x) = x_2^2$. Show that $x^* = 0$ is a point of attraction of $\mathscr{I}: x^{k+1} = Gx^k$ and that $R_1(\mathscr{I}, x^*) = 0$, but that $Q_1(\mathscr{I}, x^*) > 0$ in any norm. (Voigt [1969].)

E 10.1-7. Define $G: R^1 \to R^1$ by $Gx = \exp(-x^{-2})$, $x \neq 0$, $G(0) = 0$. Show that $x^* = 0$ is a point of attraction of $\mathscr{I}: x^{k+1} = Gx^k$ and that
$$O_R(\mathscr{I}, x^*) = O_Q(\mathscr{I}, x^*) = +\infty.$$

E 10.1-8. Assume that $F: D \subset R^n \to R^n$ is F-differentiable at a point $x^* \in \text{int}(D)$ for which $Fx^* = 0$. Let $B \in L(R^n)$ be such that $\sigma = \rho(I - BF'(x^*)) < 1$.
(a) Show that x^* is a point of attraction of the iteration $\mathscr{I}: x^{k+1} = x^k - BFx^k$, $k = 0, 1, \ldots$, and that $R_1(\mathscr{I}, x^*) = \sigma$. (b) Show that $R_1(\mathscr{I}, x^*) = 0$ if and only if $F'(x^*)$ and B are both nonsingular and $B = JF'(x^*)^{-1}$, where the eigenvalues of $J \in L(R^n)$ are all equal to unity.

10.1. BASIC RESULTS

E 10.1-9. Let $G: D \subset R^n \to R^n$ and assume that $x^* \in \text{int}(D)$ is a fixed point of G. Suppose that, on some ball $S = S(x^*, r) \subset D$, G is G-differentiable, $G'(x^*) = 0$, and

$$\|G'(x)\| \leq \alpha \|x - x^*\|^\gamma, \qquad \gamma > 0, \quad \forall x \in S.$$

Show that x^* is a point of attraction of $\mathscr{I}: x^{k+1} = Gx^k$ and that

$$O_R(\mathscr{I}, x^*) \geq O_Q(\mathscr{I}, x^*) \geq 1 + \gamma.$$

E 10.1-10. Define $G: R^1 \to R^1$ by $Gx = x^p$ for some $p > 1$. Show that $x^* = 0$ is a point of attraction of $\mathscr{I}: x^{k+1} = Gx^k$, and that $O_R(\mathscr{I}, x^*) = O_Q(\mathscr{I}, x^*) = p$. (This shows that, in general, one-step iterations can have an arbitrary order of convergence.)

E 10.1-11. Define $G: R^2 \to R^2$ by $g_1(x) \equiv 0, g_2(x) = x_1^2$. Show that $G''(0) \neq 0$, but that $x^* = 0$ is a point of attraction of $\mathscr{I}: x^{k+1} = Gx^k$, such that

$$O_R(\mathscr{I}, x^*) = O_Q(\mathscr{I}, x^*) = \infty.$$

E 10.1-12. (a) Define $G: R^2 \to R^2$ by $g_1(x) \equiv 0, g_2(x) = x_1 + x_2 x_1 + x_2^m$ for some $m \geq 3$. Show that $x^* = 0$ is a point of attraction of $\mathscr{I}: x^{k+1} = Gx^k$ and that $O_Q(\mathscr{I}, x^*) \geq m - \epsilon, \epsilon > 0$, although $G'(x^*) \neq 0$ and $G''(x^*) h \neq 0, \forall h \neq 0$, [This shows that $G'(x^*) = 0$ is not a necessary condition for convergence of higher order, and also that the condition that $G''(x^*)$ be one-to-one is not sufficient to conclude that $O_Q(\mathscr{I}, x^*) \leq 2$.]

(b) Replace g_2 by $g_2(x) = x_1 x_2 + x_2^m$ and draw the same conclusions, although now $G'(x^*) = 0$. (Voigt [1969].)

E 10.1-13. Let $G: D \subset R^n \to R^n$ and assume that G is twice F-differentiable at $x^* \in \text{int}(D)$. Show that $G''(x^*)$ is not one-to-one [as a map from R^n to $L(R^n)$] if and only if the Hessian matrices $H_1(x^*), \ldots, H_n(x^*)$ are all singular and have a common null vector.

E 10.1-14. Consider the mapping $G: R^2 \to R^2$, $g_1(x) = g_2(x) = x_1^2 - x_2^2$. Show that each Hessian of G is nonsingular, but there exist $h \neq 0$ such that $G''(x) hh = 0$.

E 10.1-15. Assume that $F: D \subset R^n \to R^n$ is F-differentiable at a point $x^* \in \text{int}(D)$ for which $Fx^* = 0$. Define $G: D \subset R^n \to R^n$ by $Gx = x - \omega Fx$ for some real parameter ω. Show that $\rho(G'(x^*)) < 1$ if and only if one of the following conditions holds:

(a) The eigenvalues $\lambda_1, \ldots, \lambda_n$ of $F'(x^*)$ satisfy $\text{Re } \lambda_i > 0, i = 1, \ldots, n$, and $0 < \omega < (2 \text{ Re } \lambda_i)/|\lambda_i|^2, i = 1, \ldots, n$.

(b) $\text{Re } \lambda_i < 0, i = 1, \ldots, n$, and $(2 \text{ Re } \lambda_i)/|\lambda_i|^2 < \omega < 0, i = 1, \ldots, n$.

E 10.1-16. Consider the special case of **E 10.1-15** in which $0 < \lambda_1 \leq \cdots \leq \lambda_n$. Show that $\rho(G'(x^*))$ is minimized, as a function of ω, for $\omega = 2/(\lambda_1 + \lambda_n)$.

E 10.1-17. Let $B \in L(R^n)$ with $\rho(B) > 1$. Formulate necessary and sufficient conditions in terms of a given x and the eigensystem of B in order that $\lim_{k\to\infty} B^k x = 0$.

E 10.1-18. Let $A \in L(R^n)$ be symmetric, positive definite. For the computation of the square root $B = A^{1/2}$ of A, consider the processes $X^{k+1} = G(X^k)$, $k = 0, 1,...$, where (a) $G(X) = X + \alpha(A - X^2)$, (b) $G(X) = X + \beta(I - A^{-1}X^2)$, (c) $G(X) = (1/2)(X + AX^{-1})$.

  I. Use the differentiation formula given in **E 3.1-15** to prove that, for these processes:

  (a) $G'(B) = (I \times I) - \alpha[(I \times B) + (B \times I)]$;

  (b) $G'(B) = (I \times I) - \beta[(A^{-1} \times B) + (B^{-1} \times I)]$;

  (c) $G'(B) = (1/2)[(I \times I) - (B \times B^{-1})]$;

where $\times$ denotes the tensor product as defined in **E 3.1-15**.

  II. If $\lambda_1 \geq \cdots \geq \lambda_n$ are the eigenvalues of A, show that the eigenvalues of the three derivative operators $G'(B)$ are:

  (a) $\mu_{ij} = 1 - \alpha(\sqrt{\lambda_i} + \sqrt{\lambda_j})$;

  (b) $\mu_{ij} = 1 - \beta(\sqrt{\lambda_j}/\lambda_i + 1/\sqrt{\lambda_i})$; $\quad i,j = 1,...,n$.

  (c) $\mu_{ij} = \tfrac{1}{2}[1 - (\lambda_i/\lambda_j)^{1/2}]$.

Hence, for $\alpha = 1/(2\sqrt{\lambda_1})$ and $\beta = \lambda_n/\sqrt{\lambda_1}$, the processes (a) and (b), respectively, are locally convergent, while, in the case of (c), the μ_{ij} can be arbitrarily large in modulus if A is ill-conditioned. Show, however, that if $X^0 A = AX^0$, then (c) is locally and R-superlinearly convergent. (Liebl [1965].)

10.2. NEWTON'S METHOD AND SOME OF ITS MODIFICATIONS

We now specialize the results of the previous section to iterations of the form

$$x^{k+1} = x^k - A(x^k)^{-1} F x^k, \quad k = 0, 1,..., \tag{1}$$

where A is a mapping from a suitable subset of R^n into $L(R^n)$. The most important special case of (1) is, of course, Newton's method, in which $A(x) = F'(x)$; but, before considering Newton's method in detail, we begin with a basic lemma concerning (1) which will be useful throughout the rest of the chapter.

10.2. NEWTON'S METHOD AND SOME OF ITS MODIFICATIONS

10.2.1. Suppose that $F: D \subset R^n \to R^n$ is F-differentiable at a point $x^* \in \text{int}(D)$ for which $Fx^* = 0$. Let $A: S_0 \to L(R^n)$ be defined on an open neighborhood $S_0 \subset D$ of x^* and continuous at x^*, and assume that $A(x^*)$ is nonsingular. Then there exists a ball $S = \bar{S}(x^*, \delta) \subset S_0$, $\delta > 0$, on which the mapping

$$G: S \to R^n, \qquad Gx = x - A(x)^{-1} Fx, \qquad \forall x \in S,$$

is well-defined; moreover, G is F-differentiable at x^*, and

$$G'(x^*) = I - A(x^*)^{-1} F'(x^*). \tag{2}$$

Proof. Set $\beta = \|A(x^*)^{-1}\|$, let ϵ satisfy $0 < \epsilon < (2\beta)^{-1}$ and choose $\delta > 0$ so that $S = \bar{S}(x^*, \delta) \subset S_0$ and

$$\|A(x) - A(x^*)\| \leqslant \epsilon, \qquad \forall x \in S. \tag{3}$$

Then, by the perturbation lemma **2.3.3**, $A(x)^{-1}$ exists for all $x \in S$, and

$$\|A(x)^{-1}\| \leqslant \beta/(1 - \beta\epsilon) < 2\beta, \qquad \forall x \in S. \tag{4}$$

Therefore, the mapping G is well-defined on S.

Since F is F-differentiable at x^*, we may assume that δ was chosen sufficiently small that

$$\|Fx - Fx^* - F'(x^*)(x - x^*)\| \leqslant \epsilon \|x - x^*\|, \qquad \forall x \in S, \tag{5}$$

also holds. Now, clearly, $x^* = Gx^*$, and, therefore, using (3)–(5), we obtain the estimate

$$\|Gx - Gx^* - [I - A(x^*)^{-1} F'(x^*)](x - x^*)\|$$
$$= \|A(x^*)^{-1} F'(x^*)(x - x^*) - A(x)^{-1} Fx\|$$
$$\leqslant \|-A(x)^{-1}[Fx - Fx^* - F'(x^*)(x - x^*)]\|$$
$$+ \|A(x)^{-1}[A(x^*) - A(x)] A(x^*)^{-1} F'(x^*)(x - x^*)\|$$
$$\leqslant [2\beta\epsilon + 2\beta^2\epsilon \|F'(x^*)\|] \|x - x^*\|, \qquad \forall x \in S, \tag{6}$$

which, since ϵ is arbitrary and $\|F'(x^*)\|$ and β are fixed quantities, shows that G is F-differentiable at x^* and that (2) holds. ∎

It follows from **10.2.1** and **10.1.4** that if, in addition to the conditions of **10.2.1**, $\rho(I - A(x^*)^{-1} F'(x^*)) = \sigma < 1$, then x^* is a point of attraction of the iteration $\mathscr{I}$ defined by (1), and $R_1(\mathscr{I}, x^*) = \sigma$. In particular, if F is G-differentiable in a neighborhood of x^* and

$A(x) = F'(x)$, then **10.2.1** and **10.1.6** show that the Newton iteration has both R and Q superlinear convergence. This is stated precisely in the following basic theorem, which also shows that additional regularity assumptions on F yield higher-order convergence.

10.2.2. Newton Attraction Theorem. Assume that $F: D \subset R^n \to R^n$ is G-differentiable on an open neighborhood $S_0 \subset D$ of a point $x^* \in D$ for which $Fx^* = 0$, and that F' is continuous at x^* and $F'(x^*)$ nonsingular. Then x^* is a point of attraction of the Newton iteration

$$\mathscr{I}: \quad x^{k+1} = x^k - F'(x^k)^{-1} Fx^k, \qquad k = 0, 1, \ldots, \tag{7}$$

and $R_1(\mathscr{I}, x^*) = Q_1(\mathscr{I}, x^*) = 0$. If, in addition, there are constants $\alpha < +\infty$ and $p \in (0, 1]$ such that

$$\|F'(x) - F'(x^*)\| \leqslant \alpha \|x - x^*\|^p, \qquad \forall x \in S_0, \tag{8}$$

then $O_R(\mathscr{I}, x^*) \geqslant O_Q(\mathscr{I}, x^*) \geqslant 1 + p$. Finally, if F is continuously differentiable on S_0 and the second F-derivative of F exists at x^* and satisfies

$$F''(x^*) hh \neq 0, \qquad \forall h \in R^n, \quad h \neq 0, \tag{9}$$

then $O_R(\mathscr{I}, x^*) = O_Q(\mathscr{I}, x^*) = 2$.

Proof. By **3.2.8**, F is F-differentiable at x^*, and, thus, with $A(x) = F'(x)$ for $x \in S_0$, it follows from **10.2.1** that the Newton iteration function $Gx = x - F'(x)^{-1} Fx$ is well-defined on some ball $S = \bar{S}(x^*, \delta) \subset S_0$, $\delta > 0$, and that $G'(x^*) = 0$. Then **10.1.6** shows that $R_1(\mathscr{I}, x^*) = Q_1(\mathscr{I}, x^*) = 0$.

Suppose, next, that (8) holds. Then, by **3.2.5**,

$$\|Fx - Fx^* - F'(x^*)(x - x^*)\| \leqslant \alpha \|x - x^*\|^{p+1}, \qquad \forall x \in S,$$

and therefore from the first inequality of (6) we obtain, using (4),

$$\|Gx - x^*\| \leqslant 4\beta\alpha \|x - x^*\|^{p+1}.$$

It then follows immediately from **10.1.6** that

$$O_R(\mathscr{I}, x^*) \geqslant O_Q(\mathscr{I}, x^*) \geqslant 1 + p.$$

Finally, assume that F is continuously differentiable on S_0 and that (9) holds. Then, by **2.3.3**, $F'(x)^{-1}$ is also continuous on S, and, because

$$F'(x)^{-1} F''(x^*) hh \neq 0, \qquad \forall h \in R^n, \quad h \neq 0, \quad \forall x \in S,$$

10.2. NEWTON'S METHOD AND SOME OF ITS MODIFICATIONS

and S is compact, if follows that there exists a constant $c > 0$ such that

$$\|F'(x)^{-1}F''(x^*)\,hh\,\| \geq c\,\|\,h\,\|^2, \qquad \forall h \in R^n, \quad \forall x \in S. \tag{10}$$

Now, choose $\delta' > 0$ such that $S' = \bar{S}(x^*, \delta') \subset S$ as well as

$$\|F'(x) - F'(x^*) - F''(x^*)(x - x^*)\| \leq c(8\beta)^{-1}\,\|\,x - x^*\,\|, \qquad \forall x \in S,$$

and set

$$R(x) = Fx - Fx^* - F'(x^*)(x - x^*) - \tfrac{1}{2}F''(x^*)(x - x^*)(x - x^*), \qquad \forall x \in S.$$

Then

$$R'(x) = F'(x) - F'(x^*) - F''(x^*)(x - x^*)$$

is continuous on S', and, hence,

$$\|R(x)\| = \|R(x) - R(x^*)\| = \left\|\int_0^1 R'(x^* + t(x - x^*))\,dt\right\|$$

$$\leq \int_0^1 c(8\beta)^{-1}t\,\|\,x - x^*\,\|^2\,dt \leq c(16\beta)^{-1}\,\|\,x - x^*\,\|^2, \qquad \forall x \in S'.$$

Hence, for any $x \in S'$,

$$\|Gx - x^*\|$$
$$= \|F'(x)^{-1}\{-\tfrac{1}{2}F''(x^*)(x - x^*)(x - x^*) + R(x)$$
$$\quad - [F'(x) - F'(x^*) - F''(x^*)(x - x^*)](x - x^*)\}\|$$
$$\geq \tfrac{1}{2}\|F'(x)^{-1}F''(x^*)(x - x^*)(x - x^*)\| - \tfrac{1}{8}c\,\|\,x - x^*\,\|^2 - \tfrac{1}{4}c\,\|\,x - x^*\,\|^2$$
$$\geq \tfrac{1}{8}c\,\|\,x - x^*\,\|^2,$$

and the result follows from **10.1.6**. ∎

In the remainder of this section, we consider some common modifications of Newton's method. One of the simplest is the following iteration, discussed in **7.1**,

$$x^{k+1} = x^k - \omega[F'(x^k) + \lambda I]^{-1}Fx^k, \qquad k = 0, 1, \ldots, \tag{11}$$

where ω and λ are fixed constants. Clearly, if $\omega = 1$ and $\lambda = 0$ then (11) reduces to Newton's method:

10.2.3. Assume that $F: D \subset R^n \to R^n$ is G-differentiable on an open neighborhood $S_0 \subset D$ of a point $x^* \in D$ for which $Fx^* = 0$, and that

F' is continuous at x^* and $F'(x^*)$ nonsingular. Let $\mu_1,...,\mu_n$ be the eigenvalues of $F'(x^*)$ and

$$\beta = \min\{|\mu_i|^2/(2\operatorname{Re}\mu_i) \mid \operatorname{Re}\mu_i > 0\}, \quad \eta = \min\{|\mu_i|^2/(-2\operatorname{Re}\mu_i) \mid \operatorname{Re}\mu_i < 0\},$$

where $\beta = +\infty$ or $\eta = +\infty$ if there is no eigenvalue with $\operatorname{Re}\mu_i > 0$ or $\operatorname{Re}\mu_i < 0$, respectively. Then x^* is a point of attraction of the process (11) for any (λ, ω) from the set

$$\Lambda = \{(\lambda, \omega) \mid 0 < \omega < 2, -\beta < \lambda/(2-\omega) < \eta\}.$$

Proof. For any fixed $(\lambda, \omega) \in \Lambda$, define $A: S_0 \to L(R^n)$ by $A(x) = (1/\omega)[F'(x) + \lambda I]$, $x \in S_0$. Then A is clearly continuous at x^*. Suppose that $A(x^*)$ is singular; then $\mu_j + \lambda = 0$ for some j, and, hence, $\lambda \neq 0$. If $\lambda > 0$, then

$$\lambda/(2-\omega) < \eta \leq \mu_j^2/(-2\operatorname{Re}\mu_j) = \lambda/2$$

represents a contradiction, and, similarly, $\lambda < 0$ is excluded. Thus, $A(x^*)$ must be nonsingular, and, by **10.2.1**, the mapping

$$Gx = x - \omega[F'(x) + \lambda I]^{-1}Fx$$

is well-defined on some ball $S = \bar{S}(x^*, \delta) \subset S_0$, is F-differentiable at x^*, and

$$G'(x^*) = I - \omega[F'(x^*) + \lambda I]^{-1}F'(x^*).$$

Thus, the eigenvalues of $G'(x^*)$ are

$$\nu_i = [\lambda + (1-\omega)\mu_i]/(\lambda + \mu_i), \quad i = 1,...,n. \qquad (12)$$

The condition $|\nu_i|^2 < 1$ is equivalent with

$$\lambda^2 + (1-\omega)^2|\mu_i|^2 + 2(1-\omega)\lambda\operatorname{Re}\mu_i < \lambda^2 + |\mu_i|^2 + 2\lambda\operatorname{Re}\mu_i$$

or

$$-(2-\omega)\omega|\mu_i|^2 < 2\lambda\omega\operatorname{Re}\mu_i, \quad i = 1,...,n,$$

which is clearly satisfied for $(\lambda, \omega) < \Lambda$. Hence, $\rho(G'(x^*)) < 1$, and **10.1.3** applies. ∎

Note that the iteration $\mathscr{I}$ defined by (11) does not, in general, share the superlinear rate of convergence of Newton's method. In fact, by (12), the condition $R_1(\mathscr{I}, x^*) = \max_i|\nu_i| = 0$ is equivalent with $|\lambda + (1-\omega)\mu_i| = 0$, $i = 1,...,n$, which holds if and only if either $\lambda = 0$ and $\omega = 1$, or $\mu_i = (\omega - 1)^{-1}\lambda$ for all i. The first case is, of

10.2. NEWTON'S METHOD AND SOME OF ITS MODIFICATIONS

course, Newton's method. Note also that the set Λ given in the formulation of **10.2.3** is not the best possible. For necessary and sufficient conditions on λ and ω so that $\rho(G'(x')) < 1$, see **E 10.2-3**.

We next consider the following modification of Newton's method, discussed in **7.1**, in which a simplified Newton step is composed with a Newton step:

$$x^{k+1} = x^k - F'(x^k)^{-1}[Fx^k + F(x^k - F'(x^k)^{-1}Fx^k)], \quad k = 0, 1, \ldots. \quad (13)$$

Although the simplified Newton iteration exhibits superlinear convergence only under very restrictive conditions (see **E 10.2-1**), the combined iteration (13) has at least cubic convergence.

10.2.4. Let $F: D \subset R^n \to R^n$ be F-differentiable in an open ball $S = S(x^*, \delta) \subset D$ and satisfy

$$\| F'(x) - F'(x^*) \| \leqslant \gamma \| x - x^* \|, \quad \forall x \in S.$$

Assume, further, that $Fx^* = 0$ and that $F'(x^*)$ is nonsingular. Then x^* is a point of attraction of the iteration $\mathscr{I}$ defined by (13), and

$$O_R(\mathscr{I}, x^*) \geqslant O_Q(\mathscr{I}, x^*) \geqslant 3.$$

Proof. It follows from the proof of **10.2.2** that the Newton function $Nx = x - F'(x)^{-1}Fx$ is well defined and satisfies an estimate of the form

$$\| Nx - x^* \| \leqslant \eta \| x - x^* \|^2, \quad \forall x \in S_1,$$

on some ball $S_1 = \bar{S}(x^*, \delta_1) \subset S$. Consequently, the mapping

$$Gx = Nx - F'(x)^{-1}F(Nx)$$

is also well defined on $S_2 = \bar{S}(x^*, \delta_2) \subset S_1$, where $\delta_2^2 \leqslant \delta_1/\eta$. Therefore, if $\| F'(x)^{-1} \| \leqslant \beta$ for all $x \in S_2$, then

$$\| Gx - x^* \|$$
$$\leqslant \| F'(x)^{-1} \| \, \| F'(x) \, Nx - F(Nx) - F'(x) \, x^* \|$$
$$\leqslant \beta \| F(Nx) - Fx^* - F'(x^*)(Nx - x^*) \| + \beta \| F'(x^*) - F'(x) \| \, \| Nx - x^* \|$$
$$\leqslant \tfrac{1}{2}\beta\gamma \| Nx - x^* \|^2 + \beta\gamma \| x - x^* \| \, \| Nx - x^* \|$$
$$\leqslant [\beta\gamma\eta \, (\tfrac{1}{2}\eta\delta_2 + 1)] \| x - x^* \|^3$$

for all $x \in S_2$. Hence, the result follows from **10.1.6**. ∎

NOTES AND REMARKS

NR 10.2-1. For one-dimensional equations, the quadratic convergence of Newton's method was established by Cauchy [1829]. For equations in R^n, a point-of-attraction theorem was given by Runge [1899], who also stressed the quadratic convergence. Independently, a result for $n = 2$ was given by Blutel [1910]. For convergence results not assuming the existence of a solution, see Chapter 12 and **NR 12.6-1**.

Traub [1964], as well as other authors, shows the quadratic convergence of Newton's method by differentiating the Newton function explicitly; this requires considerably more smoothness of F than used in **10.2.2** (see **E 10.2-6**). In fact, **10.2.2** can be slightly strengthened in that F does not have to be G-differentiable in an neighborhood of x^*, but only the Jacobian matrix needs to exist (and still be continuous at x^*). The use of only Hölder continuity of the derivative in **10.2.2** is a result of Schmidt [1968] (see also **NR 11.2-4**).

Conditions for precisely second-order convergence seem not to have been given previously.

NR 10.2-2. Theorem **10.2.4** is due to Traub [1964], although, again, the differentiability assumptions here are weaker. More generally, the corresponding m-step method,

$$x^{k,0} = x^k, \quad x^{k,i} = x^{k,i-1} - F'(x^k)^{-1} F x^{k,i-1}, \quad i = 1,\ldots, m+1, \quad x^{k+1} = x^{k,m+1},$$

discussed in **7.1**, in which m simplified Newton steps are taken between each Newton step, has convergence of order $m + 2$, or, more precisely, satisfies $O_R(\mathscr{I}, x^*) \geq O_Q(\mathscr{I}, x^*) \geq m + 2$. We sketch the proof, which is a simple modification of **10.2.4**. In fact, we already saw in **10.2.4** that

$$\| x^{k,2} - x^* \| \leq \mu \| x^k - x^* \|^3.$$

Then

$$\| x^{k,3} - x^* \| \leq \| F'(x^k)^{-1} \| \{\| F x^{k,2} - F x^* - F'(x^*)(x^{k,2} - x^*)\|$$
$$+ \|[F'(x^*) - F'(x^k)](x^{k,2} - x^*)\|\}$$
$$\leq \tfrac{1}{2}\beta\gamma \| x^{k,2} - x^* \|^2 + \beta\gamma \| x^k - x^* \| \| x^{k,2} - x^* \|$$
$$\leq \mu_1 \| x^k - x^* \|^4.$$

Proceeding in this way, one shows easily that

$$\| x^{k,m+1} - x^* \| \leq \mu_{m-2} \| x^k - x^* \|^{m+2}.$$

This result is, in essence, a special case of an even more general result of Šamanski [1967a] which will be discussed in **NR 11.2-5**.

10.2. NEWTON'S METHOD AND SOME OF ITS MODIFICATIONS

NR 10.2-3. High-order iterative processes may be generated by the composition of two lower-order processes (see Traub [1964] for an extended discussion). Let G_1 and G_2 satisfy $\| G_i x - x^* \| \leqslant \gamma_i \| x - x^* \|^{p_i}$, $i = 1, 2$, in a neighborhood of x^*. Then the composite function $G = G_1 \circ G_2$ satisfies

$$\| Gx - x^* \| \leqslant \gamma_1 \| G_2 x - x^* \|^{p_1} \leqslant \gamma_1 \gamma_2 \| x - x^* \|^{p_1 p_2},$$

so that the process G has order at least $p_1 p_2$. For example, the Newton function may be composed with itself to give an iteration of fourth order; however, one is simply doubling the work at each step, so that nothing is gained. The interest in **10.2.4**, and the extension discussed in **NR 10.2-2**, is that high-order methods may be generated which require only one evaluation of F' and no higher derivatives.

NR 10.2-4. In the proof of **10.2.2**, we obtained the error estimate

$$\| x^{k+1} - x^* \| \leqslant 4\beta\alpha \| x^k - x^* \|^{p+1}. \tag{14}$$

This estimate is unimportant from the standpoint of computing an actual error bound since, even if the constants were known, x^* is not. It is important from a theoretical viewpoint, however, since it indicates how the convergence proceeds close to x^*; in particular, the larger β or α, the worse is the convergence.

This situation is typical of point-of-attraction results, in that they are virtually never useful in ascertaining for a particular x^0 that the resulting sequence of iterates will converge. Rather, the proper interpretation of **10.2.2** is that, provided F satisfies certain conditions, the Newton iterates *must* converge, and at a certain rate, whenever x^0 (or any iterate x^k) is sufficiently close to x^*. Hence, **10.2.2** gives certain intrinsic properties of the Newton iteration rather than providing a tool for ascertaining convergence.

On the other hand, results like **10.2.2** provide the possibility of obtaining *exclusion regions* for solutions. Consider the special case of (14) in which $p = 1$, (that is, F' is Lipschitz-continuous). Then **10.2.2** guarantees that the Newton sequence $\{x^k\}$ converges to x^* provided $\| x^0 - x^* \| < 1/4\beta\alpha$. Hence, if the iterates do not converge, it follows that there is no solution in the ball $S(x^0, 1/4\beta\alpha)$, and, if estimates for α and β can be obtained, this gives an exclusion region for solutions of $Fx = 0$. This idea has not been developed to any extent as a practical tool.

NR 10.2-5. The theorems of **10.2**, as well as those of later sections, assume that at the solution x^* the derivative $F'(x^*)$ is nonsingular. For Newton's method, Rall [1966] gave a point-of-attraction theorem which assumes only that $F'(x)$ is nonsingular for points $x \neq x^*$ in a neighborhood of x^*, but not for x^* itself. Cavanagh [1970] pointed out that Rall's proof contains a flaw, and then proved the following corrected result:

Suppose that $F: D \subset R^n \to R^n$ is twice F-differentiable on an open neighborhood $S \subset D$ of a solution x^* of $Fx = 0$, and that on S, F'' is Lipschitz continuous, while $F'(x)$ is nonsingular for all $x \neq x^*$, $x \in S$. Assume, further, that if $N \subset R^n$ denotes the null space of $F'(x^*)$, there is a $\gamma > 0$ such that

$$\|F''(x^*)\,uv\| \geq \gamma \|u\|\,\|v\|, \quad \forall u \in N, \ v \in R^n.$$

Then x^* is a point of attraction of Newton's method. Moreover, if M is the orthogonal complement of N and P_N and P_M are the corresponding (natural) projections of R^n onto N and M, respectively, then the sequence $\{P_M(x^k - x^*)\}$ converges R-quadratically to zero, while the R_1-factor of $\{P_N(x^k - x^*)\}$ is, at least $1/2$.

On the other hand, Stepleman [1969] has considered a modified Newton iteration which, under certain conditions, preserves the quadratic convergence even when F is not differentiable at x^*.

EXERCISES

E 10.2-1. Apply **E 10.1-8** to conclude that the simplified Newton iteration $x^{k+1} = x^k - F'(x^0)^{-1} Fx^k$ has R-superlinear convergence if and only if $F'(x^*)^{-1}$ exists and $F'(x^0) = F'(x^*)\,J$, where all eigenvalues of J are equal to unity.

E 10.2-2. Consider the Newton iteration applied to the mapping $f: R^1 \to R^1$ defined by $f(t) = \exp(-t^{-2})$, $t \neq 0$, $f(0) = 0$. Show that 0 is a point of attraction, but that $R_1(\mathscr{I}, 0) = Q_1(\mathscr{I}, 0) = 1$.

E 10.2-3. Assume that the conditions of **10.2.3** are satisfied, let μ_i, $i = 1,\ldots,n$, denote the eigenvalues of $F'(x^*)$, and set

$$\Gamma_+ = \{|\mu_i|^2/(2\,\mathrm{Re}\,\mu_i) \mid \mathrm{Re}\,\mu_i > 0\}, \qquad \Gamma_- = \{|\mu_i|^2/(-2\,\mathrm{Re}\,\mu_i) \mid \mathrm{Re}\,\mu_i < 0\}.$$

If Γ_+ is not empty, set $\beta_m = \min \Gamma_+$, $\beta_M = \max \Gamma_+$. Similarly, if Γ_- is not empty, set $\eta_m = \min \Gamma_-$ and $\eta_M = \max \Gamma_-$. Show that $[F'(x^*) + \lambda I]^{-1}$ exists and $\rho(I - \omega[F'(x^*) + \lambda I]^{-1} F'(x^*)) < 1$ if and only if the real parameters ω and λ are in the set Λ defined by one of the following conditions:

(a) If $\mathrm{Re}\,\mu_i > 0$, $i = 1,\ldots,n$, then

$$\Lambda = \{\omega > 0,\ \ 0 > \lambda > \beta_m \omega - 2\beta_m\} \cup \{\omega > 0,\ \lambda > 0,\ \lambda > \beta_M \omega - 2\beta_M\}$$
$$\cup \{\omega < 0,\ \lambda < \beta_M \omega - 2\beta_M\}.$$

(b) If $\mathrm{Re}\,\mu_i < 0$, $i = 1,\ldots,n$, then

$$\Lambda = \{\omega > 0,\ 0 < \lambda < -\omega \eta_m + 2\eta_m\} \cup \{\omega > 0,\ \lambda < 0,\ \lambda < -\omega \eta_M + 2\eta_M\}$$
$$\cup \{\omega < 0,\ \lambda > -\omega \eta_M + 2\eta_M\}.$$

10.2. NEWTON'S METHOD AND SOME OF ITS MODIFICATIONS

(c) If there exist μ_i and μ_j such that $\operatorname{Re} \mu_i > 0$ and $\operatorname{Re} \mu_j < 0$, then
$$\Lambda = \{0 < \omega < 2, \ \omega\beta_m - 2\beta_m < \lambda < -\omega\eta_m + 2\eta_m\}.$$

(d) If $\operatorname{Re} \mu_i \geqslant 0$, $i = 1,\dots, n$, and $\operatorname{Re} \mu_i = 0$ for some i, then
$$\Lambda = \{0 < \omega < 2, \ \lambda > \omega\beta_m - 2\beta_m\}.$$

(e) If $\operatorname{Re} \mu_i \leqslant 0$, $i = 1,\dots, n$, and $\operatorname{Re} \mu_i = 0$ for some i, then
$$\Lambda = \{0 < \omega < 2, \ \lambda < -\omega\eta_m + 2\eta_m\}.$$

(f) If $\operatorname{Re} \mu_i = 0$, $i = 1,\dots, n$, then
$$\Lambda = \{0 < \omega < 2, \ -\infty < \lambda < +\infty\}.$$

Show also that Λ is necessarily empty if $F'(x^*)$ is singular.

E 10.2-4. Define $f: R^1 \to R^1$ by $f(x) = x + x^{1+\alpha}$ for some $\alpha \in (0, 1]$. Show that $O_R(\mathcal{I}, 0) = O_Q(\mathcal{I}, 0) = 1 + \alpha$ for the Newton iteration. (This shows that the precise order of Newton's method may be any number between 1 and 2, depending on the regularity of f.)

E 10.2-5. In the context of **10.2.1**, assume that the mapping $A: S_0 \to L(R^n)$ is G-differentiable. Compute $G'(x)$ explicitly in a neighborhood of x^* and conclude from this, again, that $G'(x^*) = I - A(x^*)^{-1} F'(x^*)$.

E 10.2-6. Assume that the conditions of **10.2.2** hold and, in addition, that F is three times continuously differentiable in D_0. Show that the Newton function $Gx = x - F'(x)^{-1} Fx$ is twice F-differentiable in a neighborhood S of x^*, that
$$G'(x) = F'(x)^{-1} F''(x)[F'(x)^{-1} Fx], \quad \forall x \in S,$$
and that $G''(x^*) = F'(x^*)^{-1} F''(x^*)$. Hence, conclude the final statement of **10.2.2** by means of **10.1.7**.

E 10.2-7. For $f: R^1 \to R^1$, assume that f is three times continuously differentiable on R^1 and that, for some x^*, $f(x^*) = f''(x^*) = 0$ and $f'(x^*) \neq 0$. Show that $O_R(\mathcal{I}, x^*) \geqslant O_Q(\mathcal{I}, x^*) \geqslant 3$ for the Newton iteration.

E 10.2-8. Suppose that $x^* \in R^1$ is a root of multiplicity p of $f: R^1 \to R^1$. Under suitable differentiability assumptions on f, show that
$$Q_1(\mathcal{I}, x^*) = R_1(\mathcal{I}, x^*) = (p-1)/p$$
for the Newton iteration.

E 10.2-9. Let $G: D \subset R^n \to R^n$ be F-differentiable in an open neighborhood of $x^* = Gx^*$, and $\rho(G'(x^*)) < 1$. Show that x^* is a point of attraction for the Newton iteration applied to $Fx = x - Gx$.

E 10.2-10. (*Gauss–Newton Convergence*). Let $F: D \subset R^n \to R^m$, $m \geq n$, be two times F-differentiable in a neighborhood of a point x^* for which $F'(x^*)^T Fx^* = 0$, and assume that rank $F'(x^*) = n$. Show that the Gauss–Newton iteration function (Section **8.5**),

$$Gx = x - [F'(x)^T F'(x)]^{-1} F'(x)^T Fx,$$

is well-defined in a neighborhood of x^*, that G is F-differentiable at x^*, and that

$$G'(x^*) = -[F'(x^*)^T F'(x^*)]^{-1} F''(x^*) Fx^*.$$

Conclude, in particular, that $R_1(\mathscr{I}, x^*) = Q_1(\mathscr{I}, x^*) = 0$ if $Fx^* = 0$, that is, if the equations are consistent.

E 10.2-11. Define $F: R^1 \to R^2$ by $f_1(x) = 1 + a^2 x^2$, $f_2(x) = x$. Show that $x^* = 0$ is a point of attraction of the Gauss–Newton iteration of **E 10.2-10** if and only if $2a^2 < 1$, and that $R_1(\mathscr{I}, x^*) = 0$ if and only if $a = 0$.

E 10.2-12. Let $\phi: R^n \to R^n$ be continuously differentiable and let $A \in L(R^n)$. Assume that one of the following sets of conditions holds:

(a) A is symmetric, positive definite and $\phi'(x)$ is symmetric, positive semidefinite for all $x \in R^n$.

(b) A is an M-matrix and $\phi'(x)$ is nonnegative and diagonal for all $x \in R^n$.

Then, by **4.4.1** or **5.4.1**, respectively, $Fx \equiv Ax + \phi x = 0$ has a unique solution x^*. Show that x^* is a point of attraction for the Newton iteration and that the convergence is Q-superlinear.

E 10.2-13. Assume that $F: D \subset R^n \to R^n$ has a Lipschitz-continuous derivative in an open neighborhood $S_0 \subset D$ of a point x^* for which $Fx^* = 0$, and $F'(x^*)$ is nonsingular. Show that there is a $\delta > 0$ so that $S = S(x^*, \delta) \subset S_0$ and

$$\|F(x - F'(x)^{-1} Fx)\| \leq \|Fx\|, \qquad \forall x \in S.$$

Conclude from this that whenever the Newton iterates (7) converge to x^* there is a k_0 so that $\|Fx^{k+1}\| \leq \|Fx^k\|$ for all $k \geq k_0$.

10.3. GENERALIZED LINEAR ITERATIONS

We now apply the results of the previous sections to several of the generalized linear iterations described in **7.4**. We begin with a result about the general process

$$x^{k+1} = x^k - [I + \cdots + H(x^k)^{m-1}] B(x^k)^{-1} Fx^k, \qquad k = 0, 1, \ldots, \tag{1}$$

10.3. GENERALIZED LINEAR ITERATIONS

where B and H are defined by

$$F'(x) = B(x) - C(x), \quad H(x) = B(x)^{-1}C(x). \quad (2)$$

10.3.1. Let $F: D \subset R^n \to R^n$ be G-differentiable in an open neighborhood $S_0 \subset D$ of a point $x^* \in D$ at which F' is continuous and $Fx^* = 0$. Suppose that $B: S_0 \to L(R^n)$ is continuous at x^*, that $B(x^*)$ is nonsingular, and that $\rho(H(x^*)) < 1$. Then, for any $m \geq 1$, x^* is a point of attraction of the iteration $\mathscr{I}$ defined by (1) and (2) and $R_1(\mathscr{I}, x^*) = \rho(H(x^*)^m)$.

Proof. Since $B(x^*)$ is nonsingular and B is continuous at x^*, it follows from **2.3.3** that $B(x)^{-1}$ exists on some ball $S = S(x^*, \delta) \subset S_0$ and is continuous at x^*. Hence, H also exists in S and is continuous at x^*.

Now consider the identity

$$I - H(x^*)^m = [I - H(x^*)][I + \cdots + H(x^*)^{m-1}]. \quad (3)$$

Since $\rho(H(x^*)) < 1$, the matrix $I - H(x^*)^m$ is nonsingular, and, therefore, the second term on the right side of (3) is also nonsingular. It follows from the continuity of H at x^* that the mapping

$$A(x) = B(x)[I + H(x) + \cdots + H(x)^{m-1}]^{-1}$$

is well defined on some ball $S_1 = S(x^*, \delta_1) \subset S$ and continuous at x^*. Clearly, $A(x^*)^{-1}$ exists, and thus **10.2.1** ensures that the mapping $Gx = x - A(x)^{-1}Fx$ is well defined on a neighborhood of x^*, and, moreover, that G is F-differentiable at x^* and that

$$G'(x^*) = I - A(x^*)^{-1}F'(x^*).$$

But

$$I - A(x^*)^{-1}F'(x^*) = I - [I + \cdots + H(x^*)^{m-1}]B(x^*)^{-1}B(x^*)[I - H(x^*)]$$
$$= H(x^*)^m$$

so that the result now follows from **10.1.3–10.1.4**. ∎

This result applies to any of the generalized linear methods of Chapter 7 which can be written in the form (1). As an example, we give here the corresponding attraction theorem for the iteration (7.4.12).

10.3.2. Newton–SOR Theorem. Suppose that the conditions of **10.3.1** for $F: D \subset R^n \to R^n$ and x^* are satisfied. Let

$$F'(x) = D(x) - L(x) - U(x) \quad (4)$$

be the decomposition of $F'(x)$ into its diagonal, strictly lower-, and strictly upper-triangular parts, and assume that $D(x^*)$ is nonsingular. Consider the Newton–SOR iteration

$$\mathscr{I}: \quad x^{k+1} = x^k - \omega[I + \cdots + H_\omega(x^k)^{m-1}][D(x^k) - \omega L(x^k)]^{-1} F x^k,$$
$$k = 0, 1, \ldots, \quad (5)$$

where $\omega > 0$, $m \geq 1$, and

$$H_\omega(x) = [D(x) - \omega L(x)]^{-1}[(1 - \omega) D(x) + \omega U(x)]. \quad (6)$$

If $\rho(H_\omega(x^*)) < 1$, then x^* is a point of attraction of $\mathscr{I}$, and $R_1(\mathscr{I}, x^*) = \rho(H_\omega(x^*)^m)$.

Proof. In order to apply **10.3.1**, set

$$B(x) = (1/\omega)[D(x) - \omega L(x)], \quad C(x) = (1/\omega)[(1 - \omega) D(x) + \omega U(x)].$$

It is immediate that $F'(x) = B(x) - C(x)$, and, because of the continuity of F' at x^*, both B and C are continuous at x^*. Since $L(x^*)$ is strictly lower triangular the nonsingularity of $B(x^*)$ follows directly from that of $D(x^*)$. Thus, all conditions of **10.3.1** are satisfied. ∎

We may interpret **10.3.2** as follows: $H(x^*)$ is simply the SOR iteration matrix for a linear system $F'(x^*)x = b$. Hence, for $m = 1$, the asymptotic rate of convergence of the process (5) is precisely the same as that of the SOR iteration applied to $F'(x^*)x = b$. It is then to be expected that the m-step Newton–SOR process (5) is m-times as fast as the one-step process, since for the corresponding linear problem one iteration of the m-step process amounts to m SOR steps. An analogous interpretation applies to the general method (1).

Specific results similar to **10.3.2** can also be derived from **10.3.1** for various other methods, such as the Newton–Jacobi process; in this connection, see **E 10.3-2** and **E 10.3-3**. Note, however, that **10.3.1** does not apply, for example, to the general Newton–SOR process in which the relaxation factor ω or the number of sweeps m of the secondary iteration are allowed to vary with k. For results along this line, we refer to Theorem **11.1.5**, which also contains a slightly different proof of **10.3.1** not utilizing **10.2.1**. Finally, the important question of the choice of ω so as to maximize the rate of convergence is discussed in the appendix **(10.5)** to this chapter.

We consider next the one-step SOR–Newton process. As described in Section **7.4**, this iteration is given by

$$x_i^{k+1} = x_i^k - \frac{\omega f_i(x_1^{k+1}, \ldots, x_{i-1}^{k+1}, x_i^k, \ldots, x_n^k)}{\partial_i f_i(x_1^{k+1}, \ldots, x_{i-1}^{k+1}, x_i^k, \ldots, x_n^k)}, \quad k = 0, 1, \ldots, \quad i = 1, 2, \ldots, n \quad (7)$$

10.3. GENERALIZED LINEAR ITERATIONS

where, as usual, $f_1,\ldots,f_n$ are the components of $F: D \subset R^n \to R^n$. Clearly, (7) may be written in the form $x^{k+1} = Gx^k$, although now the mapping G becomes rather complicated. In fact, the simplest representation is to define the components $g_1, g_2,\ldots,g_n$ of G recursively by

$$g_i(x) = x_i - \frac{\omega f_i(g_1(x),\ldots, g_{i-1}(x), x_i,\ldots, x_n)}{\partial_i f_i(g_1(x),\ldots, g_{i-1}(x), x_i,\ldots, x_n)}, \qquad i = 1,\ldots, n. \qquad (8)$$

10.3.3. SOR-Newton Theorem.

Given $F: D \subset R^n \to R^n$, assume that $Fx^* = 0$ for some $x^* \in \text{int}(D)$ and that F is continuously differentiable on an open neighborhood $S_0 \subset D$ of x^*. Let $D(x)$, $-L(x)$, and $-U(x)$ be the diagonal, strictly lower-, and strictly upper-triangular parts of $F'(x)$, and assume that $D(x^*)$ is nonsingular and that $\sigma = \rho(H_\omega(x^*)) < 1$, where $H_\omega(x)$ is defined by (6) and $\omega > 0$. Then the mapping G with components given by (8) is well defined on an open ball $S = S(x^*, \delta)$ in S_0, x^* is a point of attraction of the iteration $\mathscr{I}: x^{k+1} = Gx^k$, $k = 0, 1,\ldots$, and $R_1(\mathscr{I}, x^*) = \sigma$.

Proof. Define the sets

$$D_i = \{x \in S_0 \mid \partial_i f_i(x) \neq 0\}, \qquad i = 1,\ldots, n,$$

and

$$S_i = \{x \in S_{i-1} \mid \gamma^i(x) \in D_i\}, \qquad i = 1,\ldots, n,$$

where $\gamma^1(x) = x$ and

$$\gamma^i(x) = (g_1(x),\ldots, g_{i-1}(x), x_i,\ldots, x_n)^T, \qquad i = 2,\ldots, n.$$

Clearly, $S_1 = D_1$ and, by (8), each g_i is well defined on S_i, $i = 1,\ldots, n$. Thus, G is well defined on S_n, and, since $x_i^* = g_i(x^*)$, $i = 1,\ldots, n$, we have $Gx^* = x^* \in S_n$. From the continuity of $\partial_i f_i$ on S_0 and from $\partial_i f_i(x^*) \neq 0$, it follows that each D_i is open. In particular, S_1 is open, and, since g_1 is continuous on S_1, there is an open neighborhood $S_1' \subset S_1$ of x^* so that $\gamma^2(x) \in D_2$ for all $x \in S_1'$. An obvious induction now shows that S_n contains an open neighborhood S of x^*. Therefore, it suffices to show that $G: S \to R^n$ is F-differentiable at x^* and that $G'(x^*) = H_\omega(x^*)$; then the result will follow from **10.1.3–10.1.4**.

Set

$$r_i(x) = f_i(x) - \underbrace{f_i(x^*)}_{=0} - f_i'(x^*)(x - x^*), \qquad \forall x \in S_0;$$

then (8) can be rewritten as

$$g_i(x) - g_i(x^*) = x_i - x_i^* - \omega \frac{f_i'(x^*)(\gamma^i(x) - x^*) + r_i(\gamma^i(x))}{\partial_i f_i(\gamma^i(x))}, \qquad \forall x \in S. \qquad (9)$$

Now, let $S' = \bar{S}(x^*, \delta') \subset S$, $\delta' > 0$, be a closed ball in S. By the continuity of $\partial_i f_i$ and **3.2.5**, there exists a constant η so that

$$|r_i(x)| \leq \eta \|x - x^*\|, \qquad \forall x \in S'.$$

Therefore, we obtain from (9) for $i = 1$, again by using the continuity of $\partial_1 f_1$, that

$$|g_1(x) - g_1(x^*)| \leq c_1 \|x - x^*\|, \qquad \forall x \in S',$$

with some constant c_1. Assume there exist constants $c_1, \ldots, c_{i-1}$ such that

$$|g_j(x) - g_j(x^*)| \leq c_j \|x - x^*\|, \qquad \forall x \in S', \tag{10}$$

for $j = 1, \ldots, i - 1$. Then, clearly, there is a constant c_i' so that

$$\|\gamma^i(x) - x^*\| \leq c_i' \|x - x^*\|, \qquad \forall x \in S'. \tag{11}$$

Consequently, it follows from (9), using the continuity of $\partial_i f_i$ and γ^i, as well as the Lipschitz continuity of r_i, that there is a c_i so that (10) holds for $j = i$. Hence, by induction, (10) and (11) are valid for all $i, j = 1, \ldots, n$.

Observe now that (9) implies that

$$\partial_i f_i(x^*)[g_i(x) - g_i(x^*)] = \partial_i f_i(x^*)(x_i - x_i^*) - \omega f_i'(x^*)(\gamma^i(x) - x^*) - q_i(x), \tag{12}$$

for $x \in S$, $i = 1, \ldots, n$, where

$$q_i(x) = [\partial_i f_i(\gamma^i(x)) - \partial_i f_i(x^*)][g_i(x) - g_i(x^*) - (x_i - x_i^*)] - \omega r_i(\gamma^i(x)). \tag{13}$$

Evidently, (12) is equivalent with

$$\partial_i f_i(x^*)(g_i(x) - g_i(x^*)) + \omega \sum_{j=1}^{i-1} \partial_j f_i(x^*)(g_j(x) - g_j(x^*))$$

$$= \partial_i f_i(x^*)(x_i - x_i^*) - \omega \sum_{j=i}^{n} \partial_j f_i(x^*)(x_j - x_j^*) - q_i(x), \qquad i = 1, \ldots, n,$$

which, in turn, is equivalent with

$$[D(x^*) - \omega L(x^*)](Gx - Gx^*) = [(1 - \omega) D(x^*) + \omega U(x^*)](x - x^*) - Q(x).$$

where $Q(x) = (q_1(x), \ldots, q_n(x))^T$. Now, it follows from (9), (10), (13), and the continuity of $\partial_i f_i$ that

$$\lim_{x \to x^*} [|q_i(x)|/\|x - x^*\|] = 0, \qquad i = 1, \ldots, n.$$

10.3. GENERALIZED LINEAR ITERATIONS

Therefore

$$\lim_{x \to x^*} \frac{1}{\|x - x^*\|} \| Gx - Gx^* - H_\omega(x^*)(x - x^*) \|$$

$$= \lim_{x \to x^*} \frac{1}{\|x - x^*\|} \|(D(x^*) - \omega L(x^*))^{-1} Q(x)\|$$

$$= 0,$$

which shows that $G'(x^*) = H_\omega(x^*)$. ∎

We consider next the nonlinear SOR process (7.4.26):

Solve $f_i(x_1^{k+1}, \ldots, x_{i-1}^{k+1}, x_i, x_{i+1}^k, \ldots, x_n^k) = 0$ for x_i

Set $x_i^{k+1} = x_i^k + \omega(x_i - x_i^k)$, $i = 1, \ldots, n$, $k = 0, 1, \ldots$.

In this case, the iterates are not defined explicitly; nevertheless, we may still apply **10.1.4** to a suitable (but probably unknown) function G in order to obtain the rate of convergence. For the following more general result, recall from **5.2.2** that if $\hat{G}: D \times D \subset R^n \times R^n \to R^n$, then $\partial_1 \hat{G}(x, y)$ denotes the partial derivative of $\hat{G}$ with respect to the first vector variable, while $\partial_2 \hat{G}(x, y)$ is the derivative with respect to the second vector variable.

10.3.4. Suppose that $\hat{G}: D \times D \subset R^n \times R^n \to R^n$ has continuous partial derivatives $\partial_1 \hat{G}$ and $\partial_2 \hat{G}$ on an open neighborhood $S_0 \subset D$ of a point $x^* \in D$ at which $\hat{G}(x^*, x^*) = 0$. Assume, further, that $\partial_1 \hat{G}(x^*, x^*)$ is nonsingular and that

$$\sigma = \rho(-\partial_1 \hat{G}(x^*, x^*)^{-1} \partial_2 \hat{G}(x^*, x^*)) < 1.$$

Then, there is an open ball $S = S(x^*, \delta) \subset S_0$ such that, for any $y \in S$, the equation $\hat{G}(x, y) = 0$ has a unique solution $x = Gy$ in S. Therefore, the sequence

$$x^{k+1} = Gx^k, \quad k = 0, 1, \ldots, \tag{14}$$

is well-defined for any $x^0 \in S$ and satisfies

$$\hat{G}(x^{k+1}, x^k) = 0, \quad k = 0, 1, \ldots .$$

Moreover, x^* is a point of attraction of the iteration $\mathscr{I}$ defined by (14) and $R_1(\mathscr{I}, x^*) = \sigma$.

Proof. It follows from the implicit function theorem **5.2.4** that there are open neighborhoods S_1 and S_2 of x^* such that, for any $y \in S_2$, the equation $\hat{G}(x, y) = 0$ has a unique solution $x = Gy$ in S_1; in particular, $x^* = Gx^*$. Theorem **5.2.4** also shows that the mapping $G: S_2 \to S_1$ is F-differentiable at x^* and that

$$G'(x^*) = -\partial_1 \hat{G}(x^*, x^*)^{-1} \partial_2 \hat{G}(x^*, x^*).$$

Now, as in the proof of **10.1.3**, given $\epsilon > 0$ with $\sigma + 2\epsilon < 1$, there is a norm and a $\delta > 0$ so that $S = S(x^*, \delta) \subset S_1 \cap S_2$ and

$$\| Gx - x^* \| \leq (\sigma + 2\epsilon) \| x - x^* \|, \quad \forall x \in S;$$

hence, $GS \subset S$. Consequently, the first statement is proved, and, by construction, for any $x^0 \in S$, the sequence (14) satisfies $\hat{G}(x^{k+1}, x^k) = 0$, $k = 0, 1,\ldots$. The remaining statements follow immediately from **10.1.3** and **10.1.4**. ∎

In order to apply **10.3.4** to the nonlinear SOR process, we note that the mapping $\hat{G}$ may be defined in this case by

$$\begin{cases} \hat{G}(x, y) = (\hat{g}_1(x, y),\ldots, \hat{g}_n(x, y))^T \\ \hat{g}_i(x, y) = f_i(x_1,\ldots, x_{i-1}, y_i + [1/\omega][x_i - y_i], y_{i+1},\ldots, y_n), \quad i = 1,\ldots, n, \end{cases} \quad (15)$$

for some $\omega > 0$; clearly, the sequence $\{x^k\}$ satisfies the SOR prescription if and only if $\hat{G}(x^{k+1}, x^k) = 0$, $k = 0, 1,\ldots$.

10.3.5. Nonlinear SOR Theorem. Let $F: D \subset R^n \to R^n$ be continuously differentiable in an open neighborhood $S_0 \subset D$ of a point $x^* \in D$ for which $Fx^* = 0$. Consider again the decomposition (4) of $F'(x)$ into its diagonal, strictly lower-, and strictly upper-triangular parts, and suppose that $D(x^*)$ is nonsingular and $\rho(H_\omega(x^*)) < 1$, where $H_\omega(x)$ is defined by (6) and $\omega > 0$. Then there exists an open ball $S = S(x^*, \delta)$ in S_0 such that, for any $x^0 \in S$, there is a unique sequence $\{x^k\} \subset S$ which satisfies the nonlinear SOR prescription. Moreover, $\lim_{k \to \infty} x^k = x^*$ and $R_1(\mathscr{I}, x^*) = \rho(H_\omega(x^*))$.

Proof. Because of the continuous differentiability of F on S_0, it follows that the mapping $\hat{G}$ defined by (15) has continuous partial derivatives on some set $S' \times S'$, where S' is an open subset of S_0 of the form

$$S' = \{x \in S_0 \mid |x_i - x_i^*| < \delta', \quad i = 1,\ldots, n\}.$$

Moreover, a straightforward computation shows that

$$\partial_1 \hat{G}(x^*, x^*) = \omega^{-1}[D(x^*) - \omega L(x^*)],$$
$$\partial_2 \hat{G}(x^*, x^*) = \omega^{-1}[(1 - \omega) D(x^*) + \omega U(x^*)],$$

10.3. GENERALIZED LINEAR ITERATIONS

and, since $D(x^*)$ is nonsingular $\partial_1 \hat{G}(x^*, x^*)$ is also nonsingular. But then

$$H_\omega(x^*) = \partial_1 \hat{G}(x^*, x^*)^{-1} \partial_2 \hat{G}(x^*, x^*),$$

and the result follows directly from **10.3.4**. ∎

The nonlinear SOR iteration may be considered as the limiting form of an m-step SOR-Newton iteration as $m \to \infty$. Since the one-step method and the infinite-step method have the same asymptotic rate of convergence, it is reasonable that the m-step method also has the R-convergence factor $\rho(H_\omega(x^*))$. This is indeed true (see **E 10.3.5**), and, therefore, it does not enhance the asymptotic rate of convergence to take more than one Newton step. This is in contrast to the m-step Newton-SOR iteration in which the R_1-convergence factor decreases geometrically with the number of secondary iterations.

We summarize these results in Table 10.1.

TABLE 10.1

Method	R-Convergence factor
Nonlinear SOR	$\rho(H_\omega(x^*))$
m-Step SOR–Newton	$\rho(H_\omega(x^*))$
m-Step Newton–SOR	$[\rho(H_\omega(x^*))]^m$

Even when it is not possible to ascertain in advance that $\rho(H_\omega(x^*)) < 1$, these results nevertheless provide useful *a priori* information. On the other hand, it is sometimes quite easy to guarantee $\rho(H_\omega(x^*)) < 1$ even though x^* is unknown. We give a simple example of this in **10.3.7**, which, in turn, is a corollary of the next result.

10.3.6. Suppose that $F : D \subset R^n \to R^n$ is continuously differentiable on an open neighborhood $S_0 \subset D$ of a point $x^* \in D$ for which $Fx^* = 0$. Assume, further, that $F'(x^*)$ is an M-matrix. Then x^* is a point of attraction of any one of the SOR-type methods considered in **10.3.2**, **10.3.3**, and **10.3.5**, provided that $\omega \in (0, 1]$.

Proof. Consider the decomposition (4) of $F'(x)$ into its diagonal, lower-, and upper-triangular parts. By **2.4.8**, $D(x^*)$ is nonnegative and invertible, and, by **2.4.7**, $L(x^*)$ and $U(x^*)$ are also nonnegative. Hence, for any $\omega > 0$, **2.4.6** ensures that $D(x^*) - \omega L(x^*)$ has a nonnegative inverse, and, therefore,

$$F'(x^*) = (1/\omega)[D(x^*) - \omega L(x^*)] - (1/\omega)[(1 - \omega) D(x^*) + \omega U(x^*)]$$

is a regular splitting (see Definition **2.4.16**) of $F'(x^*)$ for any $\omega \in (0, 1]$. Consequently, **2.4.17** shows that $\rho(H_\omega(x^*)) < 1$, where $H_\omega(x^*)$ is defined by (6), and, thus, **10.3.2**, **10.3.3**, or **10.3.5** apply to give the result. ∎

As an immediate corollary, **10.3.6** gives the following theorem.

10.3.7. Let $A \in L(R^n)$ be an M-matrix and $\phi: R^n \to R^n$ a continuously differentiable mapping such that $\phi'(x)$ is diagonal and nonnegative for all $x \in R^n$. Then, for any $\omega \in (0, 1]$, the unique solution x^* of $Ax + \phi x = 0$ is a point of attraction of any of the SOR-type methods of this section.

Proof. Set $Fx \equiv Ax + \phi x$, $x \in R^n$; then **5.4.1** ensures that $Fx = 0$ has a unique solution x^* and **2.4.11** shows that $F'(x^*) = A + \phi'(x^*)$ is an M-matrix. The result then follows directly from **10.3.6**. ∎

We complete this section with some point-of-attraction results for the Peaceman–Rachford iterations discussed in **7.4**. Consider, first, the one-step Newton–Peaceman–Rachford iteration (7.4.25):

$$x^{k+1} = x^k - 2\alpha[V(x^k) + \alpha I]^{-1}[H(x^k) + \alpha I]^{-1}Fx^k, \quad k = 0, 1, \ldots. \quad (16)$$

10.3.8. Newton–Peaceman–Rachford Theorem. Let $F: D \subset R^n \to R^n$ be F-differentiable on an open neighborhood $S_0 \subset D$ of a point x^* for which $Fx^* = 0$, and let $H, V: S_0 \to L(R^n)$ be such that $F'(x) = H(x) + V(x)$, for all $x \in S_0$. Suppose that H and V are continuous at x^*, that $H(x^*)$ and $V(x^*)$ are positive semidefinite, and that at least one of them is positive definite. Then, for any $\alpha > 0$, x^* is a point of attraction of the iteration $\mathscr{I}$ defined by (16). Moreover, $R_1(\mathscr{I}, x^*) = \rho(B) < 1$, where

$$B = [V(x^*) + \alpha I]^{-1}[H(x^*) - \alpha I][H(x^*) + \alpha I]^{-1}[V(x^*) - \alpha I].$$

Proof. Define $G: S_0 \to R^n$ by $Gx = x - A(x)^{-1}Fx$, where

$$A(x) = (2\alpha)^{-1}[H(x) + \alpha I][V(x) + \alpha I];$$

then (16) is the iteration $x^{k+1} = Gx^k$, $k = 0, 1, \ldots$. Clearly, A is continuous at x^*, and, since $H(x^*) + \alpha I$ and $V(x^*) + \alpha I$ are positive definite and, hence, invertible, we see that $A(x^*)^{-1}$ exists. It then follows immediately from **10.2.1** that G is well defined in a neighborhood $S \subset S_0$ of x^*, that G is F-differentiable at x^*, and that

$$G'(x^*) = I - A(x^*)^{-1}F'(x^*).$$

10.3. GENERALIZED LINEAR ITERATIONS

Now, recall that, by (7.4.19),

$$A(x^*) - (2\alpha)^{-1}[H(x^*) - \alpha I][V(x^*) - \alpha I] = H(x^*) + V(x^*) = F'(x^*)$$

so that, since $H(x^*) - \alpha I$ and $[H(x^*) + \alpha I]^{-1}$ commute,

$$G'(x^*) = A(x^*)^{-1}[A(x^*) - F'(x^*)] = B.$$

Therefore, to complete the proof, it suffices, by **10.1.4**, to prove that $\rho(B) < 1$. Set $\hat{H} = H(x^*)$, $\hat{V} = V(x^*)$, $S = (\hat{H} - \alpha I)(\hat{H} + \alpha I)^{-1}$, and $T = (\hat{V} - \alpha I)(\hat{V} + \alpha I)^{-1}$. Then

$$ST = (\hat{V} + \alpha I) B (\hat{V} + \alpha I)^{-1},$$

so that ST is similar to B. Hence, it suffices to show that $\|ST\| < 1$.

Let $U \in L(R^n)$ be any positive semidefinite matrix, and, for arbitrary $x \neq 0$, set $y = (\alpha I + U)^{-1} x$. Then, under the Euclidean norm,

$$\frac{\|(U - \alpha I)(U + \alpha I)^{-1} x\|^2}{\|x\|^2} = \frac{\|(U - \alpha I) y\|^2}{\|(U + \alpha I) y\|^2} = \frac{y^T(U - \alpha I)^T(U - \alpha I) y}{y^T(U + \alpha I)^T(U + \alpha I) y}$$

$$= \frac{\|Uy\|^2 - 2\alpha y^T U y + \alpha^2 \|y\|^2}{\|Uy\|^2 + 2\alpha y^T U y + \alpha^2 \|y\|^2} \leqslant 1, \qquad (17)$$

so that $\|(U - \alpha I)(U + \alpha I)^{-1}\| \leqslant 1$. Moreover, if $y^T U y > 0$, then strict inequality holds in (17); hence, if $x^T U x > 0$, for all $x \in R^n$, $x \neq 0$, it follows that $\|(U - \alpha I)(U + \alpha I)^{-1}\| < 1$, since the value of the norm is assumed for some $x \in R^n$. But this shows that $\|S\| \|T\| < 1$. ∎

Note that **10.3.9** could be rephrased to omit the hypotheses that $H(x^*)$ and $V(x^*)$ are positive semidefinite and $\alpha > 0$, by assuming that $[H(x^*) + \alpha I]^{-1}$ and $[V(x^*) + \alpha I]^{-1}$ exist and $\rho(B) < 1$. The following result, which will cover the nonlinear Peaceman–Rachford and one-step Peaceman–Rachford–Newton iterations (7.4.35) and (7.4.36), is stated in this way.

10.3.9. Let $F: D \subset R^n \to R^n$ and assume that, on an open neighborhood $S_0 \subset D$ of a point $x^* \in D$ for which $Fx^* = 0$, there exist F-differentiable mappings K_i, $H_i : S_0 \to R^n$, $i = 1, 2$, such that

$$Fx = K_1 x - H_1 x = K_2 x - H_2 x, \qquad \forall x \in S_0.$$

Assume that K_i', H_i', $i = 1, 2$, are continuous at x^*, that $K_1'(x^*)$ and $K_2'(x^*)$ are nonsingular, and that

$$\sigma = \rho\big(K_2'(x^*)^{-1} H_2'(x^*) K_1'(x^*)^{-1} H_1'(x^*)\big) < 1.$$

Then x^* is a point of attraction of the implicit iteration

$$\mathscr{I}_1: \quad K_1 x^{k+\frac{1}{2}} = H_1 x^k, \qquad K_2 x^{k+1} = H_2 x^{k+\frac{1}{2}}, \qquad k = 0, 1, ...,$$

as well as of the corresponding composite iteration

$$\mathscr{I}_2: \quad x^{k+\frac{1}{2}} = x^k - K_1'(x^k)^{-1} F x^k, \qquad x^{k+1} = x^{k+\frac{1}{2}} - K_2'(x^{k+\frac{1}{2}})^{-1} F x^{k+\frac{1}{2}}.$$

Moreover, $R_1(\mathscr{I}_1, x^*) = R_1(\mathscr{I}_2, x^*) = \sigma$.

Proof. Since K_1' is continuous at x^*, the inverse function theorem 5.2.1 shows that K_1 is locally invertible at x^*; that is, there is a ball $\hat{S}_1 = S(x^*, \hat{\delta}_1) \subset S_0$ such that the restriction $\hat{K}_1$ of K_1 to $\hat{S}_1$ is one-to-one and that, also, $K_1 \hat{S}_1$ contains an open ball about $K_1 x^*$. Since $H_1 x^* = K_1 x^*$ and H_1 is continuous at x^*, there is also a ball $S_1 = S(x^*, \delta_1)$ such that $H_1 x \in K_1 \hat{S}_1$, for all $x \in S_1$. Therefore, the composite mapping $G_1 = \hat{K}_1^{-1} H_1$ is well defined on S_1, and $G_1 x^* = \hat{K}_1^{-1} K_1 x^* = x^*$. Moreover, 5.2.1 also ensures that $\hat{K}_1^{-1}$ is F-differentiable at $K_1 x^*$ and that $(\hat{K}_1^{-1})'(K_1 x^*) = K_1'(x^*)^{-1}$. Hence, it follows from the chain rule 3.1.7 that G_1 is F-differentiable at x^* and that

$$G_1'(x^*) = K_1'(x^*)^{-1} H_1'(x^*).$$

In precisely the same fashion, it is seen that there is a ball $S_2 = S(x^*, \delta_2)$ such that the restriction $\hat{K}_2$ of K_2 to S_2 is one-to-one, and that the mapping $G_2 = \hat{K}_2^{-1} H_2$ is well defined on S_2, F-differentiable at x^*, and $G_2'(x^*) = K_2'(x^*)^{-1} H_2'(x^*)$.

Now, G_1 is continuous at x^*, and, hence, there is still another ball $S = S(x^*, \delta) \subset S_1$ such that $G_1 S \subset S_2$. Therefore, the composition $G = G_2 \circ G_1$ is well defined on S, and, since $x^* = G_1 x^*$, another application of 3.1.7 yields

$$G'(x^*) = G_2'(x^*) G_1'(x^*) = K_2'(x^*)^{-1} H_2'(x^*) K_1'(x^*)^{-1} H_1'(x^*).$$

By hypothesis, $\rho(G'(x^*)) = \sigma < 1$, and, hence, 10.1.3 ensures that x^* is a point of attraction of the iteration $x^{k+1} = G x^k$, $k = 0, 1, ...$; but, by construction, this iteration is equivalent to $x^{k+\frac{1}{2}} = G_1 x^k$, $x^{k+1} = G_2 x^{k+\frac{1}{2}}$, that is, to the iteration $\mathscr{I}_1$. Now, $R_1(\mathscr{I}, x^*) = \sigma$ is a consequence of 10.1.4.

For the iteration $\mathscr{I}_2$, consider the mappings

$$G_1 x = x - K_1'(x)^{-1} F x, \qquad G_2 x = x - K_2'(x)^{-1} F x.$$

By hypothesis, $K_1'(x^*)^{-1}$ and $K_2'(x^*)^{-1}$ exist and K_1' and K_2' are continuous at x^*. Hence, 10.2.1 ensures that G_1 and G_2 are well defined on

10.3. GENERALIZED LINEAR ITERATIONS

balls $S_i = S(x^*, \delta_i)$, $i = 1, 2$, respectively, and F-differentiable at x^*, and that

$$G_i'(x^*) = I - K_i'(x^*)^{-1}F'(x^*) = K_i'(x^*)^{-1}H_i'(x^*), \qquad i = 1, 2.$$

Clearly, $G_i x^* = x^*$, $i = 1, 2$, and the proof now proceeds precisely as for $\mathscr{I}_1$. ∎

Theorem **10.3.9** applies to the nonlinear Peaceman–Rachford iteration (7.4.35),

$$\alpha x^{k+\frac{1}{2}} + F_H x^{k+\frac{1}{2}} = x^k - F_V x^k, \qquad \alpha x^{k+1} + F_V x^{k+1} = \alpha x^{k+\frac{1}{2}} - F_H x^{k+\frac{1}{2}},$$
$$k = 0, 1, \ldots,$$

as well as the one-step Peaceman–Rachford–Newton iteration (7.4.36),

$$x^{k+\frac{1}{2}} = x^k - [\alpha I + F_H'(x^k)]^{-1} F x^k, \qquad x^{k+1} = x^{k+\frac{1}{2}} - [\alpha I + F_V'(x^{k+\frac{1}{2}})]^{-1} F x^{k+\frac{1}{2}},$$
$$k = 0, 1, \ldots,$$

by setting

$$K_1 x = \alpha x + F_H x, \quad K_2 x = \alpha I + F_V x, \quad H_1 x = x - F_V x, \quad H_2 x = x - F_H x.$$

In order to apply **10.3.9**, it is necessary to guarantee that $[\alpha I + F_H'(x^*)]^{-1}$ and $[\alpha I + F_V'(x^*)]^{-1}$ exist and that

$$\rho([\alpha I + F_V'(x^*)]^{-1}[F_H'(x^*) - \alpha I][F_H'(x^*) + \alpha I]^{-1}[F_V'(x^*) - \alpha I]) < 1. \tag{18}$$

One way to ensure this is to assume that $F_H'(x^*)$ and $F_V'(x^*)$ are both positive semidefinite, one is positive definite, and $\alpha > 0$. It then follows, as in the proof of **10.3.6**, that (18) holds.

We also note that if $F = F_H + F_V$, then $F'(x) = F_H'(x) + F_V'(x)$. Hence, for this choice of the mapping H and V of **10.3.8**, the R_1 factor of the one-step Newton–Peaceman–Rachford method is also given by the quantity of (18).

NOTES AND REMARKS

NR 10.3-1. Theorems **10.3.3**–**10.3.5** were first given in Ortega and Rockoff [1966] under somewhat stronger differentiability assumptions; in particular, **10.3.3** was proved under the assumption that F is twice-continuously differentiable in a neighborhood of x^*. In that case, the proof of **10.3.3** may be simplified by applying **10.3.4** (see E **10.3-4**). Theorems **10.3.1**, **10.3.2**, and **10.3.9** were proved in Ortega and Rheinboldt [1970a]. Theorem **10.3.8** is new.

NR 10.3-2. The hypothesis that $F'(x^*)$ is an M-matrix in **10.3.6** may, of course, be replaced by any condition which ensures that the corresponding SOR matrix satisfies $\rho(H_\omega(x^*)) < 1$. For example, if $F'(x^*)$ is symmetric, positive definite, then $\rho(H_\omega(x^*)) < 1$ for any $\omega \in (0, 2)$; see, for example, Varga [1962, p. 77]. In particular, the conclusion of **10.3.7** holds for $\omega \in (0, 2)$, if A is symmetric, positive definite and $\phi'(x)$ is symmetric, positive semidefinite for all x. In this case, the existence of a solution is assured by **4.4.1**. More generally, the point-of-attraction results of this section, including those of the Peaceman–Rachford method, apply to any of the problems discussed in **4.4** for which it is shown that $F'(x^*)$ is positive definite.

NR 10.3-3. Theorems **10.3.2**, **10.3.3**, and **10.3.5** extend immediately to the block SOR methods discussed in **NR 7.4-10**, provided, of course, that the block SOR matrix corresponding to (6) has spectral radius less than one.

NR 10.3-4. If U is symmetric, positive semidefinite, the fact that

$$\|(U - \alpha I)(U + \alpha I)^{-1}\| \leqslant 1$$

is a standard result of the linear theory (see, for example, Varga [1962, p. 213]). The proof used in **10.3.8** for nonsymmetric U is taken from Kellogg [1964].

EXERCISES

E 10.3-1. Apply **10.3.1** to the m-step block Newton–SOR iteration in which the iteration function G is defined by (5) and (6), where (4) is the decomposition of $F'(x)$ into block diagonal, strictly lower, and strictly upper block triangular parts with the same block sizes for all x. (See **NR 7.4-10**.)

E 10.3-2. Consider the m-step Newton–Jacobi iteration

$$\mathscr{J}: \quad x^{k+1} = x^k - [I + \cdots + H(x^k)^{m-1}] D(x^k)^{-1} F x^k, \quad k = 0, 1, \ldots,$$

where $F'(x) = D(x) - L(x) - U(x)$ is the decomposition into diagonal, lower-, and upper-triangular parts, and $H(x) = D(x)^{-1}[L(x) + U(x)]$. Apply **10.3.1** to determine a result analogous to **10.3.2**. Moreover, under the conditions of **10.3.6** or **10.3.7**, show that x^* is a point of attraction.

E 10.3-3. Consider the one-step Jacobi–Newton iteration

$$x_i^{k+1} = x_i^k - f_i(x^k)/\partial_i f_i(x^k), \quad i = 1,\ldots, n; \quad k = 0, 1,\ldots .$$

Observe that this is identical to the one-step Newton–Jacobi iteration, and, hence, conclude the rate of convergence.

10.3. GENERALIZED LINEAR ITERATIONS

E 10.3-4. Let $F: R^n \to R^n$ and assume that F is twice-continuously differentiable on
$$S = \{x \mid |x_i - x_i^*| < \delta, \quad i = 1,\ldots, n\},$$
where $Fx^* = 0$. Suppose that $D(x^*)$ is nonsingular, and that $\rho(H_\omega(x^*)) < 1$. Define $\hat{G}: S \times S \subset R^n \times R^n \to R^n$ by
$$\hat{g}_i(x, y) = \partial_i f_i(x_1,\ldots, x_{i-1}, y_i,\ldots, y_n)(x_i - y_i) + \omega f_i(x_1,\ldots, x_{i-1}, y_i,\ldots, y_n),$$
for $i = 1,\ldots, n$. Prove the conclusion of **10.3.3** for the one-step SOR–Newton iteration by applying **10.3.4** to $\hat{G}$.

E 10.3-5. Let $F: R^n \to R^n$ satisfy the conditions of **E 10.3-4**. Show that the m-step SOR–Newton iteration has the same R_1-factor as the one-step iteration for any integer $m \geq 1$.

E 10.3-6. Let $F: R^n \to R^n$ be twice-continuously differentiable on R^n and let $Fx^* = 0$. Consider the iteration $\mathscr{I}$, discussed in **NR 7.4-9**:
$$x_i^{k+1} = x_i^k - \frac{\omega}{\partial_i f_i(x^{k,i})}\left[\sum_{j=1}^{i-1} \partial_j f_i(x^{k,i})(x_j^{k+1} - x_j^k) - f_i(x^{k,i})\right], \quad i = 1,\ldots, n,$$
where $x^{k,i} = (x_1^{k+1},\ldots, x_{i-1}^{k+1}, x_i^k,\ldots, x_n^k)^T$. Assume that $D(x^*)^{-1}$ exists and $\rho(H) < 1$, where
$$H = [D(x^*) - 2\omega L(x^*)]^{-1}[(1 - \omega) D(x^*) - \omega L(x^*) + \omega U(x^*)].$$
Show that x^* is a point of attraction and $R_1(\mathscr{I}, x^*) = \rho(H)$.

E 10.3-7. Let $G: R^n \to R^n$ be continuously differentiable in a neighborhood of a fixed point x^* and consider the iteration
$$x_i^{k+1} = g_i(x_1^{k+1},\ldots, x_{i-1}^{k+1}, x_i^k,\ldots, x_n^k), \quad k = 0, 1,\ldots; \quad i = 1,\ldots, n.$$
Show that x^* is a point of attraction provided that
$$\rho\{(I - L)^{-1}[G'(x^*) - L]\} < 1,$$
where L is the strictly lower-triangular part of $G'(x^*)$.

E 10.3-8. Let $F: D \subset R^n \to R^n$ be continuously differentiable in an open neighborhood of a solution x^* of $Fx^* = 0$. With the notation of **E 10.3-2**, assume that $D(x^*)$ is nonsingular and that $\rho(H(x^*)) < 1$. Show that if z is a fixed vector sufficiently close to x^*, then x^* is a point of attraction of the *Jacobi–Regula-Falsi* iteration:
$$\mathscr{I}: \quad x_i^{k+1} = x_i^k - f_i(x^k)/h_i(x^k, z), \quad k = 0, 1,\ldots; \quad i = 1,\ldots, n,$$

where

$$h_i(x, z) = \begin{cases} [f_i(x + (z_i - x_i) e^i) - f_i(x)]/(z_i - x_i), & \text{for } x_i \neq z_i, \\ \partial_i f_i(x), & \text{for } x_i = z_i. \end{cases}$$

Compute $R_1(\mathscr{I}, x^*)$ and show that, in general, this R_1 is different from the R-factor for the Jacobi–Newton process of **E 10.3-3**. Formulate the corresponding SOR–Regula–Falsi iteration and prove an analogous result (Voigt [1969]).

10.4. CONTINUATION METHODS

In this section, we apply some of the point-of-attraction results to the continuation methods discussed in **7.5**. Let us first assume that the homotopy defining the continuation process is given in the form

$$H(x, t) = x - G(x, t) = 0, \quad t \in [0, 1], \tag{1}$$

where $G: D \times [0, 1] \subset R^{n+1} \to R^n$, and suppose that there exists a continuous path $x: [0, 1] \to R^n$ such that $H(x(t), t) = 0$, for all $t \in [0, 1]$. If the initial point $x^0 = x(0)$ is explicitly known, then, as discussed in **7.5**, we may consider "moving along" this solution curve by using the sequence of iterations

$$x^{i,k+1} = G(x^{i,k}, t_i), \quad k = 0, 1, \ldots, m_i - 1, \quad i = 1, 2, \ldots, N - 1, \tag{2a}$$

$$x^{1,0} = x^0, \quad x^{i+1,0} = x^{i,m_i}, \tag{2b}$$

$$x^{N,k+1} = G(x^{N,k}, 1), \quad k = 0, 1, \ldots, \tag{2c}$$

where

$$0 = t_0 < t_1 < t_2 < \cdots < t_N = 1 \tag{3}$$

is some partition of $[0, 1]$. The problem now is to give sufficient conditions which ensure that the entire sequence $\{x^{i,k}\}$ is well defined and that $\lim_{k \to \infty} x^{N,k} = x(1)$.

10.4.1. Suppose that $G: D \times [0, 1] \subset R^n \times R^1 \to R^n$ is F-differentiable with respect to the first variable and that $\partial_1 G$ is continuous on $D \times [0, 1]$. Assume, further, that (1) has a continuous solution $x: [0, 1] \to \text{int}(D)$ with known initial point $x^0 = x(0)$, and that $\sigma_t = \rho(\partial_1 G(x(t), t)) < 1$, for all $t \in [0, 1]$. Then there exists a partition (3) of $[0, 1]$ and integers $m_1, \ldots, m_{N-1}$ such that the entire sequence $\{x^{i,k}\}$ defined by (2) remains in D and $\lim_{k \to \infty} x^{N,k} = x(1)$.

Proof. Because $C = \{x \in R^n \mid x = x(t), t \in [0, 1]\} \subset \text{int}(D)$ is compact,

10.4. CONTINUATION METHODS

there exists a compact subset $D_0 \subset D$ such that $C \subset \text{int}(D_0)$. Moreover, for each $t \in [0, 1]$, there exists, by **2.2.8**, a norm $\|\cdot\|_t$ such that $\|\partial_1 G(x(t), t)\|_t \leq \sigma_t + \epsilon$. Let $\epsilon > 0$ be such that $\sigma_t + 3\epsilon < 1$. Since $\partial_1 G$ is uniformly continuous on $D_0 \times [0, 1]$ under any norm, we can choose $\delta_1 = \delta_1(t) > 0$ such that

$$\|\partial_1 G(x, t_1) - \partial_1 G(y, t_2)\|_t \leq \epsilon,$$

$\forall x, y \in D_0, \quad \|x - y\|_t \leq \delta_1, \quad \forall t_1, t_2 \in [0, 1], \quad |t_1 - t_2| < \delta_1.$

Again by uniform continuity, there exists $\delta_2 = \delta_2(t) > 0$ for which

$$\|x(t_1) - x(t_2)\|_t \leq \delta_1, \quad \forall t_1, t_2 \in [0, 1], \quad |t_1 - t_2| \leq \delta_2.$$

Let $\delta' \leq \min(\delta_1, \delta_2)$ be such that $x \in D_0$ whenever $\|x - x(s)\|_t \leq \delta'$ for some $s \in [0, 1]$. Then, by the triangle inequality, we have

$$\|\partial_1 G(x(s), s)\|_t \leq \sigma_t + 2\epsilon, \quad \forall s \in [0, 1], \quad |s - t| \leq \delta',$$

so that, by the mean value theorem **3.2.5**,

$$\|G(x, s) - x(s)\|_t = \|G(x, s) - G(x(s), s)\|_t$$
$$\leq \{(\sigma_t + 2\epsilon) + \sup_{0 \leq \theta \leq 1} \|\partial_1 G(x(s), s) - \partial_1 G(x(s) + \theta(x - x(s)), s)\|_t\} \|x - x(s)\|_t$$
$$\leq (\sigma_t + 3\epsilon) \|x - x(s)\|_t$$

for every $s \in [0, 1]$ such that $|s - t| \leq \delta'$ and every $x \in R^n$ such that $\|x - x(s)\|_t \leq \delta'$. Therefore, **10.1.2** ensures that, for any $s \in [0, 1]$ with $|s - t| \leq \delta'$, the iteration

$$x^{k+1} = G(x^k, s), \quad k = 0, 1, \ldots, \tag{4}$$

starting from any x^0 with $\|x^0 - x(s)\|_t \leq \delta'$, remains in D_0 and converges to $x(s)$. By the norm-equivalence theorem **2.2.1**, it follows that, for any fixed norm, independent of t, there also exists a $\delta = \delta(t) > 0$ such that the iteration (4) remains in D_0 and converges to $x(s)$ for any $s \in [0, 1]$ with $|s - t| \leq \delta$ and x^0 with $\|x^0 - x(s)\| \leq \delta$.

Since this result holds for any t from the compact interval $[0, 1]$, there exists a covering of $[0, 1]$ by finitely many intervals $\{t \mid |t - \bar{t}_j| \leq \delta(\bar{t}_j)\}$, $j = 1, \ldots, M$, such that if $\delta_0 = \min_j \delta(\bar{t}_j)$, then the iterates (4) remain in D_0 and converge to $x(s)$ for any x^0 with $\|x^0 - x(s)\| \leq \delta_0$ and any $s \in [0, 1]$.

Now, let a partittion (3) be chosen such that

$$\max_{0 \leq i \leq N-1} \|x(t_{i+1}) - x(t_i)\| \leq \delta_0' < \delta_0,$$

and suppose that we have already obtained $x^{i,0}$ such that

$$\| x^{i,0} - x(t_{i-1}) \| \leqslant \delta_0 - \delta_0'.$$

This holds for $i = 1$, since $x^{1,0} = x^0 = x(0)$. Then

$$\| x^{i,0} - x(t_i) \| \leqslant \| x^{i,0} - x(t_{i-1}) \| + \| x(t_{i-1}) - x(t_i) \| \leqslant \delta_0$$

shows that the iteration (2a) remains in D_0 and converges to $x(t_i)$. Thus, we can select an integer m_i for which $\| x^{i+1,0} - x(t_i) \| \leqslant \delta_0 - \delta_0'$, where $x^{i+1,0} = x^{i,m_i}$. Therefore, the process (2) can be continued up to $i = N$, all $x^{i,k}$ are in D_0, and the iteration (2c) converges to $x(1)$. ∎

We turn now to the use of Newton's method for moving along a solution curve $x = x(t)$ of the equation

$$H(x, t) = 0, \quad t \in [0, 1], \tag{5}$$

which need no longer be of the special form (1). Thus, instead of (2), we now consider the sequence of Newton iterations (7.5.10)–(7.5.11):

$$\left. \begin{array}{l} x^{i,k+1} = x^{i,k} - \partial_1 H(x^{i,k}, t_i)^{-1} H(x^{i,k}, t_i), \\ x^{1,0} = x^0, \quad x^{i+1,0} = x^{i,m_i}, \end{array} \right\} \quad \begin{array}{l} k = 0, \ldots, m_i - 1, \\ i = 1, \ldots, N - 1, \end{array} \tag{6a}$$
$$\tag{6b}$$

$$x^{N,k+1} = x^{N,k} - \partial_1 H(x^{N,k}, 1)^{-1} H(x^{N,k}, 1), \quad k = 0, 1, \ldots, \tag{6c}$$

where $\{t_i\}$ is again a partition (3) of $[0, 1]$. The following result is, in essence, a corollary of **10.4.1**. However, in analogy with **10.2.2**, a direct proof allows weaker assumptions on H.

10.4.2. Suppose that $H: D \times [0, 1] \subset R^n \times R^1 \to R^n$ is F-differentiable with respect to the first variable and that $\partial_1 H$ is continuous on $D \times [0, 1]$. Assume, further, that there exists a continuous solution $x: [0, 1] \to \text{int}(D)$ of (5) and that $\partial_1 H(x(t), t)$ is nonsingular for all $t \in [0, 1]$. Then there exists a partition (3) of $[0, 1]$ and integers $m_1, \ldots, m_{N-1}$ such that the entire sequence $\{x^{i,k}\}$ defined by (6) remains in D and $\lim_{k \to \infty} x^{N,k} = x(1)$.

Proof. By **2.3.3**, $\partial_1 H(x(t), t)^{-1}$ is continuous for $t \in [0, 1]$, and, since $[0, 1]$ is compact, there exists a $\beta < +\infty$ such that

$$\| \partial_1 H(x(t), t)^{-1} \| \leqslant \beta, \quad \forall t \in [0, 1]. \tag{7}$$

With C defined as in the proof of **10.4.1**, let $D_0 \subset D$ be any compact set such that $C \subset \text{int}(D_0)$; then $\partial_1 H$ is uniformly continuous on $D_0 \times [0, 1]$,

10.4. CONTINUATION METHODS

and, hence, for $\epsilon \in (0, \tfrac{1}{2}\beta)$, there is $\delta > 0$ for which $\bar{S}(x(t), \delta) \subset D_0$, for all $t \in [0, 1]$, and

$$\| \partial_1 H(x, t) - \partial_1 H(y, t) \| \leq \epsilon, \qquad \forall x, y \in D_0, \quad \| x - y \| \leq \delta, \quad t \in [0, 1]. \tag{8}$$

Therefore, the perturbation lemma **2.3.2** ensures the existence of $\partial_1 H(x, t)^{-1}$ for each $t \in [0,1]$ and $x \in \bar{S}(x(t), \delta)$. Moreover, we have

$$\| \partial_1 H(x, t)^{-1} \| \leq \beta/(1 - \beta\epsilon), \qquad \forall x \in \bar{S}(x(t), \delta), \quad t \in [0, 1]. \tag{9}$$

For any fixed $t \in [0, 1]$, consider, now, the Newton process

$$x^{k+1} = x^k - \partial_1 H(x^k, t)^{-1} H(x^k, t), \qquad k = 0, 1, \ldots, \tag{10}$$

with $x^0 \in \bar{S}(x(t), \delta)$. Then we claim that

$$\| x^k - x(t) \| \leq \alpha^k \delta, \qquad k = 0, 1, \ldots, \quad \alpha = \beta\epsilon/(1 - \beta\epsilon) < 1. \tag{11}$$

In fact, (11) is true by assumption for $k = 0$, and, if it holds for some $k \geq 0$, then $x^k \in \bar{S}(x(t), \delta)$, and, hence, by (8), (9), and **3.2.12**,

$$\begin{aligned}
\| x^{k+1} - x(t) \| &= \| x^k - x(t) - \partial_1 H(x^k, t)^{-1} H(x^k, t) \| \\
&\leq \| \partial_1 H(x^k, t)^{-1} \| \, \| H(x(t), t) - H(x^k, t) - \partial_1 H(x^k, t)(x(t) - x^k) \| \\
&\leq [\beta/(1 - \beta\epsilon)] \, \epsilon \, \| x(t) - x^k \| \leq \alpha^{k+1} \delta. \tag{12}
\end{aligned}$$

Therefore, the Newton sequence (10) remains in $\bar{S}(x(t), \delta)$ and converges to $x(t)$.

Now, choose a partition (3) for which

$$\| x(t_{i+1}) - x(t_i) \| \leq \delta' < \delta, \qquad i = 0, \ldots, N - 1,$$

and let $m_i \equiv m \geq 1$, $i = 1, \ldots, N - 1$, be such that $\alpha^m \leq 1 - (\delta'/\delta)$. Suppose that $x^{i,0} \in S(x(t_{i-1}), \delta - \delta')$, which is certainly true for $i = 1$; then again by the triangle inequality, $\| x^{i,0} - x(t_i) \| \leq \delta$, and, thus, by (11),

$$\| x^{i+1,0} - x(t_i) \| = \| x^{i,m} - x(t_i) \| \leq \alpha^m \delta = \delta - \delta',$$

which proves that the process (6) can be continued up to $i = N$, that all $x^{i,k}$ are in D_0, and that the final iteration (6c) converges to $x(1)$. ∎

As a special result of this type, let us consider the homotopy (7.5.3), namely,

$$H(x, t) = Fx + (t - 1) Fx^0, \qquad t \in [0, 1], \tag{13}$$

and for it the process (7.5.12) which follows from (6) for $m_i \equiv m = 1$.

10.4.3. Suppose that $F: R^n \to R^n$ is norm-coercive and twice-continuously differentiable on R^n, and that $F'(x)$ is nonsingular for all $x \in R^n$. Then, for any $x^0 \in R^n$, there is an integer $N_0 \geq 1$ such that, for any $N \geq N_0$, the combined process

$$x^{k+1} = x^k - F'(x^k)^{-1}[Fx^k + \frac{k}{N-1}Fx^0], \quad k = 0, 1, \ldots, N-1,$$
$$x^{k+1} = x^k - F'(x^k)^{-1}Fx^k, \quad k = N, N+1, \ldots, \quad (14)$$

converges to the unique solution x^* of $Fx = 0$ in R^n.

Proof. By **7.5.1**, there exists a unique, continuously differentiable mapping $x: [0, 1] \to R^n$ for which $H(x(t), t) = 0$, for all $t \in [0, 1]$, where H is given by (13), and

$$x'(t) = -F'(x(t))^{-1}Fx^0, \quad t \in [0, 1], \quad x(0) = x^0. \quad (15)$$

Let $D_0 \subset R^n$ be a convex, compact set such that

$$\{x \mid x = x(t), t \in [0, 1]\} \subset \mathrm{int}(D_0).$$

Then
$$\|F'(x)^{-1}\| \leq \beta < +\infty, \quad \|F''(x)\| \leq \gamma < \infty, \quad \forall x \in D_0,$$

since F'' and, by **2.3.3**, $F'(\cdot)^{-1}$ are continuous on D_0.

The proof now follows that of **10.4.2**. Set $\delta = (2\beta\gamma)^{-1}$ and $\epsilon = \gamma\delta$. Since $\partial_1 H(x, t) = F'(x)$ and, by **3.3.5**,

$$\|F'(x) - F'(y)\| \leq \gamma \|x - y\|, \quad \forall x, y \in D_0,$$

the inequality (8) is valid for this choice of ϵ and δ. Moreover,

$$\|\partial_1 H(x, t)^{-1}\| \leq \beta, \quad \forall x \in D_0, \quad t \in [0, 1],$$

so that (11) holds with $\alpha = \beta\epsilon = \frac{1}{2}$.

Next, choose $N_0 \geq 4\beta^2\gamma \|Fx^0\|$ and, for any fixed $N \geq N_0$, set $\delta' = \beta \|Fx^0\|/N$. Then (15) implies, using **3.2.7**, that

$$\|x(t_{i+1}) - x(t_i)\| = \left\| \int_0^1 x'(t_i + s(t_{i+1} - t_i))(t_{i+1} - t_i) \, ds \right\|$$
$$\leq \delta' \leq (4\beta\gamma)^{-1} < \delta. \quad (16)$$

Furthermore, with this choice of δ', we have

$$1 - (\delta'/\delta) \geq 1 - \frac{2\beta\gamma}{4\beta\gamma} = \frac{1}{2},$$

10.4. CONTINUATION METHODS

so that $m_i = 1$, $i = 1,\ldots, N$, is permissible. But for this choice of m_i and the homotopy (13), the iteration (6) is precisely (14). ∎

Note that norm-coerciveness is used in **10.4.3** only to ensure the existence of the curve $x(t)$; any other assumption, such as the uniform boundedness of $F'(x)^{-1}$, which guarantees this existence may be used as well.

We end this section with a similar result about the process (7.5.18),

$$x^{k+1} = x^k - hF'(x^k)^{-1}Fx^0, \quad k = 0, 1,\ldots, N-1, \quad h = 1/N, \quad (17a)$$

which is obtained by applying Euler's method to the differential equation (15). Since (17a) only provides an approximation x^N to the solution $x^* = x(1)$ of $Fx = 0$, it is natural to continue the process with Newton's method starting from x^N:

$$x^{k+1} = x^k - F'(x^k)^{-1}Fx^k, \quad k = N, N+1,\ldots. \quad (17b)$$

10.4.4. Let $F: R^n \to R^n$ be twice-continuously differentiable on R^n and satisfy $\|F'(x)^{-1}\| \leq \beta$, for all $x \in R^n$. Then, for any $x^0 \in R^n$, there exists an $N_0 \geq 1$ such that, for any $N \geq N_0$, the combined process (17) converges to the unique solution x^* of $Fx = 0$ in R^n.

Proof. From **7.5.1**, it again follows that the differential equation (15) has a unique, continuously differentiable solution $x: [0, 1] \to R^n$ and that

$$H(x(t), t) = Fx(t) + (t-1)Fx^0 = 0, \quad \forall t \in [0, 1]. \quad (18)$$

By **10.2.2**, $x(1)$ is a point of attraction of the Newton iteration (17b). Let $\delta > 0$ be chosen such that, for any $x^N \in S(x(1), \delta)$, this iteration converges to $x(1)$, and set $r = \beta \|Fx^0\| + \delta$, as well as $S = \bar{S}(x^0, r)$. Then, by (15),

$$\|x(t) - x^0\| = \left\| t \int_0^1 x'(st)\, ds \right\| \leq \beta \|Fx^0\| t < r, \quad \forall t \in [0, 1],$$

and, for any $N \geq 1$,

$$\|x^{k+1} - x^0\| \leq \sum_{j=0}^{k} \|x^{j+1} - x^j\| \leq (k+1)\beta \|Fx^0\|/N < r, \quad k = 0, 1,\ldots, N-1,$$

which shows that not only the entire curve $x = x(t)$, $t \in [0, 1]$, but also all the iterates $\{x^k\}$ given by (17a) are in S for any N.

Set $\gamma = \max_{x \in S} \|F''(x)\|$, and for arbitrary but fixed N consider the partition $t_i = ih$, $i = 0, 1, \ldots, N$, $h = 1/N$. By **3.2.12** as well as (18), (16), and (15), it follows that

$$\| x(t_{k+1}) - x(t_k) - hx'(t_k) \| \leq \| F'(x(t_k))^{-1} \| \, \| hFx^0 - F'(x(t_k))[x(t_k) - x(t_{k+1})] \|$$

$$\leq \beta \| F(x(t_{k+1})) - F(x(t_k)) - F'(x(t_k))[x(t_{k+1}) - x(t_k)] \|$$

$$\leq \tfrac{1}{2}\beta\gamma \| x(t_{k+1}) - x(t_k) \|^2 \leq \tfrac{1}{2}\beta^3\gamma \| Fx^0 \|^2 h^2,$$

and, hence, that

$$\| x(t_{k+1}) - x^{k+1} \| \leq \| x(t_k) - x^k \| + \| x(t_{k+1}) - x(t_k) - hx'(t_k) \|$$

$$+ h \| F'(x(t_k))^{-1} - F'(x^k)^{-1} \| \, \| Fx^0 \|$$

$$\leq \| x(t_k) - x^k \| + \tfrac{1}{2}\beta^3\gamma \| Fx^0 \|^2 h^2 + h\beta^2 \| Fx^0 \| \gamma \| x(t_k) - x^k \|$$

With $\epsilon_k = \| x(t_k) - x^k \|$, $k = 0, 1, \ldots, N$, we therefore have

$$\epsilon_{k+1} \leq (1 + \eta_0 h)\epsilon_k + \eta_1 h^2, \quad k = 0, 1, \ldots, N-1, \tag{19}$$

where

$$\eta_0 = \beta^2\gamma \| Fx^0 \|, \quad \eta_1 = \tfrac{1}{2}\beta^3\gamma \| Fx^0 \|^2.$$

We show next that

$$\epsilon_k \leq (\eta_1 h/\eta_0)[\exp(\eta_0 t_k) - 1], \quad k = 0, 1, \ldots, N. \tag{20}$$

In fact, (20) clearly holds for $k = 0$ since $\epsilon_0 = 0$, and, if it is valid for some $k \leq N - 1$, then

$$\epsilon_{k+1} \leq (1 + \eta_0 h)\epsilon_k + \eta_1 h^2 \leq (\eta_1 h/\eta_0)[(1 + \eta_0 h)\exp(\eta_0 t_k) - 1]$$

$$\leq (\eta_1 h/\eta_0)[\exp(\eta_0 t_{k+1}) - 1],$$

since

$$\exp(\eta_0 t_{k+1}) = \exp(\eta_0 t_k)\exp(\eta_0 h) \geq (1 + \eta_0 h)\exp(\eta_0 t_k).$$

Therefore,

$$\| x(1) - x^N \| \leq ch, \quad c = (\eta_1/\eta_0)[\exp(\eta_0) - 1],$$

and, if we choose $N \geq c/\delta$, then, clearly, $x^N \in S(x(1), \delta)$, so that the Newton process (17b) converges to $x(1)$. ∎

NOTES AND REMARKS

NR 10.4-1. For a discussion of references relating to continuation methods, see **NR 7.5-1–NR 7.5-5**.

NR 10.4-2. Theorem **10.4.1** is a simplified version of a result of Avila [1970]. It extends to Banach spaces if $\rho(\partial_1 G(x(t), t)) < 1$ is replaced by a corresponding norm condition.

NR 10.4-3. Theorem **10.4.2** is also due to Avila [1970] and extends to Banach spaces provided $\partial_1 H$ is assumed to have a bounded, linear inverse. A related result using the Newton theorem **12.6.2** was given by Sidlovskaya [1958].

NR 10.4-4. Theorem **10.4.4** was proved by Meyer [1968] in a more general form. Related results are given by Kleinmichel [1968] and Bosarge [1968]. Again, with obvious modifications, the result can be extended to Banach spaces.

NR 10.4-5. The statement of **10.4.2** can be modified to allow $m_i = 1$, $i = 1,..., N - 1$. In this case, choose δ' in the proof so that $\alpha \leq 1 - \delta'/\delta$. This illustrates the natural balance between the number of intermediate Newton steps and the mesh size. Similarly, the same procedure may be used in **10.4.3**, and that theorem, in fact, holds as stated if F is only once-continuously differentiable. However, the proof of **10.4.3** as given exhibits an estimate for the mesh size, namely, $N_0 \geq 4\beta^2 \gamma \| Fx^0 \|$, which is not available unless F' is at least Lipschitz continuous.

10.5. APPENDIX. COMPARISON THEOREMS AND OPTIMAL ω FOR SOR METHODS

The results of Section **10.3** did not deal with the important question of the behavior of the rate of convergence of the SOR methods as a function of ω. In this appendix, we shall review, without proof, two well-known comparison theorems for spectral radii and show how these may be applied to nonlinear equations within the setting of **10.3**.

10.5.1. Let $A = B_1 - C_1 = B_2 - C_2$ be two regular splittings of $A \in L(R^n)$ and assume that A is invertible and $A^{-1} \geq 0$. If $C_2 \geq C_1$, then

$$\rho(B_1^{-1} C_1) \leq \rho(B_2^{-1} C_2) < 1. \tag{1}$$

Moreover, if $C_2 \neq C_1$ and $A^{-1} > 0$ (that is, if all elements of A^{-1} are positive), then strict inequality holds in (1).

The proof of **10.5.1** may be found in Varga [1962, p. 90]. We note that the proof given there extends verbatim to weak regular splittings (Definition **2.4.15**).

As a consequence of **10.5.1**, we prove the following comparison theorem (see Varga [1962, p. 92]) for the SOR matrix

$$H = (D - \omega L)^{-1}[(1 - \omega) D + \omega U]. \tag{2}$$

10.5.2. Let A be an M-matrix with diagonal, strictly lower-, and strictly upper-triangular parts D, $-L$, and $-U$, respectively, and let H be given by (2). If $0 < \omega_1 < \omega_2 \leq 1$, then

$$\rho(H_{\omega_1}) \leq \rho(H_{\omega_2}) < 1. \tag{3}$$

Moreover, if A is irreducible, then strict inequality holds in (3).

Proof. As in **10.3.6**,

$$A = (1/\omega)[D - \omega L] - (1/\omega)[(1-\omega)D + \omega U]$$

is a regular splitting for any $\omega \in (0, 1]$. Moreover, whenever

$$0 < \omega_1 \leq \omega_2 \leq 1,$$

then

$$(1/\omega_1)[(1-\omega_1)D + \omega_1 U] \geq (1/\omega_2)[(1-\omega_2)D + \omega_2 U],$$

and equality is not possible since D has positive diagonal entries. It is known (Varga [1962, p. 84]) that an irreducible M-matrix has a positive inverse. Hence, the result follows from **10.5.1**. ∎

In **10.3.6**, we have shown that x^* is a point of attraction of any of the SOR-type methods provided that $F'(x^*)$ is an M-matrix. By means of **10.5.2**, we may now supplement that result with a comparison theorem on the rate of convergence.

10.5.3. Let $F: D \subset R^n \to R^n$ and assume that F is continuously differentiable in a neighborhood of a solution x^* of $Fx = 0$, and that $F'(x^*)$ is an irreducible M-matrix. Let $\mathscr{I}_1$ denote the SOR–Newton iteration (10.3.7) with relaxation parameter $\omega_1 \in (0, 1)$, and $\mathscr{I}_2$ the same iteration with factor $\omega_2 \in (\omega_1, 1]$. Then x^* is a point of attraction of either iteration, and $R_1(\mathscr{I}_2, x^*) < R_1(\mathscr{I}_1, x^*) < 1$.

The proof of **10.5.3** is an immediate consequence of **10.3.3**, **10.3.6**, and **10.5.2**. The important point is simply that **10.5.2** shows that $\rho(H_\omega)$ is a strictly isotone function of ω in the interval $(0, 1]$. In particular, $\omega = 1$ maximizes the rate of convergence over all ω in $(0, 1]$. Theorem **10.5.3** also holds verbatim for the m-step Newton–SOR and the nonlinear SOR iterations, since, as **10.3.2** and **10.3.5** show, their rate of convergence depends on the spectral radius of precisely the same matrix as the SOR–Newton iteration.

Theorem **10.5.3** applies immediately to the problem $Fx \equiv Ax + \phi x$ under the conditions of **10.3.7**. Here, it is also of interest to compare

10.5. APPENDIX. COMPARISON THEOREMS

the rate of convergence of the nonlinear problem with that of the linear problem. This we may do by means of the following result, which is a consequence of Theorems 2.8 and 3.3 and Lemma 2.3 in Varga [1962].

10.5.4. Let $A \in L(R^n)$ be an M-matrix and $\hat{D} \in L(R^n)$ any nonnegative diagonal matrix. Set $\hat{A} = A + \hat{D}$. If H_ω and $\hat{H}_\omega$ are the SOR matrices for A and $\hat{A}$ for some $\omega \in (0, 1]$, then

$$\rho(\hat{H}_\omega) \leqslant \rho(H_\omega) < 1. \tag{4}$$

Moreover, if A is irreducible and $\hat{D} \neq 0$, then strict inequality holds in (4). Similarly, if $B = D^{-1}(L + U)$ and $\hat{B} = (D + \hat{D})^{-1}(L + U)$ are the Jacobi matrices for A and $\hat{A}$, then

$$\rho(\hat{B}) \leqslant \rho(B) < 1, \tag{5}$$

where, again, strict inequality holds if A is irreducible and $\hat{D} \neq 0$.

By means of **10.5.4**, we conclude that, under the conditions of **10.3.7**, the rate of convergence of the SOR–Newton (and also of the Newton–SOR, as well as of the nonlinear SOR) iteration for any $\omega \in (0, 1]$ is at least as fast as that of the SOR iteration, with the same ω, applied to a linear problem with coefficient matrix A. Moreover, if A is irreducible and $\phi'(x^*) \neq 0$, the rate of convergence is faster for the nonlinear problem.

Comparison results between different iterations are also sometimes possible, as the following example shows.

10.5.5. Assume that $F: R^n \to R^n$ satisfies the conditions of **10.5.3**, and denote by $\mathscr{I}_1$ the SOR–Newton iteration with $\omega = 1$, and by $\mathscr{I}_2$ the Jacobi–Newton iteration (7.4.30). Then $R_1(\mathscr{I}_1, x^*) < R_1(\mathscr{I}_2, x^*)$.

The proof of **10.5.5** is an immediate consequence of **10.3.2**, **E 10.3–2**, and the Stein–Rosenberg theorem (see Varga [1962, p. 70]).

So far, we have considered only underrelaxation, that is, $\omega \leqslant 1$, but for many linear problems it is known that overrelaxation improves the the rate of convergence of the SOR iteration considerably. We begin with a brief review of the relevant notation and concepts of the linear theory; for further details and extensions of this theory, see Forsythe and Wasow [1960, Chapter 22] and Varga [1962, Chapter 4].

Again let $A = D - L - U$ be the decomposition of $A \in L(R^n)$ into diagonal, strictly lower-triangular, and strictly upper-triangular parts. Let

$$B = D^{-1}(L + U) \tag{6}$$

be the Jacobi iteration matrix. Then A is 2-*cyclic* (or, has *property A*) if there is a permutation matrix P such that

$$PBP^{\mathrm{T}} = \begin{pmatrix} 0 & B_1 \\ B_2 & 0 \end{pmatrix},$$

where the diagonal blocks are square. Moreover, A is *consistently ordered* if the eigenvalues of

$$\alpha D^{-1}L + (1/\alpha) D^{-1}U$$

are independent of α for $\alpha \neq 0$. For example, it is known (see Varga [1962, Th. 6.4]) that the matrices (1.1.8) and (1.2.7) of the discrete boundary value problems (1.1.7) and (1.2.6) for $u'' = f(u)$ and $\Delta u = f(u)$, respectively, are 2-cyclic and consistently ordered.

We next state a basic result of the linear theory; for its proof, see Varga [1962, p. 111].

10.5.6. Let $A \in L(R^n)$ have nonzero diagonal elements and suppose that A is 2-cyclic and consistently ordered. Let B be the Jacobi matrix (6) and assume that all eigenvalues of B^2 lie in the interval $[0, 1)$. Set

$$\omega_L = 2/[1 + (1 - \mu^2)^{1/2}], \qquad \mu = \rho(B). \tag{7}$$

Then the spectral radius of the SOR matrix H_ω of (2) satisfies

$$\rho(H_\omega) = \tfrac{1}{4}\{\omega\mu + [\omega^2\mu^2 - 4(\omega - 1)]^{1/2}\}^2, \qquad 0 < \omega \leqslant \omega_L, \tag{8}$$

$$\rho(H_\omega) = \omega - 1, \qquad \omega_L \leqslant \omega \leqslant 2, \tag{9}$$

and

$$\rho(H_{\omega_L}) = \min_{0 \leqslant \omega \leqslant 2} \rho(H_\omega).$$

Note that the formulas (8) and (9) show that the graph of $\rho(H_\omega)$ as a function of ω has the form given in Fig. 10.1.

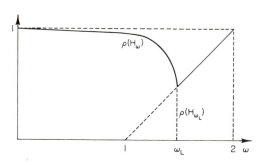

FIGURE 10.1

10.5. APPENDIX. COMPARISON THEOREMS

On the basis of **10.5.6**, we may conclude the existence of an ω which minimizes the R_1-convergence factor of any of the SOR-type methods of **10.3** provided that $F'(x^*)$ satisfies the conditions of **10.5.6**. We state explicitly a result of Ortega and Rockoff [1966] for equations of the form $Ax + \phi x = 0$.

10.5.7. Let $A \in L(R^n)$ be a 2-cyclic, consistently ordered Stieltjes matrix (see **2.4.7**), and let $\phi: D \subset R^n \to R^n$ be continuously differentiable in a neighborhood of a solution x^* of the equation $Ax + \phi x = 0$. Assume that $\phi'(x^*)$ is a nonnegative diagonal matrix. Then x^* is a point of attraction of the SOR–Newton iteration for any $\omega \in (0, 2)$, and there exists an $\omega_N \leqslant \omega_L$ which minimizes the R_1 convergence factor. Here, ω_L is the optimal ω, given by (7), for the linear problem with coefficient matrix A. Moreover, for any $\omega \in (0, 2)$, the R-rate of convergence of the SOR–Newton iteration is at least as fast as that of the SOR iteration applied to the linear problem with coefficient matrix A.

Proof. Set $\hat{D} = \phi'(x^*)$ and $\hat{A} = \hat{D} + A$. Clearly, $\hat{A}$ is also symmetric positive definite, 2-cyclic, and consistently ordered. Let B and $\hat{B}$ be the Jacobi matrices defined in **10.5.4**, and note that

$$B = D^{-1/2}[D^{-1/2}(L + U) D^{-1/2}] D^{1/2},$$

so that B is similar to a symmetric matrix, and, hence, has real eigenvalues. Therefore, by **10.5.4**, the eigenvalues of B^2 lie in $[0, 1)$. Similarly, by (5), the eigenvalues of $\hat{B}^2$ lie in $[0, \mu] \subset [0, 1)$, where $\mu = \rho(B)$. Theorem **10.5.6** now applies to both A and $\hat{A}$. Indeed, if ω_L and ω_N are the optimal ω given by **10.5.6** for A and $\hat{A}$, respectively, then

$$\omega_L = 2/[1 + (1 - \mu^2)^{1/2}] \geqslant 2/[1 + (1 - \hat{\mu}^2)^{1/2}] = \omega_N, \qquad (10)$$

where $\hat{\mu} = \rho(\hat{B})$. The result now follows from **10.3.3**. Finally, to prove the last statement, we simply note that the explicit representation (8)–(9) shows that $\rho(H_\omega)$ is an isotone function of μ for fixed ω. Hence,

$$\rho(H_\omega(x^*)) \leqslant \rho(H_\omega), \qquad \forall \omega \in (0, 2), \qquad (11)$$

where equality necessarily holds in (11) for $\omega \in [\omega_L, 2)$. ∎

Note that if A is also irreducible, then B is irreducible, and **10.5.4** shows that strict inequality holds in (5). It follows that strict inequality holds in (10) and also in (11) for $\omega \in (0, \omega_L]$. Hence, in this case, the graphs of $\rho(H_\omega)$ and $\rho(H_\omega(x^*))$ have the form shown in Fig. **10.2.** In

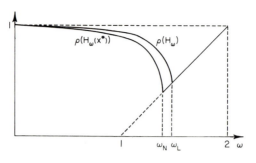

FIGURE 10.2

particular, since $\omega_N \leqslant \omega_L$, the optimal ω for the linear problem is a reasonable candidate for an approximation to ω_N.

Theorem **10.5.7** also applies verbatim, of course, to the nonlinear SOR and the m-step Newton–SOR iterations.

Chapter 11 / **MULTISTEP METHODS AND ADDITIONAL ONE-STEP METHODS**

11.1. INTRODUCTION AND FIRST RESULTS

In the previous chapter, we obtained point-of-attraction and rate-of-convergence results for stationary one-step methods $x^{k+1} = Gx^k$, $k = 0, 1,\dots$. We now consider corresponding results for multistep methods as well as for one-step iterations which are either nonstationary or in which x^{k+1} depends upon x^k in such a complex manner that the techniques of the previous chapter do not apply. The main examples of multistep methods to be analyzed here are the secant-type methods described in Section 7.2. Modified Newton iterations of the form

$$x^{k+1} = x^k - \omega_k[F'(x^k) + \lambda_k I]^{-1}Fx^k, \quad k = 0, 1,\dots, \qquad (1)$$

as described in **7.1**, or variable-step generalized linear iterations

$$x^{k+1} = x^k - [I + H(x^k) + \cdots + H(x^k)^{m_k-1}]B(x^k)^{-1}Fx^k, \quad k = 0, 1,\dots, \qquad (2)$$

discussed in **7.4**, are examples of nonstationary one-step methods, provided the parameters are given *a priori*, as, for example, if $m_k = k + 1$. Alternatively, (1) and (2) may be stationary one-step methods if ω_k, λ_k, and m_k are given functions of x^k. Finally, certain Steffensen iterations of **7.2** are stationary one-step methods for which the complex dependence of x^{k+1} upon x^k precludes the use of the results of Chapter 10.

As discussed in **7.6**, the most general iterative process may be written in the form

$$x^{k+1} = G_k(x^k, x^{k-1},\ldots, x^0,\ldots, x^{-p+1}), \qquad k = 0, 1,\ldots, \tag{3}$$

where $G_k : D_k \subset (R^n)^{k+p} \to R^n$, $k = 0, 1,\ldots,$ are given mappings, and there are p initial points $x^{-p+1},\ldots, x^0$. For these general methods, Definition **10.1.1** of a point of attraction can be extended as follows.

11.1.1. Definition. A point $x^* \in R^n$ is a point of attraction of the iterative process (3) if there exists an open neighborhood S of x^* such that for any p starting points $x^{-p+1},\ldots, x^0 \in S$ the entire sequence $\{x^k\}$ of iterates generated by (3) is well defined and converges to x^*. ∎

In most practical cases, the iteration (3) will have the form

$$\left.\begin{aligned} x^{k+1} &= G(x^k, h^k), \\ h^k &= g_k(x^k,\ldots, x^{-p+1}), \end{aligned}\right\} \quad k = 0, 1,\ldots, \tag{4}$$

where $G: D \times D_h \subset R^n \times R^m \to R^n$ and $g_k: D_k \subset (R^n)^{k+p} \to D_h \subset R^m$, $k = 0, 1,\ldots$. This is no less general than (3), since for $G(x, h) \equiv h$ and $g_k \equiv G_k$, (4) reduces to (3). However, the structure of (4) will allow us to concentrate primarily upon the mapping G and to pay somewhat less attention to the g_k. As we shall see, this dichotomy is convenient, and sufficient to handle a wide class of methods. In fact, since (4) is fairly difficult to manipulate, we shall in most cases disregard for the sake of convenience the dependence of h^k upon $x^k,\ldots, x^{-p+1}$, and phrase the majority of the results in this chapter in terms of sequences defined by

$$x^{k+1} = G(x^k, h^k), \qquad k = 0, 1,\ldots . \tag{5}$$

Here, $\{h^k\}$ is a sequence in R^m which may be treated as an abbreviation for $\{g_k(x^k,\ldots, x^{-p+1})\}$ or, alternatively, as a sequence of parameter vectors. In any case, we shall assume that the h^k are given, and, consequently, we shall not obtain point-of-attraction results for the process (4), but only results for the particular sequence (5). Corresponding point-of-attraction results under particular conditions on the mappings g_k are left as exercises.

We begin with the following simple lemma, which is the analog of **10.1.2**.

11.1.2. Let $G: D \times D_h \subset R^n \times R^m \to R^n$ and suppose that there are

11.1. INTRODUCTION AND FIRST RESULTS

sets $S = S(x^*, \delta) \subset D$ and $D_h' \subset D_h$, as well as a constant $\alpha < 1$, such that

$$\| G(x, h) - x^* \| \leq \alpha \| x - x^* \|, \quad \forall x \in S, \quad \forall h \in D_h'. \tag{6}$$

Then, for any $x^0 \in S$ and any sequence $\{h^k\} \subset D_h'$, the iterates $\{x^k\}$ generated by (5) remain in S and converge to x^*. Moreover,

$$R_1\{x^k\} \leq Q_1\{x^k\} \leq \alpha. \tag{7}$$

Proof. The proof is immediate. A simple induction shows that

$$\| x^{k+1} - x^* \| = \| G(x^k, h^k) - x^* \| \leq \alpha \| x^k - x^* \| \leq \cdots \leq \alpha^{k+1} \| x^0 - x^* \|,$$

and, hence, that all x^k remain in S and converge to x. The first inequality in (7) is given by **9.3.1**, while the second follows from (6) and the definition of Q_1. ∎

The estimate (6) may be achieved in a variety of ways. As a first possibility, we prove an extension of Theorem **10.1.3**.

11.1.3. Definition. A collection of mappings $G_h \colon D \subset R^n \to R^n$, where the parameter vector h varies over some set $D_h \subset R^m$, is *uniformly differentiable* at $x \in \mathrm{int}(D)$ if each G_h, $h \in D_h$, is F-differentiable at x and if for any $\epsilon > 0$ there exists a $\delta = \delta(\epsilon) > 0$, independent of h, such that $S(x, \delta) \subset D$ and

$$\| G_h y - G_h x - G_h'(x)(y - x) \| \leq \epsilon \| y - x \|, \quad \forall y \in S(x, \delta), \quad \forall h \in D_h. \ \blacksquare \tag{8}$$

11.1.4. Generalized Ostrowski Theorem. For $G \colon D \times D_h \subset R^n \times R^m \to R^n$ and $x^* \in \mathrm{int}(D)$ such that $x^* = G(x^*, h)$ for all $h \in D_h$, assume that the collection of mappings

$$G_h \colon D \subset R^n \to R^n, \quad G_h x = G(x, h), \quad x \in D, \quad h \in D_h, \tag{9}$$

is uniformly differentiable at x for all $h \in D_h$, and that

$$G_h'(x^*) = H^{q(h)}, \quad \forall h \in D_h,$$

where $H \in L(R^n)$ satisfies $\rho(H) < 1$ and $q(h)$ is a positive integer. Then there is an open neighborhood S of x^* such that for any $x^0 \in S$ and any sequence $\{h^k\} \subset D_h$ the iterates $\{x^k\}$ given by (5) are well defined and converge to x^*. Moreover,

$$R_1\{x^k\} \leq \rho(H^m), \quad m = \liminf_{k \to \infty} q(h^k).$$

Proof. For given $\epsilon > 0$, **2.2.8** ensures the existence of a norm on R^n for which

$$\|H\| \leqslant \sigma + \epsilon, \qquad \sigma = \rho(H).$$

In this norm, the uniform differentiability of the collection of mappings G_h allows us to choose a $\delta = \delta(\epsilon) > 0$ so that $S = S(x^*, \delta) \subset D$ and

$$\|G(x, h) - x^*\| \leqslant \|G_h x - G_h x^* - G_h'(x^*)(x - x^*)\| + \|G_h'(x^*)(x - x^*)\|$$
$$\leqslant \epsilon \|x - x^*\| + \|H^{q(h)}\| \|x - x^*\|$$
$$\leqslant [\epsilon + (\sigma + \epsilon)^{q(h)}] \|x - x^*\|,$$

whenever $x \in S$ and $h \in D_h$. Since $\sigma < 1$, we may assume that ϵ satisfies $\sigma + 2\epsilon < 1$. Then $q(h) \geqslant 1$ implies that $\epsilon + (\sigma + \epsilon)^{q(h)} \leqslant \sigma + 2\epsilon < 1$, and it follows immediately from **11.1.2** that $\lim_{k \to \infty} x^k = x^*$ as well as

$$R_1\{x^k\} \leqslant \limsup_{k \to \infty}[\epsilon + (\sigma + \epsilon)^{q(h^k)}] \leqslant \epsilon + (\sigma + \epsilon)^m.$$

But $\epsilon > 0$ was arbitrary, and, since $R_1\{x^k\}$ is independent of the norm (see **9.2.2**), we see that $R_1\{x^k\} \leqslant \sigma^m$. ∎

Note that in the special case when D_h consists of precisely one point h and $q(h) = 1$, **11.1.4** reduces to **10.1.3** and that for $D_h = \{1, 2, 3, ...\}$ in R^1 and $h^k = k$, the iteration (5) is simply the one-step nonstationary process

$$x^{k+1} = G_k x^k, \qquad k = 0, 1, ..., \qquad (10)$$

where $G_k x = G(x, k)$.

Theorem **11.1.4** has somewhat limited applicability because of the requirement that the derivatives $G_h'(x^*)$ all be powers of one and the same matrix H. But, before considering other means of satisfying the basic estimate (6), we give one useful application of **11.1.4** to the iteration (2); this extends Theorem **10.3.1** which dealt with the corresponding stationary process.

11.1.5. Let $F: D \subset R^n \to R^n$ be G-differentiable on an open neighborhood $S_0 \subset D$ of a point $x^* \in D$ at which F' is continuous and $Fx^* = 0$. Suppose that $F'(x) = B(x) - C(x)$, where $B, C: S_0 \to L(R^n)$ and B is continuous on S_0, $B(x^*)$ is nonsingular, and $\rho(H(x^*)) < 1$ with $H(x) = B(x)^{-1}C(x)$. Then there exists an open neighborhood S of x^* such that for any $x^0 \in S$ and any sequence of positive integers m_k, $k = 0, 1,...$, the iterates $\{x^k\}$ given by (2) are well-defined and converge to x^*. Moreover,

$$R_1\{x^k\} \leqslant \rho[H(x^*)]^{m'}, \qquad m' = \liminf_{k \to \infty} m_k;$$

11.1. INTRODUCTION AND FIRST RESULTS

in particular, if $\lim_{k \to \infty} m_k = +\infty$, then the rate of convergence is R-superlinear.

Proof. It follows as in the proof of **10.3.1** that there is an $r > 0$ such that $S = S(x^*, r) \subset S_0$ and $B(x)^{-1}$ exists on S and is continuous at x^*. Therefore, the mapping $G \colon S \times N \to R^n$, where N is the set of positive integers and

$$G(x, h) = x - A_h(x)Fx, \qquad A_h(x) = (I + \cdots + H(x)^{h-1})B(x)^{-1},$$

is well-defined for all $x \in S$ and $h \in N$, and, clearly, $x^* = G(x^*, h)$ for all $h \in N$. Moreover, since $\rho(H(x^*)) < 1$, **2.2.8** ensures that there is a norm on R^n such that $\| H(x^*)\| < 1$, and, then, by the continuity of H at x^*, we may choose $0 < r_1 \leqslant r$ and $\lambda < 1$ such that $\|H(x)\| \leqslant \lambda$, for all $x \in S_1 = \bar{S}(x^*, r_1)$. Therefore, $I - H(x)$ and $F'(x)$ are nonsingular for all $x \in S_1$, and, using the matrix identity

$$(I + H + \cdots + H^m)(I - H) = I - H^{m+1},$$

we have

$$A_h(x)F'(x) = (I - H(x)^h)(I - H(x))^{-1} B(x)^{-1} F'(x) = I - H(x)^h, \quad \forall x \in S_1. \quad (11)$$

Consequently, with $G_h x = G(x, h)$,

$$\| G_h x - G_h x^* - H(x^*)^h(x - x^*)\| = \| -A_h(x)Fx + A_h(x^*)F'(x^*)(x - x^*)\|$$
$$\leqslant \| A_h(x)[Fx - Fx^* - F'(x^*)(x - x^*)]\|$$
$$+ \| A_h(x)(F'(x^*) - F'(x))(x - x^*)\|$$
$$+ \|[A_h(x)F'(x) - A_h(x^*)F'(x^*)](x - x^*)\|. \quad (12)$$

Now, let

$$\beta = \sup\{\| B(x)^{-1} \| \mid x \in S_1\};$$

the existence of such a $\beta < \infty$ is a consequence of the construction of S by means of **2.3.2**. Then

$$\| A_h(x)\| \leqslant (1 + \lambda + \cdots + \lambda^{h-1}) \| B(x)^{-1}\| \leqslant \beta/(1 - \lambda), \qquad \forall x \in S_1.$$

Hence, the F-differentiability of F and the continuity of F' at x^* ensure that for given $\epsilon > 0$ we can choose $0 < r_2 \leqslant r_1$ such that each one of the first two terms of (12) is bounded by $[\beta\epsilon/(1 - \lambda)] \| x - x^* \|$ whenever $x \in S_2 = \bar{S}(x^*, r_2)$.

In order to bound the third term of (12), we claim that

$$\| H(x)^h - H(x^*)^h \| \leqslant h\lambda^{h-1} \| H(x) - H(x^*)\|, \qquad \forall x \in S_1. \quad (13)$$

In fact, this is trivially valid for $h = 1$, and, if it holds for some $h = k$, then

$$\| H(x)^{k+1} - H(x^*)^{k+1} \| \leq \| H(x)^k[H(x) - H(x^*)] \| + \| [H(x)^k - H(x^*)^k]H(x^*) \|$$
$$\leq \lambda^k \| H(x) - H(x^*) \| + k\lambda^k \| H(x) - H(x^*) \|$$
$$= (k+1)\lambda^k \| H(x) - H(x^*) \|,$$

so that (13) follows by induction. Since $\lambda < 1$, the set $\{k\lambda^{k-1} \mid k = 1, 2,...\}$ is bounded, and, therefore, by the continuity of H, we may assume that $0 < \delta \leq r_2$ has been chosen sufficiently small that $\| H(x)^h - H(x^*)^h \| \leq \epsilon$ for all $x \in S(x^*, \delta)$ and all $h \in N$. Combining these estimates, we obtain from (12) that

$$\| G_h x - G_h x^* - H(x^*)^h(x - x^*) \| \leq \{[2\beta/(1-\lambda)] + 1\} \epsilon \| x - x^* \|, \qquad \forall x \in S(x^*, \delta).$$

This shows that the collection of mappings $\{G_h\}$ is uniformly differentiable at x^* for all $h \in N$, and, moreover, that $G_h'(x^*) = H(x^*)^h$. The result now follows immediately from **11.1.4** after setting the mapping q in that theorem equal to $q(k) = k$, $k \in N$. ∎

As a special case of **11.1.5**, we may take $m_k = k + 1$, so that the iteration is

$$x^{k+1} = x^k - [I + H(x^k) + \cdots + H(x^k)^k] B(x^k)^{-1} F x^k, \qquad k = 0, 1,...;$$

that is, precisely $k + 1$ steps of the secondary linear iteration are taken to obtain x^{k+1}. In this case, $\lim_{k \to \infty} m_k = \infty$, and **11.1.5** ensures that the rate of convergence is R-superlinear.

Note that **11.1.5** applies, in particular, to the Newton–SOR iteration under the conditions of **10.3.2**. The precise formulation of this result is left to **E 11.1-8**.

NOTES AND REMARKS

NR 11.1-1. The results of this section, including Definitions **11.1.1** and **11.1.3**, appear to be new in this form. However, if each G_k is F-differentiable at a point x^* for which $x^* = G_k x^*$, then the iteration $x^{k+1} = G_k x^k$, $k = 0, 1,...$, may be written as

$$z^{k+1} = A_k z^k + \phi_k(z^k), \qquad k = 0, 1,..., \qquad (14)$$

where $A_k = G_k'(x^*)$, $z^k = x^k - x^*$, and

$$\phi_k(x - x^*) = G_k x - G_k x^* - G_k'(x^*)(x - x^*).$$

11.1. INTRODUCTION AND FIRST RESULTS

The question of when the solution $\{z^k\}$ of the difference equation (14) satisfies $\lim_{k\to\infty} z^k = 0$ has been studied by several authors including Li [1934], Hahn [1958], and Smith [1966], and Theorem **11.1.4** is, in essence, a special case of these results. For further discussion of (14) in connection with iterative processes, see Cavanagh [1970].

NR 11.1-2. Another approach to the analysis of sequential m-step methods of the form

$$\mathscr{I}: \quad x^{k+1} = G(x^k,\dots, x^{k-m+1}), \qquad k = 0, 1,\dots, \tag{15}$$

where $G: (R^n)^m \to R^n$, is to treat them as one-step methods on $(R^n)^m$. That is, define a mapping $\hat{G}: (R^n)^m \to (R^n)^m$ by

$$\hat{G}(y^1,\dots, y^m) = (G(y^1,\dots, y^m), y^1,\dots, y^{m-1}).$$

Then, with $z^k = (x^k,\dots, x^{k-m+1}) \in (R^n)^m$, the iteration (15) is equivalent with $z^{k+1} = \hat{G}z^k$, $k = 0, 1,\dots$. Assuming that G is F-differentiable at $z^* = (x^*,\dots, x^*)$, Theorems **10.1.3–10.1.4** may now be applied to conclude that x^* is a point of attraction of (15) provided that $\rho(H) < 1$, where H is the block matrix

$$H = \begin{pmatrix} H_1 & H_2 & \cdots & H_m \\ I & 0 & \cdots & 0 \\ & \cdot & \cdot & \vdots \\ \mathbf{O} & & I & 0 \end{pmatrix}, \tag{16}$$

with $H_i = \partial_i G(x^*,\dots, x^*)$, $i = 1,\dots, m$; moreover, $R_1(\mathscr{I}, x^*) = \rho(H)$. (See **E 11.1-7** for the corresponding linear result.) It is easy to see (**E 11.1-7**) that a simple sufficient condition for $\rho(H) < 1$ is that $\sum_{i=1}^{m} \alpha_i < 1$, with $\alpha_i = \|H_i\|$, or, more generally, that all roots of the polynomial $\lambda^m = \alpha_1\lambda^{m-1} + \cdots + \alpha_m$ are less than unity in modulus. It is *not*, in general, sufficient that

$$\sum_{i=1}^{m} \rho(H_i) < 1, \tag{17}$$

although, by means of a result of Poljak [1964a], (17) does imply that $\rho(H) < 1$ provided that $H_1,\dots, H_m$ are mutually commutative.

The above results have been given by Voigt [1969], who then applies them to iterations such as the Jacobi–secant or SOR–secant methods discussed in **7.4**. For example, if $F: D \subset R^n \to R^n$ is twice-continuously differentiable in a neighborhood of a solution x^*, and $\rho(H_\omega(x^*)) < 1$, where $H_\omega(x^*)$ is given by (10.3.6), then x^* is a point of attraction of the SOR–secant method (7.4.31). Moreover, the R_1 factor of the iteration is again $\rho(H_\omega(x^*))$, so that the asymptotic rate of convergence is the same as that of the SOR–Newton method (10.3.7).

EXERCISES

E 11.1-1. Assume that G satisfies the conditions of **11.1.2** and that $g_k : D_k \subset (R^n)^{k+p} \to R^n$, $k = 0, 1,...$, is a sequence of mappings such that $S^{k+p} \subset D_k$ and $g_k(S^{k+p}) \subset D_h{'}$, $k = 0, 1,...$, where S and $D_h{'}$ are as in **11.1.2**. Show that x^* is a point of attraction of the iteration

$$\mathscr{I}: \quad x^{k+1} = G(x^k, g_k(x^k,..., x^{-p+1})), \quad k = 0, 1,... .$$

E 11.1-2. Assume that the conditions of **11.1.4** hold for G and that $g_k : D_k \subset (R^n)^{k+p} \to R^n$ satisfies $g_k(D_k) \subset D_h$, $k = 0, 1,...$. Show that x^* is a point of attraction of the iteration $\mathscr{I}$ of **E 11.1-1**.

E 11.1-3. Let $G_k : D \subset R^n \to R^n$, $k = 0, 1,...$, be a sequence of mappings which satisfy $G_k x^* = x^*$, $k = 0, 1,...$, for some $x^* \in \text{int}(D)$ and

$$\beta \| x - x^* \|^p \leqslant \| G_k x - x^* \| \leqslant \alpha \| x - x^* \|^p, \quad k = 0, 1,..., \quad \forall x \in S(x^*, \delta) \subset D,$$

where $\beta > 0$ and $p > 1$. Show that $O_R(\mathscr{I}, x^*) = O_Q(\mathscr{I}, x^*) = p$, where $\mathscr{I}$ is the process $x^{k+1} = G_k x^k$, $k = 0, 1,...$.

E 11.1-4. Let $G_k : D \subset R^n \to R^n$ satisfy $G_k x^* = x^*$, and suppose that there is a $\delta > 0$ such that $S(x^*, \delta) \subset D$, and that each G_k is G-differentiable on $S(x^*, \delta)$ with $\| G_k{'}(x) \| \leqslant \alpha < 1$, $x \in S(x^*, \delta)$, $k = 0, 1,...$. Show that x^* is a point of attraction for the iteration (10).

E 11.1-5. Let $G_k : D \subset R^n \to R^n$ be uniformly differentiable at $x^* \in \text{int}(D)$, where $x^* = G_k x^*$, $k = 0, 1,...$, and suppose that

$$\| G_k{'}(x^*) \| \leqslant \alpha < 1, \quad k = 0, 1,... . \tag{18}$$

Show that x^* is a point of attraction of (10). Show also, by the example $G_k x = Bx$, k even, $G_k x = B^T x$, k odd, where

$$B = \begin{pmatrix} 0 & 2 \\ 0 & 0 \end{pmatrix},$$

that the condition (18) cannot, in general, be replaced by $\rho(G_k{'}(x^*)) \leqslant \alpha < 1$, $k = 0, 1,...$.

E 11.1-6. Formulate and prove a result corresponding to **E 10.1-1** for sequences $\{G_k\}$ and $\{\varphi_k\}$.

E 11.1-7. Let $H_i \in L(R^n)$, $i = 1,..., n$, and assume that $\sum_{i=1}^{m} \| H_i \| < 1$. Show that the equation $x = H_1 x + \cdots + H_m x + b$ has a unique solution x^* and that the sequence

$$\mathscr{I}: \quad x^{k+1} = H_1 x^k + \cdots + H_m x^{k-m+1} + b, \quad k = 0, 1,...,$$

11.2. CONSISTENT APPROXIMATIONS

converges to x^* for any $x^0,\ldots, x^{-m+1} \in R^n$. Conclude also that

$$R_1(\mathscr{I}, x^*) = \rho(H) < 1,$$

where $H \in L(R^{mn})$ is defined by (16).

E 11.1-8. Formulate and prove, using the conditions of **10.3.2**, the special case of **11.1.5** for variable-step Newton–SOR processes.

11.2. CONSISTENT APPROXIMATIONS

Both of the methods (11.1.1) and (11.1.2) are of the general form

$$x^{k+1} = x^k - J(x^k, h^k)^{-1} F x^k, \qquad k = 0, 1,\ldots, \tag{1}$$

with some mapping J from $R^n \times R^m$ into $L(R^n)$. More importantly, (1) encompasses a wide class of other methods such as, for example, the discrete Newton, the two-point secant, and the Steffensen iterations of **7.1** and **7.2**. We shall devote this section to a discussion of the general process (1).

Most of the concrete processes of the form (1) are related to Newton's method by the property that when $\|h\| \to 0$, then $J(x, h)$ tends to $F'(x)$. This observation is the basis for the following definition.

11.2.1. Definition. Let $F: D \subset R^n \to R^n$ be G-differentiable on $D_0 \subset D$ and $J: D_J \times D_h \subset R^n \times R^m \to L(R^n)$. Then J is a *consistent approximation* to F' on $D_0 \subset D_J$ if $0 \in R^m$ is a limit point of D_h and

$$\lim_{h \to 0; h \in D_h} J(x, h) = F'(x), \quad \text{uniformly for} \quad x \in D_0. \tag{2}$$

If there are constants c and $r > 0$ such that

$$\|F'(x) - J(x, h)\| \leqslant c \|h\|, \qquad \forall x \in D_0, \quad h \in D_h \cap S(0, r), \tag{3}$$

then J is a *strongly consistent approximation* to F' on D_0. ∎

The basis of the results in this section is the following estimation lemma.

11.2.2. Assume that $F: D \subset R^n \to R^n$ is G-differentiable in an open neighborhood $S_0 \subset D$ of a point $x^* \in D$ for which $Fx^* = 0$, and that F' is continuous at x^* and $F'(x^*)$ is nonsingular. Let $J: D_J \times D_h \subset R^n \times R^m$

$\to L(R^n)$ be a consistent approximation to F' on S_0. Then there exist $\delta > 0$ and $r > 0$ such that the mapping

$$G(x, h) = x - J(x, h)^{-1} Fx \tag{4}$$

is well defined for all $x \in S = S(x^*, \delta)$, $h \in D_h' = D_h \cap S(0, r)$, and satisfies

$$\| x^* - G(x, h) \| \leqslant \omega(x, h) \| x - x^* \|, \qquad \forall x \in S, \quad h \in D_h', \tag{5}$$

where

$$\omega(x, h) \to 0 \quad \text{as} \quad x \to x^* \quad \text{and} \quad h \to 0, \qquad h \in D_h'. \tag{6}$$

Moreover, if J is a strongly consistent approximation to F' on S_0 and if

$$\| F'(x) - F'(x^*) \| \leqslant \gamma \| x - x^* \|, \qquad \forall x \in S_0, \tag{7}$$

then there are constants α_1, α_2 so that

$$\| x^* - G(x, h) \| \leqslant \alpha_1 \| x - x^* \|^2 + \alpha_2 \| h \| \| x - x^* \|, \qquad \forall x \in S, \quad h \in D_h'. \tag{8}$$

Proof. Set $\beta = \| F'(x^*)^{-1} \|$ and let $\epsilon \in (0, \tfrac{1}{2}\beta^{-1})$. Since J is a consistent approximation on S_0, there is an $r > 0$ such that D_h' is not empty and

$$\| F'(x) - J(x, h) \| \leqslant \tfrac{1}{2}\epsilon, \qquad \forall x \in S_0, \quad h \in D_h'.$$

Moreover, by the continuity of F' at x^*, there is a $\delta > 0$ such that $S = S(x^*, \delta) \subset S_0$ and

$$\| F'(x) - F'(x^*) \| \leqslant \tfrac{1}{2}\epsilon, \qquad \forall x \in S.$$

Hence,

$$\| F'(x^*) - J(x, h) \| \leqslant \epsilon, \qquad \forall x \in S, \quad h \in D_h',$$

and, by the perturbation lemma **2.3.2**, it follows that $J(x, h)^{-1}$ exists and satisfies

$$\| J(x, h)^{-1} \| \leqslant \eta = \beta/(1 - \beta\epsilon), \qquad \forall x \in S, \quad h \in D_h'.$$

Hence, G is well defined on $S \times D_h'$ and

$$\| G(x, h) - x^* \| = \| J(x, h)^{-1} [J(x, h)(x - x^*) - Fx] \|$$
$$\leqslant \eta [\| J(x, h) - F'(x) \| + \| F'(x) - F'(x^*) \|] \| x - x^* \|$$
$$+ \eta \| Fx - Fx^* - F'(x^*)(x - x^*) \|,$$

which shows that (5) holds with

$$\omega(x, h) = \eta[\| J(x, h) - F'(x) \| + \| F'(x) - F'(x^*) \| + q(x)], \tag{9}$$

11.2. CONSISTENT APPROXIMATIONS

where

$$q(x) = \frac{\|Fx - Fx^* - F'(x^*)(x - x^*)\|}{\|x - x^*\|}, \quad \text{for } x \neq x^*, \quad q(x^*) = 0.$$

In order to conclude that (6) holds, we need only note that the continuity of F' at x^* ensures that both $q(x) \to 0$ and $F'(x) - F'(x^*) \to 0$ as $x \to x^*$, while the uniform convergence requirement in Definition **11.2.1** implies that $J(x, h) - F'(x) \to 0$ as $h \to 0$ and $x \to x^*$.

Now, suppose that (7) holds. Then, by **3.2.5**,

$$\|q(x)\| \leqslant \gamma \|x - x^*\|, \quad \forall x \in S, \tag{10}$$

and (8) follows immediately from (5) and (9) with $\alpha_1 = 2\gamma\eta$ and $\alpha_2 = \eta c$, where c is the constant in (3). ∎

As a first application of **11.2.2**, we give the following simple result, which shows that the rate of convergence of the sequence (1) is superlinear when $\lim_{k \to \infty} h^k = 0$.

11.2.3. Assume that $F: D \subset R^n \to R^n$ is G-differentiable in an open neighborhood $S_0 \subset D$ of $x^* \in D$, where $Fx^* = 0$, and that F' is continuous at x^* and $F'(x^*)$ nonsingular. Let $J: D_J \times D_h \subset R^n \times R^m \to L(R^n)$ be a consistent approximation to F' on S_0. Then there exists a ball $S_1 = S(x^*, \delta_1) \subset S_0$ and an $r_1 > 0$ such that for any $x^0 \in S_1$ and any sequence $\{h^k\} \subset D_h \cap S(0, r_1)$ the iterates $\{x^k\}$ given by (1) remain in S_1 and converge to x^*. Moreover, if $\lim_{k \to \infty} h^k = 0$, then $R_1\{x^k\} = Q_1\{x^k\} = 0$.

Proof. Let $\delta > 0$ and $r > 0$ be the constants obtained in **11.2.2**. Then, for given $\alpha \in (0, 1)$, (6) ensures that we can choose $\delta_1 \leqslant \delta$ and $r_1 \leqslant r$ such that

$$\omega(x, h) \leqslant \alpha, \quad \forall x \in S(x^*, \delta_1), \quad h \in D_h \cap S(0, r_1).$$

Hence, the existence and convergence of $\{x^k\}$ follows from **11.1.2**. Moreover, if $\lim_{k \to \infty} h^k = 0$, then **11.1.2** together with (5) and (6) show that

$$R_1\{x^k\} \leqslant Q_1\{x^k\} \leqslant \limsup_{k \to \infty} \omega(x^k, h^k) = 0. \quad \blacksquare$$

In order to apply **11.2.2** or **11.2.3** to concrete iterations, it is, of course, necessary to ensure that the corresponding J is a consistent approximation. One possibility arises from the modified Newton iteration (11.1.1).

11.2.4. Let F and x^* satisfy the conditions of **11.2.3**. Then there exist constants $1 > c_1 > 0$, $c_2 > 0$, and a ball $S_1 = S(x^*, \delta_1) \subset S_0$ such that, for any $x^0 \in S_1$ and any sequences $\{\omega_k\}, \{\lambda_k\}$ satisfying

$$1 - c_1 \leqslant \omega_k \leqslant 1 + c_1, \quad -c_2 \leqslant \lambda_k \leqslant c_2, \quad k = 0, 1,...,$$

the iterates

$$x^{k+1} = x^k - \omega_k[F'(x^k) + \lambda_k I]^{-1} F x^k, \quad k = 0, 1,...,$$

remain in S_1 and converge to x^*. Moreover, if $\lim_{k\to\infty} \omega_k = 1$ and $\lim_{k\to\infty} \lambda_k = 0$, then $R_1\{x^k\} = Q_1\{x^k\} = 0$.

Proof. Define $J: S_0 \times D_h \subset R^n \times R^2 \to L(R^n)$ by

$$J(x, h) = (1 - h_1)^{-1}[F'(x) + h_2 I],$$

where $D_h = \{h \in R^2 \mid h_1 \neq 1\}$. Since F' is continuous at x^*, there is, for given $\eta > 0$, a $\delta > 0$ so that $\|F'(x)\| \leqslant \|F'(x^*)\| + \eta = \eta_1$, for all $x \in S(x^*, \delta) \subset D_0$. Therefore,

$$\| J(x, h) - F'(x) \| = \left\| \frac{h_1}{1 - h_1} F'(x) + \frac{h_2}{1 - h_1} I \right\| \leqslant \frac{|h_1|\eta_1 + |h_2|}{|1 - h_1|},$$

which shows that J is a consistent approximation to F' on $S(x^*, \delta)$. The result now follows from **11.2.3**. ∎

Our interest in the remainder of this section will center on consistent approximations which arise as difference approximations to derivatives. Consider, for example, the matrix $J(x, h) \in L(R^n)$ whose components are defined by

$$[J(x, h)]_{i,j} = \begin{cases} (1/h_{ij})\left[f_i\left(x + \beta \sum_{k=1}^{j-1} h_{ik} e^k + h_{ij} e^j\right) - f_i\left(x + \beta \sum_{k=1}^{j-1} h_{ik} e^k\right)\right], \\ \quad \text{if } h_{ij} \neq 0, \\ \partial_j f_i\left(x + \beta \sum_{k=1}^{j-1} h_{ik} e^k\right), \quad \text{if } h_{ij} = 0, \end{cases} \quad (11)$$

where $\beta \in [0, 1]$ and, as usual, $e^1, ..., e^n$ denote the coordinate vectors. If $\beta = 1$, then (11) corresponds to the approximation (7.1.15) to $\partial_j f_i(x)$, while, for $\beta = 0$, (11) corresponds to (7.1.16). If F is F-differentiable in a neighborhood of a point x, then, clearly, $J(x, h) \to F'(x)$ as $h \to 0$. The next lemma shows that, under nominal conditions, the limit may be achieved uniformly so that J is, in fact, a consistent approximation to F'.

11.2. CONSISTENT APPROXIMATIONS

11.2.5. Assume that $F: D \subset R^n \to R^n$ is continuously differentiable on the open set D. Then, for any compact set $D_0 \subset D$, there exists an $r > 0$ such that, with $D_h = \{h \in R^{n^2} \mid |h_{ij}| \leqslant r, i, j = 1,..., n\}$, the mapping $J: D_0 \times D_h \subset R^n \times R^{n^2} \to L(R^n)$, given by (11), is well-defined for any $\beta \in [0, 1]$, and is a consistent approximation to F' on D_0. Moreover, if

$$\|F'(x) - F'(y)\| \leqslant \gamma \|x - y\|, \qquad \forall x, y \in D, \tag{12}$$

then J is a strongly consistent approximation to F' on D_0.

Proof. We shall prove the result only for the l_1-norm; for an arbitrary norm, it then follows easily from the norm-equivalence theorem **2.2.1**.

Since D_0 is compact and D is open, there exists a $\delta > 0$ such that the compact set $D_1 = \{x \mid \|x - y\|_1 \leqslant \delta, \text{ for some } y \in D_0\}$ is in D. Clearly, F' is uniformly continuous on D_1, and, consequently, for given $\epsilon > 0$, there is a $\delta_1 \in (0, \delta)$ such that

$$|\partial_j f_i(x) - \partial_j f_i(y)| \leqslant \epsilon, \qquad i, j = 1,..., n, \quad \forall x, y \in D_1, \quad \|x - y\|_1 \leqslant \delta_1.$$

Set $r = \delta_1/n$ and $\Delta_{ij}(h) = \beta \sum_{k=1}^{j-1} h_{ik} e^k$. Then, for any $h \in D_h$,

$$\|\Delta_{ij}(h) + h_{ij} e^j\|_1 \leqslant nr \leqslant \delta_1 < \delta, \qquad i, j = 1,..., n,$$

which shows that $x + \Delta_{ij}(h) + h_{ij} e^j \in D_1$ whenever $x \in D_0$. Therefore, it follows from the mean-value theorem **3.2.12** that

$$|(1/h_{ij})[f_i(x + \Delta_{ij}(h) + h_{ij} e^j) - f_i(x + \Delta_{ij}(h))] - \partial_j f_i(x)|$$
$$\leqslant |(1/h_{ij})[f_i(x + \Delta_{ij}(h) + h_{ij} e^j) - f_i(x + \Delta_{ij}(h))] - \partial_j f_i(x + \Delta_{ij}(h))|$$
$$+ |\partial_j f_i(x + \Delta_{ij}(h)) - \partial_j f_i(x)| \leqslant 2\epsilon, \tag{13}$$

and, hence, that

$$\|F'(x) - J(x, h)\|_1 \leqslant 2n\epsilon, \qquad \forall x \in D_0, \quad h \in D_h.$$

Since ϵ was arbitrary, this proves that J is a consistent approximation to F' on D_0. If (12) also holds, then, clearly,

$$|\partial_j f_i(x) - \partial_j f_i(y)| \leqslant \gamma_1 \|x - y\|_1, \qquad \forall x, y \in D,$$

and **3.2.12** shows that the right-hand side of the estimate (13) may be replaced by $\gamma_1[|h_{ij}| + |\Delta_{ij}(h)|] \leqslant \gamma_1 \sum_{k=1}^n |h_{ik}|$. Hence,

$$\|F'(x) - J(x, h)\|_1 \leqslant \gamma_1 \sum_{i,j=1}^h |h_{ij}| = \gamma_1 \|h\|_1, \qquad \forall x \in D_0,$$

which proves that J is a strongly consistent approximation on D_0. ∎

Note that the restriction of D_h in 11.2.5 to points h that are suitably small is necessary to ensure that J is well defined. In case F is defined on all of R^n, we may take $D_h = R^{n^2}$.

As an immediate corollary of 11.2.3 and 11.2.5, we have the following result.

11.2.6. Discrete Newton Theorem. Assume that $F: R^n \to R^n$ is continuously differentiable, and that there is a solution x^* of $Fx = 0$ at which $F'(x^*)$ is nonsingular. Define $J: R^n \times R^{n^2} \to L(R^n)$ by (11). Then there exist $r_1 > 0$ and $\delta_1 > 0$ so that, for any $x^0 \in S(x^*, \delta_1)$ and any sequence $\{h^k\} \subset S(0, r_1) \subset R^{n^2}$, the iterates $\{x^k\}$ given by (1) are well-defined and converge to x^*. Moreover, if $\lim_{k \to \infty} h^k = 0$, then $R_1\{x^k\} = Q_1\{x^k\} = 0$.

So far in this section, we have been concerned only with obtaining superlinear convergence. In order to obtain higher-order convergence, it is generally necessary to introduce three additional conditions: F is sufficiently smooth; J is a strongly consistent approximation; and the rate of decrease of h^k is sufficiently rapid.

If F' satisfies the Lipschitz condition (7) and J is a strongly consistent approximation, then the basic estimate (8) shows that

$$\| x^{k+1} - x^* \| \leqslant \alpha_1 \| x^k - x^* \|^2 + \alpha_2 \| h^k \| \| x^k - x^* \|.$$

If α_2 were zero, this would show at least quadratic convergence for the sequence $\{x^k\}$, but, if $\alpha_2 \neq 0$, then the behavior of the h^k as $k \to \infty$ plays the dominant role in assessing the rate of convergence. The next result indicates two possible requirements on the h^k in order to achieve higher-order convergence.

11.2.7. Assume that $F: D \subset R^n \to R^n$ is G-differentiable in an open neighborhood $S_0 \subset D$ of $x^* \in D$, where $Fx^* = 0$, and that the Lipschitz condition (7) holds and $F'(x^*)$ is nonsingular. Let $J: D_J \times D_h \subset R^n \times R^m \to L(R^n)$ be a strongly consistent approximation to F' on S_0. Suppose that for some $\{h^k\} \subset D_h$ the iterates $\{x^k\}$ given by (1) are well defined and converge to x^*. If, in addition, the condition

$$\| h^k \| \leqslant \beta_1 \| Fx^k \|, \quad \forall k \geqslant k_0, \tag{14}$$

holds, then $O_R\{x^k\} \geqslant O_Q\{x^k\} \geqslant 2$, while, if

$$\| h^k \| \leqslant \beta_2 \| x^k - x^{k-1} \|, \quad \forall k \geqslant k_0, \tag{15}$$

then $O_R\{x^k\} \geqslant \tfrac{1}{2}(1 + \sqrt{5})$.

11.2. CONSISTENT APPROXIMATIONS   361

Proof. By **11.2.2**, there exist $\delta > 0$, $r > 0$ such that, for any $x \in S = S(x^*, \delta) \subset S_0$ and $h \in D_h' = D_h \cap S(0, r)$, the estimate (8) holds. Assume that (14) holds. Then $\lim_{k \to \infty} x^k = x^*$ implies that $\lim_{k \to \infty} Fx^k = 0$, and, hence, $x^k \in S$, $h^k \in D_h'$, for all $k \geq k_1 \geq k_0$. Therefore, by (8),

$$\| x^{k+1} - x^* \| \leq \alpha_1 \| x^k - x^* \|^2 + \alpha_2 \beta_1 \| Fx^k \| \| x^k - x^* \|, \qquad \forall k \geq k_1.$$

But

$$\| Fx^k \| \leq \| Fx^k - Fx^* - F'(x^*)(x^k - x^*) \| + \| F'(x^*)(x^* - x^k) \|$$
$$\leq [\epsilon_k + \| F'(x^*) \|] \| x^k - x^* \|,$$

where $\lim_{k \to \infty} \epsilon_k = 0$, and the result is a direct consequence of **9.3.3**. Similarly, if (15) holds, then

$$\| x^{k+1} - x^* \| \leq \alpha_1 \| x^k - x^* \|^2 + \alpha_2 \beta_2 \| x^k - x^{k-1} \| \| x^k - x^* \|$$
$$\leq (\alpha_1 + \alpha_2 \beta_2) \| x^k - x^* \|^2 + \alpha_2 \beta_2 \| x^{k-1} - x^* \| \| x^k - x^* \|,$$
$$\forall k \geq k_1. \qquad (16)$$

It then follows from **9.2.9** (with $m = 1$) that $O_R\{x^k\} \geq \tau$, where $\tau = \frac{1}{2}(1 + \sqrt{5})$ is the positive root of $t^2 - t - 1 = 0$. ∎

The estimates (14) and (15) can be satisfied in a variety of ways. For example, if the h^k are regarded as parameter vectors, (14) and (15) may be considered as constraints on the choice of the h^k, since the quantities $\| Fx^k \|$ or $\| x^k - x^{k-1} \|$ are easily computed.

All of the results of this section have so far assumed that the sequence $\{h^k\}$ satisfied certain assumptions but was otherwise not precisely specified. As a consequence, we have not been able to consider point-of-attraction theorems. This we do in the next theorem for two simple explicit choices of h^k which arise in the secant and Steffensen methods.

11.2.8. Assume that $F: D \subset R^n \to R^n$ is G-differentiable in an open neighborhood $S_0 \subset D$ of $x^* \in D$, where $Fx^* = 0$, and that F' is continuous at x^* and $F'(x^*)$ is nonsingular. Let $J: D \times D_h \subset R^n \times R^n \to L(R^n)$ be a consistent approximation to F' on S_0, with $0 \in \text{int}(D_h)$. Then x^* is a point of attraction of either one of the iterations

$$\mathscr{I}_1: \quad x^{k+1} = x^k - J(x^k, Fx^k)^{-1} Fx^k, \qquad k = 0, 1, \ldots, \qquad (17)$$

or

$$\mathscr{I}_2: \quad x^{k+1} = x^k - J(x^k, x^{k-1} - x^k)^{-1} Fx^k, \qquad k = 0, 1, \ldots, \qquad (18)$$

and $R_1(\mathscr{I}_i, x^*) = Q_1(\mathscr{I}_i, x^*) = 0$, $i = 1, 2$. Moreover, if F' satisfies the Lipschitz condition (7) and J is a strongly consistent approximation, then $O_R(\mathscr{I}_1, x^*) \geq O_Q(\mathscr{I}_1, x^*) \geq 2$, while $O_R(\mathscr{I}_2, x^*) \geq (1 + \sqrt{5})/2$.

Proof. Consider first the iteration $\mathscr{I}_2$. By **11.2.2** and the argument of **11.2.3**, we may choose $\alpha < 1$, $\delta_1 > 0$, and $r_1 > 0$ such that $S(0, r_1) \subset D_h$ and

$$\| x - J(x, y - x)^{-1} F x - x^* \| \leqslant \alpha \| x - x^* \|, \tag{19}$$

provided $x \in S_1 = S(x^*, \delta_1)$ and $\| x - y \| \leqslant r_1$. Clearly, we may assume that $\delta_1 \leqslant \tfrac{1}{2} r_1$, so that (19) will hold if $x, y \in S_1 = S(x^*, \delta_1)$. It then follows immediately from (19) that whenever $x^0, x^{-1} \in S_1$, the sequence (18) is well-defined, lies in S_1, and converges to x^*. Hence, x^* is a point of attraction. Similarly, the continuity of F at x^* implies that we may assume that δ_1 is so small that $\|Fx\| \leqslant r_1$ whenever $x \in S_1$. It follows that x^* is a point of attraction for $\mathscr{I}_1$. The rate-of-convergence statements are then immediate consequences of **11.2.3** and **11.2.7**. ∎

Particular instances of (17)–(18) are the two-point secant methods and corresponding Steffensen methods described in **7.2**. Consider the special case of the matrix $J(x, h)$ of (11) in which $h_{ij} = h_j$, $i, j = 1,\ldots, n$. That is, the columns J_j of $J(x, h)$ are defined by

$$J_j(x, h) = \begin{cases} (1/h_j)\left[F\left(x + \beta \sum_{k=1}^{j-1} h_k e^k + h_j e^j\right) - F\left(x + \beta \sum_{k=1}^{j-1} h_k e^k\right)\right], & \text{if } h_j \neq 0, \\ F'\left(x + \beta \sum_{k=1}^{j-1} h_k e^k\right) e^j, & \text{if } h_j = 0, \end{cases} \tag{20}$$

for $j = 1,\ldots, n$, where again $\beta \in [0, 1]$. Here, we may consider the parameter vector h to lie in R^n rather than R^{n^2}, so that $J: D_J \times D_h \subset R^n \times R^n \to L(R^n)$. With this J and $\beta = 0$, the iteration (18) becomes the two-point secant method defined by (7.2.20, 19), while, if $\beta = 1$, (18) is the two-point method (7.2.20, 22). (Note that we have now allowed for the possibility that $h_j = 0$.) Similarly, (17) becomes one of the Steffensen methods (7.2.31, 19) or (7.2.31, 22), depending on whether $\beta = 0$ or 1.

As a consequence of **11.2.5** and **11.2.8**, we now have the following concrete result.

11.2.9. Two-Point Secant–Steffensen Theorem. Assume that $F: D \subset R^n \to R^n$ is continuously differentiable in an open neighborhood $S_0 \subset D$ of $x^* \in D$, where $Fx^* = 0$ and $F'(x^*)$ is nonsingular. Then, for any $\beta \in [0, 1]$, there exist $\delta > 0$ and $r > 0$ so that the mapping $J: D_J \times D_h \subset R^n \times R^n \to L(R^n)$ given by (20) is well defined for $D_J = \bar{S}(x^*, \delta) \subset S_0$ and $D_h = \{h \in R^n \mid \|h\| \leqslant r\}$, and x^* is a point of attraction of either one of the iterations (17) or (18). Moreover

11.2. CONSISTENT APPROXIMATIONS

$R_1(\mathscr{I}_i, x^*) = Q_1(\mathscr{I}_i, x^*) = 0$, $i = 1, 2$, and if the Lipschitz condition (12) holds, then

$$O_R(\mathscr{I}_1, x^*) \geq O_Q(\mathscr{I}_1, x^*) \geq 2, \qquad O_R(\mathscr{I}_2, x^*) \geq \tfrac{1}{2}(1 + \sqrt{5}).$$

Proof. From **11.2.5**, it follows that J is well defined on $D_J \times D_h$ and is a consistent approximation to F' on D_J. The result is then a direct consequence of **11.2.8**. ∎

NOTES AND REMARKS

NR 11.2-1. The definition **11.2.1** of consistent approximations and the results **11.2.2**–**11.2.4**, **11.2.7**, and **11.2.8** extend immediately to Banach spaces.

NR 11.2-2. The result of **11.2.9** for the approximation J of (20) with $\beta = 0$ was obtained by Korganoff [1961] for the two-point secant method (18) and by Wegge [1966] for the Steffensen iteration (17). Both of these results were based on a direct analysis using considerable differentiability conditions. A similar result for (17), (20) with $\beta = 0$ was also given by Maergoiz [1967]. Independently, Schmidt [1966a] obtained essentially the result of **11.2.9** for $\beta = 1$ and for both the secant and Steffensen methods as a special case of a more general treatment using divided differences (see **NR 7.2-6** and **NR 11.2-3**) in Banach spaces. (See also **NR 11.2-6** and Schmidt and Schwetlick [1968].) However, in earlier papers, Schmidt [1961, 1963a] had already obtained the $(1 + \sqrt{5})/2$ order of convergence of the secant method (18), (20), with $\beta = 1$, working in the setting of Chapter 12; that is, the existence of a solution was not assumed. Chen [1964] then considered the analogous Steffensen method in the same setting also using Schmidt's divided differences on a Banach space; quadratic convergence was shown but under the restrictive assumption that $\|F'(x^*)\| < 1$. In concurrent work, Ul'm [1964c] did not assume that $\|F'(x^*)\| < 1$, but made the even more restrictive assumption that the divided-difference operator was symmetric (see **NR 7.2-7**). This condition was weakened somewhat in Ul'm [1965a] and dispensed with completely in Johnson and Scholz [1968]. (See also **NR 12.6-4**.)

The superlinear convergence of a particular discrete Newton method for a gradient operator F was obtained by Goldstein and Price [1967].

NR 11.2-3. Consistent approximations were introduced by Ortega [1967] and are related to, but more general than, divided differences. Recall from **NR 7.2-6** that a divided-difference operator $J: R^n \times R^n \to L(R^n)$, as defined by Schmidt [1961], satisfies

$$J(x, h) h = F(x + h) - Fx, \qquad (21)$$

$$\| J(x, h) - J(x + h, k)\| \leq a \|h + k\| + b(\|h\| + \|k\|), \qquad (22)$$

where a and b are certain constants, and the vectors x, h, and k in (22) are suitably restricted to compact sets.

Now, (22) is essentially our condition for a strongly consistent approximation. Indeed, note first that if (21) and (22) hold for all $x \in \bar{S}(z, \delta)$ and $h, k \in \bar{S}(0, r)$, then F has a Lipschitz-continuous F-derivative on $\bar{S}(z, \delta - r)$; this follows from

$$\| F(x+h) - Fx - J(x, 0) h \| = \|[J(x, h) - J(x, 0)] h \| \leqslant (a+b) \| h \|^2,$$

which shows that $F'(x) = J(x, 0)$, and

$$\| F'(x) - F'(y) \| = \| J(x, 0) - J(y, 0) \| = \| J(x, 0) - J(x, y-x) \|$$
$$+ \| J(x, y-x) - J(x+y-x, 0) \|$$
$$\leqslant 2(a+b) \| y - x \|.$$

Moreover,

$$\| F'(x) h - J(x, h) h \| = \| F(x+h) - Fx - F'(x) h \| \leqslant (a+b) \| h \|$$

implies that J is a strongly consistent approximation to F'. Conversely, if J is a strongly consistent approximation to F' and F' is Lipschitz continuous, then

$$\| J(x, h) - J(x+h, k) \| \leqslant \| J(x, h) - F'(x) \| + \| F'(x+h) - F'(x) \|$$
$$+ \| F'(x+h) - J(x+h, k) \|$$
$$\leqslant c \| h \| + \gamma \| h \| + c \| k \|$$
$$\leqslant c \| h \| + \gamma \| h + k \| + \gamma \| k \| + c \| k \|,$$

so that (22) holds with $a = \gamma$ and $b = c + \gamma$. Therefore, all of our results which require that J be a strongly consistent approximation and F' be Lipschitz continuous can be rephrased in terms of the condition (22), provided, of course, that $J: R^n \times R^n \to L(R^n)$. However, we do not need the restrictive condition (21), nor do we require that D_h be a subset of R^n.

NR 11.2-4. Schmidt [1968] extended his earlier point-of-attraction results (Schmidt [1966a], see **NR 11.2-2**) by means of a theorem for iterations of the form

$$x^{k+1} = x^k - A(x^k, ..., x^{k-m})^{-1} F(x^k, ..., x^{k-m}), \quad k = 0, 1, ...,$$

in Banach spaces. We shall state the result only for the iteration

$$x^{k+1} = x^k - A(x^k, x^{k-1})^{-1} Fx^k, \quad k = 0, 1, ..., \quad (23)$$

where $F: R^n \to R^n$ and $A: R^n \times R^n \to L(R^n)$.

Assume that $Fx^* = 0$, $A(x^*, x^*)^{-1}$ exists, and, for some ball $S = S(x^*, r)$, the following estimates hold for all $x, y \in S$:

$$\| A(x^*, x^*) - A(x, y) \| \leqslant \sum_{i=1}^{p} a_i \| x - x^* \|^{\alpha_{i1}} \| y - x^* \|^{\alpha_{i2}}, \quad (24)$$

$$\| Fx - Fx^* - A(x, y)(x - x^*) \| \leqslant \sum_{i=1}^{p} b_i \| x - x^* \|^{\beta_{i1}} \| y - x^* \|^{\beta_{i2}}, \quad (25)$$

11.2. CONSISTENT APPROXIMATIONS

where all constants are nonnegative. Set

$$a = \min_{1 \leq i \leq p} \{\alpha_{i1} + \alpha_{i2} \mid a_i > 0\}, \qquad a = 0 \quad \text{if} \quad a_1 = \cdots = a_p = 0,$$

$$b = \min_{1 \leq i \leq p} \{\beta_{i1} + \beta_{i2} \mid b_i > 0\},$$

and assume that $b > 1$ and $b - a \geq 1$. Then x^* is a point of attraction for (23), and, for some constant c,

$$\| x^{k+1} - x^* \| \leq c \sum_{i=1}^{p} b_i \| x^k - x^* \|^{\beta_{i1}} \| x^{k-1} - x^* \|^{\beta_{i2}}, \qquad k = 1, 2, \ldots.$$

This theorem contains the earlier results of Schmidt [1966a] on the secant and Steffensen methods (see **NR 11.2-2**) as well as a point-of-attraction result for Newton's method using only Hölder-continuity of the derivative (see **NR 10.2-1**). Although Schmidt applied the result to the secant and Steffensen methods only under his divided-difference assumption, it is of interest to note that the more general theorem **11.2.7** can also be recovered in this setting (see **E 11.2-14**).

NR 11.2-5. Šamanskii [1967a] has extended the result of **10.2.4** (and **NR 10.2-2**) by replacing the derivative in Newton's method by the operator $J: R^n \times R^1 \to L(R^n)$ with the components

$$[J(x, h)]_{ij} = \begin{cases} [f_i(x + he^j) - f_i(x)]/h, & \text{if } h \neq 0, \\ \partial_j f_i(x), & \text{if } h = 0, \end{cases} \qquad i, j = 1, \ldots, n. \qquad (26)$$

More specifically, he shows that, under suitable conditions, the iterates

$$x^{k+1} = x^{k,m}, \quad x^{k,i+1} = x^{k,i} - J(x^k, h^k)^{-1} F x^{k,i}, \quad i = 0, 1, \ldots, m-1, \quad x^{k,0} = x^k,$$

converge to a solution x^*, and, if $\mid h_k \mid \leq c \, \| F x^k \|$, then the order of convergence is $m + 1$. The special case $m = 1$ is covered in **E 11.2-6**.

NR 11.2-6. It is important to note that both the discrete Newton theorem **11.2.6** and the secant–Steffensen theorem **11.2.9** require the difference approximation J of (11) to be defined for $h_{ij} = 0$. It is easy to see that **11.2.5** remains valid if J is defined only for $h_{ij} \neq 0$, provided that D_h is restricted in the same way. This allows the corresponding modification of **11.2.6**, but, in order to obtain a convergence result for, say, the two-point secant method, it would be necessary to guarantee that $x_i^{k-1} - x_i^k \neq 0$, $k = 1, 2, \ldots$; $i = 1, \ldots, n$. No meaningful results in this direction are known.

NR 11.2-7. In analogy to the exact rate-of-convergence results of **10.1.7** and **10.2.2**, Voigt [1969] has given the following theorem for the two-point secant method: Assume that the conditions of **11.2.9** hold for F, and that, in addition, F is twice-continuously differentiable on S_0 and satisfies

$$\|F''(x) - F''(x^*)\| \leq c\|x - x^*\|, \qquad \forall x \in S_0.$$

Suppose, further, that

$$F'(x^*)^{-1}\left(\partial_i^2 f_1(x^*),\ldots,\partial_i^2 f_n(x^*)\right)^T > 0, \qquad i = 1,\ldots,n, \tag{27}$$

and

$$\beta F'(x^*)^{-1}\left(\partial_i\partial_j f_1(x^*),\ldots,\partial_i\partial_j f_n(x^*)\right)^T \geq 0, \qquad i,j = 1,\ldots,n, \quad i < j, \tag{28}$$

where the strict inequality in (27) means that all elements of the vector are positive. Then $O_R(\mathscr{I}_2, x^*) = \frac{1}{2}(1 + \sqrt{5})$ for the iteration $\mathscr{I}_2$ of (18), (20).

Note that (28) is automatically satisfied if $\beta = 0$, and that, for $n = 1$, (27) reduces to $F''(x^*) \neq 0$. [The signs in (27) and (28) may be replaced by $<$ and $\leq$, respectively.] Note also that it is easy to see that (27) and (28) are satisfied by (1.2.5) when, for example, $f(s, t, u) = e^u$.

Voigt has also shown that (27) and (28) are necessary to preclude higher-order convergence in the sense that if the quantities in both (27) and (28) are zero, then $O_R(\mathscr{I}_2, x^*) \geq 2$. Note that this requires that $F''(x^*) = 0$ in one dimension but not in higher dimensions.

NR 11.2-8. As mentioned in **NR 7.2-11**, one may consider the *regula falsi* type iteration

$$x^{k+1} = x^k - J(x^k, \bar{x} - x^k)^{-1} F x^k, \qquad k = 0, 1,\ldots, \tag{29}$$

where $\bar{x}$ is held fixed, and J is defined by (11) or, more generally, is a consistent approximation to F'. Voigt [1969] has given a point-of-attraction theorem for this iteration (see **E 11.2-13**) and has also shown that the rate of convergence can, in general, be no faster than R-linear.

NR 11.2-9. The superlinear convergence of the iteration (11.1.2) when $\lim_{k\to\infty} m_k = \infty$ may also be obtained by the methods of this section by showing that the mapping $J: R^n \times [1, \infty) \to L(R^n)$ defined by

$$J(x, h) = B(x)[I + H(x) + \cdots + H(x)^{k-1}]^{-1}$$

is a consistent approximation to F' on some ball about x^*.

EXERCISES

E 11.2-1. Assume that $F: D \subset R^n \to R^n$ satisfies the conditions of **11.2.2**. Let $J: D_J \times D_h \subset R^n \times R^m \to L(R^n)$ and assume that

$$\|F'(x^*) - J(x, h)\| \leq \alpha < \tfrac{1}{2}\|F'(x^*)^{-1}\|, \qquad \forall x \in S(x^*, \delta), \quad h \in D_h.$$

11.2. CONSISTENT APPROXIMATIONS

Show that there is a $\delta_1 \leq \delta$ such that, for any $x^0 \in S(x^*, \delta_1)$ and any sequence $\{h^k\} \subset D_h$, the sequence (1) is well defined and converges to x^*. If, in addition, $\lim_{k \to \infty} J(x^k, h^k) = F'(x^*)$, show that $R_1\{x^k\} = Q_1\{x^k\} = 0$.

E 11.2-2. Assume that F and J satisfy the conditions of **11.2.3** and that $g: D_h \subset (R^n)^p \to R^m$ is continuous at $(x^*, \dots, x^*) \subset D_h$, $g(x^*, x^*, \dots, x^*) = 0$, and D_h contains a neighborhood of 0. Show that x^* is a point of attraction of the iteration

$$\mathscr{I}: \quad x^{k+1} = x^k - J(x^k, g(x^k, \dots, x^{k-p+1}))^{-1} F x^k, \qquad k = 0, 1, \dots,$$

and that $R_1(\mathscr{I}, x^*) = Q_1(\mathscr{I}, x^*) = 0$.

E 11.2-3. Assume that F and J satisfy the conditions of **11.2.7** and that $g: R^n \to R^m$ is such that $\|g(x)\| \leq \eta \|Fx\|$ for all x in some neighborhood of x^*. Show that x^* is a point of attraction of the iteration

$$\mathscr{I}: \quad x^{k+1} = x^k - J(x^k, g(x^k))^{-1} F x^k, \qquad k = 0, 1, \dots,$$

and that $O_R(\mathscr{I}, x^*) \geq O_Q(\mathscr{I}, x^*) \geq 2$.

E 11.2-4. Assume that the conditions of **11.2.7** hold for F and J and that $g: R^n \times R^n \to R^m$ satisfies $\|g(x, y)\| \leq \eta \|x - y\|$ for all x, y in some neighborhood of x^*. Show that x^* is a point of attraction of the iteration

$$\mathscr{I}: \quad x^{k+1} = x^k - J(x^k, g(x^k, x^{k-1}))^{-1} F x^k, \qquad k = 0, 1, \dots,$$

and that $O_R(\mathscr{I}, x^*) \geq (1 + \sqrt{5})/2$.

E 11.2-5. Prove **11.2.5** for an arbitrary norm.

E 11.2-6. Assume that $F: D \subset R^n \to R^n$ satisfies the conditions of **11.2.9** as well as (12), and let $J: R^n \times R^1 \to L(R^n)$ be defined by (26). Show that the sequence (1) converges to x^* provided that $\|x^0 - x^*\|$ and $|h^k|$, $k = 0, 1, \dots$, are suitably small. If, in addition, there is a constant c and a k_0 such that either $|h^k| \leq c \|x^{k+1} - x^k\|$, or $|h^k| \leq c \|Fx^k\|$, for all $k \geq k_0$, then

$$O_R\{x^k\} \geq O_Q\{x^k\} \geq 2.$$

[Note that the first condition may be obtained from the second by

$$|h^k| \leq c \|Fx^k\| = c \|J(x^k, h^k)(x^{k+1} - x^k)\|$$

if one has a uniform bound on $J(x^k, h^k)$.] (Šamanskii[1966a])

E 11.2-7. Assume that F satisfies (12). Show that the mapping J of **11.2.4** is a strongly consistent approximation to F' on some ball $S(x^*, \delta)$, if D_h is restricted to

$$D_h = \{h \in R^2 \mid |1 - h_1| \geq c > 0\}.$$

E 11.2-8. Suppose that $F: D \subset R^n \to R^n$ satisfies the conditions of **11.2.3**, and, in addition, that $\|F'(x) - F'(x^*)\| \leq K\|x - x^*\|^p$, $p \in (0, 1]$. Assume that $\{A_k\} \subset L(R^n)$ and that the sequence $x^{k+1} = x^k - A_k^{-1} F x^k$, $k = 0, 1, \ldots$, satisfies $\lim_{k \to \infty} x^k = x^*$ and $\|A_k - F'(x^*)\| \leq c\|x^k - x^*\|^q$, $q \in (0, 1]$. Show that $O_R\{x^k\} \geq O_Q\{x^k\} \geq 1 + \min(p, q)$.

E 11.2-9. Let F satisfy the conditions of **11.2.7**, and let $A_k : R^n \to L(R^n)$, $k = 0, 1, \ldots$, be a given sequence of operators such that

$$\|A_k(x)^{-1}\| \leq \beta, \quad \|A_k(x) - F'(x)\| \leq \alpha_k(x) \leq (2\beta)^{-1}, \quad \forall x \in S(x^*, \delta), \ k = 0, 1, \ldots,$$

where the mappings $\alpha_k : R^n \to R^1$ satisfy $\lim_{k \to \infty} \alpha_k(x) = 0$ uniformly for all $x \in S(x^*, \delta)$. Show that x^* is a point of attraction of the iteration

$$\mathscr{I}: \quad x^{k+1} = x^k - A_k(x^k)^{-1} F x^k, \quad k = 0, 1, \ldots,$$

and $Q_1(\mathscr{I}, x^*) = 0$.

E 11.2-10. Let $F: R^n \to R^n$ be F-differentiable. Define

$$J: R^n \times \{h \in R^n \mid h_i \neq 0, i = 1, \ldots, n\} \to L(R^n)$$

by

$$J(x, h) = \operatorname{diag}\left(\frac{f_1(x) - f_1(x + h_1 e^1)}{h_1}, \ldots, \frac{f_n(x) - f_n(x + h_n e^n)}{h_n}\right).$$

Show that $\lim_{h \to 0} J(x, h) = F'(x)$ if and only if $F'(x)$ is diagonal. Hence, J is a consistent approximation to F' only on those sets where $F'(x)$ is diagonal.

E 11.2-11. Replace J in **E 11.2-10** by

$$J(x, h) = \operatorname{diag}\left(\frac{f_1(x + h) - f_1(x)}{h_1}, \ldots, \frac{f_n(x + h) - f_n(x)}{h_n}\right)$$

and obtain the same conclusion.

Note. With J defined either as in **E 11.2-10** or **E 11.2-11**, the process (18) may be considered as a naive generalization of the one-dimensional secant method to n dimensions. Both exercises show that these generalizations do not preserve the intrinsic properties of the one-dimensional method. Note, however, that, for the J of **E 11.2-10**, (18) is simply the Jacobi-secant process discussed in **7.4**.

E 11.2-12. Let $F: R^n \to R^n$ be F-differentiable at a solution x^* and assume that $J: R^n \times R^m \to L(R^n)$ is such that for some fixed $h \in R^m$ and $\delta > 0$, $J(x, h)^{-1}$ is defined and bounded for all $x \in S(x^*, \delta)$. Show that the iteration $x^{k+1} = x^k - J(x^k, h)^{-1} F x^k$ is R-superlinearly convergent at x^* if and only if the eigenvalues of $J(x^*, h)^{-1} F'(x^*)$ are all equal to unity.

E 11.2-13. Let F and J satisfy the conditions of **11.2.4**. Show that there is a $\delta > 0$ such that, for any $\bar{x} \in S(x^*, \delta)$, x^* is a point of attraction of the iteration (29). (Voigt [1969])

E 11.2-14. Assume that $F: D \subset R^n \to R^n$ satisfies (12) and that, for $J: D \times D_h \subset R^n \times R^n \to L(R^n)$,

$$\|F'(x) - J(x, h)\| \leq c \|h\|, \quad \forall x \in D, \quad h \in S(0, 2r) \subset D_h.$$

Set $A(x, y) = J(x, y - x)$ and assume that $S(x^*, r) \subset D$. Show that the estimates (24) and (25) hold with $p = 2$, $a_1 = \gamma + c$, $a_2 = c$, $\alpha_{11} = \alpha_{22} = 1$, $\alpha_{12} = \alpha_{21} = 0$, $\beta_{11} = 2$, $\beta_{12} = 0$, and $\beta_{21} = \beta_{22} = 1$.

11.3. THE GENERAL SECANT METHOD

In the previous section, we obtained point-of-attraction results for some particular two-point secant and related Steffensen methods. We now consider the general secant method, which, as described in **7.2**, is defined by

$$x^{k+1} = x^k - J(x^k, H_k)^{-1} F x^k, \quad k = 0, 1, \ldots, \tag{1}$$

where

$$J(x, H) = (F(x + He^1) - Fx, \ldots, F(x + He^n) - Fx) H^{-1}, \tag{2}$$

and, with given auxiliary points $x^{k,i}$, $i = 1, \ldots, n$,

$$H_k = (x^{k,1} - x^k, \ldots, x^{k,n} - x^k). \tag{3}$$

Note that, as with the approximation (11.2.11), J depends on the n^2 parameters h_{ij} which are the elements of H. Here, as in the previous section, we could consider J to be defined on a subset of $R^n \times R^{n^2}$; but it is now more convenient to associate R^{n^2} with $L(R^n)$, that is, to set $J: D_J \times D_h \subset R^n \times L(R^n) \to L(R^n)$.

We will show that J is a consistent approximation to F' provided that H is restricted to the following subsets of $L(R^n)$:

11.3.1. Definition. For given $\sigma > 0$, set

$$K(\sigma) = \Big\{ H = (h^1, \ldots, h^n) \in L(R^n) \mid$$

$$h^j \neq 0, \quad j = 1, \ldots, n; \left| \det \left(\frac{h^1}{\|h^1\|}, \ldots, \frac{h^n}{\|h^n\|} \right) \right| \geq \sigma \Big\}. \tag{4}$$

Then a family of matrices $Q \subset L(R^n)$ is *uniformly nonsingular if* $Q \subset K(\sigma)$ for some $\sigma > 0$. ∎

Note that if $K(\sigma)$ is not empty, then all matrices in $K(\sigma)$ are, of course, nonsingular. Note also that $K(\sigma)$ may be empty if σ is too large but is always nonempty for sufficiently small σ (see **E 11.3.2**).

Clearly, any set of nonsingular diagonal matrices is uniformly nonsingular. More generally, the following lemma gives useful criteria for the uniform nonsingularity of certain subsets of $L(R^n)$.

11.3.2. Let $Q \subset L(R^n)$ be a collection of nonsingular matrices. Then Q is uniformly nonsingular if and only if there exists a constant α such that

$$\|(h^1/\|h^1\|, \ldots, h^n/\|h^n\|)^{-1}\| \leqslant \alpha, \qquad \forall H \in Q. \tag{5}$$

Moreover, if there exists a constant β such that $\|H\|\|H^{-1}\| \leqslant \beta$, for all $H \in Q$, then Q is uniformly nonsingular.

Proof. Assume that $Q \subset K(\sigma)$ for some $\sigma > 0$ and set

$$K_1 = \{A = (a^1, \ldots, a^n) \in K(\sigma) \mid \|a^j\| = 1, \; j = 1, \ldots, n\}.$$

Clearly, K_1 is bounded. Moreover, if $\{A_k\} \subset K_1$ is a convergent sequence with $\lim_{k \to \infty} A_k = A \in L(R^n)$, then $\|a^j\| = 1, j = 1, \ldots, n$, and, since the determinant is a continuous function of its elements, it follows that $|\det A| \geqslant \sigma$. Hence, $A \in K_1$, so that K_1 is closed. The mapping $f: K_1 \to R^1$, $f(A) = \|A^{-1}\|$, is continuous, and, therefore, there exists a constant α so that $\|A^{-1}\| \leqslant \alpha$ for all $A \in K_1$. But, for any $H \in Q$, the matrix

$$\hat{H} = (h^1/\|h^1\|, \ldots, h^n/\|h^n\|)$$

is contained in K_1, and, hence, $\|\hat{H}^{-1}\| \leqslant \alpha$.

Conversely, suppose that (5) holds. The set

$$S = \{A \in L(R^n) \mid \|A\| \leqslant \alpha\}$$

is compact, and, again by the continuity of the determinant, there exists a constant $\tau > 0$ such that $|\det A| \leqslant \tau$ for all $A \in S$. But $\hat{H}^{-1} \in S$ for all $H \in Q$, and thus

$$|\det \hat{H}| = 1/|\det \hat{H}^{-1}| \geqslant 1/\tau, \qquad \forall H \in Q,$$

so that $Q \subset K(1/\tau)$.

For the proof of the last part, observe that $\hat{H} = HD^{-1}$, with $D = \text{diag}(\|h^1\|, \ldots, \|h^n\|)$. Now, in the l_1-norm, $\|D\|_1 = \max_j \|h^j\|_1 = \|H\|_1$, and, therefore, by **2.2.5**, there is a constant c such that $\|D\| \leqslant c\|H\|$. Consequently,

$$\|\hat{H}^{-1}\| \leqslant c\|H\|\|H^{-1}\| \leqslant c\alpha, \qquad \forall H \in Q,$$

which, by the first part, implies that Q is uniformly nonsingular. ∎

11.3. THE GENERAL SECANT METHOD

Note that the converse of the second statement of **11.3.2** does not hold, as **E 11.3-1** shows.

We now prove that the mapping J of (2) is a consistent approximation to F' provided that H is restricted to a uniformly nonsingular set.

11.3.3. Let $F: D \subset R^n \to R^n$ be continuously differentiable on the open set D and let $Q \subset L(R^n)$ be a uniformly nonsingular family such that the null matrix is a limit point of Q. Then, for any compact subset D_0 of D, there is an $r > 0$ such that the mapping J given by (2) is well-defined on $D_0 \times D_h$ with $D_h = \{H \in Q \mid \|H\| \leq r\}$, and is a consistent approximation to F' on D_0. Moreover, if

$$\|F'(x) - F'(y)\| \leq \gamma \|x - y\|, \quad \forall x, y \in D, \tag{6}$$

then J is a strongly consistent approximation.

Proof. As in the proof of **11.2.5**, choose $\delta > 0$ such that

$$D_1 = \{x \mid \|x - y\| \leq \delta, \text{ for some } y \in D_0\} \subset D.$$

Then F' is uniformly continuous on D_1, and, for given $\epsilon > 0$, **3.2.5** ensures that we can choose $r \in (0, \delta)$ such that

$$\Gamma(h) = F(x + h) - Fx - F'(x)h$$

satisfies $\|\Gamma(h)\| \leq \epsilon \|h\|$ whenever $x \in D_0$ and $\|h\| \leq r$.

Now let β be a constant such that

$$\|A\| \leq \beta \max_i \|a^i\|, \quad \forall A = (a^1, ..., a^n) \in L(R^n);$$

β depends only on the norm, and $\beta = 1$ for the l_1-norm (see **E 2.2-2**). Moreover, let α be the constant such that (5) holds for all $H \in Q$. Then, for any $x \in D_0$ and $H \in Q$ such that $\|h^i\| \leq r$, $i = 1, ..., n$, we obtain

$$\|J(x, H) - F'(x)\| = \|(\Gamma(h^1), ..., \Gamma(h^n)) H^{-1}\|$$

$$= \left\|(\Gamma(h^1), ..., \Gamma(h^n))[\text{diag}(\|h^1\|, ..., \|h^n\|)]^{-1} \left(\frac{h^1}{\|h^1\|}, ..., \frac{h^n}{\|h^n\|}\right)^{-1}\right\|$$

$$\leq \alpha \left\|\left(\frac{\Gamma(h^1)}{\|h^1\|}, ..., \frac{\Gamma(h^n)}{\|h^n\|}\right)\right\| \leq \alpha\beta\epsilon. \tag{7}$$

Hence, J is a consistent approximation to F' on D_0. If, in addition, (6) holds, then **3.2.12** shows that $\|\Gamma(h)\| \leq \frac{1}{2}\gamma\|h\|^2$, so that (7) becomes

$$\|J(x, H) - F'(x)\| \leq \frac{1}{2}\alpha\beta\gamma \max_i \|h^i\|. \quad \blacksquare$$

We note that, as in **11.2.5**, if F is defined on all of R^n, then the restriction in **11.3.3** that $\|H\| \leq r$ may be removed and we may take $D_h = Q$.

By means of **11.3.3**, the consistent approximation theorem **11.2.3** now applies to the general secant iteration (1)–(2) provided that the matrices $\{H_k\}$ form a uniformly nonsingular set.

11.3.4. General Secant Theorem. Let $F: D \subset R^n \to R^n$ be continuously differentiable on the open set D, and assume that there exists an $x^* \in D$ for which $Fx^* = 0$ and $F'(x^*)$ is nonsingular. Assume, further, that $\sigma > 0$ is chosen so that $K(\sigma)$ is not empty. Then there is a constant $r_1 > 0$ such that, for any sequence $\{H_k\} \subset K(\sigma)$ with $\|H_k\| \leq r_1$, $k = 0, 1, \ldots$, the sequence (1)–(2) is well defined and $\lim_{k \to \infty} x^k = x^*$, provided that $\|x^0 - x^*\|$ is suitably small. Moreover, if $\lim_{k \to \infty} H_k = 0$, then $R_1\{x^k\} = Q_1\{x^k\} = 0$.

Proof. Note, first, that $K(\sigma)$ is invariant under scaling; that is, if $H \in K(\sigma)$, then $cH \in K(\sigma)$ for any $c \neq 0$. Hence, the null matrix is a limit point of $K(\sigma)$. Now, choose $\delta > 0$ so that $S = \bar{S}(x^*, \delta) \subset D$. Then **11.3.3** shows that there is an $r > 0$ such that J is well defined on $S \times D_h$ with $D_h = \{H \in K(\sigma) \mid \|H\| \leq r\}$, and is a consistent approximation to F' on S. The result now follows directly from **11.2.3**. ∎

Theorem **11.3.4** does not specify the selection mechanism for the auxiliary points $x^{k,1}, \ldots, x^{k,n}$ which, by (3), determine the H_k in the general secant method. We study next some specific examples.

Consider, first, the choice of auxiliary points

$$x^{k,i} = x^k + (x_i^{k-1} - x_i^k) e^i, \quad i = 1, \ldots, n, \quad k = 1, 2, \ldots,$$

discussed in Section **7.2**. In this case,

$$H_k = \mathrm{diag}(x_1^{k-1} - x_1^k, \ldots, x_n^{k-1} - x_n^k), \quad k = 1, 2, \ldots,$$

and we may restrict the set D_h over which J is defined to

$$D_h = \{H \in L(R^n) \mid H \text{ nonsingular and diagonal}\}.$$

It follows from **E 11.3-2** that $D_h \subset K(\sigma)$ for some $\sigma > 0$, and, consequently, **11.3.4** applies provided that the $\|H_k\|$ are sufficiently small, that $x_i^k \neq x_i^{k-1}$, $k = 0, 1, \ldots$, $i = 1, \ldots, n$, and that the mapping F satisfies the given conditions. Note, however, that we have already obtained a result for this choice of auxiliary points in Theorem **11.2.9**

11.3. THE GENERAL SECANT METHOD

(with $\beta = 0$). Our real interest in **11.3.4** is the possibility of applying it to other secant methods.

We consider next the $(n + 1)$-point sequential secant method (7.2.24)

$$x^{k+1} = x^k - J(x^k, H_k)^{-1} F x^k, \quad H_k = (x^{k-1} - x^k, \ldots, x^{k-n} - x^k), \quad k = 0, 1, \ldots, \quad (8)$$

and the related Steffensen methods (7.2.33) and (7.2.34),

$$x^{k+1} = x^k - J(x^k, H_k)^{-1} F x^k, \quad H_k = (F x^k, \ldots, F x^{k-n+1}), \quad k = 0, 1, \ldots, \quad (9)$$

$$x^{k+1} = x^k - J(x^k, H_k)^{-1} F x^k, \quad H_k = (G x^k - x^k, \ldots, G^n x^k - x^k), \quad Gx = x - Fx. \quad (10)$$

Note that the iteration (10) is a stationary one-step method for the fixed-point equation $x = Gx$, while (9) is, of course, a multistep method.

11.3.5. The $(n + 1)$-Point Secant–Steffensen Theorem. Let F satisfy the conditions of **11.3.4**. Then there exist constants $\sigma > 0$ and $r > 0$ so that if the sequence $\{x^k\}$ of (8) is well defined, $\| x^0 - x^* \|$ is sufficiently small, and the corresponding H_k all lie in the set $\{H \in K(\sigma) \mid \| H \| \leqslant r\}$, then $\lim_{k \to \infty} x^k = x^*$ and $R_1\{x^k\} = Q_1\{x^k\} = 0$. Precisely the same statement also applies to each of the iterations (9) and (10). If, in addition, the Lipschitz condition (6) holds, then the sequence (10) satisfies $O_R\{x^k\} \geqslant O_Q\{x^k\} \geqslant 2$, while the sequences of (8) and (9) satisfy, respectively, $O_R\{x^k\} \geqslant \tau_i$, $i = 1, 2$, where τ_1 is the positive root of $t^{n+1} - t^n - 1 = 0$ and τ_2 is the positive root of $t^n - t^{n-1} - 1 = 0$.

Proof. The convergence statements follow immediately from **11.3.4**, as does the superlinear rate of convergence, since $\lim_{k \to \infty} x^k = x^*$ implies that $\lim_{k \to \infty} H_k = 0$ in all three cases. Now, suppose that F' satisfies (6). Then **11.3.3** shows that J is a strongly consistent approximation to F' on some ball $S_1 = S(x^*, \delta_1) \subset S$. Hence, by **11.2.2**, the sequences of (8)–(10) will all satisfy estimates of the form

$$\| x^{k+1} - x^* \| \leqslant \alpha_1 \| x^k - x^* \|^2 + \alpha_2 \| x^k - x^* \| \| H_k \|, \quad \forall k \geqslant k_0. \quad (11)$$

Consider now the H_k of (8). By **2.2.6**, there exists a constant β such that

$$\| H_k \| \leqslant \beta \sum_{j=1}^n \| x^k - x^{k-j} \| \leqslant n\beta \| x^k - x^* \| + \beta \sum_{j=1}^n \| x^{k-j} - x^* \|,$$

so that (11) becomes

$$\| x^{k+1} - x^* \| \leqslant (\alpha_1 + n\beta\alpha_2) \| x^k - x^* \|^2 + \alpha_2 \beta \| x^k - x^* \| \sum_{j=1}^n \| x^{k-j} - x^* \|.$$

Hence, the order of convergence statement is a direct consequence of **9.2.9**. Similarly, for (9), the H_k satisfy

$$\|H_k\| \leqslant \beta \sum_{j=0}^{n-1} \|Fx^{k-j}\|, \qquad k = 0, 1, \ldots . \tag{12}$$

Since F' is continuous on S_1, there is a constant η such that $\|F'(x)\| \leqslant \eta$ for all $x \in S_1$, and, hence, the mean-value theorem **3.2.3** ensures that

$$\|Fx - Fy\| \leqslant \eta \|x - y\|, \qquad \forall x, y \in S_1 . \tag{13}$$

Since $\lim_{k \to \infty} x^k = x^*$, we may assume that the k_0 of (11) has been chosen sufficiently large that $x^k \in S_1$, $k \geqslant k_0 - n$. Hence, by (13),

$$\|Fx^k\| = \|Fx^k - Fx^*\| \leqslant \eta \|x^k - x^*\|, \qquad \forall k \geqslant k_0 - n,$$

and combining this with (11) and (12) yields

$$\|x^{k+1} - x^*\| \leqslant (\alpha_1 + \alpha_2 \beta \eta) \|x^k - x^*\|^2 + \alpha_2 \beta \eta \|x^k - x^*\| \sum_{j=0}^{n-1} \|x^{k-j} - x^*\|,$$

$$\forall k \geqslant k_0 .$$

The result then follows again from **9.2.9**.

Finally, consider the iteration (10), and set $\delta_2 = \delta_1/(1 + \eta)^n$, where η is the constant of (13). We will show that, for any $x \in S_2 = S(x^*, \delta_2)$, $Gx = x - Fx$ satisfies

$$\|G^i x - x\| \leqslant [(1 + \eta)^i - 1] \|x - x^*\|, \tag{14}$$

for $i = 1, \ldots, n$. For $i = 1$, (14) follows immediately from (13); hence, assume that (14) holds for $i = 1, \ldots, p < n$. Then, by choice of δ_2, we have $G^p x \in S_1$, and, therefore, (13) shows that

$$\|G^{p+1} x - x\| = \|G^p x - F(G^p x) - x\| \leqslant (1 + \eta) \|G^p x - x\| + \|Fx - Fx^*\|$$

$$\leqslant \{(1 + \eta)[(1 + \eta)^p - 1] + \eta\} \|x - x^*\|,$$

which implies that (14) holds for $i = p + 1$, and, therefore, for $i = 1, \ldots, n$. Thus, whenever $x^k \in S_2$, the matrix H_k of (10) satisfies

$$\|H_k\| \leqslant \beta \sum_{j=1}^{n} \|G^j x^k - x^k\| \leqslant \beta n[(1 + \eta)^n - 1] \|x^k - x^*\|.$$

The rate-of-convergence statement then follows immediately from (11) and **9.2.9**. ∎

11.3. THE GENERAL SECANT METHOD

Theorem **11.3.5** is a weak result in that it assumes that the whole sequence $\{x^k\}$ is well-defined, that the corresponding sequence $\{H_k\}$ remains in some $K(\sigma)$, and that all $\| H_k \|$ are suitably small. Indeed, the convergence statement of **11.3.5** is of little interest, although the corresponding rate-of-convergence results are.

It is important to note, however, that **11.3.5** will also hold whenever the H_k remain in any set $D_h{}'$ for which J is a strongly consistent approximation to F' on $D_J \times D_h{}'$. While **11.3.3** shows that J is a consistent approximation to F' provided that H is allowed to vary in some set $K(\sigma)$, this is only a sufficient, and by no means a necessary, restriction on H. In fact, consider the set

$$D_h{}' = \{H \in L(R^n) \mid H = DP, \quad D \text{ diagonal and nonsingular}\},$$

where P is the matrix $P = (e^1, e^1 + e^2,..., e^1 + \cdots + e^n)$. Clearly, H is nonsingular whenever $H \in D_h{}'$, but it is easy to see (**E 11.3-3**) that

$$\inf\{|\det(h^1/\| h^1 \|,..., h^n/\| h^n \|)| \mid H \in D_h{}'\} = 0,$$

which shows that $D_h{}'$ is not contained in any $K(\sigma)$. However, **7.2.7** shows that if J is defined by (2) with H restricted to lie in $D_h{}'$, then J may be written as

$$J(x, H) = \left(d_1^{-1}[F(x + d_1 e^1) - Fx],..., d_n^{-1}\left[F\left(x + \sum_{j=1}^{n} d_j e^j\right) - F\left(x + \sum_{j=1}^{n-1} d_j e^j\right)\right]\right)$$

and this is precisely the J defined by (11.2.11) whenever $d_i \neq 0$, $i = 1,..., n$. Then a simple modification of **11.2.5** shows that J is a consistent approximation to F'.

The above observation suggests that we might be able to prove that J is a consistent approximation to F' if D_h were allowed to be all of the set $\{H \in L(R^n) \mid H^{-1} \text{ exists}\}$. The following result shows that this is not possible except in the trivial case in which F is an affine operator.

11.3.6. Let $F: D \subset R^n \to R^n$, $n \geqslant 2$, be twice F-differentiable at a point $x \in D$, and suppose that $F''(x) \neq 0$. Let J be defined by (2). Then, given any constants $\gamma > 0$ and $\eta > 0$, there exists a nonsingular $H \in L(R^n)$ such that $\| H \| \leqslant \gamma$ and $\| F'(x) - J(x, H) \| \geqslant \eta$.

Proof. We prove the result only for the l_1-norm; the proof for a general norm is an easy consequence of the norm-equivalence theorem **2.2.1**. Therefore, throughout the proof, $\| \cdot \|$ shall denote the l_1-norm.

Let

$$Q_k = (\partial_i \partial_j f_k(x)), \quad k = 1,..., n,$$

be the Hessian matrices of $F''(x)$. By assumption, not all Q_i are zero, and we may assume that $Q_1 \neq 0$. Hence, since, by **3.3.4**, Q_1 is symmetric, there exists a $u \in R^n$ such that

$$|u^T Q_1 u| = \lambda > 0 \quad \text{and} \quad \|u\| = 1. \tag{15}$$

Now, let c be the constant in the relation between the l_1- and l_2-matrix norms, that is, $\|A\| \leq c\|A\|_2$, for all $A \in L(R^n)$, and choose

$$0 < \epsilon < \frac{\lambda}{12c^2}.$$

For this ϵ, **3.3.12** shows that there is a $\delta > 0$ such that, for any $H = (h^1, \ldots, h^n) \in L(R^n)$ with $\|H\| \leq \delta$, we have

$$(F(x + h^1) - Fx, \ldots, F(x + h^n) - Fx) = F'(x)H + \tfrac{1}{2}P + R, \tag{16}$$

where

$$P = (F''(x) h^1 h^1, \ldots, F''(x) h^n h^n) = (p^1, \ldots, p^n),$$

and $\|R\| \leq \epsilon \|H\|^2$.

Now, select a particular H such that h^1 is proportional to the u of (15), and, moreover, that $\|h^i\| = \|h^1\| < \min(\delta, \gamma)$, $i = 1, \ldots, n$, and that $h^1, \ldots, h^n$ are mutually orthogonal. Clearly, H is nonsingular, $\|H\| = \|h^1\| < \delta$, and, by **2.2.3**,

$$\|H\| \|H^{-1}\| \leq c^2 \|H\|_2 \|H^{-1}\|_2 = c^2. \tag{17}$$

For this fixed H, define next the matrices

$$H_t = (h^1, \tfrac{1}{2} h^1 + t h^2, h^3, \ldots, h^n) = H + [\tfrac{1}{2} h^1 + (t-1) h^2](e^2)^T,$$

for $t \in (0, \tfrac{1}{2}]$. Then, by the Sherman–Morrison formula (2.3.14), we have

$$H_t^{-1} = H^{-1} - \frac{H^{-1}[\tfrac{1}{2} h^1 + (t-1) h^2](e^2)^T H^{-1}}{1 + (e^2)^T H^{-1}[\tfrac{1}{2} h^1 + (t-1) h^2]}$$

$$= H^{-1} - \frac{1}{t}\left[\tfrac{1}{2} e^1 + (t-1) e^2\right](e^2)^T H^{-1}, \quad \forall t \in \left(0, \tfrac{1}{2}\right],$$

and, hence,

$$\|H_t^{-1}\| \leq \frac{1}{t}\|(\tfrac{1}{2} e^1 - e^2)(e^2)^T H^{-1}\| + \|H^{-1} - e^2(e^2)^T H^{-1}\| \leq \frac{3}{2t}\|H^{-1}\| + b. \tag{18}$$

11.3. THE GENERAL SECANT METHOD

The matrix P of (16) may be represented by means of the Hessian matrices Q_i as $P = ((h^j)^T Q_i h^j)$, and, hence, if $h_t{}^i$ denotes the ith column of H_t, then

$$P_t \equiv ((h_t{}^j)^T Q_i h_t{}^j) = P + [\tfrac{1}{4} p^1 + (t^2 - 1) p^2 + tq](e^2)^T,$$

where

$$q = ((h^1)^T Q_1 h^2, ..., (h^1)^T Q_n h^2)^T.$$

By choice of h^1, we see, from (15) and the general norm relation $\| AB \| \geqslant \| A^{-1} \|^{-1} \| B \|$, that

$$\| p^1(e^2)^T H^{-1} \| \geqslant \frac{\| p^1(e^2)^T \|}{\| H \|} = \frac{\| p^1 \|}{\| H \|} \geqslant \frac{|(h^1)^T Q_1 h^1|}{\| H \|} = \lambda \frac{\| h^1 \|^2}{\| H \|} = \lambda \| H \|;$$

hence,

$$\| P_t H_t^{-1} \| = \| PH^{-1} + [-(4t)^{-1} p^1 + (t-1) p^2 + q](e^2)^T H^{-1} \|$$
$$\geqslant \lambda (4t)^{-1} \| H \| - a, \tag{19}$$

where

$$\| PH^{-1} + [(t-1) p^2 + q](e^2)^T H^{-1} \| \leqslant a, \quad \forall t \in [0, \tfrac{1}{2}].$$

Now, clearly, $\| H_t \| = \| H \| < \delta$ for all $t \in (0, \tfrac{1}{2}]$, and (16) applied to H_t yields, using the estimates (18) and (19),

$$\| J(x, H_t) - F'(x) \| = \| \tfrac{1}{2} P_t H_t^{-1} + R_t H_t^{-1} \| \geqslant \frac{\lambda \| H \| - 4ta}{8t} - \epsilon \| H_t \|^2 \| H_t^{-1} \|$$
$$\geqslant \left[\frac{\lambda \| H \| - 4ta}{12 \| H^{-1} \| + 8tb} - \epsilon \| H \|^2 \right] \left(\frac{3 \| H^{-1} \| + 2tb}{2t} \right). \tag{20}$$

But (17) and the choice of ϵ show that

$$\lambda \| H \|/(12 \| H^{-1} \|) - \epsilon \| H \|^2 > 0,$$

and, hence, the right-hand side of (20) tends to $+\infty$ as $t \to 0$. It follows that there is a $t > 0$ such that $\| J(x, H_t) - F'(x) \| \geqslant \eta$. ∎

Although **11.3.4** shows that one cannot expect the operator J of (2) to be a consistent approximation unless H is restricted, the question still remains as to whether the $(n + 1)$-point secant method (8) does really fail. We end this section with the following example.

Define $F: R^2 \to R^2$ by

$$f_1(x) = x_1, \quad f_2(x) = x_1{}^2 - \tfrac{1}{2} x_2,$$

and let

$$x^{-2} = (0, -\alpha)^T, \quad x^{-1} = (-\alpha, 2\alpha^2)^T, \quad x^0 = (\alpha, 2\alpha^2)^T, \quad 0 < \alpha < 1.$$

Then
$$H_0 = -\begin{pmatrix} 2\alpha & \alpha \\ 0 & \alpha + 2\alpha^2 \end{pmatrix},$$

and the initial points x^0, x^{-1}, and x^{-2} are "well placed" in the sense that $H_0 \in K(\sigma)$, for some σ, as $\alpha \to 0$.

The first iterate x^1 of (1) is then

$$x^1 = \begin{pmatrix} \alpha \\ 2\alpha^2 \end{pmatrix} - \begin{pmatrix} 2\alpha & \alpha \\ 0 & \alpha + 2\alpha^2 \end{pmatrix}\begin{pmatrix} 2\alpha & \alpha \\ 0 & -\alpha/2 \end{pmatrix}^{-1}\begin{pmatrix} \alpha \\ 0 \end{pmatrix} = \begin{pmatrix} 0 \\ 2\alpha^2 \end{pmatrix}.$$

Hence, x^1 is considerably closer to $x^* = 0$ than any of the initial points, but x^{-1}, x^0, and x^1 are now collinear and H_1 is singular.

NOTES AND REMARKS

NR 11.3-1. The treatment in this section is essentially new, but parts of it are closely related to results in the literature. The convergence statement in **11.3.5** for the $(n+1)$-point sequential secant method (8) was first given by Bittner [1959] and the rate-of-convergence statement by Tornheim [1964] (under slightly different conditions). The results of **11.3.5** for the Steffensen methods (9) and (10) are new, but (10) has also been treated by Krawcyzk [1966]. Theorem **11.3.6** was suggested by a result of Tornheim [1964] which states that when F is twice-continuously differentiable in a neighborhood of the solution x^*, then, in any smaller neighborhood of x^*, there exist distinct points $x^0, \ldots, x^n$ such that $Fx^0, \ldots, Fx^n$ are not in general position. This, in turn, is suggested by the geometric observation that secant planes need not tend to the tangent plane.

NR 11.3-2. Note that if we dropped, in the example following **11.3.6**, the points x^{-1} or x^0 instead of x^{-2}, then we could continue the iteration. In other words, although x^{-1}, x^0, and x^1 are collinear, the points in either one of the sets $\{x^{-2}, x^0, x^1\}$ or $\{x^{-2}, x^{-1}, x^1\}$ are in general position. This suggests the following procedure, essentially due to Bittner [1959]. Given the new approximation x^{k+1}, define the matrices

$$H_{k,j} = (x^{k+1} - x^k, \ldots, x^{k+1} - x^{k-j+2}, x^{k+1} - x^{k-j}, \ldots, x^{k+1} - x^{k-n}), \quad j = 1, \ldots, n,$$

and select the index j^0 for which

$$|\det(h^1/\|h^1\|, \ldots, h^n/\|h^n\|)|$$

is maximal. If σ_{k+1} is the maximal value and $\sigma_{k+1} \geq \sigma_k$, then form

$$x^{k+2} = x^{k+1} - J(x^{k+1}, H_{k,j_0})^{-1} Fx^{k+1},$$

11.3. THE GENERAL SECANT METHOD

while if $\sigma_{k+1} < \sigma_k$, then set

$$x^{k+2} = x^{k+1} - J(x^k, H_k)^{-1} F x^{k+1};$$

that is, again use the J evaluated at the previous step. In the limiting case, J is held fixed throughout the whole process and the iteration degenerates into a fixed-chord method. Note that the evaluation of n determinants at each step is computationally unattractive.

NR 11.3-3. We can attempt a comparison of the relative efficiencies of the various secant and Steffensen methods on the basis of the rate-of-convergence results. As discussed in **7.2**, the method (11.2.18, 20) requires $n + 1$ or n evaluations of F with $\beta = 0$ or 1, respectively. Hence, since **11.2.9** shows their orders of convergence to be identical, $\beta = 1$ would be preferable. For the corresponding Steffensen methods (11.2.17), precisely the same number of evaluations are required, and **11.2.9** furnishes no basis for a choice.

The comparison of the two-point secant methods with the $(n + 1)$-point sequential method (8) is more difficult. The two-point method has guaranteed local convergence of order 1.61..., but, with $\beta = 1$, at the expense of n evaluations of F. The estimate of the order of convergence of (8) given in **11.3.5** decreases rapidly with n (for example, if $n = 100$, the largest root of $t^{n+1} - t^n - 1 = 0$ is 1.03...) and there is no guaranteed local convergence. However, the requirement of only one evaluation of F per step is still attractive.

The same difficulty results in comparing the Steffensen methods (11.2.17, 20) and (9). However, the comparison of (11.2.17, 20) with (10) seems clear, since both require $n + 1$ evaluations of F per stage, but (11.2.17, 20) has guaranteed local quadratic convergence while (10) does not. Hence, there would seem to be little reason to prefer (10).

NR 11.3-4. Voigt [1969] has considered the *regula falsi* iteration (1)–(2) in which

$$H_k = (x^{-1} - x^k, \ldots, x^{-n} - x^k), \quad k = 0, 1, \ldots; \tag{21}$$

that is, the n auxiliary points of (3) are now held constant (see **NR 7.2-11**). Voigt shows that if the $n + 1$ points $x^{-1}, \ldots, x^{-n}, x^*$ are in general position and the points $x^{-i}, i = 0, \ldots, n$, are sufficiently close to x^*, then, under the conditions of **11.3.4** for F, x^* is a point of attraction of the iteration (1), (2), (21).

EXERCISES

E 11.3-1. Let $H = (h^1, h^2) \in L(R^2)$ be nonsingular and assume that $\|H\|_1 = \|h^1\|_1$. Define

$$Q = \{H_t \in L(R^n) \mid H_t = (h^1, th^2), \quad \forall t \in (0, 1]\}.$$

Show that Q is uniformly nonsingular but that $\|H_t\| \|H_t^{-1}\| \to \infty$ as $t \to 0$.

E 11.3-2. Let $K(\sigma)$ be defined by (4). Use the Hadamard inequality

$$|\det H| \leq \prod_{i=1}^{n} \|h^i\|_2$$

to show that, for a given norm, there is a σ_0 such that $K(\sigma)$ is empty for all $\sigma > \sigma_0$. Conversely, for a given norm, show that there is always a $\sigma_0 > 0$ such that $K(\sigma)$ is not empty for all $\sigma \leq \sigma_0$ and includes at least all matrices of the form PD, where P is orthogonal and D is diagonal and nonsingular.

E 11.3-3. Show that

$$\inf\left\{\left\|\det\left(\frac{h_1 e^1}{\|h_1 e^1\|},\ldots,\frac{h_1 e^1 + \cdots + h_n e^n}{\|h_1 e^1 + \cdots + h_n e^n\|}\right)\right\| \; h_i \neq 0, \; i = 1,\ldots,n\right\} = 0.$$

E 11.3-4. Assume that F satisfies the conditions of **11.3.4**. Let P_k be a sequence of orthogonal matrices and define $H_k = \|x^k - x^{k-1}\| P_k$, $k = 0, 1, \ldots$. Show that if x^0 and x^{-1} are chosen sufficiently close to x^*, then the sequence (1)–(2) converges to x^* and $Q_1\{x^k\} = 0$. In addition, if (6) holds for all x, y in a neighborhood of x^*, then $O_R\{x^k\} \geq (1 + \sqrt{5})/2$. (Robinson [1966])

E 11.3-5. Prove **11.3.6** for an arbitrary norm.

E 11.3-6. Let $A = (a^1,\ldots,a^n) \in L(R^n)$ be nonsingular, and set

$$\sigma = |\det A| \Big/ \prod_{i=1}^{n} \|a^i\|.$$

Show that, for any nonsingular diagonal matrix $D \in L(R^n)$, $AD \in K(\sigma)$, where $K(\sigma)$ is defined by (4). (Voigt [1969])

Part V / SEMILOCAL AND GLOBAL CONVERGENCE

In the previous part we obtained convergence and rate-of-convergence results for various iterative processes under the assumption that a solution x^* of the system $Fx = 0$ existed and that the initial approximation x^0 is sufficiently close to x^*. Those results provide a means for characterizing and comparing different iterative processes in terms of their asymptotic convergence behavior and are very useful for this theoretical purpose. However, they do not give answers to three very important questions: (1) Can the existence of solutions be ascertained directly from the convergence of the iterative process? (2) For a given initial approximation x^0, can criteria be given which ensure that the iterative process converges starting from this particular x^0? (3) If the iteration is stopped after the kth step, how can an estimate for the error vector $x^k - x^*$ be found? This last question is closely related to the important problem of deciding when to stop the iteration in order to achieve a prescribed accuracy. In fact, if a computable estimate for the error vector is available, then it may—at least in principle—be used as a criterion for terminating the iteration.

In the next chapter we begin an attack on these questions by considering in more detail the contraction-mapping theorem **5.1.3** and many of its variants. These results are then extended to allow a closer study of Newton's method, as well as of related iterative processes. While Chapter 12 is based on the use of some norm on R^n, in Chapter 13 we consider R^n as a partially ordered linear space and obtain results showing monotonic convergence of the iterates. Finally, in Chapter 14 we discuss the convergence of methods for the minimization of functionals on R^n.

Chapter 12 / CONTRACTIONS AND NONLINEAR MAJORANTS

12.1. SOME GENERALIZATIONS OF THE CONTRACTION THEOREM

In Chapter 5 we proved the contraction-mapping theorem **5.1.3** and subsequently used it as a tool for deriving other existence theorems. While these other existence results of Chapter 5 are generally nonconstructive in character, the contraction theorem itself is by nature constructive. Indeed, the proof of **5.1.3** showed that the sequence

$$x^{k+1} = Gx^k, \quad k = 0, 1, \ldots, \tag{1}$$

converges to a fixed point of G. In this chapter we shall examine several variations of the contraction theorem with particular emphasis on the convergence of the iteration (1). We begin with a generalization of **5.1.3** which sets the stage for subsequent developments and also serves as a review of that theorem.

Recall that G^p denotes the pth power of the operator G defined recursively by $G^0 = I$ and $G^p x = G(G^{p-1} x)$, $p \geqslant 1$.

12.1.1. Suppose that $G \colon D \subset R^n \to R^n$ maps a closed set $D_0 \subset D$ into itself and that for some integer $p \geqslant 1$

$$\| G^{kp} x - G^{kp} y \| \leqslant \alpha_k \| x - y \|, \quad \forall x, y \in D_0, \quad k = 1, 2, \ldots, \tag{2}$$

where $\beta = \sum_{k=1}^{\infty} \alpha_k < +\infty$. Then G has a unique fixed point $x^* \in D_0$, and for any $x^0 \in D_0$ the sequence (1) converges to x^*. Moreover, the error estimate

$$\| x^k - x^* \| \leqslant \beta \| x^k - x^{k-p} \|, \qquad k = 1, 2, \ldots, \tag{3}$$

holds.

Proof. Let $z^* \in D_0$ be any fixed point of G^q for some $q \geqslant 1$; then $G^{2q}z^* = G^q(G^q z^*) = G^q z^* = z^*$, and, by induction, $G^{kq}z^* = z^*$ for all $k \geqslant 1$. Thus, if $x^*, y^* \in D_0$ are any two fixed points of G^p, then

$$\| x^* - y^* \| = \| G^{kp} x^* - G^{kp} y^* \| \leqslant \alpha_k \| x^* - y^* \|, \tag{4a}$$
$$\| x^* - Gx^* \| = \| G^{kp} x^* - G^{kp}(Gx^*) \| \leqslant \alpha_k \| x^* - Gx^* \|, \tag{4b}$$
$$k = 1, 2, \ldots.$$

Since $\lim_{k \to \infty} \alpha_k = 0$, it follows from (4a) that $x^* = y^*$ and from (4b) that $x^* = Gx^*$. Hence, both G^p and G have exactly the same fixed points in D_0, and there exists at most one such point.

Now, observe that, because $GD_0 \subset D_0$, the sequence $\{x^k\}$ of (1) is well-defined and contained in D_0, provided only that $x^0 \in D_0$. For fixed $0 \leqslant i \leqslant p - 1$, we consider the subsequence $y^k \equiv y^{i,k} = x^{i+kp}$, $k = 0, 1, \ldots$. Then $y^{k+1} = G^p y^k$, $k \geqslant 0$, and, hence,

$$\| y^{k+j} - y^{k+j-1} \| = \| G^{jp} y^k - G^{jp} y^{k-1} \| \leqslant \alpha_j \| y^k - y^{k-1} \|, \qquad \forall j \geqslant 1$$

which implies that

$$\| y^{k+m} - y^k \| \leqslant \sum_{j=1}^{m} \| y^{k+j} - y^{k+j-1} \| \leqslant \left(\sum_{j=1}^{m} \alpha_j \right) \| y^k - y^{k-1} \|$$
$$\leqslant \beta \| y^k - y^{k-1} \| \leqslant \beta \alpha_{k-1} \| y^1 - y^0 \|, \qquad \forall k, m \geqslant 1. \tag{5}$$

Consequently, since $\lim_{k \to \infty} \alpha_k = 0$, $\{y^{i,k}\}$ is a Cauchy sequence, and, since D_0 is closed, it converges to a limit $\bar{y}^i \in D_0$. From

$$\| \bar{y}^i - G^p \bar{y}^i \| \leqslant \| \bar{y}^i - y^{i,k+1} \| + \| G^p y^{i,k} - G^p \bar{y}^i \|$$
$$\leqslant \| \bar{y}^i - y^{i,k+1} \| + \alpha_1 \| y^{i,k} - \bar{y}^i \|, \qquad k \geqslant 0,$$

it now follows that $\bar{y}^i = G^p \bar{y}^i$, $i = 0, 1, \ldots, p - 1$, and, hence, by the first part of the proof, $\bar{y}^0 = \cdots = \bar{y}^{p-1} = x^*$ and $x^* = Gx^*$. Clearly, $\lim_{k \to \infty} y^{i,k} = x^*$, $i = 0, \ldots, p - 1$, is equivalent to $\lim_{k \to \infty} x^k = x^*$.

Finally, the error estimate (3) follows from (5) for $m \to \infty$. ∎

If, for some $p \geqslant 1$, G^p satisfies

$$\| G^p x - G^p y \| \leqslant \alpha \| x - y \|, \qquad x, y \in D_0, \quad \alpha < 1, \tag{6}$$

12.1. SOME GENERALIZATIONS OF THE CONTRACTION THEOREM

then, evidently, (2) holds with $\alpha_k = \alpha^k$ and $\beta = \alpha/(1 - \alpha)$. Note that when, in addition to (6), G is also Lipschitz-continuous,

$$\| Gx - Gy \| \leq \gamma \| x - y \|, \quad x, y \in D_0,$$

then the conditions of **12.1.1** are satisfied for $p = 1$ (see **E 12.1-1**). However, it is possible that (2)—or even (6)—holds even though G is not even continuous on D_0 (see **E 12.1-2**).

The special case $p = 1$ of (6) again gives the contraction condition **5.1.2**, and thus we obtain as an immediate corollary of **12.1.1** the contraction-mapping theorem **5.1.3**—but this time with an error estimate.

12.1.2. Contraction-Mapping Theorem. Suppose that $G: D \subset R^n \to R^n$ maps a closed set $D_0 \subset D$ into itself and that

$$\| Gx - Gy \| \leq \alpha \| x - y \|, \quad \forall x, y \in D_0, \tag{7}$$

for some $\alpha < 1$. Then, for any $x^0 \in D_0$, the sequence (1) converges to the unique fixed point x^* of G in D_0 and

$$\| x^k - x^* \| \leq [\alpha/(1 - \alpha)] \| x^k - x^{k-1} \|, \quad k = 1, 2, \ldots . \tag{8}$$

Note that (8) provides a computable error estimate; that is, if the contraction constant α is known, then the actual error $x^k - x^*$ after the kth step of the iteration can be bounded in terms of the last step $x^k - x^{k-1}$. This is a possibility not available in the attraction theorems of Part IV. Note also that **12.1.2** requires not only that G is contractive on D_0 but that it maps D_0 into itself. [It may be mentioned that one trivial modification of **12.1.2** permits weakening this requirement by assuming that for some x^0 the entire sequence (1) remains in D_0.] A criterion for determining a suitable D_0 in terms of the initial point x^0 is given in **12.2.3**. (See also **12.4.4**.) However, in the important special case that $D_0 = D = R^n$, so that G is a contraction on all of R^n, **12.1.2** provides a *global convergence theorem*, that is, for any $x^0 \in R^n$, the sequence (1) converges to the unique fixed point of G in R^n. As an example of how this condition might be satisfied, we give the following simple result.

12.1.3. Let $A \in L(R^n)$ be symmetric, positive definite, and assume that $\phi: R^n \to R^n$ is continuously differentiable, that $\phi'(x)$ is symmetric, positive semidefinite for all $x \in R^n$, and that there is a constant $\beta < +\infty$ for which

$$\| \phi'(x) \|_2 \leq \beta, \quad \forall x \in R^n. \tag{9}$$

Then the equation $Ax + \phi x = 0$ has a unique solution x^*, and for any $x^0 \in R^n$ the sequence $\{x^k\}$ defined by the Picard iteration

$$(A + \gamma I) x^{k+1} = \gamma x^k - \phi x^k, \quad k = 0, 1, \ldots, \quad \gamma = \beta/2,$$

converges to x^*.

Proof. Define $G: R^n \to R^n$ by

$$Gx = (A + \gamma I)^{-1} [\gamma x - \phi x], \quad \forall x \in R^n.$$

Clearly, G is well defined and has a fixed point x^* if and only if $Ax^* + \phi x^* = 0$; moreover, G is continuously differentiable on R^n and, since $\gamma = \beta/2$,

$$\| G'(x) \|_2 = \|(A + \gamma I)^{-1} [\gamma I - \phi'(x)]\|_2 \leqslant \gamma/(\gamma + \lambda) < 1, \quad \forall x \in R^n,$$

where $\lambda > 0$ is the minimal eigenvalue of A. Therefore, the mean-value theorem **3.2.5** ensures that G is a contraction on R^n, and the result follows from **12.1.2**. ∎

Note that the existence and uniqueness statement in **12.1.3** is of little interest to us since this has already been proved in **4.4.1** without the additional condition (9). Note also that **12.1.3** has immediate application to the discrete analogs (1.1.7) and (1.2.5) of the differential equations $u'' = f(u)$ and $\Delta u = f(u)$ provided that $0 \leqslant f'(u) \leqslant \beta$, for all $u \in R^1$ (see **E 12.1-5**).

Theorem **12.1.3** illustrates the general observation that it is usually difficult to obtain global contractions unless rather stringent conditions, such as (9), are imposed. Many of the remaining results in this chapter are attempts to circumvent the contraction condition, but in such a way that convergence of the iterates is maintained; indeed, **12.1.1** is already an example of this type. We next consider two additional possibilities of weakening the contraction hypothesis which, although almost self-evident, are sometimes useful.

12.1.4. Suppose that $G: R^n \to R^n$ has the property that for any compact set $C \in R^n$ there is a constant $\alpha_c < 1$ so that

$$\| Gx - Gy \| \leqslant \alpha_c \| x - y \|, \quad \forall x, y \in C. \tag{10}$$

If G has a (necessarily unique) fixed point x^*, then for any $x^0 \in R^n$ the iterates (1) converge to x^*.

12.1. SOME GENERALIZATIONS OF THE CONTRACTION THEOREM

Proof. For given x^0, set $C = \bar{S}(x^*, \|x^* - x^0\|)$. Then (10) shows that G maps C into itself, and, hence, **12.1.2** applies. ∎

We note that the condition of **12.1.4** is not sufficient to ensure the existence of a fixed point (see **E 12.1-7**).

The next observation concerns a mapping G which, in analogy to matrix theory, is "similar" to a contraction.

12.1.5. Let $G: R^n \to R^n$, and suppose there is a homeomorphism $T: R^n \to R^n$ such that $T^{-1}GT$ is a contraction on all of R^n. Then G has a unique fixed point x^*, and for any $x^0 \in R^n$, the iterates (1) converge to x^*.

The proof of **12.1.5** is immediate from **12.1.2** and the following general principle.

12.1.6. Given $G: D \subset R^n \to R^n$ and a homeomorphism $T: D \to D$ from D onto itself, then G has precisely the same number of fixed points in D as $T^{-1}GT$, and for any $x^0 \in D$ the iterates (1) remain in D and converge if and only if the sequence $y^{k+1} = T^{-1}GTy^k$, $k = 0, 1, \ldots$, $y^0 = T^{-1}x^0$, remains in D and converges.

Proof. Clearly, if $x^* \in D$ is a fixed point of G, then $y^* = T^{-1}x^*$ is a fixed point of $T^{-1}GT$, and conversely. Moreover, since the sequences $\{y^k\}$ and $\{x^k\}$ are connected by $x^k = Ty^k$, $k = 0, 1, \ldots$, they must have the same convergence behavior. ∎

As an example of the use of these results, we consider the nonlinear Peaceman–Rachford method (7.4.35):

$$\mu x^{k+\frac{1}{2}} + Hx^{k+\frac{1}{2}} = \mu x^k - Vx^k,$$
$$\mu x^{k+1} + Vx^{k+1} = \mu x^{k+\frac{1}{2}} - Hx^{k+\frac{1}{2}}, \qquad k = 0, 1, \ldots . \qquad (11)$$

Recall from Definition **5.4.2** that a mapping $F: R^n \to R^n$ is uniformly monotone on R^n if there is a $\gamma > 0$ so that

$$(x - y)^T(Fx - Fy) \geq \gamma \|x - y\|^2, \qquad \forall x, y \in R^n, \qquad (12)$$

and monotone if (12) holds with $\gamma = 0$.

12.1.7. Global Peaceman–Rachford Theorem. Assume that the mappings $H, V: R^n \to R^n$ are monotone and that at least one of them is uniformly monotone. Assume further that on each compact set of R^n both H and V are Lipschitz continuous. Then the equation

$Hx + Vx = 0$ has a unique solution x^*, and for any $x^0 \in R^n$ and any $\mu > 0$, the sequence $\{x^k\}$ of (11) is well-defined and converges to x^*.

Proof. The mappings $H + V$, $H + \mu I$, and $V + \mu I$ are all continuous and uniformly monotone on R^n, and, therefore, by **6.4.4**, they are all homeomorphisms of R^n onto itself. In particular, $Hx + Vx = 0$ has a unique solution x^*, and the iterates (11) are well-defined and satisfy (1) with

$$Gx \equiv (\mu I + V)^{-1}(\mu I - H)(\mu I + H)^{-1}(\mu I - V)x, \quad x \in R^n. \qquad (13)$$

From $Hx^* + Vx^* = 0$, it follows that $(\mu I + H)^{-1}(\mu I - V)x^* = x^*$, as well as $(\mu I + V)^{-1}(\mu I - H)x^* = x^*$; hence, x^* is a fixed point of G.

Now, suppose that H is uniformly monotone on R^n with constant $\gamma > 0$ and let $C \subset R^n$ be any compact set. Then $\hat{C} = (H + \mu I)^{-1}(C)$ is also compact, and we denote the Lipschitz constant of H on $\hat{C}$ by η. For arbitrary $x, y \in C$, $x \neq y$, set $u = (\mu I + H)^{-1}x$, $v = (\mu I + H)^{-1}y$; then, in the l_2-norm, the mapping $G_1 = (\mu I - H)(\mu I + H)^{-1}$ satisfies

$$\left[\frac{\|G_1 x - G_1 y\|}{\|x - y\|}\right]^2 = \frac{\mu^2 \|u - v\|^2 - 2\mu(u - v)^T(Hu - Hv) + \|Hu - Hv\|^2}{\mu^2 \|u - v\|^2 + 2\mu(u - v)^T(Hu - Hv) + \|Hu - Hv\|^2}$$

$$\leqslant \frac{(\mu^2 - 2\mu\gamma)\|u - v\|^2 + \|Hu - Hv\|^2}{(\mu^2 + 2\mu\gamma)\|u - v\|^2 + \|Hu - Hv\|^2}$$

$$\leqslant \frac{\mu^2 - 2\mu\gamma + \eta^2}{\mu^2 + 2\mu\gamma + \eta^2} < 1,$$

where the next to last inequality follows from the isotonicity of functions of the form $(a + t)/(b + t)$ when $b - a \geqslant 0$. This shows that G_1 is a contraction on C, and if V is only monotone, the same proof shows that the mapping $G_2 = (\mu I - V)(\mu I + V)^{-1}$ satisfies $\|G_2 x - G_2 y\| \leqslant \|x - y\|$ for all $x, y \in R^n$. Hence, the composite mapping $G_1 G_2$ is contractive on any compact set, and the same is true provided that H is monotone and V is uniformly monotone. Since $G = (\mu I + V)^{-1} G_1 G_2 (\mu I + V)$, and G has the fixed point x^*, it follows from **12.1.6** that $G_1 G_2$ has the fixed point $y^* = (\mu I + V)x^*$, and, thus, **12.1.4** implies that $y^{k+1} = G_1 G_2 y^k$, $y^0 = (\mu I + V)x^0$, $k = 0, 1, \ldots$, converges to y^*. But then, by **12.1.6**, $\{x^k\}$ converges to x^*. ∎

We note that Theorem **12.1.7** has immediate application to equations of the form $Ax + \phi x = 0$; see **E 12.1-8**.

So far, we have only considered the one-step iteration (1). We conclude this section with a contraction-type theorem for two-step processes of the form

$$x^{k+1} = G(x^k, x^{k-1}), \quad k = 1, 2, \ldots. \qquad (14)$$

12.1. SOME GENERALIZATIONS OF THE CONTRACTION THEOREM

12.1.8. Suppose that $G: D \times D \subset R^n \times R^n \to R^n$ satisfies on some closed set $D_0 \subset D$

$$\| G(x, y) - G(y, z) \| \leq \alpha \| x - y \| + \beta \| y - z \|, \qquad \forall x, y, z \in D_0, \quad (15)$$

where $\alpha + \beta < 1$, and there exist $x^0, x^1 \in D_0$ such that the iterates $\{x^k\}$ defined by (14) remain in D_0. Then $\lim_{k \to \infty} x^k = x^*$, where x^* is the unique fixed point of the operator $\hat{G}: D \subset R^n \to R^n$, $\hat{G}x = G(x, x)$. Moreover, there exist constants $\bar{\gamma}_1, \bar{\gamma}_2 > 0$ for which

$$\| x^k - x^* \| \leq \bar{\gamma}_1 t_1^k + \bar{\gamma}_2 t_2^k, \qquad k = 0, 1, \dots, \quad (16)$$

where t_1 and t_2 are the two roots of $t^2 - \alpha t - \beta = 0$.

Proof. By assumption, the sequence $s_k = \| x^{k+1} - x^k \|$, $k = 0, 1, \dots$, is well-defined, and from (14) and (15) it follows that

$$s_{k+1} \leq \alpha s_k + \beta s_{k-1}, \qquad k = 1, 2, \dots .$$

The linear difference equation $\sigma_{k+1} = \alpha \sigma_k + \beta \sigma_{k-1}$ has the general solution $\sigma_k = \gamma_1 t_1^k + \gamma_2 t_2^k$, as can be verified directly, and it is a simple calculation to show that

$$-1 < t_2 < 0 < t_1 < 1. \quad (17)$$

We determine the constants γ_1, γ_2 such that

$$\sigma_0 = \gamma_1 + \gamma_2 = s_0, \qquad \sigma_1 = \gamma_1 t_1 + \gamma_2 t_2 = s_1.$$

Since the determinant of this linear system is $t_2 - t_1 \neq 0$, the constants γ_1, γ_2 are uniquely determined. With these constants, we now obtain by induction that $s_k \leq \sigma_k$, $k = 0, 1, \dots$. In fact, if this is true up to some $k \geq 1$, then

$$s_{k+1} \leq \alpha s_k + \beta s_{k-1} \leq \alpha \sigma_k + \beta \sigma_{k-1} = \sigma_{k+1}.$$

From (17) and

$$\| x^{k+p} - x^k \| \leq \sum_{j=k}^{k+p-1} \| x^{j+1} - x^j \| \leq \gamma_1 \sum_{j=k}^{k+p-1} t_1^j + \gamma_2 \sum_{j=k}^{k+p-1} t_2^j$$

$$\leq \frac{\gamma_1}{1 - t_1} t_1^k + \frac{\gamma_2}{1 - t_2} t_2^k, \qquad k \geq 0, \quad p > 0, \quad (18)$$

it follows that $\{x^k\}$ is a Cauchy sequence and, hence, that $\lim_{k \to \infty} x^k = x^* \in D_0$ exists. Now,

$$\| \hat{G}x^* - x^{k+1} \| \leq \| G(x^*, x^*) - G(x^*, x^k) \| + \| G(x^*, x^k) - G(x^k, x^{k-1}) \|$$

$$\leq \beta \| x^* - x^k \| + \alpha \| x^* - x^k \| + \beta \| x^k - x^{k-1} \|,$$

and the convergence of the right-hand side to zero shows that $\hat{G}x^* = x^*$. The uniqueness of the fixed point is a consequence of

$$\|x^* - y^*\| \leqslant \|G(x^*, x^*) - G(x^*, y^*)\| + \|G(x^*, y^*) - G(y^*, y^*)\|$$
$$\leqslant \beta \|x^* - y^*\| + \alpha \|x^* - y^*\| < \|x^* - y^*\|, \qquad (19)$$

and, finally, the error estimate (16) follows from (18) when $p \to \infty$. ∎

Note that an estimate analogous to (19) shows that the operator $\hat{G}$ is a contraction on D_0. Hence, the uniqueness of x^* could have been concluded from that fact.

NOTES AND REMARKS

NR 12.1-1. All theorems in this section except for **12.1.3** and **12.1.7** hold verbatim for operators which map a complete metric space into itself, and the proofs carry over word for word.

NR 12.1-2. The use of a contraction condition for proving the convergence of successive approximations goes back well into the last century. Based probably on ideas of Cauchy and Liouville, Picard used it in his classical and well-known proof of the existence and uniqueness of solutions of initial value problems for ordinary differential equations. For a discussion and a historical summary of the contraction principle in connection with this problem, and with integral equations, see Wouk [1964]. As mentioned in **NR 5.1-1**, the first abstract formulation of the contraction-mapping theorem appears to be due to Banach [1922]. Since then, there have been numerous generalizations of the basic theorem. The special case $p = 1$ of **12.1.1** was given by Weissinger [1952]. (We note that this result was attributed by Wouk [1964] to Cacciopoli [1931], but this paper of Cacciopoli does not seem to contain a theorem of this type.) Kolmogoroff and Fomin [1954] gave the somewhat different extension in which, for some $p > 1$, G^p satisfies (6); in suitable Banach spaces, this result is particularly useful for Volterra integral equations. Diaz [1964] then gave a thorough analysis of the Kolmogoroff–Fomin theorem and pointed out various possible modifications. The combined version **12.1.1** of the Weissinger and Kolmogoroff–Fomin theorems appears to be new.

For various other generalizations, also see the other sections of Chapter 12, the beginning part of Chapter 13, and the Notes and Remarks of these sections.

NR 12.1-3. Applications of the contraction theorem are so numerous in the literature that it would be impossible to give a meaningful survey here. For a number of examples, see, for instance, Berezin and Zhidkov [1959], Collatz [1964], Kantorovich and Akilov [1959], and Keller [1968].

12.1. SOME GENERALIZATIONS OF THE CONTRACTION THEOREM

NR 12.1-4. The idea of using the iteration of Theorem **12.1.3** with a suitable γ appears in various forms in the literature; it was, for example, used by Douglas [1961] in connection with a discussion of ADI methods.

NR 12.1-5. Theorems **12.1.5** and **12.1.6** are due, in essence, to Chu and Diaz [1964, 1965], who used changes of norms to study Volterra-type integral equations and certain functional equations.

NR 12.1-6. Theorem **12.1.7** is due to Kellogg [1969], and Theorem **12.1.4** is contained in Kellogg's proof. Contraction-type convergence theorems for various other ADI methods of the form discussed in **7.4** have been given by Caspar [1969].

NR 12.1-7. Theorem **12.1.8** is due to Weinitschke [1964], who also considers analogous results for m-step methods. Weinitschke also points out that in many cases the combined iteration

$$x^{k+1} = \mu[\lambda \hat{G} x^k + (1 - \lambda) G(x^k, x^{k-1})] + (1 - \mu) x^k,$$
$$k = 1, 2, \ldots, \quad 0 < \lambda, \mu \leq 1,$$

shows a considerable improvement in the convergence over the one-step process $x^{k+1} = \hat{G} x^k$ and the two-step process $x^{k+1} = G(x^k, x^{k-1})$. It is readily seen that when $G(x, y)$ satisfies the condition (15), then also

$$H(x, y) = \mu[\lambda \hat{G} x + (1 - \lambda) G(x, y)] + (1 - \mu) x$$

obeys the inequality

$$\| H(x, y) - H(y, z) \| \leq \hat{\alpha} \| x - y \| + \hat{\beta} \| y - z \|$$

where now

$$\hat{\alpha} = \mu \alpha + \mu \lambda \beta + (1 - \mu), \quad \hat{\beta} = \mu (1 - \lambda) \beta.$$

Clearly, μ and λ can be chosen such that, again, $\hat{\alpha} + \hat{\beta} < 1$, and we obtain the best convergence if the corresponding roots t_1 and t_2 are minimal.

EXERCISES

E 12.1-1. Suppose that $G: D \subset R^n \to R^n$ maps a closed set $D_0 \subset D$ into itself, is Lipschitz-continuous on D_0, and satisfies for some fixed $q \geq 1$ and $\alpha < 1$

$$\| G^q x - G^q y \| \leq \alpha \| x - y \|, \quad \forall x, y \in D_0.$$

Show that the conditions of **12.1.1** are satisfied with $p = 1$.

E 12.1-2. Show that the discontinuous function $g: [0, 2] \subset R^1 \to [0, 2]$ defined by $g(x) = 0$ for $x \in [0, 1]$ and $g(x) = 1$ for $x \in (1, 2]$ satisfies the conditions of **12.1.1**.

E 12.1-3. Let $G: R^n \to R^n$ be the affine operator $Gx = Hx + b$ for some $H \in L(R^n)$ and $b \in R^n$. Show that the following statements are equivalent:
  (a) G is a contraction in some norm.
  (b) For some integer $m > 1$, G^m is a contraction in some norm.
  (c) $\rho(H) < 1$.

E 12.1-4. Let $K: [0, 1] \times [0, 1] \times R^1 \to R^1$ satisfy

$$|K(s, t, u) - K(s, t, v)| \leq \eta |u - v|, \quad \forall s, t \in [0, 1], \quad u, v \in R^1,$$

where $\eta < 1$, and let $\gamma_1, \ldots, \gamma_n$ be positive constants such that $\sum_{j=1}^{n} \gamma_j = 1$. Show that the discrete integral equation

$$x_i = \psi(t_i) + \sum_{j=1}^{n} \gamma_j K(t_i, t_j, x_j), \quad i = 1, \ldots, n,$$

where $t_1, \ldots, t_n \in [0, 1]$ and $\psi: [0, 1] \to R^1$ is a given function, has a unique solution x^*, and that the iterates

$$x_i^{k+1} = \psi(t_i) + \sum_{j=1}^{n} \gamma_j K(t_i, t_j, x_j^k), \quad i = 1, \ldots, n, \quad k = 0, 1, \ldots,$$

converge to x^* for any x^0. Apply this result to the discrete Hammerstein equation (1.3.10–11), which arises from the two-point boundary value problem (1.3.9), under the assumptions that

$$|f(t, u) - f(t, v)| \leq \eta_0 |u - v|, \quad \forall u, v \in R^1, \quad t \in [0, 1],$$

where $\eta_0 < 8$, $t_j = jh$, $\gamma_j = h$, $j = 1, \ldots, n$, and $h = (n + 1)^{-1}$.

Hint. By **E 1.3-2**, $B = -hA^{-1}$, where A and B are the matrices of (1.1.8) and (1.3.10), respectively. Hence, the vector $z \in R^n$ with components $z_i = \sum_{j=1}^{n} b_{ij}$ can be obtained as the solution of $Az = -he$, $e = (1, \ldots, 1)^T$.

E 12.1-5. Consider the two-point boundary value problem (1.1.2) and the corresponding discrete problem $Ax + \phi x = 0$, where $A \in L(R^n)$ and $\phi: R^n \to R^n$ are defined by (1.1.8) and (1.1.9). Suppose that $\partial_2 f(s, u)$ exists, is continuous in u for each fixed $s \in [0, 1]$, and satisfies

$$0 \leq \partial_2 f(s, u) \leq \beta, \quad \forall s \in [0, 1], \quad u \in R^1.$$

12.2. APPROXIMATE CONTRACTIONS AND SEQUENCES

Show that the iterates

$$(A + \gamma I) x^{k+1} = \gamma x^k - \phi x^k, \quad k = 0, 1, \ldots, \quad \gamma = h^2 \beta/2,$$

converge to the unique solution of $Ax + \phi x = 0$ for any x^0. State and prove the corresponding result for the discrete analog (1.2.5) of the boundary value problem (1.2.1).

E 12.1-6. Consider the two-point boundary value problem

$$y'' = f(t, y, y'), \quad 0 < t < 1,$$
$$\alpha_0 y(0) + \alpha_1 y'(0) = \alpha, \quad \beta_0 y(1) + \beta_1 y'(1) = \beta,$$

where f is continuous on $S = \{(t, u, v) \in R^3 \mid t \in [0, 1], u, v \in R^1\}$, and $|f(t, u, v) - f(t, w, z)| \leq \lambda_1 |u - w| + \lambda_2 |v - z|$ on S for all $t \in [0, 1]$. For the discretized problem

$$x_{i+1} - 2x_i + x_{i-1} = h^2 f\bigl(ih, x_i, (2h)^{-1}(x_{i+1} - x_{i-1})\bigr), \quad i = 1, \ldots, n,$$
$$\alpha_0 x_0 + \alpha_1 (1/h)(x_1 - x_0) = \alpha, \quad \beta_0 x_{n+1} + \beta_1 (1/h)(x_{n+1} - x_n) = \beta,$$

with $h = 1/(n + 1)$, give conditions which ensure that the contraction mapping theorem applies (Berezin and Zhidkov [1959]).

E 12.1-7. Show that the function $G: R^1 \to R^1$ defined by $Gx = \ln(1 + e^x)$ is contractive on compact sets but has no fixed point.

E 12.1-8. Let $\phi: R^n \to R^n$ be continuously differentiable on R^n and assume that $\phi'(x)$ is positive semidefinite for all $x \in R^n$. Let $A_1, A_2 \in L(R^n)$ be positive definite, set $A = A_1 + A_2$, and for some $\beta \in [0, 1]$ define $H, V: R^n \to R^n$ by

$$Hx = A_1 x + \beta \phi x, \quad Vx = A_2 x + (1 - \beta) \phi x, \quad x \in R^n.$$

Show that for any $\mu > 0$ and any $x^0 \in R^n$ the iterates of (11) converge to the unique solution of $Ax + \phi x = 0$.

12.2. APPROXIMATE CONTRACTIONS AND SEQUENCES OF CONTRACTIONS

In the previous section we considered the iterative process

$$x^{k+1} = Gx^k, \quad k = 0, 1, \ldots, \qquad (1)$$

under various assumptions on G. Because of rounding or discretization error in the evaluation of G, an approximate sequence $\{y^k\}$ is, in general, produced in place of the exact sequence $\{x^k\}$. The question is then, what can be said about the behavior of $\{y^k\}$ if we know that the exact sequence $\{x^k\}$ converges? Closely related to this problem is the study of non-stationary processes of the form

$$y^{k+1} = G_k y^k, \quad k = 0, 1, ..., \tag{2}$$

where the G_k are all contractions.

The following theorem contains the basic result about the relation of the approximate sequence $\{y^k\}$ to the exact sequence (1).

12.2.1. Let $G: D \subset R^n \to R^n$ be a contraction on $D_1 \subset D$ (with constant α) and $D_0 \subset D_1$ a closed set such that $GD_0 \subset D_0$. (Then, by **12.1.2**, the sequence (1) starting from any $x^0 \in D_0$ converges to the unique fixed point x^* of G in D_0.) Let $\{y^k\} \subset D_1$ be any sequence and set $\epsilon_k = \| Gy^k - y^{k+1} \|$, $k = 0, 1, \ldots$. Then

$$\| y^{k+1} - x^* \| \leq [1/(1-\alpha)][\alpha \| y^{k+1} - y^k \| + \epsilon_k], \quad k = 0, 1, ..., \tag{3}$$

$$\| y^{k+1} - x^* \| \leq \| x^{k+1} - x^* \| + \sum_{j=0}^{k} \alpha^{k-j} \epsilon_j + \alpha^{k+1} \| x^0 - y^0 \|, \quad k = 0, 1, ..., \tag{4}$$

and

$$\lim_{k \to \infty} y^k = x^* \quad \text{if and only if} \quad \lim_{k \to \infty} \epsilon_k = 0 \tag{5}$$

Proof. The estimate (3) follows from

$$\| y^{k+1} - x^* \| \leq \| y^{k+1} - Gy^k \| + \| Gy^k - Gy^{k+1} \| + \| Gy^{k+1} - Gx^* \|$$
$$\leq \epsilon_k + \alpha \| y^k - y^{k+1} \| + \alpha \| y^{k+1} - x^* \|,$$

while (4) is obtained from

$$\| x^{k+1} - y^{k+1} \| \leq \| Gx^k - Gy^k \| + \| Gy^k - y^{k+1} \|$$
$$\leq \alpha \| x^k - y^k \| + \epsilon_k \leq \cdots \leq \sum_{j=0}^{k} \alpha^{k-j} \epsilon_j + \alpha^{k+1} \| x^0 - y^0 \|,$$

together with

$$\| y^{k+1} - x^* \| \leq \| y^{k+1} - x^{k+1} \| + \| x^{k+1} - x^* \|.$$

Suppose, now, that $\lim_{k \to \infty} \epsilon_k = 0$, and, more specifically, that for given $\epsilon > 0$, $\epsilon_k \leq \epsilon$ for $k \geq k_0$. Then with $\gamma_k = \sum_{j=0}^{k} \alpha^{k-j} \epsilon_j$ we have

$$\gamma_k \leq \alpha^{k-k_0} \gamma_{k_0} + \sum_{j=k_0+1}^{k} \alpha^{k-j} \epsilon_j \leq \alpha^{k-k_0} \gamma_{k_0} + \epsilon[\alpha^{k_0+1}/(1-\alpha)]$$

12.2. APPROXIMATE CONTRACTIONS AND SEQUENCES

which shows that $\lim_{k \to \infty} \gamma_k = 0$; thus, by (4), $\lim_{k \to \infty} y^k = x^*$. Conversely, if $\lim_{k \to \infty} y^k = x^*$, then

$$0 \leq \epsilon_k = \| Gy^k - y^{k+1} \| \leq \| Gy^k - Gx^* \| + \| x^* - y^{k+1} \|$$
$$\leq \alpha \| y^k - x^* \| + \| x^* - y^{k+1} \|,$$

which implies that $\lim_{k \to \infty} \epsilon_k = 0$. ∎

Note that no assumptions were made about $\{y^k\}$ except that the sequence is contained in the domain $D_1 \subset D$ in which G is a contraction. In particular, the y^k need not lie in D_0, nor need the ϵ_k be small. For the estimates to be useful, we shall, of course, interpret the sequence $\{y^k\}$ as an approximation to the exact sequence (1). Note that in the special case $x^k = y^k$, that is, $\epsilon_k = 0$, the estimate (3) reduces to the error estimate (12.1.8) of the contraction-mapping theorem **12.1.2**. Finally, note that (3) and (4) play different roles: (4) relates the exact sequence (1) to the sequence $\{y^k\}$ and is useful for such theoretical purposes as proving (5). On the other hand, (3) is a computable estimate which may be used, for example, to terminate a computation.

Theorem **12.2.1** can be applied to nonstationary processes of the form (2), as the next result shows.

12.2.2. Let $G: D \subset R^n \to R^n$ be a contraction on a closed set $D_0 \subset D$ with $GD_0 \subset D_0$. Suppose that $G_k : D_0 \subset R^n \to R^n$, $k = 0, 1,...$, are any mappings such that $G_k D_0 \subset D_0$ for all $k \geq 0$ and

$$\lim_{k \to \infty} \| G_k x - Gx \| = 0, \quad \text{uniformly for} \quad x \in D_0. \tag{6}$$

Then the sequence $\{y^k\}$ defined by (2) converges to the only fixed point of G in D_0.

The proof is immediate; in fact, (6) implies that the sequence $\{\epsilon_k\} = \{\| G_k y^k - Gy^k \|\}$ converges to zero, and, hence, the result is a direct consequence of the last part of **12.2.1**.

Note that (6) cannot, in general, be weakened to pointwise convergence as the following simple example on R^1 shows:

$$Gx = 0, \quad G_k x = e^x/(k+1), \quad k = 0, 1,... .$$

With more stringent conditions on the G_k a stronger result may be obtained. We first prove the following lemma.

12.2.3. Suppose that $G: D \subset R^n \to R^n$ satisfies

$$\|Gx - Gy\| \leq \alpha \|x - y\| + \gamma, \qquad \forall x, y \in D,$$

where $\alpha < 1$ and $\gamma \geq 0$. If there are two points $y^0, y^1 \in D$ such that $D_0 = \bar{S}(y^1, \delta) \subset D$ where

$$\delta = [1/(1 - \alpha)](\alpha \|y^1 - y^0\| + \|Gy^0 - y^1\| + \gamma),$$

then $GD_0 \subset D_0$.

Proof. If $x \in D_0$, then

$$\begin{aligned}\|Gx - y^1\| &\leq \|Gx - Gy^0\| + \|Gy^0 - y^1\| \\ &\leq \alpha \|x - y^0\| + \gamma + \|Gy^0 - y^1\| \\ &\leq \alpha \|x - y^1\| + [\alpha \|y^1 - y^0\| + \|Gy^0 - y^1\| + \gamma] \\ &\leq \alpha \delta + (1 - \alpha) \delta = \delta. \quad \blacksquare\end{aligned}$$

Note that if $x^0 = y^0 = y^1$ and $\gamma = 0$, then $GD_0 \subset D_0$ provided that

$$\bar{S}(x^0, [1 - \alpha]^{-1} \|Gx^0 - x^0\|) \subset D.$$

This is a condition which is frequently invoked in conjunction with the contraction-mapping theorem **12.1.2**, and, on the basis of **12.2.3**, the following three results will be phrased in terms of the initial iterates satisfying similar conditions.

12.2.4. Let $G: D \subset R^n \to R^n$ be a contraction on D with constant α, and suppose that for some $\gamma \geq 0$ the operators $G_k: D \subset R^n \to R^n$, $k = 0, 1, \ldots$, satisfy

$$\|G_k x - G_k y\| \leq \alpha \|x - y\| + \gamma, \qquad \forall x, y \in D. \tag{7}$$

Assume, further, that there exists a $y^0 \in D$ such that $S = \bar{S}(G_0 y^0, r) \subset D$, where

$$r = [1/(1 - \alpha)][\alpha \|G_0 y^0 - y^0\| + \delta + \gamma],$$

and

$$\|G_k y^0 - G_0 y^0\| \leq \delta, \qquad \forall k \geq 0, \qquad \|Gy^0 - G_0 y^0\| \leq \gamma + \delta. \tag{8}$$

Then the sequences $\{x^k\}$ and $\{y^k\}$ defined by (1) and (2) with $x^0 = y^0$ remain in S, and $\lim_{k \to \infty} x^k = x^*$ where x^* is the unique fixed point of G in S. In addition, the error estimates (3) and (4) hold with $\epsilon_k = \|G_k y^k - G y^k\|$.

12.2. APPROXIMATE CONTRACTIONS AND SEQUENCES

Moreover, if $\gamma = 0$, each G_k has a unique fixed point z^k in S, and the following four statements are equivalent: (a) $\lim_{k \to \infty} y^k = x^*$, (b) $\lim_{k \to \infty} z^k = x^*$, (c) $\lim_{k \to \infty} \epsilon_k = 0$, (d) $\lim_{k \to \infty} \| G_k x^* - G x^* \| = 0$.

Proof. By (8) and **12.2.3**, it follows that $GS \subset S$ and $G_k S \subset S$, $k \geqslant 0$. Thus, $\{x^k\}$, $\{y^k\} \subset S$ and **12.1.2** ensures the existence of x^* and the convergence of $\{x^k\}$ to x^*. The error estimates (3) and (4) now apply.

If $\gamma = 0$, then each G_k is a contraction, and since $G_k S \subset S$ there exist, by **12.1.2**, unique z^k such that $z^k = G_k z^k$, $k \geqslant 0$. Now,

$$\| y^{k+1} - x^* \| \leqslant \| G_k y^k - G_k x^* \| + \| G_k x^* - G x^* \|$$
$$\leqslant \alpha \| y^k - x^* \| + \| G_k x^* - G x^* \|$$
$$\leqslant \cdots \leqslant \sum_{j=0}^{k} \alpha^{k-j} \| G_k x^* - x^* \| + \alpha^{k+1} \| y^0 - x^* \|.$$

As in the proof of the last part of **12.2.1**, this shows that (d) implies (a). From

$$\| z^k - x^* \| \leqslant \| G_k z^k - G_k x^* \| + \| G_k x^* - G_k y^k \| + \| G_k y^k - G x^* \|,$$

we obtain

$$\| z^k - x^* \| \leqslant [1/(1 - \alpha)][\alpha \| x^* - y^k \| + \| y^{k+1} - x^* \|],$$

so that (a) implies (b). Next

$$\| G_k x^* - x^* \| \leqslant \| G_k x^* - G_k z^k \| + \| G_k z^k - G x^* \|$$
$$\leqslant \alpha \| x^* - z^k \| + \| z^k - x^* \|,$$

and (b) implies (d). Finally, (a) and (c) are equivalent by **12.2.1**. ∎

As a first corollary, we obtain a result about the influence of an error in the evaluation of G upon the iteration (1) under the assumption that this error does not exceed a certain magnitude.

12.2.5. Let $G: D \subset R^n \to R^n$ be a contraction on D with constant α, and $\tilde{G}: D \subset R^n \to R^n$ another mapping for which

$$\| \tilde{G} x - G x \| \leqslant \epsilon, \quad \forall x \in D.$$

Suppose that for some $y^0 \in D$, $S = \bar{S}(\tilde{G} y^0, r) \subset D$ where

$$r = [1/(1 - \alpha)][\alpha \| \tilde{G} y^0 - y^0 \| + 2\epsilon].$$

Then the sequence $y^{k+1} = \hat{G}y^k$, $k = 0, 1, \ldots$, remains in S and

$$\| y^{k+1} - x^* \| \leq [\alpha/(1-\alpha)] \| y^k - y^{k+1} \| + \epsilon/(1-\alpha),$$

where x^* is the unique fixed point of G in S.

Moreover, if $\{x^k\}$ is the sequence defined by (1) with $x^0 = y^0$, then

$$\| y^{k+1} - x^* \| \leq \| x^{k+1} - x^* \| + \epsilon/(1-\alpha).$$

The proof is an immediate consequence of the first part of **12.2.4**, since

$$\| \hat{G}x - \hat{G}y \| \leq \| \hat{G}x - Gx \| + \| \hat{G}y - Gy \| + \| Gx - Gy \|$$
$$\leq \alpha \| x - y \| + 2\epsilon, \qquad \forall x \in D.$$

As a corollary along the lines of **12.2.2**, we obtain:

12.2.6. Let $G_k : D \subset R^n \to R^n$, $k = 0, 1, \ldots$, be contractions on D with a uniform contraction constant α. Suppose that for some $G : D \subset R^n \to R^n$

$$\lim_{k \to \infty} G_k x = Gx, \qquad \forall x \in D. \tag{9}$$

If $S = \bar{S}(G_0 y^0, r) \subset D$ where $r = [1/(1-\alpha)][\alpha \| G_0 y^0 - y^0 \| + \delta]$ and $\| G_0 y^0 - G_k y^0 \| \leq \delta$ for $k = 0, 1, \ldots$, then the sequences $\{x^k\}$ and $\{y^k\}$ defined by (1) and (2), respectively, remain in S and converge to the only fixed point x^* of G in S. Moreover, the error estimates (3) and (4) hold with $\epsilon_k = \| G_k y^k - Gy^k \|$.

Proof. From

$$\| Gx - Gy \| \leq \| Gx - G_k x \| + \| G_k x - G_k y \| + \| G_k y - Gy \|$$

and (9) it follows that G itself is a contraction on D with constant α. Thus, **12.2.4** applies and ensures that the sequences $\{x^k\}$, $\{y^k\}$ remain in S and that $\{x^k\}$ converges to x^*. Moreover, the estimates (3) and (4) are valid. By (9) the statement (d) of **12.2.4** is true, and therefore we have $\lim_{k \to \infty} y^k = x^*$. ∎

NOTES AND REMARKS

NR 12.2-1. As all contraction-type theorems, the results of this section hold verbatim for operators on complete metric spaces, and the proofs also remain word for word the same. In fact, most of these results were originally proved in so-called pseudometric spaces; that is, spaces in which distances are measured

12.2. APPROXIMATE CONTRACTIONS AND SEQUENCES

in terms of elements of a partially ordered topological space (see, for example, Collatz [1964]).

NR 12.2-2. Many authors have contributed to the problem area discussed in this section; in particular, Ehrmann [1959a], Gardner [1965], Ostrowski [1967c], Schmidt [1960], Urabe [1956, 1962], and Warga [1952]. Our presentation here follows Ortega and Rheinboldt [1967b].

NR 12.2-3. The basic theorem **12.2.1** represents a minor extension of a result of Ostrowski [1967c]. In the proof, the so-called Toeplitz lemma is used in the following form (see **E 12.2-1** for a more general statement):
If the sequence $\{\alpha_k\} \subset R^1$ converges to zero, then for any α with $|\alpha| < 1$ the sequence $\beta_k = \sum_{j=0}^{k} \alpha^{k-j}\alpha_j$, $k = 0, 1,...$, also converges to zero.
When pseudometric spaces are used, this lemma extends only under suitable additional conditions. In particular, we shall make use of one such generalization in the proof of **13.1.2**. For some related comments, see also Ortega and Rheinboldt [1967b].

NR 12.2-4. A domain condition of the form used in **12.2.3** appears to have been given first by Weissinger [1952].

NR 12.2-5. Iterations of the general form $x^{k+1} = G_k x^k$, $k = 0, 1,...$, have been studied by several authors, and, in particular, Theorem **12.2.6** represents a principal result of Ehrmann [1959a]. Schmidt [1960] considered the implicit process $x^k = G_k x^k$, $k = 0, 1,...$, and Theorem **12.2.4** contains one of Schmidt's results. Both Theorems **12.2.3** and **12.2.4** were first proved in Ortega and Rheinboldt [1967b].

NR 12.2-6. Theorem **12.2.5** is a frequently cited result of Urabe [1956]. Its interpretation is that if at each step of the iteration a roundoff error not exceeding $\epsilon > 0$ is made, then the ball of limiting accuracy has radius $\epsilon/(1 - \alpha)$ where α is the contraction constant. For a further discussion of the use of this and similar theorems in the study of rounding error, see Lancaster [1966].

EXERCISES

E 12.2-1. Prove the "general Toeplitz lemma": If the sequence $\{\alpha_k\} \subset R^1$ converges to zero and the coefficients $\gamma_{ik} \in R^1$, $i = 0, 1,...$, $k = 0, 1,..., i$, satisfy $\lim_{i \to \infty} \gamma_{ik} = 0$ for each k, and $\sum_{k=0}^{i} |\gamma_{ik}| \leq c$ for all $i = 0, 1,...$, then the sequence $\beta_i = \sum_{k=0}^{i} \gamma_{ik} \alpha_k$ converges to zero.

E 12.2-2. Let $G: D \subset R^n \to R^n$ be a contraction on $D_0 \subset D$. Show that for any $x^0 \in D_0$ such that $\bar{S}(Gx^0, \alpha[1 - \alpha]^{-1} \| Gx^0 - x^0 \|) \subset D_0$ the sequence $\{G^k x^0\}$ remains in D_0 and converges to the only fixed point of G in D_0.

E 12.2-3. Let $G: D \subset R^n \to R^n$ be a contraction on D with constant $\alpha < 1$, and suppose that all $G_k: D \subset R^n \to R^n$, $k = 0, 1,...$, map D into itself. Assume, further, that $S = \bar{S}(G_0 y^0, r) \subset D$ with

$$r = (1-\alpha)^{-1}(\alpha \| G_0 y^0 - y^0 \| + \| Gy^0 - G_0 y^0 \|).$$

Show that G has a unique fixed point x^* in S and that the sequence (2) satisfies (3) and (4).

E 12.2-4. Suppose that $G_k: D \subset R^n \to R^n$, $k = 0, 1,...$, are contractions on D with the same constant $\alpha < 1$, and have (unique) fixed points $y^k \in D$. Let $G: D \subset R^n \to R^n$ be such that

$$\|(Gx - Gy) - (G_k x - G_k y)\| \leq \beta \|x - y\|, \qquad \forall x, y \in D,$$

with $\beta < 1 - \alpha$. Finally, assume that $S = \bar{S}(y^1, r) \subset D$ with

$$r = [1 - (\alpha + \beta)]^{-1}(\beta \| y^1 - y^0 \| + \| G_1 y^0 - Gy^0 \|).$$

Show that G has a unique fixed point x^* in S, that

$$\| y^{k+1} - x^* \| \leq [1 - (\alpha + \beta)]^{-1}(\beta \| y^{k+1} - y^k \| + \| G_{k+1} y^k - Gy^k \|), \quad k = 0, 1,...,$$

and, moreover, that the following four statements are equivalent:

(i) $\epsilon_k = \| y^{k+1} - Gy^k \| \to 0$, (ii) $\| G_{k+1} y^k - Gy^k \| \to 0$, (iii) $y^k \to x^*$, (iv) $\| G_k x^* - Gx^* \| \to 0$ (Ortega and Rheinboldt [1967b]).

12.3. ITERATED CONTRACTIONS AND NONEXPANSIONS

In the previous two sections we have examined various extensions of the basic contraction theorem **12.1.2**. In the present section we continue this program, and consider first a different type of extension by restricting the set of points in which the contraction condition has to hold.

12.3.1. Definition. The mapping $G: D \subset R^n \to R^n$ is an *iterated contraction* on the set $D_0 \subset D$ if there is an $\alpha < 1$ such that

$$\| G(Gx) - Gx \| \leq \alpha \| Gx - x \| \tag{1}$$

whenever x and Gx are in D_0. ∎

Clearly, if G is a contraction on D_0, then it is also an iterated contraction, but the converse is not necessarily true as the simple one-dimensional example $Gx = x^2$ shows. Indeed, in this case G is a contraction on any closed interval $[a, b] \subset (-\frac{1}{2}, \frac{1}{2})$, but an iterated contraction on any $[a, b] \subset (-1, +1)$. Note, also, that an iterated contraction need not be continuous nor need its fixed points be unique (**E 12.3-1**).

12.3. ITERATED CONTRACTIONS AND NONEXPANSIONS

In spite of this lack of properties, iterated contractions turn out to be very useful in the study of certain iterative processes, and in Sections 12.4–12.6 we shall extend this concept to a more general form, which has wide applicability. For the present, we state only the following simple result, which is an immediate corollary of the more general theorem **12.4.3**; a direct proof along the lines of **12.1.1** is also easily given (**E 12.3-2**).

12.3.2. Suppose that $G: D \subset R^n \to R^n$ is an iterated contraction on the closed set $D_0 \subset D$ and that for some $x^0 \in D_0$ the sequence

$$x^{k+1} = Gx^k, \qquad k = 0, 1, \ldots, \tag{2}$$

remains in D_0. Then, $\lim_{k \to \infty} x^k = x^* \in D_0$, and the estimate

$$\| x^k - x^* \| \leqslant [\alpha/(1-\alpha)] \| x^k - x^{k-1} \|, \qquad k = 0, 1, \ldots, \tag{3}$$

holds. Moreover, if G is continuous at x^*, then $x^* = Gx^*$.

A simple, but typical, application of **12.3.2** is the following convergence result.

12.3.3. Let $F: D \subset R^n \to R^n$ be F-differentiable on D, and

$$\| F'(x) - F'(y) \| \leqslant \gamma, \qquad \forall x, y \in D.$$

Suppose that the mapping $A: D \subset R^n \to L(R^n)$ satisfies

$$\| A(x)^{-1} \| \leqslant \beta, \qquad \| F'(x) - A(x) \| \leqslant \delta, \qquad \forall x \in D,$$

where $\alpha = \beta(\gamma + \delta) < 1$, and that there is an $x^0 \in D$ for which $S = \bar{S}(x^0, r) \subset D$ with $r \geqslant \beta \| Fx^0 \|/(1-\alpha)$. Then the iterates

$$x^{k+1} = x^k - A(x^k)^{-1} Fx^k, \qquad k = 0, 1, \ldots,$$

remain in S and converge to the only solution x^* of $Fx = 0$ in S. Moreover, the error estimate (3) holds.

Proof. Set $Gx = x - A(x)^{-1} Fx$; then for any $x \in S$ such that $Gx \in S$ we find, with the help of **3.2.5**, that

$$\| G(Gx) - Gx \|$$
$$= \| A(Gx)^{-1} F(Gx) \| \leqslant \beta \| F(Gx) - Fx - A(x)(Gx - x) \|$$
$$\leqslant \beta \| F(Gx) - Fx - F'(x)(Gx - x) \| + \beta \| F'(x) - A(x) \| \| Gx - x \|$$
$$\leqslant \beta(\gamma + \delta) \| Gx - x \| = \alpha \| Gx - x \|.$$

This shows that G is an iterated contraction on S. Now, if $x^0,..., x^k \in S$, then

$$\| x^{k+1} - x^0 \| \leq \sum_{j=0}^{k} \| x^{j+1} - x^j \| \leq \sum_{j=0}^{k} \alpha^j \| x^1 - x^0 \| \leq \beta \| Fx^0 \|/(1 - \alpha) \leq r; \quad (4)$$

hence, $x^{k+1} \in S$, and, by induction, $\{x^k\} \subset S$. The convergence of $\{x^k\}$ to some $x^* \in S$ and the error estimate follow directly from **12.3.2**. Since

$$\| A(x) \| \leq \| A(x) - F'(x) \| + \| F'(x) - F'(x^*) \| + \| F'(x^*) \| \leq \delta + \gamma + \| F'(x^*) \|,$$

we have

$$\| Fx^k \| \leq \eta \| x^{k+1} - x^k \|, \quad k = 0, 1,..., \quad \eta = \delta + \gamma + \| F'(x^*) \|.$$

Hence, $\lim_{k \to \infty} Fx^k = 0$, and because, by **3.1.6**, F is continuous at x^*, it follows that $Fx^* = 0$. For the uniqueness assume that $Fy^* = 0$, $x^* \neq y^* \in S$. Then, using **3.2.5**, we obtain the contradiction

$$\| x^* - y^* \| \leq \beta \| A(x^*)(x^* - y^*) \|$$

$$\leq \beta \| F'(x^*)(x^* - y^*) - Fx^* - Fy^* \| + \beta \| F'(x^*) - A(x^*) \| \| x^* - y^* \|$$

$$\leq \beta(\gamma + \delta) \| x^* - y^* \| < \| x^* - y^* \|. \quad \blacksquare$$

There are many possible variations of results of this type; we have included several as exercises, and in later sections we shall also derive several extensions. Of some theoretical as well as practical interest is the following modification of **12.3.3** which does not follow from **12.3.2** but can be proved analogously.

12.3.4. Suppose that the conditions of **12.3.3** hold. Then for any sequence $\{z^k\} \subset S$ the iterates $x^{k+1} = x^k - A(z^k)^{-1} Fx^k$, $k = 0, 1,...$, remain in S and converge to x^*. Moreover, the error estimate (3) holds.

Proof. If $x^0,..., x^k \in S$, then, again by **3.2.5**,

$$\| x^{k+1} - x^k \| = \| A(z^k)^{-1} Fx^k \| \leq \beta \| Fx^k - Fx^{k-1} - A(z^{k-1})(x^k - x^{k-1}) \|$$

$$\leq \beta \{ \| Fx^k - Fx^{k-1} - F'(z^{k-1})(x^k - x^{k-1}) \|$$

$$+ \| F'(z^{k-1}) - A(z^{k-1}) \| \| x^k - x^{k-1} \| \}$$

$$\leq \beta(\gamma + \delta) \| x^k - x^{k-1} \| \leq \cdots \leq \alpha^k \| x^1 - x^0 \|,$$

which, as in (4), proves that $x^{k+1} \in S$ and, thus, by induction that $\{x^k\} \subset S$.

12.3. ITERATED CONTRACTIONS AND NONEXPANSIONS

Now,

$$\| x^{k+p} - x^k \| \leq \sum_{j=k}^{k+p-1} \| x^{j+1} - x^j \| \leq \frac{\alpha}{1-\alpha} \| x^k - x^{k-1} \|$$

$$\leq \frac{\alpha^k}{1-\alpha} \| x^1 - x^0 \| \qquad (5)$$

shows that $\{x^k\}$ is a Cauchy sequence. Since x^* was the only solution of $Fx = 0$ in S, we see that $\lim_{k \to \infty} x^k = x^*$, and the error estimate (3) follows from (5) for $p \to \infty$. ∎

For the applications the z^k will usually be chosen from among the prior iterates. More precisely, z^k may be held constant for several steps, so that $A(z^k)$ is not changed at each stage of the iteration. Note that for $F'(x) = A(x)$ we obtain from **12.3.3** and **12.3.4** convergence results about Newton's method and one of its variants.

We return to the general result **12.3.2**. An interesting extension is the following theorem, which—in analogy to the strict nonexpansivity definition **5.1.1**—weakens the condition (1) to only strict inequality.

12.3.5. Suppose that $G: D \subset R^n \to R^n$ maps $D_0 \subset D$ into itself, GD_0 is compact, and

$$\| G(Gx) - Gx \| < \| Gx - x \|, \qquad \forall x \in D_0, \quad x \neq Gx. \qquad (6)$$

Assume, further, that G is continuous on D_0 and has at most one fixed point in D_0. Then a fixed point x^* exists, and for any $x^0 \in D_0$ the sequence (2) converges to x^*.

Proof. For any $x^0 \in D_0$ the sequence (2) is well defined, $x^k \in GD_0$ for $k \geq 1$, and, since GD_0 is compact, $\{x^k\}$ has limit points in GD_0. Suppose that $x^* \in GD_0$ is such a limit point and that $\lim_{i \to \infty} x^{k_i} = x^*$. If $x^* \neq Gx^*$, then the mapping

$$r(x) = \| G(Gx) - Gx \| / \| Gx - x \|$$

is well defined and continuous in some neighborhood of x^* and by (6) we have $r(x^*) < 1$. Hence, for given $\alpha \in (r(x^*), 1)$, there exists a $\delta > 0$ so that $r(x) \leq \alpha$ for all $x \in S(x^*, \delta) \cap D_0$. Consequently, there is an index $j = j(\delta)$ so that $r(x^{k_i}) \leq \alpha$ for $i \geq j$; that is,

$$\| Gx^{k_i+1} - Gx^{k_i} \| \leq \alpha \| x^{k_i+1} - x^{k_i} \|, \qquad i \geq j.$$

Now, (6) implies that $\| x^{k+1} - x^k \| < \| x^k - x^{k-1} \|$ for all k. Hence,

$$\| x^{k_{i+1}+1} - x^{k_{i+1}} \| < \| x^{k_i+1} - x^{k_{i+1}-1} \| < \cdots < \| Gx^{k_i+1} - Gx^{k_i} \|$$
$$\leq \alpha \| x^{k_i+1} - x^{k_i} \| \leq \cdots \leq \alpha^{i-j+1} \| x^{k_j+1} - x^{k_j} \|,$$

which shows that $\lim_{i \to \infty} (x^{k_i+1} - x^{k_i}) = 0$. But then, by continuity of G, the right-hand side of

$$\| Gx^* - x^* \| \leq \| Gx^* - Gx^{k_i} \| + \| x^{k_i+1} - x^{k_i} \| + \| x^{k_i} - x^* \|$$

tends to zero, and, thus, $x^* = Gx^*$, which is a contradiction. Therefore, every limit point of $\{x^k\}$ is a fixed point of G, and, since G has at most one fixed point in D_0, it follows that the whole sequence $\{x^k\}$ must converge. ∎

The condition (6) is, of course, satisfied if G is strictly nonexpansive. In that case, the second half of the assumptions of **12.3.5** are automatically satisfied, and, hence, we have the following immediate corollary of that theorem.

12.3.6. Edelstein Theorem. Suppose that $G: D \subset R^n \to R^n$ maps $D_0 \subset D$ into itself, GD_0 is compact, and G is strictly nonexpansive on D_0. Then for any $x^0 \in D_0$ the sequence (2) converges to the unique fixed point of G in D_0.

Note that in both **12.3.5** and **12.3.6** the continuity of G implies that the compactness of GD_0 can be replaced by the stronger condition that D_0 itself is compact. It is important to recall, however, that when G is a contraction neither one of these compactness assumptions is necessary.

The simple one-dimensional example $Gx = -x$, $D_0 = [-1, 1]$ shows that **12.3.6** need not remain valid if G is only nonexpansive, since the sequence (2) will not converge to $x^* = 0$ unless $x^0 = x^*$. But we can prove the following convergence result for a modified iteration obtained by a convex combination of G and the identity.

12.3.7. Suppose that under the Euclidean norm $G: D \subset R^n \to R^n$ is nonexpansive on the closed, convex set $D_0 \subset D$. Assume, further, that $GD_0 \subset D_0$ and that D_0 contains a fixed point of G. Then for any $\omega \in (0, 1)$ and $x^0 \in D_0$ the iteration

$$x^{k+1} = \omega x^k + (1 - \omega) Gx^k, \quad k = 0, 1, \ldots, \tag{7}$$

converges to a fixed point of G in D_0.

12.3. ITERATED CONTRACTIONS AND NONEXPANSIONS

Proof. The convexity of D_0 ensures that the sequence (7) is well-defined and remains in D_0. If x^* is a fixed point of G in D_0, then, in the l_2-norm,

$$\|x^{k+1} - x^*\|^2 = \omega^2 \|x^k - x^*\|^2 + (1-\omega)^2 \|Gx^k - x^*\|^2$$
$$+ 2\omega(1-\omega)(Gx^k - x^*)^T (x^k - x^*) \quad (8)$$

and

$$\|x^k - Gx^k\|^2 = \|x^k - x^*\|^2 + \|Gx^k - x^*\|^2 - 2(Gx^k - x^*)^T (x^k - x^*). \quad (9)$$

After multiplication of (9) by $\omega(1-\omega)$ and subsequent addition to (8), this leads to

$$\|x^{k+1} - x^*\|^2 + \omega(1-\omega)\|x^k - Gx^k\|^2$$
$$= \omega\|x^k - x^*\|^2 + (1-\omega)\|Gx^k - Gx^*\|^2 \leq \|x^k - x^*\|^2. \quad (10)$$

Therefore, for any $m \geq 0$,

$$\omega(1-\omega) \sum_{k=0}^{m} \|x^k - Gx^k\|^2 \leq \sum_{k=0}^{m} [\|x^k - x^*\|^2 - \|x^{k+1} - x^*\|^2]$$
$$= \|x^0 - x^*\|^2 - \|x^{m+1} - x^*\|^2 \leq \|x^0 - x^*\|^2,$$

which proves that for $m \to \infty$ the series on the left converges, and, in particular, that $\lim_{k \to \infty} \|x^k - Gx^k\| = 0$.

Since

$$\|x^{k+1} - x^*\| = \|\omega(x^k - x^*) + (1-\omega)(Gx^k - Gx^*)\|$$
$$\leq \|x^k - x^*\| \leq \|x^j - x^*\| \leq \|x^0 - x^*\|, \quad \forall k \geq 0, j \leq k, \quad (11)$$

it follows that the sequence $\{x^k\}$ is bounded and, hence, has a convergent subsequence $\{x^{k_i}\}$ which, by the closedness of D_0, must have its limit point y^* in D_0. Then (7) shows that

$$\lim_{i \to \infty}(x^{k_i+1} - y^*) = \lim_{i \to \infty}(x^{k_i} - y^*) + (1-\omega) \lim_{i \to \infty}(Gx^{k_i} - x^{k_i}) = 0,$$

or, by the continuity of G, that $y^* = Gy^*$. Therefore, (11) holds with y^* instead of x^*, and, accordingly, the whole sequence $\{x^k\}$ must converge to the fixed point y^*. ∎

Note that if the set D_0 is also bounded, then **5.1.4** (and also the Brouwer fixed-point theorem **6.3.2**) guarantees the existence of a fixed point. Note, also, that (7) is equivalent to

$$\hat{x}^k = Gx^k, \qquad x^{k+1} = x^k + \omega(\hat{x}^k - x^k), \qquad k = 0, 1, \ldots;$$

that is, (7) may be considered as an "underrelaxed" version of the basic iteration (2). Finally, note that the operator $\hat{G} = \omega I + (1 - \omega) G$ is not necessarily contractive under the conditions of **12.3.7** as the simple example $Gx = x$ in R^1 shows.

NOTES AND REMARKS

NR 12.3-1. Relations of the form (1) and (6) were used by Goldstein [1967] (see also Cheney and Goldstein [1959]), but the concept of an iterated contraction was independently developed and extended by Rheinboldt [1968].

NR 12.3-2. Theorems **12.3.3** and **12.3.4** represent general versions of a variety of theorems in the literature; see, for example, Ben-Israel [1965, 1966], Bartle [1955], and Pereyra [1967a]. In particular, Theorem **12.3.4** is essentially due to Bartle (see also **E 12.3-3–E 12.3-4**). Ben-Israel considered the generalized Newton iteration

$$x^{k+1} = x^k - B(x^k) F x^k, \qquad k = 0, 1, \ldots, \tag{12}$$

where now $F: D \subset R^n \to R^m$, and $B(x) = F'(x)^+$ is a generalized inverse of the derivative (see **NR 8.5-4**). A theorem similar to **12.3.3** can then be phrased as follows:

Let $F: D \subset R^n \to R^m$ be F-differentiable on $S = \bar{S}(x^0, r) \subset D$ and

$$\|Fy - Fx - F'(y)(x-y)\| \leqslant \beta \|x - y\|, \qquad \forall x, y \in S.$$

Let $B: S \subset R^n \to L(R^n, R^m)$ be such that

$$\|(B(x) - B(y)) Fy\| \leqslant \delta \|x - y\|, \qquad \|B(x) F'(x) - I\| \leqslant \tau, \qquad \|B(x)\| \leqslant \eta,$$

for all $x, y \in S$. If $\alpha = \eta\beta + \delta + \tau < 1$ and $r \geqslant \eta \|Fx^0\|/(1 - \alpha)$, then the iterates (12) remain in S and converge to a solution $x^* \in S$ of $B(x) Fx = 0$.

NR 12.3-3. For additional results along the line of **12.3.4** when $A(x) = F'(x)$, see Dennis [1969].

NR 12.3-4. Theorem **12.3.5** is an apparently new generalization of the earlier theorem **12.3.6** of Edelstein [1962]. A different generalization also using the condition (6) is given by Goldstein [1967] (see **E 12.3-6**).

NR 12.3-5. In Kantorovich and Akilov [1959], the following interesting proof of the existence part of **12.3.6** is attributed to a communication from M. Krein: The functional $g(x) = \|Gx - x\|$ is continuous on the compact set GD_0 and, hence, $x^* \in GD_0$ exists such that $g(x^*) = \min\{g(x) \mid x \in GD_0\}$. If $g(x^*) > 0$, then $g(Gx^*) < g(x^*)$ gives a contradiction to the construction of x^*, so that $Gx^* = x^*$.

Clearly, this proof also applies in the case of **12.3.5**.

12.3. ITERATED CONTRACTIONS AND NONEXPANSIONS

NR 12.3-6. Theorem **12.3.7** was first proved by Browder and Petryshyn [1966] for nonexpansive operators which map a closed, convex set in a uniformly convex Banach space into itself. Other authors who contributed to this problem are Krasnoselskii [1955], Schaefer [1957], Zarantonello [1960], and Opial [1967a]. See also Diaz and Metcalf [1969].

EXERCISES

E 12.3-1. Show that the discontinuous function $g\colon [0, 1] \subset R^1 \to R^1$ defined by $g(x) = 0$, $x \in [0, \frac{1}{2})$, $g(x) = 1$, $x \in [\frac{1}{2}, 1]$, is an iterated contraction and also has more than one fixed point.

E 12.3-2. Give a direct proof of **12.3.2**.

E 12.3-3. Suppose that $F\colon D \subset R^n \to R^n$ is F-differentiable on $S = \bar{S}(x^0, r) \subset D$ and that $\|F'(x)^{-1}\| \leq \beta$ for all $x \in S$. Assume, further, that either one of the following two conditions holds:

(a) $\|Fy - Fx - F'(z)(y - x)\| \leq \gamma \|y - x\|$, $\forall x, y, z \in S$; $\alpha = \beta\gamma < 1$, $r \geq \beta \|Fx^0\|/(1 - \alpha)$
(b) $\|F'(x) - F'(x^0)\| \leq \gamma$, $\forall x \in S$; $\alpha = 2\gamma\beta < 1$, $r \geq \beta \|Fx^0\|/(1 - \alpha)$.

Show that the Newton iterates starting at x^0 remain in S and converge to the only solution x^* of $Fx = 0$ in S, and, moreover, that (3) holds. Show also that for any sequence $\{z^k\} \subset S$ the same is true for the iterates

$$x^{k+1} = x^k - F'(z^k)^{-1} Fx^k, \qquad k = 0, 1, \ldots. \tag{13}$$

E 12.3-4. Suppose $F\colon D \subset R^n \to R^n$ is F-differentiable on $S = \bar{S}(x^0, r) \subset D$ and that $\|F'(x^0)^{-1}\| \leq \beta$. Assume, further, that one of the following three conditions holds:

(a) $\|Fy - Fx - F'(z)(y - x)\| \leq \gamma \|y - x\|$, $\forall x, y, z \in S$; $\alpha = 3\beta\gamma < 1$, $r \geq \beta \|Fx^0\|/(1 - \alpha)$.
(b) $\|F'(y) - F'(x)\| \leq \gamma$, $\forall x, y \in S$; $\alpha = 2\beta\gamma < 1$, $r \geq \beta \|Fx^0\|/(1 - \alpha)$.
(c) $\|F'(x) - F'(x^0)\| \leq \gamma$, $\forall x \in S$; $\alpha = 3\beta\gamma < 1$, $r \geq \beta \|Fx^0\|/(1 - \alpha)$.

Show that, again, the Newton iterates starting at x^0 remain in S and converge to the only solution of $Fx = 0$ in S, and that the same is true for the iterates (13).

E 12.3-5. State and prove convergence theorems analogous to those in **E 12.3-3** and Theorem **12.3.4** for iterations of the form

$$x^{k+1} = x^k - A_k^{-1} Fx^k, \qquad k = 0, 1, \ldots,$$

where $A_k \in L(R^n)$, $k = 0, 1, \ldots$, is a given sequence of nonsingular matrices.

E 12.3-6. Let $G: D \subset R^n \to R^n$ be nonexpansive on $D_0 \subset D$, where $GD_0 \subset D_0$ and GD_0 is compact. Moreover, suppose that (6) holds. Show that for given $x^0 \in D_0$ the limit point of any convergent subsequence of $\{G^k x^0\}$ is a fixed point of G, and, based upon this, that the whole sequence converges to a fixed point of G. (Goldstein [1967])

E 12.3-7. Suppose that $F: D \subset R^n \to R^n$ satisfies

$$1 \leqslant \gamma = \sup\{\|Fx - Fy\|_2 / \|x - y\|_2 \mid x, y \in D, \ x \neq y\}.$$

(a) If, for some $\mu < 1$,

$$(x - y)^T (Fx - Fy) \leqslant \mu \|x - y\|_2^2, \qquad \forall x, y \in D,$$

show that $Gx = (1 - \alpha) x + \alpha Fx$ is a contraction on D for any $0 < \alpha < 2(1 - \mu)/(1 - 2\mu + \gamma^2)$.

(b) If, for some $\mu > 1$,

$$(x - y)^T (Fx - Fy) \geqslant \mu \|x - y\|_2^2, \qquad \forall x, y \in D,$$

show that G is a contraction for $2(1 - \mu)/(1 - 2\mu + \gamma^2) < \alpha < 0$. (Zarantonello [1960])

E 12.3-8. Let $F: D \subset R^n \to R^n$ and set

$$\mu(x, y) = \frac{(x - y)^T (Fx - Fy)}{\|x - y\|_2^2}, \qquad \gamma(x, y) = \frac{\|Fx - Fy\|_2}{\|x - y\|_2}, \qquad x, y \in D, \ x \neq y.$$

If $\mu(x, y) \leqslant \bar{\mu} < 1$ for all $x, y \in D$, show that $Gx = (1 - \alpha) x + \alpha Fx$ is a contraction on D for any

$$0 < \alpha < \inf \left\{ \frac{2[1 - \mu(x, y)]}{1 - 2\mu(x, y) + \gamma(x, y)^2} \ \middle| \ x, y \in D, \ x \neq y \right\}.$$

E 12.3-9. Suppose that $F: D \subset R^n \to R^n$ satisfies for some $\mu > 0$ and $\gamma < \infty$

$$(y - x)^T (Fy - Fx) \geqslant \mu \|y - x\|^2, \qquad \|Fy - Fx\| \leqslant \gamma \|y - x\|, \qquad \forall x, y \in D.$$

(a) Show that $Gx = x - \omega Fx$ is a contraction on D with contraction constant $q_\omega = (\gamma^2 \omega^2 - 2\mu\omega + 1)^{1/2}$ provided that $0 < \omega < 2\mu/\gamma^2$. When is q_ω minimal? (Kolomy [1964])

(b) Suppose that $S = \bar{S}(x^0, r) \subset D$, with $r \geqslant [\mu^{-1} + (\mu^{-2} - \gamma^{-2})]^{1/2} \|Fx^0\|$. Show that $x^{k+1} = x^k - (\mu/\gamma^2) Fx^k$, $k = 0, 1, \ldots$, converge to the only solution of $Fx = 0$ in S. (Linkov [1964b])

E 12.3-10. Let $B \in L(R^n)$, and assume that $\rho(B) \leqslant 1$ and 1 is not an eigenvalue of B. Show that $\rho\{\omega I + (1 - \omega) B\} < 1$ for all $\omega \in (0, 1)$.

12.4. NONLINEAR MAJORANTS

In previous sections we considered various modifications of the contraction-mapping theorem. We now continue this study by introducing more general nonlinear estimates for the differences $Gy - Gx$ or $G^2x - Gx$. In this section we develop the basic setting, and then in subsequent sections we extend these ideas to more general situations.

12.4.1. Definition. Let $\{x^k\}$ be any sequence in R^n. Then a sequence $\{t_k\} \subset [0, \infty) \subset R^1$ for which

$$\| x^{k+1} - x^k \| \leqslant t_{k+1} - t_k, \qquad k = 0, 1, \ldots, \qquad (1)$$

holds is a *majorizing sequence* for $\{x^k\}$. ∎

Note that any majorizing sequence is necessarily monotonically increasing.

The following simple result will be a frequently used lemma.

12.4.2. Let $\{t_k\} \subset R^1$ be a majorizing sequence for $\{x^k\} \subset R^n$, and suppose that $\lim_{k \to \infty} t_k = t^* < \infty$ exists. Then $x^* = \lim_{k \to \infty} x^k$ exists and

$$\| x^* - x^k \| \leqslant t^* - t_k, \qquad k = 0, 1, \ldots. \qquad (2)$$

Proof. The estimate

$$\| x^{k+m} - x^k \| \leqslant \sum_{j=k}^{k+m-1} \| x^{j+1} - x^j \| \leqslant \sum_{j=k}^{k+m-1} (t_{j+1} - t_j) = t_{k+m} - t_k \qquad (3)$$

shows that $\{x^k\}$ is a Cauchy sequence, and the error estimate (2) follows from (3) as $m \to \infty$. ∎

In the sequel, majorizing sequences will arise as solutions of certain nonlinear difference equations which in turn are based on estimates for $G(Gx) - Gx$. The idea is given in the following extension of **12.3.2**.

12.4.3. For $G: D \subset R^n \to R^n$, suppose that there exists an isotone function $\varphi: [0, \infty) \subset R^1 \to [0, \infty)$ such that on some set $D_0 \subset D$

$$\| G^2x - Gx \| \leqslant \varphi(\| Gx - x \|), \qquad \text{if} \quad x, Gx \in D_0. \qquad (4)$$

Assume, further, that for some $x^0 \in D_0$ the iterates $x^k = G^k x^0$, $k = 1, 2, \ldots$, remain in D_0 and that the sequence $\{t_k\}$ defined by

$$t_{k+1} = t_k + \varphi(t_k - t_{k-1}), \qquad t_0 = 0, \qquad t_1 \geqslant \| Gx^0 - x^0 \|, \qquad k = 1, 2, \ldots, \qquad (5)$$

converges to $t^* < +\infty$. Then $\lim_{k\to\infty} x^k = x^*$ exists and the estimate (2) holds. Moreover, if $x^* \in D$ and G is continuous at x^*, then $x^* = Gx^*$.

Proof. We show by induction that $\{t_k\}$ majorizes $\{x^k\}$. By assumption, $\|x^1 - x^0\| \leq t_1 - t_0$, and if $\|x^j - x^{j-1}\| \leq t_j - t_{j-1}$ for $j = 1,\ldots, k$, then (4) and (5) together with the isotonicity of φ imply that

$$\|x^{k+1} - x^k\| = \|G^2 x^{k-1} - Gx^{k-1}\| \leq \varphi(\|x^k - x^{k-1}\|)$$
$$\leq \varphi(t_k - t_{k-1}) = t_{k+1} - t_k. \tag{6}$$

The result now follows from **12.4.2**. ∎

Note that **12.4.3** contains the iterated contraction theorem **12.3.2** and, hence, also the contraction theorem **12.1.2**. In this case the majorant function φ is simply $\varphi(t) = \alpha t$ with $\alpha < 1$. Hence, the difference equation (5) becomes

$$t_{k+1} - t_k = \alpha(t_k - t_{k-1}), \qquad t_0 = 0, \qquad t_1 = \|Gx^0 - x^0\|,$$

with solution

$$t_k = \left(\sum_{j=0}^{k-1} \alpha^j\right) \|Gx^0 - x^0\|,$$

so that

$$\lim_{k\to\infty} t_k = t^* = [1/(1-\alpha)] \|Gx^0 - x^0\|.$$

The condition $\{x^k\} \subset D_0$ is, of course, satisfied if $GD_0 \subset D_0$. Alternatively, it is possible to give a condition on the initial iterates analogous to **12.2.3**.

12.4.4. Assume that the conditions of **12.4.3** hold with the exception that only $x^0,\ldots, x^m$ are assumed to lie in D_0. Then $x^k \in D_0$, $k = m+1,\ldots$, if either one of the following relations holds:

$$\bar{S}(x^m, t^* - t_m) \subset D_0, \tag{7}$$

$$S(x^m, t^* - t_m) \subset D_0, \quad \text{and} \quad t_k < t^*, \quad \forall k \geq m. \tag{8}$$

Proof. We proceed by induction. Suppose that $x^j \in D_0$, $j = 0,\ldots, k$, for some $k \geq m$. Then it follows from (6) that

$$\|x^{j+1} - x^j\| \leq t_{j+1} - t_j, \qquad j = 0, 1,\ldots, k,$$

and therefore

$$\|x^{k+1} - x^m\| \leq \sum_{j=m}^{k} \|x^{j+1} - x^j\| \leq \sum_{j=m}^{k} (t_{j+1} - t_j) = t_{k+1} - t_m.$$

12.4. NONLINEAR MAJORANTS

Hence, $x^{k+1} \in \bar{S}(x^m, t_{k+1} - t_m) \subset \bar{S}(x^m, t^* - t_m)$ when $t_{k+1} \leqslant t^*$, or $x^{k+1} \in S(x^m, t^* - t_m)$ if $t_{k+1} < t^*$. This completes the induction. ∎

As an application of **12.4.3**, we prove a result similar to **12.3.3**.

12.4.5. Let $F: D \subset R^n \to R^n$ be F-differentiable on a convex set $D_0 \subset D$ and

$$\|F'(x) - F'(y)\| \leqslant \gamma \|x - y\|, \qquad \forall x, y \in D_0.$$

Assume that the mapping $A: D_0 \subset R^n \to L(R^n)$ satisfies

$$\|A(x)^{-1}\| \leqslant \beta, \qquad \|F'(x) - A(x)\| \leqslant \delta, \qquad \forall x \in D_0,$$

where $\beta\delta < 1$. If $x^0 \in D_0$ is such that

$$\|A(x^0)^{-1} F x^0\| \leqslant \eta, \qquad \alpha = \tfrac{1}{2}\beta\gamma\eta + \beta\delta < 1,$$

as well as $S = \bar{S}(x^0, \eta/[1 - \alpha]) \subset D_0$, then the iterates

$$x^{k+1} = x^k - A(x^k)^{-1} F x^k, \qquad k = 0, 1, \ldots,$$

remain in S and converge to a solution x^* of $Fx = 0$.

Proof. Again, set $Gx = x - A(x)^{-1} Fx$; then, whenever $x, Gx \in D_0$, we have, using **3.2.12**,

$$\|G^2 x - Gx\| = \|A(Gx)^{-1} F(Gx)\|$$
$$\leqslant \beta \|F(Gx) - Fx - F'(x)(Gx - x)\| + \beta \|(F'(x) - A(x))(Gx - x)\|$$
$$\leqslant \tfrac{1}{2}\beta\gamma \|Gx - x\|^2 + \beta\delta \|Gx - x\|. \qquad (9)$$

Hence, G satisfies (4) with $\varphi(t) = \tfrac{1}{2}\beta\gamma t^2 + \beta\delta t$. Consider, now, the difference equation (5) with initial values $t_0 = 0$ and $t_1 = \eta$. Clearly, $t_2 - t_1 = \alpha\eta$, and it follows easily by induction that

$$t_{k+1} - t_k \leqslant \alpha^k \eta, \qquad k = 1, 2, \ldots. \qquad (10)$$

Thus, $t_k \leqslant \eta \sum_{j=0}^{k-1} \alpha^j$ and $\lim_{k \to \infty} t_k = t^* \leqslant \eta/(1 - \alpha)$ exists. Now, **12.4.4** ensures that $\{x^k\} \subset \bar{S}(x^0, t^*) \subset D_0$, and the convergence statement follows from **12.4.3**. Finally,

$$\|Fx^k\| = \|A(x^k)(x^{k+1} - x^k)\|$$
$$\leqslant \|(A(x^k) - F'(x^k))(x^{k+1} - x^k)\| + \|F'(x^0)(x^{k+1} - x^k)\|$$
$$\quad + \|(F'(x^k) - F'(x^0))(x^{k+1} - x^k)\|$$
$$\leqslant (\delta + \|F'(x^0)\| + \gamma t^*) \|x^{k+1} - x^k\|,$$

so that $\lim_{k \to \infty} Fx^k = 0$, and, by the continuity of F, $Fx^* = 0$. ∎

In contrast to Theorem **12.3.3**, we assumed here that F' is Lipschitz-continuous, and we obtained a slightly better convergence condition. Moreover, **12.4.5** could have been strengthened by assuming only that $\bar{S}(x^0, t^*) \subset D_0$; but t^* is not, in general, known and hence this improvement has only theoretical value. For the same reason, the error estimate (2) is only theoretically useful in this case. However, it is easy to obtain computable estimates, and, in fact, from (9) and (10) it follows that for any $i \geq k$ we have

$$\| x^{i+1} - x^i \| \leq (\tfrac{1}{2}\beta\gamma \| x^i - x^{i-1} \| + \beta\delta) \| x^i - x^{i-1} \|$$
$$\leq (\tfrac{1}{2}\beta\gamma\alpha^{i-1}\eta + \beta\delta) \| x^i - x^{i-1} \| \leq \alpha \| x^i - x^{i-1} \|$$
$$\leq \cdots \leq \alpha^{i-k+1} \| x^k - x^{k-1} \|,$$

so that

$$\| x^{k+m} - x^k \| \leq \left(\sum_{j=0}^{m-1} \alpha^j \right) \| x^{k+1} - x^k \|$$
$$\leq [1/(1-\alpha)][\tfrac{1}{2}\beta\gamma\alpha^{k-1}\eta + \beta\delta] \| x^k - x^{k-1} \|,$$

which, for $m \to \infty$, leads to

$$\| x^* - x^k \| \leq [1/(1-\alpha)][\tfrac{1}{2}\beta\gamma\alpha^{k-1}\eta + \beta\delta] \| x^k - x^{k-1} \|.$$

This estimate is obviously very crude, and, in particular, it does not reduce to an estimate exhibiting the expected quadratic convergence when $A(x) = F'(x)$. A more precise error estimate is included in the following specialization of **12.4.5** to Newton's method.

12.4.6. Newton–Mysovskii Theorem. Suppose that $F: D \subset R^n \to R^n$ is F-differentiable on a convex set $D_0 \subset D$ and that for each $x \in D_0$ $F'(x)$ is nonsingular and satisfies

$$\| F'(x) - F'(y) \| \leq \gamma \| x - y \|, \qquad \| F'(x)^{-1} \| \leq \beta, \qquad \forall x, y \in D_0.$$

If $x^0 \in D_0$ is such that $\| F'(x^0)^{-1} Fx^0 \| \leq \eta$ and $\alpha = \tfrac{1}{2}\beta\gamma\eta < 1$, as well as $\bar{S}(x^0, r_0) \subset D_0$ where

$$r_0 = \eta \sum_{j=0}^{\infty} \alpha^{2^j - 1},$$

then the Newton iterates

$$x^{k+1} = x^k - F'(x^k)^{-1} Fx^k, \qquad k = 0, 1, \ldots,$$

remain in $\bar{S}(x^0, r_0)$ and converge to a solution x^* of $Fx = 0$. Moreover,

$$\| x^* - x^k \| \leq \epsilon_k \| x^k - x^{k-1} \|^2, \qquad k = 1, 2, \ldots, \tag{11}$$

12.4. NONLINEAR MAJORANTS

where

$$\epsilon_k = (\alpha/\eta) \sum_{j=0}^{\infty} (\alpha^{2^k})^{2^j-1} \leq \alpha[\eta(1-\alpha^{2^k})]^{-1}.$$

Proof. The estimate (9) leads in this case to the difference equation

$$t_{k+1} - t_k = \tfrac{1}{2}\beta\gamma(t_k - t_{k-1})^2, \quad t_0 = 0, \quad t_1 = \eta, \quad k = 1,\ldots.$$

By induction we will show that

$$t_{k+1} - t_k \leq \eta\alpha^{2^k-1}, \quad k = 0, 1,\ldots; \tag{12}$$

in fact, this is clearly correct for $k = 0$, and if it holds for $k = j - 1$, then

$$t_{j+1} - t_j \leq \tfrac{1}{2}\beta\gamma(\eta\alpha^{2^{j-1}-1})^2 = \tfrac{1}{2}\beta\gamma\eta^2\alpha^{2^j-2} = \eta\alpha^{2^j-1}.$$

Therefore,

$$t^* = \lim_{k\to\infty} t_k = \lim_{k\to\infty} \sum_{j=0}^{k-1} (t_{j+1} - t_j) \leq \eta \sum_{j=0}^{\infty} \alpha^{2^j-1} = r_0,$$

and the convergence statement follows from **12.4.4** and **12.4.3** precisely as in **12.4.5**.

In order to obtain the error estimate (11), note that by (9), since $\delta = 0$,

$$\|x^{k+1} - x^k\| \leq \alpha_0 \|x^k - x^{k-1}\|^2 \leq \alpha_0(t_k - t_{k-1})^2 \leq \alpha_0\eta^2\alpha^{2^k-2},$$

where $\alpha_0 = \tfrac{1}{2}\beta\gamma$. Therefore,

$$\|x^{k+m} - x^k\| \leq \sum_{j=k}^{k+m-1} \|x^{j+1} - x^j\| \leq \sum_{j=1}^{m} \alpha_0^{2^j-1} \|x^k - x^{k-1}\|^{2^j}$$

$$\leq \|x^k - x^{k-1}\|^2 \sum_{j=1}^{\infty} \alpha_0^{2^j-1}(t_k - t_{k-1})^{2^j-2}$$

$$\leq \alpha[\eta(1 - \alpha^{2^k})]^{-1}\|x^k - x^{k-1}\|^2,$$

since, because of (12),

$$\sum_{j=1}^{m} \alpha_0^{2^j-1}(t_k - t_{k-1})^{2^j-2} \leq \sum_{j=1}^{m} \alpha_0^{2^j-1}[\alpha^{2^{k-1}}/\alpha_0]^{2^j-2} = \alpha_0 \sum_{j=0}^{m-1} (\alpha^{2^k})^{2^j-1}$$

$$\leq \epsilon_k \leq (\alpha/\eta) \sum_{j=0}^{\infty} (\alpha^{2^k})^j = \alpha[\eta(1-\alpha^{2^k})]^{-1}. \quad \blacksquare$$

There are many possible technical variations of Theorem **12.4.5** and, hence, also of **12.4.6**; for two of them, we refer to **E 12.4-2–E 12.4-3**.

NOTES AND REMARKS

NR 12.4-1. Based on ideas contained in the majorant proof of Kantorovich [1949] for Newton's method (see **NR 12.6-1**), the concept of majorizing sequences was first explicitly defined by Ortega [1968] and Rheinboldt [1968]. Geometrically, the estimate

$$\| x^k - x^0 \| \leq \sum_{j=0}^{k-1} \| x^{j+1} - x^j \| \leq t_k - t_0$$

represents an upper bound for the "pathlength" $\sum_{j=0}^{k-1} \| x^{j+1} - x^j \|$ traversed by the iterates, and, hence, the convergence of $\{t_k\}$ states that this pathlength is finite as k tends to infinity.

NR 12.4-2. For a discussion of **12.4.3** and the more general results of Section **12.5**, see **NR 12.5-1**.

NR 12.4-3. Theorem **12.4.5** is essentially contained in Kantorovich and Akilov [1959]. Related results are due to Kolomy [1963], Poljak [1964b], and Zinčenko [1963a, b] (see **E 12.4-3**). Theorem **12.4.6** is a frequently cited result of Mysovskii [1950] and is also presented in Kantorovich and Akilov [1959]. Both **12.4.5** and **12.4.6**, together with their proofs, hold in an arbitrary Banach space provided that the inverses of $A(x)$ and $F'(x)$ are assumed to be bounded linear operators defined on the entire range space. For a further discussion of Newton's method, see **NR 12.6-1**.

EXERCISES

E 12.4-1. Let $\varphi: [0, \infty) \to [0, \infty)$ be continuous and isotone. Suppose that for some $t_1 > 0$ the solution of

$$t_{k+1} - t_k = \varphi(t_k - t_{k-1}), \quad t_0 = 0, \quad k = 1,...,$$

satisfies $\lim t_k = t^* < \infty$. Show that $\varphi(0) = 0$, and, in addition, that one of the following hold:

(a) If φ has a fixed point $s^* > 0$, then $t_1 < s^*$ and $\varphi(s) < s$ for all $s \in [0, s^*)$.
(b) If φ has no positive fixed point, then $\varphi(s) < s$ for all $s > 0$.

Show also that these conditions are not sufficient to imply that $\{t_k\}$ converges.

E 12.4-2. Let $F: D \subset R^n \to R^n$ be F-differentiable in $S = \bar{S}(x^0, r) \subset D$ and $\| F'(x) - F'(y)\| \leq \gamma \| x - y \|$, for all $x, y \in S$. Suppose that $B: S \to L(R^n)$ satisfies $\| B(x)\| \leq \beta$, $\| B(x) F'(x) - I \| \leq \delta$, and $\|(B(x) - B(y)) Fy \| \leq \eta \| x - y \|$, for all $x, y \in S$. Show that if $\alpha = \frac{1}{2}\beta\gamma \| B(x^0) Fx^0 \| + \eta + \delta < 1$ and $r \geq \| B(x^0) Fx^0 \|/(1 - \alpha)$, then the iterates

$$x^{k+1} = x^k - B(x^k) Fx^k, \quad k = 0, 1, \ldots,$$

remain in S and converge to the only solution x^* of $B(x) Fx = 0$ in $S(x^0, r')$, where $r' = \min\{r, (\beta\gamma)^{-1}[1 - \eta - \delta]\}$.

E 12.4-3. Let $F: D \subset R^n \to R^n$ be continuous and $K: D \subset R^n \to R^n$ be F-differentiable on $S = \bar{S}(x^0, r) \subset D$. Moreover, suppose that

$$\| K'(x) - K'(y) \| \leq \gamma \| x - y \|, \quad \|(Fx - Kx) - (Fy - Ky)\| \leq \delta \| y - x \|,$$

for all $x, y \in S$, and that $K'(x)$ is nonsingular and $\| K'(x)^{-1} \| \leq \beta$ for all $x \in S$. Show that if $\| K'(x^0)^{-1} Fx^0 \| \leq \eta$, $\alpha = \frac{1}{2}\beta\gamma\eta + \beta\delta < 1$, and $r \geq \eta/(1 - \alpha)$, then the iterates

$$x^{k+1} = x^k - K'(x^k)^{-1} Fx^k, \quad k = 0, 1, \ldots,$$

remain in S and converge to a solution of $Fx = 0$. (Zinčenko [1963a])

E 12.4-4. Under the conditions of **12.4.6**, show that

$$\| x^* - x^k \| \leq \eta \alpha^{2^k - 1} \sum_{i=0}^{\infty} (\alpha^{2^k})^i \leq \eta \alpha^{2^k - 1}/(1 - \alpha^{2^k}).$$

Conclude from this that the R-order of $\{x^k\}$ is at least two.

E 12.4-5. Consider the discrete integral equation of **E 12.1-4**. Formulate and prove a corollary of **12.4.6** for Newton's method applied to this equation. Do the same for the discrete two-point boundary value problem of **E 12.1-5**.

12.5. MORE GENERAL MAJORANTS

In place of the nonlinear estimate (12.4.4) used in the previous section, we shall now consider more general estimates involving a dependence upon the initial data. More specifically, we will suppose that an inequality of the form

$$\| G^2 x - Gx \| \leq \varphi(\| Gx - x \|, \| Gx - x^0 \|, \| x - x^0 \|) \tag{1}$$

holds, where x^0 is a given point. We will see that, for certain problems, this leads to sharper results than those possible with the estimate used in the previous section.

We begin by observing that **12.4.3** extends immediately to this more general setting:

12.5.1. Let $G: D \subset R^n \to R^n$ and $\varphi: J_1 \times J_2 \times J_3 \subset R^3 \to [0, \infty) \subset R^1$, where each J_i is an interval of the form $[0, \alpha]$, $[0, \alpha)$, or $[0, \infty)$ and φ is isotone (see **2.4.3**) in each variable. Suppose that there is a set $D_0 \subset D$ and an $x^0 \in D_0$ such that (1) holds whenever $x, Gx \in D_0$, and that with $t_0 = 0$, $t_1 \geqslant \|x^0 - Gx^0\|$ the solution of the difference equation

$$t_{k+1} - t_k = \varphi(t_k - t_{k-1}, t_k, t_{k-1}), \qquad k = 1, 2, \ldots, \qquad (2)$$

exists and converges to $t^* < +\infty$. Finally, assume either that $\bar{S}(x^0, t^*) \subset D_0$ or that $S(x^0, t^*) \subset D_0$ and $t_k < t^*$ for all $k \geqslant 0$. Then the iterates $x^{k+1} = Gx^k$, $k = 0, 1, \ldots$, are well-defined, lie in $\bar{S}(x^0, t^*)$, converge to some $x^* \in \bar{S}(x^0, t^*)$, and satisfy

$$\|x^* - x^k\| \leqslant t^* - t_k, \qquad k = 0, 1, \ldots. \qquad (3)$$

If $x^* \in D$ and G is continuous at x^*, then $x^* = Gx^*$.

Proof. As in the proof of **12.4.3**, we show that $\{x^k\}$ is majorized by $\{t_k\}$. Assume that $x^j \in \bar{S}(x^0, t^*) \subset D_0$ and $\|x^j - x^{j-1}\| \leqslant t_j - t_{j-1}$ for $j = 1, 2, \ldots, k$ and some $k \geqslant 1$. This is certainly satisfied for $k = 1$, since $\|x^1 - x^0\| \leqslant t_1 \leqslant t^*$. Evidently, x^{k+1} is well defined, and, using

$$\|x^j - x^0\| \leqslant \sum_{i=1}^{j} \|x^i - x^{i-1}\| \leqslant \sum_{i=1}^{j}(t_i - t_{i-1}) = t_j, \qquad 1 \leqslant j \leqslant k, \qquad (4)$$

and the isotonicity of φ, we have

$$\|x^{k+1} - x^k\| = \|G^2 x^{k-1} - Gx^{k-1}\| \leqslant \varphi(\|x^k - x^{k-1}\|, \|x^k - x^0\|, \|x^{k-1} - x^0\|)$$
$$\leqslant \varphi(t_k - t_{k-1}, t_k, t_{k-1}) = t_{k+1} - t_k.$$

Hence, by (4), $\|x^{k+1} - x^0\| \leqslant t_{k+1} \leqslant t^*$, and the induction step is proved. The proof is analogous if only $S(x^0, t^*) \subset D_0$ but $t_k < t^*$ for all k. The result now follows from **12.4.2**. ∎

The analysis of the difference equation (2) is considerably simplified in the special case when there exists a "first integral." By this, we mean that there exists a mapping $\psi: J \subset R^1 \to R^1$ such that if $\{t_k\}$ satisfies

$$t_{k+1} = \psi(t_k), \qquad t_0 = 0, \qquad k = 0, 1, \ldots, \qquad (5)$$

then $\{t_k\}$ also satisfies (2). A sufficient condition for this is given by the following lemma, whose proof is immediate.

12.5.2. Let $\varphi: J_1 \times J_2 \times J_3 \subset R^3 \to R^1$ and $\psi: J \subset R^1 \to R^1$, where the

12.5. MORE GENERAL MAJORANTS

intervals J_i, J are as defined in **12.5.1**. Suppose that $J \subset J_1 \cap J_2 \cap J_3$ and

$$\psi(s) - \psi(t) = \varphi(s - t, s, t), \qquad \forall s, t \in J, \quad s \geq t. \tag{6}$$

If the sequence $\{t_k\}$ generated by (5) remains in J, then $\{t_k\}$ satisfies the difference equation (2) with initial conditions $t_0 = 0$, $t_1 = \psi(0)$.

We consider next a uniqueness theorem based on the condition (6). First, we quote the following result which is geometrically evident and will be proved in a more general context in Chapter 13. (See **13.2.2**.)

12.5.3. Kantorovich Lemma. Let $\psi: [t_0, s_0] \subset R^1 \to R^1$ be isotone, and $t_0 \leq \psi(t_0)$, $s_0 \geq \psi(s_0)$. Then the sequences $t_{k+1} = \psi(t_k)$, $s_{k+1} = \psi(s_k)$, $k = 0, 1, \ldots$, are monotonically increasing and decreasing, respectively, and

$$\lim_{k \to \infty} t_k \equiv t^* \leq s^* \equiv \lim_{k \to \infty} s_k.$$

Moreover, if ψ is continuous on $[t_0, s_0]$, then t^* and s^* are, respectively, the smallest and largest fixed points of ψ in $[t_0, s_0]$.

12.5.4. Suppose that the conditions of **12.5.1** and **12.5.2** hold except that, instead of (1),

$$\| Gx - Gy \| \leq \varphi(\| x - y \|, \| x - x^0 \|, \| y - x^0 \|), \qquad \forall x, y \in D_0. \tag{7}$$

Moreover, let $\psi(0) = \| Gx^0 - x^0 \|$, and suppose that $t^* = \lim_{k \to \infty} t_k = \psi(t^*) \in J$. Then x^* is the only possible fixed point of G in $\bar{S}(x^0, t^*)$.

In addition, if, instead of (7),

$$\| Gx - Gy \| \leq \varphi(\| x - y \|, \| y - x^0 \|), \qquad \forall x, y \in D_0, \tag{8}$$

and if ψ is continuous and has a fixed point $t^{**} > t^*$, $t^{**} \in J$, such that $\psi(t) < t$ for $t^* < t < t^{**}$, then x^* is the only possible fixed point of G in $D_0 \cap S(x^0, t^{**})$.

Proof. Suppose that (7) holds and that $y^* = Gy^* \in \bar{S}(x^0, t^*)$. Assume that $\| y^* - x^j \| \leq t^* - t_j$ for $j = 0, 1, \ldots, k$ and some $k \geq 0$. For $k = 0$ this is clearly correct, since $\| y^* - x^0 \| \leq t^* = t^* - t_0$. Since $\{t_k\}$ majorizes $\{x^k\}$, we have, by (6) and (7),

$$\| y^* - x^{k+1} \| = \| Gy^* - Gx^k \| \leq \varphi(\| y^* - x^k \|, \| y^* - x^0 \|, \| x^k - x^0 \|)$$
$$\leq \varphi(t^* - t_k, t^*, t_k) = t^* - t_{k+1}.$$

This completes the induction step, and, since $\lim_{k \to \infty} t_k = t^*$, it follows that $x^* = \lim_{k \to \infty} x^k = y^*$.

Now consider the second case. If $y^* = Gy^* \in D_0 \cap S(x^0, t^{**})$, then—in view of the first part of the proof—it suffices to assume that $s_0 = \|y^* - x^0\| \in (t^*, t^{**})$. By induction we see that $\|y^* - x^k\| \leq s_k - t_k$ where $s_{k+1} = \psi(s_k)$; in fact, by (8),

$$\|y^* - x^{k+1}\| \leq \varphi(\|y^* - x^k\|, \|x^k - x^0\|) \leq \varphi(s_k - t_k, t_k)$$
$$= \psi(s_k) - \psi(t_k) = s_{k+1} - t_{k+1}.$$

Now, clearly, (6) and the nonnegativity of φ imply that ψ is isotone. Also, by assumption, $t^* = \psi(t^*)$ and $\psi(s_0) < s_0$, since $s_0 \in (t^*, t^{**})$. Hence, 12.5.3 implies that $\lim_{k \to \infty} s_k = t^*$ so that, again, $x^* = \lim_{k \to \infty} x^k = y^*$. ∎

As a first application of these theorems we prove now:

12.5.5. Assume that $G: D \subset R^n \to R^n$ is F-differentiable on a convex set $D_0 \subset D$ and that

$$\|G'(y) - G'(x)\| \leq \gamma \|y - x\|, \quad \forall x, y \in D_0.$$

Suppose that there is an $x^0 \in D_0$ such that $\|G'(x^0)\| \leq \delta < 1$ and $\alpha = \gamma\eta/(1-\delta)^2 \leq \tfrac{1}{2}$, where $\eta = \|x^0 - Gx^0\|$. Set

$$t^* = [(1-\delta)/\gamma][1 - (1-2\alpha)^{1/2}], \qquad t^{**} = [(1-\delta)/\gamma][1 + (1-2\alpha)^{1/2}]$$

and assume that $\bar{S}(x^0, t^*) \subset D_0$. Then the iterates $x^{k+1} = Gx^k$, $k = 0, 1,...$, remain in $\bar{S}(x^0, t^*)$ and converge to the only fixed point x^* of G in $D_0 \cap S(x^0, t^{**})$. Moreover, the error extimate (3) holds where the sequence $\{t_k\}$ is generated by

$$t_{k+1} = \tfrac{1}{2}\gamma t_k^2 + \delta t_k + \eta, \qquad t_0 = 0, \qquad k = 0, 1,.... \qquad (9)$$

Proof. For all $x, y \in D_0$, we have by the mean-value theorem 3.2.12

$$\|Gx - Gy\| \leq \|Gx - Gy - G'(y)(x-y)\|$$
$$+ \|(G'(y) - G'(x^0))(x-y)\| + \|G'(x^0)(x-y)\|$$
$$\leq \tfrac{1}{2}\gamma \|x - y\|^2 + \gamma \|y - x^0\| \|x - y\| + \delta \|x - y\|$$
$$\equiv \varphi(\|x - y\|, \|y - x^0\|),$$

where $\varphi(s, t) = \tfrac{1}{2}\gamma s^2 + \gamma st + \delta s$. It is easy to verify that $\varphi(s - t, t) = \psi(s) - \psi(t)$, where $\psi(t) = \tfrac{1}{2}\gamma t^2 + \delta t + \eta$, that ψ is continuous and isotone on $[0, \infty)$, has the two fixed points t^* and t^{**}, and that $\psi(t) < t$ for $t \in (t^*, t^{**})$ unless $\alpha = \tfrac{1}{2}$ in which case $t^* = t^{**}$. From 12.5.2 and

12.5. MORE GENERAL MAJORANTS

12.5.3 it follows that the sequence generated by (9) satisfies $t_{k+1} - t_k = \varphi(t_k - t_{k-1}, t_{k-1})$ and converges to t^*. Hence, **12.5.1** applies and shows that the sequence $\{x^k\}$ remains in $\bar{S}(x^0, t^*)$, converges to x^*, and satisfies (3). Moreover, G is continuous on $\bar{S}(x^0, t^*)$, and, hence, $x^* = Gx^*$. Finally, the uniqueness statement is a direct consequence of **12.5.4**. ∎

This theorem has immediate application to chord methods of the form

$$x^{k+1} = x^k - A^{-1}Fx^k, \quad k = 0, 1, \ldots .$$

(See **E 12.5-1**). In particular, in the next section, we shall use a result of this type for the simplified Newton method in which case $A = F'(x^0)$.

Note that the error estimate (3) is computable in the case of **12.5.5** since t^* is known explicitly. This is in contrast to the situation in **12.4.5** or **12.4.6**, where t^* is not known. Note also that the use of a nonlinear majorant function produces in **12.5.5** only a slightly better result than the contraction theorem **12.1.2**. In fact, under the hypotheses of **12.5.5**, $x \in \bar{S}(x^0, t^*)$ implies that

$$\| Gx - x^0 \| \leqslant \| Gx - Gx^0 - G'(x^0)(x - x^0) \| + \| Gx^0 - x^0 \| + \| G'(x^0)(x - x^0) \|$$

$$\leqslant \tfrac{1}{2}\gamma t^{*2} + \delta t^* + \eta = \psi(t^*) = t^*,$$

so that G maps $\bar{S}(x^0, t^*)$ into itself. Moreover,

$$\| G'(x) \| \leqslant \| G'(x) - G'(x^0) \| + \| G'(x^0) \| \leqslant \gamma \| x - x^0 \| + \delta,$$

so that for $x, y \in \bar{S}(x^0, t^*)$

$$\| Gx - Gy \| \leqslant (\gamma t^* + \delta) \| x - y \|. \tag{10}$$

Hence, if $\alpha < \tfrac{1}{2}$, then $\gamma t^* + \delta < 1$, and G is a contraction on $\bar{S}(x^0, t^*)$. Moreover, (10) shows that G is a strict nonexpansion in $S(x^0, [1 - \delta]/\gamma)$, and thus we have uniqueness in $S(x^0, [1 - \delta]/\gamma) \cap D_0$. Since $(1 - \delta)/\gamma = \tfrac{1}{2}(t^* + t^{**})$, **12.5.5** provides in this case a sharper uniqueness result. On the other hand, if $\alpha = \tfrac{1}{2}$, then $t^* = t^{**}$ and $\gamma t^* + \delta = 1$, so that in this case the contraction theorem does not apply; however, (10) now gives uniqueness in $S(x^0, t^{**}) = S(x^0, t^*)$.

NOTES AND REMARKS

NR 12.5-1. The basic majorant theorems **12.5.1** and **12.4.3** are due to Rheinboldt [1968], and, in fact, the presentation of Sections **12.4–12.6** follows this article. There are essentially three concepts entering into the analysis. The first is the idea of "majorization" due to Kantorovich [1949]; that is, the comparison of a higher-dimensional iterative process with another such process in one dimension such that the convergence of the process in R^1 implies that of the original iteration. (In this connection, see also **NR 12.6-1**.) The second concept is the use of nonlinear functions on the right-hand side of (12.4.4) and (1). Such nonlinear estimates were first used by Schröder [1956b, 1957] (see also the presentation of these results in Collatz [1964]), but in all of Schröder's estimates, the comparison process was ultimately assumed to be of the form $s_{k+1} = \psi(s_k)$, $k = 0, 1,\ldots$. The more general form of the estimates leading to the nonlinear difference equation (2) is due to Rheinboldt [1968]. Finally, the third concept is that of using the iterated difference $G^2x - Gx$ rather than $Gy - Gx$. As mentioned in **NR 12.3-1**, such iterated differences were used by Goldstein [1967] in connection with nonexpansions. Their importance in connection with the convergence of Newton-type processes was recognized by Ortega [1968] and Rheinboldt [1968].

NR 12.5-2. The uniqueness theorem **12.5.4** is due to Rheinboldt [1968], and generalizes a result of Schröder [1956b]. Theorem **12.5.5** is also due to Rheinboldt [1968].

EXERCISES

E 12.5-1. Let $F: D \subset R^n \to R^n$ be F-differentiable on $S = \bar{S}(x^0, r) \subset D$ and $\|F'(x) - F'(y)\| \leq \gamma \|x - y\|$, for all $x, y \in S$. Suppose that $A \in L(R^n)$ is nonsingular and $\|I - A^{-1}F'(x^0)\| \leq \delta < 1$. Show that if $\|A^{-1}Fx^0\| \leq \eta$ and $\alpha = \gamma\eta \|A^{-1}\|/(1 - \delta)^2 \leq \frac{1}{2}$ as well as $r \geq (1 - \delta)[1 - (1 - 2\alpha)^{1/2}]/(\gamma \|A^{-1}\|)$, then the iterates

$$x^{k+1} = x^k - A^{-1}Fx^k, \qquad k = 0, 1, \ldots,$$

remain in S and converge to the only solution of $Fx = 0$ in $S(x^0, r')$, where $r' = \min\{r, (1 - \delta)[1 + (1 - 2\alpha)^{1/2}]/(\gamma \|A^{-1}\|)\}$.

E 12.5-2. Let $F: D \subset R^n \to R^n$ be F-differentiable on the Euclidean ball $S = \bar{S}(x^0, r) \subset D$ and satisfy $\|F'(x) - F'(y)\|_2 \leq \gamma \|x - y\|_2$ for all $x, y \in S$. Suppose that

$$dh^T h \geq h^T F'(x^0) h \geq ch^T h, \qquad \forall h \in R^n,$$

with $d \geq c > 0$. Show that if $\alpha = c^2\gamma \|Fx^0\|_2 \leq \frac{1}{2}$ and $r \geq c[1 - (1 - 2\alpha)^{1/2}]/\gamma$, then the sequence

$$x^{k+1} = x^k - [2/(c + d)]Fx^k, \qquad k = 0, 1, \ldots,$$

remains in S and converges to the only solution of $Fx = 0$ in $S(x^0, r')$, where $r' = \min\{r, d[1 + (1 - 2\alpha)^{1/2}]/\gamma\}$. (Linkov [1964a])

12.6. NEWTON'S METHOD AND RELATED ITERATIONS

In this section we consider the application of the previous theory to the important special case of Newton's method, and to some related generalizations. We begin with a corollary of **12.5.5** for the simplified Newton method

$$x^{k+1} = x^k - F'(x^0)^{-1} Fx^k, \qquad k = 0, 1, \ldots . \tag{1}$$

12.6.1. Assume that $F: D \subset R^n \to R^n$ is F-differentiable on a convex set $D_0 \subset D$ and that

$$\|F'(x) - F'(y)\| \leq \gamma \|x - y\|, \qquad \forall x, y \in D_0. \tag{2}$$

Suppose that there exists an $x^0 \in D_0$ such that $\|F'(x^0)^{-1}\| \leq \beta$ and $\alpha = \beta\gamma\eta \leq \frac{1}{2}$, where $\eta \geq \|F'(x^0)^{-1} Fx^0\|$. Set

$$t^* = (\beta\gamma)^{-1}[1 - (1 - 2\alpha)^{1/2}], \qquad t^{**} = (\beta\gamma)^{-1}[1 + (1 - 2\alpha)^{1/2}], \tag{3}$$

and assume that $\bar{S}(x^0, t^*) \subset D_0$. Then the iterates (1) are well-defined, remain in $\bar{S}(x^0, t^*)$, and converge to a solution x^* of $Fx = 0$ which is unique in $S(x^0, t^{**}) \cap D_0$.

Proof. Define $G: D_0 \subset R^n \to R^n$, $Gx = x - F'(x^0)^{-1} Fx$; then $G'(x) = I - F'(x^0)^{-1} F'(x)$, so that

$$\|G'(x) - G'(y)\| = \|F'(x^0)^{-1}[F'(x) - F'(y)]\| \leq \beta\gamma \|x - y\|, \qquad \forall x, y \in D_0,$$

and $G'(x^0) = 0$. The result now follows directly from **12.5.5**. ∎

It turns out that the hypotheses of this theorem are also sufficient to prove the convergence of Newton's method itself.

12.6.2. Newton–Kantorovich Theorem Assume that the conditions of **12.6.1** hold. Then the Newton iterates

$$x^{k+1} = x^k - F'(x^k)^{-1} Fx^k, \qquad k = 0, 1, \ldots, \tag{4}$$

are well-defined, remain in $\bar{S}(x^0, t^*)$, and converge to a solution x^* of $Fx = 0$ which is unique in $S(x^0, t^{**}) \cap D_0$. Moreover, the error estimate

$$\|x^* - x^k\| \leq (\beta\gamma 2^k)^{-1} (2\alpha)^{2^k}, \qquad k = 0, 1, \ldots, \tag{5}$$

holds.

Proof. Set $D_1 = S(x^0, (\beta\gamma)^{-1}) \cap D_0$; then for $x \in D_1$ we have

$$\|F'(x) - F'(x^0)\| \leq \gamma \|x - x^0\| < 1/\beta.$$

Thus, the perturbation lemma **2.3.2** shows that $F'(x)$ is nonsingular for all $x \in D_1$, and that

$$\|F'(x)^{-1}\| \leq \beta/(1 - \beta\gamma \|x - x^0\|), \qquad \forall x \in D_1. \tag{6}$$

In particular, if $\alpha < \tfrac{1}{2}$, then $t^* < (\beta\gamma)^{-1}$, so that $Gx = x - F'(x)^{-1} Fx$ is defined on $\bar{S}(x^0, t^*)$; if $\alpha = \tfrac{1}{2}$, then $t^* = (\beta\gamma)^{-1}$ and G is defined on $S(x^0, t^*)$. In either case, if x, $Gx \in S(x^0, t^*)$, then it follows with the aid of **3.2.12** and (6) that

$$\|G^2x - Gx\| = \|F'(Gx)^{-1}F(Gx)\| = \|F'(Gx)^{-1}[F(Gx) - Fx - F'(x)(Gx - x)]\|$$

$$\leq \tfrac{1}{2}\beta\gamma \|Gx - x\|^2 /(1 - \beta\gamma \|Gx - x^0\|)$$

$$= \varphi(\|Gx - x\|, \|Gx - x^0\|)$$

where

$$\varphi(s, t) = \tfrac{1}{2}\beta\gamma s^2 / (1 - \beta\gamma t).$$

In order to apply **12.5.1**, we therefore have to consider the difference equation

$$t_{k+1} - t_k = \varphi(t_k - t_{k-1}, t_k), \qquad t_0 = 0, \qquad t_1 = \eta, \qquad k = 1, 2, \ldots. \tag{7}$$

We show now that (7) has a "first integral"

$$t_{k+1} = \psi(t_k), \qquad t_0 = 0, \qquad k = 0, 1, \ldots, \tag{8}$$

where

$$\psi(t) = (\tfrac{1}{2}\beta\gamma t^2 - \eta)/(\beta\gamma t - 1).$$

In fact, multiplying (8) by $(1 - \beta\gamma t_k)$ we see after a simple manipulation that

$$\tfrac{1}{2}\beta\gamma(t_{k+1} - t_k)^2 = \tfrac{1}{2}\beta\gamma t_{k+1}^2 - t_{k+1} + \eta, \qquad k = 0, 1, \ldots,$$

and, hence, that

$$t_{k+1} - t_k = \frac{\tfrac{1}{2}\beta\gamma t_k^2 - t_k + \eta}{1 - \beta\gamma t_k} = \frac{\tfrac{1}{2}\beta\gamma(t_k - t_{k-1})^2}{1 - \beta\gamma t_k} = \varphi(t_k - t_{k-1}, t_k).$$

Clearly, (8) also implies that $t_1 = \eta$.

Now, suppose that $\alpha < \tfrac{1}{2}$; then ψ is isotone on $[0, t^*] \subset [0, (\beta\gamma)^{-1})$ and t^* is the smallest fixed point of ψ. Hence, **12.5.3** and the continuity of ψ

12.6. NEWTON'S METHOD AND RELATED ITERATIONS

show that $\lim_{k\to\infty} t_k = t^*$. Therefore, **12.5.1** now ensures that all x^k are well-defined, remain in $\bar{S}(x^0, t^*)$, and converge to x^*. Moreover, since $\bar{S}(x^0, t^*) \subset S(x^0, (\beta\gamma)^{-1})$, G is continuous at x^* so that $x^* = Gx^*$. This, in turn, implies that $F'(x^*)^{-1} Fx^* = 0$, and thus $Fx^* = 0$.

If $\alpha = \frac{1}{2}$, then $t^{**} = t^* = (\beta\gamma)^{-1}$ and $\psi(t) = \frac{1}{2}t + \eta$. Clearly, $t_k < t^*$ for all $k \geq 0$, so that **12.5.1** again ensures that $\lim_{k\to\infty} x^k = x^* \in \bar{S}(x^0, t^*)$. To show that x^* solves $Fx = 0$, note that

$$\|Fx^k\| = \|F'(x^k)(x^{k+1} - x^k)\|$$
$$\leq \|[F'(x^k) - F'(x^0)](x^{k+1} - x^k)\| + \|F'(x^0)(x^{k+1} - x^k)\|$$
$$\leq [\gamma t^* + \|F'(x^0)\|] \|x^{k+1} - x^k\|;$$

hence, by the continuity of F, we have $Fx^* = \lim_{k\to\infty} Fx^k = 0$. The uniqueness statement has already been shown in **12.6.1**.

In order to obtain the error estimate (5), we show first that

$$t_{k+1} - t_k \leq \eta 2^{-k}, \quad k = 0, 1, \ldots, \tag{9}$$

which immediately implies that

$$t_{k+1} = \sum_{j=0}^{k} (t_{j+1} - t_j) \leq \eta \sum_{j=0}^{k} 2^{-j} = 2[1 - (1/2^{k+1})]\eta, \tag{10}$$

and, consequently, that

$$1/(1 - \beta\gamma t_{k+1}) \leq 1/[1 - 2\alpha(1 - 2^{-k-1})] \leq 2^{k+1}.$$

Clearly, (9) holds for $k = 0$, and the general induction step follows from

$$t_{k+2} - t_{k+1} = \frac{\frac{1}{2}\beta\gamma(t_{k+1} - t_k)^2}{1 - \beta\gamma t_{k+1}} \leq \frac{1}{2}\beta\gamma\eta^2 2^{-2k} 2^{k+1} \leq \frac{1}{2^{k+1}}\eta.$$

Now, we can show that

$$t^* - t_k \leq (\beta\gamma 2^k)^{-1} (2\alpha)^{2^k}, \quad k = 0, 1, \ldots.$$

For $k = 0$ this follows from $1 - (1 - 2\alpha)^{1/2} \leq 2\alpha$, and the general induction step is a consequence of

$$t^* - t_{k+1} = \frac{t^* - \beta\gamma t^* t_k + \frac{1}{2}\beta\gamma t_k^2 - \eta}{1 - \beta\gamma t_k} = \frac{\frac{1}{2}\beta\gamma(t^* - t_k)^2}{1 - \beta\gamma t_k} \leq 2^{k-1}\beta\gamma(t^* - t_k)^2$$

$$\leq \frac{2^{k-1}\beta\gamma(2\alpha)^{2^{k+1}}}{\beta^2\gamma^2 2^{2k}} = \frac{1}{\beta\gamma 2^{k+1}}(2\alpha)^{2^{k+1}}.$$

Hence, (5) holds. ∎

Note that the function ψ is simply the Newton iteration function for the polynomial $p(t) = \frac{1}{2}\beta\gamma t^2 - t + \eta$ with roots t^* and t^{**}. Hence, the sequence $\{x^k\}$ is majorized by the Newton sequence $\{t_k\}$ for p.

Note also that an error estimate of the form

$$\|x^* - x^k\| \leqslant c\|x^k - x^{k-1}\|^2, \quad k \geqslant 1,$$

is possible if $\alpha < \frac{1}{2}$. In fact, in that case the proof shows that $\|F'(x^k)^{-1}\| \leqslant \beta/(1 - \beta\gamma t^*)$ for all $x \in \bar{S}(x^0, t^*)$. Hence, as soon as

$$[\beta\gamma/(1 - \beta\gamma t^*)]\|Fx^k\| < 2,$$

the estimate (12.4.11) of **12.4.6** applies.

Theorem **12.6.2** can be extended to Newton-related processes of the form

$$x^{k+1} = x^k - A(x^k)^{-1}Fx^k, \quad k = 0, 1, \ldots. \qquad (11)$$

Under the conditions of **12.6.4**, the difference equation (7) will take the form

$$t_{k+1} - t_k = [1/(1 - p_4 t_k)][p_1(t_k - t_{k-1}) + (p_2 + p_3 t_{k-1})](t_k - t_{k-1}),$$
$$k = 1, 2, \ldots, \qquad (12)$$

and we first prove the following lemma which gives sufficient conditions that the solution of (12) converges.

12.6.3. Assume that $p_i \geqslant 0, i = 1, \ldots, 4, p_1 > 0, p_2 < 1, p_3 + p_4 = 2p_1$, and $0 \leqslant \eta \leqslant (1 - p_2)^2(4p_1)^{-1}$. Then the sequence $\{t_k\}$ of (12) with initial values $t_0 = 0$ and $t_1 = \eta$ is strictly increasing unless $\eta = 0$, and

$$\lim_{k \to \infty} t_k = t^* = (2p_1)^{-1}\{(1 - p_2) - [(1 - p_2)^2 - 4p_1\eta]^{1/2}\}.$$

Proof. Set

$$u(t) = p_1 t^2 - (1 - p_2)t + \eta, \quad v(t) = 1 - p_4 t, \quad \psi(t) = t + u(t)/v(t).$$

Then, if $\{t_k\} \subset [0, 1/p_4)$ satisfies

$$t_{k+1} = \psi(t_k), \quad t_0 = 0, \quad k = 0, 1, \ldots, \qquad (13)$$

it follows that

$$t_{k+1} - t_k = [1/v(t_k)]\{u(t_k) - u(t_{k-1}) - u'(t_{k-1})(t_k - t_{k-1})$$
$$+ [u'(t_{k-1}) + v(t_{k-1})](t_k - t_{k-1})\}, \qquad (14)$$

12.6. NEWTON'S METHOD AND RELATED ITERATIONS

and, hence, that (13) is a "first integral" of (12). Now, $v(t) > 0$ for $0 \leqslant t < t^*$, since either $p_4 = 0$ and $v(t) = 1$, or $p_4 > 0$ and

$$v(t) = 1 - p_4 t > 1 - p_4 t^* = (2p_1)^{-1}\{p_2 p_4 + p_3 + p_4[(1-p_2)^2 - 4p_1\eta]^{1/2}\} \geqslant 0.$$

Moreover, $v(t^*) > 0$ unless $p_2 = p_3 = 0$ and $\eta = (4p_1)^{-1}$; in this case, $t^* = (2p_1)^{-1}$, and, by L'Hospital's rule, $u(t^*)/v(t^*) = 0$. Since t^* is the smallest root of $u(t) = 0$, it therefore follows in all cases that t^* is the smallest fixed point of ψ and that $\psi(t) > t$ for $0 \leqslant t < t^*$. Furthermore, a computation analogous to (14) shows that

$$t^* - \psi(t) = [1/v(t)][p_1(t^* - t)^2 + (p_2 + p_3 t)(t^* - t)] > 0, \qquad 0 \leqslant t < t^*.$$

If $0 < t_1 = \eta < t^*$, we obtain by induction that $t_k < t_{k+1} < t^*$ for all $k \geqslant 0$. Hence, because ψ is continuous and has no fixed point in $[0, t^*)$, it follows that $\lim_{k\to\infty} t_k = t^*$. ∎

Using this lemma, we can now prove the following extension of the Newton–Kantorovich theorem 12.6.2.

12.6.4. Suppose that $F: D \subset R^n \to R^n$ is F-differentiable on the convex set $D_0 \subset D$ and

$$\|F'(x) - F'(y)\| \leqslant \gamma \|x - y\|, \qquad x, y \in D_0.$$

Let $A: D_0 \subset R^n \to L(R^n)$ and $x^0 \in D_0$ be such that with $\delta_0, \delta_1 \geqslant 0$

$$\left.\begin{array}{l} \|A(x) - A(x^0)\| \leqslant \mu \|x - x^0\|, \\ \|F'(x) - A(x)\| \leqslant \delta_0 + \delta_1 \|x - x^0\|, \end{array}\right\} \forall x \in D_0.$$

Assume, moreover, that $A(x^0)$ is nonsingular and $\|A(x^0)^{-1} Fx^0\| \leqslant \eta$, $\|A(x^0)^{-1}\| \leqslant \beta$, and that $\beta\delta_0 < 1$ and $\alpha = \sigma\beta\gamma\eta/(1 - \beta\delta_0)^2 \leqslant \frac{1}{2}$ where $\sigma = \max(1, [\mu + \delta_1]/\gamma)$. Set

$$t^* = \frac{1 - (1 - 2\alpha)^{1/2}}{\alpha} \frac{\eta}{1 - \beta\delta_0}, \qquad t^{**} = \frac{1 + (1 - 2\alpha/\sigma)^{1/2}}{\alpha} \frac{\sigma\eta}{1 - \beta\delta_0}. \qquad (15)$$

If $\bar{S}(x^0, t^*) \subset D_0$, then the sequence $\{x^k\}$ defined by

$$x^{k+1} = x^k - A(x^k)^{-1} Fx^k, \qquad k = 0, 1, \dots, \qquad (16)$$

remains in $S(x^0, t^*)$ and converges to a solution x^* of $Fx = 0$ which is unique in $D_0 \cap S(x^0, t^{**})$.

Proof. For $x \in S(x^0, t^*)$, we have

$$\| A(x) - A(x^0) \| \leq \mu \| x - x^0 \| < \mu t^* \leq \sigma \gamma t^* \leq (1 - \beta \delta_0)/\beta \leq 1/\beta.$$

Hence, by the perturbation lemma **2.3.2**, $A(x)$ is nonsingular and

$$\| A(x)^{-1} \| \leq \beta/(1 - \beta \mu \| x - x^0 \|), \qquad \forall x \in S(x^0, t^*).$$

Therefore, $Gx = x - A(x)^{-1} Fx$ is defined for $x \in S(x^0, t^*)$, and if $x, Gx \in S(x^0, t^*)$, then

$$\| G^2 x - Gx \| = \| -A(Gx)^{-1} F(Gx) \|$$

$$\leq \frac{\beta}{1 - \beta\mu \| Gx - x^0 \|} [\| F(Gx) - Fx - F'(x)(Gx - x) \|$$

$$+ \|(F'(x) - A(x))(Gx - x)\|]$$

$$\leq \frac{\beta}{1 - \beta\mu \| Gx - x^0 \|} [\tfrac{1}{2} \gamma \| Gx - x \| + \delta_0$$

$$+ \delta_1 \| x - x^0 \|] \| Gx - x \|.$$

Therefore, since $\beta\gamma \leq \beta\gamma\sigma$ and $\beta\delta_1 \leq \beta\gamma\sigma - \beta\mu$, we have, whenever $x, Gx \in S(x^0, t^*)$,

$$\| G^2 x - Gx \| \leq \varphi(\| Gx - x \|, \| Gx - x^0 \|, \| x - x^0 \|),$$

where

$$\varphi(u, v, w) = [1/(1 - \beta\mu v)][\tfrac{1}{2}\beta\gamma\sigma u + \beta\delta_0 + \beta(\sigma\gamma - \mu)w]u.$$

The case $\eta = 0$ may be excluded since otherwise $Fx^0 = 0$. Therefore, **12.6.3** shows that the solution of the difference equation

$$t_{k+1} - t_k = \varphi(t_k - t_{k-1}, t_k, t_{k-1}), \qquad k = 0, 1, \ldots, \quad t_0 = 0, \quad t_1 = \eta,$$

satisfies $\lim_{k \to \infty} t_k = t^*$ and $t_k < t^*$, $k \geq 0$, where t^* is given by (15). Now, **12.5.1** ensures that $\{x^k\} \subset S(x^*, t^*)$ and the convergence of the sequence (16) to $x^* \in \bar{S}(x^0, t^*)$. Finally, from

$$\| Fx^k \| \leq \|[A(x^k) - A(x^0)](x^{k+1} - x^k)\| + \| A(x^0)(x^{k+1} - x^k) \|$$

$$\leq (\mu t^* + \| A(x^0)\|)\| x^{k+1} - x^k \|,$$

it follows that $Fx^* = 0$. The uniqueness is a consequence of **12.5.5** applied to the simplified process $x^{k+1} = x^k - A(x^0)^{-1} Fx^k, k = 0, 1, \ldots$. ∎

In the special case $A(x) = F'(x)$, we have $\mu = \gamma$, $\delta_0 = \delta_1 = 0$, and $\sigma = 1$, so that **12.6.4** reduces to the Newton–Kantorovich theorem **12.6.2** except for the rate of convergence estimate (5).

12.6. NEWTON'S METHOD AND RELATED ITERATIONS

As an application of this result, we consider the one-step Newton–SOR process

$$x^{k+1} = x^k - \omega[D(x^k) - \omega L(x^k)]^{-1} F x^k, \qquad k = 0, 1, \ldots, \tag{17}$$

discussed in Section 7.4. Here, as usual, $D(x)$, $-L(x)$, and $-U(x)$ are the diagonal, strictly lower-, and strictly upper-triangular parts of $F'(x)$, respectively.

12.6.5. Assume that $F: D \subset R^n \to R^n$ is F-differentiable and satisfies on a convex set $D_0 \subset D$

$$\|F'(x) - F'(y)\|_1 \leq \gamma \|x - y\|_1, \qquad \forall x, y \in D_0,$$

where we use the l_1-norm. Suppose that for some $x^0 \in D_0$ the following estimates hold:

$$\|[D(x^0) - \omega L(x^0)]^{-1}\|_1 \leq \beta/\omega, \qquad \|U(x^0)\|_1 \leq \delta, \qquad \|D(x^0)\|_1 \leq \delta,$$

$$\|[D(x^0) - \omega L(x^0)]^{-1} F x^0\|_1 \leq \eta/\omega,$$

where, with $\tau = 1 + |1 - \omega^{-1}|$,

$$\theta = \beta \tau \delta < 1, \qquad \alpha = 2\tau \beta \gamma \eta / (1 - \theta)^2 \leq \tfrac{1}{2}.$$

Set

$$t^* = \frac{(1-\theta)}{2\tau \beta \gamma} [1 - (1 - 2\alpha)^{1/2}], \qquad t^{**} = \frac{(1-\theta)}{\beta \gamma} \left\{ 1 + \left[1 - \frac{\beta \gamma}{(1-\theta)^2} \right]^{1/2} \right\},$$

and assume that $\bar{S}(x^0, t^*) \subset D_0$. Then the Newton–SOR iterates (17) are well defined and converge to a solution $x^* \in \bar{S}(x^0, t^*)$ of $Fx = 0$ which is unique in $D_0 \cap S(x^0, t^{**})$.

Proof. We note, first, that because of the choice of norm $D(x)$, $D(x) - L(x)$, and $U(x)$ are all Lipschitz continuous on D_0 with the same constant γ as F'. Set

$$A(x) = \omega^{-1} D(x) - L(x) = D(x) - L(x) + (\omega^{-1} - 1) D(x).$$

Then

$$\|A(x) - A(x^0)\|_1 \leq \|[D(x) - L(x)] - [D(x^0) - L(x^0)]\|_1$$
$$+ |\omega^{-1} - 1| \|D(x) - D(x^0)\|_1 \leq \tau \gamma \|x - x^0\|_1,$$

and

$$\|F'(x) - A(x)\|_1 = \|(\omega^{-1} - 1) D(x) + U(x)\|_1 \leq |1 - \omega^{-1}| \|D(x)\|_1 + \|U(x)\|_1$$
$$\leq |\omega^{-1} - 1| [\|D(x^0)\|_1 + \|D(x) - D(x^0)\|_1]$$
$$+ \|U(x^0)\|_1 + \|U(x) - U(x^0)\|_1$$
$$\leq \tau \delta + \tau \gamma \|x - x^0\|_1.$$

Hence, we can apply **12.6.4** with $\mu = \tau\gamma$, $\delta_0 = \tau\delta$, $\delta_1 = \tau\gamma$, $\sigma = 2\tau$, and the result follows immediately. ∎

We note that **12.6.5** holds as stated also in the l_∞-norm, or, more generally, in any norm with the property that $\|C\| \leq \|B\|$ if C is a submatrix of B.

NOTES AND REMARKS

NR 12.6-1. The convergence analysis of Newton's method has a long history. Point-of-attraction results dating back to the last century have already been discussed in **NR 10.2-1**, but Fine [1916] appears to have been the first to prove the convergence of Newton's method in n dimensions without assuming the existence of a solution. His conditions are similar to those of the Mysovskii theorem **12.4.6**, and, in particular, the derivative $F'(x)$ is assumed to be invertible on some suitable ball S. In the same year, Bennett [1916] formulated related results for operators on infinite dimensional spaces but for the proofs he referred to Fine. The article of Fine appears to have been overlooked, and twenty years later Ostrowski [1936] presented, independently, new convergence theorems and also discussed error estimates. Concurrently, Willers [1938] also proved similar convergence theorems. Although both these authors observed that their results extend immediately to the case of a general n, they themselves presented them only for $n = 2$ and $n = 3$, respectively. Bussmann [1940], in an unpublished dissertation, proved these results and some extensions for general n; Bussmann's theorems are quoted by Rehbock [1942]. Then Kantorovich [1948a, b] gave his now-famous convergence results for Newton's method in Banach spaces. The main result is **12.6.2**, and a year later Kantorovich [1949] also presented a new proof using for the first time the majorant principle (see also Kantorovich [1951a, b] and Kantorovich and Akilov [1959]). Briefly, an operator equation $x = Gx$ on a Banach space X is said to be majorized by the real equation $t = \varphi(t)$ if $\|Gx^0 - x^0\| \leq \varphi(t^0)$ and $\|G'(x)\| \leq \varphi'(t)$ whenever $\|x - x^0\| \leq t - t^0$. Using this assumption, the convergence of the process $x^{k+1} = Gx^k$ on X is deduced from that of $t_{k+1} = \varphi(t_k)$ on R^1. Although this approach proves to be effective for the study of Newton's method itself, it rests essentially on the requirement that the majorizing process have the same form as the underlying process, and this, in turn, is a severe limitation when it comes to the study of more general iterations. This observation led to the new convergence proof for Newton's method by Ortega [1968] and the general convergence theory for Newton-related processes by Rheinboldt [1968]. The proof of **12.6.2** is taken from these two papers, while **12.6.4** was first given in the latter. Both of these results, and their proofs, hold in Banach spaces provided

12.6. NEWTON'S METHOD AND RELATED ITERATIONS

$F'(x)^{-1}$ and $A(x)^{-1}$ are bounded linear operators defined on the entire range space. For another generalization of the Kantorovich majorant concept, see Altman [1961c].

NR 12.6-2. Newton's method has been applied to a variety of functional equations, and it would be impossible to list all relevant publications here. For a survey of applications to integral equations, see Anselone [1964] and, in particular, the articles by Moore [1964] and Noble [1964]. For two-point boundary value problems, see, for example, the books by Henrici [1962], Keller [1968], and Lee [1968]. For elliptic boundary value problems, see Bellman and Kalaba [1965] and Collatz [1964]. We also note that the Newton–Kantorovich theorem or related results have frequently been employed as tools for obtaining existence theorems; see, for example, Moser [1966] and Mamedov [1965].

NR 12.6-3. The application of **12.6.2**, or related theorems, to obtain error bounds requires the calculation of various quantities, such as the Lipschitz constant of (2), and is usually no easy task. For studies in this direction and also computer programs see, for example, Rall [1969], Lohr and Rall [1967], and Pereyra [1967a]. We note, in this connection, that it is sometimes convenient to regard x^0 in **12.6.2** as the final computed approximation to x^*, so that (5) becomes the error estimate $\|x^* - x^0\| \leqslant 2\alpha/(\beta\gamma)$. In other words, one computes with the Newton iteration until the iterates satisfy some nominal convergence criterion such as $\|Fx^k\| \leqslant \epsilon$ or $\|x^k - x^{k-1}\| \leqslant \epsilon$, then renumbers the approximate solution x^k as x^0, and uses **12.6.2** to compute a final rigorous error bound. For additional comments along this line and also a study of the effects of rounding error in the use of Newton's method, see Lancaster [1966].

NR 12.6-4. Theorems analogous to **12.6.2** have also been given for various secant and Steffensen-type methods by, among others, Bittner [1963], Chen [1964], Johnson and Scholz [1968], Koppel [1966], Schmidt [1963a], and Ul'm [1964c, 1965a]. (Also see **NR 11.2-2** for additional discussion.) However, the majorization approach discussed in **12.4–12.5** fails, in general, to yield these results.
Cavanagh [1970] observed that, instead of one difference equation for the majorizing sequence $\{t_k\}$ of $\{x^k\}$, a system of nonlinear difference equations in several variables is now needed to majorize such processes. The kth value of any of these variables bounds some quantity occurring at the kth step of the iteration; moreover, one of these variables, or a combination of several of them, provides an estimate of the form

$$\|x^{k+1} - x^k\| \leqslant s_k, \qquad k = 0, 1, \ldots .$$

Thus, the sequence

$$t_{k+1} = s_k + t_k, \qquad t_0 = 0, \qquad t_1 = s_0, \qquad k = 0, 1, \ldots,$$

again represents a majorizing sequence of $\{x^k\}$ in the sense of **12.4.1**, and the problem is to find suitable intitial conditions for the system of difference equations which ensure that the sum $\sum_{k=0}^{\infty} s_k$ and, hence, also $\lim_{k\to\infty} t_k$ remains finite. By **12.4.2**, we can then conclude that $\{x^k\}$ converges.

For example, for the two-point secant method

$$x^{k+1} = x^k - J(x^k, x^{k-1} - x^k)^{-1} Fx^k, \quad k = 0, 1, ...,$$

Schmidt [1963a] proved, under a divided-difference condition on J (see **NR 7.2-6**), a convergence theorem which is, in essence, based upon the system of difference equations

$$s_{k+1} = \frac{\eta_k}{1 - \eta_k} s_k, \quad s_0 = s_1 \geq 0,$$

$$\eta_{k+1} = \frac{\eta_k}{1 - \eta_k} \frac{\eta_{k-1}}{1 - \eta_{k-1}}, \quad \eta_0 = \frac{1}{2}, \quad 0 \leq \eta_1 \leq \frac{2}{7}, \quad k = 0, 1, ...$$

Here, η_k bounds the change of $J(x^k, x^{k-1} - x^k)$ when k increases to $k+1$, while s_k has the property indicated above. For details we refer to Cavanagh [1970] where a corresponding convergence result of Johnson and Scholz [1968] for a Steffensen iteration is also proved in terms of this general majorization approach.

The variety of systems of difference equations which can occur in conjunction with different processes appears to prevent the formulation of a comprehensive theory of such general majorizations.

EXERCISES

E 12.6-1. Under the assumptions of **12.6.2**, show that the R-order of $\{x^k\}$ is at least two provided that $\alpha < \frac{1}{2}$.

E 12.6-2. Consider the quadratic polynomial $Fx = (\gamma/2) x^2 - (1/\beta) x + \eta/\beta$ in one dimension. Take $x^0 = 0$ and show that the uniqueness statement of **12.6.1** is sharp.

E 12.6-3. Apply **12.6.2** to the discrete two-point boundary value problem of **E 12.1-5**.

E 12.6-4. Let p_i^0, $i = 1, ..., 4$, and η^0 denote a set of coefficients and initial values satisfying the conditions of **12.6.3**, and $\{t_k^0\}$ the corresponding solution of the difference equation (12). Show that if $0 \leq p_i \leq p_i^0$, $i = 1, ..., 4$, and $0 \leq \eta \leq \eta^0$ is another set of coefficients and initial values, and $\{t_k\}$ the corresponding solution of (12) with $t_0 = 0$, then $\lim_{k\to\infty} t_k \leq \lim_{k\to\infty} t_k^0$.

12.6. NEWTON'S METHOD AND RELATED ITERATIONS 431

E 12.6-5. In 12.6.4, replace the condition on $A(x) - F'(x)$ by $\|A(x^0) - F'(x)\| \leq \delta_0$. Show that $\|F'(x) - A(x)\| \leq \delta_0 + (\gamma + \mu)\|x - x^0\|$ for all $x \in D_0$, and, hence, that the result remains valid if $\sigma = 1 + 2\mu/\gamma$.

E 12.6-6. Let $F: D \subset R^n \to R^n$ be continuous, $K: D \subset R^n \to R^n$ be F-differentiable on $S = \bar{S}(x^0, r) \subset D$, and $\|K'(x) - K'(y)\| \leq \gamma\|x - y\|$, as well as $\|(Fx - Kx) - (Fy - Ky)\| \leq \delta\|x - y\|$, for all $x, y \in S$. Assume that $\|K'(x^0)^{-1}\| \leq \beta$, $\|K'(x^0)^{-1}Fx^0\| \leq \eta$, and $\beta\delta < 1$, $\alpha = \beta\gamma\eta/(1 - \beta\delta)^2 \leq \tfrac{1}{2}$, $r \geq \eta[1 - (1 - 2\alpha)^{1/2}]/[\alpha(1 - \beta\delta)]$. Then the iterates

$$x^{k+1} = x^k - K'(x^0)^{-1}Fx^k \quad \text{and} \quad x^{k+1} = x^k - K'(x^k)^{-1}Fx^k, \quad k = 0, 1, \ldots,$$

remain in S and converge to the only solution of $Fx = 0$ in $S(x^0, r')$, where $r' = \min\{r, \eta[1 + (1 - 2\alpha)^{1/2}]/[\alpha(1 - \beta\delta)]\}$. (Zincenko [1963b])

E 12.6-7. Let $F: D \subset R^n \to R^n$ be F-differentiable on $S = \bar{S}(x^0, r) \subset D$ and $\|F'(x) - F'(y)\| \leq \gamma\|x - y\|$, for all $x, y \in S$. Let $P: L(R^n) \to L(R^n)$ be a linear operator such that [under the induced norm on $L(R^n)$] $\|P\| \leq 1$ and $\|I - P\| \leq 1$. Assume, further, that $\|(PF'(x^0))^{-1}\| \leq \beta$, $\|F'(x^0)\| \leq \delta$, $\|(PF'(x^0))^{-1}Fx^0\| \leq \eta$, $\beta\delta < 1$, $\alpha = 2\beta\gamma\eta/(1 - \beta\delta)^2 \leq \tfrac{1}{2}$, and

$$r \geq \eta[1 - (1 - 2\alpha)^{1/2}]/[\alpha(1 - \beta\delta)].$$

Then the iterates

$$x^{k+1} = x^k - (PF'(x^k))^{-1}Fx^k, \quad k = 0, 1, \ldots,$$

remain in S and converge to the only solution of $Fx = 0$ in $S(x^0, r')$, where $r' = \min\{r, \eta[1 + (1 - 2\alpha)^{1/2}]/[\alpha(1 - \beta\delta)]\}$. Apply this result to the case when $PA = \text{diag } A$ for all $A \in L(R^n)$, and the norm is monotonic. (Bryan [1964])

Chapter 13 / CONVERGENCE UNDER PARTIAL ORDERING

13.1. CONTRACTIONS UNDER PARTIAL ORDERING

In our discussions in Chapter 12 we have measured the convergence of the iterative sequence $\{x^k\}$ to its limit x^* in terms of a norm on R^n. In some sense, this means that we concern ourselves only with the worst convergence behavior of the component sequences $\{x_i^k\}$, $i = 1,..., n$, and when it comes to error estimates this has certain disadvantages. In order to obtain a measure of the convergence behavior in terms of the individual components, it is convenient to use the absolute value of vectors on R^n:

$$|x| = (|x_1|, |x_2|,..., |x_n|)^T, \quad x \in R^n. \tag{1}$$

Recall from Section **2.4** that the natural partial ordering on R^n is defined by

$$x \leqslant y, \quad x, y \in R^n \quad \text{if and only if} \quad x_i \leqslant y_i, \quad i = 1,..., n. \tag{2}$$

The properties of this partial ordering are given in **2.4.1**. Recall also that, as given in **2.4.2**, the absolute value (1) has the following properties:

$$|x| \geqslant 0, \quad \forall x \in R^n \quad \text{and} \quad |x| = 0 \quad \text{if and only if} \quad x = 0; \tag{3a}$$

$$|\alpha x| = |\alpha| |x|, \quad \forall x \in R^n, \quad \alpha \in R^1; \tag{3b}$$

$$|x + y| \leqslant |x| + |y|, \quad \forall x, y \in R^n. \tag{3c}$$

13.1. CONTRACTIONS UNDER PARTIAL ORDERING

The concept of a contraction in terms of this partial ordering may now be defined as follows.

13.1.1. Definition. An operator $G: D \subset R^n \to R^n$ is called a *P-contraction* on a set $D_0 \subset D$ if there exists a linear operator $P \in L(R^n)$ with the properties

$$P \geq 0, \quad \rho(P) < 1 \qquad (4)$$

such that

$$|Gx - Gy| \leq P |x - y|, \quad \forall x, y \in D_0. \qquad (5)$$

The contraction-mapping theorem **12.1.2** extends in a natural way to P-contractions. Recall that, from **2.4.5**,

$$(I - P)^{-1} = \sum_{i=0}^{\infty} P^i \geq 0, \quad \sum_{i=0}^{k} P^i \leq (I-P)^{-1}, \quad \forall k \geq 1. \qquad (6)$$

13.1.2. Suppose that $G: D \subset R^n \to R^n$ is a P-contraction on the closed set $D_0 \subset D$ such that $GD_0 \subset D_0$. Then for any $x^0 \in D_0$ the sequence

$$x^{k+1} = Gx^k, \quad k = 0, 1, \ldots, \qquad (7)$$

converges to the only fixed point of G in D_0 and the error estimate

$$|x^k - x^*| \leq (I - P)^{-1} P |x^k - x^{k-1}|, \quad k = 1, 2, \ldots, \qquad (8)$$

holds.

Proof. The proof follows that of the contraction-mapping theorem. From (6) we find that

$$|x^{k+m} - x^k| \leq \sum_{j=1}^{m} |x^{k+j} - x^{k+j-1}| \leq \sum_{j=1}^{m} P^j |x^k - x^{k-1}|$$

$$\leq (I - P)^{-1} P |x^k - x^{k-1}|$$

$$\leq (I - P)^{-1} P^k |x^1 - x^0|, \quad k, m \geq 0. \qquad (9)$$

Hence, $\{x^k\}$ is a Cauchy sequence which therefore converges to some $x^* \in D_0$. Since

$$|x^* - Gx^*| \leq |x^* - x^{k+1}| + |Gx^k - Gx^*| \leq |x^* - x^{k+1}| + P |x^k - x^*|,$$

we see that $x^* = Gx^*$. The error estimate (8) is a direct consequence of (9) for $m \to \infty$, and, finally, if $y^* \in D_0$ is another fixed point of G, then

$$|x^* - y^*| = |Gx^* - Gy^*| \leq P |x^* - y^*|$$

implies that $(I - P) \mid x^* - y^* \mid \leqslant 0$, or, because of $(I - P)^{-1} \geqslant 0$, that $\mid x^* - y^* \mid \leqslant 0$; hence, $x^* = y^*$. ∎

For a typical application of this theorem, assume that the components g_i, $i = 1,..., n$, of the mapping $G: R^n \to R^n$ satisfy

$$\mid g_i(y_1,..., y_n) - g_i(x_1,..., x_n) \mid \leqslant \sum_{j=1}^{n} p_{ij} \mid y_j - x_j \mid, \quad i = 1,..., n,$$

where $p_{ij} \geqslant 0$, $i, j = 1,..., n$. With $P = (p_{ij}) \in L(R^n)$, this is equivalent with

$$\mid Gy - Gx \mid \leqslant P \mid x - y \mid.$$

Thus, if $\rho(P) < 1$, then for any $x^0 \in R^n$ the iterates (7) converge to the unique fixed point x^* of G in R^n, and we have the componentwise error estimate (8).

As a more interesting application, we obtain a global convergence theorem for the nonlinear Jacobi iteration discussed in **7.4**. Recall that the definitions of an M-matrix, a diagonal mapping, and an isotone mapping are given in **2.4.7**, **1.1.1**, and **2.4.3**, respectively.

13.1.3. Global Jacobi Theorem. Let $A \in L(R^n)$ be an M-matrix, assume that $\phi: R^n \to R^n$ is a continuous, diagonal, and isotone mapping, and set $Fx = Ax + \phi x$. Then for any $\omega \in (0, 1]$ and any $x^0 \in R^n$ the Jacobi sequence $\{x^k\}$ given by

$$\begin{cases} \text{Solve} \quad a_{ii} x_i + \varphi_i(x_i) + \sum_{j=1; j \neq i}^{n} a_{ij} x_j^k = 0 \quad \text{for} \quad x_i, \\ \text{Set} \quad x_i^{k+1} = x_i^k + \omega(x_i - x_i^k), \quad i = 1,..., n, \quad k = 0, 1,..., \end{cases} \tag{10}$$

is well-defined and converges to the unique solution x^* of $Fx = 0$.

Proof. Let D be the diagonal part of $A = (a_{ij})$, set $B = D - A$, and

$$r_i(t) = a_{ii} t + \varphi_i(t), \quad i = 1,..., n, \quad t \in R^1. \tag{11}$$

Since, by **2.4.8**, $a_{ii} > 0$, and, since the φ_i are isotone, it follows that each r_i is one-to-one and maps R^1 onto R^1. In particular, then, the sequence (10) is well defined and the operator $D + \phi$ has an inverse defined on all of R^n. Set

$$G: \quad R^n \to R^n, \quad Gx = (1 - \omega) x + \omega(D + \phi)^{-1} Bx. \tag{12}$$

13.1. CONTRACTIONS UNDER PARTIAL ORDERING

Then x^* is a fixed point of G if and only if $Fx^* = 0$. Moreover, since (10) can be written in the form

$$(D + \phi)[x^k + (1/\omega)(x^{k+1} - x^k)] - Bx^k = 0, \quad k = 0, 1,\ldots,$$

we see that the iterates x^k satisfy $x^{k+1} = Gx^k$, $k = 0, 1,\ldots$. In order to show that G is a P-contraction, note that, because of the isotonicity of the φ_i,

$$|t_1 - t_2| \leq |t_1 - t_2 + (1/a_{ii})[\varphi_i(t_1) - \varphi_i(t_2)]| = (1/a_{ii})|r_i(t_1) - r_i(t_2)|,$$

for all $t_1, t_2 \in R^1$ and $i = 1,\ldots, n$. Consequently, for arbitrary $s_1, s_2 \in R^1$ and $t_1 = r_i^{-1}(s_1)$, $t_2 = r_i^{-1}(s_2)$, we obtain

$$|r_i^{-1}(s_1) - r_i^{-1}(s_2)| \leq (1/a_{ii})|s_1 - s_2|, \quad i = 1,\ldots, n, \quad (13)$$

or

$$|(D + \phi)^{-1} x - (D + \phi)^{-1} y| \leq D^{-1} |x - y|, \quad \forall x, y \in R^n.$$

This, in turn, implies that

$$|Gx - Gy| \leq |(1 - \omega)(x - y)| + \omega |(D + \phi)^{-1} Bx - (D + \phi)^{-1} By|$$
$$\leq [(1 - \omega)I + \omega D^{-1}B] |x - y| = P |x - y|, \quad \forall x, y \in R^n,$$

where $P = (1 - \omega) I + \omega D^{-1}B \geq 0$. Since

$$A = (1/\omega) D - (1/\omega)[(1 - \omega) D + \omega B]$$

is a regular splitting of A, and $A^{-1} \geq 0$, it follows from **2.4.18** that $\rho(P) < 1$. Hence, **13.1.2** applies. ∎

The corresponding result for the nonlinear SOR iteration also holds. This will be a consequence of the following theorem for the implicit process

$$x^k = G_k(x^k, x^{k-1}), \quad k = 1, 2,\ldots, \quad (14)$$

where the G_k are now operators with domain in $R^n \times R^n$.

13.1.4. Suppose that the mappings $G_k : D \times D \subset R^n \times R^n \to R^n$, $k = 1, 2,\ldots$, satisfy on some set $D_0 \subset D$ the conditions

$$|G_k(x, z) - G_k(y, z)| \leq Q |x - y|, \quad (15)$$
$$|G_k(z, x) - G_k(z, y)| \leq R |x - y|, \quad \forall x, y, z \in D_0, \quad (16)$$

where both Q, $R \in L(R^n)$ are nonnegative, and $\rho(Q) < 1$, as well as $\rho(P) < 1$ with $P = (I - Q)^{-1}R$. For given $G: D \subset R^n \to R^n$ set

$$H_k: D \subset R^n \to R^n, \qquad H_k x = |G_k(x, x) - Gx|, \qquad k = 1, 2,\ldots,$$

and assume that

$$\lim_{k \to \infty} H_k x = 0, \qquad \forall x \in D_0. \tag{17}$$

Moreover, let $x^0 \in D_0$ be such that $x = G_1(x, x^0)$ has a solution $x^1 \in D_0$ and

$$S = \{x \in R^n \mid |x - x^1| \leq u = (I - P)^{-1}(P \mid x^1 - x^0 \mid + 2(I - Q)^{-1} v)\} \subset D_0,$$

where $v \geq H_k(x^1)$, $k = 1, 2,\ldots$. Then the equations

$$x = G_k(x, x^{k-1}), \qquad k = 1, 2,\ldots, \tag{18}$$

have unique solutions $x^k \in S$, and $\lim_{k \to \infty} x^k = x^*$ where $x^* \in S$ is the unique fixed point of G in D_0.

Proof. For $x, y \in D_0$ and $k \geq 1$ we obtain

$$|Gx - Gy| \leq |Gx - G_k(x, x)| + |G_k(x, x) - G_k(x, y)|$$
$$+ |G_k(x, y) - G_k(y, y)| + |G_k(y, y) - Gy|$$
$$\leq (Q + R) |x - y| + H_k x + H_k y.$$

Since k can be taken arbitrarily large, it follows from (17) that

$$|Gx - Gy| \leq T |x - y|, \qquad \forall x, y \in D_0,$$

with $T = Q + R$. Clearly, $I - T = (I - Q)(I - P)$, and since $\rho(P) < 1$ and $\rho(Q) < 1$, as well as $P, Q \geq 0$, we see that

$$(I - T)^{-1} = (I - P)^{-1}(I - Q)^{-1} \geq 0,$$

and, hence, by **2.4.5**, that $\rho(T) < 1$. Thus, G is a T-contraction on D_0. In order to show that G maps S into itself, let $x \in S$. Then

$$|Gx - x^1| \leq |Gx - Gx^1| + |Gx^1 - G_1(x^1, x^1)| + |G_1(x^1, x^1) - G_1(x^1, x^0)|$$
$$\leq T |x - x^1| + H_1(x^1) + R |x^1 - x^0| \leq Tu + v + R |x^1 - x^0|$$
$$= [(Q + R)(I - P)^{-1} P + R] |x^1 - x^0|$$
$$+ [2(Q + R)(I - P)^{-1}(I - Q)^{-1} + I]v$$
$$\leq (I - P)^{-1}[P | x^1 - x^0 | + 2(I - Q)^{-1}v] = u,$$

13.1. CONTRACTIONS UNDER PARTIAL ORDERING

since a simple calculation shows that

$$(Q + R)(I - P)^{-1}(I - Q)^{-1} + I = (I - P)^{-1}(I - Q)^{-1},$$
$$(Q + R)(I - P)^{-1}P + R = (I - P)^{-1}P. \tag{19}$$

Therefore, by **13.1.2**, G has a fixed point x^* in S which is unique in D_0.

We prove next that $x, y \in S$ implies that $G_k(x, y) \in S$ or, in other words, that for each fixed $y \in S$, $G_k(\cdot, y)$ maps S into itself. Let $x, y \in S$; then, again using the identities (19), we find that

$$|G_k(x, y) - x^1| \leq |G_k(x, y) - G_k(x, x^1)| + |G_k(x, x^1) - G_k(x^1, x^1)|$$
$$+ |G_k(x^1, x^1) - Gx^1| + |Gx^1 - G_1(x^1, x^1)|$$
$$+ |G_1(x^1, x^1) - G_1(x^1, x^0)|$$
$$\leq R|y - x^1| + Q|x - x^1| + H_k x^1 + H_1 x^1 + R|x^1 - x^0|$$
$$\leq [(Q + R)(I - P)^{-1}P + R]|x^1 - x^0|$$
$$+ [(Q + R)(I - P)^{-1}(I - Q)^{-1} + I]2v$$
$$\leq (I - P)^{-1}[P|x^1 - x^0| + 2(I - Q)^{-1}v] = u.$$

Thus by (15) and **13.1.2** the solution x^k of (18) exists in S and is unique in D_0.

Finally,

$$|x^k - x^*| \leq |G_k(x^k, x^{k-1}) - G_k(x^*, x^{k-1})| + |G_k(x^*, x^{k-1}) - G_k(x^*, x^*)|$$
$$+ |G_k(x^*, x^*) - Gx^*| \leq Q|x^k - x^*| + R|x^{k-1} - x^*| + H_k x^*,$$

or

$$|x^k - x^*| \leq P|x^{k-1} - x^*| + (I - Q)^{-1}H_k x^*.$$

For abbreviation, set $u^k = |x^k - x^*|$, $v^k = (I - Q)^{-1}H_k x^*$. Then, by (17), $\lim_{k \to \infty} v^k = 0$, and moreover,

$$u^k \leq P u^{k-1} + v^k \leq \cdots \leq P^k u^0 + \sum_{j=1}^{k} P^{k-j} v^j.$$

By **2.2.8** we can choose a norm such that $\| P \| < 1$. Let $\| v^k \| \leq \epsilon$ for $k \geq k_0$; then

$$\| u^k \| \leq \sum_{j=0}^{k_0} \| P^{k-j} v^j \| + \sum_{j=k_0+1}^{k} \| P^{k-j} v^j \| + \| P^k u^0 \|$$
$$\leq \| P \|^{k-k_0} \| u^{k_0} \| + \epsilon \frac{\| P^{k_0+1} \|}{1 - \| P \|} + \| P \|^k \| u^0 \|,$$

which shows that $\lim_{k \to \infty} u^k = \lim_{k \to \infty} |x^k - x^*| = 0$. ∎

Note that when the operators G_k do not depend on the first variable, so that $G_k(x, y) \equiv G_k y$, then we are considering the iteration

$$x^k = G_k x^{k-1}, \quad k = 1, 2, \ldots. \tag{20}$$

In that case, $Q = 0$ and $P = R$, and **13.1.4** states, essentially, that when the G_k are R-contractions and $\lim_{k \to \infty} G_k x = Gx$ for each $x \in D_0$, then the x^k will converge to x^*.

We complete this section by applying **13.1.4** to the nonlinear SOR iteration

$$\begin{cases} \text{Solve} & a_{ii} x_i + \varphi_i(x_i) + \sum_{j=1}^{i-1} a_{ij} x_j^{k+1} + \sum_{j=i+1}^{n} a_{ij} x_j^{k} = 0 \quad \text{for } x_i, \\ \text{Set} & x_i^{k+1} = x_i^{k} + \omega(x_i - x_i^{k}), \quad i = 1, \ldots, n, \quad k = 0, 1, \ldots, \end{cases} \tag{21}$$

for the equation $Ax + \phi x = 0$.

13.1.5. Global SOR Theorem. Under the conditions of **13.1.3**, the sequence $\{x^k\}$ given by (21) is well defined for any $x^0 \in R^n$ and $\omega \in (0, 1]$, and $\{x^k\}$ converges to the unique solution x^* of $Fx = 0$.

Proof. Let r_i again be given by (11), and define the components g_i^0 of $G_0 : R^n \times R^n \to R^n$ by

$$g_i^0(x, y) = (1 - \omega) y_i + \omega r_i^{-1}(-\sum_{j=1}^{i-1} a_{ij} x_j - \sum_{j=i+1}^{n} a_{ij} y_j), \quad i = 1, \ldots, n.$$

Then the sequence $\{x^k\}$ of (21) satisfies $x^{k+1} = G_0(x^{k+1}, x^k), k = 0, 1, \ldots$. In order to apply **13.1.4**, observe first that

$$G_0(x, x) = (1 - \omega) x + \omega(D + \phi)^{-1} Bx = Gx,$$

where D is again the diagonal part of A, $B = D - A$, and G was given by (12). Recall that x^* is a fixed point of G if and only if $Fx^* = 0$. Since in this case $G_k \equiv G_0$ for $k \geq 1$, the condition (17) of **13.1.4** is automatically satisfied. Next, the application of (13) yields

$$| g_i^0(x, z) - g_i^0(y, z) |$$

$$= \omega \left| r_i^{-1}(-\sum_{j=1}^{i-1} a_{ij} x_j - \sum_{j=i+1}^{n} a_{ij} z_j) - r_i^{-1}(-\sum_{j=1}^{i-1} a_{ij} y_j - \sum_{j=i+1}^{n} a_{ij} z_j) \right|$$

$$\leq \frac{\omega}{a_{ii}} \left| -\sum_{j=1}^{i-1} a_{ij}(x_j - y_j) \right| \leq \frac{\omega}{a_{ii}} \sum_{j=1}^{i-1} (-a_{ij}) | x_j - y_j |,$$

13.1. CONTRACTIONS UNDER PARTIAL ORDERING

for $i = 1,\ldots, n$, since $a_{ij} \leqslant 0$ for $i \neq j$. Therefore,

$$| G_0(x, z) - G_0(y, z) | \leqslant \omega D^{-1} L \,|\, x - y \,| \equiv Q \,|\, x - y \,|, \qquad \forall x, y, z \in R^n,$$

where $-L$ and $-U$ denote the strictly lower- and strictly upper-triangular parts of A. Similarly, it follows that

$$| g_i^0(z, x) - g_i^0(z, y) | \leqslant (1 - \omega) \,|\, x_i - y_i \,|$$
$$+ \omega(1/a_{ii}) \sum_{j=i+1}^{n} (-a_{ij}) \,|\, x_j - y_j \,|, \qquad i = 1,\ldots, n,$$

so that

$$| G_0(z, x) - G_0(z, y) | \leqslant [(1 - \omega) I + \omega D^{-1} U] \,|\, x - y \,|$$
$$\equiv R \,|\, x - y \,|, \qquad \forall x, y, z \in R^n.$$

Hence, the conditions (15) and (16) hold with $Q = \omega D^{-1} L \geqslant 0$ and $R = (1 - \omega) I + \omega D^{-1} U \geqslant 0$. Since A is an M-matrix and

$$A = (1/\omega)(D - \omega L) - (1/\omega)[(1 - \omega) D + \omega U]$$

is a regular splitting, **2.4.18** shows that $\rho(P) < 1$, where $P = (I - Q)^{-1} R$, and, clearly, $\rho(Q) = 0$. Thus, all conditions of **13.1.4** are satisfied. ∎

NOTES AND REMARKS

NR 13.1-1. Iterative processes in partially ordered spaces appear to have been studied first by Kantorovich [1939]; this followed earlier work by Kantorovich [1937] on the topological properties of partially ordered linear spaces. If the spaces are infinite-dimensional, it is these topological properties, and, more specifically, the connection between the topology and the partial ordering which plays a critical role in the study of convergent processes. Kantorovich, in his 1939 paper, used what is now called the "order convergence"; in this connection, see also Birkhoff [1948].

NR 13.1-2. Schröder [1956a] first proved a generalized contraction theorem in spaces metricized by elements of a partially ordered linear space. Such spaces are called pseudometric by Collatz [1964]. Schröder avoided the topological difficulties in partially ordered spaces by defining the convergence of sequences by means of a number of explicit axioms. Later, Bohl [1964, 1967] and also Vandergraft [1967] analyzed in more detail the various topologies on partially ordered linear spaces which have the desired properties needed for the study of the convergence behavior of iterative processes. In finite-dimensional spaces the topological problems mentioned above disappear and all results simplify considerably. This permits us to give the easy proof of the P-contraction theorem **13.1.2** which is a special case of the more general theorem of Schröder [1956a].

NR 13.1-3. Theorem **13.1.4** is due to Ortega and Rheinboldt [1967a]. The last section of this article also mentions some of the topological problems connected with extending the theorem to infinite-dimensional spaces.

NR 13.1-4. Global convergence results for the nonlinear SOR and Jacobi iterations were first proved by Bers [1953] for discrete analogs of the mildly nonlinear boundary value problem:

$$\Delta u = f(u, u_s, u_t) \quad \text{in domain;} \quad u = g \quad \text{on boundary.}$$

Theorems **13.1.3** and **13.1.5** are essentially special cases of Bers' results, although Bers assumed that $\omega = 1$ and that ϕ was differentiable but did not require the assumption that ϕ is diagonal; the results as given in the text appeared in Ortega and Rheinboldt [1970a].

NR 13.1-5. Both **13.1.2** and **13.1.4** also hold if more general partial orderings on R^n are used. In particular, we could consider the orderings

$$x \leqslant y \quad \text{if} \quad \sum_{j=1}^{n} c_{ij} x_j \leqslant \sum_{j=1}^{n} c_{ij} y_j, \quad i = 1,\ldots, n,$$

where $C = (c_{ij})$ is a nonsingular matrix. (See also **NR 2.4-3**.)

EXERCISES

E 13.1-1. Extend all the results of **12.2** to P-contractions.

E 13.1-2. Under the conditions of **13.1.3**, show that $F: R^n \to R^n$ is a one-to-one mapping from R^n onto itself.

E 13.1-3. Apply **13.1.3** and **13.1.5** to the equations (1.1.7) and (1.2.5) under the assumptions of **4.4.2** and **4.4.3**, respectively.

E 13.1-4. Under the conditions of **13.1.2**, show that

$$|G^k x - G^k y| \leqslant P^k |x - y|, \quad \forall x, y \in D_0, \quad k = 1, 2,\ldots.$$

Hence, conclude that $\| G^k x - G^k y \| \leqslant \| P^k \| \| x - y \|$ in the l_∞-norm, and then prove the convergence of the sequence (7) by means of **12.1.1**.

E 13.1-5. Suppose that $G: R^n \to R^n$ is continuous and

$$|G^2 x - Gx| \leqslant P |Gx - x|, \quad \forall x \in R^n,$$

where $P \in L(R^n)$ satisfies (4). Show that the sequence (7) converges to a fixed point x^* of G and that (8) holds.

E 13.1-6. Suppose that $G: R^n \to R^n$ satisfies

$$|g_i(x_1,\ldots, x_i, z_{i+1},\ldots, z_n) - g_i(y_1,\ldots, y_i, z_{i+1},\ldots, z_n)| \leq \sum_{j=1}^{i} q_{ij} |x_j - y_j|$$

$$|g_i(z_1,\ldots, z_i, x_{i+1},\ldots, x_n) - g_i(z_1,\ldots, z_i, y_{i+1},\ldots, y_n)| \leq \sum_{j=i+1}^{n} r_{ij} |x_j - y_j|$$

for $i = 1,\ldots, n$ and all $x, y, z \in R^n$. Give conditions on the q_{ij} and r_{ij} which permit the application of **13.1.4** to the Gauss–Seidel process

$$x_i^{k+1} = g_i(x_1^{k+1},\ldots, x_i^{k+1}, x_{i+1}^k,\ldots, x_n^k), \quad i = 1,\ldots, n, \quad k = 0, 1,\ldots.$$

13.2. MONOTONE CONVERGENCE

In this section we consider basic results about monotone convergence of iterative processes. For abbreviation, throughout the remainder of this chapter, we shall use the notation $x^k \downarrow x^*$, $k \to \infty$, to mean that

$$x^0 \geq x^1 \geq \cdots \geq x^k \geq x^{k+1} \geq \cdots \geq x^*, \quad \lim_{k\to\infty} x^k = x^*;$$

$x^k \uparrow x^*$, $k \to \infty$, is defined analogously. In either case, the sequence $\{x^k\}$ is said to converge monotonically to x^*. In addition, we define for any $x, y \in R^n$ such that $x \leq y$ the *order interval*

$$\langle x, y \rangle = \{u \in R^n \mid x \leq u \leq y\}.$$

A first result provides a way of bounding solutions of operator equations.

13.2.1. Suppose that $K, H: D \subset R^n \to R^n$ are isotone mappings and that $x^0 \leq y^0$, $\langle x^0, y^0 \rangle \subset D$. Consider the iterations

$$x^{k+1} = Kx^k - Hy^k, \quad y^{k+1} = Ky^k - Hx^k, \quad k = 0, 1,\ldots,$$

and assume that $x^0 \leq x^1$, $y^0 \geq y^1$; then there exist points $x^0 \leq x^* \leq y^* \leq y^0$ such that $x^k \uparrow x^*$ and $y^k \downarrow y^*$ as $k \to \infty$. Moreover, any fixed point u of the operator $Gx = Kx - Hx$ in $\langle x^0, y^0 \rangle$ is contained in $\langle x^*, y^* \rangle$.

Proof. We show by induction that $x^k \leq y^k$, $x^{k+1} \geq x^k$, and $y^{k+1} \leq y^k$

imply that $x^{k+1} \leqslant y^{k+1}$, $x^{k+2} \geqslant x^{k+1}$, and $y^{k+2} \leqslant y^{k+1}$. In fact, because of the isotonicity, we have

$$x^{k+1} = Kx^k - Hy^k \leqslant Ky^k - Hx^k = y^{k+1},$$
$$x^{k+2} = Kx^{k+1} - Hy^{k+1} \geqslant Kx^k - Hy^k = x^{k+1},$$
$$y^{k+2} = Ky^{k+1} - Hx^{k+1} \leqslant Ky^k - Hx^k = y^{k+1},$$

so that for all $k \geqslant 0$

$$x^0 \leqslant x^1 \cdots \leqslant x^k \leqslant x^{k+1} \leqslant y^{k+1} \leqslant y^k \leqslant \cdots \leqslant y^1 \leqslant y^0.$$

Hence, as monotone sequences, $\{x_i^k\}$ and $\{y_i^k\}$ have limits x_i^* and y_i^*, so that also the vector sequences $\{x^k\}$ and $\{y^k\}$ have limits x^* and y^* for which, clearly, $x^* \leqslant y^*$. Finally, if $x^0 \leqslant u \leqslant y^0$ and $u = Gu$, then we see by induction that $x^k \leqslant u \leqslant y^k$ for all k. In fact,

$$x^{k+1} = Kx^k - Hy^k \leqslant Ku - Hu = u = Ku - Hu \leqslant Ky^k - Hx^k = y^{k+1}.$$

The induction is complete, and it follows that $x^* \leqslant u \leqslant y^*$. ∎

In general, x^* and y^* are not fixed points of G, but there is an important special case when they have this property.

13.2.2. Kantorovich Lemma. Let $G: D \subset R^n \to R^n$ be isotone on D, and suppose that $x^0 \leqslant y^0$, $\langle x^0, y^0 \rangle \subset D$, $x^0 \leqslant Gx^0$, $y^0 \geqslant Gy^0$. Then the sequences

$$x^{k+1} = Gx^k, \quad y^{k+1} = Gy^k, \quad k = 0, 1, \ldots, \tag{1}$$

satisfy $x^k \uparrow x^*$, $k \to \infty$, $y^k \downarrow y^*$, $k \to \infty$, and $x^* \leqslant y^*$. Moreover, if G is continuous on $\langle x^0, y^0 \rangle$, then $x^* = Gx^*$, $y^* = Gy^*$, and any fixed point $u \in \langle x^0, y^0 \rangle$ of G is contained in $\langle x^*, y^* \rangle$.

The proof is a direct consequence of **13.2.1** with $K \equiv G$ and $H \equiv 0$. Clearly, if G is continuous on $\langle x^0, y^0 \rangle$, then it follows from (1) that $x^* = Gx^*$ and $y^* = Gy^*$.

Evidently, the most desirable case is when $x^* = y^*$, since then x^* is the only fixed point of G in $\langle x^0, y^0 \rangle$, the sequences $\{x^k\}$ and $\{y^k\}$ both converge to it, and we have the upper and lower bounds $x^k \leqslant x^* \leqslant y^k$, $k = 0, 1, \ldots$.

We turn now to a different approach to constructing sequences which converge monotonically to the solution of a system $Fx = 0$. The following theorem is basic to our subsequent discussions. Recall from **2.4.4** that $B \in L(R^n)$ is a subinverse of $A \in L(R^n)$ if $BA \leqslant I$ and $AB \leqslant I$.

13.2. MONOTONE CONVERGENCE

13.2.3. Let $F: D \subset R^n \to R^n$ and suppose that

$$x^0 \leqslant y^0, \quad \langle x^0, y^0 \rangle \subset D, \quad Fx^0 \leqslant 0 \leqslant Fy^0. \tag{2}$$

Assume, further, that there exists a mapping $A: \langle x^0, y^0 \rangle \to L(R^n)$ such that

$$Fy - Fx \leqslant A(y)(y - x), \quad x^0 \leqslant x \leqslant y \leqslant y^0. \tag{3}$$

If $P_k : \langle x^0, y^0 \rangle \to L(R^n)$, $k = 0, 1, \ldots$, are any mappings such that $P_k(x)$ is a nonnegative subinverse of $A(x)$ for all $x \in \langle x^0, y^0 \rangle$, then the iterates

$$y^{k+1} = y^k - P_k(y^k) Fy^k, \quad k = 0, 1, \ldots, \tag{4}$$

are well-defined and $y^k \downarrow y^*$, $k \to \infty$, with $y^* \in \langle x^0, y^0 \rangle$. Any solution of $Fx = 0$ in $\langle x^0, y^0 \rangle$ is contained in $\langle x^0, y^* \rangle$, and if F is continuous at y^* and there exists a nonsingular $P \in L(R^n)$ such that

$$P_k(y^k) \geqslant P \geqslant 0, \quad \forall k \geqslant k_0, \tag{5}$$

then $Fy^* = 0$.

Proof. We prove by induction that

$$y^0 \geqslant y^{k-1} \geqslant y^k \geqslant x^0, \quad Fy^k \geqslant 0.$$

If this is correct for some $k \geqslant 0$, then $P_k(y^k) \geqslant 0$ and $Fy^k \geqslant 0$ together imply that $y^{k+1} \leqslant y^k$. Now, using (4) and (3), together with the fact that $P_k(x)$ is a subinverse of $A(x)$, we find that, for any $x \in \langle x^0, y^k \rangle$,

$$x - P_k(y^k) Fx = y^{k+1} - (y^k - x) + P_k(y^k)(Fy^k - Fx)$$
$$\leqslant y^{k+1} - [I - P_k(y^k) A(y^k)](y^k - x) \leqslant y^{k+1}; \tag{6}$$

hence, in particular, $x^0 \leqslant x^0 - P_k(y^k) Fx^0 \leqslant y^{k+1}$. Similarly, we obtain

$$Fy^{k+1} \geqslant Fy^k + A(y^k)(y^{k+1} - y^k) = [I - A(y^k) P_k(y^k)] Fy^k \geqslant 0.$$

This completes the induction. Now, $\{y^k\}$, as a bounded, monotonically decreasing sequence, has a limit $y^* \geqslant x^0$. Suppose that $z \in \langle x^0, y^0 \rangle$ is a solution of $Fx = 0$. Then (6) for $k = 0$ implies that

$$z = z - P_0(y^0) Fz \leqslant y^1,$$

and, by induction, we see that $z \leqslant y^k$ for all $k \geqslant 0$; hence, $z \leqslant y^*$. If now (5) holds, then for $k \geqslant k_0$,

$$y^k - y^{k+1} = P_k(y^k) Fy^k \geqslant P Fy^k \geqslant 0.$$

But $\lim_{k\to\infty}(y^k - y^{k+1}) = 0$, so that $\lim_{k\to\infty} PFy^k = 0$. The continuity of F at y^* and the nonsingularity of P then imply that $Fy^* = 0$. ∎

Note that some condition such as (5) is necessary in order to ensure that $Fy^* = 0$; in fact, the sequence $P_k(x) \equiv 0$ for all $x \in \langle x^0, y^0 \rangle$, $k \geq 0$, satisfies the remaining hypotheses of **13.2.3**, and in this case $y^k = y^0$ for all k. However, (5) may be replaced by a requirement such as

$$\| P_k(y^k)^{-1} \| \leq \alpha, \qquad \forall k \geq k_0$$

(see **E 13.2-1**).

We note also that there are other versions of **13.2.3** corresponding to different sign configurations. We indicate these versions schematically in Table **13.1**, where the first column represents Theorem **13.2.3** as stated.

TABLE 13.1

$x^0 \leq y^0$	$x^0 \geq y^0$	$x^0 \leq y^0$	$x^0 \geq y^0$
$Fx^0 \leq 0 \leq Fy^0$	$Fx^0 \geq 0 \geq Fy^0$	$Fx^0 \geq 0 \geq Fy^0$	$Fx^0 \leq 0 \leq Fy^0$
$Fy - Fx \leq A(y)(y-x)$	$Fy - Fx \geq A(y)(y-x)$	$Fy - Fx \leq A(y)(y-x)$	$Fy - Fx \geq A(y)(y-x)$
$P_k(x) \geq 0$	$P_k(x) \leq 0$	$P_k(x) \leq 0$	$P_k(x) \geq 0$
$y^{k+1} \leq y^k$	$y^{k+1} \geq y^k$	$y^{k+1} \leq y^k$	$y^{k+1} \geq y^k$

We consider next the construction of an additional, monotonically increasing sequence starting at x^0.

13.2.4. Given $F: D \subset R^n \to R^n$, let the points $x^0, y^0 \in D$ satisfy (2), and suppose that for the mapping $A: \langle x^0, y^0 \rangle \to L(R^n)$ the condition (3) holds as well as

$$A(x) \leq A(y) \quad \text{if} \quad x^0 \leq x \leq y \leq y^0. \tag{7}$$

Let the sequence $\{y^k\}$ and y^* be defined as in **13.2.3**, and assume that $Q_k \in L(R^n)$, $k = 0, 1,\ldots$, are nonnegative subinverses of $A(y^k)$. Then the sequence

$$x^{k+1} = x^k - Q_k F x^k, \qquad k = 0, 1,\ldots, \tag{8}$$

is well-defined, and $x^k \uparrow x^*$, $k \to \infty$, with $x^* \in \langle x^0, y^* \rangle$. The interval $\langle x^*, y^* \rangle$ contains all solutions of $Fx = 0$ in $\langle x^0, y^0 \rangle$. If F is continuous at x^* and if there is a nonsingular $Q \in R^n$ for which $Q_k \geq Q \geq 0$, for all $k \geq k_0$, then $Fx^* = 0$.

13.2. MONOTONE CONVERGENCE

Proof. We show by induction that

$$x^0 \leq x^{k-1} \leq x^k \leq y^k, \qquad Fx^k \leq 0.$$

Suppose this holds for some $k \geq 0$; then $Fx^k \leq 0$ and $Q_k \geq 0$ imply $x^{k+1} \geq x^k$, and because of $Fy^k \geq 0$, (3), and the fact that Q_k is a subinverse of $A(y^k)$, we have

$$y^k \geq y^k - Q_k Fy^k = x^{k+1} + (y^k - x^k) + Q_k(Fx^k - Fy^k)$$
$$\geq x^{k+1} + (I - Q_k A(y^k))(y^k - x^k) \geq x^{k+1}.$$

Therefore, using (7),

$$Fx^{k+1} \leq Fx^k + A(x^{k+1})(x^{k+1} - x^k) \leq (I - A(y^k)Q_k)Fx^k \leq 0,$$

and, finally,

$$x^{k+1} \leq x^{k+1} - P_k(y^k)Fx^{k+1} = y^{k+1} - (y^k - x^{k+1}) + P_k(y^k)(Fy^k - Fx^{k+1})$$
$$\leq y^{k+1} - [I - P_k(y^k)A(y^k)](y^k - x^{k+1}) \leq y^{k+1}. \tag{9}$$

This completes the induction, and all conclusions of **13.2.4** follow in a manner analogous to **13.2.3**. ∎

If the points y^*, x^* of **13.2.3-4** are solutions of $Fx = 0$, we shall call them the *maximal* and *minimal* solution of $Fx = 0$ in $\langle x^0, y^0 \rangle$. The case of most interest is, of course, again when $x^* = y^*$, because then the sequences $\{y^k\}$ and $\{x^k\}$ constitute upper and lower bounds for x^*; that is, we have the componentwise estimates

$$x_i^k \leq x_i^* \leq y_i^k, \qquad k = 0, 1, \ldots, \quad i = 1, 2, \ldots, n, \tag{10}$$

and (10) provides a termination criterion for the iteration.

The following result gives a sufficient condition that $x^* = y^*$.

13.2.5. Let $F: D \subset R^n \to R^n$, and suppose that the points $x^0, y^0 \in D$ satisfy (2). In addition, assume there is a mapping $B: \langle x^0, y^0 \rangle \to L(R^n)$ such that

$$Fy - Fx \geq B(x)(y - x) \qquad \text{if} \qquad x^0 \leq x \leq y \leq y^0, \tag{11}$$

where, for all $x \in \langle x^0, y^0 \rangle$, $B(x)$ is nonsingular and $B(x)^{-1} \geq 0$. If $Fx = 0$ has either a maximal or a minimal solution in $\langle x^0, y^0 \rangle$, then no other solutions exist in $\langle x^0, y^0 \rangle$.

Proof. Suppose x^* is a minimal solution in $\langle x^0, y^0 \rangle$ and $z^* \in \langle x^0, y^0 \rangle$ is any other solution of $Fx = 0$. Then

$$0 = Fz^* - Fx^* \geqslant B(x^*)(z^* - x^*),$$

and, since $B(x^*)^{-1} \geqslant 0$, it follows that $z^* \leqslant x^*$; therefore, $z^* = x^*$. The proof is similar if a maximal solution exists. ∎

We conclude this section by noting that the Kantorovich lemma **13.2.2** is also a special case of **13.2.3-4**. In fact, if G satisfies the conditions of **3.2.2**, then $Fx \equiv x - Gx$ has the property

$$Fy - Fx = y - x - (Gy - Gx) \leqslant y - x \quad \text{if} \quad x^0 \leqslant x \leqslant y \leqslant y^0.$$

Hence, all conditions of **13.2.3** and **13.2.4** hold for F with $P_k(x) \equiv P \equiv A(x) \equiv Q_k \equiv Q = I$.

NOTES AND REMARKS

NR 13.2-1. Theorem **13.2.1** is a special case of a more general result of Schröder [1959]; here, we have followed the presentation by Collatz [1964]. Extensions of Schröder's results were given by Albrecht [1961, 1962] and Bohl [1967].

NR 13.2-2. The Kantorovich lemma **13.2.2** was proved by Kantorovich [1939] in more general partially ordered linear spaces (see **NR 13.1-1**).

NR 13.2-3. Theorem **13.2.3–13.2.5** were given by Ortega and Rheinboldt [1967a]. Related results are due to Baluev [1952, 1956] and Slugin [1955, 1958a–c], who were motivated by the Chaplygin method for differential equations (see, for example, Berezin and Zhidkov [1959] for a discussion of Chaplygin's method). For some modifications of **13.2.3-4**, see Yoshiaki [1968].

EXERCISES

E 13.2-1. Replace (5) by $\| P_k(y^k)^{-1} \| \leqslant \alpha$ for $k \geqslant k_0$ and show that **13.2.3** remains valid.

E 13.2-2. Give a direct proof of **13.2.2**.

E 13.2-3. State and prove the variations of **13.2.3** indicated in Table **13.1**.

E 13.2-4. Let $F: L(R^n) \to L(R^n)$ be defined by $FX = AX - I$ where $A \in L(R^n)$ is nonsingular and $A^{-1} \geqslant 0$. Consider the Schultz iteration

$$X_{k+1} = X_k - X_k F X_k, \quad k = 0, 1, \ldots.$$

If X_0 is a nonnegative, nonsingular subinverse of A, show that all X_k are nonnegative, nonsingular subinverses of A and that $X_k \uparrow A^{-1}$, as $k \to \infty$. (Albrecht [1961])

E 13.2-5. Consider the mapping $G: R^n \to R^n$ with components

$$g_i(x) = 1 + \sum_{j=1}^{n} \alpha_j x_i x_j, \qquad i = 1,\ldots, n,$$

where $\alpha_j \geq 0$, and $\sum_{j=1}^{n} \alpha_j \leq \frac{1}{4}$. Show that G is isotone on $D = \{x \in R^n \mid x \geq 0\}$. Find a suitable interval on which **13.2.2** applies, and show that in any such interval G has a unique fixed point.

E 13.2-6. Under the conditions of **13.2.1**, show that

$$G\langle x^k, y^k\rangle \subset \langle x^{k+1}, y^{k+1}\rangle \subset \langle x^k, y^k\rangle, \qquad k = 0, 1,\ldots.$$

Conclude, by the Brouwer fixed-point theorem **6.3.2**, that if G is continuous on $\langle x^0, y^0\rangle$, then G has a fixed point in $\langle x^*, y^*\rangle$.

E 13.2-7. Suppose that the mappings H_1, $H_2: D \times D \subset R^n \to R^n$ are isotone in the first vector variable and antitone in the second one, and that

$$H_1(x, x) \leq H_2(x, x), \qquad \forall x \in D.$$

If

$$x^0 \leq y^0, \qquad \langle x^0, y^0\rangle \subset D, \qquad x^0 \leq H_1(x^0, y^0), \qquad H_2(y^0, x^0) \leq y^0,$$

then the sequences

$$x^{k+1} = H_1(x^k, y^k), \qquad y^{k+1} = H_2(y^k, x^k), \qquad k = 0, 1,\ldots,$$

satisfy $x^k \uparrow x^*$, $y^k \downarrow y^*$, $k \to \infty$, and $x^0 \leq x^* \leq y^* \leq y^0$. Moreover, if $G: D \subset R^n \to R^n$ is any continuous mapping such that

$$H_1(x, x) \leq Gx \leq H_2(x, x), \qquad \forall x \in D,$$

then G has a fixed point in $\langle x^*, y^*\rangle$, and any fixed point of G in $\langle x^0, y^0\rangle$ is contained in $\langle x^*, y^*\rangle$. (Schröder [1960a])

13.3. CONVEXITY AND NEWTON'S METHOD

The Kantorovich lemma provides one way of obtaining the mapping A of **13.2.3**. A more interesting possibility arises by extending the concept of a convex function.

13.3.1. Definition. A mapping $F: D \subset R^n \to R^m$ is *order-convex* on a convex subset $D_0 \subset D$ if

$$F(\lambda x + (1 - \lambda) y) \leqslant \lambda Fx + (1 - \lambda) Fy, \tag{1}$$

whenever $x, y \in D_0$ are comparable ($x \leqslant y$ or $y \leqslant x$) and $\lambda \in (0, 1)$. If (1) holds for all $x, y \in D_0$ and $\lambda \in (0, 1)$, then F is *convex* on D_0. ∎

Clearly, if $F: D \subset R^n \to R^1$, then the above definition of convexity reduces to that previously given in **3.4.1**. More generally, if $F: D \subset R^n \to R^m$ has components $f_1, \ldots, f_m$, then it follows immediately that F is (order) convex on $D_0 \subset D$ if and only if each f_i is (order) convex on D_0. Therefore, **3.4.3** extends to the present setting to show that if $F: D \subset R^n \to R^m$ is convex on an open convex set $D_0 \subset D$, then F is continuous on D_0. The differential characterizations of **3.4** also extend in a natural way. Although the proofs are essentially identical, we repeat them here for convenience.

13.3.2. Let $F: D \subset R^n \to R^m$ be G-differentiable on the convex set $D_0 \subset D$. Then the following statements are equivalent:

$$F \text{ is order-convex on } D_0; \tag{2}$$

$$Fy - Fx \geqslant F'(x)(y - x), \text{ for all comparable } x, y \in D_0; \tag{3}$$

$$[F'(y) - F'(x)](y - x) \geqslant 0, \text{ for all comparable } x, y \in D_0. \tag{4}$$

Similarly, F is convex on D_0 if and only if (3) and (4) hold for all $x, y \in D_0$. If F is twice G-differentiable in D_0, then F is order-convex on D_0 if and only if

$$F''(x) hh \geqslant 0, \tag{5}$$

for all $x \in D_0$ and $h \in R^n$, $h \geqslant 0$; F is convex on D_0 if and only if (5) holds for all $x \in D_0$, and $h \in R^n$.

Proof. Suppose, first, that (3) holds, and for given comparable $x, y \in D_0$ and $\lambda \in (0, 1)$ set $z = \lambda x + (1 - \lambda) y$. Then z is comparable with x and y, and, hence,

$$Fx - Fz \geqslant F'(z)(x - z), \quad Fy - Fz \geqslant F'(z)(y - z).$$

Multiplying these inequalities by λ and $(1 - \lambda)$, respectively, and adding, we see that

$$\lambda Fx + (1 - \lambda) Fy - Fz \geqslant F'(z)[\lambda x + (1 - \lambda) y - z] = 0,$$

13.3. CONVEXITY AND NEWTON'S METHOD

so that F is order-convex. Conversely, if (2) holds and $x, y \in D_0$ are comparable, then, for any $t \in (0, 1)$, x and $x + t(y - x)$ are comparable, so that

$$Fy - Fx \geq (1/t)\{F(x + t[y - x]) - Fx\}.$$

Hence, in view of the G-differentiability of F, (3) follows as $t \to 0$. To show the equivalence of (3) and (4), note first that if (3) holds, then addition of

$$Fy - Fx \geq F'(x)(y - x), \qquad Fx - Fy \geq F'(y)(x - y)$$

immediately gives (4). Conversely, if (4) holds, then

$$[f_i'(y) - f_i'(x)](y - x) \geq 0, \quad i = 1, \ldots, m, \quad \text{all comparable} \ x, y \in D_0,$$

where $f_1, \ldots, f_m$ are the components of F. Now, for given comparable $x, y \in D_0$, the mean-value theorem **3.2.2** shows that there are $t_i \in (0, 1)$ such that

$$f_i(y) - f_i(x) = f_i'(z^i)(y - x), \quad i = 1, \ldots, m,$$

where $z^i = x + t_i(y - x)$. But each z^i is comparable with x and y, so that

$$[f_i'(z^i) - f_i'(x)](y - x) = (1/t_i)[f_i'(z^i) - f_i'(x)](z^i - x) \geq 0, \quad i = 1, \ldots, m,$$

and, hence,

$$f_i(y) - f_i(x) = f_i'(z^i)(y - x) \geq f_i'(x)(y - x), \quad i = 1, \ldots, m.$$

This, in turn, shows that (3) holds. Finally, if F is twice G-differentiable on D_0 and (5) holds, then $f_i''(x) hh \geq 0$, $i = 1, \ldots, m$, for all $h \geq 0$ and $x \in D_0$. Consequently, by the mean-value theorem **3.3.10**,

$$f_i(y) - f_i(x) - f_i'(x)(y - x) = \tfrac{1}{2} f_i''(x + t_i[y - x])(y - x)(y - x) \geq 0,$$

$$i = 1, \ldots, m,$$

whenever x and y are comparable. Therefore, (3) holds and F is order-convex. Conversely, if F is order-convex, then (4) holds, and for any $x \in D_0$, $h \geq 0$, and $t > 0$, $x + th$ and x are comparable, so that

$$(1/t)[F'(x + th) h - F'(x)h] \geq 0.$$

Hence, (5) follows as $t \to 0$. The proofs for the convex case proceed analogously. ∎

Note that if F' is isotone ($x \leqslant y$ implies that $F'(x) \leqslant F'(y)$), then (4) is satisfied, and, hence, F is order-convex (but not necessarily convex). Moreover, if the second G-derivative exists on D_0, then a sufficient (but not necessary) condition that F be order-convex is that

$$f_i''(x) \geqslant 0, \quad i = 1,\ldots, m, \quad \forall x \in D_0, \qquad (6)$$

where, again, $f_1,\ldots, f_m$ are the components of F. Note that if F'' is continuous on D_0, then the mean-value theorem **3.3.7** and (6) imply that if $x \leqslant y$, then

$$F'(y) - F'(x) = \int_0^1 F''(x + t[y - x])(y - x)\, dt \geqslant 0,$$

so that F' is isotone on D_0. On the other hand, if F is twice G-differentiable on D_0, then **13.3.2** shows that F is convex on D_0 if and only if each matrix $f_i''(x)$, $i = 1,\ldots, m$, $x \in D_0$, is positive semidefinite. Hence, a simple example of an order-convex but not convex function is the quadratic functional $f: R^n \to R^1$ defined by $f(x) = x^{\mathrm{T}} Ax$, where $A \geqslant 0$ is not positive semidefinite.

We now apply the convergence results of the previous section to order-convex mappings.

13.3.3. For $F: D \subset R^n \to R^n$, assume that

$$x^0 \leqslant y^0, \quad \langle x^0, y^0 \rangle \subset D, \quad Fx^0 \leqslant 0 \leqslant Fy^0, \qquad (7)$$

and that F is G-differentiable and order-convex on $\langle x^0, y^0 \rangle$. Assume, further, that the mappings $P_k : \langle x^0, y^0 \rangle \to L(R^n)$ are such that, for each $x \in \langle x^0, y^0 \rangle$, $P_k(x)$ is a nonnegative subinverse of $F'(x)$. Then the sequence

$$y^{k+1} = y^k - P_k(y^k)Fy^k, \quad k = 0, 1,\ldots,$$

is well-defined and satisfies $y^k \downarrow y^*$, $k \to \infty$, with $y^* \in \langle x^0, y^0 \rangle$. Moreover, if F' is isotone on $\langle x^0, y^0 \rangle$, then the subsidiary sequence

$$x^{k+1} = x^k - Q_k Fx^k, \quad k = 0, 1,\ldots,$$

where the Q_k are nonnegative subinverses of $F'(y^k)$, has the property that $x^k \uparrow x^*$, $k \to \infty$, $x^* \in \langle x^0, y^* \rangle$. Any solution of $Fx = 0$ in $\langle x^0, y^0 \rangle$ is contained in $\langle x^*, y^* \rangle$, and if F is continuous at y^* or x^*, and if there is a nonsingular $P \in L(R^n)$ such that

$$P_k(y^k) \geqslant P \geqslant 0, \quad \forall k \geqslant k_0,$$

13.3. CONVEXITY AND NEWTON'S METHOD

or a nonsingular $Q \in L(R^n)$ such that

$$Q_k \geq Q \geq 0, \quad \forall k \geq k_1,$$

then $Fy^* = 0$ or $Fx^* = 0$, respectively.

Proof. Since F is order-convex and G-differentiable on the convex set $\langle x^0, y^0 \rangle$, **13.3.2** shows that

$$Fy - Fx \leq F'(y)(y - x), \quad x^0 \leq x \leq y \leq y^0.$$

Hence, **13.2.3** and **13.2.4** apply with $A(x) = F'(x)$, $x \in \langle x^0, y^0 \rangle$. ∎

As a special case of **13.3.3**, we have the following important result for Newton's method.

13.3.4. Monotone Newton Theorem. Let $F: D \subset R^n \to R^n$, and assume that there exist $x^0, y^0 \in D$ such that (7) holds. Suppose that F is continuous, G-differentiable, and order-convex on $\langle x^0, y^0 \rangle$, and that, for each $x \in \langle x^0, y^0 \rangle$, $F'(x)^{-1}$ exists and is nonnegative. Then the Newton iterates

$$y^{k+1} = y^k - F'(y^k)^{-1} Fy^k, \quad k = 0, 1, \ldots, \tag{8}$$

satisfy $y^k \downarrow y^* \in \langle x^0, y^0 \rangle$ as $k \to \infty$, and if F' is either continuous at y^* or isotone on $\langle x^0, y^0 \rangle$, then y^* is the unique solution of $Fx = 0$ in $\langle x^0, y^0 \rangle$. Moreover, if F' is isotone on $\langle x^0, y^0 \rangle$, then the sequence

$$x^{k+1} = x^k - F'(y^k)^{-1} Fx^k, \quad k = 0, 1, \ldots, \tag{9}$$

satisfies $x^k \uparrow y^*$ as $k \to \infty$. Finally, if, in addition,

$$\| F'(x) - F'(y) \| \leq \gamma \| x - y \|, \quad \forall x, y \in \langle x^0, y^0 \rangle, \tag{10}$$

then there is a constant c such that

$$\| y^{k+1} - x^{k+1} \| \leq c \| x^k - y^k \|^2, \quad k = 0, 1, \ldots. \tag{11}$$

Proof. Since $P_k(x) = F'(x)^{-1} \geq 0$, $x \in \langle x^0, y^0 \rangle$, the first part of **13.3.3** applies and shows that $y^k \downarrow y^*$, $k \to \infty$, $y^* \in \langle x^0, y^0 \rangle$. In order to show that $Fy^* = 0$, suppose, first, that F' is isotone on $\langle x^0, y^0 \rangle$. Then $F'(y^k) \leq F'(y^0)$, so that the nonnegativity of the inverses implies that

$$P_k(y^k) = F'(y^k)^{-1} \geq F'(y^0)^{-1} = P \geq 0, \quad k = 0, 1, \ldots.$$

On the other hand, if F' is continuous at y^*, then there exists a matrix E and an integer k_0 such that $P = F'(y^*)^{-1} - E \geq 0$ is nonsingular and

$F'(y^k)^{-1} \geqslant P$ for $k \geqslant k_0$. Hence, in both cases **13.2.3** shows that $Fy^* = 0$. The uniqueness of y^* in $\langle x^0, y^0 \rangle$ follows immediately from **13.2.5** with $B(x) = F'(x)$, whereas $x^k \uparrow y^*$, $k \to \infty$, is a consequence of **13.3.3** with $Q = F'(y^0)^{-1}$. Finally, to prove (11), suppose that (10) holds. Then F' is continuous on $\langle x^0, y^0 \rangle$ and, since $F'(x)$ is nonsingular, there is a β such that $\|F'(x)^{-1}\| \leqslant \beta$, $x \in \langle x^0, y^0 \rangle$. It then follows by the mean-value theorem **3.2.12** that

$$\| y^{k+1} - x^{k+1} \| = \| y^k - x^k - F'(y^k)^{-1}(Fy^k - Fx^k)\|$$
$$\leqslant \beta \| F'(y^k)(y^k - x^k) - (Fy^k - Fx^k)\| \leqslant \tfrac{1}{2}\beta\gamma \| y^k - x^k \|^2. \quad\blacksquare$$

Note that if Gaussian elimination is used to solve the linear systems implied by (8), then it requires very little additional work to obtain the subsidiary sequence $\{x^k\}$. Moreover, the estimate (11) shows that the intervals $\langle x^k, y^k \rangle$ "converge quadratically" to x^*, so that use of the sequence $\{x^k\}$ does not negate the quadratic convergence of Newton's method itself; indeed, it is possible to show (see **E 13.3-7**) that the convergence of $\{x^k\}$ to x^* is quadratic. Note finally that, corresponding to Table **13.1**, there are three other natural versions of **13.3.4**, depending upon the signs of $Fy - Fx - F'(x)(y - x)$ and $F'(x)^{-1}$ (**E 13.3-6**).

We turn next to the crucial question of condition (7); that is, the question of finding suitable starting points. In this connection the following lemma is sometimes useful.

13.3.5. Let $F: D \subset R^n \to R^n$ be order-convex and G-differentiable on the convex set $D_0 \subset D$ and suppose that there is a nonnegative $C \in L(R^n)$ such that $F'(x) C \geqslant I$, $x \in D_0$. If $Fy^0 \geqslant 0$ and $x^0 = y^0 - CFy^0 \in D_0$, then $Fx^0 \leqslant 0$. Similarly, if $Fx^0 \leqslant 0$ and $y^0 = x^0 - CFx^0 \in D_0$, then $Fy^0 \geqslant 0$.

Proof. Assume that $Fy^0 \geqslant 0$ and $x^0 = y^0 - CFy^0 \in D_0$; then $x^0 \leqslant y^0$ and, by **13.3.2**,

$$Fx^0 \leqslant Fy^0 + F'(x^0)(x^0 - y^0) = [I - F'(x^0)C]Fy^0 \leqslant 0.$$

Similarly, if $Fx^0 \leqslant 0$ and $y^0 = x^0 - CFy^0 \in D_0$, then, again, **13.3.2** shows that

$$Fy^0 \geqslant Fx^0 + F'(x^0)(y^0 - x^0) = [I - F'(x^0)C]Fx^0 \geqslant 0. \quad\blacksquare$$

Under the conditions of **13.3.5**, the other end-point of a suitable interval $\langle x^0, y^0 \rangle$ may be computed if one knows one end-point. One possibility of obtaining the first point is by taking one Newton step.

13.3. CONVEXITY AND NEWTON'S METHOD

13.3.6. Let $F: D \subset R^n \to R^n$ be convex and G-differentiable on the convex set D. Assume that, for some $x \in D$, $F'(x)^{-1}$ exists and that $y^0 = x - F'(x)^{-1} Fx \in D$. Then $Fy^0 \geq 0$.

The proof is self-evident since, by **13.3.2**,

$$Fy^0 \geq Fx + F'(x)(y^0 - x) = 0.$$

As a consequence of **13.3.4** and **13.3.6** we have the following global theorem.

13.3.7. Global Newton Theorem. Suppose that $F: R^n \to R^n$ is continuous, G-differentiable, and convex on R^n, and that $F'(x)$ is nonsingular and $F'(x)^{-1} \geq 0$, for all $x \in R^n$. Assume, further, that $Fx = 0$ has a solution x^* and that F' is either isotone or continuous on R^n. Then x^* is unique, and for any $y^0 \in R^n$ the Newton iterates (8) converge to x^* and

$$y^k \geq y^{k+1} \geq x^*, \quad k = 1, 2, \dots. \tag{12}$$

Proof. For arbitrary $y^0 \in R^n$, **13.3.6** shows that $Fy^1 \geq 0$. Moreover, by **13.3.2** we have

$$0 = Fx^* \geq Fy^1 + F'(y^1)(x^* - y^1),$$

which implies that

$$x^* \leq y^1 - F'(y^1)^{-1} Fy^1 \leq y^1.$$

Hence, **13.3.4** applies, with x^0 set equal to x^* and y^1 taken as the upper end-point, and shows convergence to a solution y^*. But x^* is the unique solution since, if y^* is any other solution, then

$$F'(x^*)(x^* - y^*) \geq 0 = Fx^* - Fy^* \geq F'(y^*)(x^* - y^*),$$

which, since $F'(y)^{-1} \geq 0$, implies that $x^* \leq y^*$ as well as $y^* \leq x^*$. ∎

Theorem **13.3.7** may, of course, be modified in an obvious way to replace the assumption of the existence of x^* by an assumption, as in **13.3.5**, that $F'(x) C \geq I$ for some nonnegative $C \in L(R^n)$ and all $x \in R^n$. This will, in turn, guarantee the existence of a solution as well as of a starting point x^0 for the subsidiary sequence.

We conclude this section by applying **13.3.7** to the equation $Ax + \phi x = 0$.

13.3.8. Let $A \in L(R^n)$ be an M-matrix and set $Fx = Ax + \phi x$, where $\phi: R^n \to R^n$ is continuously differentiable, diagonal, isotone, and convex on R^n. Then for any $y^0 \in R^n$ the Newton iterates (8) converge to the unique solution x^* of $Fx = 0$, and (12) holds.

Proof. The conditions of **13.1.3** are satisfied, and, hence, $Fx = 0$ has a unique solution in R^n. Clearly, F is convex, and, since ϕ is isotone and diagonal, it follows that $\phi'(x)$ is nonnegative and diagonal for all $x \in R^n$. Thus, by **2.4.11**, $F'(x)$ is an M-matrix for all x. Moreover, because ϕ is diagonal, **13.3.2** ensures that ϕ' is isotone. The result then follows directly from **13.3.7**. ∎

We note that the convexity assumption on ϕ in **13.3.8** cannot, in general, be removed (see **E 13.3-11**). Note also that **13.3.8** has immediate application to discrete analogs of boundary value problems for $u'' = f(u)$ or $\Delta u = f(u)$ if f is isotone and convex and, in particular, if $f(u) = e^u$. (See **E 13.3-12**.)

NOTES AND REMARKS

NR 13.3-1. The concept of order-convexity was used by Ortega and Rheinboldt [1967a], where Theorems **13.3.2** and **13.3.3** are also proved.

NR 13.3-2. Various extensions of the definition **3.4.1** of strict or uniform convexity of a functional to mappings $F: R^n \to R^m$ are possible. For example, one could define F to be strictly (uniformly) convex if each component functional is strictly (uniformly) convex. A somewhat different definition of strict convexity has been given and applied by Stepleman [1969].

NR 13.3-3. Theorem **13.3.4**, without the rate-of-convergence statement (11), dates back at least to Baluev [1952], who proved the result in more general spaces (see **NR 13.2-3**). More recently, Vandergraft [1967] has also considered the monotone convergence of Newton's method in partially ordered linear spaces; his results apply, in particular, to the quasilinearization technique of Kalaba [1959] for differential equations.

NR 13.3-4. Lemmas **13.3.5–6** are only one possible way of finding suitable initial points x^0, y^0. For other ways, also see **13.4.7** and Schmidt [1964].

EXERCISES

E 13.3-1. Assume that $F: R^n \to R^n$ is convex and $A \in L(R^n)$. Show that $G = FA$ is convex.

13.3. CONVEXITY AND NEWTON'S METHOD

E 13.3-2. Write out the proofs for the convex case of **13.3.2**.

E 13.3-3. Assume that $F: R^n \to R^m$ is twice G-differentiable and F' is isotone. Show that $F''(x) \geq 0$ for all x (that is, (6) holds for all $x \in R^n$).

E 13.3-4. Suppose that $F: R^n \to R^n$ satisfies
$$Fy - Fx \leq A(y)(y - x), \qquad \forall x, y \in D,$$
where D is convex and A is an arbitrary mapping from D to $L(R^n)$. Show that F is convex on D.

E 13.3-5. Let
$$A = \begin{pmatrix} 1 & -1 \\ 0 & 1 \end{pmatrix}.$$
Show that $h^T A h \geq 0$ for all $h \geq 0$. Conclude from this that if $F: R^n \to R^1$, then $F''(x) \geq 0$ is not a necessary condition for order-convexity.

E 13.3-6. Write down a table for **13.3.4** corresponding to Table **13.1**. Prove your assertions.

E 13.3-7. Under the conditions of **13.3.4**, show that $O_R\{x^k\} \geq 2$.

E 13.3-8. Assume that F satisfies the conditions of **13.3.4**. Show that the simplified Newton iterates $y^{k+1} = y^k - F'(y^0)^{-1} F y^k$ satisfy $y^k \downarrow x^*$, $k \to \infty$.

E 13.3-9. If $F: R^n \to R^n$ is convex on $\langle x^0, y^0 \rangle$ and $F'(x)^{-1} \geq 0$ for all $x \in \langle x^0, y^0 \rangle$, then $Fx = 0$ has at most one solution in $\langle x^0, y^0 \rangle$.

E 13.3-10. Assume that $F: D \subset R^n \to R^n$ is continuous, G-differentiable, and order-convex on $\langle x^0, y^0 \rangle$, where x^0, y^0 satisfy (7). Assume, further, that there is a nonnegative, nonsingular $C \in L(R^n)$ which is a subinverse of $F'(x)$ for every $x \in \langle x^0, y^0 \rangle$. Show that there exists a sequence $\{y^k\} \subset \langle x^0, y^0 \rangle$ which satisfies
$$F'(y^k)(y^{k+1} - y^k) + F y^k = 0, \qquad y^k \downarrow y^*, \quad k \to \infty, \quad F x^* = 0.$$
(Vandergraft [1967])

E 13.3-11. Define $A \in L(R^2)$ and $\phi: R^2 \to R^2$ by
$$A = \begin{pmatrix} a - 1 & -1 \\ -1 & a - 1 \end{pmatrix}, \qquad x = \begin{pmatrix} x_1 + \sin x_1 \\ x_2 + \sin x_2 \end{pmatrix},$$
where a is the positive root of $t^2 - 3t - 2 = 0$. Show that all the conditions of **13.3.8**, except the convexity of ϕ, are satisfied, but that the Newton iterates starting from $x^0 = (\pi, \pi)^T$ satisfy $x^{2k} = x^0$ for $k = 1, 2,\ldots$.

E 13.3-12. Apply **13.3.8** to the equations (1.1.7) and (1.2.5), where $f(s, u) = e^u$ or $f(s, t, u) = e^u$, respectively.

13.4. NEWTON–SOR ITERATIONS

In this section we apply the previous convergence results to an analysis of the Newton–SOR methods of **7.4**. This, in turn, will provide us with a nontrivial example of the use of the concept of subinverses. More generally, we shall consider a generalized linear iteration of the form

$$y^{k+1} = y^k - [I + \cdots + H_k(y^k)^{m_k-1}] B_k(y^k)^{-1} Fy^k, \quad k = 0, 1, \ldots. \quad (1)$$

As discussed in **7.4**, this process is derived by using Newton's method as primary iteration and by applying to the kth Newton step,

$$F'(y^k)(y - y^k) + Fy^k = 0,$$

m_k steps of the secondary linear process defined by the splitting

$$F'(x) = B_k(x) - C_k(x), \quad H_k(x) = B_k(x)^{-1} C_k(x). \quad (2)$$

In connection with (1), it will be of interest to know when the matrix

$$P_k = (I + \cdots + H^k) B^{-1}, \quad H = B^{-1}C, \quad (3)$$

is a subinverse of $B - C$, and we first give the following lemma. Recall from Definition **2.4.16** that $A = B - C$ is a weak regular splitting of $A \in L(R^n)$ if $B^{-1} \geq 0$, $B^{-1}C \geq 0$, and $CB^{-1} \geq 0$.

13.4.1. Let $A = B - C$ be a weak regular splitting. Then for any integer $k \geq 0$ the matrix P_k defined by (3) is a subinverse of A. Moreover, if A is nonsingular and $A^{-1} \geq 0$, then P_k is also nonsingular and $A = P_k^{-1} - (P_k^{-1} - A)$ is a weak regular splitting.

Proof. Since $H \geq 0$, we obtain by the identity (2.4.10)

$$\begin{aligned} P_k A &= (I + H + \cdots + H^k) B^{-1}(B - C) \\ &= (I + \cdots + H^k)(I - H) = I - H^{k+1} \leq I, \end{aligned} \quad (4)$$

and, similarly, since $CB^{-1} \geq 0$,

$$AP_k = (B - C)(I + \cdots + H^k) B^{-1} = B(I - H^{k+1}) B^{-1} = I - (CB^{-1})^{k+1} \leq I.$$

Hence, P_k is a subinverse of A. If $A^{-1} \geq 0$, then, by **2.4.17**, $\rho(H) < 1$, and therefore $\rho(H^{k+1}) < 1$. Thus, **2.3.1** shows that $(I - H^{k+1})^{-1}$ exists, and it follows from (4) that P_k^{-1} exists. Since $P_k \geq 0$, **2.4.16** implies that $A = P_k^{-1} - (P_k^{-1} - A)$ is a weak regular splitting. ∎

We now return to the iteration (1).

13.4. NEWTON–SOR ITERATIONS

13.4.2. For $F: D \subset R^n \to R^n$, assume that

$$x^0 \leq y^0, \quad \langle x^0, y^0 \rangle \subset D, \quad Fx^0 \leq 0 \leq Fy^0. \tag{5}$$

Suppose that F is continuous, G-differentiable, and order-convex on $\langle x^0, y^0 \rangle$, and that $F'(x) = B_k(x) - C_k(x)$, $k = 0, 1, \ldots$, is for each $x \in \langle x^0, y^0 \rangle$ a sequence of weak regular splittings. Then, for any sequence of integers $\{m_k\}$, $m_k \geq 1$, the sequence $\{y^k\}$ given by (1) with $H_k(x) = B_k(x)^{-1} C_k(x)$ is well defined and satisfies $y^k \downarrow y^*$, $k \to \infty$, for some $y^* \in \langle x^0, y^0 \rangle$. If, in addition, there is a nonsingular $B \in L(R^n)$ so that

$$B_k(y^k)^{-1} \geq B \geq 0, \quad \forall k \geq k_0, \tag{6}$$

then $Fy^* = 0$. Finally, if F' is isotone on $\langle x^0, y^0 \rangle$, then the subsidiary sequence

$$x^{k+1} = x^k - [I + \cdots + H_k(y^k)^{m_k - 1}] B_k(y^k)^{-1} Fx^k, \quad k = 0, 1, \ldots, \tag{7}$$

satisfies $x^k \uparrow x^*$, $k \to \infty$, $x^* \in \langle x^0, y^* \rangle$, and (6) implies again that $Fx^* = 0$.

Proof. By **13.4.1**

$$P_k(x) = [I + \cdots + H_k(x)^{m_k - 1}] B_k(x)^{-1} \tag{8}$$

is, for any $x \in \langle x^0, y^0 \rangle$ and $m_k \geq 1$, a subinverse of $F'(x)$ and, since $H_k(x) \geq 0$, we have $P_k(x) \geq 0$. Thus, $y^k \downarrow y^*$, $k \to \infty$, $y^* \in \langle x^0, y^0 \rangle$, follows directly from **13.3.3**, and if F' is isotone, then $x^k \uparrow x^*$, $k \to \infty$, $x^* \in \langle x^0, y^* \rangle$, is also a consequence of the same theorem. Now, if (6) holds, then, since $H_k(y^k) \geq 0$, we see that $P_k(y^k) \geq B_k(y^k)^{-1} \geq B \geq 0$ for all $k \geq k_0$ and, hence, again by **13.3.3**, that $Fy^* = 0$ as well as $Fx^* = 0$. ∎

We now specialize this result to the Newton–SOR iteration (7.4.12) under more stringent assumptions on F. Recall that the general Newton–SOR iteration is defined by (1) and (2), where

$$B_k(x) = (1/\omega_k)[D(x) - \omega_k L(x)], \quad C_k(x) = (1/\omega_k)[(1 - \omega_k) D(x) + \omega_k U(x)], \tag{9}$$

and

$$F'(x) = D(x) - L(x) - U(x) \tag{10}$$

is the partition of $F'(x)$ into diagonal, strictly lower-, and strictly upper-triangular parts, respectively.

13.4.3. Monotone Newton–SOR Theorem. For $F: D \subset R^n \to R^n$, assume that there exist $x^0, y^0 \in D$ for which (5) holds, and that F is continuous, G-differentiable and order-convex on $\langle x^0, y^0 \rangle$. Moreover, suppose that $F'(x)$ is an M-matrix for each $x \in \langle x^0, y^0 \rangle$. Then for any sequence of integers $m_k \geq 1$ and parameters $\omega_k \in (0, 1]$ the Newton–SOR iterates defined by (1), (2), (9), and (10) satisfy $y^k \downarrow y^*$, $k \to \infty$, $y^* \in \langle x^0, y^0 \rangle$. If, in addition, $\omega_k \geq \omega > 0$, $k \geq k_0$, and F' is either continuous or isotone on $\langle x^0, y^0 \rangle$, then y^* is the unique solution of $Fy = 0$ in $\langle x^0, y^0 \rangle$. Finally, if F' is isotone on $\langle x^0, y^0 \rangle$ and $\omega_k \geq \omega > 0$, $k \geq k_0$, then the sequence $\{x^k\}$ defined by (7), (2), and (9) satisfies $x^k \uparrow y^*$, $k \to \infty$.

Proof. By **2.4.8**, $D(x)^{-1} \geq 0$ for any $x \in \langle x^0, y^0 \rangle$, and, hence, we have $\omega_k D(x)^{-1} L(x) \geq 0$. Therefore, suppressing in the following equations the explicit dependence of all matrices on x, we have

$$B_k^{-1} = \omega_k(I - \omega_k D^{-1}L)^{-1} D^{-1} = \omega_k \sum_{i=0}^{n-1} (\omega_k D^{-1}L)^i D^{-1} \geq 0, \quad (11)$$

and, thus,

$$H_k = B_k^{-1} C_k = (I - \omega_k D^{-1}L)^{-1} [(1 - \omega_k) I + \omega_k D^{-1} U] \geq 0.$$

Similarly,

$$C_k B_k^{-1} = [(1 - \omega_k) I + \omega_k U D^{-1}] D D^{-1}(I - \omega_k L D^{-1})^{-1} \geq 0,$$

so that $F'(x) = B_k(x) - C_k(x)$ is a weak regular splitting for all $x \in \langle x^0, y^0 \rangle$ and $k = 0, 1, \ldots$. The first part of the result now follows from **13.4.2**. Next, if F' is isotone on $\langle x^0, y^0 \rangle$, it follows that $D(y^k) \leq D(y^0)$ for all k, so that $D(y^k)^{-1} \geq D(y^0)^{-1}$. Hence, from (11), $B_k(y^k)^{-1} \geq \omega D(y^0)^{-1}$ for all k, and we may take $B = \omega D(y_0)^{-1}$ in **13.4.2**. On the other hand, if F' is continuous at y^*, then $\lim_{k \to \infty} D(y^k)^{-1} = D(y^*)^{-1}$. Hence, there exists a matrix E and an integer k_0 such that $B = \omega D(y^*)^{-1} - E \geq 0$ is nonsingular and $\omega D(y^k)^{-1} \geq B$ for $k \geq k_0$. Thus, **13.4.2** again ensures that $Fy^* = 0$. The uniqueness of y^* in $\langle x^0, y^0 \rangle$ follows immediately from **13.2.5** and **13.3.2**. Finally, if F' is isotone on $\langle x^0, y^0 \rangle$, then the convergence of $\{x^k\}$ is a consequence of **13.4.2**. ∎

We know from **10.3.2** that with increasing m_k the Newton–SOR iterates have an increasing asymptotic rate of convergence, tending for $m_k \to \infty$ to the superlinear convergence of the Newton iterates. Under the assumptions of the previous theorem, we can give the following comparison result for different Newton–SOR processes and Newton's method.

13.4. NEWTON-SOR ITERATIONS

13.4.4. Let $F: D \subset R^n \to R^n$ and $x^0, y^0 \in D$ satisfy (5). Suppose, further, that F is continuous and G-differentiable with isotone F' on $\langle x^0, y^0 \rangle$, and that $F'(x)$ is an M-matrix for each $x \in \langle x^0, y^0 \rangle$. Consider any two Newton–SOR sequences $\{y^k\}$ and $\{\hat{y}^k\}$ defined by (1), (2), and (9), with $0 < \hat{\omega}_k = \omega_k \leqslant 1$, $1 \leqslant \hat{m}_k \leqslant m_k$, and $\hat{y}^0 = y^0$, respectively, as well as the corresponding Newton sequence $\{u^k\}$ starting at y^0. Then

$$u^k \leqslant y^k \leqslant \hat{y}^k, \quad k = 0, 1, \ldots. \tag{12}$$

Proof. All conditions of **13.3.4** and **13.4.3** are satisfied; hence, the three sequences are well defined, contained in $\langle x^0, y^0 \rangle$, and monotonically decreasing. Since F' is isotone, D, $-L$, and $-U$ have the same property, and we see that for $x^0 \leqslant x \leqslant y \leqslant y^0$

$$D(y)^{-1}L(y) \leqslant D(x)^{-1}L(x), \quad [I - \omega_k D(y)^{-1} L(y)]^{-1} \leqslant [I - \omega_k D(x)^{-1} L(x)]^{-1},$$

and

$$(1/\omega_k)[(1 - \omega_k) I + \omega_k D(y)^{-1} U(y)] \leqslant (1/\omega_k)[(1 - \omega_k) I + \omega_k D(x)^{-1} U(x)].$$

Therefore,

$$B_k(y)^{-1} \leqslant B_k(x)^{-1}, \quad H_k(y) \leqslant H_k(x), \quad x^0 \leqslant x \leqslant y \leqslant y^0,$$

and if $P_k(x)$ and $\hat{P}_k(x)$ are defined by (8) with m_k and $\hat{m}_k$, respectively, then

$$\hat{P}_k(y) \leqslant P_k(y) \leqslant P_k(x), \quad x^0 \leqslant x \leqslant y \leqslant y^0. \tag{13}$$

Here, the first of the inequalities in (13) is an immediate consequence of $H_k(y) \geqslant 0$. Moreover, we have

$$P_k(x) \leqslant [I - H_k(x)]^{-1} B_k(x)^{-1} = [B_k(x) - C_k(x)]^{-1} = F'(x)^{-1}.$$

Now, (12) follows easily by induction. In fact, if it holds for some $k \geqslant 0$, then (13), together with the order-convexity of F and the facts that $Fy^k \geqslant 0$ and $P_k(y)$ is a subinverse of $F'(y)$, show that

$$\hat{y}^{k+1} - y^{k+1} = \hat{y}^k - y^k + [P_k(y^k) - \hat{P}_k(\hat{y}^k)] Fy^k - \hat{P}_k(\hat{y}^k)[F\hat{y}^k - Fy^k]$$
$$\geqslant [I - \hat{P}_k(\hat{y}^k) F'(\hat{y}^k)](\hat{y}^k - y^k) \geqslant 0.$$

In a similar fashion, we obtain

$$y^{k+1} - u^{k+1} = y^k - u^k + [F'(u^k)^{-1} - P_k(y^k)] Fu^k - P_k(y^k)[Fy^k - Fu^k]$$
$$\geqslant [I - P_k(y^k) F'(y^k)](y^k - u^k) \geqslant 0. \quad \blacksquare$$

As an application of the previous results, we again consider the problem $Ax + \phi x = 0$.

13.4.5. Let $A \in L(R^n)$ be an M-matrix and set $Fx = Ax + \phi x$, where $\phi: R^n \to R^n$ is continuously differentiable, diagonal, isotone, and convex. Assume that there exist points x^0, y^0 for which (5) holds. Then all of the conclusions of both **13.4.3** and **13.4.4** hold, where $x^* = y^*$ is the unique solution of $Fx = 0$.

The proof is immediate since F is convex, F' is isotone, and, by **2.4.11**, $F'(x)$ is an M-matrix for all $x \in R^n$. Hence, **13.4.3** and **13.4.4** both apply and, by **13.1.3**, $y^* = x^*$.

Given the conditions on A and ϕ in **13.4.5**, it is still necessary to find suitable starting points x^0 and y^0. One possibility is to apply **13.3.6** and then **13.3.5** with $C = A^{-1}$. Other possibilities which require weaker conditions on A and ϕ are given in the following lemma.

13.4.6. Set $Fx = Ax + \phi x$, where $A \in L(R^n)$ is nonsingular, $A^{-1} \geqslant 0$, and $\phi: R^n \to R^n$.

(a) If $-a \leqslant \phi x \leqslant a$ for some $a \geqslant 0$ and all $x \in R^n$, then $Fx^0 \leqslant 0 \leqslant Fy^0$ for $y^0 = A^{-1}a$, $x^0 = -y^0$.

(b) If $\phi(0) \leqslant 0$ and $\phi x \geqslant \phi(0)$ for all $x \geqslant 0$, then $Fy^0 \geqslant 0$ for $y^0 = -A^{-1}\phi(0)$.

(c) If ϕ is isotone, then $Fx^0 \leqslant 0 \leqslant Fy^0$ for $y^0 = A^{-1}|\phi(0)|$ and $x^0 = -y^0$.

Proof. (a) $Fx^0 = Ax^0 + \phi x^0 = -a + \phi x^0 \leqslant 0 \leqslant a + \phi y^0 = Fy^0$.

(b) $Fy^0 = Ay^0 + \phi y^0 = -\phi(0) + \phi y^0 \geqslant 0$.

(c) Clearly, $y^0 \geqslant 0$ and hence $x^0 \leqslant 0$ and $\phi x^0 \leqslant \phi(0) \leqslant \phi(y^0)$. Thus,

$$Fx^0 = -|\phi(0)| + \phi x^0 \leqslant -|\phi(0)| + \phi(0) \leqslant 0 \leqslant |\phi(0)| + \phi(0)$$
$$\leqslant |\phi(0)| + \phi y^0 = Fy^0. \blacksquare$$

Note that, in each case of **13.4.6**, as with **13.3.5** and **13.3.6**, the evaluation of at least one of the two points requires the solution of a linear system of equations.

While **13.4.5**, together with **13.4.6**, provides us with an effective means of approximating the unique solution of $Ax + \phi x = 0$ and also gives guaranteed two-sided error estimates

$$x^k \leqslant x^* \leqslant y^k, \quad k = 0, 1, \ldots,$$

13.4. NEWTON–SOR ITERATIONS

it is of interest that, at least for the one-step Newton–SOR iteration with $\omega = 1$, we are able to prove global convergence. This will be a consequence of the following, more general result.

13.4.7. Let $F: R^n \to R^n$ be continuously differentiable and convex, and assume that there exists a mapping $B: R^n \to L(R^n)$ such that $B(x)$ is nonsingular for all $x \in R^n$ and

$$0 \leqslant B(x)^{-1} \leqslant C_0, \qquad (14)$$
$$0 \leqslant B(x) - F'(x) \leqslant C_1, \qquad \forall x \in R^n, \qquad (15)$$

where $\rho(C) < 1$ with $C = C_0 C_1$. Then for any $x^0 \in R^n$ the iterates

$$x^{k+1} = x^k - B(x^k)^{-1} F x^k, \qquad k = 0, 1, \dots,$$

converge to the unique solution x^* of $Fx = 0$.

Proof. Observe, first, that

$$0 \leqslant I - B(x)^{-1} F'(x) = B(x)^{-1} [B(x) - F'(x)] \equiv H(x) \leqslant C, \qquad \forall x \in R^n.$$

Therefore, **2.4.5** and **2.4.9** imply that $I - H(x)$ and, hence, $F'(x)$ are nonsingular and

$$0 \leqslant F'(x)^{-1} = (I - H(x))^{-1} B(x)^{-1} \leqslant (I - C)^{-1} C_0, \qquad \forall x \in R^n.$$

It follows that $\| F'(x)^{-1} \|$ is uniformly bounded in any norm and, by the Hadamard theorem **5.3.10**, this shows that $Fx = 0$ has a unique solution x^*.

Clearly, the sequence $\{x^k\}$ is well defined, and, by convexity of F, we have

$$x^{k+1} - x^* = x^k - x^* - B(x^k)^{-1} (Fx^k - Fx^*) \geqslant [I - B(x^k)^{-1} F'(x^k)](x^k - x^*)$$
$$= H(x^k)(x^k - x^*) \geqslant \left[\prod_{j=0}^{k} H(x^j) \right] (x^0 - x^*) \geqslant -C^k | x^0 - x^* |.$$

Since $\rho(C) < 1$, the right-hand side tends to zero and, hence, $\{x^k\}$ is bounded below; that is, we have

$$x^k \geqslant w, \qquad k = 0, 1, \dots, \qquad (16)$$

for some $w \in R^n$. Similarly, we obtain

$$x^{k+1} - x^k = B(x^k)^{-1}(-Fx^k) \leqslant B(x^k)^{-1} [-Fx^{k-1} - F'(x^{k-1})(x^k - x^{k-1})]$$
$$= B(x^k)^{-1} [B(x^{k-1}) - F'(x^{k-1})](x^k - x^{k-1}) \equiv K_k(x^k - x^{k-1}),$$

where
$$0 \leqslant K_k = B(x^k)^{-1}[B(x^{k-1}) - F'(x^{k-1})] \leqslant C, \quad k = 1, 2, \ldots.$$

It follows that
$$x^{k+1} - x^k \leqslant \left(\prod_{j=1}^{k} K_j\right)(x^1 - x^0) \leqslant C^k \mid x^1 - x^0 \mid,$$

and, in particular, that for any $k > m \geqslant 1$

$$x^k - x^m \leqslant \left[\sum_{j=m}^{k-1} C^j\right] \mid x^1 - x^0 \mid \leqslant C^m(I - C)^{-1} \mid x^1 - x^0 \mid, \quad (17)$$

since $C \geqslant 0$. For fixed m, (17) shows that $\{x^k\}$ is bounded above, which, together with (16), implies that $\{x^k\}$ is bounded. Hence, the set Ω of limit points of $\{x^k\}$ is compact and nonempty. Let v be any point of Ω and $\{x^{k_i}\}$ a subsequence of $\{x^k\}$ converging to v. Then, given $\epsilon > 0$, there exists an i_0 so that $x^{k_i} \leqslant v + \frac{1}{2}\epsilon e$, $e = (1,\ldots, 1)^\mathrm{T}$, for $i \geqslant i_0$, and, hence, if i_0 was chosen sufficiently large,

$$x^k \leqslant x^m + C^m(I - C)^{-1} \mid x^1 - x^0 \mid \leqslant v + \epsilon e, \quad \forall k \geqslant m = k_{i_0}.$$

Since $\epsilon > 0$ was arbitrary, this means that any other limit point $u \in \Omega$ of $\{x^k\}$ must satisfy $u \leqslant v$. But $v \in \Omega$ was arbitrary and, hence, Ω can contain only one point v, so that $\lim_{k\to\infty} x^k = v$. To show that $v = x^*$, note that (15) implies that

$$F'(x^k) \leqslant B(x^k) \leqslant C_1 + F'(x^k)$$

so that, by continuity of F', $\{B(x^k)\}$ is bounded. Consequently,

$$Fv = \lim_{k\to\infty} Fx^k = \lim_{k\to\infty} B(x^k)(x^{k+1} - x^k) = 0,$$

and, thus, $v = x^*$. ∎

As an almost immediate corollary, we have the previously announced result for the equation $Ax + \phi x = 0$.

13.4.8. Global Newton–SOR Theorem. Suppose A and ϕ satisfy the conditions of **13.4.5**, and $A = D - L - U$ is the splitting of A into its diagonal, lower-, and upper-triangular parts. Then for any $x^0 \in R^n$ the one-step Newton–SOR iterates

$$x^{k+1} = x^k - [D + \phi'(x^k) - L]^{-1} Fx^k, \quad k = 0, 1, \ldots, \quad (18)$$

converge to the unique solution x^* of $Fx = 0$.

13.4. NEWTON-SOR ITERATIONS

Proof. To apply Theorem **13.4.7**, set $B(x) = D + \phi'(x) - L$. Then **2.4.10** and **2.4.11** imply that $B(x)$ is an M-matrix and that

$$0 \leq B(x)^{-1} \leq (D - L)^{-1}, \quad \forall x \in R^n.$$

Moreover, we have

$$0 \leq B(x) - F'(x) = U.$$

Since $A = (D - L) - U$ is a regular splitting, **2.4.17** ensures that $\rho((D - L)^{-1} U) < 1$, and, hence, the result follows directly from **13.4.7**. ∎

As in **13.3.8**, the convexity assumption on ϕ in **13.4.8** cannot be removed (see **E 13.4-5**).

We note that under the conditions of **13.4.8** we have the option of first obtaining suitable starting points and applying **13.4.5** in order to obtain two-sided error estimates

$$x^k \leq x^* \leq y^k, \quad k = 0, 1,\ldots, \tag{19}$$

or to start with an arbitrary point. In the latter case, we are still assured of convergence, but no longer, in general, will the convergence be monotonic.

NOTES AND REMARKS

NR 13.4-1. Theorems **13.4.1–13.4.6** were given in Ortega and Rheinboldt [1967a] as generalizations of results of Greenspan and Parter [1965], who treated discrete analogs of elliptic boundary value problems.

NR 13.4-2. Theorem **13.4.8** is due to Greenspan and Parter [1965], and the more general **13.4.7**, whose proof is a modification of that used by Greenspan and Parter, was given in Ortega and Rheinboldt [1970a]. It is of interest to note, however, that **13.4.7** is not sufficiently general to handle, under the conditions of **13.4.8**, m-step Newton–SOR methods nor even the one-step method if $\omega \neq 1$.

NR 13.4-3. Theorems **13.4.5** and **13.4.8** have immediate application to boundary value problems for $\Delta u = f(u)$ provided f is isotone and convex (see **E 13.4-4**). More generally, it is shown in Ortega and Rheinboldt [1967a] that corresponding results may be given for $\Delta u = f(u, u_s, u_t)$ provided, in particular, that f is a convex functional on R^3.

EXERCISES

E 13.4-1. Suppose that $A \in L(R^n)$ has a positive row or column. Show that the zero matrix is the only nonnegative subinverse of A.

E 13.4-2. Give an example to show that $AB \geq I$, $BA \geq I$, $B \geq 0$, and B nonsingular, do not imply that $A \geq B^{-1}$.

E 13.4-3. Extend **13.4.3** to block splittings of $F'(x)$ (see **NR 7.4-10**).

E 13.4-4. Apply **13.4.5** and **13.4.8** to the equations (1.1.7) and (1.2.5), where $f(s, u) = e^u$ or $f(s, t, u) = e^u$, respectively.

E 13.4-5. Let $A \in L(R^2)$ and $\phi: R^2 \to R^2$ be given by

$$A = \begin{pmatrix} 1 & -1 \\ -1 & 2 \end{pmatrix}, \quad \phi x = \begin{pmatrix} 2x_1 + 2\sin x_1 \\ x_2 + \sin x_2 \end{pmatrix}.$$

Show that A and ϕ satisfy all conditions of **13.4.8** except for convexity of ϕ, but that if $x^0 = (\pi, \pi)^T$, then $x^{2k} = x^0$, $k = 1, 2, \ldots$.

E 13.4-6. Apply **13.4.7** to prove again the global Newton theorem **13.3.7** under the more restrictive assumptions that $F: R^n \to R^n$ is continuously differentiable and convex and that $0 \leq F'(x)^{-1} \leq C$ for some $C \in L(R^n)$ and all $x \in R^n$.

13.5. M-FUNCTIONS AND NONLINEAR SOR PROCESSES

In this section we consider monotone and global convergence theorems for the nonlinear SOR process

$$\begin{cases} \text{Solve } f_i(x_1^{k+1}, \ldots, x_{i-1}^{k+1}, x_i, x_{i+1}^k, \ldots, x_n^k) = b_i & \text{for } x_i; \\ \text{Set } x_i^{k+1} = x_i^k + \omega(x_i - x_i^k); \quad i = 1, \ldots, n, \quad k = 0, 1, \ldots. \end{cases} \quad (1)$$

discussed in **7.4**, as well as the nonlinear Jacobi process

$$\begin{cases} \text{Solve } f_i(x_1^k, \ldots, x_{i-1}^k, x_i, x_{i+1}^k, \ldots, x_n^k) = b_i & \text{for } x_i; \\ \text{Set } x_i^{k+1} = x_i^k + \omega(x_i - x_i^k); \quad i = 1, \ldots, n, \quad k = 0, 1, \ldots. \end{cases} \quad (2)$$

provided, in each case, that $\omega \in (0, 1]$. The following definition delineates the class of functions F of interest.

13.5.1. Definition. A mapping $F: R^n \to R^n$ is *diagonally isotone* if, for any $x \in R^n$, the n functions

$$\psi_{ii}: R^1 \to R^1, \quad \psi_{ii}(t) = f_i(x + te^i), \quad i = 1, \ldots, n, \quad (3)$$

13.5. M-FUNCTIONS AND NONLINEAR SOR PROCESSES

are isotone. The function F is *strictly diagonally isotone* if, for any $x \in R^n$, the ψ_{ii}, $i = 1,\ldots, n$, are strictly isotone, and, finally, F is *off-diagonally antitone* if, for any $x \in R^n$, the functions

$$\psi_{ij}: R^1 \to R^1, \qquad \psi_{ij}(t) = f_i(x + te^j), \qquad i \neq j, \quad i,j = 1,\ldots, n, \qquad (4)$$

are antitone. ∎

13.5.2. Let $F: R^n \to R^n$ be continuous, off-diagonally antitone, and strictly diagonally isotone, and suppose that for some $b \in R^n$ there exist points $x^0, y^0 \in R^n$ such that

$$x^0 \leqslant y^0, \qquad Fx^0 \leqslant b \leqslant Fy^0. \qquad (5)$$

Then, for any $\omega \in (0, 1]$, the SOR iterates $\{y^k\}$ and $\{x^k\}$ given by (1) and starting from y^0 and x^0, respectively, are uniquely defined and satisfy

$$x^k \uparrow x^*, \qquad y^k \downarrow y^*, \quad k \to \infty, \qquad x^* \leqslant y^*, \qquad Fx^* = Fy^* = b. \qquad (6)$$

The corresponding result holds for the Jacobi iteration (2).

Proof. We give the proof only for the SOR iteration; the analogous proof for the Jacobi process is left to **E 13.5-1**.

As induction hypothesis, suppose that for some $k \geqslant 0$ and $i \geqslant 1$

$$x^0 \leqslant x^k \leqslant y^k \leqslant y^0, \qquad Fx^k \leqslant b \leqslant Fy^k, \qquad (7)$$

$$x_j^k \leqslant x_j^{k+1} \leqslant y_j^{k+1} \leqslant y_j^k, \qquad j = 1,\ldots, i-1, \qquad (8)$$

where for $i = 1$ the set of j satisfying (8) is empty. Clearly, (7) and (8) hold for $k = 0$, $i = 1$. By the off-diagonal antitonicity, it now follows that the functions

$$\alpha(s) = f_i(x_1^{k+1},\ldots, x_{i-1}^{k+1}, s, x_{i+1}^k,\ldots, x_n^k),$$

$$\beta(s) = f_i(y_1^{k+1},\ldots, y_{i-1}^{k+1}, s, y_{i+1}^k,\ldots, y_n^k),$$

satisfy

$$\beta(s) \leqslant \alpha(s), \qquad \forall s \in R^1, \qquad (9)$$

and

$$\beta(x_i^k) \leqslant \alpha(x_i^k) \leqslant f_i(x^k) \leqslant b_i \leqslant f_i(y^k) \leqslant \beta(y_i^k) \leqslant \alpha(y_i^k). \qquad (10)$$

By the continuity and strict isotonicity of α and β, (10) implies the existence of unique $\hat{y}_i^k$ and $\hat{x}_i^k$ for which

$$\beta(\hat{y}_i^k) = b_i = \alpha(\hat{x}_i^k), \qquad x_i^k \leqslant \hat{x}_i^k \leqslant \hat{y}_i^k \leqslant y_i^k,$$

where $\hat{x}_i^k \leqslant \hat{y}_i^k$ is a consequence of (9). Since $\omega \in (0, 1]$, we therefore have

$$y_i^k \geqslant y_i^{k+1} = y_i^k + \omega(\hat{y}_i^k - y_i^k) \geqslant \hat{y}_i^k \geqslant \hat{x}_i^k \geqslant x_i^{k+1}$$
$$= x_i^k + \omega(\hat{x}_i^k - x_i^k) \geqslant x_i^k,$$

which shows that (8) holds for $i = 1,\ldots, n$, and, hence, that $x^k \leqslant x^{k+1} \leqslant y^{k+1} \leqslant y^k$. But then we obtain

$$f_i(y^{k+1}) \geqslant f_i(y_1^{k+1},\ldots, y_i^{k+1}, y_{i+1}^k, \ldots, y_n^k)$$
$$\geqslant f_i(y_1^{k+1},\ldots, y_{i-1}^{k+1}, \hat{y}_i^k, y_{i+1}^k, \ldots, y_n^k) = b_i,$$

and, similarly, $f_i(x^{k+1}) \leqslant b_i$, $i = 1,\ldots, n$. This completes the induction and, hence, the proof of (7). Clearly, the limits $x^* \leqslant y^*$ exist, and, since $\omega > 0$, we have

$$\lim_{k \to \infty} \hat{x}_i^k = (1/\omega) \lim_{k \to \infty} (x_i^{k+1} - x_i^k) + \lim_{k \to \infty} x_i^k = x_i^*, \qquad i = 1,\ldots, n,$$

and, similarly, $\lim_{k \to \infty} \hat{y}^k = y^*$. Therefore, it follows from the definition (1) of the SOR process and the continuity of F that $Fx^* = Fy^* = b$. ∎

Theorem 13.5.2 has application to the equation $Ax + \phi x = 0$. Before stating a specific corollary, however, we introduce some additional terminology.

13.5.3. Definition. A mapping $F: R^n \to R^n$ is *inverse isotone* if $Fx \leqslant Fy$ for any $x, y \in R^n$ implies that $x \leqslant y$. ∎

13.5.4. A mapping $F: R^n \to R^n$ is inverse isotone if and only if F is one-to-one and $F^{-1}: FR^n \subset R^n \to R^n$ is isotone.

Proof. If F is inverse isotone, then $Fx = Fy$ implies both $x \leqslant y$ and $x \geqslant y$ so that $x = y$, and F is one-to-one. Moreover, if $u = Fx$, $v = Fy$, and $u \leqslant v$, then $F^{-1}u = x \leqslant y = F^{-1}v$, so that F^{-1} is isotone. Similarly, if F is one-to-one and F^{-1} is isotone on FR^n, then $v = Fy \geqslant Fx = u$ implies that $y = F^{-1}v \geqslant F^{-1}u = x$. ∎

Another interesting property of inverse isotone functions is given by the following result.

13.5.5. A continuous inverse isotone function $F: R^n \to R^n$ with $FR^n = R^n$ is a homeomorphism from R^n onto itself.

13.5. M-FUNCTIONS AND NONLINEAR SOR PROCESSES

Proof. Because of **13.5.4** it suffices to show that $F^{-1}: R^n \to R^n$ is continuous. Let $\{y^k\} \subset R^n$ be such that $\lim_{k \to \infty} y^k = y$. Then $\{y^k\}$ is bounded, and if, say, $u \leqslant y^k \leqslant v$, then it follows from the inverse isotonicity that $F^{-1}u \leqslant x^k = F^{-1}y^k \leqslant F^{-1}v$ for all $k \geqslant 0$, so that $\{x^k\}$ is bounded. If x is any limit point of $\{x^k\}$, and if $\lim_{i \to \infty} x^{k_i} = x$, then, by the continuity of F, we have $Fx = \lim_{i \to \infty} Fx^{k_i} = \lim_{i \to \infty} y^{k_i} = y$, or $x = F^{-1}y$. Hence, $\{x^k\}$ has the unique limit point $F^{-1}y$, which shows that $\lim_{k \to \infty} F^{-1}y^k = F^{-1}y$ and, therefore, that F^{-1} is continuous. ∎

As an application of the previous results, we return to the equation $Ax + \phi x = 0$.

13.5.6. Let $A \in L(R^n)$ be an M-matrix, $\phi: R^n \to R^n$ a continuous, isotone, and diagonal mapping, and set $Fx = Ax + \phi x$, $x \in R^n$. Then F is inverse isotone and a homeomorphism of R^n onto itself. Moreover, for any $b \in R^n$, set

$$y^0 = A^{-1} \mid \phi(0) - b \mid, \qquad x^0 = -y^0; \tag{11}$$

then the SOR iterates (1) for any $\omega \in (0, 1]$ and starting from x^0 and y^0, respectively, satisfy

$$x^k \uparrow x^*, \qquad y^k \downarrow x^*, \qquad k \to \infty, \tag{12}$$

where x^* is the unique solution of $Fx = b$. Precisely the same convergence result holds for the Jacobi iterates (2).

Proof. We show first that F is inverse isotone. Suppose that $Fx \leqslant Fy$ for some $x, y \in R^n$ for which $x \not\leqslant y$, and set $S = \{1 \leqslant j \leqslant n \mid x_j > y_j\}$. Then, by the isotonicity of ϕ and the fact that the off-diagonal elements of A are nonpositive, we obtain

$$0 \leqslant f_j(y) - f_j(x) = \sum_{k=1}^{n} a_{jk}(y_k - x_k) + \varphi_j(y_j) - \varphi_j(x_j)$$

$$\leqslant \sum_{k \in S} a_{jk}(y_k - x_k), \qquad j \in S. \tag{13}$$

But by **2.4.10** it follows that the submatrix $(a_{jk} \mid j, k \in S)$ is also an M-matrix, so that (13) shows that $y_j \geqslant x_j$ for all $j \in S$. This is a contradiction, and F is inverse isotone.

Now, let $b \in R^n$ be arbitrary and $x^0 \leqslant y^0$ given by (11); then **13.4.6(c)** shows that $Fx^0 \leqslant b \leqslant Fy^0$. Clearly, F satisfies the conditions of **13.5.2** (see **2.4.8**), and, hence, the SOR iterates satisfy (6). But, by **13.5.4**, F is one-to-one, so that $x^* = y^*$. Therefore, we have also shown that F is onto R^n, and **13.5.5** ensures that F is a homeomorphism.

The convergence proof for the Jacobi iterates is also an immediate consequence of **13.5.2**. ∎

Note that we invoked **13.4.6** for the particular choice of points (11), but any other choice for which (5) is satisfied would, of course, yield the same result. Note, also, that we have already shown in **13.1.3** that $A + \phi$ is one-to-one and onto; however, **13.5.6** gives a different proof of that result, and, in conjunction with **13.5.5**, the stronger statement that F is a homeomorphism. On the other hand, **13.1.3-5** showed the global convergence of the Jacobi and SOR iterates.

We show next that global convergence is also a consequence of the present setting under somewhat stronger conditions on the function F of **13.5.2**.

In analogy to the concept of an M-matrix, we now introduce the following class of functions.

13.5.7. Definition. A mapping $F: R^n \to R^n$ is an *M-function* if F is inverse isotone and off-diagonally antitone. ∎

Clearly, an affine function $Ax + b$, $A \in L(R^n)$, $b \in R^n$, is an M-function if and only if A is an M-matrix (**E 13.5-4**). Moreover, **13.5.6** shows that $A + \phi$, under the conditions of that theorem, is an M-function.

The diagonal elements of an M-matrix are necessarily positive. For M-functions, this property extends as follows.

13.5.8. An M-function $F: R^n \to R^n$ is strictly diagonally isotone. Moreover, if $FR^n = R^n$, then for any $x \in R^n$ and $1 \leqslant i \leqslant n$

$$\lim_{t \to +\infty} f_i(x + te^i) = +\infty, \qquad \lim_{t \to -\infty} f_i(x + te^i) = -\infty. \tag{14}$$

Proof. Suppose that for some $x \in R^n$ there exists an index i and numbers $t > s$ such that

$$f_i(x + se^i) \geqslant f_i(x + te^i).$$

By the off-diagonal antitonicity, we have

$$f_j(x + se^i) \geqslant f_j(x + te^i), \qquad i \neq j, \quad j = 1,...,n,$$

and, hence, altogether,

$$F(x + se^i) \geqslant F(x + te^i),$$

which, by the inverse isotonicity, leads to the contradiction $s \geqslant t$. Thus,

13.5. M-FUNCTIONS AND NONLINEAR SOR PROCESSES

F is strictly diagonally isotone. Now, assume $FR^n = R^n$, and suppose that the first condition of (14) does not hold; that is, for some $x \in R^n$ and some index i, there is sequence $\{t_k\} \subset R^1$ with $\lim_{k\to\infty} t_k = +\infty$ such that

$$f_i(x + t_k e^i) \leqslant a_i < +\infty, \qquad k = 0, 1, \ldots\,.$$

If $t_k \geqslant t$, $k = 0, 1, \ldots$, then, again by the off-diagonal antitonicity,

$$f_j(x + t_k e^i) \leqslant f_j(x + t e^i) \equiv a_j < +\infty, \qquad j \neq i, \quad j = 1, \ldots, n, \quad k = 0, 1, \ldots,$$

or

$$F(x + t_k e^i) \leqslant a = (a_1, \ldots, a_n)^T, \qquad k = 0, 1, \ldots\,.$$

Since $FR^n = R^n$, there is a $y \in R^n$ so that $Fy = a$ and, hence, by the inverse isotonicity,

$$x + t_k e^i \leqslant y, \qquad k = 0, 1, \ldots\,.$$

This shows that $\{t_k\}$ is bounded above which is a contradiction. The proof for the second condition of (14) is analogous. ∎

By **13.5.8** any continuous M-function satisfies the conditions of **13.5.2**. Moreover, with the additional assumption that F is onto, we are able to show global convergence of the SOR or Jacobi iterates.

13.5.9. Let $F\colon R^n \to R^n$ be a continuous M-function from R^n onto itself. Then for any $b \in R^n$, any starting point $x^0 \in R^n$, and any $\omega \in (0, 1]$, the SOR iterates (1), as well as the Jacobi iterates (2), converge to the unique solution x^* of $Fx = b$.

Proof. We prove the result again only for the SOR iteration. For given $x^0, b \in R^n$, define

$$\begin{aligned} u^0 &= F^{-1}(\max[f_1(x^0), b_1], \ldots, \max[f_n(x^0), b_n]), \\ v^0 &= F^{-1}(\min[f_1(x^0), b_1], \ldots, \min[f_n(x^0), b_n]). \end{aligned} \qquad (15)$$

Then, by the inverse isotonicity,

$$Fu^0 \geqslant b \geqslant Fv^0, \qquad u^0 \geqslant x^0 \geqslant v^0, \qquad u^0 \geqslant x^* \geqslant v^0.$$

Let $\{u^k\}$, $\{v^k\}$, and $\{x^k\}$ denote the SOR sequences starting from u^0, v^0, and x^0, respectively, each of them formed with the same $\omega \in (0, 1]$. By

13.5.8 and the continuity of F, the solutions $\hat{u}_i^k$, $\hat{v}_i^k$, and $\hat{x}_i^k$ of the equations

$$\left.\begin{array}{l} f_i(u_1^{k+1},...,u_{i-1}^{k+1},\hat{u}_i^k,u_{i+1}^k,...,u_n^k) = b_i \\ f_i(v_1^{k+1},...,v_{i-1}^{k+1},\hat{v}_i^k,v_{i+1}^k,...,v_n^k) = b_i \\ f_i(x_1^{k+1},...,x_{i-1}^{k+1},\hat{x}_i^k,x_{i+1}^k,...,x_n^k) = b_i \end{array}\right\} \quad i=1,...,n, \quad k=0,1,...,$$

exist and are unique, and, therefore, the three SOR sequences are well-defined. Moreover, by **13.5.2** we have

$$v^0 \leqslant v^k \leqslant v^{k+1} \leqslant \lim_{k \to \infty} v^k = x^* = \lim_{k \to \infty} u^k \leqslant u^{k+1} \leqslant u^k \leqslant u^0, \tag{16}$$

$$Fv^k \leqslant b \leqslant Fu^k, \quad k=0,1,....$$

Suppose that for some $k \geqslant 0$ and $i \geqslant 1$

$$v^k \leqslant x^k \leqslant u^k, \quad v_j^{k+1} \leqslant x_j^{k+1} \leqslant u_j^{k+1}, \quad j=1,...,i-1, \tag{17}$$

which is valid for $k = 0$, and, vacuously, for $i = 1$. Then

$$f_i(u_1^{k+1},...,u_{i-1}^{k+1},\hat{u}_i^k,u_{i+1}^k,...,u_n^k) = b_i = f_i(x_1^{k+1},...,x_{i-1}^{k+1},\hat{x}_i^k,x_{i+1}^k,...,x_n^k)$$
$$\geqslant f_i(u_1^{k+1},...,u_{i-1}^{k+1},\hat{x}_i^k,u_{i+1}^k,...,u_n^k),$$

together with the strict diagonal isotonicity of F, implies that $\hat{u}_i^k \geqslant \hat{x}_i^k$. Similarly, we find that $\hat{v}_i^k \leqslant \hat{x}_i^k$. Hence, because of $\omega \in (0, 1]$, it follows that

$$v_i^{k+1} = v_i^k + \omega(\hat{v}_i^k - v_i^k) \leqslant x_i^k + \omega(\hat{x}_i^k - x_i^k)$$
$$= x_i^{k+1} \leqslant u_i^k + \omega(\hat{u}_i^k - u_i^k) = u_i^{k+1}.$$

This completes the induction, and (17) and (16) together now imply that $\lim_{k \to \infty} x^k = x^*$. ∎

As immediate corollaries of **13.5.9** together with **13.5.6**, we again obtain the global convergence theorems **13.1.3** and **13.1.5** for the Jacobi and SOR iterations applied to the equation $Ax + \phi x = 0$.

NOTES AND REMARKS

NR 13.5-1. Theorem **13.5.2** for the special case $Fx \equiv Ax + \phi x$, where A and ϕ are as in **13.5.6**, was given in Ortega and Rheinboldt [1970a] and represented an improvement on an earlier result (Ortega and Rheinboldt [1967a]) in which ϕ

13.5. M-FUNCTIONS AND NONLINEAR SOR PROCESSES

was assumed to be continuously differentiable and convex. For a certain class of mappings arising in the theory of thermal networks, Birkhoff and Kellogg [1966] proved a result similar to **13.5.2** for the Jacobi process, and Porsching [1969] extended it to the SOR process. Porsching also gave a comparison result between the SOR and Jacobi iterates. In our terms, a somewhat more general result, also allowing comparison for different ω, may be stated as follows.

Suppose that the conditions of **13.5.2** hold and $0 < \omega \leqslant \bar{\omega} \leqslant 1$. Let $\{y^k\}$ and $\{\bar{y}^k\}$ be the SOR iterates (1) with ω and $\bar{\omega}$, respectively, and with $y^0 = \bar{y}^0$, and let $\{v^k\}$ and $\{\bar{v}^k\}$ denote the corresponding Jacobi sequences. Then

$$y^k \geqslant \bar{y}^k \geqslant y^*, \qquad v^k \geqslant \bar{v}^k \geqslant y^*, \qquad v^k \geqslant y^k, \qquad k = 0, 1, \dots,$$

where y^* is the maximal solution of $Fx = b$ in $\langle x^0, y^0 \rangle$.

For a proof of this, and also of the corresponding result with the sequences starting from x^0, and y^* replaced by x^*, see Rheinboldt [1969b].

NR 13.5-2. The concept of an inverse isotone operator on a partially ordered linear space is due to Collatz [1952], who used the terminology "operator of monotone kind." Since then, a considerable literature dealing with conditions which ensure inverse isotonicity has developed, especially for differential operators. For a summary, see Collatz [1964] and, for some central results, Schröder [1962, 1966].

NR 13.5-3. Theorem **13.5.5** is a special case of the more general result that if $F: R^n \to R^n$ is continuous and inverse isotone, then F^{-1} is continuous on FR^n. The proof follows that of **13.5.5** after noting that, by the domain-invariance theorem (see **NR 6.2-1**), FR^n is open, and, hence, u and v may be chosen as vertices of a hypercube $S \subset FR^n$ such that $u \leqslant y^k \leqslant v$ for all $k \geqslant k_0$.

NR 13.5-4. The concept of an M-function was introduced by Ortega in an unpublished note and developed by Rheinboldt [1969b]. In particular, Theorems **13.5.8** and **13.5.9**, as well as additional results with application to nonlinear network problems and boundary value problems for $u'' = f(t, u, u')$ are given in that paper. For the class of mappings arising in the theory of thermal networks, mentioned in **NR 13.5-1**, Porsching [1969] has also proved the global convergence of the SOR process. Since Rheinboldt has shown these particular mappings to be M-functions, this convergence result is subsumed under **13.5.9**.

EXERCISES

E 13.5-1. Prove **13.5.2**, **13.5.6**, and **13.5.9** for the Jacobi process (2).

E 13.5-2. Given $F, G: R^n \to R^n$, show that FG is inverse isotone if both F and G are inverse isotone and that G is inverse isotone if F is isotone and FG is inverse isotone.

E 13.5-3. An inverse isotone function $F: R^n \to R^n$ is strictly inverse isotone if $Fx < Fy$ implies that $x < y$. Show that a continuous inverse isotone function is strictly inverse isotone. (Schröder [1962])

E 13.5-4. Show that the affine function $Ax + b$, $A \in L(R^n)$, $b \in R^n$, is an M-function if and only if A is an M-matrix.

E 13.5-5. Give a one-dimensional example of an M-function which is not onto.

E 13.5-6. Let $G: R^n \to R^n$ be an M-function onto R^n, and set $F = G^{-1}$. Show that (14) holds.

E 13.5-7. Apply **13.5.6** to the equations (1.1.7) and (1.2.5) under the assumptions of **4.4.2** and **4.4.3**, respectively.

Chapter 14 / **CONVERGENCE OF MINIMIZATION METHODS**

14.1. INTRODUCTION AND CONVERGENCE OF SEQUENCES

In this chapter we shall consider the convergence of iterative methods of the general form

$$x^{k+1} = x^k - \omega_k \alpha_k p^k, \quad k = 0, 1, \dots, \qquad (1)$$

for finding minimizers or critical points of a given functional g: $D \subset R^n \to R^1$. Here, as in Chapter 8, p^k is a "direction" vector, α_k the basic steplength, and ω_k a relaxation parameter.

Our attention will be restricted to descent methods, that is, to methods for which the sequence (1) satisfies

$$g(x^{k+1}) \leqslant g(x^k), \quad k = 0, 1, \dots. \qquad (2)$$

Many of the algorithms discussed in Chapter 8 were, indeed, constructed so that this condition is valid. For others it will be necessary to restrict the parameter ω_k in (1) appropriately and, in any case, to verify that (2) holds.

The verification of (2) is always the first step of the convergence analysis, and if g is bounded below, then, clearly, (2) implies the convergence of the sequence $\{g(x^k)\}$. It is important to realize, however, that this by itself implies nothing about the validity of the final convergence statement:

$$\lim_{k \to \infty} x^k = x^*, \quad g'(x^*)^T = 0. \qquad (3)$$

Indeed, our analysis of (3) will be based upon consideration of two additional fundamental intermediate propositions, namely,

$$\lim_{k \to \infty} g'(x^k) p^k / \| p^k \| = 0, \tag{4}$$

and

$$\lim_{k \to \infty} g'(x^k)^T = 0. \tag{5}$$

In Chapter 8 we saw that there is considerable independence between the selection of p^k and α_k. The importance of the technical condition (4) is that this independence is maintained in the convergence analysis and that, in fact, the validity of (4) depends only upon the steplength algorithm and on very mild conditions for g itself. Section **14.2** is devoted exclusively to the analysis of various steplength algorithms with the verification of (4) as the primary consideration.

For a very important class of methods—the gradient-related methods to be treated in **14.3** and parts of **14.4** and **14.5**—an inequality of the form

$$g'(x^k) p^k \geq c \| g'(x^k)^T \| \| p^k \|, \quad c > 0, \quad \forall k \geq k_0, \tag{6}$$

holds. This ensures that (4) implies (5). The validity of (6) is primarily a property of the direction vectors p^k; intuitively, we may say that the vectors $g'(x^k)^T$ and p^k "are bounded away from orthogonality." On the other hand, the inequality (6) does not hold for such important methods as univariate relaxation (Gauss–Seidel) to be considered in **14.6**. In that case the deduction of (5) from (4) is still based strongly on the properties of the p^k but also requires additional assumptions on g and, in addition, frequently, the validity of another intermediate proposition:

$$\lim_{k \to \infty}(x^{k+1} - x^k) = 0. \tag{7}$$

The verification of (7), in turn, is sometimes immediate for certain steplength algorithms and sometimes requires additional assumptions on g.

Finally, (3) is usually a consequence of only (2) and (5), together with suitable assumptions on g. Under weaker assumptions on g it may still be valid provided that (7) also holds. The remainder of this first section is devoted to the question of deducing (3), as well as to a basic rate-of-convergence result. We begin with two definitions.

14.1.1. Definition. A functional $g: D \subset R^n \to R^1$ is *hemivariate* on a set $D_0 \subset D$ if it is not constant on any line segment of D_0, that is, if there exist no distinct points $x, y \in D_0$ such that $(1 - t)x + ty \in D_0$ and $g([1 - t]x + ty) = g(x)$ for all $t \in [0, 1]$. ∎

14.1. INTRODUCTION AND CONVERGENCE OF SEQUENCES

Clearly, any strictly quasiconvex functional (Definition 4.2.5) is hemivariate on convex subsets of D. (See also E 14.1-4.)

The following definition strengthens the descent condition (2) inherent in all the methods (1) considered here.

14.1.2. Definition. Given $g: D \subset R^n \to R^1$, a sequence $\{x^k\}$ in some subset $D_0 \subset D$ is *strongly downward* in D_0 if

$$(1-t)x^k + tx^{k+1} \in D_0, \qquad \forall t \in [0, 1], \tag{8}$$

and

$$g(x^k) \geq g([1-t]x^k + tx^{k+1}) \geq g(x^{k+1}), \qquad \forall t \in [0, 1]. \quad \blacksquare \tag{9}$$

With these definitions we can now prove the following result about (7).

14.1.3. Suppose that the functional $g: D \subset R^n \to R^1$ is continuous and hemivariate on a compact set $D_0 \subset D$. Then every strongly downward sequence $\{x^k\} \subset D_0$ satisfies $\lim_{k \to \infty} (x^k - x^{k+1}) = 0$.

Proof. Suppose there exists a subsequence $\{x^{k_i}\}$ such that

$$\| x^{k_i+1} - x^{k_i} \| \geq \epsilon > 0, \qquad \forall i \geq 0.$$

Because of the compactness of D_0 it is no restriction to assume that $\lim_{i \to \infty} x^{k_i} = x^*$ and $\lim_{i \to \infty} x^{k_i+1} = x^{**}$. Then $\| x^* - x^{**} \| \geq \epsilon > 0$, and, of course, $x^*, x^{**} \in D_0$. By assumption, we have $g(x^k) \geq g(x^{k+1})$, and, since g is bounded below on D_0, it follows that

$$\lim_{k \to \infty} (g(x^k) - g(x^{k+1})) = 0,$$

and, therefore, that $g(x^*) = g(x^{**})$. Finally, we obtain from (8) and the closedness of D_0 that $(1-t)x^* + tx^{**} \in D_0$ for $t \in [0, 1]$, and hence, from (9), by continuity of g, that

$$g(x^*) = g([1-t]x^* + tx^{**}) = g(x^{**}), \qquad \forall t \in [0, 1].$$

Thus, g is constant on the line segment from x^* to x^{**} which contradicts the hemivariateness of g. $\blacksquare$

We turn now to the final convergence statement (3) and prove the following simple but basic result.

14.1.4. Let $g: D \subset R^n \to R^1$ be continuously differentiable on a compact set $D_0 \subset D$ and suppose that $\{x^k\} \subset D_0$ is any sequence which satisfies $\lim_{k\to\infty} g'(x^k)^T = 0$. Then the set $\Omega = \{x \in D_0 \mid g'(x)^T = 0\}$ of critical points of g in D_0 is not empty and

$$\lim_{k\to\infty} [\inf_{x\in\Omega} \| x^k - x \|] = 0. \tag{10}$$

In particular, if Ω consists of a single point x^*, then $\lim_{k\to\infty} x^k = x^*$ and $g'(x^*)^T = 0$.

Proof. Because of the compactness of D_0, $\{x^k\}$ has convergent subsequences, and if $\lim_{i\to\infty} x^{k_i} = x$ then, by continuity of g', $g'(x)^T = 0$ so that $x \in \Omega$. Now, let $\delta_k = \inf_{x\in\Omega} \| x^k - x \|$ and suppose that $\lim_{i\to\infty} \delta_{k_i} = \delta$. Then $\{x^{k_i}\}$ has a convergent subsequence, and, since this subsequence must have its limit point in Ω, it follows that $\delta = 0$, which proves (10). ∎

The primary importance of the previous result is that when Ω consists of only one point the sequence converges. If, however, Ω consists of finitely many points, convergence is still assured provided that (7) holds.

14.1.5. Let $g: D \subset R^n \to R^1$ be continuously differentiable on the compact set $D_0 \subset D$, and suppose that the set Ω of critical points of g in D_0 is finite. Let $\{x^k\} \subset D_0$ be any sequence for which $\lim_{k\to\infty} (x^k - x^{k+1}) = 0$ and $\lim_{k\to\infty} g'(x^k)^T = 0$. Then $\lim_{k\to\infty} x^k = x^*$ and $g'(x^*)^T = 0$.

Proof. Let Λ be the set of limit points of $\{x^k\}$. As in **14.1.4**, any limit point is also a critical point of g so that $\Lambda \subset \Omega$; that is, Λ is finite. Suppose that $\Lambda = \{z^1, ..., z^m\}$ with $m > 1$; then

$$\delta = \min\{\| z^i - z^j \| \mid i \neq j,\ i, j = 1, ..., m\} > 0,$$

and we can choose $k_0 \geq 0$ such that $x^k \in \bigcup_{i=1}^{m} S(z^i, \delta/4)$ and $\| x^k - x^{k+1} \| \leq \delta/4$ for all $k \geq k_0$. But then $x^{k_1} \in S(z^1, \delta/4)$ for some $k_1 \geq k_0$ implies that

$$\| z^i - x^{k_1+1} \| \geq \| z^i - z^1 \| - (\| z^1 - x^{k_1} \| + \| x^{k_1} - x^{k_1+1} \|)$$
$$\geq \delta - 2\delta/4 = \delta/2, \quad i \geq 2,$$

and, hence, necessarily, that $x^{k_1+1} \in S(z^1, \delta/4)$. By induction, therefore, $x^k \in S(z^1, \delta/4)$ for all $k \geq k_1$, which contradicts the fact that $z^2, ..., z^m$ are limit points of $\{x^k\}$; therefore, $m = 1$. ∎

We end this section with a basic rate-of-convergence result. Recall from Definition **9.2.1** that the R_1-convergence factor of a sequence $\{x^k\}$ which converges to x^* is defined by

$$R_1\{x^k\} = \limsup_{k\to\infty} \| x^k - x^* \|^{1/k},$$

14.1. INTRODUCTION AND CONVERGENCE OF SEQUENCES

and that the sequence is said to have at least an R-linear rate of convergence if $R_1\{x^k\} < 1$.

14.1.6. Let $g: D \subset R^n \to R^1$ be G-differentiable on an open set $D_0 \subset D$ and suppose that $\{x^k\} \subset D_0$ converges to $x^* \in D_0$. Assume that $g'(x^*)^T = 0$, that g has a second F-derivative at x^* and the Hessian matrix $H_g(x^*)$ is invertible, and that there is an $\eta > 0$ and a k_0 for which

$$g(x^k) - g(x^{k+1}) \geq \eta \| g'(x^k)^T \|^2, \quad \forall k \geq k_0. \tag{11}$$

Then $R_1\{x^k\} < 1$.

Proof. Set $\alpha = \| H_g(x^*)^{-1} \|^{-1}$. Then for given $\epsilon \in (0, \alpha)$ we may choose $\delta > 0$ so that $S \equiv S(x^*, \delta) \subset D_0$ and

$$\| g'(x)^T - H_g(x^*)(x - x^*) \| \leq \epsilon \| x - x^* \|, \quad \forall x \in S.$$

Hence

$$\| g'(x)^T \| \geq \| H_g(x^*)(x - x^*) \| - \| g'(x)^T - H_g(x^*)(x - x^*) \|$$
$$\geq (\alpha - \epsilon) \| x - x^* \|, \quad \forall x \in S.$$

Consequently, in view of (11), there is a $k_1 \geq k_0$ so that with $\gamma_0 = [\eta(\alpha - \epsilon)^2]^{-1}$

$$\| x^k - x^* \|^2 \leq (\alpha - \epsilon)^{-2} \| g'(x^k)^T \|^2 \leq \gamma_0 [g(x^k) - g(x^{k+1})], \quad \forall k \geq k_1. \tag{12}$$

Now note that the mean-value theorem **3.3.12** ensures that

$$g(x^k) - g(x^*) \leq \gamma_1 \| x^k - x^* \|^2, \quad \forall k \geq k_2,$$

where $\gamma_1 = \frac{1}{2} \| g''(x^*) \| + \epsilon$, and $k_2 \geq k_1$ is sufficiently large. Hence, we obtain, using (11) and (12),

$$0 \leq g(x^{k+1}) - g(x^*) \leq g(x^k) - g(x^*) - \eta \| g'(x^k)^T \|^2$$
$$\leq g(x^k) - g(x^*) - (1/\gamma_0) \| x^k - x^* \|^2 \leq \lambda [g(x^k) - g(x^*)], \quad \forall k \geq k_2, \tag{13}$$

where $\lambda = 1 - (\gamma_0 \gamma_1)^{-1}$. Clearly, $0 \leq \lambda < 1$, and, by **9.3.1**, (13) shows that

$$R_1\{[g(x^k) - g(x^*)]^{1/2}\} \leq Q_1\{[g(x^k) - g(x^*)]^{1/2}\} \leq \lambda^{1/2} < 1.$$

Therefore, by (12), and remembering that $g(x^k) - g(x^{k+1}) \leq g(x^k) - g(x^*)$,

$$R_1\{x^k\} \leq R_1\{[g(x^k) - g(x^*)]^{1/2}\} \leq \lambda^{1/2} < 1. \quad \blacksquare$$

NOTES AND REMARKS

NR 14.1-1. Although contained in the literature in an implicit form (see, for example, Goldstein [1967] or Ostrowski [1966]), the explicit use of the basic relation (4) is due to Elkin [1968], as are **14.1.1–14.1.3** (although we have used different terminology).

NR 14.1-2. Theorems 14.1.4 and 14.1.5 are from Ostrowski [1966]. In fact, 14.1.5 is a special case of his more general result that the set of limit points of any bounded sequence $\{x^k\} \subset R^n$ satisfying $\lim_{k\to\infty}(x^k - x^{k+1}) = 0$ is a closed, connected set.

NR 14.1-3. If the set of critical points of g is not finite, Ostrowski [1967a] has proved the following theorem: Let $g: D \subset R^n \to R^1$ be twice F-differentiable in an open set $D_0 \subset D$, and let $\{x^k\} \subset D_0$ satisfy $\lim_{k\to\infty}(x^{k+1} - x^k) = 0$ and $\lim_{k\to\infty} g'(x^k)^T = 0$. If $\{x^k\}$ has a limit point x^* for which $H_g(x^*)$ is nonsingular, then $\lim_{k\to\infty} x^k = x^*$.

This result states, in particular, that, under the conditions on g and $\{x^k\}$, if $\{x^k\}$ does not converge, then $H_g(x)$ is singular at every limit point. Ostrowski [1967a] has actually shown a stronger result—namely, that if g is four times continuously differentiable on D_0, then rank $H_g(x) \leqslant n - 2$ at every limit point.

EXERCISES

E 14.1-1. Let $g: D \subset R^1 \to R^1$ be continuously differentiable, strictly convex, and bounded below on a convex compact set $D_0 \subset D$. Show that any sequence $\{x^k\} \subset D_0$ with $g(x^k) \geqslant g(x^{k+1})$ and $\lim_{k\to\infty} g'(x^k)(x^k - x^{k+1}) = 0$ satisfies $\lim_{k\to\infty}(x^k - x^{k+1}) = 0$. By means of the examples

$$D_0 = [-1, 1] \subset R^1, \quad g(x) = x^2, \quad x^k = (-1)^k [\tfrac{1}{2} + 2^{-k}]$$

and

$$D_0 = [-1, 1] \subset R^1, \quad g(x) = e^x, \quad x^k = 2^{-k},$$

show also that the condition $\lim_{k\to\infty} g'(x^k)(x^k - x^{k+1}) = 0$ cannot, in general, be omitted, and that the limit of $\{x^k\}$ need not be a local minimizer.

E 14.1-2. Show by means of the example $g: R^1 \to R^1$, $g(x) = \exp(-x^2)$, $x^k = \sum_{j=1}^k j^{-1}$, that Theorem **14.1.5** need not hold if D_0 is not compact.

E 14.1-3. Show that the following functionals are hemivariate:

(a) $g: R^2 \to R^1, g(x_1, x_2) = a^2(x_1^2 - x_2)^2 + (1 - x_1)^2$. (Rosenbrock [1960])
(b) $g: R^4 \to R^1, g(x_1, x_2, x_3, x_4) = (x_1 + 10x_2)^2 + 5(x_3 - x_4)^2 + (x_2 - 2x_3)^4 + 10(x_1 - x_4)^4$. (Powell [1962])

14.2. STEPLENGTH ANALYSIS

E 14.1-4. Suppose that $g: D \subset R^n \to R^1$ is quasiconvex on the convex set $D_0 \subset D$. Show that g is hemivariate if and only if it is strictly quasiconvex.

E 14.1-5. Give an example of a compact sequence $\{x^k\} \subset R^1$ such that $\lim_{k \to \infty}(x^k - x^{k+1}) = 0$ but that $\lim_{k \to \infty} x^k$ does not exist.

14.2. STEPLENGTH ANALYSIS

In this section we shall prove that the basic proposition

$$\lim_{k \to \infty} g'(x^k) \, p^k / \| p^k \| = 0 \tag{1}$$

is valid for certain steplength algorithms and arbitrary directions $p^k \neq 0$.

Even simple one-dimensional examples (see **E 14.1-1**) show that the descent condition

$$g(x^k) \geqslant g(x^{k+1}) \tag{2}$$

by itself does not in general imply (1), even if strict inequality holds for all $k \geqslant 0$; only if g is decreased "sufficiently" at each step can (1) be concluded. As a means of measuring the sufficiency of this decrease, we introduce the following class of functions.

14.2.1. Definition. A mapping $\sigma: [0, \infty) \to [0, \infty)$ is a *forcing function* (*F-function*) if for any sequence $\{t_k\} \subset [0, \infty)$

$$\lim_{k \to \infty} \sigma(t_k) = 0 \quad \text{implies} \quad \lim_{k \to \infty} t_k = 0. \quad \blacksquare \tag{3}$$

Note that any nondecreasing function $\sigma: [0, \infty) \to [0, \infty)$ such that $\sigma(0) = 0$ and $\sigma(t) > 0$ for $t > 0$ is necessarily an *F*-function. Recall that a function of this type has already been used in the definition **4.3.5** of a uniformly connected functional.

All subsequent results about the validity of (1) reduce to the following simple *principle of sufficient decrease*.

14.2.2. Suppose that $g: D \subset R^n \to R^1$ is G-differentiable and bounded below on some set $D_0 \subset D$ and that the vectors

$$x^{k+1} = x^k - \omega_k \alpha_k p^k, \quad p^k \neq 0, \quad k = 0, 1, \ldots, \tag{4}$$

remain in D_0. If

$$g(x^k) - g(x^{k+1}) \geqslant \sigma(|\, g'(x^k) \, p^k \,|/\| p^k \|), \quad k \geqslant 0, \tag{5}$$

for some *F*-function σ, then $\lim_{k \to \infty} g'(x^k) \, p^k / \| p^k \| = 0$.

Proof. Since g is bounded below on D_0 and (5) implies that $g(x^k) \geqslant g(x^{k+1})$, it follows that $\lim_{k \to \infty} (g(x^k) - g(x^{k+1})) = 0$, and, hence, by the definition of F-functions, that (1) holds. ∎

Our aim is now to derive estimates of the form (5) for the various steplength algorithms discussed in Chapter 8. However, in order to apply 14.2.2, it will be necessary to guarantee that the sequence (4), generated by the particular algorithm under consideration, remains in D_0. For most of the algorithms, we shall be able to prove that $\{x^k\}$ is strongly downward, that is,

$$g(x^k + t[x^{k+1} - x^k]) \leqslant g(x^k), \quad \forall t \in [0, 1], \quad k = 0, 1, \ldots, \quad (6)$$

so that $\{x^k\} \subset L^0(g(x^0))$. Here, as in Chapter 8,

$$L(g(x^0)) = \{x \in D \mid g(x) \leqslant g(x^0)\}, \quad (7)$$

and $L^0(g(x^0))$ is the path-connected component of the level set $L(g(x^0))$ which contains x^0 itself. [Whenever there is no danger of confusion, we shall write L and L^0 in place of $L(g(x^0))$ and $L^0(g(x^0))$, respectively.]

The verification of (6) will usually be carried out by showing that for any $t \in (0, 1]$ there is a $\hat{t} \in (0, t)$ such that

$$g(x^k - t\omega_k \alpha_k p^k) - g(x^k) = -t_k \omega_k \alpha_k g'(x^k - \hat{t}\omega_k \alpha_k p^k) p^k < 0. \quad (8)$$

The equality in (8) is, of course, obtained by the mean-value theorem, while the final inequality will follow from properties of the algorithm. It is not possible, however, to apply the mean-value theorem unless we first know that

$$[x^k, x^k - t\omega_k \alpha_k p^k] \subset D. \quad (9)$$

The following fundamental lemma allows us to ensure this condition.

14.2.3. Let $g: D \subset R^n \to R^1$ be continuous on the open set D and G-differentiable on the compact set $L^0 = L^0(g(x^0))$ for some $x^0 \in D$. Then for any $x \in L^0$ and $p \in R^n$ with $g'(x) p > 0$ there is an $\alpha^* > 0$ such that $g(x) = g(x - \alpha^* p)$ and $[x, x - \alpha^* p] \subset L^0$. In particular, if $\eta > 0$ is any number with the property that

$$g(x - \alpha p) < g(x), \quad \forall x - \alpha p \in (x, x - \eta p] \cap L^0,$$

then $[x, x - \eta p] \subset L^0$.

14.2. STEPLENGTH ANALYSIS

Proof. Set $\alpha^* = \sup J$ where

$$J = \{\alpha > 0 \mid [x, x - \alpha p] \subset D \text{ and } g(x - \beta p) < g(x), \ \forall \beta \in (0, \alpha]\}.$$

It follows from **8.2.1** that J is not empty, and, hence, α^* is well-defined. Moreover, by the compactness of L^0, $\alpha^* < +\infty$ and $[x, x - \alpha^* p] \subset L^0$. Now, suppose that $g(x - \alpha^* p) < g(x)$. Then, since D is open and g is continuous, we may choose a $\delta > 0$ such that $x - \alpha p \in D$ and $g(x - \alpha p) < g(x)$ for $\alpha \in [\alpha^*, \alpha^* + \delta]$. This contradicts the definition of α^*, and, hence, $g(x) = g(x - \alpha^* p)$. The last statement follows immediately since, clearly, $\eta < \alpha^*$. ∎

We begin our steplength analyses by considering the majorization principle discussed in **8.3.(d)** In the following theorem, as in all subsequent results, the sequence $\{p^k\}$ is assumed to be given and essentially arbitrary. Moreover, recall that $g'(x) \in L(R^n, R^1)$ is a row vector and that the norm on $L(R^n, R^1)$ is defined by (2.2.9). Hence, under any norm on R^n, $|g'(x)h| \leqslant \|g'(x)\| \|h\|$, for all $x \in D$, $h \in R^n$.

14.2.4. Let $g: D \subset R^n \to R^1$ be continuously differentiable on the open set D. Assume that $L^0 = L^0(g(x^0))$ is compact, and that

$$\|g'(x) - g'(y)\| \leqslant \gamma \|x - y\|, \qquad \forall x, y \in L^0.$$

Consider the iteration (4) with any sequence $\{p^k\}$ of nonzero vectors, and suppose that when $x^k \in D$, the sign of p^k is normalized so that $g'(x^k) p^k \geqslant 0$. Assume that the steplength and the relaxation parameter satisfy

$$\alpha_k = (\gamma \|p^k\|)^{-1} g'(x^k) p^k / \|p^k\|, \qquad \epsilon \leqslant \omega_k \leqslant 2 - \epsilon, \qquad k = 0, 1, \ldots,$$

where $\epsilon \in (0, 1]$ is given. Then the sequence $\{x^k\}$ remains in L^0, $\lim_{k \to \infty} (x^k - x^{k+1}) = 0$, and $\lim_{k \to \infty} g'(x^k) p^k / \|p^k\| = 0$.

Proof. We proceed by induction and assume that $x^k \in L^0$. If $g'(x^k) p^k = 0$, then $x^{k+1} = x^k$; hence, assume that $g'(x^k) p^k > 0$. It follows from **8.3.1** that

$$g(x^k) - g(x^k - tp^k) \geqslant tg'(x^k) p^k - \tfrac{1}{2}\gamma t^2 \|p^k\|^2, \tag{10}$$

whenever $[x^k, x^k - tp^k] \subset D$, and, hence, **14.2.3** with $\eta = (2 - \epsilon) \alpha_k$ ensures that $x^{k+1} \in L^0$.

Now, by (10),

$$g(x^k) - g(x^{k+1}) \geq \omega_k \alpha_k g'(x^k) p^k - \frac{\gamma}{2}(\omega_k \alpha_k \|p^k\|)^2$$

$$= \frac{\omega_k}{\gamma}\left[g'(x^k)\frac{p^k}{\|p^k\|}\right]^2 - \frac{\omega_k^2}{2}\left[g'(x^k)\frac{p^k}{\|p^k\|}\right]^2$$

$$= \frac{1}{2\gamma}(2\omega_k - \omega_k^2)\left[g'(x^k)\frac{p^k}{\|p^k\|}\right]^2$$

$$\geq \frac{1}{2\gamma}\epsilon(2-\epsilon)\left[g'(x^k)\frac{p^k}{\|p^k\|}\right]^2,$$

since $2\omega_k - \omega_k^2 = 1 - (1-\omega_k)^2 \geq 1 - (1-\epsilon)^2 = \epsilon(2-\epsilon)$. Clearly, $\sigma(t) \equiv \frac{1}{2}(\epsilon/\gamma)(2-\epsilon)t^2$ is an F-function and it follows from **14.2.2** that (1) holds. Finally, from

$$\|x^k - x^{k+1}\| = \omega_k \alpha_k \|p^k\| = \omega_k \gamma^{-1} g'(x^k) p^k / \|p^k\|,$$

we obtain $\lim_{k\to\infty}(x^k - x^{k+1}) = 0$. ∎

The proof of **14.2.4** exhibits a particularly simple forcing function, namely, ct^2. In most of the remaining results, more complex F-functions will be needed, and the majority of these are obtained in one way or the other from the following function.

14.2.5. Definition. Let $g: D \subset R^n \to R^1$ be continuously differentiable, and assume that on some $D_0 \subset D$

$$\alpha = \sup\{\|g'(x) - g'(y)\| \mid x, y \in D_0\} > 0.$$

Then the mapping $\delta: [0, \infty) \to [0, \infty)$ defined by

$$\delta(t) = \begin{cases} \inf\{\|x-y\| \mid x, y \in D_0, \ \|g'(x) - g'(y)\| \geq t\}, & t \in [0, \alpha), \\ \lim_{s \to \alpha^-} \delta(s), & t \in [\alpha, +\infty), \end{cases}$$

is the *reverse modulus of continuity* of $g': D \subset R^n \to L(R^n, R^1)$ on D_0. ∎

Note that δ is always well defined and isotone, and that $\delta(0) = 0$. We wish to preclude, however, the possibility that δ is identically zero.

14.2.6. Assume that $g: D \subset R^n \to R^1$ has a uniformly continuous derivative on $D_0 \subset D$ and that the quantity α of **14.2.5** is positive. Then $\delta(t) > 0$ for all $t > 0$. Hence, δ is an F-function.

14.2. STEPLENGTH ANALYSIS

Proof. If $\delta(t) = 0$ for some $t > 0$, then given any $\epsilon > 0$ there exist $x, y \in D_0$ so that $\| g'(x) - g'(y) \| \geq t$ and $\| x - y \| \leq \epsilon$. This contradicts the uniform continuity. ∎

The principle of sufficient decrease (5) is essentially equivalent to the requirement that $g'(x^{k+1}) p^k$ be "sufficiently smaller" than $g'(x^k) p^k$. This is made precise by the Altman principle of **8.3.(b)** For fixed $\mu \in [0, 1)$, α_k is defined to be the smallest nonnegative solution of the equation

$$g'(x^k - \alpha p^k) p^k = \mu g'(x^k) p^k. \tag{11}$$

Recall that for $\mu = 0$ this reduces to the Curry principle of **8.3.(b)**. Although for $\mu > 0$ the Altman principle has little interest as a computational procedure, the following result provides a tool for the analysis of other algorithms.

14.2.7. Suppose that $g: D \subset R^n \to R^1$ is continuously differentiable on the open set D and that $L^0 = L^0(g(x^0))$ is compact. Let $\mu \in [0, 1)$ and $\epsilon \in (0, 1]$ be given, and consider the iteration (4) where $\{p^k\}$ is any sequence such that $g'(x^k) p^k \geq 0$, $p^k \neq 0$, as well as $\epsilon \leq \omega_k \leq 1$, and

$$\alpha_k = \min\{\alpha \geq 0 \mid g'(x^k - \alpha p^k) p^k = \mu g'(x^k) p^k\}. \tag{12}$$

Then $\{x^k\} \subset L^0$, $\{x^k\}$ is strongly downward in L^0, and

$$\lim_{k \to \infty} g'(x^k) p^k / \| p^k \| = 0.$$

Proof. We proceed again by induction and assume that $x^k \in L^0$. If $g'(x^k) p^k = 0$, then $\alpha_k = 0$ and, hence, $x^{k+1} = x^k$. Suppose, therefore, that $g'(x^k) p^k > 0$. Then **14.2.3** guarantees the existence of a $t_k > 0$ such that $[x^k, x^k - t_k p^k] \subset L^0$ and $g(x^k) = g(x^k - t_k p^k)$. Hence, by the mean-value theorem **3.2.2**, there is an $\hat{t} \in (0, t_k)$ for which

$$g'(x^k - \hat{t} p^k) p^k = 0. \tag{13}$$

Therefore, by the continuity of g', (11) has solutions in $(0, t_k)$, and, since $g'(x^k) p^k > 0$, there must be a smallest solution $\alpha_k > 0$. Since $\omega_k \leq 1$, this shows that x^{k+1} is well defined and that $x^{k+1} \in L^0$.

Now, clearly,

$$g'(x^k - \alpha p^k) p^k > \mu g'(x^k) p^k, \qquad \forall \alpha \in [0, \omega_k \alpha_k), \tag{14}$$

so that $g(x^k - \alpha p^k)$ is monotone decreasing on $[0, \omega_k \alpha_k]$. Hence,

$$g(x^k) \geq g([1 - t] x^k + t x^{k+1}) \geq g(x^{k+1}), \qquad \forall t \in [0, 1]$$

and $\{x^k\}$ is strongly downward in L^0.

For the proof of the last statement, suppose first that $\mu > 0$. Then, again by the mean-value theorem and (14), it follows that

$$g(x^k) - g(x^{k+1}) = \omega_k \alpha_k g'(x^k - \alpha p^k) p^k \geq \alpha_k \epsilon \mu g'(x^k) p^k. \tag{15}$$

It is no restriction to assume that $g'(x) \not\equiv 0$ for $x \in L^0$, since otherwise the statement of the theorem is trivial. Then **14.2.6** shows that the reverse-modulus-of-continuity function δ of g' on L^0 is an F-function. From

$$(1 - \mu) g'(x^k) \frac{p^k}{\|p^k\|} = g'(x^k) \frac{p^k}{\|p^k\|} - g'(x^k - \alpha_k p^k) \frac{p^k}{\|p^k\|}$$

$$\leq \| g'(x^k) - g'(x^k - \alpha_k p^k)\|,$$

it then follows by the definition of δ that

$$\alpha_k \| p^k \| \geq \delta([1 - \mu] g'(x^k) p^k / \| p^k \|).$$

Hence, the inequality (15) can be continued to

$$g(x^k) - g(x^{k+1}) \geq \alpha_k \| p^k \| \epsilon \mu g'(x^k) \frac{p^k}{\|p^k\|} \geq \sigma \left(g'(x^k) \frac{p^k}{\|p^k\|} \right), \tag{16}$$

where $\sigma(t) \equiv \mu \epsilon t \delta([1 - \mu] t)$, $t \geq 0$. It is immediate that σ is also an F-function, and the result follows from **14.2.2**. This completes the proof for $\mu > 0$.

Now, let $\mu = 0$ and consider besides the steplength α_k the corresponding steplength $\bar{\alpha}_k$ obtained from (12) for $\mu = \frac{1}{2}$. If we set $\bar{x}^{k+1} = x^k - \omega_k \bar{\alpha}_k p^k$, then, clearly, $g(\bar{x}^{k+1}) \geq g(x^{k+1})$. Moreover, the estimate (16) holds with $\bar{x}^{k+1}$ in place of x^{k+1} for $\sigma(t) = \frac{1}{2} \epsilon t \delta(t/2)$, and, therefore, with this F-function,

$$g(x^k) - g(x^{k+1}) \geq g(x^k) - g(\bar{x}^{k+1}) \geq \sigma\big(g'(x^k) p^k / \| p^k \|\big). \tag{17}$$

Hence, the result again follows from **14.2.2**. ∎

The reasoning of the last part of this proof is particularly noteworthy and shall be called the *"comparison principle."* Suppose that we have two different steplength algorithms I and II, and that at a point x^k the application of I and II yields x_{I}^{k+1} and x_{II}^{k+1}, respectively. If

$$g(x^k) - g(x_{\text{I}}^{k+1}) \geq \sigma(|g'(x^k) p^k| / \| p^k \|),$$

where σ is an F-function, then it suffices to show that $g(x_{\text{I}}^{k+1}) \geq g(x_{\text{II}}^{k+1})$, in order to obtain

$$g(x^k) - g(x_{\text{II}}^{k+1}) \geq g(x^k) - g(x_{\text{I}}^{k+1}) \geq \sigma(|g'(x^k) p^k| / \| p^k \|).$$

14.2. STEPLENGTH ANALYSIS

Theorem **14.2.7** allows only relaxation factors $\omega_k \leqslant 1$; for a corresponding result which permits $\omega_k > 1$ we need stronger conditions on g.

14.2.8. Let $g: D \subset R^n \to R^1$ be twice-continuously differentiable on the open set D and assume that $L^0 = L^0(g(x^0))$ is compact. Suppose, further, that

$$\eta_0 \| h \|^2 \leqslant g''(x) hh \leqslant \eta_1 \| h \|^2, \qquad \forall x \in L^0, \quad h \in R^n, \tag{18}$$

where $0 < \eta_0 \leqslant \eta_1$. Given $\mu \in [0, 1)$ and $\epsilon \in (0, 1)$, consider the iteration (4) with $g'(x^k) p^k \geqslant 0$, $p^k \neq 0$, α_k defined by (12), and

$$1 \leqslant \omega_k \leqslant \bar{\omega} = 1 + (\eta_0/\eta_1)^{1/2} (1 - \epsilon).$$

Then $\{x^k\} \subset L^0$, $\lim_{k\to\infty} g'(x^k) p^k / \| p^k \| = 0$, and $\lim_{k\to\infty}(x^k - x^{k+1}) = 0$.

Proof. We proceed again by induction and suppose that $x^k \in L^0$ and that $g'(x^k) p^k > 0$. [The case $g'(x^k) p^k = 0$ is trivial.] We have already shown in **14.2.7** that α_k is well defined, that $[x^k, x^k - \alpha_k p^k] \subset L^0$, and that $g(x^k - \alpha_k x^k) < g(x^k)$; hence, by continuity of g, there exists an $\alpha \in (\alpha_k, \bar{\omega}\alpha_k]$ so that $[x^k, x^k - \alpha p^k] \subset L^0$. For any such α it then follows from the mean-value theorem **3.3.11**, using the abbreviations $z^k = x^k - \alpha_k p^k$ and $y^k = x^k - \beta p^k$ for some $\beta \in (0, \epsilon\alpha_k)$, that

$$g(x^k) - g(x^k - \alpha p^k)$$
$$= [g(x^k) - g(y^k)] + [g(y^k) - g(z^k)] - [g(x^k - \alpha p^k) - g(z^k)]$$
$$= g(x^k) - g(y^k) + (\alpha_k - \beta) g'(z^k) p^k$$
$$\quad + (\alpha_k - \beta)^2 \int_0^1 (1 - t) g''(z^k + t[\alpha_k - \beta] p^k) p^k p^k \, dt$$
$$\quad - (\alpha_k - \alpha) g'(z^k) p^k - (\alpha_k - \alpha)^2 \int_0^1 (1-t) g''(z^k + t[\alpha_k - \alpha] p^k) p^k p^k \, dt$$
$$\geqslant g(x^k) - g(y^k) + (\alpha - \beta) g'(z^k) p^k$$
$$\quad + \tfrac{1}{2}(\alpha_k - \beta)^2 \eta_0 \| p^k \|^2 - \tfrac{1}{2}(\alpha_k - \alpha)^2 \eta_1 \| p^k \|^2$$
$$\geqslant g(x^k) - g(x^k - \beta p^k) + (1 - \epsilon) \alpha_k g'(x^k - \alpha_k p^k) p^k$$
$$\quad + \tfrac{1}{2}\alpha_k^2 \| p^k \|^2 [(1 - \epsilon)^2 \eta_0 - (\eta_0/\eta_1)(1 - \epsilon)^2 \eta_1]$$
$$= g(x^k) - g(x^k - \beta p^k) + (1 - \epsilon) \mu \alpha_k g'(x^k) p^k > 0. \tag{19}$$

Thus, Lemma **14.2.3** implies that $[x^k, x^k - \bar{\omega}\alpha_k] \subset L^0$ and, in particular, that $x^{k+1} \in L^0$.

Observe now that

$$(1 - \mu) g'(x^k) p^k = g'(x^k) p^k - g'(x^k - \alpha_k p^k) p^k$$
$$= \alpha_k \int_0^1 g''(x^k - t\alpha_k p^k) p^k p^k \, dt, \qquad (20)$$

and, hence, that

$$\alpha_k \eta_1 \| p^k \| \geq (1 - \mu) g'(x^k) p^k / \| p^k \| \geq \eta_0 \alpha_k \| p^k \| \geq \eta_0 \| x^k - x^{k+1} \|. \qquad (21)$$

Thus, for $\mu > 0$ we get from (19) that

$$g(x^k) - g(x^{k+1}) \geq (1 - \epsilon) \mu \alpha_k \| p^k \| g'(x^k) p^k / \| p^k \|$$
$$\geq (1 - \epsilon) \mu [(1 - \mu)/\eta_1][g'(x^k) p^k / \| p^k \|]^2,$$

which by Lemma **14.2.2**, shows that (1) holds.

For $\mu = 0$ we again apply the comparison principle. Let $\bar{\alpha}_k \in (0, \alpha_k)$ be the steplength (12) for $\mu = \tfrac{1}{2}$, and set $\beta = \epsilon \bar{\alpha}_k \in (0, \epsilon \alpha_k)$. Then (19) together with (21), applied to $\bar{\alpha}_k$, implies that for some $\tilde{\beta} \in (0, \beta)$

$$g(x^k) - g(x^{k+1}) \geq g(x^k) - g(x^k - \beta p^k) = \epsilon \bar{\alpha}_k g'(x^k - \tilde{\beta} p^k) p^k$$
$$> \tfrac{1}{2} \epsilon \bar{\alpha}_k g'(x^k) p^k \geq \tfrac{1}{4} \epsilon (1/\eta_1)[g'(x^k) p^k / \| p^k \|]^2,$$

and, hence, again by **14.2.2**, that (1) holds.

The last statement, $\lim_{k \to \infty}(x^k - x^{k+1}) = 0$, is a direct consequence of (1) and (21). ∎

As already discussed in **8.3**, a common steplength choice is

$$\alpha_k = g'(x^k) p^k / g''(x^k) p^k p^k, \qquad (22)$$

which is the result of taking precisely one Newton step toward the solution of

$$g'(x^k - \alpha p^k) p^k = 0. \qquad (23)$$

We next analyze this steplength, again permitting certain relaxation factors ω_k.

14.2.9. Let $g: D \subset R^n \to R^1$ be twice-continuously differentiable on the open set D, and suppose that $L^0 = L^0(g(x^0))$ is compact and that (18) holds. Consider the iteration (4) with $g'(x^k) p^k \geq 0$, $p^k \neq 0$, α_k defined by (22), and with $0 < \epsilon \leq \omega_k \leq \bar{\omega}_k = (2/\gamma_k) - \epsilon$, where $\epsilon \in (0, 1]$ and

$$\gamma_k = \sup \left\{ \frac{g''(x^k - \alpha p^k) p^k p^k}{g''(x^k) p^k p^k} \,\bigg|\, \alpha > 0, \; g(x^k - \beta p^k) < g(x^k), \; \forall \beta \in (0, \alpha] \right\}. \qquad (24)$$

14.2. STEPLENGTH ANALYSIS

Then $\{x^k\} \subset L^0$, $\lim_{k\to\infty} g'(x^k) p^k / \| p^k \| = 0$, and $\lim_{k\to\infty} (x^k - x^{k+1}) = 0$.

Proof. Suppose that $x^k \in L^0$; then, by (18), α_k and γ_k are both well-defined. We may assume again that $g'(x^k) p^k > 0$ and, hence, that $\alpha_k > 0$. Let $[x^k, x^k - \alpha p^k] \subset L^0$, $\alpha \in (0, \bar{\omega}\alpha_k)$; then, by the mean-value theorem **3.3.10**, with some $\tilde{\alpha} \in (0, \alpha)$,

$$g(x^k) - g(x^k - \alpha p^k)$$
$$= \alpha g'(x^k) p^k - \tfrac{1}{2}\alpha^2 g''(x^k - \tilde{\alpha}p^k) p^k p^k$$
$$\geq \alpha g'(x^k) p^k \{1 - \tfrac{1}{2}[(2/\gamma_k) - \epsilon] g''(x^k - \tilde{\alpha}p^k) p^k p^k / g''(x^k) p^k p^k\}$$
$$\geq \tfrac{1}{2}\epsilon \alpha g'(x^k) p^k > 0, \tag{25}$$

which, by Lemma **14.2.3**, implies that $[x^k, x^k - \bar{\omega}\alpha_k p^k] \subset L^0$ and, in particular, that $x^{k+1} \in L^0$. It then follows immediately from (25) that

$$g(x^k) - g(x^{k+1}) \geq \tfrac{1}{2}\epsilon \omega_k \alpha_k g'(x^k) p^k \geq \tfrac{1}{2}\epsilon^2 [g'(x^k) p^k]^2 / g''(x^k) p^k p^k$$
$$\geq \tfrac{1}{2}\epsilon^2 (1/\eta_1) [g'(x^k) p^k / \| p^k \|]^2,$$

so that **14.2.2** implies that (1) holds. The final statement is now a consequence of

$$\| x^k - x^{k+1} \| = \alpha_k \omega_k \| p^k \| \leq [(2 - \epsilon)/\eta_0] g'(x^k) p^k / \| p^k \|. \blacksquare$$

Note that $1 \leq \gamma_k \leq \eta_1/\eta_0$ and, hence, $1/\gamma_k \geq \eta_0/\eta_1$. Thus, a conservative choice for ω_k is always

$$0 < \epsilon \leq \omega_k \leq 2(\eta_0/\eta_1) - \epsilon.$$

If $g''(x^k - \alpha p^k) p^k p^k$ is nonincreasing in α, then we have, of course, that $\gamma_k = 1$, and hence $\epsilon \leq \omega_k \leq 2 - \epsilon$. This applies, in particular, for quadratic functionals g.

Note, further, that in the SOR–Newton method (8.3.9) the p^k are coordinate directions, and, thus, $g''(x) p^k p^k$ is a diagonal element of the Hessian matrix H_g of g. Therefore, if we set

$$C_i = \sup\{\partial_i^2 g(x) \mid x \in L^0\}, \quad c_i = \inf\{\partial_i^2 g(x) \mid x \in L^0\}, \quad i = 1,\dots,n,$$

then in this case $\gamma^k \leq C_i/c_i$, which gives the permissible range

$$\epsilon \leq \omega_k \leq 2[\max_i (c_i/C_i)] - \epsilon \tag{26}$$

for ω_k.

We turn next to the minimization principles for obtaining the steplength. As discussed in **8.3.(a)**, there are three possibilities for these minimizations, and these are the subjects of the next three results. Together with $L^0(g(x^0))$, we will now need to consider also the connected components $L^0(g(x^k))$ containing x^k of the level sets $L(g(x^k))$, $k = 1, 2,...$

14.2.10. Assume that $g: D \subset R^n \to R^1$ is continuously differentiable on the open set D and that $L^0 = L^0(g(x^0))$ is compact. Consider the iteration

$$x^{k+1} = x^k - \alpha_k p^k, \qquad k = 0, 1,..., \qquad (27)$$

where $p^k \neq 0$ and α_k is chosen so that

$$g(x^k - \alpha_k p^k) = \min\{g(x^k - \alpha p^k) \mid [x^k, x^k - \alpha p^k] \subset L^0(g(x^k))\}. \qquad (28)$$

Then $\{x^k\} \subset L^0$, $\{x^k\}$ is strongly downward, and $\lim_{k \to \infty} g'(x^k) p^k / \| p^k \| = 0$.

Proof. If $x^k \in L^0(g(x^0))$, then $L^0(g(x^k))$ is also compact, and there is at least one α_k so that (28) holds. Hence, $x^{k+1} \in L^0(g(x^k))$. We may assume, without loss of generality, that $g'(x^k) p^k \geq 0$. Now, let $\bar{x}^{k+1}$ be obtained by the Curry principle (12) (with $\mu = 0$). Then, clearly, $g(x^{k+1}) \leq g(\bar{x}^{k+1})$ and, consequently, by (17),

$$g(x^k) - g(x^{k+1}) \geq \sigma(g'(x^k) p^k / \| p^k \|),$$

where $\sigma(t) = \frac{1}{2}\epsilon t \delta(t/2)$ is the F-function of **14.2.7**. Therefore, **14.2.2** shows that (1) holds. Finally, since $[x^k, x^{k+1}] \subset L^0(g(x^k))$, we have

$$g(x^k) \geq g([1 - t] x^k + t x^{k+1}) \geq g(x^{k+1}), \qquad \forall t \in [0, 1],$$

so that $\{x^k\}$ is strongly downward. ∎

As immediate corollaries of this result, we conclude by the comparison principle that (1) holds for either one of the other minimization possibilities of **8.3.(a)**, that is, for the minimization over the whole component $L^0(g(x^k))$ or over the whole level set $L(g(x^k))$. Clearly, in either case, the sequence $\{x^k\}$ need not be strongly downward. We summarize this in the following two statements.

14.2.11. Let g satisfy the conditions of **14.2.10**, and consider the iteration (27) where $p^k \neq 0$ and α_k is chosen so that

$$g(x^k - \alpha_k p^k) = \min\{g(x^k - \alpha p^k) \mid x^k - \alpha p^k \in L^0(g(x^k))\}. \qquad (29)$$

Then $\{x^k\} \subset L^0(g(x^0))$ and $\lim_{k \to \infty} g'(x^k) p^k / \| p^k \| = 0$.

14.2. STEPLENGTH ANALYSIS

14.2.12. Let g satisfy the conditions of **14.2.10** and assume that the whole level set $L = L(g(x^0))$ is compact. Consider the iteration (27) with $p^k \neq 0$ and α_k chosen so that

$$g(x^k - \alpha_k p^k) = \min\{g(x^k - \alpha p^k) \mid x^k - \alpha p^k \in L(g(x^k))\}. \tag{30}$$

Then $\{x^k\} \subset L(g(x^0))$ and $\lim_{k \to \infty} g'(x^k) p^k / \| p^k \| = 0$.

Under the assumptions of **14.2.10**, we are not, in general, able to underrelax the steplength, since it is possible that $g(x^k - \alpha p^k) = g(x^k)$ for some $\alpha \in (0, \alpha_k)$ (Fig. **14.1**), and, hence, the estimate (5) could not hold. If g is quasiconvex, this situation cannot occur.

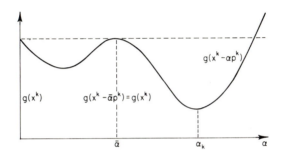

FIGURE 14.1

14.2.13. Let $g: D \subset R^n \to R^1$ be continuously differentiable and quasiconvex on the open, convex set D, and assume that the level set $L(g(x^0))$ is compact. Consider the iteration

$$x^{k+1} = x^k - \omega_k \alpha_k p^k, \quad k = 0, 1, \ldots,$$

where $g'(x^k) p^k \geq 0$, $p^k \neq 0$, $0 < \epsilon \leq \omega_k \leq 1$, and α_k is defined by (28). Then $\{x^k\} \subset L^0$, $\lim_{k \to \infty} g'(x^k) p^k / \| p^k \| = 0$, and $\{x^k\}$ is strongly downward.

Proof. Let $\bar{\alpha}_k$ be the Curry steplength of **14.2.7**. Since D is open, it follows that $g'(x^k - \bar{\alpha}_k p^k) p^k = 0$ and, hence, by definition of $\bar{\alpha}_k$, that $\alpha_k \geq \bar{\alpha}_k$. Let $g'(x^k) p^k > 0$; then $\bar{\alpha}_k > 0$, and we have the inequality (17), namely

$$g(x^k) - g(x^k - \omega_k \bar{\alpha}_k p^k) \geq \sigma\big(g'(x^k) p^k / \| p^k \|\big),$$

where $\sigma(t) = \tfrac{1}{2} \epsilon t \delta(t/2)$. On the other hand, it follows from the quasiconvexity that

$$g(x^k - \omega_k \alpha_k p^k) \leq g(x^k - \omega_k \bar{\alpha}_k p^k)$$

and, hence, that

$$g(x^k) - g(x^{k+1}) \geq \sigma(g'(x^k) p^k / \| p^k \|).$$

Therefore, by **14.2.2**, (1) holds, and the quasiconvexity of g implies that $\{x^k\}$ is strongly downward. ∎

In all our discussions so far about steplength algorithms, we had to ensure, on the one hand, that g is sufficiently decreased, and, on the other hand, that $\| x^k - x^{k+1} \|$ is not too small. These two requirements specify a certain range of "permissible" steplengths and instead of determining α_k by a definite steplength algorithm we can consider prescribing only a permissible range of steplength and of choosing α_k more or less arbitrarily in this range.

One possibility is the Goldstein principle discussed in **8.3.(e)**:

$$\begin{cases} \text{If } g'(x^k) p^k = 0, \text{ set } \alpha_k = 0. \text{ Otherwise, let } \alpha_k > 0 \text{ be such that} \\ \mu_1 \alpha_k g'(x^k) p^k \leq g(x^k) - g(x^k - \alpha_k p^k) \leq \mu_2 \alpha_k g'(x^k) p^k, \quad (31) \\ \text{where } 0 < \mu_1 \leq \mu_2 < 1 \text{ are fixed numbers.} \end{cases}$$

14.2.14. Let $g: D \subset R^n \to R^1$ be continuously differentiable on the open, convex set D, and assume that the level set $L = L(g(x^0))$ is compact. Consider the iteration

$$x^{k+1} = x^k - \alpha_k p^k, \quad k = 0, 1, \dots,$$

where $g'(x^k) p^k \geq 0$, $p^k \neq 0$, and $\alpha_k \geq 0$ is any number satisfying (31) with fixed $0 < \mu_1 \leq \mu_2 < 1$. Then $\{x^k\} \subset L$ and $\lim_{k \to \infty} g'(x^k) p^k / \| p^k \| = 0$.

Proof. Suppose that $x^k \in L$ and $g'(x^k) p^k > 0$. It follows easily from **14.2.3** that α_k satisfying (31) exist. For any such α_k we have $g(x^k) - g(x^{k+1}) > 0$ and, hence, $x^{k+1} \in L$. By the convexity of D we also see that $x^k - \alpha p^k \in D$ for all $\alpha \in [0, \alpha_k]$.

Let ω be the modulus of continuity of g' on some compact convex set $D_0 \subset D$ containing L. Then **3.1.11** shows that ω is well-defined and continuous on $[0, \infty)$, and, hence, $r(s) = \int_0^1 \omega(st) dt$ is well-defined for all $s \geq 0$. Then, by **8.3.1**,

$$g(x^k) - g(x^{k+1}) \geq \alpha_k g'(x^k) p^k - \alpha_k \| p^k \| r(\alpha_k \| p^k \|),$$

so that, by (31),

$$\mu_2 \alpha_k \| p^k \| g'(x^k) p^k / \| p^k \| \geq \alpha_k \| p^k \| g'(x^k) p^k / \| p^k \| - \alpha_k \| p^k \| r(\alpha_k \| p^k \|),$$

14.2. STEPLENGTH ANALYSIS

which implies that

$$\hat{r}(\alpha_k \| p^k \|) \geq r(\alpha_k \| p^k \|) \geq (1 - \mu_2) g'(x^k) p^k / \| p^k \|, \quad (32)$$

where $\hat{r}: [0, \infty) \to [0, \infty)$ is any strictly isotone function such that $\hat{r}(t) \geq r(t)$ for all $t \geq 0$. Thus, (31) and (32) together show that

$$g(x^k) - g(x^{k+1}) \geq \mu_1 \alpha_k \| p^k \| g'(x^k) p^k / \| p^k \|$$
$$\geq (\mu_1 g'(x^k) p^k / \| p^k \|) \hat{r}^{-1}((1 - \mu_2) g'(x^k) p^k / \| p^k \|).$$

Since $\hat{r}^{-1}$ exists and is an F-function, the function $\sigma(t) = \mu_1 t \hat{r}^{-1}([1 - \mu_2]t)$ is also an F-function, and **14.2.2** applies. ∎

The Goldstein algorithm does not provide constructive techniques for finding α_k but, as discussed in **8.3.(e)**, it forms the basis of a search procedure. One such procedure is the following.

Goldstein–Armijo Algorithm: Let σ be a fixed F-function and $\mu \in (0, 1)$ and $q > 1$ given constants.

I. *Basic steplength selection.* If $g'(x^k) p^k = 0$, set $\alpha_k = 0$. Otherwise, let $\alpha_k \geq 0$ be *any* real number such that $\alpha_k \| p^k \| \geq \sigma(g'(x^k) p^k / \| p^k \|)$.

II. *Choice of relaxation factor.* If $x^k - \alpha_k p^k \in D$ and

$$g(x^k) - g(x^k - \alpha_k p^k) \geq \mu \alpha_k g'(x^k) p^k, \quad (33)$$

then set $\omega_k = 1$. If $x^k - \alpha_k p^k \notin D$ or if (33) is not satisfied, let ω_k be the largest number in the sequence $\{q^{-j}\}_1^\infty$ such that $x^k - \omega_k \alpha_k p^k \in D$, and

$$g(x^k) - g(x^k - \omega_k \alpha_k p^k) \geq \mu \omega_k \alpha_k g'(x^k) p^k. \quad (34)$$

Note that if D is open, then $x^k - q^{-j} \alpha_k p^k \in D$ for sufficiently large j, and it follows from **8.3.2** that (34) can be satisfied.

14.2.15. Let $g: D \subset R^n \to R^1$ be continuously differentiable on the open convex set D, and assume that $L = L(g(x^0))$ is compact. Consider the iteration (4), where $g'(x^k) p^k \geq 0$, $p^k \neq 0$, and α_k, ω_k are selected according to the Goldstein–Armijo algorithm. Then $\{x^k\} \subset L$ and $\lim_{k \to \infty} g'(x^k) p^k / \| p^k \| = 0$.

Proof. If $x^k \in L$ and $g'(x^k) p^k > 0$, then (33) and (34) together imply that $g(x^k) - g(x^{k+1}) > 0$, and, hence, that $x^{k+1} \in L$.

If (33) holds, then

$$g(x^k) - g(x^{k+1}) \geq \mu\sigma\big(g'(x^k) p^k/\| p^k \|\big) g'(x^k) p^k/\| p^k \|, \tag{35}$$

which has the form of the basic estimate in **14.2.2**. Suppose, therefore, that (34) applies. Then, by definition of ω_k, either $x^k - q\omega_k\alpha_k p^k \notin D$ or

$$g(x^k) - g(x^k - q\omega_k\alpha_k p^k) < \mu q\omega_k\alpha_k g'(x^k) p^k. \tag{36}$$

Let δ again denote the reverse modulus of continuity of g' on L. By **14.2.6**, δ is an F-function unless $g'(x) = 0$ for all $x \in L$ which can be excluded. Suppose that $x^k - q\omega_k\alpha_k p^k \notin D$ and consider the Curry step

$$\bar{\alpha}_k = \min\{\alpha > 0 \mid g'(x^k - \alpha p^k) p^k = 0\}.$$

Then **14.2.7** implies that $x^k - \bar{\alpha}_k p^k \in L$ and, from

$$g'(x^k) p^k/\| p^k \| = g'(x^k) p^k/\| p^k \| - g'(x^k - \bar{\alpha}_k p^k) p^k/\| p^k \|$$
$$\leq \| g'(x^k) - g'(x^k - \bar{\alpha}_k p^k)\|,$$

it follows that

$$\bar{\alpha}_k \| p^k \| \geq \delta\big(g'(x^k) p^k/\| p^k \|\big).$$

Since $x^k - q\omega_k\alpha_k p^k \notin D$ and, by the convexity of D, $[x^k, x^k - \bar{\alpha}_k p^k] \subset D$, we see that $q\omega_k\alpha_k > \bar{\alpha}_k$ and, hence, that

$$g(x^k) - g(x^{k+1}) \geq \mu\omega_k\alpha_k \| p^k \| g'(x^k) \frac{p^k}{\| p^k \|}$$
$$\geq \frac{\mu}{q} g'(x^k) \frac{p^k}{\| p^k \|} \delta\left(g'(x^k) \frac{p^k}{\| p^k \|}\right), \tag{37}$$

which again is of the form of the estimate in **14.2.2**.

Finally, suppose that $x^k - q\omega_k\alpha_k p^k \in D$, but that (36) holds. If $x^k - q\omega_k\alpha_k p^k \notin L$, then again $q\omega_k\alpha_k > \bar{\alpha}_k$, and thus we obtain again (37). If, on the other hand, $x^k - q\omega_k\alpha_k p^k \in L$, then there exists an $\tilde{\alpha} \in (0, q\omega_k\alpha_k)$ such that

$$g'(x^k - \tilde{\alpha} p^k) p^k = \frac{1}{q\omega_k\alpha_k} [g(x^k) - g(x^k - q\omega_k\alpha_k p^k)] < \mu g'(x^k) p^k,$$

or

$$(1 - \mu) g'(x^k) \frac{p^k}{\| p^k \|} \leq (g'(x^k) - g'(x^k - \tilde{\alpha} p^k)) \frac{p^k}{\| p^k \|}$$
$$\leq \| g'(x^k) - g'(x^k - \tilde{\alpha} p^k)\|.$$

14.2. STEPLENGTH ANALYSIS

This implies that

$$q\omega_k \alpha_k \| p^k \| \geq \tilde{\alpha} \| p^k \| \geq \delta((1-\mu) g'(x^k) p^k / \| p^k \|)$$

and, hence, that

$$g(x^k) - g(x^{k+1}) \geq \frac{1}{q} \mu g'(x^k) \frac{p^k}{\| p^k \|} \delta\left([1-\mu] g'(x^k) \frac{p^k}{\| p^k \|}\right). \tag{38}$$

Altogether, (35), (37), and (38) show that

$$g(x^k) - g(x^{k+1}) \geq \hat{\sigma}(g'(x^k) p^k / \| p^k \|),$$

where $\hat{\sigma}$ is the F-function defined by

$$\hat{\sigma}(t) = \min\{\mu t \sigma(t), (1/q) \mu t \delta(t), (1/q) \mu t \delta([1-\mu] t)\}.$$

Thus, 14.2.2 implies that (1) holds. ∎

NOTES AND REMARKS

NR 14.2-1. Theorem **14.2.4** is essentially contained in Ostrowski [1966; Th. 27.1]. The remaining results of this section are taken from Elkin [1968], although several of the steplength algorithms had been analyzed previously in conjuction with specific directions. In particular, Curry [1944] and Altman [1966a] treated the principles named after them in conjunction with the gradient direction, and Goldstein [1967] has also analyzed the Curry principle as well as the minimization principle (12), again for the gradient direction. Theorem **14.2.15** represents an extension of results of Armijo [1966] and Goldstein [1967], as well as of Elkin [1968].

NR 14.2-2. The results of this section are, in essence, one-dimensional, and therefore extend immediately to general Banach spaces. However, the condition that $L^0(g(x^0))$ [or $L(g(x^0))$] be compact is then too stringent and has to be replaced by the condition that $L^0(g(x^0))$ is closed and bounded together with the assumption that g is bounded below.

EXERCISES

E 14.2-1. Let $g: D \subset R^n \to R^1$ be continuously differentiable on the open set D, and assume that $L^0 = L^0(g(x^0))$ is compact and that for some $\lambda \in (0, 1]$

$$\| g'(x) - g'(y) \| \leq \gamma \| x - y \|^\lambda, \quad \forall x, y \in L^0.$$

Consider the iteration $x^{k+1} = x^k - \omega_k \alpha_k p^k$ with $g'(x^k) p^k \geqslant 0$, $p^k \neq 0$, and

$$\alpha_k = \frac{1}{\|p^k\|} \left[\frac{(1+\lambda)}{2\gamma} g'(x^k) \frac{p^k}{\|p^k\|} \right]^{1/\lambda}, \quad (1+\lambda)\epsilon \leqslant \omega_k^\lambda \leqslant (1+\lambda)(1-\epsilon).$$

Show that $\{x^k\} \subset L^0$, $\lim_{k\to\infty}(x^k - x^{k+1}) = 0$, and $\lim_{k\to\infty} g'(x^k) p^k / \|p^k\| = 0$.

E 14.2-2. Let $g: D \subset R^n \to R^1$ be continuously differentiable on the open set D, and bounded below on the closed set $L_0(g(x^0))$. Consider the iteration

$$x^{k+1} = x^k - \omega_k \alpha_k p^k, \quad \text{with} \quad g'(x^k) p^k \geqslant 0, \quad \|p^k\| = 1, \quad 0 < \epsilon \leqslant \omega_k \leqslant 1,$$

$$\alpha_k = \sup\{\alpha \in [0, \mu_2 g'(x^k) p^k] \mid g'(x^k - \beta p^k) p^k \geqslant \mu_1 g'(x^k) p^k \text{ for } \beta \in [0, \alpha]\},$$

where $0 \leqslant \mu_1 < \mu_2$. Show that

$$\{x^k\} \subset L^0, \quad \lim_{k\to\infty} g'(x^k) p^k = 0, \quad \text{and} \quad \lim_{k\to\infty}(x^k - x^{k+1}) = 0.$$

(Elkin [1968])

E 14.2-3. Suppose that $g: R^n \to R^1$ is continuously differentiable on R^n and g is bounded below on some level set L. Consider $x^{k+1} = x^k - \alpha_k p^k$, $k = 0, 1, \ldots$, $x^0 \in L$, where $p^k = g'(x^k)^T$, and $\alpha_k \in (0, 1]$ is chosen such that

$$g(x^k) - g(x^k - \alpha_k p^k) \geqslant \tfrac{1}{2} \|p^k\|^2.$$

Show that the sequence $\{x^k\}$ exists, that $\lim_{k\to\infty} g'(x^k)^T = 0$, and that $\lim_{k\to\infty}(x^k - x^{k+1}) = 0$.

14.3. GRADIENT AND GRADIENT-RELATED METHODS

In the previous section we considered the iteration

$$x^{k+1} = x^k - \omega_k \alpha_k p^k, \quad k = 0, 1, \ldots, \tag{1}$$

and for various steplength algorithms we showed that, under suitable conditions,

$$\lim_{k\to\infty} g'(x^k) p^k / \|p^k\| = 0. \tag{2}$$

As discussed in **14.1**, these results are only intermediate, and complete convergence theorems depend, of course, upon the choice of the directions p^k as well as the steplength algorithm. In this and the next three sections, we obtain such complete convergence theorems for various algorithms, and we now begin by considering the following general class of directions. Recall from Definition **14.2.1** that $\sigma: [0,\infty) \to [0,\infty)$ is an *F*-function if $\lim_{k\to\infty} \sigma(t_k) = 0$ implies that $\lim_{k\to\infty} t_k = 0$.

14.3. GRADIENT AND GRADIENT-RELATED METHODS

14.3.1. Definition. Let $g: D \subset R^n \to R^1$ be G-differentiable on D and let $\{x^k\} \subset D$ be a given sequence. Then a sequence $\{p^k\} \subset R^n$ of nonzero vectors is *gradient-related* to $\{x^k\}$ if there is an F-function σ such that

$$|g'(x^k) p^k / \| p^k \| | \geq \sigma(\| g'(x^k)\|), \quad k = 0, 1, \ldots . \quad \blacksquare \qquad (3)$$

Note that the concept of gradient-relatedness is independent of the choice of norm on R^n, since a change of norm may always be subsumed in the F-function σ (see **E 14.3-1**).

If $\{x^k\} \subset D$ satisfies (2) for a gradient-related sequence $\{p^k\}$, then, clearly, (3) and the definition of an F-function imply that

$$\lim_{k \to \infty} g'(x^k)^T = 0. \qquad (4)$$

In order to conclude that the iterates $\{x^k\}$ themselves converge to a critical point of g, we need additional conditions on g or the sequence $\{x^k\}$. Results of this type were proved in **14.1**, and we now combine these with the steplength analyses of the previous section in order to obtain some general results. Recall that $L^0(g(x^0))$ denotes the connected component containing x^0 of the level set $L(g(x^0)) = \{x \in D \mid g(x) \leq g(x^0)\}$.

14.3.2. Gradient-Related Theorem. Assume that $g: D \subset R^n \to R^1$ is continuously differentiable on the open set D and that there is an $x^0 \in D$ such that $L^0 = L^0(g(x^0))$ is compact. Assume that g has a unique critical point x^* in L^0, and consider the iteration (1) where ω_k, α_k are chosen by any steplength algorithm such that $x^k \in L^0$ for all k and (2) holds. Assume, finally, that $\{p^k\}$ is gradient-related to $\{x^k\}$. Then $\lim_{k \to \infty} x^k = x^*$.

The proof is immediate since, as indicated above, (2) and the gradient-relatedness of $\{p^k\}$ imply that (4) holds, and the result then follows from **14.1.4**. Note that **14.3.2** applies to any of the steplength algorithms analyzed in the previous section provided the conditions of the relevant theorems of that section are satisfied.

The definition of gradient-related directions is motivated in part by the gradient iteration

$$x^{k+1} = x^k - \omega_k \alpha_k g'(x^k)^T, \quad k = 0, 1, \ldots, \qquad (5)$$

although, as we shall see in the following two sections, the concept has much wider applicability. In the case of (5), $p^k = g'(x^k)^T$ so that in the l_2-norm

$$g'(x^k) p^k = \| g'(x^k)^T \|_2 \| p^k \|_2, \quad k = 0, 1, \ldots, \qquad (6)$$

while in any other norm it follows from the norm-equivalence theorem **2.2.1** that there is a constant $c > 0$ such that

$$g'(x^k) p^k \geq c \, \| g'(x^k)^T \| \, \| p^k \|, \quad k = 0, 1, \ldots. \tag{7}$$

Hence, in the first case the forcing function σ of **14.3.1** is simply $\sigma(t) = t$, while for (7) $\sigma(t) = ct$.

In conjunction with the steplength algorithms of the previous section, we can now give the following result for the iteration (5).

14.3.3. Gradient Theorem. Assume that $g: D \subset R^n \to R^1$, $x^0 \in D$, and $x^* \in D$ satisfy the conditions of **14.3.2**. Then the iterates (5) are well defined and $\lim_{k\to\infty} x^k = x^*$ provided that ω_k, α_k are chosen by any of the following steplength algorithms: Curry–Altman (as defined in Theorem **14.2.7**), minimization (Theorems **14.2.10–11**, with $\omega_k = 1$), and, if D is also convex, Goldstein (Theorem **14.2.14**), and Goldstein–Armijo (Theorem **14.2.15**). More generally, the result holds for any steplength algorithm such that $\{x^k\} \subset L^0$ and (2) holds with $p^k = g'(x^k)^T$.

Proof. Consider, first, the Curry–Altman algorithm and assume that $x^0, \ldots, x^k$ are well-defined and lie in L^0. Then, with $p^k = g'(x^k)^T$, the first part of the proof of **14.2.7** shows that x^{k+1} is well defined and lies in L^0. Hence, by induction, the whole sequence $\{x^k\}$ of (5) is well defined and lies in L^0. If $g'(x^j)^T = 0$ for some j, then $x^j = x^*$, and the proof is complete. Otherwise, the sequence $p^k = g'(x^k)^T$, $k = 0, 1, \ldots$, satisfies (7) and is gradient-related to $\{x^k\}$. Therefore, since **14.2.7** shows that (2) holds, the result follows from **14.3.2**. The proof for any of the other steplength algorithms proceeds in precisely the same fashion using the relevant theorem from **14.2**. ∎

We note that **14.3.3** is not a direct corollary of **14.2.7** (or the other theorems of **14.2**) since in **14.2.7** we assumed the prior existence of the direction sequence $\{p^k\}$, while for iterations like (5) the sequence $\{p^k\}$ is defined in terms of the sequence $\{x^k\}$. Hence, a repetition of the inductive argument of **14.2.7**, as just done in **14.3.3**, is necessary in order to ensure the existence of the p^k.

In case g has more than one critical point in L^0, we may still conclude convergence by means of **14.1.5** provided that the number of critical points is finite and that

$$\lim_{k\to\infty}(x^{k+1} - x^k) = 0. \tag{8}$$

As shown by **14.1.3**, (8) will be satisfied for any strongly downward

14.3. GRADIENT AND GRADIENT-RELATED METHODS

sequence (Definition 14.1.2) provided that g is hemivariate (Definition 14.1.1). Hence, we have the following companion result to 14.3.2.

14.3.4. Assume that the conditions of 14.3.2 hold except that the uniqueness of x^* is replaced by the assumption that g is hemivariate and has only finitely many critical points in L^0, and that the sequence $\{x^k\}$ is strongly downward. Then $\{x^k\}$ converges to a critical point of g.

Note that 14.3.4 applies immediately to the Curry–Altman and restricted minimization algorithms since 14.2.7 and 14.2.10, respectively, show that for these two steplength algorithms the sequence $\{x^k\}$ is strongly downward. Note also that a modification of 14.3.3 along the lines of 14.3.4 is immediate for these steplength algorithms (see E 14.3-6).

Neither 14.3.2 nor 14.3.3 apply directly to the overrelaxed Curry–Altman, the one-step Newton, or the approximate minimization steplength algorithms studied in Theorems 14.2.8, 14.2.9, and 14.2.4, respectively, since these all require more stringent conditions on g. We next give two theorems utilizing these algorithms for general gradient-related methods.

14.3.5. Assume that $g: D \subset R^n \to R^1$ is continuously differentiable on the open set D, that $L^0 = L^0(g(x^0))$ is compact, and that

$$\| g'(x) - g'(y) \| \leqslant \gamma \| x - y \|, \qquad \forall x, y \in L^0. \tag{9}$$

Consider the iteration (1) where ω_k, α_k are chosen as in 14.2.4 and the p^k are nonzero vectors satisfying (7) for some $c > 0$. If g has only finitely many critical points in L^0, then the sequence $\{x^k\}$ converges to a critical point x^*. Moreover, if g has a second F-derivative at x^* and the Hessian matrix $H_g(x^*)$ is nonsingular, then the rate of convergence is at least R-linear.

Proof. By 14.2.4 we have that $\{x^k\} \subset L^0$ and that (2) holds, so that (7) implies (4). But 14.2.4 also shows that (8) is valid, and the convergence result again follows from 14.1.5. For the rate of convergence, we note that in the proof of 14.2.4 we obtained the inequality

$$g(x^k) - g(x^{k+1}) \geqslant \tfrac{1}{2}(\epsilon/\gamma)(2 - \epsilon)[g'(x^k) p^k / \| p^k \|]^2, \qquad k = 0, 1, \ldots.$$

Therefore, (7) implies that

$$g(x^k) - g(x^{k+1}) \geqslant \tfrac{1}{2}[\epsilon(2 - \epsilon) c^2/\gamma] \| g'(x^k)^{\mathrm{T}} \|^2, \qquad k = 0, 1, \ldots,$$

and the result follows from 14.1.6. ∎

14.3.6. Let $g: D \subset R^n \to R^1$ be twice-continuously differentiable on the open convex set D, assume that $L^0 = L^0(g(x^0))$ is compact, and suppose that $g''(x)$ is positive definite for all $x \in D$. Consider the iteration (1) with ω_k, α_k defined by either the overrelaxed Curry–Altman algorithm (Theorem **14.2.8**) or the one-step Newton algorithm (Theorem **14.2.9**), and assume that the p^k are nonzero vectors satisfying (7) for some $c > 0$. Then $\lim_{k \to \infty} x^k = x^*$, where x^* is the unique critical point of g in L^0, and the rate of convergence is at least R-linear.

Proof. From **14.2.8** or **14.2.9**, respectively, it follows that the iterates (1) remain in L^0 and satisfy (2). Moreover, since L^0 is compact and D is open, **4.1.3**, **4.2.2**, and **4.2.9** show that g has a unique critical point in L^0, and (7) implies that the sequence $\{p^k\}$ is gradient-related. The convergence then follows from **14.3.2**. For the rate of convergence, observe that we showed in the proof of **14.2.8** and **14.2.9** that

$$g(x^k) - g(x^{k+1}) \geqslant \beta [g'(x^k) p^k / \| p^k \|]^2, \qquad k = 0, 1, \ldots, \qquad (10)$$

where $\beta = \epsilon(4\eta_1)^{-1}$ in **14.2.8** and $\beta = \epsilon^2(2\eta_1)^{-1}$ in **14.2.9**. Hence, by (7),

$$g(x^k) - g(x^{k+1}) \geqslant \beta c^2 \| g'(x^k)^T \|^2, \qquad k = 0, 1, \ldots, \qquad (11)$$

and **14.1.6** again applies, since the Hessian matrix $H_g(x^*)$ is non-singular. ∎

We note that both **14.3.5** and **14.3.6** give rate-of-convergence results. The key to these results is that the steplength algorithm yields an inequality of the form (10) for some $\beta > 0$ so that under the assumption (7) on the directions p^k the inequality (11) holds and **14.1.6** is applicable. The inequality (11) is indeed valid for all of the steplength algorithms analyzed in **14.2** under the assumptions of those theorems. The verification of this for the Goldstein, Goldstein–Armijo, and minimization algorithms is left to **E 14.3-8**. In the sequel, we shall need the following result for the Curry–Altman algorithm.

14.3.7. Assume that g, x^0, and x^* satisfy the conditions of **14.3.2** and, in addition, that g is twice-continuously differentiable in a neighborhood of x^*, and $H_g(x^*)$ is nonsingular. Let $\{p^k\}$ be a sequence of nonzero vectors satisfying (7), and consider the iteration (1) with ω_k, α_k chosen by the Curry–Altman algorithm of Theorem **14.2.7**. Then the convergence of $\{x^k\}$ to x^* is at least R-linear.

Proof. The convergence of $\{x^k\}$ to x^* is a consequence of **14.3.2**. For

14.3. GRADIENT AND GRADIENT-RELATED METHODS

the rate of convergence, observe that we showed in the proof of **14.2.7** that

$$g(x^k) - g(x^{k+1}) \geq \epsilon \mu_0 \alpha_k g'(x^k) p^k, \tag{12}$$

where $\mu_0 = \mu$ if $\mu > 0$, and $\mu_0 = \frac{1}{2}$ otherwise. Let $\delta > 0$ and k_0 be such that g is twice-continuously differentiable on $S = \bar{S}(x^*, \delta) \subset D$ and $x^k \in S$ for all $k \geq k_0$. Then, by the mean-value theorem **3.3.5** and the definitions of α_k and μ_0, we have, proceeding as in **14.2.7**,

$$\begin{aligned}
(1 - \mu_0) g'(x^k) p^k &\leq g'(x^k) p^k - g'(x^k - \alpha_k p^k) p^k \\
&\leq \| g'(x^k) - g'(x^k - \alpha_k p^k)\| \, \| p^k \| \\
&\leq \gamma_1 \alpha_k \| p^k \|^2, \quad \forall k \geq k_0, \tag{13}
\end{aligned}$$

where $\gamma_1 = \max\{\| g''(x)\| \mid x \in S\}$. Hence, combining (12), (13), and (7), we see that

$$g(x^k) - g(x^{k+1}) \geq \epsilon(\mu_0/\gamma_1)(1 - \mu_0) c^2 \| g'(x^k)^T \|^2, \quad \forall k \geq k_0,$$

and the result follows from **14.1.6**. ∎

NOTES AND REMARKS

NR 14.3-1. Although Cauchy [1847] proposed the gradient method, he gave no convergence theorems. The first such results appear to be due to Temple [1939] for quadratic functions and to Curry [1944] for general functionals. Since then a large number of authors, including Kantorovich [1948b], Crockett and Chernoff [1955], Lumiste [1955], Vainberg [1960], Goldstein [1962], Polyak [1963], Nashed [1964, 1965], Glazman [1964], Altman [1966a], Armijo [1966], Blum [1966], Braess [1966], and Ljubic [1968], have contributed to this problem. Many of the above papers deal with the gradient iteration using different steplength algorithms, and, in particular, **14.3.3** includes several of these results. On the other hand, many of these authors treat the problem in either Hilbert or Banach space. See, for example, Goldstein [1967] and Poljak [1966].

NR 14.3-2. The hypothesis of general directions p^k which satisfy (7) is due to Ostrowski [1966], who also proved, in essence, **14.3.5**. However, estimates of the form (7) had been obtained previously, but implicitly, in the analysis of such concrete methods as that of Newton. On the other hand, the use of gradient-related directions independently of any particular steplength is due to Elkin [1968], who recognized the essential independence of the steplength analysis.

NR 14.3-3. The convergence statements of **14.3.6** are due to Elkin [1968], while the rate-of-convergence statements of this theorem, as well as **14.3.5** and **14.3.7**, extend and modify results of Ostrowski [1966, 1967a].

NR 14.3-4. Definition **14.3.1** of gradient-relatedness can be generalized by replacing (3) with

$$|g'(x^k) p^k| \geq \sigma_k(\|g'(x^k)^T\|), \qquad k = 0, 1, \ldots,$$

where $\sigma_k : [0, \infty) \to [0, \infty)$, $k = 0, 1, \ldots$, is a sequence of functions such that $\lim_{k \to \infty} \sigma_k(t_k) = 0$ implies that $\lim_{k \to \infty} t_k = 0$. All results of this and subsequent sections pertaining to gradient-relatedness then remain the same word for word.

EXERCISES

E 14.3-1. Suppose that $\{p^k\} \subset R^n$ is gradient-related to $\{x^k\}$ in some norm. Show that $\{p^k\}$ is gradient-related to $\{x^k\}$ in any norm.

E 14.3-2. Let $g: D \subset R^n \to R^1$ be G-differentiable and assume that $\{x^k\} \subset D$, $\{p^k\} \subset R^n$, $p^k \neq 0$, satisfy

$$g'(x^k) p^k = [\alpha_k \|g'(x^k)\| - \beta_k] \|p^k\|,$$

where $0 < \alpha \leq |\alpha_k|$ and $\lim_{k \to \infty} \beta_k = 0$. Show that $\{p^k\}$ is gradient-related to $\{x^k\}$ under the extended definition of **NR 14.3-4**. (Altman [1966a])

E 14.3-3. Let $g: D \subset R^n \to R^1$ be G-differentiable and suppose that $\{p^k\} \subset R^n$ is gradient-related to $\{x^k\} \subset D$. Show that the sequence $q^k = (|g'(x^k) p^k|/\|p^k\|) p^k$ is also gradient-related to $\{x^k\}$.

E 14.3-4. Let $g: D \subset R^n \to R^1$ be continuously differentiable on the open, convex set D, and $x^0 \in D$ such that $L(g(x^0))$ is compact. Assume that g has only finitely many critical points in L, and consider the Jacobi process

$$x^{k+1} = x^k - \omega_k \alpha_k p^k,$$

where ω_k, α_k are determined by the Goldstein–Armijo algorithm of **14.2.15** and p^k is determined as follows. Let

$$\alpha_{j,k} = \min\{\alpha \geq 0 \mid x^k - \hat{e}^j \in D, \; g'(x^k - \alpha \hat{e}^j) \hat{e}^j = 0\}, \qquad j = 1, \ldots, n,$$

where $\hat{e}^j = \operatorname{sgn}(g'(x^k) e^j) e^j$, and set $p^k = \sum_{j=1}^n \alpha_{j,k} \hat{e}^j$. Show that $\lim_{k \to \infty} x^k = x^*$, where $g'(x^*)^T = 0$. (Elkin [1968])

E 14.3-5. Show that the steplength algorithms analyzed in Theorems **14.2.7**, **14.2.10**, and **14.2.11** may be used in **14.3.2**, and that the algorithms of **14.2.12**, **14.2.14**, and **14.2.15** may be used if the whole level set $L = L(g(x^0))$ is assumed to be compact and x^* unique in L.

E 14.3-6. Assume that $g: D \subset R^n \to R^1$ satisfies the conditions of **14.3.4**, and ω_k, α_k in the iteration (5) are obtained by the Curry–Altman algorithm of **14.2.7**. Show that $\{x^k\}$ converges to a critical point of g. Show the same for the

14.4. NEWTON-TYPE METHODS

iteration $x^{k+1} = x^k - \alpha_k g'(x^k)^T$, where α_k is determined by the minimization algorithm of **14.2.10**.

E 14.3-7. Show that **14.3.7** remains valid if the Curry–Altman algorithm is replaced by the Goldstein algorithm (Theorem **14.2.14**), the Goldstein–Armijo algorithm (Theorem **14.2.15**), or the minimization algorithm (Theorem **14.2.10**).

14.4. NEWTON-TYPE METHODS

As we saw in Chapter 8, the Newton method does not, in general, decrease the value of the functional at each point, so that the results presented in this chapter do not apply directly. However, as also discussed in Chapter 8, we can ensure that $g(x^k) \geq g(x^{k+1})$ by means of adding a factor α_k in order to obtain the damped Newton method:

$$x^{k+1} = x^k - \alpha_k H_g(x^k)^{-1} g'(x^k)^T, \qquad k = 0, 1, \ldots, \tag{1}$$

where $H_g(x)$ is again the Hessian matrix of g at x. Our approach will be to show the Newton directions

$$p^k = H_g(x^k)^{-1} g'(x^k)^T \tag{2}$$

are gradient-related and then to apply the general results of the previous section. We begin with the following basic lemma.

14.4.1. Let $g: D \subset R^n \to R^1$ be G-differentiable on a compact set $D_0 \subset D$ and assume that $A: D_0 \to L(R^n)$ is a continuous mapping such that $A(x)$ is positive definite for each $x \in D_0$. Set $p(x) = A(x)^{-1} g'(x)^T$, $x \in D_0$. Then there is a constant $c > 0$ such that

$$g'(x) p(x) \geq c \, \| g'(x)^T \| \, \| p(x) \|, \qquad \forall x \in D_0. \tag{3}$$

Proof. We shall work in the l_2-norm; for an arbitrary norm the result follows easily from the norm-equivalence theorem **2.2.1**. Since A is continuous on the compact set D_0, it follows that there are constants $\mu_2 \geq \mu_1 > 0$ such that

$$\| A(x) \| \leq \mu_2, \qquad h^T A(x) h \geq \mu_1 h^T h, \qquad \forall x \in D_0, \ h \in R^n.$$

Moreover, by the Cauchy–Schwarz inequality, we have

$$\| A(x) h \| \geq \mu_1 \| h \|, \qquad \forall x \in D_0, \ h \in R^n.$$

Hence,

$$\| h \| = \| A(x) A(x)^{-1} h \| \geq \mu_1 \| A(x)^{-1} h \|, \qquad \forall x \in D_0, \ h \in R^n, \tag{4}$$

and, similarly,

$$\|h\| \leq \mu_2 \|A(x)^{-1} h\|, \qquad \forall x \in D_0, \quad h \in R^n.$$

Therefore, with $\hat{h} = A(x)^{-1}h$, we see that

$$h^T A(x)^{-1} h = \hat{h}^T A(x)^T \hat{h} = \hat{h}^T A(x) \hat{h} \geq \mu_1 \|\hat{h}\|^2 \geq \gamma \|h\|^2,$$

where we have set $\gamma = \mu_1/\mu_2^2$. Therefore, using (4),

$$g'(x) p(x) \geq \gamma \|g'(x)^T\|^2 \geq \gamma \|g'(x)^T\| \mu_1 \|p(x)\|,$$

so that (3) holds with $c = \mu_1^2/\mu_2^2$. ∎

We consider now the iteration

$$x^{k+1} = x^k - \alpha_k A(x^k)^{-1} g'(x^k)^T, \qquad k = 0, 1, \ldots. \tag{5}$$

If A satisfies the conditions of **14.4.1** and $\{x^k\} \subset D_0$, then (3) shows that the directions

$$p^k = A(x^k)^{-1} g'(x^k)^T, \qquad k = 0, 1, \ldots,$$

are gradient-related to $\{x^k\}$. Hence, convergence results for (5) follow immediately from the gradient-related theorem **14.3.2** provided $\{\alpha_k\}$ is chosen under the conditions of that theorem. We summarize this in the next result.

14.4.2. Assume that $g: D \subset R^n \to R^1$ is continuously differentiable on the open set D, that there is an $x^0 \in D$ such that $L^0 = L^0(g(x^0))$ is compact, and that g has a unique critical point x^* in L^0. Suppose that $A: L^0 \to L(R^n)$ is continuous and $A(x)$ is positive definite for all $x \in L^0$. Then the iterates (5) converge to x^* for any sequence of α_k such that $\{x^k\} \subset L^0$ and

$$\lim_{k \to \infty} g'(x^k) p^k / \|p^k\| = 0, \tag{6}$$

where $p^k = A(x^k)^{-1} g'(x^k)^T$.

Under the conditions of **14.4.2**, admissible choices for α_k are summarized in **14.3.3**, and, as in that theorem, relaxation factors ω_k may also be used in conjunction with (5).

There are numerous variations of **14.4.2** corresponding, in particular, to those of **14.3.4**. Rather than repeat these variations, some of which are stated in **E 14.4-3/4**, we shall give a few samples of concrete theorems. We stress that, especially in the choice of steplength, these results are

14.4. NEWTON-TYPE METHODS

meant to be only illustrative, and, again, several variations are given in the exercises.

14.4.3. Global Damped Newton Theorem. Assume that $g\colon R^n \to R^1$ is uniformly convex and twice-continuously differentiable. Then, given any x^0, there exists a sequence $\{\alpha_k\}$ so that the iterates (1) converge to the unique critical point x^* of g. Moreover, there is a k_0 depending on x^0 so that we may take $\alpha_k = 1$ for $k \geqslant k_0$; hence, in this case the rate convergence of $\{x^k\}$ to x^* is at least (R- or Q-) superlinear.

Proof. Theorem **4.3.9** guarantees that g has a unique critical point x^*, and, by **4.3.7**, for any x^0 the level set $L = L(g(x^0))$ is compact. For the determination of α_k, we may use any of the steplength algorithms of Section **14.2**. For example, consider the minimization algorithm of **14.2.12**. Proceeding precisely as in **14.3.3**, an inductive argument shows that all x^k are well defined and lie in L, and **14.2.12** proves that (6) holds. Hence, the convergence follows from **14.4.2**, since $L = L^0$. For the final statement, let $S = \bar{S}(x^*, \delta)$, $\delta > 0$, be a ball such that the Newton iterates themselves converge to x^* whenever $x^k \in S$; the existence of this ball, as well as the superlinear rate of convergence, is a consequence of **10.2.2**. Hence, the convergence of the sequence (1) ensures that $x^{k_0} \in S$ for some k_0, and from this point on we may choose $\alpha_k = 1$. ∎

As indicated in the proof, any of the steplength algorithms of **14.2** may be used to obtain the α_k of (1); see **E 14.4-5**. Although we chose to use the minimization principle, a sometimes computationally useful procedure is the Goldstein–Armijo algorithm: Choose α_k as the first number in the sequence $\{q^j\colon j = 0, 1,...\}$ such that

$$g(x^k) - g(x^k - \alpha_k p^k) \geqslant \mu \alpha_k g'(x^k) p^k, \tag{7}$$

where μ, $q \in (0, 1)$ are fixed and p^k is the Newton direction. (We deviate now from the notation of Theorem **14.2.15**.) This is advantageous when g is defined only on a subset D, and the result of a Newton step itself is outside of D. Results for this steplength procedure in the context of **14.4.2–3** are given in **E 14.4-6–7**. In addition, sufficient conditions that the α_k in (1) may be taken as unity for all k are given in **E 14.4-8**.

Theorem **14.4.3** could, of course, have been phrased in a local rather than global form (see **E 14.4-5**). We do this in the following result for the damped Gauss–Newton process

$$x^{k+1} = x^k - \alpha_k [F'(x^k)^T F'(x^k)]^{-1} F'(x^k)^T Fx^k, \quad k = 0, 1,..., \tag{8}$$

discussed in Section **8.5**.

14.4.4. Damped Gauss–Newton Theorem Let $F: D \subset R^n \to R^m$, $m \geq n$, be continuously differentiable on the open set D, define $g: D \subset R^n \to R^1$ by $g(x) = \frac{1}{2}(Fx)^T Fx$, and suppose that there is an $x^0 \in D$ so that $L^0 = L^0(g(x^0))$ is compact. Assume, further, that rank $F'(x) = n$ for all $x \in L^0$ and that the equation $F'(x)^T Fx = 0$ has a unique solution x^* in L^0. Then there exist α_k, $k = 0, 1,\ldots$, such that the sequence (8) remains in L^0 and $\lim_{k \to \infty} x^k = x^*$.

Proof. By assumption, the mapping

$$A: L^0 \to L(R^n), \qquad A(x) = F'(x)^T F'(x), \qquad x \in L^0,$$

is continuous, and, since rank $F'(x) = n$, $A(x)$ is positive definite for all $x \in L^0$. Moreover, g is continuously differentiable and $g'(x)^T = F'(x)^T Fx$. Now, choose some admissible steplength algorithm for α_k, say, the Curry–Altman algorithm of **14.2.7**. Then, proceeding as in **14.3.3**, an inductive argument shows that all x^k lie in L^0 and (6) holds, so that **14.2.2** applies. ∎

In both **14.4.3** and **14.4.4**, the matrix $A(x)$ was symmetric. We give one last example of the use of **14.4.2** in which $A(x)$ is not symmetric but still positive definite. Consider the Newton–SOR iteration

$$x^{k+1} = x^k - \alpha_k \omega [D(x^k) - \omega L(x^k)]^{-1} g'(x^k)^T, \tag{9}$$

discussed in **7.4**, with the addition of the damping factor α_k. Here, $D(x)$ and $-L(x)$ are the diagonal and strictly lower-triangular parts of $H_g(x)$.

14.4.5. Damped Newton–SOR Theorem. Let $g: D \subset R^n \to R^1$ be twice-continuously differentiable on the open convex set D, and suppose that there is an $x^0 \in D$ such that $L^0 = L^0(g(x^0))$ is compact. Assume that $g''(x)$ is positive definite for all $x \in D$, that $\omega \in (0, 2)$, and that the factors α_k in (9) are chosen by the minimization algorithm of Theorem **14.2.10**. Then $\{x^k\} \subset L^0$, $\lim_{k \to \infty} x^k = x^*$, where x^* is the unique critical point of g in L^0, and the rate of convergence is at least R-linear.

Proof. Set

$$A(x) = (1/\omega)[D(x) - \omega L(x)], \qquad x \in L^0;$$

then

$$A(x) + A(x)^T = (2/\omega) D(x) - L(x) - L(x)^T = [(2/\omega) - 1] D(x) + H_g(x).$$

Consequently, since $\omega \in (0, 2)$ and

$$h^T A(x) h = \tfrac{1}{2} h^T [A(x) + A(x)^T] h,$$

14.4. NEWTON-TYPE METHODS

it follows that $A(x)$ is positive definite for all $x \in L^0$. As in **14.3.6**, g has a unique critical point in L^0, and it follows as in **14.4.4** that $\lim_{k \to \infty} x^k = x^*$. Finally, the positive definiteness of $g''(x)$ on L^0 implies that the minimization and Curry principles for α_k are identical, so that the rate of convergence is a consequence of **14.3.7**. ∎

Note that we chose to state **14.4.5** in terms of a specific algorithm for the α_k. This allows the rate-of-convergence statement, but convergence alone could also have been obtained, of course, for other steplength algorithms.

Up to now, we have ensured that the iterations under consideration become descent methods by the addition of the damping factors α_k. An alternative approach is the use of factors λ_k in the Levenberg fashion as discussed in **8.5**:

$$x^{k+1} = x^k - [A(x^k) + \lambda_k I]^{-1} g'(x^k)^T, \quad k = 0, 1, \dots. \tag{10}$$

For suitable choice of λ_k this is a descent method, as the following result shows.

14.4.6. Let $g: D \subset R^n \to R^1$ be twice-continuously differentiable on the open set D, and suppose that there is an $x^0 \in D$ such that $L^0 = L^0(g(x^0))$ is compact and that g has only finitely many critical points in L^0. Assume that $A: L^0 \to L(R^n)$ is a continuous mapping such that $A(x)$ is symmetric for all $x \in L^0$, and that

$$\mu_0 h^T h \leq h^T A(x) h \leq \mu_1 h^T h, \quad g''(x) hh \leq \gamma_1 h^T h, \quad \forall x \in L^0, \; h \in R^n, \tag{11}$$

where μ_0, μ_1, and γ_1 need not be positive. Consider the process (10) with

$$\tfrac{1}{2}\hat{\gamma}_1 - \mu_0 < \eta_0 \leq \lambda_k \leq \eta_1 < +\infty, \quad \hat{\gamma}_1 = \max(0, \gamma_1), \quad k = 0, 1, \dots. \tag{12}$$

Then $\{x^k\} \subset L^0$, $\lim_{k \to \infty} x^k = x^*$, where $g'(x^*)^T = 0$, and if $H_g(x^*)$ is nonsingular, then the convergence is at least R-linear.

Proof. Observe, first, that $\mu_0 + \eta_0 > 0$ and that, for any $\lambda \in [\eta_0, \eta_1]$ and $x \in L^0$

$$(\mu_0 + \eta_0) h^T h \leq h^T [A(x) + \lambda I] h \leq (\mu_1 + \eta_1) h^T h. \tag{13}$$

Therefore, if $x^k \in L^0$, then $p^k = [A(x^k) + \lambda_k I]^{-1} g'(x^k)^T$ is well-defined, and we have

$$g'(x^k) p^k \geq (\mu_1 + \eta_1)^{-1} \| g'(x^k)^T \|_2^2$$

in the l_2-norm. Hence, $g'(x^k) p^k > 0$ unless $g'(x^k)^T = 0$, that is, unless the process terminates at x^k. Now, let $\alpha \in (0, 1]$ be such that $[x^k, x^k - \alpha p^k] \subset L^0$. Then, by the mean-value theorem **3.3.10**, there is an $\hat{\alpha} \in (0, \alpha)$ so that

$$g(x^k) - g(x^k - \alpha p^k) = \alpha g'(x^k) p^k - \tfrac{1}{2}\alpha^2 g''(x^k - \hat{\alpha} p^k) p^k p^k$$
$$\geq \alpha[\mu_0 + \eta_0 - \tfrac{1}{2}\hat{\gamma}_1] \| p^k \|_2^2 > 0, \qquad (14)$$

since $g'(x^k) p^k = (p^k)^T [A(x^k) + \lambda_k I] p^k$. It follows from **14.2.3** that $x^{k+1} = x^k - p^k \in L^0$ and, by induction, that $\{x^k\} \subset L^0$.

Since (13) implies that

$$(\mu_1 + \eta_1)^{-1} \| g'(x^k)^T \|_2 \leq \| p^k \|_2 \leq (\mu_0 + \eta_0)^{-1} \| g'(x^k)^T \|_2, \qquad (15)$$

we have from (14) that

$$g(x^k) - g(x^{k+1}) \geq (\mu_1 + \eta_1)^{-2}(\mu_0 + \eta_0 - \tfrac{1}{2}\hat{\gamma}_1) \| g'(x^k)^T \|_2^2. \qquad (16)$$

Since L^0 is compact, the left side converges to zero and, hence, $\lim_{k \to \infty} g'(x^k)^T = 0$. Therefore, since $p^k = x^k - x^{k+1}$, (15) shows that $\lim_{k \to \infty} (x^{k+1} - x^k) = 0$, and the result follows from **14.1.5** and **14.1.6**. ∎

Note that we could have shown that the directions $\{p^k\}$ in **14.4.6** are gradient-related to $\{x^k\}$ (see **E 14.4-10**). However, in this case it is simpler to obtain the necessary final inequality (16) directly.

Note also that **14.4.6** has immediate application to the cases $A(x) = F'(x)^T F'(x)$ and $A(x) = H_g(x)$ (see **E 14.4-11**). In particular, if $0 < \tfrac{1}{2}\gamma_1 < \mu_0$, then $\lambda_k = 0$ is permitted in the iteration (10), and, in this case, **14.4.6** provides results on the Newton and Gauss–Newton processes without damping factors.

NOTES AND REMARKS

NR 14.4-1. There are various results in the literature similar to **14.4.2**. For example, Nashed [1964] considers the iteration

$$x^{k+1} = x^k - \alpha_k A_k^{-1} g'(x^k)^T, \qquad k = 0, 1, \ldots, \qquad (17)$$

for a sequence of symmetric linear operators A_k which satisfy

$$\mu_1 h^T h \leq h^T A_k h \leq \mu_2 h^T h, \qquad \forall h \in R^n, \qquad (18)$$

where $0 < \mu_1 \leq \mu_2$. However, Nashed also assumes the very stringent requirement that each A_k commutes with $H_g(x)$ for each x. On the other hand, it is easy to show that (18) implies that the sequence $p^k = A_k^{-1} g'(x^k)^T$, $k = 0, 1, \ldots$, is gradient-related to $\{x^k\}$, and convergence theorems along the lines of **14.4.2**

14.4. NEWTON-TYPE METHODS

immediately follow for (17) (see **E 14.4-2**). Jakovlev [1964] has also used the condition (18) in connection with (17), where α_k is chosen by the minimization principle, and obtained results similar to **14.4.2**.

NR 14.4-2. The idea of the damping factor α_k in (1) had been suggested by Gleyzal [1959] and Crockett and Chernoff [1955], but no convergence was proved. Theorem **14.4.3** itself is implicit in the work of Jakovlev [1964], who also states that if α_k is obtained by the minimization principle, then $\lim_{k \to \infty} \alpha_k = 1$ and the iteration (1) maintains superlinear convergence. Similar results are given by Goldstein [1967], who used his own form of steplength algorithm and showed that $\alpha_k = 1$ for $k \geq k_0$, and by Elkin [1968], who analyzed the procedure described after **14.4.3** (see **E 14.4-6** and also **E 14.4-7**).

NR 14.4-3. Theorem **14.4.4** appears to be new, although Hartley [1961] gives a partial convergence theorem for (8) if α_k is chosen by the minimization principle. See also Tornheim [1963], who shows that the Gauss–Newton directions p^k satisfy $g'(x^k) p^k > 0$ if $g'(x^k)^T \neq 0$, and Braess [1966].

NR 14.4-4. Theorem **14.4.5** is due to Stepleman [1969]. Theorem **14.4.6** is apparently new, although the introduction of the parameter λ_k in (10) dates back to Levenberg [1944], and has also been studied by Marquardt [1963].

EXERCISES

E 14.4-1. Let $g: D \subset R^n \to R^1$ be G-differentiable, assume that $\{x^k\} \subset D$ is an arbitrary sequence, and suppose that $\{B_k\} \subset L(R^n)$ is a sequence of matrices such that for some $\beta < \infty$ and $\mu > 0$

$$\| B_k \| \leq \beta, \qquad h^T B_k h \geq \mu h^T h, \qquad \forall h \in R^n, \quad k = 0, 1, \ldots.$$

Show that the sequence $p^k = B_k g'(x^k)^T$, $k = 0, 1, \ldots$, is gradient-related to $\{x^k\}$. State and prove a convergence theorem analogous to **14.4.2**.

E 14.4-2. Let g and $\{x^k\}$ be as in **E 14.4-1**, and $\{A_k\} \subset L(R^n)$ a sequence of symmetric matrices such that

$$\mu_1 h^T h \leq h^T A_k h \leq \mu_2 h^T h, \qquad \forall h \in R^n, \quad k = 0, 1, \ldots,$$

where $0 < \mu_1 \leq \mu_2$. Show that the sequence $p^k = A_k^{-1} g'(x^k)^T$, $k = 0, 1, \ldots$, is gradient related to $\{x^k\}$. State and prove a convergence theorem analogous to **14.4.2**.

E 14.4-3. Assume that $g: D \subset R^n \to R^1$, x^0, and $A: L^0 \to L(R^n)$ satisfy the conditions of **14.4.2**. Show that the iterates (5) converge to x^* provided α_k is chosen by any of the steplength algorithms specified in **14.3.3**.

E 14.4-4. Replace the uniqueness assumption in **14.4.2** by: g is hemivariate and has at most finitely many critical points in L^0, and $\{x^k\}$ is strongly downward. Show that the result remains valid.

E 14.4-5. Prove the convergence statement of **14.4.3** with α_k chosen by any of the steplength algorithms analyzed in **14.2**. Also state and prove the local form of **14.4.3** in which $L^0 = L^0(g(x^0))$ is compact, g is twice-continuously differentiable, and $g''(x)$ positive definite on L^0.

E 14.4-6. Let $g: D \subset R^n \to R^1$ be continuously differentiable on the open, convex set D, and $x^0 \in D_0$ such that $L^0 = L^0(g(x^0))$ is compact. Assume that g has only finitely many critical points in L^0 and let $A: L^0 \to L(R^n)$ be continuous and $A(x)$ symmetric, positive definite for all $x \in L_0$. Consider the process (5) with α_k chosen as the first number in the sequence $\{q^j : j = 0, 1,...\}$ such that

$$x^k - \alpha_k p^k \in D, \qquad g(x^k) - g(x^k - \alpha_k p^k) \geq \mu \alpha_k g'(x^k) p^k,$$

where $\mu, q \in (0, 1)$ are fixed and $p^k = A(x^k)^{-1} g'(x^k)^T$. Show that $\lim_{k \to \infty} x^k = x^*$, where $g'(x^*)^T = 0$. (Elkin [1968])

E 14.4-7. Let $g: R^n \to R^1$ satisfy the conditions of **14.4.3**, and let the α_k in (1) be chosen by the procedure of **E 14.4-6** with $A(x) = H_g(x)$ and with $\mu \in (0, q)$. Show there is a k_0 so that $\alpha_k = 1$ for all $k \geq k_0$.

E 14.4-8. Let $g: R^n \to R^1$ satisfy the conditions of **14.4.3**, and for some x^0 suppose that

$$\mu_1 h^T h \leq g''(x) hh \leq \mu_2 h^T h, \qquad \forall h \in R^n, \quad x \in L(g(x^0)),$$

where $\mu_2/\mu_1 \leq \sqrt{2}$. Show that the Newton iterates starting from x^0 satisfy $g(x^{k+1}) \leq g(x^k)$, $k = 0, 1,...$.

E 14.4-9. Let $F: D \subset R^n \to R^n$ be differentiable, and set $g(x) = \|Fx\|_2$. Let $P: D \subset R^n \to R^n$ be a continuous mapping which satisfies

$$(Fx)^T F'(x) Px = g(x)^2 \|Px\|_2 \|F'(x) Fx\|_2 \leq \eta g(x)^2, \qquad \forall x \in D.$$

Show that, for any $\{x^k\} \subset D$ such that $Fx^k \neq 0$, the sequence $p^k = Px^k$, $k = 0, 1,...$, is gradient-related to $\{x^k\}$.

E 14.4-10. Show that the directions $\{p^k\}$ of **14.4.6** are gradient-related to $\{x^k\}$.

E 14.4-11. Apply **14.4.6** to the iterations

$$x^{k+1} = x^k - [H_g(x^k) + \lambda_k I]^{-1} g'(x^k)^T$$

and

$$x^{k+1} = x^k - [F'(x^k)^T F'(x^k) + \lambda_k I]^{-1} F'(x^k)^T Fx^k.$$

14.5. CONJUGATE-DIRECTION METHODS

We now apply the results of the previous sections to conjugate-direction methods, and first consider the Daniel algorithm discussed in 8.4:

$$\begin{cases} x^{k+1} = x^k - \alpha_k p^k, \quad g(x^{k+1}) = \min\{g(x^k - \alpha p^k) \mid x^k - \alpha p^k \in D\}, \\ p^0 = g'(x^0)^T \\ p^{k+1} = g'(x^{k+1})^T - \beta_k p^k, \quad \beta_k = \dfrac{g''(x^{k+1}) g'(x^{k+1})^T p^k}{g''(x^{k+1}) p^k p^k}. \end{cases} \quad (1)$$

14.5.1. Let $g: D \subset R^n \to R^1$ be twice-continuously differentiable on the open convex set D, and assume that $g''(x)$ is positive definite for all $x \in D$. Suppose, further, that there is an x^0 in D so that the level set $L = L(g(x^0))$ is compact. Then the iterates (1) are well defined, lie in L, and satisfy $\lim_{k \to \infty} x^k = x^*$ where x^* is the unique critical point of g in L.

Proof. As in **14.3.6**, g has a unique critical point in L. Moreover, L is convex, and the minimization algorithms of **14.2.10**–**14.2.12** are identical. Using **14.2.10**, an induction argument as in the proof of **14.3.3** shows that all x^k are well-defined, lie in L, and satisfy

$$\lim_{k \to \infty} g'(x^k) p^k / \| p^k \| = 0, \quad (2)$$

provided that no $p^k = 0$. However, since the steplength algorithm implies that $g'(x^k) p^{k-1} = 0$, $k \geq 1$, we have

$$g'(x^k) p^k = g'(x^k) g'(x^k)^T - \beta_{k-1} g'(x^k) p^{k-1} = \| g'(x^k)^T \|_2^2. \quad (3)$$

Thus, $p^k = 0$ implies that $x^k = x^*$ in which case the result is proved. Hence, we may assume that $p^k \neq 0$ for all k, and, by **14.3.2**, it suffices to show that the sequence $\{p^k\}$ is gradient-related to $\{x^k\}$.

Since L is compact, g'' continuous, and $g''(x)$ positive definite on L, there exist constants $0 < \mu_0 \leq \mu_1$ such that

$$\mu_0 h^T h \leq g''(x) hh \leq \mu_1 h^T h, \quad \forall x \in L, \quad h \in R^n. \quad (4)$$

By definition of β_k and the symmetry of $g''(x^k)$, we have that

$$g''(x^k) p^k p^k = g''(x^k) g'(x^k)^T g'(x^k)^T$$
$$+ \beta_{k-1}^2 g''(x^k) p^{k-1} p^{k-1} - 2\beta_{k-1} g''(x^k) g'(x^k)^T p^{k-1}$$
$$= g''(x^k) g'(x^k)^T g'(x^k)^T - \beta_{k-1}^2 g''(x^k) p^{k-1} p^{k-1},$$

and, hence, it follows from (4) that

$$\mu_1 \| g'(x^k)^T \|_2^2 \geq g''(x^k) g'(x^k)^T g'(x^k)^T = g''(x^k) p^k p^k + \beta_{k-1}^2 g''(x^k) p^{k-1} p^{k-1}$$
$$\geq g''(x^k) p^k p^k \geq \mu_0 \| p^k \|_2^2.$$

Therefore, (3) shows that

$$g'(x^k) p^k = \| g'(x^k)^T \|_2^2 \geq \| g'(x^k)^T \|_2 (\mu_0/\mu_1)^{1/2} \| p^k \|_2,$$

and $\{p^k\}$ is gradient-related to $\{x^k\}$. ∎

As a corollary of **14.5.1**, we obtain the following global convergence theorem.

14.5.2. Assume that $g: R^n \to R^1$ is twice-continuously differentiable, $g''(x)$ is positive definite for all $x \in R^n$, and

$$\lim_{\|x\| \to \infty} g(x) = +\infty. \tag{5}$$

Then, for any $x^0 \in R^n$, the iterates (1) are well defined and converge to the unique critical point of g in R^n.

The proof is immediate since **4.3.2** shows that for any x^0 the level set $L(g(x^0))$ is compact, and **4.2.9** guarantees the uniqueness of critical points. Hence, **14.5.1** applies.

We turn now to the Fletcher–Reeves algorithm discussed in Section **8.4** in the following modified form:

$$\begin{cases} x^{k+1} = x^k - \alpha_k p^k, \\ p^{k+1} = g'(x^{k+1})^T - \beta_k p^k, \quad p^0 = g'(x^0)^T; \\ \beta_k = \begin{cases} 0, & \text{if } k+1 = 0 (\bmod m), \quad m > 0, \\ \dfrac{g'(x^{k+1}) g'(x^{k+1})^T}{g'(x^k) g'(x^k)^T}, & \text{otherwise.} \end{cases} \end{cases} \tag{6}$$

The definition of β_k is equivalent to a "restart" procedure after every m steps; usually, m is chosen to be equal to n.

The proof of **14.5.1** was accomplished by showing that the directions $\{p^k\}$ of (1) were gradient-related to $\{x^k\}$. This proof procedure does not quite work for the iteration (6), but, by a related argument, we can prove the following result. Recall that $L^0 = L^0(g(x^0))$ denotes the connected component containing x^0 of the level set $L(g(x^0))$ and that a functional is hemivariate if it is not constant on line segments (Definition **14.1.1**).

14.5. CONJUGATE-DIRECTION METHODS

14.5.3. Assume that $g: D \subset R^n \to R^1$ is continuously differentiable on the open set D, that there is an $x^0 \in D$ so that L^0 is compact, and that g is hemivariate and has only finitely many critical points in L^0. Then the iterates (6), where α_k is chosen by the restricted minimization algorithm of **14.2.10**, are well-defined, lie in L^0, and converge to a critical point of g.

Proof. Assume that $x^k \in L^0$. Then **14.2.10** shows that x^{k+1} is well-defined and $x^{k+1} \in L^0$. Clearly, p^{k+1} is well defined and, precisely as in **14.5.1**, $p^k \neq 0$ unless x^k is a critical point, in which case the proof is complete. Otherwise, it follows by induction that $\{x^k\} \subset L^0$, and **14.2.10** shows, further, that $\{x^k\}$ is strongly downward and

$$\lim_{k \to \infty} g'(x^k) p^k / \| p^k \| = 0. \tag{7}$$

Moreover, by **14.1.3**, $\lim_{k \to \infty} (x^{k+1} - x^k) = 0$.

For given $k \geq 0$, let k_0 denote the largest integer not exceeding k which is congruent to zero modulo m, and set $\gamma_j = \| g'(x^j)^T \|_2^2, j = 0, 1, \ldots$. Then

$$p^k = \sum_{j=k_0}^{k} (\gamma_k / \gamma_j) g'(x^j)^T,$$

and with

$$\eta = \max_{x \in L^0} \| g'(x)^T \|_2, \qquad \gamma_{j_k} = \min_{k_0 \leq j \leq k} \gamma_j,$$

we have

$$\| p^k \|_2 \leq (k - k_0 + 1) \gamma_k \eta / \gamma_{j_k} \leq (m + 1) \gamma_k \eta / \gamma_{j_k}.$$

Now note that the steplength algorithm implies that $g'(x^j) p^{j-1} = 0$, and, hence,

$$g'(x^j) p^j = g'(x^j) g'(x^j)^T - \beta_{j-1} g'(x^j) p^{j-1} = \gamma_j, \qquad j = 0, 1, \ldots.$$

Therefore,

$$g'(x^k) \frac{p_k}{\| p^k \|_2} = \frac{\gamma_k}{\| p^k \|_2} \geq \frac{1}{(m+1) \eta} \gamma_{j_k},$$

which, by (7), shows that

$$\lim_{k \to \infty} g'(x^{j_k})^T = 0.$$

Since g' is uniformly continuous on L^0 and $\lim_{k \to \infty} \| x^k - x^{k+1} \| = 0$, there exists for any given $\epsilon > 0$ an index $k' \geq 0$ such that

$$\| g'(x^k)^T - g'(x^{k+1})^T \| \leq \epsilon, \qquad k \geq k'.$$

Therefore,

$$\| g'(x^k)^T \| \leq \| g'(x^{j_k})^T \| + \| g'(x^{j_k})^T - g'(x^k)^T \| \leq \| g'(x^{j_k})^T \| + m\epsilon, \quad \forall k \geq k',$$

which proves that $\lim_{k \to \infty} g'(x^k)^T = 0$. The result now follows from **14.1.5**. ∎

Note that the choice of steplength algorithm in **14.5.3** is not restricted to the minimization algorithm of **14.2.10**. All that is required is that $\{x^k\}$ be strongly downward, $g'(x^k) p^{k-1} \geq 0$, $p^{k-1} \neq 0$, and, of course, (7). In particular, the theorem also holds for the Curry–Altman algorithm of **14.2.7**.

As a global convergence theorem corresponding to **14.5.2**, we obtain the following corollary of **14.5.3**.

14.5.4. Assume that $g: R^n \to R^1$ satisfies the conditions of **14.5.2**. Then, for any x^0 the iterates (6) with α_k satisfying

$$g(x^{k+1}) = \min\{g(x^k - \alpha p^k) \mid \alpha \in R^1\} \qquad (8)$$

converge to the unique critical point of g in R^n.

Again the proof is immediate since g is strictly convex and, hence, hemivariate, has a unique critical point, and, as in **14.5.2**, L^0 is compact for any x^0.

NOTES AND REMARKS

NR 14.5-1. Theorem **14.5.1** is due to Daniel [1967a], who considers the method in Hilbert space. Theorem **14.5.3** was given in Ortega and Rheinboldt [1970a] in a somewhat more restricted form.

NR 14.5-2. The rate of convergence of both conjugate-gradient methods discussed in this section is still open. It is easy to see (**E 14.5-2**) that, under the conditions of **14.5.1**, the convergence of (1) is at least R-linear, but the fact that (1) converges in finitely many steps for quadratic functionals strongly indicates a superlinear rate of convergence in general. (Indeed, various authors erroneously assume that finite convergence in the quadratic case automatically implies quadratic convergence in general.) In fact, Daniel [1967b] announced, under essentially the conditions of **14.5.1**, that the order of convergence of (1) is at least $2^{1/n}$; unfortunately, the proof was incorrect. (Note added in proof: A corrected result showing superlinear convergence will appear in *SIAM J. Num. Anal.* [1970].)

14.6. UNIVARIATE RELAXATION AND RELATED PROCESSES

EXERCISES

E 14.5-1. Assume that $g: R^n \to R^1$ is twice-continuously differentiable and uniformly convex on R^n. Show that the conclusions of **14.5.2** hold.

E 14.5-2. Under the conditions of **14.5.1**, show that the rate of convergence of $\{x^k\}$ is at least R-linear.

E 14.5-3. Suppose that $F: R^n \to R^n$ is continuously differentiable and $F'(x)$ is symmetric and satisfies for some $c > 0$

$$h^T F'(x) h \geq c h^T h, \quad \forall x, h \in R^n.$$

Show that either of the iterations (1) or (6), (8), with $g'(x)^T$ replaced by Fx, are globally convergent to the unique solution of $Fx = 0$.

E 14.5-4. Apply **E 14.5-3** to $Fx \equiv Ax + \phi x$ under the conditions of **4.4.1**(a).

E 14.5-5. Show that either (1) or (6), (8) is globally convergent for the functional g defined in **4.4.4**. Using **4.4.6**, obtain global convergence theorems for the discrete analog (4.4.7) of the Plateau problem (4.4.16).

14.6. UNIVARIATE RELAXATION AND RELATED PROCESSES

In the previous three sections we have obtained convergence results by showing that the directions p^k in the general iteration

$$x^{k+1} = x^k - \omega_k \alpha_k p^k, \quad k = 0, 1, \ldots, \qquad (1)$$

were gradient-related. In this final section, we consider primarily another class of direction algorithms in which the sequence $\{p^k\}$ consists of only finitely many distinct vectors which are repeated in some manner. The following definition delineates the two basic classes of sequences of this type which will be of interest to us.

14.6.1. Definition. A sequence $\{p^k\} \subset R^n$ containing only finitely many distinct vectors is *essentially periodic* if there is an integer $m \geq n$ and an index k_0 such that for any $k \geq k_0$ the m vectors $p^{k+1}, \ldots, p^{k+m}$ span R^n. The sequence is *free-steering* if for any $k \geq 0$ the vectors p^j, $j = k, k+1, \ldots,$ span R^n. ∎

The prototype essentially-periodic-sequence is given by the coordinate vectors e^i, $i = 1, \ldots, n$, repeated periodically; that is,

$$p^k = e^{k(\bmod n)+1}, \quad k = 0, 1, \ldots. \qquad (2)$$

Recall that this is the sequence of direction vectors of the cyclic univariate relaxation methods discussed in Section **8.2**. Similarly, the vectors e^i, $i = 1,..., n$, give rise to a free-steering sequence provided that each e^i appears in the sequence $\{p^k\}$ infinitely often; the same is, of course, true for any linearly independent set $q^1,..., q^n$. Finally, note that any essentially periodic sequence is free steering, but the converse is not true (**E 14.6-1**).

In addition to these sequences $\{p^k\}$ containing only finitely many different vectors, we shall also consider sequences which may have infinitely many different members, but which show a behavior similar to that of the essentially periodic sequences.

14.6.2. Definition. A sequence $\{p^k\} \subset R^n$, $p^k \neq 0$, is *uniformly linearly independent* if there exist a constant $\gamma > 0$ and indices $m \geq n$, $k_0 \geq 0$, such that for each $k \geq k_0$

$$\max\{|\,x^T p^j\,|/\|\,x\,\|\,\|\,p^j\,\|\,|\,j = k+1,..., k+m\} \geq \gamma, \quad \forall x \in R^n,\ x \neq 0. \quad \blacksquare \quad (3)$$

The connection between essentially periodic and uniformly linearly independent sequences is contained in the following lemma.

14.6.3. Let $\{p^k\} \subset R^n$, $p^k \neq 0$, be a sequence containing only finitely many distinct vectors. Then $\{p^k\}$ is essentially periodic if and only if it is uniformly linearly independent.

Proof. Let $\{p^k\}$ be uniformly linearly independent. If for some $k \geq k_0$ the vectors $p^{k+1},..., p^{k+m}$ span only a linear subspace $V \subset R^n$ of dimension less than n, then $x^T p^j = 0$, $j = k+1,..., k+m$, for any $x \neq 0$ in the orthogonal complement of V; this contradicts (3). Conversely, let $\{p^k\}$ be essentially periodic, and denote the distinct vectors of $\{p^k\}$ by $q^1,..., q^r$. Since these vectors span R^n, $\|\,x\,\|' = \max\{|\,x^T q^j\,|/\|\,q^j\,\|\,|\,j = 1,..., r\}$ defines a norm on R^n, and hence by **2.2.1** there exists a constant $\gamma > 0$ such that $\|\,x\,\|' \geq \gamma \|\,x\,\|$ for all $x \in R^n$; that is, (3) holds. $\blacksquare$

In connection with the general process (1), the basic result for uniformly linearly independent, and, hence, also for essentially periodic sequences is the following.

14.6.4. Let $g: D \subset R^n \to R^1$ be continuously differentiable on the compact set $D_0 \subset D$, and assume that $\{p^k\} \subset R^n$, $p^k \neq 0$, is a uniformly linearly independent sequence. Further, let $\{x^k\} \subset D_0$ be any sequence such that

$$\lim_{k \to \infty} \|\,x^{k+1} - x^k\,\| = 0 \quad (4)$$

14.6. UNIVARIATE RELAXATION AND RELATED PROCESSES

and
$$\lim_{k \to \infty} g'(x^k) p^k / \| p^k \| = 0. \tag{5}$$

Then $\lim g'(x^k)^T = 0$.

Proof. Let ω denote the modulus of continuity of g' on D_0. Since the case $g'(x) \equiv 0$, $x \in D_0$, is trivial, we may assume that $\omega(t) > 0$ for $t > 0$. For given $\epsilon > 0$, let k_0 be such that

$$\| x^{k+1} - x^k \| \leq \epsilon/m, \qquad | g'(x^k) p^k |/\| p^k \| \leq \omega(\epsilon), \qquad \forall k \geq k_0.$$

Then $\| x^{k+j} - x^k \| \leq \epsilon$ if $j \leq m$, and, therefore,

$$\| g'(x^{k+j}) - g'(x^k) \| \leq \omega(\epsilon), \qquad j = 1, 2, \ldots, m.$$

This implies that

$$| g'(x^k) p^{k+j} |/\| p^{k+j} \| \leq \| g'(x^k) - g'(x^{k+j}) \| + | g'(x^{k+j}) p^{k+j} |/\| p^{k+j} \| \leq 2\omega(\epsilon)$$

and, thus, by (3),

$$2\omega(\epsilon) \geq \max_{j=1,\ldots,m} | g'(x^k) p^{k+j} |/\| p^{k+j} \| \geq \gamma \| g'(x^k) \|.$$

Since $\epsilon \to 0$ implies that $\omega(\epsilon) \to 0$, it follows that $\lim_{k \to \infty} g'(x^k)^T = 0$. ∎

Note that if g has only finitely many critical points in D_0, then **14.1.5** ensures that the sequence $\{x^k\}$ of **14.6.4** converges to a critical point of g. We use this in the following corollary. Here, as usual, L^0 denotes the connected component of $L(g(x^0))$ containing x^0.

14.6.5. Assume that $g: D \subset R^n \to R^1$ is continuously differentiable on the open set D and that there is an $x^0 \in D$ so that L^0 is compact. Suppose, in addition, that g is hemivariate on L^0 and has only finitely many critical points in L^0. Let $q^1, \ldots, q^n$ be given linearly independent vectors. Then the iterates (1) with

$$p^k = \operatorname{sgn}(g'(x^k) \hat{q}^k) \hat{q}^k, \qquad \hat{q}^k = q^{k(\bmod n)+1}, \tag{6}$$

$$0 < \epsilon \leq \omega_k \leq 1, \tag{7}$$

and α_k defined by the Curry–Altman algorithm of **14.2.7**, are well-defined, lie in L^0, and converge to a critical point of g.

Proof. Precisely as in **14.3.3**, it follows that $\{x^k\}$ is well defined and lies in L^0, and **14.2.7** also shows that (5) holds, and that $\{x^k\}$ is strongly downward. Hence, **14.1.3** ensures that (4) is valid. Clearly, the sequence $\{p^k\}$ is essentially periodic, and hence by **14.6.3** uniformly linearly independent. Therefore, **14.6.4** applies, and then **14.1.5** shows the convergence of $\{x^k\}$. ∎

The use of the Curry–Altman steplength algorithm in **14.6.5** is not essential, and any other steplength algorithm for which $\{x^k\}$ is strongly downward or $\lim_{k\to\infty} \| x^k - x^{k-1} \| = 0$ is permissible provided, of course, that (5) holds. In particular, the restricted minimization algorithm of **14.2.10** may be applied (**E 14.6-4**).

There are numerous possible modifications and corollaries of **14.6.5**. Some of these are given in the exercises, and here we give only the following two global results. Recall that a strictly quasiconvex functional is defined in **4.2.5.**

14.6.6. Global Essentially-Periodic Theorem. Assume that $g\colon R^n \to R^1$ is continuously differentiable, strictly quasiconvex, and satisfies

$$\lim_{\|x\|\to\infty} g(x) = +\infty. \tag{8}$$

Then for any $x^0 \in R^n$ the sequence (1) with p^k and ω_k satisfying (6) and (7), and α_k such that

$$g(x^k - \alpha_k p^k) = \min\{g(x^k - \alpha p^k) \mid -\infty < \alpha < +\infty\}, \tag{9}$$

is well defined and converges to the unique (local and global) minimizer of g.

Proof. By **4.2.7** and **4.3.3**, g has a unique minimizer x^* which is also the unique critical point; moreover, **4.3.2** shows that for any x^0 the level set $L(g(x^0))$ is compact. The strict quasiconvexity implies that g is hemivariate and also that the α_k determined by (9) are precisely those of the Curry algorithm. Hence, **14.6.5** applies. ∎

The usual choice of the vectors q^k of **14.6.5**, and, hence, also of **14.6.6**, is $q^i = e^i$, $i = 1,\ldots, n$. In this case, we have the following result.

14.6.7. Global SOR Theorem. Assume that $F\colon R^n \to R^n$ is continuously differentiable, that $F'(x)$ is symmetric for all $x \in R^n$, and that there is a constant $c > 0$ for which

$$h^\mathrm{T} F'(x) h \geqslant c h^\mathrm{T} h, \qquad \forall x, h \in R^n. \tag{10}$$

14.6. UNIVARIATE RELAXATION AND RELATED PROCESSES

Then, for any $x^0 \in R^n$ and $\omega \in (0, 1]$ the nonlinear SOR iterates

$$\begin{cases} \text{Solve} & f_i(x_1^{k+1},\ldots, x_{i-1}^{k+1}, x_i, x_{i+1}^k,\ldots, x_n^k) = 0 \quad \text{for } x_i; \\ \text{Set} & x_i^{k+1} = x_i^k + \omega(x_i - x_i^k), \quad k = 0, 1,\ldots, \quad i = 1,\ldots, n, \end{cases} \quad (11)$$

are uniquely defined and converge to the only solution of $Fx = 0$ in R^n.

Proof. By the symmetry principle **4.1.6**, there is a $g: R^n \to R^1$ such that $g'(x) = (Fx)^T$ for all $x \in R^n$. Because of (10) and **3.4.6**, g is uniformly convex, and **4.3.6** shows that (8) holds. After a renumbering of the iterates, the algorithm (11) is that of **14.6.6** with $q^i = e^i$, $i = 1,\ldots, n$, and, hence, the result follows. ∎

We conclude with the following result for free-steering sequences.

14.6.8. Let $g: D \subset R^n \to R^1$ be continuously differentiable on the open set D and uniformly convex on the compact, convex set $D_0 \subset D$. Let $\{p^k\}$ be a free-steering sequence of nonzero vectors and assume that the iterates (1) exist, lie in D_0, and satisfy $g(x^{k+1}) \leq g(x^k)$, $k = 0, 1,\ldots$, as well as (5). Then $\{x^k\}$ converges to the unique global minimizer of g on D_0.

Proof. Let Λ be the set of limit points of $\{x^k\}$, and suppose that $g'(x) \neq 0$ for some $x \in \Lambda$. Since $\{p^k\}$ is free-steering, the set

$$\{| g'(x) p^k |/\| p^k \| \mid k = 0, 1,\ldots\}$$

has a least positive element α. Let ω denote the modulus of continuity of g' on D_0 and choose $r > 0$ so that $\omega(r) < \alpha/2$. Then, by (5), there is a k_0 so that

$$| g'(x^k) p^k |/\| p^k \| \leq \omega(r), \quad \forall k \geq k_0,$$

and, therefore, whenever $\| x^k - x \| < r$ for $k \geq k_0$, we have

$$| g'(x) p^k |/\| p^k \| \leq \| g'(x) - g'(x^k)\| + | g'(x^k) p^k |/\| p^k \| \leq 2\omega(r) < \alpha.$$

By the definition of α, this implies that $g'(x) p^k = 0$.

If $\Lambda \subset S(x, r)$, then $x^k \in S(x, r)$ for all $k \geq k_1 \geq k_0$, and, hence, $g'(x) p^k = 0$, $k \geq k_1$. But the free-steering property would then imply that $g'(x) = 0$, and therefore $\Lambda_1 = \{y \in \Lambda \mid \| y - x \| \geq r\}$ cannot be empty. Let $y \in \Lambda_1$. Then, from the fact that $\{g(x^k)\}$ converges, it follows that $g(x) = g(y)$. Hence, **3.4.4** shows that

$$g'(x)(x - y) \geq c \| x - y \|^2 \geq cr^2 \equiv 2\gamma, \quad (12)$$

where $c > 0$ is the constant of uniform convexity. Therefore, whenever $\|x^k - y\| < \gamma/\|g'(x)\|$, we have

$$|g'(x)(x - x^k)| \geq |g'(x)(x - y)| - |g'(x)(y - x^k)| > \gamma.$$

Since $y \in \Lambda_1$ was arbitrary, it follows that we can choose $k_2 \geq k_1$ so that either

$$\|x^k - x\| < r, \quad \text{or} \quad |g'(x)(x - x^k)| > \gamma, \quad \forall k \geq k_2, \quad (13)$$

and, since Λ_1 is not empty, the latter relation holds for some $k \geq k_2$. If $\|x^{k+1} - x\| < r$, then

$$|g'(x)(x - x^{k+1})| = |g'(x)(x - x^k)| > \gamma,$$

because $g'(x)(x^k - x^{k+1}) = \omega_k \alpha_k g'(x) p^k = 0$. On the other hand, if $\|x^{k+1} - x\| > r$, then (13) implies that, again, $|g'(x)(x - x^{k+1})| > \gamma$. Therefore,

$$|g'(x)(x - x^{k+j})| > \gamma, \quad j = 0, 1, \ldots,$$

which contradicts the assumption that $x \in \Lambda$. Hence Λ_1 must be empty, and, as noted previously, this implies that $g'(x) = 0$; that is, $g'(x) = 0$ for every $x \in \Lambda$. But, by **4.2.8**, Λ can consist of only one point x^*, which is the unique global minimizer of g on D_0. Hence, $\lim_{k \to \infty} x^k = x^*$. ∎

NOTES AND REMARKS

NR 14.6-1. Theorems **14.6.4** and **14.6.5** as well as the concept of a uniformly linearly independent sequence are due to Elkin [1968], and were motivated by results of Schechter [1962] for univariate relaxation (see **NR 14.6-2**). A related result is also stated in Goldstein [1967, p. 33].

NR 14.6-2. Theorem **14.6.8** is due to Schechter [1962] for twice-continuously differentiable g, for α_k determined by the Curry algorithm, and for $\{p^k\}$ consisting of the coordinate directions e^i, each repeated infinitely often. The more general result as stated is due to Elkin [1968], and the proof given in the text is a modification of that of Elkin which, in turn, was a modification of Schechter's proof. More generally, Elkin gives the result for functionals which satisfy the relation

$$g(x) \geq g(y) \quad \text{implies} \quad g'(x)(x - y) \geq c \|x - y\|^2. \quad (14)$$

In analogy with **NR 4.2-4**, such functionals may be called uniformly pseudoconvex, and contain the class of continuously differentiable uniformly convex functionals. For this class of functionals, the proof of **14.6.8** remains precisely the

14.6. UNIVARIATE RELAXATION AND RELATED PROCESSES

same, since the uniform convexity was used only to ensure (14) as well as the uniqueness of critical points; the latter property is implied by (14) (see **E4.2-17**).

Elkin also considers more general "block univariate relaxation" methods in which a sequence of subspaces, rather than directions, is given, and the algorithm consists of minimizing g over these subspaces in some order. Schechter [1968] also treats related block methods.

NR 14.6-3. It is easy to prove convergence of the Rosenbrock algorithm of **8.4** by noting that the directions are uniformly linear independent (see **E 14.6-3**). One can handle the Zangwill algorithm described in **NR 8.4-7** in a similar way.

NR 14.6-4. Purely free-steering processes are probably rarely used. More frequently, the next direction vector p^k is selected by a criterion such as

$$| g'(x^k) e^m | = | \max_{1 \leq j \leq n} g'(x^k) e^j |, \qquad p^k = \text{sgn}(g'(x^k) e^m) e^m.$$

The resulting sequence is, in general, neither essentially periodic nor free steering. However, since

$$g'(x^k) p^k \geq n^{-1/2} \| g'(x^k) \|_2 ,$$

it follows that the sequence $\{p^k\}$ is gradient-related, and, hence, convergence results may be obtained by the methods of **14.3**. Results of this kind have been given by Goldstein [1967] and Elkin [1968].

NR 14.6-5. By means of **14.2.9**, convergence theorems analogous to **14.6.7** and **14.6.8** may be obtained for the SOR-Newton iteration discussed in **7.4**. Results for this iteration were first given by Schechter [1962], and considerably extended in Schechter [1968].

EXERCISES

E 14.6-1. Give an example of a free-steering sequence in R^n which is not essentially periodic.

E 14.6-2. In **14.6.4** assume that $\{p^k\}$ is essentially periodic rather than uniformly linearly independent. Give a direct proof of the result without using **14.6.3**.

E 14.6-3. Assume that $g: R^n \to R^1$ is continuously differentiable, strictly convex, and has a (unique) critical point x^*. Show that the Rosenbrock iterates (8.4.12)–(8.4.15) converge to x^*. (Elkin [1968])

E 14.6-4. Show that **14.6.5** remains valid if α_k is determined by the minimization algorithm of **14.2.10**.

E 14.6-5. Assume that $g: R^n \to R^1$ is twice-continuously differentiable and $g''(x)\, hh \geq ch^T h$ for all $x, h \in R^n$, where $c > 0$. Show that the conclusions of **14.6.6** remain valid. For given x^0 assume, in addition, that

$$g''(x)\, hh \leq dh^T h, \qquad \forall h \in R^n, \qquad x \in L(g(x^0)).$$

Show that the conclusions of **14.6.6** remain valid if

$$1 \leq \omega_k < 2d/(2d - c).$$

E 14.6-6. Assume that $g: D \subset R^n \to R^1$ is continuously differentiable on D, that L^0 is compact, that

$$\|g'(x) - g'(y)\| \leq \gamma \|x - y\|, \qquad \forall x, y \in L^0,$$

and that g has only finitely many critical points in L^0. Consider the iteration (1) with p^k given by (6), ω_k by (7), and

$$\alpha_k = \gamma^{-1} g'(x^k)\, p^k, \qquad k = 0, 1, \ldots .$$

Show that the conclusions of **14.6.5** remain valid.

E 14.6-7. Assume that $g: R^n \to R^1$ is continuously differentiable and uniformly convex. Show that the conclusions of **14.6.6** remain valid. By means of the example

$$g(x) = |x_1 - x_2| + x_1^2 + x_2^2,$$

and the starting point $x^0 = (\tfrac{1}{4}, \tfrac{1}{4})^T$, show that the differentiability is necessary. (Elkin [1968])

E 14.6-8. Apply **14.6.7** to $Fx \equiv Ax + \phi x$ under the conditions of **4.4.1(a)**.

E 14.6-9. Apply **14.6.6** to the problems of **E 14.5-5**.

E 14.6-10. Assume that $A \in L(R^n)$ is symmetric, positive definite with diagonal and strictly lower-triangular parts D and $-L$. Conclude, by means of **14.6.7**, that if $\omega \in (0, 1]$, then

$$\rho\{(D - \omega L)^{-1} [(1 - \omega) D + \omega L^T]\} < 1.$$

Use **14.2.9** to show that this also holds if $\omega \in [1, 2)$.

AN ANNOTATED LIST OF BASIC REFERENCE BOOKS

Anselone, P. (ed.), "Nonlinear Integral Equations." University of Wisconsin Press, Madison, Wisconsin, 1964.
   An anthology which contains several important articles about the application of successive approximations and Newton's method to integral equations and boundary value problems.

Apostol, T., "Mathematical Analysis." Addison-Wesley, Reading, Massachusetts, 1957.
   An excellent undergraduate analysis text which provides a good reference for background material from calculus.

Collatz, L., "Functional Analysis and Numerical Mathematics." Springer-Verlag, Berlin, 1964; translation by H. Oser, Academic Press, New York, 1966.
   A survey of the applications of functional analytic techniques to numerical analysis. It covers, in particular, iterative techniques, including numerous examples, and contains many references to German work in the field.

Dieudonné, J., "Foundations of Modern Analysis." Academic Press, New York, 1960.
   A fundamental reference for the abstract approach to analysis in general and to differentiation in Banach spaces in particular.

Forsythe, G., and Wasow, W., "Finite Difference Methods for Partial Differential Equations." Wiley, New York, 1960.
   A survey of numerical methods for partial differential equations which contains, in particular, a great deal of material on iterative processes for linear systems of equations.

Goldstein, A., "Constructive Real Analysis." Harper and Row, New York, 1967.
   An important text for the solution of nonlinear operator equations as well as constrained optimization problems.

Greenspan, D., "Introductory Numerical Analysis of Elliptic Boundary Value Problems." Harper and Row, New York, 1965.
   An introduction which also discusses nonlinear equations and, in particular, some generalized linear methods for their discrete analogues.

Householder, A., "The Theory of Matrices in Numerical Analysis." Ginn (Blaisdell), Boston, Massachusetts, 1964.

A fundamental reference for much material on linear algebra and linear iterative methods, including a detailed discussion of norms on R^n. It also contains a large bibliography.

Kantorovich, L., and Akilov, G., "Functional Analysis in Normed Spaces." Fizmatgiz, Moscow, 1959; translation by D. Brown and A. Robertson, Pergamon Press, Oxford, 1964.

An important reference to the Russian work in the field. The last part of the book treats iterative methods and presents the convergence theory of Newton's method based on Kantorovich's majorant technique.

Keller, H., "Numerical Methods for Two-point Boundary Value Problems." Ginn (Blaisdell), Boston, Massachusetts, 1968.

This book includes many results about two-point boundary value problems which complement and supplement those in the text.

Kowalik, J., and Osborne, M., "Methods for Unconstrained Optimization Problems." American Elsevier, New York, 1968.

A survey of minimization methods which includes numerical examples.

Krasnoselskii, M., "Topological Methods in the Theory of Nonlinear Integral Equations." Gostekhteoretizdat, Moscow, 1956; translation by A. Armstrong, Pergamon Press, Oxford, 1964.

An analysis of nonlinear integral equations using, in particular, the theory of degree of a mapping.

Ostrowski, A., "Solution of Equations and Systems of Equations." Academic Press, New York, 1960, second edition, 1966.

Although large parts are concerned with one-dimensional equations, this book also covers important aspects of the numerical solution of systems of equations. Of particular interest is the discussion of points of attraction and the convergence of the steepest descent method.

Rall, L., "Computational Solution of Nonlinear Operator Equations." Wiley, New York, 1969.

A textbook on iterative methods in Banach spaces with emphasis on Newton's method. It stresses some of the important computational aspects.

Schwartz, J., "Nonlinear Functional Analysis (1963/64)." Lecture Notes, Courant Inst. Math. Sci., New York University, New York, 1964.

A valuable collection of material on nonlinear functional analysis which includes, in particular, an introduction to degree theory.

Traub, J., "Iterative Methods for the Solution of Equations." Prentice-Hall, Englewood Cliffs, New Jersey, 1964.

A detailed coverage of iterative methods for one equation in one unknown.

Vainberg, M., "Variational Methods for the Study of Nonlinear Operators." Gostekhteoretizdat, Moscow, 1956; translation by A. Feinstein, Holden-Day, San Francisco, California, 1964.

An important reference for variational problems in Hilbert space.

Varga, R., "Matrix Iterative Analysis." Prentice-Hall, Englewood Cliffs, New Jersey, 1962.

A fundamental reference for the theory of linear iterative methods.

BIBLIOGRAPHY

Aalto, S. [1968]. An iterative procedure for the solution of nonlinear equations in a Banach space. *J. Math. Anal. Appl.* **24**, 686–691.

Ablow, C., and Perry, C. [1959]. Iterative solution of the Dirichlet problem for $\Delta u = u^2$, *SIAM J. Appl. Math.* **7**, 459–467.

Adachi, R. [1955]. On Newton's method for the approximate solution of simultaneous equations, *Kumamoto J. Sci. Ser. A* **2**, 259–272.

Agaev, G. [1967]. Solvability of nonlinear operator equations in Banach space, *Dokl. Akad. Nauk SSSR* **174**, 1239–1242; *Soviet Math. Dokl.* **8**, 726–730.

Ahamed, S. [1965]. Accelerated convergence of numerical solution of linear and nonlinear vector field problems, *Comput. J.* **8**, 73–76.

Akhieser, N. [1955]. "The Calculus of Variations." Gos. Izdat. Tehn.-Teor. Lit., Moscow; translation by A. Frink, Ginn (Blaisdell), Boston, Massachusetts, 1962.

Akilov, G. (see Kantorovich, L.).

Albrecht, J. [1961]. Bemerkungen zum Iterationsverfahren von Schulz zur Matrixinversion, *Z. Angew. Math. Mech.* **41**, 262–263.

Albrecht, J. [1962]. Fehlerschranken und Konvergenzbeschleunigung bei einer monotonen oder alternienden Iterationsfolge, *Numer. Math.* **4**, 196–208.

Alexandroff, P. and Hopf, H. [1935]. "Topologie." Springer-Verlag, Berlin; reprinted by Chelsea, New York, 1965.

Allen, B. [1966]. An investigation into direct numerical methods for solving some calculus of variations problems, *Comput. J.* **9**, 205–210.

Altman, M. [1955]. A generalization of Newton's method, *Bull. Acad. Polon. Sci. Ser. Sci. Math. Astronom. Phys.* **3**, 189–193.

Altman, M. [1957a]. A fixed point theorem in Hilbert space, *Bull. Acad. Polon. Sci. Ser. Sci. Math. Astronom. Phys.* **5**, 19–22.

Altman, M. [1957b]. A fixed point theorem in Banach space. *Bull. Acad. Polon. Sci. Ser. Sci. Math. Astronom. Phys.* **5**, 89–92.

Altman, M. [1957c]. On the approximate solution of nonlinear functional equations, *Bull. Acad. Polon. Sci. Ser. Sci. Math. Astronom. Phys.* **5**, 457–460, 461–465.

Altman, M. [1957d]. On the approximate solution of operator equations in Hilbert spaces, *Bull. Acad. Polon. Sci. Ser. Sci. Math. Astronom. Phys.* **5**, 605–609, 711–715, 783–787.
Altman, M. [1957e]. On a generalization of Newton's method, *Bull. Acad. Polon. Sci. Ser. Sci. Math. Astronom. Phys.* **5**, 789–795.
Altman, M. [1957f]. Connection between the method of steepest descent and Newton's method, *Bull. Acad. Polon. Sci. Ser. Sci. Math. Astronom. Phys.* **5**, 1031–1036.
Altman, M. [1958]. On the approximate solution of nonlinear functional equations in Banach spaces, *Bull. Acad. Polon. Sci. Ser. Sci. Math. Astronom. Phys.* **6**, 19–24.
Altman, M. [1960]. Functional equations involving a parameter, *Proc. Amer. Math. Soc.* **11**, 54–61.
Altman, M. [1961a]. A generalization of Laguerre's method for functional equations, *Bull. Acad. Sci. Ser. Sci. Math. Astronom. Phys.* **9**, 581–586.
Altman, M. [1961b]. Concerning the method of tangent hyperbolas for operator equations, *Bull. Acad. Polon. Sci. Ser. Sci. Math. Astronom. Phys.* **9**, 633–637.
Altman, M. [1961c]. A general majorant principle for functional equations, *Bull. Acad. Polon. Sci. Ser. Sci. Math. Astronom. Phys.* **9**, 745–750.
Altman, M. [1961d]. Connection between gradient methods and Newton's method for functionals, *Bull. Acad. Polon. Sci. Ser. Sci. Math. Astronom. Phys.* **9**, 877–880.
Altman, M. [1966a]. Generalized gradient methods of minimizing a functional, *Bull. Acad. Polon. Sci. Ser. Sci. Math. Astronom. Phys.* **14**, 313–318.
Altman, M. [1966b]. A generalized gradient method for the conditional minimum of a functional, *Bull. Acad. Polon. Sci. Ser. Sci. Math. Astronom. Phys.* **14**, 445–451.
Altman, M. [1966c]. A generalized gradient method of minimizing a functional on a Banach space, *Mathematica (Cluj)* **8**, 15–18.
Altman, M. [1967a]. A generalized gradient method with self-fixing step size for the conditional minimum of a functional, *Bull. Acad. Polon. Sci. Ser. Sci. Math. Astronom. Phys.* **15**, 19–24.
Altman, M. [1967b]. A generalized gradient method for the conditional extremum of a functional, *Bull. Acad. Polon. Sci. Ser. Sci. Math. Astronom. Phys.* **15**, 177–183.
Ames, W. [1965]. "Nonlinear Partial Differential Equations in Engineering." Academic Press, New York.
Ames, W. (ed.) [1967]. "Nonlinear Partial Differential Equations." Academic Press, New York.
Anderson, D. [1965]. Iterative procedures for nonlinear integral equations, *J. Assoc. Comput. Mach.* **12**, 547–560.
Anonymous [1960]. *Symposium on the Numerical Treatment of Ordinary Differential Equations, Integral and Integro-differential Equations*, Birkhäuser Verlag, Basel.
Anselone, P. (ed.) [1964]. "Nonlinear Integral Equations." Univ. of Wisconsin Press, Madison, Wisconsin.
Anselone, P. [1965]. Convergence and error bounds for approximate solutions to integral and operator equations, *in* "Error in Digital Computation II" (L. Rall, ed.), pp. 231–252. Wiley, New York.
Anselone, P., and Moore, R. [1964]. Approximate solutions of integral and operator equations, *J. Math. Anal. Appl.* **9**, 268–277.
Anselone, P., and Moore, R. [1966]. An extension of the Newton-Kantorovich method for solving nonlinear equations with an application to elasticity, *J. Math. Anal. Appl.* **13**, 476–501.
Anselone, P., and Rall, L. [1968]. The solution of characteristic value-vector problems by Newton's method, *Numer. Math.* **11**, 38–45.

Antosiewicz, H. [1968]. Newton's method and boundary value problems, *J. Comput. System Sci.* **2**, 177–203.
Antosiewicz, H., and Rheinboldt, W. [1962]. Functional analysis and numerical analysis, *in* "Survey of Numerical Analysis" (J. Todd, ed.), p. 485–517. Wiley, New York.
Apostol, T. [1957]. "Mathematical Analysis," Addison-Wesley, Reading, Massachusetts.
Apslund, E. [1968]. Frechet differentiability of convex functions, *Acta Math.* **121**, 31–47.
Armijo, L. [1966]. Minimization of functions having Lipschitz-continuous first partial derivatives, *Pacific J. Math.* **16**, 1–3.
Avila, J. [1970]. Continuation Methods for Nonlinear Equations, Ph.D. Diss., Univ. of Maryland, College Park, Maryland.
Axelson, O. [1964]. Global integration of differential equations through Lobatto quadrature, *BIT* **4**, 69–86.
Baer, R. [1962a]. Note on an extremum locating algorithm, *Comput. J.* **5**, 193.
Baer, R. [1962b]. Nonlinear regression and the solution of simultaneous equations, *Comm. ACM* **5**, 397–398.
Bailey, P., and Shampine, L. [1968]. On shooting methods for two-point boundary value problems, *J. Math. Anal. Appl.* **23**, 235–249.
Bailey, P., Shampine, L., and Waltman, P. [1968]. "Nonlinear Two-Point Boundary Value Problems." Academic Press, New York.
Balakrishnan, A., and Neustadt, L. (eds.) [1964]. "Computing Methods in Optimization Problems." Academic Press, New York.
Balázs, M. (see Jankó, B.).
Baluev, A. [1952]. On the method of Chaplygin (Russian), *Dokl. Akad. Nauk SSSR* **83**, 781–784.
Baluev, A. [1956]. On the method of Chaplygin for functional equations (Russian), *Vestnik Leningrad Univ.* **13**, 27–42.
Baluev, A. [1958]. Application of semi-ordered norms in approximate solution of nonlinear equations (Russian), *Leningrad. Gos. Univ. Učen. Zap. Ser. Mat. Nauk* **33**, 18–27.
Banach, S. [1922]. Sur les opérations dans les ensembles abstraits et leur applications aux équations intégrales, *Fund. Math.* **3**, 133–181.
Bard, Y. [1968]. On a numerical instability of Davidon-like methods, *Math. Comp.* **22**, 665–666.
Barnes, J. [1965]. An algorithm for solving nonlinear equations based on the secant method, *Comput. J.* **8**, 66–72.
Bartis, M. [1968]. Certain iteration methods for solving nonlinear operator equations (Ukrainian), *Ukrain. Mat. Ž.* **20**, 104–113.
Bartle, R. [1955]. "Newton's method in Banach spaces," *Proc. Amer. Math. Soc.* **6**, 827–831.
Beale, E. [1964]. Numerical methods, *in* "Nonlinear Programming" (NATO Summer School, Menton), pp. 133–205. North-Holland Publ., Amsterdam.
Beightler, C. (see Wilde, D.).
Bellman, R. [1953]. "Stability Theory of Differential Equations." McGraw-Hill, New York.
Bellman, R. [1960]. "Introduction to Matrix Analysis." McGraw-Hill, New York.
Bellman, R. [1961]. Successive approximations and computer storage problems in ordinary differential equations, *Comm. ACM* **4**, 222–223.
Bellman, R. [1965]. A new approach to the numerical solution of a class of linear and nonlinear integral equations of Fredholm type, *Proc. Nat. Acad. Sci. USA* **54**, 1501–1503.
Bellman, R., and Kalaba, R. [1965]. "Quasilinearization and Nonlinear Boundary-value Problems." American Elsevier, New York.

Bellman, R., Juncosa, M., and Kalaba, R. [1961]. Some numerical experiments using Newton's method for nonlinear parabolic and elliptic boundary value problems, *Comm. ACM* **4**, 187–191.

Bellman, R., Kagiwada, H., and Kalaba, R. [1962a]. Orbit determination as a multi-point boundary-value problem and quasi-linearization, *Proc. Nat. Acad. Sci. USA* **48**, 1327–1329.

Bellman, R., Kagiwada, H., and Kalaba, R. [1962b]. A computational procedure for optimal system design and utilization, *Proc. Nat. Acad. Sci. USA* **48**, 1524–1528.

Bellman, R., Kagiwada, H., and Kalaba, R. [1965]. Nonlinear extrapolation and two-point boundary value problems, *Comm. ACM* **8**, 511–512.

Bellman, R., Kagiwada, H., Kalaba, R., and Vasudevan, R. [1968]. Quasilinearization and the estimation of differential operators from eigenvalues, *Comm. ACM* **11**, 255–256.

Belluce, L., and Kirk, W. [1966]. Fixed point theorems for families of contraction mappings, *Pacific J. Math.* **18**, 213–217.

Belluce, L., and Kirk, W. [1967]. Nonexpansive mappings and fixed points in Banach spaces, *Illinois J. Math.* **11**, 474–479.

Belluce, L., and Kirk, W. [1969]. Fixed point theorems for certain classes of nonexpansive mappings, *Proc. Amer. Math. Soc.* **20**, 141–146.

Bel'tjukov, B. [1965a]. Construction of rapidly converging iterative algorithms for the solution of integral equations (Russian), *Sibirsk, Mat. Ž.* **6**, 1415–1419.

Bel'tjukov, B. [1965b]. On a certain method of solution of nonlinear functional equations (Russian), *Ž. Vyčisl. Mat. i Mat. Fiz.* **5**, 927–931.

Bel'tjukov, B. [1966]. On solving nonlinear integral equations by Newton's method (Russian), *Differencial'nye Uravnenija* **2**, 1072–1083.

Ben-Israel, A. [1965]. A modified Newton-Raphson method for the solution of systems of equations, *Israel J. Math.* **3**, 94–99.

Ben-Israel, A. [1966]. A Newton-Raphson method for the solution of systems of equations, *J. Math. Anal. Appl.* **15**, 243–252.

Bennett, A. [1916]. Newton's method in general analysis, *Proc. Nat. Acad. Sci. USA* **2**, 592–598.

Berezin, I., and Zhidkov, N. [1959]. "Computing Methods." Fizmatgiz, Moscow; transl. by O. Blunn and A. Booth, Pergamon Press, Oxford, 1965.

Berge, C. [1959]. "Espaces Topologiques, Fonctions Multivoques." Dunod, Paris; translation by E. Patterson as "Topological Spaces." MacMillan, New York, 1963.

Berman, G. [1966]. Minimization by successive approximation, *SIAM J. Numer. Anal.* **3**, 123–133.

Berman, G. [1969]. Lattice approximations to the minima of functions of several variables, *J. Assoc. Comput. Mach.* **16**, 286–294.

Bers, L. [1953]. On mildly nonlinear partial difference equations of elliptic type, *J. Res. Nat. Bur. Standards Sect. B* **51**, 229–236.

Bers, L., John, F., and Schechter, M. [1964]. "Partial Differential Equations." Wiley (Interscience), New York.

Bezlyudnaya, L. [1968]. On the investigation of a general iterative method by the majorant method (Ukrainian), *Dopovidi Akad. Nauk Ukrain. RSR* **1968**, 867–869.

Birkhoff, G. D., and Kellogg, O. [1922]. Invariant points in function space, *Trans. Amer. Math. Soc.* **23**, 96–115.

Birkhoff, G. [1948]. "Lattice Theory," Colloquium Publ. Vol. 25, Am. Math. Soc., Providence, Rhode Island.

Birkhoff, G., and Diaz, J. [1956]. Nonlinear network problems, *Quart. Appl. Math.* **13**, 431–443.
Birkhoff, G., and Kellogg, R. [1966]. Solution of equilibrium equations in thermal networks, *Proc. Symp. Generalized Networks*, p. 443–452. Brooklyn Polytechnic Press, Brooklyn, New York.
Birkhoff, G., Schultz, M., and Varga, R. [1968]. Piecewise Hermite interpolation in one and two variables with applications to partial differential equations, *Numer. Math.* **11**, 232–256.
Bittner, E. [1965]. Numerical Analysis of Laplace's Equation with Nonlinear Boundary Conditions, Ph.D. Diss., Case Inst. of Techn., Cleveland, Ohio.
Bittner, L. [1959]. Eine Verallgemeinerung des Sekantenverfahrens zur näherungsweisen Berechnung der Nullstellen eines nichtlinearen Gleichungssystems, *Wiss. Z. Techn. Univ. Dresden* **9**, 1959/60, 325–329.
Bittner, L. [1963]. Mehrpunktverfahren zur Auflösung von Gleichungssystemen, *Z. Angew. Math. Mech.* **43**, 111–126.
Bittner, L. [1967]. Einige kontinuierliche Analogien von Iterationsverfahren, *in* "Funktionalanalysis, Approximationstheorie, Numerische Mathematik, ISNM 7," pp. 114–135. Birkhäuser Verlag. Basel.
Bliss, G. [1925]. "Calculus of Variations," Carus Math. Monograph No. 1. The Math. Assoc. of Am., Open Court, LaSalle, Illinois.
Block, H. [1953]. Construction of solutions and propagation of errors in nonlinear problems, *Proc. Amer. Math. Soc.* **4**, 715–722.
Blum, E. [1966]. A convergent gradient procedure in pre-Hilbert spaces, *Pacific J. Math.* **18**, 25–29.
Blum, E. [1968]. Stationary points of functionals in pre-Hilbert spaces, *J. Comput. System Sci.* **1**, 86–90.
Blundell, P. [1962]. A method for solving simultaneous polynomial equations, *Proc. IFIP Congr. 1962*, pp. 39–42. North Holland, Amsterdam.
Blutel, E. [1910]. Sur l'application de la méthode d'approximation de Newton à la résolution approchée des équations à plusieurs inconnues, *C. R. Acad. Sci. Paris* **151**, 1109–1112.
Bohl, E. [1964]. Die Theorie einer Klasse linearer Operatoren und Existenzsätze für Lösungen nichtlinearer Probleme in halbgeordneten Banachräumen, *Arch. Rational Mech. Anal.* **15**, 263–288.
Bohl, E. [1967]. Nichtlineare Aufgaben in halbgeordneten Räumen, *Numer. Math.* **10**, 220–231.
Bondarenko, P. [1961]. On the choice of round-off errors in solutions of difference equations by a convergent iteration method (Russian), *Visnik Kiiv. Univ.* **1961**, 27–31.
Bondarenko, P., [1964]. Computation algorithms for approximate solution of operator equations, *Dokl. Akad. Nauk SSSR* **154**, 754–756; *Soviet Math. Dokl.* **5**, 153–155.
Booth, A. [1949]. An application of the method of steepest descent to the solution of systems of nonlinear simultaneous equations, *Quart. J. Mech. Appl. Math.* **2**, 460–468.
Booth, R. [1967]. Random search for zeroes, *J. Math. Anal. Appl.* **20**, 239–257.
Bosarge, W. [1968]. Infinite dimensional iterative methods and applications, IBM Houston Sci. Center Rept. 320.2347. Houston, Texas.
Bosarge, W., and Falb, P. [1968]. Infinite dimensional multipoint methods and the solution of two-point boundary value problems, IBM Houston Sci. Center Rept. 320.2349. Houston, Texas
Box, M. [1965]. A new method of constrained optimization and a comparison with other methods, *Comput. J.* **8**, 42–52.

Box, M. [1966]. A comparison of several current optimization methods and the use of transformations in constrained problems, *Comput. J.* **9**, 67–77.
Braess, D. [1966]. Über Dämpfung bei Minimalisierungsverfahren, *Computing* **1**, 264–272.
Bramble, J. (ed.) [1966]. "Numerical Solution of Partial Differential Equations." Academic Press, New York.
Bramble, J., and Hubbard, B. [1962]. A theorem on error estimation for finite difference analogues of the Dirichlet problem for elliptic equations, *Contr. Diff. Eqn.* **2**, 319–340.
Bramble, J., and Hubbard, B. [1964]. On a finite difference analogue of an elliptic boundary value problem which is neither diagonally dominant nor of non-negative type, *J. Math. and Phys.* **40**, 117–132.
Brandler, F. [1966]. Numerical solution of a system of two quadratic equations by the method of smoothing planes, *Apl. Mat.* **11**, 352–361.
Brannin, F., and Wang, H. [1967]. A fast reliable iteration method for the analysis of nonlinear networks, *Proc. IEEE* **55**, 1819–1825.
Brauer, F. [1959a]. A note on uniqueness and convergence of successive approximations, *Canad. Math. Bull.* **2**, 5–8.
Brauer, F. [1959b]. Some results on uniqueness and successive approximations, *Canad. J. Math.* **11**, 527–533.
Brauer, F., and Sternberg, S. [1958]. Local uniqueness, existence in the large, and the convergence of successive approximations, *Amer. J. Math.* **80**, 421–430.
Brouwer, L. [1912]. Über Abbildungen von Mannigfaltigkeiten, *Math. Ann.* **71**, 97–115.
Browder, F. [1963a]. The solvability of nonlinear functional equations, *Duke Math. J.* **33**, 557–567.
Browder, F. [1963b]. Nonlinear elliptic boundary value problems, *Bull. Amer. Math. Soc.* **69**, 862–874.
Browder, F. [1964]. Remarks on nonlinear functional equations, *Proc. Nat. Acad. Sci. USA* **51**, 985–989.
Browder, F. [1965a]. Nonlinear elliptic boundary value problems II, *Trans. Amer. Math. Soc.* **117**, 530–550.
Browder, F. [1965b]. Nonexpansive nonlinear operators in a Banach space, *Proc. Nat. Acad. Sci. USA*, **54**, 1041–1044.
Browder, F. [1965c]. Existence and uniqueness theorems for solutions of nonlinear boundary value problems, *Proc. Sympos. Appl. Math.* **17**, 24–49.
Browder, F. [1967]. Convergence of approximants to fixed points of nonexpansive nonlinear mappings in Banach spaces, *Arch. Rational Mech. Anal.* **24**, 82–90.
Browder, F., and Petryshyn, W. [1966]. The solution by iteration of nonlinear functional equations in Banach spaces, *Bull. Amer. Math. Soc.* **72**, 571–575.
Browder, F., and Petryshyn, W. [1967]. Construction of fixed points of nonlinear mappings in Hilbert space, *J. Math. Anal. Appl.* **20**, 197–228.
Brown, K. [1966]. A Quadratically Convergent Method for Solving Simultaneous Nonlinear Equations. Ph.D. Diss., Purdue Univ., Lafayette, Indiana.
Brown, K. [1967]. Solution of simultaneous nonlinear equations, *Comm. ACM* **10**, 728–729.
Brown, K. [1969]. A quadratically convergent Newton-like method based upon Gaussian elimination, *SIAM J. Numer. Anal.* **6**, 560–569.
Brown, K., and Conte, S. [1967]. The solution of simultaneous nonlinear equations, *Proc. 22nd, Nat. Conf. Assoc. Comp. Mach.*, pp. 111–114. Thompson Book Co., Washington, D. C.
Brown, K., and Dennis, J. [1968]. On Newton-like iteration functions: general convergence theorems and a specific algorithm, *Numer. Math.* **12**, 186–191.

Broyden, C. [1965]. A class of methods for solving nonlinear simultaneous equations, *Math. Comp.* **19**, 577–593.
Broyden, C. [1967]. Quasi-Newton methods and their application to function minimization, *Math. Comp.* **21**, 368–381.
Broyden, C. [1969]. A new method of solving nonlinear simultaneous equations, *Comput. J.* **12**, 94–99.
Brumberg, V. [1962]. Numerical solution of boundary value problems by the method of steepest descent (Russian), *Bjull. Inst. Teoret. Astronom.* **8**, 269–282.
Bryan, C. [1964]. On the convergence of the method of nonlinear simultaneous displacements, *Rend. Circ. Mat. Palermo* **13**, 177–191.
Bryant, V. [1968]. A remark on a fixed-point theorem for iterated mappings, *Amer. Math. Monthly* **75**, 399–400.
Budak, B., and Gol'dman, N. [1967]. An application of Newton's method to the solution of nonlinear boundary value problems (Russian), in "Comput. Methods Programming, VI," Izdat. Moskov Univ., Moscow.
Bueckner, H. [1952]. "Die praktische Behandlung von Integralgleichungen." Springer-Verlag, Berlin.
Buehler, R. (see Shah, B.)
Bussmann, K. [1940]. Ph.D. Diss., Inst. of Techn. Braunschweig, Germany.
Cacciopoli, R. [1930]. Un teorema generale sull'esistenza di elementi uniti in una trasformazione funzionale, *Atti Accad. Naz. Lincei Rend. Cl. Sci. Fis. Mat. Natur. Ser. 6* **11**, 794–799.
Cacciopoli, R. [1931]. Sugli elementi uniti delle trasformazioni funzionali: un'osservazione sui problemi di valori ai limiti, *Atti. Accad. Naz. Lincei Rend. Cl. Sci. Fis. Mat. Natur. Ser. 6*, **13**, 498–502.
Cacciopoli, R. [1932]. Sugli elementi uniti delle trasformazioni funzionali, *Rend. Sem. Mat. Univ. Padova* **3**, 1–15.
Caspar, J. [1969]. Applications of Alternating Direction Methods to Mildly Nonlinear Problems. Ph.D. Diss., Univ. of Maryland, College Park, Maryland.
Cauchy, A. [1829]. Sur la détermination approximative des racines d'une équation algébrique ou transcendante, *Œuvres Complète (II)* **4**, 573–609. Gauthier-Villars, Paris, 1899.
Cauchy, A. [1847]. Méthode générale pour la résolution des systèms d'équations simultanes, *C. R. Acad. Sci. Paris* **25**, 536–538.
Cavanagh, R. [1970]. Difference Equations and Iterative Methods. Ph.D. Diss., Univ. of Maryland, College Park, Maryland.
Cesari, L. [1966]. The implicit function theorem in functional analysis, *Duke Math. J.* **33**, 417–440.
Chandrasekhar, S. [1950]. "Radiative Transfer." Oxford Univ. Press (Clarendon), London and New York.
Chen, K. [1964]. Generalization of Steffensen's method for operator equations, *Comment. Math. Univ. Carolinae* **5**, 47–77.
Chen, W. [1957]. Iterative processes for solving nonlinear functional equations, *Advancement in Math.* **3**, 434–444.
Cheney, E. [1966]. "Introduction to Approximation Theory." McGraw-Hill, New York.
Cheney, E., and Goldstein, A. [1959]. Proximity maps for convex sets. *Proc. Amer. Math. Soc.* **10**, 448–450.
Chernoff, H. (see Crockett, J.)
Chu, S., and Diaz, J. [1964]. A fixed point theorem for in-the-large application of the contraction principle, *Atti Accad. Sci. Torino Cl. Sci. Fis. Mat. Natur.* **99**, 351–363, 1964/65.

Chu, S., and Diaz, J. [1965]. Remarks on a generalization of Banach's principle of contraction mappings, *J. Math. Anal. Appl.* **11**, 440–446.
Ciarlet, P. [1966]. Variational Methods for Nonlinear Boundary Value Problems. Ph.D. Diss., Case Inst. of Techn., Cleveland, Ohio.
Ciarlet, P., Schultz, M., and Varga, R. [1967]. Numerical methods of high order accuracy for nonlinear boundary value problems I, *Numer. Math.* **9**, 394–430.
Coffman, C. [1964]. Asymptotic behavior of solutions of ordinary difference equations, *Trans. Amer. Math. Soc.* **110**, 22–51.
Collatz, L. [1952]. Aufgaben monotoner Art, *Arch. Math.* **3**, 365–376.
Collatz, L. [1953]. Einige Anwendungen funktionalanalytischer Methoden in der praktischen Analysis, *Z. Angew. Math. Phys.* **4**, 327–357.
Collatz, L. [1958]. Näherungsverfahren höherer Ordnung für Gleichungen in Banach Räumen, *Arch. Rational Mech. Anal.* **2**, 66–75.
Collatz, L. [1960a]. Application of the theory of monotonic operators to boundary value problems, *in* "Boundary Problems in Differential Equations" (R. Langer, ed.), pp. 35–45. Univ. of Wisconsin Press, Madison, Wisconsin.
Collatz, L. [1960b]. "The Numerical Treatment of Differential Equations." Springer-Verlag, Berlin.
Collatz, L. [1961]. Monotonie und Extremal-prinzipien beim Newtonschen Verfahren, *Numer. Math.* **3**, 99–106.
Collatz, L., [1964]. "Functional Analysis and Numerical Mathematics." Springer-Verlag, Berlin; transl. by H. Oser, Academic Press, New York, 1966.
Collatz, L. [1965]. Applications of functional analysis to error estimation, *in* "Error in Digital Computation II" (L. Rall, ed.), pp. 253–269. Univ. Wisconsin Press, Madison, Wisconsin.
Concus, P. [1967a]. Numerical solution of the nonlinear magnetostatic field equation in two dimensions, *J. Computational Phys.* **1**, 330–342.
Concus, P. [1967b]. Numerical solution of Plateau's problem, *Math. Comp.* **21**, 340–350.
Conte, S. (see Brown, K.)
Crockett, J. B., and Chernoff, H. [1955]. Gradient methods of maximization, *Pacific J. Math.* **5**, 33-50.
Cronin, J. [1964]. "Fixed Points and Topological Degree in Nonlinear Analysis," Mathem. Surveys II. Am. Math. Soc., Providence, Rhode Island.
Cryer, C. [1967]. On the numerical solution of a quasi-linear elliptic equation, *J. Assoc. Comput. Mach.* **14**, 363–375.
Curry, H. [1944]. The method of steepest descent for nonlinear minimization problems, *Quart. Appl. Math.* **2**, 258–261.
Daniel, J. [1967a]. The conjugate gradient method for linear and nonlinear operator equations, *SIAM J. Numer. Anal.* **4**, 10–26.
Daniel, J. [1967b]. Convergence of the conjugate gradient method with computationally convenient modifications, *Numer. Math.* **10**, 125–131.
Daugavet, I., and Samokis, B. [1963]. A posteriori error estimate in the numerical solution of differential equations (Russian), *Metody Vyčisl* **1**, 52–57.
Davidenko, D. [1953a]. On a new method of numerically integrating a system of nonlinear equations (Russian), *Dokl. Akad. Nauk SSSR* **88**, 601–604.
Davidenko, D. [1953b]. On the approximate solution of a system of nonlinear equations (Russian), *Ukrain. Mat. Ž.* **5**, 196-206.
Davidenko, D. [1955]. On application of the method of variation of parameters to the theory of nonlinear functional equations (Russian), *Ukrain. Mat. Ž.* **7**, 18–28.
Davidenko, D. [1965a]. An application of the method of variation of parameters to the

construction of iterative formulas of higher accuracy for the determination of the elements of the inverse matrix, *Dokl. Akad. Nauk USSR* **162**, 743–746; *Soviet Math. Dokl.* **6**, 738–742.

Davidenko, D. [1965b]. An application of the method of variation of parameters to the construction of iterative formulas of increased accuracy for numerical solutions of nonlinear integral equations, *Dokl. Akad. Nauk SSSR* **162**, 499–502; *Soviet Math. Dokl.* **6**, 702–706.

Davidon, W. [1959]. Variable metric methods for minimization, A.E.C. Res. and Develop. Rept. ANL-5990. Argonne Nat'l Lab., Argonne, Illinois.

Davidon, W. [1967]. Variance algorithm for minimization. *Comput. J.* **10**, 406–411.

Davis, J. [1966]. The Solution of Nonlinear Operator Equations with Critical Points. Ph.D. Diss., Oregon State Univ., Corvallis, Oregon.

de Figueiredo, D. [1967]. Topics in nonlinear functional analysis, Inst. Fluid Dyn. and Appl. Math., Lecture Series 48. Univ. of Maryland, College Park, Maryland.

Deist, F., and Sefor, L. [1967]. Solution of systems of nonlinear equations by parameter variation, *Comput. J.* **10**, 78–82.

Dennis, J. [1967]. On Newton's method and nonlinear simultaneous displacements, *SIAM J. Numer. Anal.* **4**, 103–108.

Dennis, J. [1968]. On Newton-like methods, *Numer. Math.* **11**, 324–330.

Dennis, J. [1969]. On the Kantorovich hypothesis for Newton's method, *SIAM J. Numer. Anal.* **6**, 493-507.

Dennis, J. (see Brown, K.)

Derendjaev, I. [1958]. A modification of Newton's method of solving nonlinear functional equations (Russian), *Perm. Gos. Univ. Učen. Zap. Mat.* **16**, 43–45.

Diaz, J. [1964]. Solution of the Singular Cauchy Problem, NATO Intern. Summer Inst. of Theory of Distributions, Lisbon, Portugal.

Diaz, J., and Metcalf, F. [1967]. On the structure of the set of subsequential limit points of successive approximations, *Bull. Amer. Math. Soc.* **73**, 516–519.

Diaz, J., and Metcalf, F. [1969]. On the set of subsequential limit points of successive approximations, *Trans. Amer. Math. Soc.* **135**, 1–27.

Diaz, J. (see Birkhoff, G., and Chu, S.)

Dieudonné, J. [1960]. "Foundations of Modern Analysis." Academic Press, New York.

D'jakonov, Ye. [1966]. The construction of iterative methods based on the use of spectrally equivalent operators (Russian), *Ž. Vyčisl. Mat. i Mat. Fiz.* **6**, 12–34.

Dolph, C., and Minty, G. [1964]. On nonlinear integral equations of Hammerstein type, *in* "Nonlinear Integral Equations" (P. Anselone, ed.), pp. 99–154. Univ. of Wisconsin Press, Madison, Wisconsin.

Douglas, J., Jr. [1961]. Alternating direction iteration for mildly nonlinear elliptic difference equations, *Numer. Math.* **3**, 92–98; **4**, 301–302, 1962.

Douglas, J., Jr. [1962]. Alternating direction methods for three space variables, *Numer. Math.* **4**, 41–63.

Draper, N., and Smith, H. [1966]. "Applied Regression Analysis." Wiley, New York.

Dressel, H. (see Schmidt, J.)

Dreyfus, S. [1962]. The numerical solution of variational problems, *J. Math. Anal. Appl.* **5**, 30–45.

Dubovik, L. [1965a]. General form of an iteration process of third order for nonlinear functional equations (Ukrainian), *in* "First Republ. Conf. of Young Researchers," Part I, pp. 219–225. Akad. Nauk. Ukrain. SSR Inst. Mat , Kiev."

Dubovik, L. [1965b]. On the convergence of the generalized method of Newton and its application to the solution of matrix equations (Ukrainian), *in* "First Republ. Math.

Conf. of Young Researchers," Part I, pp. 225-232. Akad. Nauk Ukrain. SSR Inst. Mat., Kiev.

Dück, W. [1966]. Iterative Verfahren und Abänderungsmethoden zur Inversion von Matrizen, *Wiss. Z. Karl Marx Univ. Leipzig, Math.-Natur. Reihe* **8**, 259-273.

Dulcau, J. [1963]. Résolution d'un système d'équations polynomiales, *C. R. Acad. Sci. Paris Ser. A-B* **256**, 2284-2286.

Dulley, D., and Pitteway, M. [1967]. Finding a solution of n functional equations in n unknowns, *Comm. ACM* **10**, 1967, 726.

Dunford, N., and Schwartz, J. [1958]. "Linear Operators I." Wiley (Interscience), New York.

Durand, E. [1960]. "Solutions Numériques des Équations Algébriques I, II." Masson et Cie, Paris.

Edelstein, M. [1962]. On fixed and periodic points under contractive mappings, *J. London Math. Soc.* **37**, 74-79.

Edelstein, M. [1963]. A theorem on fixed points under isometries, *Amer. Math. Monthly* **70**, 298-300.

Edmunds, D. [1967]. Remarks on nonlinear functional equations, *Math. Ann.* **174**, 233-239.

Eggleston, H. [1958]. "Convexity." Cambridge Univ. Press, Cambridge, England.

Ehrmann, H. [1959a]. Iterationsverfahren mit veränderlichen Operatoren, *Arch. Rational Mech. Anal.* **4**, 45-64.

Ehrmann, H. [1959b]. Konstruktion und Durchführung von Iterationsverfahren höherer Ordnung, *Arch. Rational Mech. Anal.* **4**, 65-88.

Ehrmann, H. [1961a]. Schranken für Schwingungsdauer und Lösung bei der freien ungedämpften Schwingung, *Z. Angew. Math. Mech.* **41**, 364-369.

Ehrmann, H. [1961b]. Ein Existenzsatz für die Lösungen gewisser Gleichungen mit Nebenbedingungen bei beschränkter Nichtlinearität, *Arch. Rational Mech. Anal.* **7**, 349-358.

Ehrmann, H. [1963]. On implicit function theorems and the existence of solutions of nonlinear equations, *Enseignement Math.* **9**, 129-176.

Ehrmann, H., and Lahmann, H. [1965]. Anwendungen des Schauderschen Fixpunktsatzes auf gewisse nichtlineare Integralgleichungen, *Enseignement Math.* **11**, 267-280.

Elkin, R. [1968]. Convergence Theorems for Gauss-Seidel and Other Minimization Algorithms. Ph.D. Diss., Univ. of Maryland, College Park, Maryland.

Emelina, L. [1965]. A two-point iteration process for solving nonlinear functional equations (Ukrainian), *in* "First Rep. Math. Conf. of Young Researchers," Part I, pp. 233-240. Akad. Nauk Ukrain. SSR Inst. Mat., Kiev.

Faddeev, D., and Faddeeva, V. [1960]. "Computational Methods of Linear Algebra." Fizmatgiz, Moscow; transl. by R. Williams, Freeman, San Francisco, California, 1963.

Falb. P., (see Bosarge, W.)

Farforovskaya, J., (see Gavurin, M.)

Feder, D. [1966]. Lens design viewed as an optimization problem, *in* "Recent Advances in Optimization Techniques" (A. Lavi and T. Vogl, ed.), pp. 5-21. Wiley, New York.

Fel'dman, I. [1966]. Several remarks on convergence of the method of successive approximations (Russian), *Bull. Akad. Stiince RSS Moldoven*, 94-96.

Feldstein, A., and Firestone, R. [1967]. Hermite interpolatory iteration theory and parallel numerical analysis, Div. Appl. Math. Report, Brown Univ., Providence, Rhode Island.

Fenchel, W. [1953]. Convex cones, sets, and functions. Dept. of Math. Report, Princeton Univ., Princeton, New Jersey.

BIBLIOGRAPHY

Fenyo, I. [1954]. Über die Lösung der im Banachschen Raume definierten nichtlinearen Gleichungen, *Acta Math. Acad. Sci. Hungar.* **5**, 85–93.
Fiacco, A., and McCormick, G. [1968]. "Nonlinear Programming." Wiley, New York.
Ficken, F. [1951]. The continuation method for functional equations, *Comm. Pure Appl. Math.* **4**, 435–456.
Filippi, S. [1967]. Untersuchungen zur numerischen Lösung von nichtlinearen Gleichungssystemen mit Hilfe der LIE-Reihen von W. Gröbner, *Elektron. Datenverarbeitung* **9**, 75–79.
Filippi, S., and Glasmacher, W. [1967]. Zum Verfahren von Davidenko, *Elektron. Datenverarbeitung* **9**, 55–58.
Fine, H. [1916]. On Newton's method of approximation, *Proc. Nat. Acad. Sci. USA* **2**, 546–552.
Firestone, R. (see Feldstein, A.)
Fleming, W. [1965]. "Functions of Several Variables." Addison-Wesley, Reading, Massachusetts.
Fletcher, R. [1965]. Function minimization without evaluating derivatives; a review, *Comput. J.* **8**, 33–41.
Fletcher, R. [1968]. Generalized inverse methods for the best least squares solution of systems of nonlinear equations, *Comput. J.* **10**, 392–399.
Fletcher, R., and Powell, M. [1963]. A rapidly convergent descent method for minimization, *Comput. J.* **6**, 163–168.
Fletcher, R., and Reeves, C. [1964]. Function minimization by conjugate gradients, *Comput. J.* **7**, 149–154.
Flood, M., and Leon, A. [1966]. A universal adaptive code for optimization (GROPE), *in* "Recent Advances in Optimization Techniques" (A. Lavi and T. Vogl, ed.), pp. 101–130. Wiley, New York.
Fomin, S. (see Kolmogoroff, A.)
Forster, P. [1967]. Existenzaussagen und Fehlerabschätzungen bei gewissen nichtlinearen Randwertaufgaben mit gewöhnlichen Differentialgleichungen, *Numer. Math.* **10**, 410–422.
Forsythe, G., and Wasow, W. [1960]. "Finite Difference Methods for Partial Differential Equations." Wiley, New York.
Fox, L. [1957]. "Numerical Solution of Two-point Boundary Value Problems." Oxford Univ. Press (Clarendon), London and New York.
Fox, L. (ed.) [1962]. "Numerical Solution of Ordinary and Partial Differential Equations." Addison-Wesley, Reading, Massachusetts.
Frank, T. [1967]. Error bounds on numerical solutions of Dirichlet problems for quasilinear equations, Los Alamos Rept. LA-3685. Los Alamos, New Mexico.
Fréchet, M. [1925]. La notion de différentielle dans l'analyse générale, *Ann. Sci. École Norm. Sup.* **42**, 293–323.
Freudenstein, F., and Roth, B. [1963]. Numerical solution of systems of nonlinear equations, *J. Assoc. Comput. Mach.* **10**, 550–556.
Frey, T. [1967]. Fixpunktsätze für Iterationen mit veränderlichen Operatoren, *Studia Sci. Math. Hungar.* **2**, 91–114.
Fridman, V. [1961]. An iteration process with minimum errors for a nonlinear operator equation, *Dokl. Akad. Nauk SSSR* **139**, 1063–1066; *Soviet Math. Dokl.* **2**, 1058–1061.
Fridrih, F. [1966]. On a certain modification of the Newton and gradient methods for solving functional equations (Russian), *Metody Vyčisl.* **3**, 22–29.
Fujii, M. [1963]. Remarks on accelerated iterative processes for numerical solution of equations, *J. Sci. Hiroshima Univ. Ser. A-1 Math.* **27**, 97–118.

Gagliardo, E. [1965]. A method that combines known methods of solution for systems of equations, *Calcolo* **2**, Supp. No. 1, 81–83.
Galanov, B. [1965]. On a general way of obtaining methods for solving nonlinear equations (Ukrainian), *Dopovidi Akad. Nauk Ukrain RSR* **1965**, 1553–1558.
Gale, D., and Nikaido, H. [1965]. The Jacobian matrix and global univalence of mappings, *Math. Ann.* **159**, 81–93.
Gantmacher, F. [1953]. "The Theory of Matrices." Gosud. Izdat. Tehn.-Teor. Lit., Moscow; transl. by K. Hirsch. Chelsea, New York, 1959.
Gardner, G. [1965]. Numerical errors in iterative processes. Div. of Appl. Math. Report, Brown Univ., Providence, Rhode Island.
Gâteaux, R. [1913]. Sur les fonctionelles continues et les fonctionelles analytiques, *C. R. Acad. Sci. Paris* **157**, 325–327.
Gâteaux, R. [1922]. Sur les fonctionelles continues et les fonctionelles analytiques, *Bull. Soc. Math. France* **50**, 1–21.
Gavurin, M. [1958]. Nonlinear functional equations and continuous analogues of iterative methods (Russian), *Izv. Vysš. Učebn. Zaved. Matematica* **6**, 18–31.
Gavurin, M. [1963]. Existence theorems for nonlinear functional equations (Russian), *Metody Vyčisl* **2**, 24–28.
Gavurin, M., and Farforovskaya, J. [1966]. An iterative method for finding the minimum of sums of squares (Russian), *Ž. Vyčisl. Mat. i. Mat. Fiz.* **6**, 1094–1097.
Gendžojan, G. [1964]. Two-sided Chaplygin approximations to the solution of a two-point boundary problem (Russian), *Izv. Akad. Nauk Armjan. SSR Ser. Mat.* **17**, 21–27.
Geraščenko, S. [1967]. The choice of the right-hand sides in the system of differential equations by the gradient method (Russian), *Differencial'nye Uravnenija* **3**, 2144–2150.
Ghinea, M. [1964]. Sur la résolution des équations opérationnelles dans les espaces de Banach, *C. R. Acad. Sci. Paris Ser. A-B* **258**, 2966–2969.
Ghinea, M. [1965]. Sur la résolution des équations opérationnelles dans les espaces de Banach, *Rev. Française Traitement Information Chiffres* **8**, 3–22.
Glasmacher, W. (see Filippi, S.)
Glazman, I. [1964]. Gradient relaxation for nonquadratic functionals, *Dokl. Akad. Nauk SSSR* **154**, 1011–1014; *Soviet Math. Dokl.* **5**, 210–213.
Glazman, I. [1965]. Relaxation on surfaces with saddle points, *Dokl. Akad. Nauk SSSR* **161**, 750–752; *Soviet Math. Dokl.* **6**, 487–490.
Glazman, I., and Senčuk, Ju. [1966a]. A direct method for minimization of certain functionals of the calculus of variations (Russian), *Teor. Funkcii Funkcional Anal. i Priložen* **2**, 7–20.
Glazman, I., and Senčuk, Ju. [1966b]. Minimization of quasi-quadratic functionals in a Hilbert space (Ukrainian), *Dopovidi Akad. Nauk Ukrain. RSR* **1966**, 981–985.
Gleyzal, A. [1959]. Solution of nonlinear equations, *Quart. J. Appl. Math.* **17**, 95–96.
Golab, S. [1966]. La comparison de la rapidité de convergence des approximations successives de la méthode de Newton avec la méthode de "regula falsi," *Mathematica (Cluj)* **8**, 45–49.
Goldfeld, S., Quandt, R., and Trotter, H. [1966]. Maximization by quadratic hill climbing, *Econometrica* **34**, 541–551.
Gol'dman, N. (see Budak, B.)
Goldstein, A. [1962]. Cauchy's method of minimization, *Numer. Math.* **4**, 146–150.
Goldstein, A. [1964]. Minimizing functionals on Hilbert space, *in* "Computing Methods in Optimization Problems" (A. Balakrishnan and L. Neustadt, eds.). Academic Press, New York.

Goldstein, A. [1965a]. On Newton's method, *Numer. Math.* **7**, 391-393.
Goldstein, A. [1965b], On steepest descent, *SIAM J. Control* **3**, 147-151.
Goldstein, A. [1966]. Minimizing functionals on normed linear spaces, *SIAM J. Control.* **4**, 81-89.
Goldstein, A. [1967]. "Constructive Real Analysis." Harper & Row, New York.
Goldstein, A., and Price, J. [1967]. An effective algorithm for minimization, *Numer. Math.* **10**, 184-189.
Goldstein, A. (see Cheney, E.)
Gotusso, L. [1965]. Su un metodo iterativo per la risoluzione di sistemi non lineari, *Ist. Lombardo Accad. Sci. Lett. Rend. A.* **99**, 933-949.
Gotusso, L, [1967]. Sull'impiego dell'integrale di Kronecker per la separazione delle radici di sistemi non lineari, *Ist. Lombardo Accad. Sci. Lett. Rend. A* **101**, 8-28.
Graves, L. (see Hildebrandt. T.)
Gray, J., and Rall, L. [1967]. NEWTON: A general purpose program for solving nonlinear systems. Math. Res. Center Rept. 790, Univ. of Wisconsin, Madison, Wisconsin.
Grebenjuk, V. [1966]. Application of the principle of majorants to a class of iteration processes (Russian), *Ukrain. Mat. Ž.* **18**, 102–106.
Greenspan, D. [1965a]. "Introductory Numerical Analysis of Elliptic Boundary Value Problems." Harper & Row, New York.
Greenspan, D. [1965b]. On approximating extremals of functionals, I, *ICC Bull.* **4**, 99–120.
Greenspan, D. (ed.) [1966]. "Numerical Solution of Nonlinear Differential Equations." Wiley, New York.
Greenspan, D. [1967]. On approximating extremals of functionals, II, *Internat. J. Engrg. Sci.* **5**, 571–588.
Greenspan, D., and Jain, P. [1967]. Application of a method for approximating extremals of functionals to compressible subsonic flow, *J. Math. Anal. Appl.* **18**, 85–111.
Greenspan, D., and Parter, S. [1965]. Mildly nonlinear elliptic partial differential equations and their numerical solution, II, *Numer. Math.* **7**, 129–147.
Greenspan, D., and Yohe, M. [1963]. On the approximate solution of $\Delta u = F(u)$, *Comm. ACM* **6**, 564–568.
Greenstadt, J. [1967]. On the relative efficiencies of gradient methods, *Math. Comp.* **21**, 360–367.
Greub, W. [1967]. "Multilinear Algebra." Springer-Verlag, Berlin.
Groschaftová, Z. [1967]. Approximate solutions of equations in Banach spaces by the Newton iterative method, I, II, *Comment. Math. Univ. Carolinae* **8**, 335–358, 469–501.
Gunn, J. [1964a]. On the two-stage iterative method of Douglas for mildly nonlinear elliptic difference equations, *Numer. Math.* **6**, 243–249.
Gunn, J. [1964b]. The numerical solution of $\nabla(a\nabla u) = f$ by a semi-explicit alternating direction method, *Numer. Math.* **6**, 181–184.
Gunn, J. [1965]. The solution of elliptic difference equations by semi-explicit iterative techniques, *SIAM J. Numer. Anal.* **2**, 24–45.
Gurr, S. [1967]. Über ein neues Matrizen-Differenzenverfahren zur Lösung von einigen nichtlinearen Randwertaufgaben der Mechanik, *Z. Angew. Math. Mech.* **47**, T47–48.
Hadamard, J. [1906]. Sur les transformations ponctuelles, *Bull. Soc. Math. France* **34**, 71–84.
Hadamard, J. [1910]. Sur quelques applications de d'indice de Kronecker, *in* "Introduction à la Théorie des Fonctions d'une Variable," by J. Tannery, pp. 437–477. Herman, Paris.
Hadeler, K. [1968]. Newton-Verfahren für inverse Eigenwertaufgaben, *Numer. Math.* **12**, 35–39.

Hahn, W. [1958]. Über die Anwendung der Methode von Liapunov auf Differenzengleichungen, *Math. Ann.* **136**, 430–441.
Hajtman, B. [1961]. On systems of equations containing only one nonlinear equation, *Magyar Tud. Akad. Mat. Fiz. Oszt. Közl.* **6**, 145–155.
Hanson, E. [1968]. On solving systems of equations using interval arithmetic, *Math. Comp.* **22**, 374–384.
Hanson, M. [1964]. Bounds for functionally convex optimal control problems, *J.Math. Anal. Appl.* **8**, 84–89.
Hardaway, R. [1968]. An algorithm for finding a solution of simultaneous nonlinear equations, *Proc. AFIPS 1968 Spring Joint Computer Conference*, pp. 105–114. Thompson Book Co., Washington, D.C.
Hart, H., and Motzkin, T. [1956]. A composite Newton-Raphson gradient method for the solution of systems of equations, *Pacific J. Math.* **6**, 691–707.
Hartley, H. [1961]. The modified Gauss-Newton method for the fitting of nonlinear regression functions of least squares, *Technometrics* **3**, 269–280.
Haselgrove, C. [1961]. Solution of nonlinear equations and of differential equations with two-point boundary conditions, *Comput. J.* **4**, 255–259.
Heinla, L. (see Tamme, E.)
Heinz, E. [1959]. An elementary theory of the degree of a mapping in n-dimensional space, *J. Math. Mech.* **8**, 231–247.
Henrici, P. [1962]. "Discrete Variable Methods for Ordinary Differential Equations." Wiley, New York.
Henrici, P. [1964]. "Elements of Numerical Analysis." Wiley, New York.
Hestenes, M. [1956]. The conjugate-gradient method for solving linear systems, *Proc. Sixth Symp. Appl. Math.*, pp. 83–102. Am. Math. Soc., Providence, Rhode Island.
Hestenes, M. [1966]. "Calculus of Variations and Optimal Control Theory." Wiley, New York.
Hestenes, M., and Stiefel, E. [1952]. Methods of conjugate gradients for solving linear systems, *J. Res. Nat. Bur. Standards* **49**, 409–436.
Hildebrandt, T., and Graves, L. [1927]. Implicit functions and their differentials in general analysis, *Trans. Amer. Math. Soc.* **29**, 127–153.
Hill, W., and Hunter, W. [1966]. A review of response surface methodology, *Technometrics* **8**, 571–590.
Hirasawa, Y. [1954]. On Newton's method in convex linear topological spaces, *Comment. Math. Univ. St. Paul* **3**, 15–27.
Holst, W. (see Witte, B.).
Holt, J. [1964]. Numerical solution of nonlinear two-point boundary value problems by finite difference methods, *Comm. ACM* **7**, 366–377.
Homma, T. [1964]. On an iterative method, *Amer. Math. Monthly* **71**, 77–78.
Homuth, H. [1967]. Eine Verallgemeinerung der Regula Falsi auf Operatorgleichungen. *Z. Angew. Math. Mech.* **47**, T51–52.
Hooke, R., and Jeeves, T. [1961]. Direct search solution of numerical and statistical problems, *J. Assoc. Comput. Mach.* **8**, 212–229.
Hooker, W., and Thompson, G. [1962]. Iterative procedure for operators, *Arch. Rational Mech. Anal.* **9**, 107–110.
Hopf, H. (see Alexandroff, P.)
Horwitz, L., and Sarachik, P. [1968]. Davidon's method in Hilbert space, *SIAM J. Appl. Math.* **6**, 676–695.
Householder, A. [1964]. "The Theory of Matrices in Numerical Analysis." Ginn (Blaisdell) Boston, Massachusetts.

Hrouda, J. [1966]. The valley algorithm for minimizing a function of several variables (Czech), *Apl. Mat.* 11, 271–277.
Hubbard, B. [1966]. Remarks on the order of convergence in the discrete Dirichlet problem, *in* "Numerical Solution of Partial Differential Equations" (J. Bramble, ed.), pp. 21–34. Academic Press, New York.
Hubbard, B. (see Bramble, J.)
Hunter, W. (see Hill, W.)
Ichida, K. (see Tsuda, T.)
Isaev, V., and Sonin, V. [1963]. On a modification of Newton's method for the numerical solution of boundary value problems (Russian), *Ž. Vyčisl. Mat. i Mat. Fiz.* 3, 1114–1116.
Ivanov, V. [1962]. Algorithms of quick descent, *Dokl. Akad. Nauk SSSR* 143, 775–778; *Soviet Math. Dokl.* 3, 476–479.
Jain, P. (see Greenspan, D.)
Jakovlev, M. [1964a]. On the solution of nonlinear equations by iterations, *Dokl. Akad. Nauk SSSR* 156, 522–524; *Soviet Math. Dokl.* 5, 697–699.
Jakovlev, M. [1964b]. On the solution of nonlinear equations by an iteration method (Russian), *Sibirsk. Mat. Ž.* 5, 1428–1430.
Jakovlev, M. [1964c]. The solution of systems of nonlinear equations by a method of differentiation with respect to a parameter (Russian), *Ž. Vyčisl. Mat. i Mat. Fiz.* 4, 146–149.
Jakovlev, M. [1965]. On certain methods of solution of nonlinear equations (Russian), *Trudy Mat. Inst. Steklov* 84, 8–40.
Jakovlev, M. [1967a]. Algorithms of minimization of strictly convex functionals (Russian), *Ž. Vyčisl. Mat. i Mat. Fiz.* 7, 429–431.
Jakovlev, M. [1967b]. On the finite differences method of solution of nonlinear boundary-value problems, *Dokl. Akad. Nauk SSSR* 172, 798–800; *Soviet Math. Dokl.* 8, 198–200.
Jankó, B. [1960]. Sur l'analogue de la méthode de Tchebycheff et de la méthode des hyperboles tangentes, *Mathematica* (Cluj) 2, 269–275.
Jankó, B. [1962a]. Sur les méthodes d'itération appliquées dans l'espace de Banach pour la résolution des équations fonctionnelles nonlinéaires, *Mathematica (Cluj)* 4, 261–266.
Jankó, B. [1962b]. On the generalized method of tangent hyperbolas (Rumanian), *Acad. R. P. Romîne Fil. Cluj Stud. Cerc. Mat.* 13, 301–308.
Jankó, B. [1962c]. Sur une nouvelle généralisation de la méthode des hyperboles tangentes pour la résolution des équations fonctionnelles nonlinéaires définies dans l'espace de Banach, *Ann. Polon. Math.* 12, 297–298, 1962/63.
Jankó, B. [1963a]. On the generalized method of Čebyšev, II (Rumanian), *Acad. R. P. Romîne Fil. Cluj Stud. Cerc. Mat.* 14, 57–62.
Jankó, B. [1963b]. On a general iterative method of order k (Rumanian), *Acad. R. P. Romîne Fil. Cluj Stud. Cerc. Mat.* 14, 63–71.
Jankó, B. [1965a]. Solution of nonlinear equations by Newton's method and the gradient method, *Apl. Mat.* 10, 230–234.
Jankó, B. [1965b]. Sur la résolution des équations opérationelles nonlinéaires, *Mathematica (Cluj)* 7, 1965, 257–262.
Jankó, B., and Balázs, M. [1966]. On the generalized Newton method in the solution of nonlinear operator equations, *An. Univ. Timişoara Ser. Sti. Mat.-Fiz.* 4, 189–193.
Jeeves, T. (see Hooke, R.)
John, F. (see Bers, L.)

Johnson, L., and Scholz, D. [1968]. On Steffensen's method, *SIAM J. Num. Anal.* **5**, 296–302.
Juncosa, M. (see Bellman, R.)
Kaazik, Y. [1957]. On approximate solution of nonlinear operator equations by iterative methods (Russian), *Uspehi Mat. Nauk* **12**, 195–199.
Kac, I., and Maergoiz, M. [1967]. The solution of nonlinear and transcendental equations in a complex region (Russian), *Ž. Vyčisl. Mat. i Mat. Fiz.* **7**, 654–661.
Kačurovskii, R. [1960]. On monotone operators and convex functionals (Russian), *Uspehi Mat. Nauk* **15**, 213–215.
Kačurovskii, R. [1962]. On monotone operators and convex functionals (Russian), *Učen. Zap. Mosk. Reg. Ped. Inst.* **110**, 231–243.
Kačurovskii, R. [1965]. Monotonic nonlinear operators in Banach spaces, *Dokl. Akad. Nauk SSSR* **163**, 559–562; *Soviet Math. Dokl.* **6**, 953–956.
Kačurovskii, R. [1966]. Nonlinear operators of bounded variation, monotone and convex operators in Banach spaces (Russian), *Uspehi Mat. Nauk* **21**, 256–257.
Kačurovskii, R. [1967]. Nonlinear equations with monotonic operators and with some other ones, *Dokl. Akad. Nauk SSSR* **173**, 515–519; *Soviet Math. Dokl.* **8**, 427–430.
Kačurovskii, R. [1968]. Three theorems on nonlinear equations with monotone operators, *Dokl. Akad. Nauk SSSR* **183**, 33–36; *Soviet Math. Dokl.* **9**, 1322–1325.
Kagiwada, H. (see Bellman, R.)
Kalaba, R. [1959]. On nonlinear differential equations, the maximum operation, and monotone convergence, *J. Math. Mech.* **8**, 519–574.
Kalaba, R. [1963]. Some aspects of quasi-linearization, *in* "Nonlinear Differential Equations and Nonlinear Mechanics," pp. 135–146. Academic Press, New York.
Kalaba, R., (see Bellman, R.)
Kalaida, O. [1964]. A new method of solution for functional equations (Ukrainian), *Visnik Kiiv. Univ.* **6**, 123–129.
Kalaida, O. [1966]. On a new method of numerical solution of functional equations (Ukrainian), *Dopovidi Akad. Nauk Ukrain. RSR* **1966**. 20–23.
Kannan, R. [1969]. Some results on fixed points, II, *Amer. Math. Monthly* **76**, 405–408.
Kantorovich, L. [1937]. Lineare halbgeordnete Räume, *Matem. Sbornik* **2**, 121–165.
Kantorovich, L. [1939]. The method of successive approximations for functional equations, *Acta Math.* **71**, 63–97.
Kantorovich, L. [1948a]. On Newton's method for functional equations (Russian), *Dokl. Akad. Nauk SSSR* **59**, 1237–1240.
Kantorovich, L. [1948b]. Functional analysis and applied mathematics (Russian), *Uspehi Mat. Nauk* **3**, 89–185; transl. by C. Benster as Nat. Bur. Standards Report 1509. Washington, D.C., 1952.
Kantorovich, L. [1949]. On Newton's method (Russian), *Trudy Mat. Inst. Steklov* **28**, 104–144.
Kantorovich, L. [1951a]. The majorant principle and Newton's method (Russian), *Dokl. Akad. Nauk SSSR* **76**, 17–20.
Kantorovich, L. [1951b]. Some further applications of the majorant principle (Russian), *Dokl. Akad. Nauk SSSR* **80**, 849–852.
Kantorovich, L. [1956]. Approximate solution of functional equations (Russian), *Uspehi Mat. Nauk* **11**, 99–116.
Kantorovich, L. [1957]. On some further applications of the Newton approximation method (Russian), *Vestnik. Leningrad Univ.* **12**, 68–103.
Kantorovich, L. and Akilov, G. [1959]. "Functional Analysis in Normed Spaces."

Fizmatgiz, Moscow; transl. by D. Brown and A. Robertson, Pergamon Press, Oxford, 1964.
Kantorovich, L., Vulich, B., and Pinsker, A. [1950]. "Functional Analysis in Partially Ordered Spaces" (Russian), Gos. Izdat. Tehn.-Teor. Lit., Moscow.
Kas'janjuk, S. (see Varjuhin, V.)
Kasriel, R., and Nashed, M. [1966]. Stability of solutions of some classes of nonlinear operator equations, *Proc. Amer. Math. Soc.* **17**, 1036–1042.
Katetov, M. [1967]. A theorem on mappings, *Comment. Math. Univ. Carolinae* **8**, 431–433.
Keller, H. [1968]. "Numerical Methods for Two-point Boundary Value Problems." Ginn (Blaisdell), Boston, Massachusetts.
Keller, H., and Reiss, E. [1958]. Iterative solutions for the nonlinear bending of circular plates, *Comm. Pure Appl. Math.* **11**, 273–292.
Kellogg, O. (see Birkhoff, G. D.)
Kellogg, R. [1964]. An alternating direction method for operator equations, *SIAM J. Appl. Math.* **12**, 848–854.
Kellogg, R. [1969]. A nonlinear alternating direction method, *Math. Comp.* **23**, 23–28.
Kellogg, R. (see Birkhoff, G.)
Kempthorne, O. (see Shah, B.)
Kenneth, P. (see McGill, R.)
Kenneth, P., and McGill, R. [1966]. Two-point boundary value problem techniques, *in* "Advances in Control Systems," Vol. 3, pp. 69–109. Academic Press, New York.
Kerner, M. [1933]. Die Differentiale in der allgemeinen Analysis, *Ann. of Math.* **34**, 546–572.
Kerr, D. [1967]. On Some Iterative Methods for Solving a Class of Nonlinear Boundary Value Problems. Ph.D. Diss., Purdue University, Lafayette, Indiana.
Kiefer, J. [1953]. Sequential minimax search for a maximum, *Proc. Amer. Math. Soc.* **4**, 503–506.
Kiefer, J. [1957]. Optimum sequential search and approximation methods under minimum regularity assumptions, *SIAM J. Appl. Math.* **5**, 105–136.
Kincaid, W. [1948]. Solution of equations by interpolation, *Ann. Math. Statist.* **19**, 207–219.
Kincaid, W. [1961]. A two-point method for the numerical solution of systems of simultaneous equations, *Quart. Appl. Math.* **18**, 313–324.
Kirk, W. [1965]. A fixed point theorem for mappings which do not increase distance, *Amer. Math. Monthly* **72**, 1004–1006.
Kirk, W. (see Belluce, L.).
Kitchen, J. [1966]. Concerning the convergence of iterates to fixed points, *Studia Math.* **27**, 247–249.
Kivistik, L. [1960a]. The method of steepest descent for nonlinear equations (Russian), *Eesti NSV Tead. Akad. Toimetised Füüs-Mat. Tehn.-tead. Seer.* **9**, 145–159.
Kivistik, L. [1960b]. On certain iteration methods for solving operator equations in Hilbert space (Russian), *Eesti NSV Tead. Akad. Toimetised Füüs-Mat. Tehn.-tead. Seer.* **9**, 229–241.
Kivistik, L. [1960c]. On a generalization of Newton's method (Russian), *Eesti NSV Tead. Akad. Toimetised Füüs-Mat. Tehn.-tead. Seer.* **9**, 301–312.
Kivistik, L. [1961]. A modification of the minimum residual iteration method for the solution of equations involving nonlinear operators, *Dokl. Akad. Nauk USSR* **136**, 22–25; *Soviet Math. Dokl.* **2**, 13–16.
Kivistik, L. [1962]. On a class of iteration processes in Hilbert space (Russian), *Tartu Riikl. Ül. Toimetised* **129**, 365–381.

Kivistik, L., and Ustaal, A. [1962]. Some convergence theorems for iteration processes with minimal residues (Russian), *Tartu Riikl. Ül. Toimetised* **129**, 382–393.
Kiyono, T. (see Tsuda, T.)
Kizner, W. [1964]. A numerical method for finding solutions of nonlinear equations, *SIAM J. Appl. Math.* **12**, 424–428.
Kleinmichel, H. [1968]. Stetige Analoga und Iterationsverfahren für nichtlineare Gleichungen in Banachräumen, *Math. Nachr.* **37**, 313–344.
Knill, R. [1965]. Fixed points of uniform contractions, *J. Math. Anal. Appl.* **12**, 449–456.
Kogan, T. [1964a]. Construction of iteration processes of high orders for systems of algebraic and transcendental equations (Russian), *Ž. Vyčisl. Mat. i Mat. Fiz.* **4**, 545–546.
Kogan, T. [1964b]. The construction of high order iteration processes for systems of algebraic and transcendental equations (Russian), *Taškent. Gos. Univ. Naučn. Trudy* **265**, 37–46.
Kogan, T. [1964c]. A modified Newton method for solving systems of equations (Russian), *Taškent. Gos. Univ. Naučn. Trudy* **265**, 64–67.
Kolmogoroff, A., and Fomin, S. [1954]. "Elements of the Theory of Functions and Functional Analysis. Vol. 1: Metric and Normed Spaces." Izdat. Moscow Univ., Moscow; transl. by L. Boron, Graylock Press, Rochester, New York, 1957.
Kolomy, J. [1963]. Contribution to the solution of nonlinear equations, *Comment. Math. Univ. Carolinae* **4**, 165–171.
Kolomy, J. [1964]. Remark to the solution of nonlinear functional equations in Banach spaces, *Comment. Math. Univ. Carolinae* **5**, 97–116.
Kolomy, J. [1965]. On the solution of functional equations with linear bounded operators, *Comment. Math. Univ. Carolinae* **6**, 141–143.
Kolomy, J. [1966]. Some existence theorems for nonlinear problems, *Comment. Math. Univ. Carolinae* **7**, 207–217.
Kolomy, J. [1967]. Solution of nonlinear functional equations in linear normed spaces, *Časopis Pěst. Mat.* **92**, 125–132.
Kolomy, J. [1968]. On the differentiability of operators and convex functionals, *Comment. Math. Univ. Carolinae* **9**, 441–454.
Koppel', H. [1966]. Convergence of the generalized method of Steffensen (Russian), *Eesti NSV Tead. Akad. Toimetised Füüs-Mat. Tehn.-tead. Seer.* **15**, 531–539.
Korganoff, A. [1961]. "Méthodes de Calcul Numérique. Vol. 1: Algèbre nonlinéaire." Dunod, Paris.
Kosolev, A. [1962]. Convergence of the method of successive approximations for quasi-linear elliptic equations, *Dokl. Akad. Nauk SSSR* **142**, 1007–1110; *Soviet Math. Dokl.* **3**, 219–222.
Kotze, W. [1964]. Iterative Solution of Equations in Linear Topological Spaces. Ph.D. Diss., McGill Univ., Montreal, Canada.
Kowalik, J., and Osborne, M. [1968]. "Methods for Unconstrained Optimization Problems." American Elsevier, New York.
Krasnoselskii, M. [1954]. Some problems of nonlinear analysis (Russian), *Uspehi Mat. Nauk* **9**, 57–114.
Krasnoselskii, M. [1955]. Two comments on the method of successive approximations (Russian), *Uspehi Mat. Nauk* **10**, 123–127.
Krasnoselskii, M. [1956]. "Topological Methods in the Theory of Nonlinear Integral Equations." Gostekhteoretizdat, Moscow; transl. by A. Armstrong, Pergamon Press, Oxford, 1964.
Krasnoselskii, M. [1962]. "Positive Solution of Operator Equations." Gos. Izdat. Fiz.

Mat., Moscow; transl. by R. Flaherty and L. Boron. P. Noordhoff, Groningen, 1964.

Krasnoselskii, M., and Rutickii, Y. [1961]. Some approximate methods of solving nonlinear operator equations based on linearization, *Dokl. Akad. Nauk SSSR* **141**, 785–788; *Soviet Math. Dokl.* **2**, 1542–1546.

Krawczyk, R. [1966]. Über ein Verfahren zur Bestimmung eines Fixpunktes bei nichtlinearen Gleichungsystemen, *Z. Angew. Math. Mech.* **46**, T67–69.

Krivonos, J. [1968]. The application of averages of functions to the solution of equations (Ukrainian), *Dopovidi Akad. Nauk Ukrain. RSR Ser. A* **1968**, 304–308.

Kronecker, L. [1869]. Über Systeme von Funktionen mehrerer Variablen, *Monatsb. Deutsch Akad. Wiss. Berlin* **1869**, 159–193; 688–698.

Kuiken, H. [1968]. Determination of the intersection points of two plane curves by means of differential equations, *Comm. ACM* **11**, 502–506.

Kulik, S. [1964]. The solution of two simultaneous equations, *Duke Math. J.* **31**, 119–122.

Kuntzmann, J. [1959]. "Méthodes Numériques Interpolation Dérivées." Dunod, Paris.

Kuo, M. [1968]. Solution of nonlinear equations, *IEEE Trans. Computers* **17**, 897–898.

Kurpel', M. [1964]. Some approximate methods of solving nonlinear equations in a coordinate Banach space (Ukrainian), *Ukrain. Mat. Ž.* **16**, 115–120.

Kurpel', M. [1965]. The convergence and error estimates of certain general iterative methods of solution of operator equations (Ukrainian), *Dopovidi Akad. Nauk Ukrain. RSR* **1965**, 1423–1427.

Kurpel', M., and Migovič, F. [1968]. An approximate solution of some nonlinear operator equations by the projective-iterative method (Ukrainian), *Dopovidi Akad. Nauk Ukrain. RSR Ser. A.* **1968**, 13–16.

Kusakin, A. [1965a]. On the convergence of some methods for the approximate solution of operator equations (Ukrainian), *Dopovidi Akad. Nauk Ukrain. RSR* **1965**, 830–834.

Kusakin, A. [1965b]. Convergence of certain iteration methods (Russian), *Azerbaidžan Gos. Univ. Učen. Zap. Ser. Fiz.-Mat. i Him Nauk* **1965**, 19–23.

Kusakin, A. [1967]. The convergence of the method of chords for the approximate solution of operator equations (Ukrainian), *Dopovidi Akad. Nauk Ukrain. RSR Ser. A* **1967**, 786–790.

Kushner, H. [1968]. On the numerical solution of degenerate linear and nonlinear elliptic boundary value problems, *SIAM J. Numer. Anal.* **5**, 664–679.

Kuznečov, I. (see Marčuk, G.)

Kwan, C. [1956]. A remark on Newton's method for the solution of nonlinear functional equations, *Advancement in Math.* **2**, 290–295.

Ladyzhenskaya, O., and Ural'tseva, N. [1964]. "Linear and Quasilinear Elliptic Equations." Nauka Press, Moscow; transl. by Scripta Technica, Academic Press, New York, 1968.

Lahaye, E. [1934]. Une méthode de résolution d'une catégorie d'équations transcendantes, *C. R. Acad. Sci. Paris* **198**, 1840–1842.

Lahaye, E. [1935]: Sur la représentation des racines systèmes d'équations transcendantes, *Deuxième Congrès National des Sciences* **1**, 141–146.

Lahaye, E. [1948]. Solution of systems of transcendental equations, *Acad. Roy. Belg. Bull. Cl. Sci.* **5**, 805–822.

Lahmann, H. (see Ehrmann, H.)

Lancaster, P. [1966]. Error Analysis for the Newton-Raphson method, *Numer. Math.* **9**, 55–68.

Lance, G. [1959]. Solution of algebraic and transcendental equations on an automatic digital computer, *J. Assoc. Comp. Mach.* **6**, 97–101.

Langlois, W. [1966]. Conditions for termination of the method of steepest descent after a finite number of iterations, *IBM J. Res. Develop.* **10**, 98–99.

Langer, R. (ed.) [1960]. "Boundary Problems in Differential Equations." Univ. of Wisconsin Press, Madison, Wisconsin.

Laptinskii, V. [1965]. On one method of successive approximations (Russian), *Dokl. Akad. Nauk BSSR* **9**, 219–221.

Lastman, G. [1968]. A modified Newton's method for solving trajectory optimization problems, *AIAA J.* **6**, 777–780.

Lavi, A., and Vogl, L. (eds.) [1966]. "Recent Advances in Optimization Techniques." Wiley, New York.

Lavrent'ev, I. [1967]. Solubility of nonlinear equations, *Dokl. Akad. Nauk SSSR* **175**, 1219–1222; *Soviet Math. Dokl.* **8**, 993–996.

Leach, E. [1961]. A note on inverse function theorems, *Proc. Amer. Math. Soc.* **12**, 694–697.

Lee, E. [1968]. "Quasilinearization and Invariant Imbedding." Academic Press, New York.

Lees, M. [1966]. Discrete methods for nonlinear two-point boundary value problems, *in* "Numerical Solution of Partial Differential Equations" (J. Bramble, ed.), pp. 59–72. Wiley, New York.

Lees, M., and Schultz, M. [1966]. A Leray-Schauder principle for A-compact mappings and the numerical solution of two-point boundary value problems, *in* "Numerical Solution of Nonlinear Differential Equations" (D. Greenspan, ed.), pp. 167–180. Wiley, New York.

Leon, A. [1966]. A comparison among eight known optimizing procedures, *in* "Recent Advances in Optimization Techniques" (A. Lavi and T. Vogl, eds.), pp. 28–46. Wiley, New York.

Leon, A. (see Flood, M.).

Leray, J. [1950]. La theorie des points fixes et ses applications en analyse, *Proc. Intern. Congr. Math. 1950*, pp. 202–208. Am. Math. Soc., Providence, Rhode Island, 1952.

Leray, J., and Schauder, J. [1934]. Topologie et équations fonctionelles, *Ann. Sci. École Norm. Sup.* **51**, 45–78.

Levenberg, K. [1944]. A method for the solution of certain nonlinear problems in least squares, *Quart. Appl. Math.* **2**, 164–168.

Levin, A. [1965]. An algorithm for the minimization of convex functions, *Dokl. Akad. Nauk SSSR* **160**, 1244–1248; *Soviet Math. Dokl.* **6**, 186–190.

Levin, A., and Strygin, V. [1962]. "On the rate of convergence of the Newton-Kantorovich method" (Russian), *Uspehi Mat. Nauk* **17**, 185–187.

Levy, M. [1920]. Sur les fonctions de lignes implicites, *Bull. Soc. Math. France* **48**, 13–27.

Lezanski, T. [1967]. Über die Methode des "schnellsten Falles" für das Minimumproblem von Funktionalen in Hilbertschen Räumen, *Studia Math.* **28**, 183–192.

Li, T. [1934]. Die Stabilitätsfrage bei Differenzengleichungen, *Acta Math.* **63**, 99–141.

Lieberstein, H. [1959]. Overrelaxation for nonlinear elliptic partial differential equations, Math. Res. Center Rept. 80, Univ. of Wisconsin, Madison, Wisconsin.

Lieberstein, H. [1960]. A numerical test case for the nonlinear overrelaxation algorithm, Math. Res. Center Rept. 122, Univ. of Wisconsin, Madison, Wisconsin.

Liebl, P. [1965]. Einige Bemerkungen zur numerischen Stabilität von Matrizeniterationen, *Apl. Mat.* **10**, 249–254.

Lika, D. [1965]. An iteration process for nonlinear functional equations (Russian), *in* "Studies in Algebra and Math. Anal.," pp. 134–139. Izdat. "Karta Moldovenjaske," Kishinev.

Lika, D. [1967]. The principle of majorants in certain iteration processes (Russian), *Mat. Issled* **2**, 26–44.
Lin, C. [1962]. On approximate methods of solution for a certain type of nonlinear differential equation, *Chinese Math. Acta* **1**, 374–379.
Lin'kov, E. [1964a]. The convergence of some iterative methods (Russian), *Moskov. Oblast. Ped. Inst. Učen. Zap.* **150**, 71–80.
Lin'kov, E. [1964b]. On the convergence of a method of the steepest descent type in Hilbert space and in L^p (Russian), *Moskov. Oblast. Ped. Inst. Učen. Zap.* **150**, 181–187.
Ljaščenko, M. [1963]. On the numerical solution of nonlinear integral equations (Ukrainian), *Dopovidi Akad. Nauk Ukrain. RSR* **1963**, 1139–1144.
Ljaščenko, M. [1964]. On the numerical solution of a class of nonlinear integro-differential equations (Ukrainian), *Dopovidi Akad. Nauk Ukrain RSR* **1964**, 3–7.
Ljubič, Y. [1966]. On the rate of convergence of a stationary gradient iteration (Russian), *Ž. Vyčisl. Mat. i Mat. Fiz.* **6**, 356–360.
Ljubič, Y. [1968]. Convergence of the process of steepest descent, *Dokl. Akad. Nauk SSSR* **179**, 1054–1057; *Soviet Math. Dokl.* **9**, 506–508.
Lohr, L., and Rall, L. [1967]. Efficient use of Newton's method, *ICC Bull.* **6**, 99–103.
Lorentz, G. [1953]. "Bernstein Polynomials." Univ. of Toronto Press, Toronto.
Lotkin, M. [1955]. The solution by iteration of nonlinear integral equations, *J. Math. and Phys.* **33**, 346–355.
Luchka, A. [1963]. "Theory and Application of the Averaging Method for Functional Corrections." Izdat. Akad. Nauk Ukrain. SSSR, Kiev, 1963; translated as "The Method of Functional Averages." Academic Press, New York, 1965.
Ludwig, R. [1952]. Verbesserung einer Iterationsfolge bei Gleichungssystemen, *Z. Angew. Math. Mech.* **32**, 232–234.
Ludwig, R. [1954]. Über Iterationsverfahren für Gleichungen und Gleichungssysteme I, II, *Z. Angew. Math. Mech.* **34**, 210–225, 404–416.
Lumiste, Y. [1955]. The method of steepest descent for nonlinear equations (Russian), *Tartu. Gos. Univ. Trudy Estest.-Mat. Fak.* **37**, 106–113.
Luxemburg, W. [1958]. On the convergence of successive approximations in the theory of ordinary differential equations I, II, III, (I) *Canad. Math. Bull.* **1**, 9–20; (II) *Nederl. Akad. Wetensch. Proc. Ser. A* **61** (*Indag. Math.*) **20**, 540–546; (III) *Nieuw Arch. Wisk.* (3) **6**, 93–98.
Lyubčenko, I. [1961]. Newton's method as a basis for approximate solution of the boundary value problem for a nonlinear ordinary differential equation of second order involving a small parameter in the term with derivative of highest order, *Dokl. Akad. Nauk SSSR* **138**, 39–42; *Soviet Math. Dokl.* **2**, 525–528.
Maddison, R. [1966]. A procedure for nonlinear least squares refinement in adverse practical conditions, *J. Assoc. Comput. Mach.* **13**, 124–134.
Madorskii, V. [1967]. A variant of the descent method for nonlinear functional equations (Russian), *Vesci Akad. Nauk BSSR Ser. Fiz.-Mat. Navuk* **3**, 121–124.
Maergoiz, M. [1967]. A method for solving systems of nonlinear algebraic and transcendental equations (Russian), *Ž. Vyčisl. Mat. i Mat. Fiz.* **7**, 869–874.
Maergoiz, M. (see Kac, I.)
Maistrov'skii, G. [1967]. A local relaxation theory for nonlinear equations, *Dokl. Akad. Nauk SSSR* **177**, 37–39; *Soviet Math. Dokl.* **8**, 1366–1369.
Mamedov, A. [1965]. On an approximate solution of nonlinear integral equations (Russian) *Izv. Akad. Nauk Azerbaidzan. SSR Ser. Fiz.-Tehn. Mat. Nauk* **1965**, 41–48.
Mancino, O. [1967]. Resolution by iteration of some nonlinear systems, *J. Assoc. Comput. Mach.* **14**, 341–350.

Mangasarian, O. [1965]. Pseudo-convex functions, *SIAM J. Control* **3**, 281–290.
Marčuk, G., and Kuznečov, I. [1968]. On optimal iteration processes, *Dokl. Akad. Nauk SSSR* **181**, 1331–1334; *Soviet Math. Dokl.* **9**, 1041–1045.
Marčuk, G., and Sarbasov, K. [1968]. A method for solving a stationary problem, *Dokl. Akad. Nauk SSSR* **182**, 42–45; *Soviet Math. Dokl.* **9**, 1105–1108.
Marquardt, D. [1963]. An algorithm for least squares estimation of nonlinear parameters, *SIAM J. Appl. Math.* **11**, 431–441.
Martos, B. [1967]. Quasi-convexity and quasi-monotonicity in nonlinear programming, *Studia Sci. Math. Hungar.* **2**, 265–273.
Maslova, N. [1968]. A method of solving relaxation equations, *Dokl. Akad. Nauk SSSR* **182**, 760–763; *Soviet Math. Dokl.* **9**, 1197–1200.
Matveev, V. [1964]. A method of approximate solution of systems of nonlinear equations (Russian), *Z. Vycisl. Mat. i Mat. Fiz.* **4**, 983–994.
McAllister, G. [1964]. Some nonlinear elliptic partial differential equations and difference equations, *SIAM J. Appl. Math.* **12**, 772–777.
McAllister, G. [1966a]. Quasilinear uniformly elliptic partial differential equations and difference equations, *SIAM J. Numer. Anal.* **3**, 13–33.
McAllister, G. [1966b]. Difference methods for a nonlinear elliptic system of partial differential equations, *Quart. Appl. Math.* **23**, 355–360.
McCormick, G. (see Fiacco, A.)
McGill, R. (see Kenneth, P.)
McGill, R., and Kenneth, P. [1964]. Solution of variational problems by means of a generalized Newton-Raphson operator, *AIAA J.* **2**, 1761–1766.
Mead, R. (see Nelder, J.)
Meeter, D. [1966]. On a theorem used in nonlinear least squares, *SIAM J. Appl. Math.* **14**, 1176–1179.
Meinardus, G. [1964]. "Approximation von Funktionen und ihre numerische Behandlung." Springer-Verlag, Berlin.
Melon, S. [1962]. On nonlinear numerical iteration processes, *Comment. Math. Univ. Carolinae* **3**, 14–22.
Mertvecova, M. [1953]. An analog of the process of tangent hyperbolas for general functional equations (Russian), *Dokl. Akad. Nauk SSSR* **88**, 611–614.
Metcalf, F. (see Diaz, J.)
Meyer, F. (see Sylvester, R.)
Meyer, G. [1968]. On solving nonlinear equations with a one-parameter operator imbedding, *SIAM J. Numer. Anal.* **5**, 739-752.
Meyers, P. [1965]. Some extensions of Banach's contraction theorem, *J. Res. Nat. Bur. Standards Sect. B* **69B**, 179–185.
Meyers, P. [1967]. A converse to Banach's contraction theorem, *J. Res. Nat. Bur. Standards Sect. B* **71B**, 73–76.
Migovic, F. (see Kurpel', M.)
Mihklin, S. [1957]. "Variational Methods in Mathematical Physics." Gos. Izdat. Tehn.-Teor. Lit., Moscow; transl. by T. Boddington, Pergamon Press, Oxford, 1964.
Minkowski, H. [1892]. Theorie der konvexen Körper, insbesondere Begründung ihres Oberflächenbegriffs, *Gesammelte Abhandlungen* **2**, 131–229. Teubner, Leipzig, 1911.
Minty, G. [1962]. Monotone (nonlinear) operators in Hilbert space, *Duke Math. J.* **29**, 341–346.
Minty, G. [1963]. Two theorems on nonlinear functional equations in Hilbert space, *Bull. Amer. Math. Soc.* **69**, 691–692.

Minty, G. [1964]. On the monotonicity of the gradient of a convex function, *Pacific J. Math.* **14**, 243–247.
Minty, G. [1965]. A theorem on maximal monotonic sets in Hilbert space, *J. Math. Anal. Appl.* **11**, 434–440.
Minty, G. [1967]. On the generalization of a direct method of the calculus of variations, *Bull. Amer. Math. Soc.* **73**, 315–321.
Minty, G. (see Dolph, C.)
Mirakov, V. [1957]. The majorant principle and the method of tangent parabolas for nonlinear functional equations (Russian), *Dokl. Akad. Nauk SSSR* **113**, 977–979.
Moore, J. [1967]. A convergent algorithm for solving polynomial equations, *J. Assoc. Comput. Mach.* **14**, 311–315.
Moore, R. H. [1964]. Newton's method and variations, *in* "Nonlinear Integral Equations" (P. Anselone, ed.), p. 65–98. Univ. of Wisconsin Press, Madison, Wisconsin.
Moore, R. H. [1966]. Differentiability and convergence for compact nonlinear operators, *J. Math. Anal. Appl.* **16**, 65–72.
Moore, R. H. [1968]. Approximations to nonlinear operator equations and Newton's method, *Numer. Math.* **12**, 23–34.
Moore, R. H. (see Anselone, P.)
Morozov, V. [1966]. Solution of functional equations by the regularization method, *Dokl. Akad. Nauk SSSR* **167**, 510–513; *Soviet Math. Dokl.* **7**, 414–417.
Morrison, D. [1962]. Multiple shooting method for two-point boundary value problems, *Comm. ACM* **5**, 613–614.
Moser, J. [1966]. A rapidly convergent iteration method and nonlinear partial differential equations I, II, *Ann. Scuola Norm. Sup. Pisa* **20**, 265–315, 499–535.
Moses, J. [1955]. Solution of systems of polynomial equations by elimination, *Comm. ACM* **9**, 634–637.
Moszyński, K. [1965]. The Newton's method for finding an approximate solution to an eigenvalue problem of ordinary linear differential equations, *Algorytmy* **3**, 7–33.
Motzkin, T. (see Hart, H.).
Motzkin, T., and Wasow, W. [1953]. On the approximation of linear elliptic differential equations with positive coefficients, *J. Math. and Phys.* **31**, 253–259.
Muir, T. [1933]. "A Treatise on the Theory of Determinants." Longman, Green, and Co., New York; republished by Dover, New York, 1960.
Myers, G. [1968]. Properties of the conjugate gradient and Davidon methods, *J. Optimization Theory Appl.* **2**, 209–219.
Mysovskii, I. [1949]. On convergence of Newton's method (Russian), *Trudy Mat. Inst. Steklov* **28**, 145–147.
Mysovskii, I. [1950]. On the convergence of L. V. Kantorovich's method of solution of functional equations and its applications (Russian), *Dokl. Akad. Nauk SSSR* **70**, 565–568.
Mysovskii, I. [1953]. On the convergence of the method of L. V. Kantorovich for the solution of nonlinear functional equations and its applications (Russian), *Vestnik Leningrad Univ.* **11**, 25–48.
Mysovskii, I. [1963]. An error bound for the numerical solution of a nonlinear integral equation, *Dokl. Akad. Nauk SSSR* **153**, 30–33; *Soviet Math. Dokl.* **4**, 1603–1607.
Nagumo, M. [1951]. A theory of degree of mappings based on infinitesimal analysis, *Amer. J. Math.* **73**, 485–496.
Nashed, M. [1964]. The convergence of the method of steepest descents for nonlinear equations with variational or quasi-variational operators, *J. Math. Mech.* **13**, 765–794.

Nashed, M. [1965]. On general iterative methods for the solutions of a class of nonlinear operator equations, *Math. Comp.* **19**, 14–24.
Nashed, M. [1967]. Supportably and weakly convex functionals with applications to approximation theory and nonlinear programming, *J. Math. Anal. Appl.* **18**, 504–521.
Nashed, M. (see Kasriel, R.)
Nečepurenko, M. [1954]. On Chebysheff's method for functional equations (Russian), *Uspehi Mat. Nauk* **9**, 163–170.
Nelder, J., and Mead, R. [1965]. A simplex method for function minimization, *Comput. J.* **7**, 308–313.
Nemyckii, V. [1960]. A method for finding all solutions of nonlinear operator equations, *Dokl. Akad. Nauk SSSR* **130**, 746–747; *Soviet Math. Dokl.* **1**, 330–331.
Neustadt, L. (see Balakrishnan, A.)
Newman, D. [1965]. Location of the maximum on unimodal surfaces, *J. Assoc. Comput. Mach.* **12**, 395–398.
Nicolovius, R. [1956]. Das Extrapolations-Verfahren von J. Albrecht für nichtlineare Aufgaben, *Z. Angew. Math. Mech.* **45**, 65–67.
Nicula, A. [1963]. The method of successive approximation in solving systems of equations, *Bul. Sti. Inst. Politehn. Cluj* **6**, 43–48.
Nikaido, H. (see Gale, D.)
Noble, B. [1964]. The numerical solution of nonlinear integral equations and related topics, *in* "Nonlinear Integral Equations" (P. Anselone, ed.), pp. 215–318. Univ. of Wisconsin Press, Madison, Wisconsin.
Opial, Z. [1967a]. Lecture notes on nonexpansive and monotone mappings in Banach spaces, Div. of Appl. Math. Lecture Notes 67-1, Brown Univ., Providence, Rhode Island.
Opial, Z. [1967b]. Weak convergence of the sequence of successive approximations for nonexpansive mappings, *Bull. Amer. Math. Soc.* **73**, 591–597.
Ortega, J. [1967]. Notes on Newton and secant methods in n dimensions, Tech. Note, IBM Federal Systems Div., Bethesda, Maryland.
Ortega, J. [1968]. The Newton-Kantorovich theorem, *Amer. Math. Monthly* **75**, 658–660.
Ortega, J., and Rheinboldt, W. [1966]. On discretization and differentiation of operators with application to Newton's method, *SIAM J. Numer. Anal.* **3**, 143–156.
Ortega, J., and Rheinboldt, W. [1967a]. Monotone iterations for nonlinear equations with application to Gauss-Seidel methods, *SIAM J. Numer. Anal.* **4**, 171–190.
Ortega, J., and Rheinboldt, W. [1967b]. On a class of approximate iterative processes, *Arch. Rational Mech. Anal.* **23**, 352–365.
Ortega, J., and Rheinboldt, W. [1970a]. Local and global convergence of generalized linear iterations, *in* "Numerical Solution of Nonlinear Problems" (J. Ortega and W. Rheinboldt, eds.), pp. 122–143. Soc. Ind. Appl. Math., Philadelphia, Pennsylvania.
Ortega, J., and Rheinboldt, W. (eds.) [1970b]. "Numerical Solution of Nonlinear Problems," Studies in Numerical Analysis II, Soc. Ind. Appl. Math., Philadelphia, Pennsylvania.
Ortega, J., and Rockoff, M. [1966]. Nonlinear difference equations and Gauss-Seidel type iterative methods, *SIAM J. Numer. Anal.* **3**, 497–513.
Osborne, M. (see Kowalik, J.)
Ostrowski, A. [1936]. Konvergenzdiskussion und Fehlerabschätzung für die Newton'sche Methode bei Gleichungssystemen, *Comment. Math. Helv.* **9**, 79–103, 1936/37.
Ostrowski, A. [1957]. Les points d'attraction et de répulsion pour l'itération dans l'espace à n dimensions, *C.R. Acad. Sci. Paris* **244**, 288–289.

Ostrowski, A. [1966], [1960]. "Solution of Equations and Systems of Equations." Academic Press, New York, 1960; second edition, 1966.

Ostrowski, A. [1967a]. Contributions to the theory of steepest descent, *Arch. Rational Mech. Anal.* **26**, 257–280.

Ostrowski, A. [1967b]. General existence criteria for the inverse of an operator, *Amer. Math. Monthly* **74**, 826–827.

Ostrowski, A. [1967c]. The round-off stability of iterations, *Z. Angew. Math. Mech.* **47**, 77–82.

Overholt, K. [1965]. An instability in the Fibonacci and golden section search methods, *BIT* **5**, 284–286.

Panov, A. [1959]. The behavior of the solutions of difference equations near a fixed point (Russian), *Izv. Vysš. Učebn. Zaved. Matematika* **12**, 174–183.

Panov, A. [1960]. A qualitative study of trajectories of difference equations in the neighborhood of a fixed point (Russian), *Izv. Vysš. Učebn. Zaved. Matematika* **14**, 166–174.

Panov, A. [1964]. Qualitative behavior of the trajectories of a system of difference equations in the neighborhood of a singularity (Russian), *Izv. Vysš. Učebn. Zaved. Matematika* **40**, 111–115.

Parmet, I., and Saibel, E. [1968]. The Newton-Raphson approximation applied to the Navier-Stokes equations with application to the base flow problem, *Z. Angew. Math. Mech.* **48**, 1–11.

Parter, S. [1965]. Mildly nonlinear elliptic partial differential equations and their numerical solution, I, *Numer. Math.* **7**, 113–128.

Parter, S. (see Greenspan, D.)

Peaceman, D., and Rachford, H. [1955]. The numerical solution of parabolic and elliptic differential equations, *SIAM J. Appl. Math.* **3**, 28–41.

Penrose, R. [1955]. A generalized inverse for matrices, *Proc. Cambridge Philos. Soc.* **51**, 406–413.

Pereyra, V. [1967a]. Iterative methods for solving nonlinear least square problems, *SIAM J. Numer. Anal.* **4**, 27–36.

Pereyra, V. [1967b]. Accelerating the convergence of discretization algorithms, *SIAM J. Numer. Anal.* **4**, 508–533.

Perron, O. [1929]. Über Stabilität und asymptotisches Verhalten der Lösungen eines Systems endlicher Differenzengleichungen, *J. Reine Angew. Math.* **161**, 41–64.

Perry, C. (see Ablow, C.)

Peteanu, V. [1964]. Simultaneous equations for which the iterative process is convergent, *Mathematica (Cluj)* **29**, 101–104.

Peteanu, V. [1965]. Sur le mode de convergence d'un proces iteratif, *Mathematica (Cluj)* **30**, 67–70.

Petry, W. [1965]. Das Iterationsverfahren zum Lösen von Randwertproblemen gewöhnlicher, nichtlinearer Differentialgleichungen zweiter Ordnung, *Math. Z.* **87**, 323–333.

Petryshyn, W. [1966a]. On nonlinear P-compact operators in Banach space with applications to constructive fixed-point theorems, *J. Math. Anal. Appl.* **15**, 228–242.

Petryshyn, W. [1966b]. On the extension and solution of nonlinear operator equations, *Illinois J. Math.* **10**, 255–274.

Petryshyn, W. [1967a]. Remarks on fixed point theorems and their extensions, *Trans. Amer. Math. Soc.* **126**, 43–54.

Petryshyn, W. [1967b]. Projection methods in nonlinear numerical functional analysis, *J. Math. Mech.* **17**, 353–372.

Petryshyn, W. [1968a]. On the approximation-solvability of nonlinear equations, *Math. Ann.* **177**, 156–164.

Petryshyn, W. [1968b]. On the iteration, projection, and projection–iteration methods in the solution of nonlinear functional equations, *J. Math. Anal. Appl.* **21**, 575–607.
Petryshyn, W. (see Browder, F.)
Pietrzykowski, T. [1963]. On a certain class of iteration methods for nonlinear equations, *Algorytmy* **1**, 21–27.
Pinsker, A. (see Kantorovich, L.)
Pitteway, M. (see Dulley, D.)
Poljak, B. [1963]. Gradient methods for minimizing functionals (Russian), *Ž. Vyčisl. Mat. i Mat. Fiz.* **3**, 643–653.
Poljak, B. [1964a]. Some methods of speeding up the convergence of iterative methods (Russian), *Ž. Vyčisl. Mat. i Mat. Fiz.* **4**, 791–803.
Poljak, B. [1964b]. Gradient methods for solving equations and inequalities (Russian), *Ž. Vyčisl. Mat. i Mat. Fiz.* **4**, 995–1005.
Poljak, B. [1966]. Existence theorems and convergence of minimizing sequences in extremum problems with restrictions, *Dokl. Akad. Nauk SSSR* **166**, 287–290; *Soviet Math. Dokl.* **7**, 72–75.
Poljak, B. [1967]. A general method of solving extremum problems, *Dokl. Akad. Nauk SSSR* **174**, 33–37; *Soviet Math. Dokl.* **8**, 593–597.
Poll', V. [1967a]. Certain methods of finding stationary points of functions of several variables (Russian), *Eesti NSV Tead. Akad. Toimetised Füüs-Mat.* **16**, 35–44.
Poll', V. [1967b]. Convergence of certain methods for finding the stationary points of functions of several variables (Russian), *Eesti NSV Tead. Akad. Toimetised Füüs-Mat.* **16**, 157–167.
Poll' V. [1967c]. On methods for finding stationary points (Russian), *Eesti NSV Tead. Akad. Toimetised Füüs-Mat.* **16**, 382–384.
Poll, V. (see Ul'm, S.)
Ponstein, J. [1967]. Seven kinds of convexity, *SIAM Rev.* **9**, 115–119.
Porsching, T. [1969]. Jacobi and Gauss-Seidel methods for nonlinear network problems, *SIAM J. Numer. Anal.* **6**, 437-449.
Powell, M. [1962]. An iterative method for finding stationary values of a function of several variables, *Comput. J.* **5**, 147–151.
Powell, M. [1964]. An efficient method for finding the minimum of a function of several variables without calculating derivatives, *Comput. J.* **7**, 155–162.
Powell, M. [1965]. A method for minimizing a sum of squares of nonlinear functions without calculating derivatives, *Comput. J.* **7**, 303–307.
Powell, M. [1966]. Minimization of functions of several variables, *in* "Numerical Analysis, An Introduction" (J. Walsh, ed.), pp. 143–157. Academic Press, New York.
Powell, M. [1968]. On the calculation of orthogonal vectors, *Comput. J.* **11**, 302–304.
Powell, M. (see Fletcher, R.)
Prager, M., and Vitasek, E. [1963]. Stability of numerical processes, *in* "Differential Equations and Their Applications," pp. 123–130. Academic Press, New York.
Price, H. [1968]. Monotone and oscillation matrices applied to finite difference approximations, *Math. Comp.* **22**, 489–516.
Price, J. (see Goldstein, A.)
Pugačev, B. [1962a]. On the acceleration of the convergence of second-degree iterative processes (Russian), *Ž. Vyčisl. Mat. i Mat. Fiz.* **2**, 703–705.
Pugačev, B. [1962b]. Remarks on the foundations of certain iteration processes (Russian), *Ž. Vyčisl. Mat. i Mat. Fiz.* **2**, 912–915.
Pugačev, B. [1963a]. On the acceleration of the convergence of iterative processes of the second degree (Russian), *Ž. Vyčisl. Mat. i Mat. Fiz.* **4**, 784–787.

Pugačev, B. [1963b]. Notes on the proofs of certain iterative processes (Russian), Ž. Vyčisl. Mat. i Mat. Fiz. 5, 1059–1064.
Puzynin, I. (see Zhidkov, N.)
Quandt, R. (see Goldfeld, S.)
Rachford, H. (see Peaceman, D.)
Rademacher, H. [1922]. Über eine funktionale Ungleichung in der Theorie der konvexen Körper, Math. Z. 13, 18–27.
Rado, T. [1951]. "On the Problem of Plateau." Chelsea, New York.
Rall, L. [1961a]. Quadratic equations in Banach spaces, Rend. Circ. Mat. Palermo 10, 314–332.
Rall, L. [1961b]. Newton's method for the characteristic value problem $Ax = \lambda Bx$, SIAM J. Appl. Math. 9, 288–293.
Rall, L. (ed.) [1965]. "Errors in Digital Computation," Vol. II. Wiley, New York.
Rall, L. [1966]. Convergence of the Newton process to multiple solutions, Numer. Math. 9, 23–37.
Rall, L. [1969]. "Computational Solution of Nonlinear Operator Equations." Wiley, New York.
Rall, L. (see Anselone, P.; Gray, J.; Lohr, L.)
Reeves, C. (see Fletcher, R.)
Rehbock, F. [1942]. Zur Konvergenz des Newtonschen Verfahrens für Gleichungssysteme, Z. Angew. Math. Mech. 22, 361–362.
Reiss, E. (see Keller, H.)
Rheinboldt, W. [1968]. A unified convergence theory for a class of iterative processes, SIAM J. Numer. Anal. 5, 42–63.
Rheinboldt, W. [1969a]. Local mapping relations and global implicit function theorems, Trans. Amer. Math. Soc. 138, 183–198.
Rheinboldt, W. [1969b]. On M-functions and their application to nonlinear Gauss-Seidel iterations and to network flows, Gesellschaft für Mathematik und Datenverarbeitung m.b.H., Tech. Rept. 22. Birlinghoven, Germany. To appear in J. Math. Anal. Appl. 32, 1970.
Rheinboldt, W. (see Antosiewicz, H.; Ortega, J.)
Rice, J. [1969]. Minimization and techniques in nonlinear approximation, in "Numerical Solution of Nonlinear Problems" (J. Ortega, W. Rheinboldt, eds.), pp. 80–98. Soc. Ind. Appl. Math., Philadelphia, Pennsylvania.
Roberts, S., and Shipman, J. [1966]. The Kantorovich theorem and two-point boundary value problems, IBM J. Res. Develop. 10, 402–406.
Roberts, S., and Shipman, J. [1967a]. Some results in two-point boundary value problems, IBM J. Res. Develop. 11, 383–388.
Roberts, S., and Shipman, J. [1967b]. Continuation in shooting methods for two-point boundary value problems, J. Math. Anal. Appl. 18, 45–58.
Roberts, S., and Shipman, J. [1968]. Justification for the continuation method in two-point boundary value problems, J. Math. Anal. Appl. 21, 23–30.
Roberts, S., Shipman J., and Roth, C. [1968]. Continuation in quasi-linearization, J. Optimization Theory Appl. 2, 157–163.
Robinson, S. [1966]. Interpolative solution of systems of nonlinear equations, SIAM J. Numer. Anal. 3, 650–658.
Rockafellar, R. [1967]. "Monotone Processes of Convex and Concave Type," Math. Mem. 77, Am. Math. Soc., Providence, Rhode Island.
Rockoff, M. (see Ortega, J.)
Rosenbloom, P. [1956]. The method of steepest descent, Sixth Symp. Appl. Math., pp. 127–176. Am. Math. Soc., Providence, Rhode Island.

Rosenbrock, H. [1960]. An automatic method for finding the greatest or least value of a function, *Comput. J.* **3**, 175–184.
Roth, B. (see Freudenstein, F.)
Roth, C. (see Roberts, S.)
Rothe, E. [1937]. Zur Theorie der topologischen Ordnung und der Vektorfelder in Banachschen Räumen, *Compositio Math.* **5**, 177–197.
Runge, C. [1899]. Separation und Approximation der Wurzeln von Gleichungen, *Enzykl. d. Mathem. Wissensch.*, Vol. 1, pp. 405–449. Teubner, Leipzig.
Rutickii, Y. (see Krasnoselskii, M.).
Saaty, T. [1967]. "Modern Nonlinear Equations." McGraw-Hill, New York.
Šafiev, R. [1963a]. On the method of tangent hyperbolas, *Dokl. Akad. Nauk SSSR* **149**, 788–791; *Soviet Math. Dokl.* **4**, 482–485.
Šafiev, R. [1963b]. On a modification of Chebyshev's method (Russian), *Ž. Vyčisl. Mat. i Mat. Fiz.* **3**, 950–953.
Šafiev, R. [1963c]. An iteration process for nonlinear operator equations (Russian), *Akad. Nauk Azerbaidžan SSR Dokl.* **19**, 3–9.
Šafiev, R. [1964]. On some iteration processes (Russian), *Ž. Vyčisl. Mat. i Mat. Fiz.* **4**, 139–143.
Šafiev, R. [1965]. Certain iteration methods for nonlinear integro-differential equations (Russian), *Izv. Akad. Nauk Azerbaidžan SSR Ser. Fiz.–Tehn. Mat. Nauk* **6**, 3–10.
Šafiev, R. [1967a]. Certain iteration methods of solution of nonlinear equations with non-differentiable operators (Russian), *Ž. Vyčisl. Mat. i Mat. Fiz.* **7**, 425–429.
Šafiev, R. [1967b]. Certain iteration methods of solution of functional equations (Russian), *in* "Func. Anal. Certain Problems of Differential Equations and Theory of Functions," pp. 173–179; Izdat. Akad. Nauk Azerbaidžan SSR, Baku.
Saibel, E. (see Parmet, I.).
Šamanskii, V. [1966a]. A realization of the Newton method on a computer (Russian), *Ukrain. Mat. Ž.* **18**, 135–140.
Šamanskii, V. [1966b]. "Methods of Numerical Solution of Boundary Value Problems on a Computer. Part II: Nonlinear Boundary Value Problems and Eigenvalue Problems for Differential Equations" (Russian), Naukova Dumka, Kiev.
Šamanskii, V. [1967a]. On a modification of the Newton method (Russian), *Ukrain. Mat. Ž.* **19**, 133–138.
Šamanskii, V. [1967b]. The application of Newton's method in the singular case (Russian), *Ž. Vyčisl. Mat. i Mat. Fiz.* **7**, 774–783.
Šamanskii, V. [1968]. The method of linearization for solving nonlinear boundary value problems (Russian), *Ukrain. Mat. Ž.* **20**, 218–227.
Samokis, B. (see Daugavet, I.)
Sapagovas, M. [1965]. On the question of the solution of quasi-linear elliptic equations by the method of finite differences (Russian), *Litovsk Mat. Sb.* **5**, 637–644.
Sarachik, P. (see Horwitz, L.).
Sarbasov, K. (see Marčuk, G.)
Sard, A. [1942]. The measure of the critical values of differentiable maps, *Bull. Amer. Math. Soc.* **48**, 883–890.
Šarkovskii, O. [1961]. Rapidly converging iterative processes (Russian), *Ukrain. Math. Ž.* **13**, 210–215.
Savenko, S. [1964]. Iteration methods for solving algebraic and transcendental equations (Russian), *Ž. Vyčisl. Mat. i Mat. Fiz.* **4**, 738–744.
Scarf, H. [1967]. The approximation of fixed points of a continuous mapping, *SIAM J. Appl. Math.* **15**, 1328–1343.

Schaefer, H. [1957]. Über die Methode der sukzessiven Approximationen, *Jber. Deutsch. Math. Verein.* **59**, 131–140.
Schauder, J. [1930]. Der Fixpunktsatz in Funktionalräumen, *Studia Math.* **2**, 171–180.
Schauder, J. (see Leray, J.)
Schechter, E. [1965]. Observations relative to the method of Chaplygin for systems of equations (Rumanian), *Studia Univ. Babes-Bolyai Ser. Math.-Phys.* **10**, 45–49.
Schechter, M. (see Bers, L.)
Schechter, S. [1962]. Iteration methods for nonlinear problems, *Trans. Amer. Math. Soc.* **104**, 179–189.
Schechter, S. [1968]. Relaxation methods for convex problems, *SIAM J. Numer. Anal.* **5**, 601–612.
Schmetterer, L. [1968]. Über ein Iterationsverfahren, *Arch. Math. (Basel)* **19**, 195–200.
Schmidt, J. [1960]. Konvergenzuntersuchungen und Fehlerabschätzungen für ein verallgemeinertes Iterationsverfahren, *Arch. Rational Mech. Anal.* **6**, 261–276.
Schmidt, J. [1961]. Die Regula Falsi für Operatoren in Banachräumen, *Z. Angew. Math. Mech.* **41**, 61–63.
Schmidt, J. [1963a]. Eine Übertragung der Regula Falsi auf Gleichungen im Banachraum I, II, *Z. Angew. Math. Mech.* **43**, 1–8, 97–110.
Schmidt, J. [1963b]. Zur Fehlerabschätzung näherungsweiser Lösungen von Gleichungen in halbgeordneten Räumen, *Arch. Math. (Basel)* **14**, 130–138.
Schmidt, J. [1963c]. Extremwertermittlung mit Funktionswerten, *Wiss. Z. Techn. Univ. Dresden* **12**, 1601–1605.
Schmidt, J. [1964]. Ausgangsvektoren für monotone Iterationen, *Numer. Math.* **6**, 78–88.
Schmidt, J. [1965a]. Konvergenzbeschleunigung bei monotonen Vektorfolgen, *Acta Math. Acad. Sci. Hungar.* **16**, 221–229.
Schmidt, J. [1965b]. Fehlerabschätzung und Konvergenzbeschleunigung zu Iterationen bei linearen Gleichungssystemen, *Apl. Mat.* **10**, 297–301.
Schmidt, J. [1966a]. Konvergenzgeschwindigkeit der Regula Falsi und des Steffensen Verfahrens im Banachraum, *Z. Angew. Math. Mech.* **46**, 146–148.
Schmidt, J. [1966b]. Asymptotische Einschliessung bei konvergenzbeschleunigenden Verfahren, *Numer. Math.* **8**, 105–113.
Schmidt, J. [1968]. Ein Konvergenzsatz für Iterationsverfahren, *Math. Nachr.* **37**, 67–83.
Schmidt, J., and Dressel, H. [1967]. Error estimations in connection with polynomial equations using the fixed point theorem of Brouwer, *Numer. Math.* **10**, 42–50.
Schmidt, J., and Schoenheinz, H. [1962]. Fehlerschranken zum Differenzenverfahren unter ausschliesslicher Benutzung verfügbarer Grössen, *Arch. Rational Mech. Anal.* **10**, 311–322.
Schmidt, J., and Schwetlick, H. [1968]. Ableitungsfreie Verfahren mit höherer Konvergenzgeschwindigkeit, *Computing* **3**, 215–226.
Schmidt, J., and Trinkaus, H. [1966]. Extremwertermittlung mit Funktionswerten bei Funktion von mehreren Veränderlichen, *Computing* **1**, 224–232.
Schoenheinz, H. (see Schmidt, J.)
Scholz, D. (see Johnson, L.)
Schröder, E. [1870]. Über unendlich viele Algorithmen zur Auflösung der Gleichungen, *Math. Ann.* **2**, 317–365.
Schröder, J. [1956a]. Das Iterationsverfahren bei allgemeinerem Abstandsbegriff, *Math. Z.* **66**, 111–116.
Schröder, J. [1956b]. Nichtlineare Majoranten beim Verfahren der schrittweisen Näherung, *Arch. Math. (Basel)* **7**, 471–484.

Schröder, J. [1956c]. Neue Fehlerabschätzungen für verschiedene Iterationsverfahren, *Z. Angew. Math. Mech.* **36**, 168–181.
Schröder, J. [1956d]. Über das Differenzenverfahren bei nichtlinearen Randwertaufgaben I, II, *Z. Angew. Math. Mech.* **36**, 319–331, 443–455.
Schröder, J. [1957]. Über das Newtonsche Verfahren, *Arch. Rational Mech. Anal.* **1**, 154–180.
Schröder, J. [1959]. Anwendung von Fixpunktsätzen bei der numerischen Behandlung nichtlinearer Gleichungen in halbgeordneten Räumen, *Arch. Rational Mech. Anal.* **4**, 177–192.
Schröder, J. [1960]. Error estimates for boundary value problems using fixed point theorems, *in* "Boundary Problems for Differential Equations" (R. Langer, ed.). Univ. of Wisconsin Press, Madison, Wisconsin.
Schröder, J. [1962]. Invers-monotone Operatoren, *Arch. Rational Mech. Anal.* **10**, 276–295.
Schröder, J. [1966]. Operator-Ungleichungen und ihre numerische Anwendung bei Randwertaufgaben, *Numer. Math.* **9**, 149–162.
Schultz, M. (see Birkhoff, G., Ciarlet, P., and Lees, M.)
Schwartz, J. [1964]. "Nonlinear Functional Analysis (1963/64)," Lecture Notes, Courant Inst. of Math. Sci., New York Univ., New York.
Schwartz, J. (see Dunford, N.)
Schwetlick, H. (see Schmidt, J.)
Seda, V. [1968]. A remark to quasi-linearization, *J. Math. Anal. Appl.* **23**, 130–138.
Sefor, L. (see Deist, F.)
Sen, R. [1966]. A modification of the Newton-Kantorovich method, *Mathematica (Cluj)* **31**, 155–161.
Senčuk, Ju. (see Glazman, I.)
Sergeev, A. [1961]. On the secant method (Russian), *Sibirsk Mat. Ž.* **11**, 282–289.
Shah, B., Buehler, R., and Kempthorne, O. [1964]. Some algorithms for minimizing a function of several variables, *SIAM J. Appl. Math.* **12**, 74–92.
Shah, M., and Syn, W. [1965]. A search technique for functional approximation, *Internat. J. Comput. Math.* **1**, 193–198.
Shampine, L. (see Bailey, P.)
Shanno, D. [1966]. A Modified Newton-Raphson Technique for Constrained Nonlinear Estimation Problems, Ph.D. Diss., Carnegie-Mellon Univ., Pittsburgh, Pennsylvania.
Shinbrot, M. [1964]. A fixed point theorem and some applications, *Arch. Rational. Mech. Anal.* **17**, 255–271.
Shipman, J. (see Roberts, S.)
Sidlovskaya, N. [1958]. Application of the method of differentiation with respect to a parameter to the solution of nonlinear equations in Banach spaces (Russian), *Leningrad Gos. Univ. Učen. Zap. Ser. Mat. Nauk* **33**, 3–17.
Simeonov, S. [1961]. On a process of successive approximation and its application to the solutions of functional equations with nonlinear operators of positive type, *Dokl. Akad. Nauk SSSR* **138**, 1033–1034; *Soviet Math. Dokl.* **2**, 790–791.
Simeonov, S. [1963]. The application of a process of successive approximations to the solution of certain types of functional equations, *Dokl. Akad. Nauk SSSR* **148**, 534–537; *Soviet Math. Dokl.* **4**, 144–147.
Simpson, R. [1968]. Approximation of the minimizing element for a class of functionals, *SIAM J. Numer. Anal.* **5**, 26–41.
Sisler, M. [1961]. On an iteration process for the solution of systems of nonlinear equations (Czech), *Časopis Pěst. Mat.* **86**, 439–461.

Sisler, M. [1964]. On an iteration process for the solution of an approximately linear system of equations (Czech), Časopis Pěst. Mat. **89**, 36–52.
Sisler, M. [1965]. Solution of a system of nonlinear equations with a functional matrix of special type (Czech), Časopis Pěst. Mat. **90**, 344–352.
Sisler, M. [1967]. Approximative Formeln für den Fehler bei Iterationsverfahren von höherer Ordnung, Apl. Mat. **12**, 1–14.
Slugin, S. [1955]. Approximate solution of operator equations by the method of Chaplygin (Russian), Dokl. Akad. Nauk SSSR **103**, 565–568.
Slugin, S. [1956]. An unrestrictedly applicable method of Chaplygin type for ordinary differential equations of nth order (Russian), Dokl. Akad. Nauk SSSR **110**, 936–939.
Slugin, S. [1957]. Solution of functional equations by iteration methods of one-sided approximations (Russian), Izv. Akad. Nauk SSSR Ser. Mat. **21**, 117–124.
Slugin, S. [1958a]. On the theory of Newton's and Chaplygin's method (Russian), Dokl. Akad. Nauk SSSR **120**, 472–474.
Slugin, S. [1958b]. A modification of the abstract analogue to Chaplygin's method (Russian), Dokl. Akad. Nauk SSSR **120**, 256–258.
Slugin, S. [1958c]. Some applications of methods of two-sided approximations, Izv. Vysš. Učebn. Zaved. Matematika **7**, 244–256.
Smith, H., (see Draper, N.)
Smith, R. [1966]. Sufficient conditions for stability of a solution of difference equations, Duke Math. J. **33**, 725–734.
Sonin, V. (see Isaev, V.)
Southwell, R. [1946]. "Relaxation Methods in Theoretical Physics." Oxford Univ. Press (Clarendon), London and New York.
Spang, H. [1962]. A review of minimization techniques for nonlinear functions, SIAM Rev. **4**, 343–365.
Späth, H. [1967]. The damped Taylor's series method for minimizing a sum of squares and for solving systems of nonlinear equations, Comm. ACM **10**, 726–728.
Spiridonov, V. [1968]. The application of the gradient relaxation method for the solution of systems of nonlinear equations (Russian), Ž. Vyčisl. Mat. i Mat. Fiz. **8**, 872–873.
Srinivasacharyulu, K. [1968]. On some nonlinear problems, Canad. J. Math. **20**, 394–397.
Stečenko, V. [1968]. On a method for accelerating the convergence of iterative processes, Dokl. Akad. Nauk SSSR **178**, 1021–1025; Soviet Math. Dokl. **9**, 241–244.
Steffensen, J. [1933]. Remarks on iteration, Skand. Aktuarietidskr. **16**, 64–72.
Stein, M. [1952]. Sufficient conditions for the convergence of Newton's method in complex Banach spaces, Proc. Amer. Math. Soc. **3**, 858–863; **13**, 1000, (1962).
Stepleman, R. [1969]. Finite Dimensional Analogues of Variational and Quasilinear Elliptic Dirichlet Problems, Ph.D. Diss. Univ. of Maryland, College Park, Maryland.
Sternberg, S. (see Brauer, F.)
Stewart, G. [1967]. A modification of Davidon's minimization method to accept difference approximations of derivatives, J. Assoc. Comput. Mach. **14**, 72–83.
Stiefel, E. (see Hestenes, M.)
Strygin, V. (see Levin, A.)
Swann, W. [1964]. Report on the development of a new direct searching method of optimization, Research Note, I.C.I. Ltd.
Sylvester, R., and Meyer, F. [1965]. Two-point boundary problems by quasilinearization, SIAM J. Appl. Math. **13**, 586–602.
Syn, W. (see Shah, M.)
Takahashi, I. [1965]. A note on the conjugate gradient method, Information Processing in Japan **5**, 45–49.

Takota, M. [1966]. A numerical method for boundary value problems of nonlinear ordinary differential equations, *Information Processing in Japan* **6**, 16–23.
Tamme, E. [1958]. A class of convergent iteration methods (Russian), *Izv. Vysš. Učebn. Zaved. Matematika* **7**, 115–121.
Tamme, E., and Heinla, L. [1959]. An approximate solution of operator equations depending on a parameter (Russian), *Izv. Vysš. Učebn. Zaved. Matematika* **8**, 229–232.
Tapia, R. [1967]. A generalization of Newton's method with application to the Euler-Lagrange equation, Ph.D. Diss., Univ. of California, Los Angeles, California.
Taylor, A. [1958]. "Introduction to Functional Analysis." Wiley, New York.
Temple, G. [1939]. The general theory of relaxation methods applied to linear systems, *Proc. Roy. Soc. Ser. A.* **169**, 476–500.
Thompson, G. (see Hooker, W.)
Thurston, G. [1965]. Newton's method applied to problems in nonlinear mechanics, *Trans. ASME Ser. E. J. Appl. Mech.* **32**, 383–388.
Todd, J. (ed.) [1962]. "Survey of Numerical Analysis." McGraw-Hill, New York.
Tornheim, L. [1963]. Convergence in nonlinear regression, *Technometrics* **5**, 513–514.
Tornheim, L. [1964]. Convergence of multipoint iterative methods, *J. Assoc. Comput. Mach.* **11**, 210–220.
Traub, J. [1964]. "Iterative Methods for the Solution of Equations." Prentice Hall, Englewood Cliffs, New Jersey.
Trinkaus, H. (see Schmidt, J.)
Trotter, H. (see Goldfeld, S.).
Tsuda, T., Ichida, K., and Kiyono, T. [1967]. Monte Carlo path-integral calculations for two-point boundary problems, *Numer. Math.* **10**, 110–116.
Tsuda, T., and Kiyono, T. [1964]. Application of the Monte Carlo method to systems of nonlinear algebraic equations, *Numer. Math.* **6**, 59–67.
Turner, L. [1960]. Solution of nonlinear systems, *Ann. New York Acad. Sci.* **86**, 817–827.
Ul'm, S. [1956]. On the convergence of certain iteration processes in Banach space (Russian), *Učen. Zap. Tartu Gos. Univ.* **42**, 135–142.
Ul'm, S. [1963a]. On a class of iteration methods in Hilbert space (Russian), *Eesti NSV Tead. Akad. Toimetised Füüs.-Mat. Tehn.-tead. Seer.* **12**, 132–140.
Ul'm, S. [1963b]. Iterative methods for solving a nonlinear equation based on linearization by the Newton interpolation formula (Russian), *Eesti NSV Tead. Akad.Toimetised Füüs-Mat. Tehn.-tead. Seer.* **12**, 384–390.
Ul'm, S. [1964a]. A majorant principle and the method of secants (Russian), *Eesti NSV Tead. Akad. Toimetised Füüs-Mat. Tehn.-tead. Seer.* **13**, 217–227.
Ul'm, S. [1964b]. Iteration methods with divided differences of the second order, *Dokl. Akad. Nauk USSR* **158**, 55–58; *Soviet Math. Dokl.* **5**, 1187–1190.
Ul'm, S. [1964c]. A generalization of Steffensen's method for solving nonlinear operator equations (Russian), *Ž. Vyčisl. Mat. i. Mat. Fiz.* **4**, 1093–1097.
Ul'm, S. [1965a]. Algorithms of the generalized Steffensen method (Russian), *Eesti NSV Tead. Akad. Toimetised Füüs-Mat. Tehn.-tead. Seer.* **14**, 435–443.
Ul'm, S. [1965b]. A class of iteration methods with convergence speed of third order (Russian), *Eesti NSV Tead. Akad. Toimetised Füüs-Mat. Tehn.-tead. Seer.* **14**, 534–539.
Ul'm, S. [1967a]. Generalized divided differences I, II (Russian), *Eesti NSV Tead. Akad. Toimetised Füüs-Mat.* **16**, 13–26, 146–156.
Ul'm, S. [1967b]. Iteration methods with successive approximation of the inverse operator (Russian), *Eesti NSV Tead. Akad. Toimetised Füüs-Mat.* **16**, 403–411.
Ul'm, S., and Poll', V. [1968]. Certain methods for solving minimum problems (Russian), *Eesti NSV Tead. Akad. Toimetised Füüs-Mat.* **17**, 151–163.

Urabe, M. [1956]. Convergence of numerical iteration in solution of equations, *J. Sci. Hiroshima Univ. Ser. A -I Math.* **19**, 479–489.

Urabe, M. [1962]. Error estimation in numerical solution of equations by iteration process, *J. Sci. Hiroshima Univ. Ser. A - I Math.* **26**, 77–91.

Ural'tseva, N. (see Ladyzhenskaya, O.)

Ustaal, A. (see Kivistik, L.)

Vainberg, M. [1956]. "Variational Methods for the Study of Nonlinear Operators." Gostekhteoretizdat, Moscow; transl. by A. Feinstein, Holden-Day, San Francisco, California, 1964.

Vainberg, M. [1960]. On the convergence of the method of steepest descent, *Dokl. Akad. Nauk SSSR* **129**, 9–12; *Soviet Math. Dokl.* **1**, 1–4.

Vainberg, M. [1961]. On the convergence of the method of steepest descent (Russian), *Sibirsk. Mat. Z.* **2**, 201–220.

Vandergraft, J. [1967]. Newton's method for convex operators in partially ordered spaces, *SIAM J. Numer. Anal.* **4**, 406–432.

Vandergraft, J. [1968]. Spectral properties of matrices which have invariant cones, *SIAM J. Appl. Math.* **16**, 1208–1222.

Varga, R. [1962]. "Matrix Iterative Analysis." Prentice Hall, Englewood Cliffs, New Jersey.

Varga, R. [1969]. Accurate numerical methods for nonlinear boundary value problems, in "Numerical Solution of Nonlinear Problems" (J. Ortega and W. Rheinboldt, eds.), pp. 99–113. Soc. Ind. Appl. Math., Philadelphia, Pennsylvania.

Varga, R. (see Birkhoff, G.; Ciarlet, P.)

Varjuhin, V., and Kas'janjuk, S. [1966]. On a certain method for solving nonlinear systems of a certain type (Russian), *Ž. Vyčisl. Mat. i Mat. Fiz.* **6**, 347–352.

Vasudevan, R. (see Bellman, R.)

Vertgeim, B. [1953]. On the solution of nonlinear functional equations (Russian), *Perm. Gos. Univ. Učen. Zap. Mat. Meh.* **103**, 160–163.

Vertgeim, B. [1965]. On certain devices for the linearization and approximate solution of nonlinear functional equations (Russian), *Sibirsk Mat. Ž.* **6**, 686–691.

Vitasek, E. (see Prager, M.)

Voevodin, V. [1961]. Application of descent methods for determination of all roots of an algebraic polynomial (Russian), *Ž. Vyčisl. Mat. i Mat. Fiz.* **1**, 187–195.

Vogl, L. (see Lavi, A.)

Voigt, R. [1969]. Rates of Convergence for Iterative Methods for Nonlinear Systems of Equations, Ph.D. Diss., Univ. of Maryland, College Park, Maryland.

Vulich, B. (see Kantorovich, L.)

Vyhandu, L. [1955]. Generalization of Newton's method for the solution of nonlinear equations (Russian), *Tartu Gos. Univ. Trudy Estest. Mat. Fak.* **37**, 114–117.

Wachspress, E. [1966]. "Iterative Solution of Elliptic Systems." Prentice Hall, Englewood, Cliffs, New Jersey.

Wall, D. [1956]. The order of an iteration formula, *Math. Comp.* **10**, 167–168.

Walsh, J. (ed.) [1966]. "Numerical Analysis, An Introduction." Academic Press, New York.

Waltman, P. (see Bailey, P.)

Wang, H. (see Brannin, F.)

Warga, J. [1952]. On a class of iterative procedures for solving normal systems of ordinary differential equations, *J. Math. and Phys.* **31**, 223–243.

Wasow, W. (see Forsythe, G.; Motzkin, T.)

Wasscher, E. [1963]. Steep 1, *Comm. ACM* **6**, 517–519.

Wegge, L. [1966]. On a discrete version of the Newton-Raphson method, *SIAM J. Numer. Anal.* 3, 134–142.
Weinitschke, H. [1964]. Über eine Klasse von Iterationsverfahren, *Numer. Math.* 6, 395–404.
Weinstock, R. [1952]. "Calculus of Variations with Applications to Physics and Engineering." McGraw Hill, New York.
Weissinger, J. [1951]. Über das Iterationsverfahren, *Z. Angew. Math. Mech.* 31, 245–246.
Weissinger, J. [1952]. Zur Theorie und Anwendung des Iterationsverfahrens, *Math. Nachr.* 8, 193–212.
Wells, M. [1965]. Function minimization, *Comm. ACM* 8, 169–170.
Wetterling, W. [1963]. Anwendung des Newtonschen Iterationsverfahrens bei der Tschebyscheff-Approximation I, II, *Math.-Tech.-Wirtschaft* 1963, 61–63, 112–115.
Whiteside, D. [1961]. Patterns of mathematical thought in the latter seventeenth century, *Arch. History Exact Sci.* 1, 179–388 (see p. 207 for a discussion of the history of the Newton-Raphson method).
Wilde, D. [1964]. "Optimum Seeking Methods." Prentice Hall, Englewood Cliffs, New York.
Wilde, D. [1966]. Objective function indistinguishability in unimodal optimization, *in* "Recent Advances in Optimization Techniques" (A. Lavi and T. Vogl, eds.), pp. 341–349. Wiley, New York.
Wilde, D. [1965]. A multi-variable dichotomous optimum seeking method, *IEEE Trans. Automatic Control* AC-10, 85–87.
Wilde, D., and Beightler, C. [1967]. "Foundations of Optimization," Prentice Hall, Englewood Cliffs, New Jersey.
Wilkinson, J. [1963]. "Rounding Errors in Algebraic Processes." Prentice Hall, Englewood Cliffs, New Jersey.
Wilkinson, J. [1965]. "The Algebraic-Eigenvalue Problem." Oxford Univ. Press (Clarendon), London and New York.
Willers, F. [1938]. Zur Konvergenz des Newtonschen Näherungsverfahrens, *Z. Angew. Math. Mech.* 18, 197–200.
Winslow, A. [1966]. Numerical solution of the quasilinear Poisson equation in a non-uniform triangle mesh, *J. Computational Phys.* 1, 149–172.
Witte, B., and Holst, W. [1964]. Two new direct minimum search procedures for functions of several variables, *Proc. 1964 Spring Joint Comp. Conf.*, pp. 195–209. Spartan Books, Baltimore, Maryland.
Wolfe, P. [1959]. The secant method for simultaneous nonlinear equations, *Comm. ACM* 2, 12–13.
Wouk, A. [1964]. Direct iteration, existence, and uniqueness, *in* "Nonlinear Integral Equations" (P. Anselone, ed.), pp. 3–34. Univ. of Wisconsin Press, Madison, Wisconsin.
Yamamuro, S. [1963]. Some fixed point theorems in locally convex linear spaces, *Yokohama Math. J.* 11, 5–12.
Yohe, M. (see Greenspan, D.)
Yoshiaki, M. [1968]. Practical monotonous iterations for nonlinear equations, *Mem. Fac. Sci. Kyushu Univ. Ser A* 22, 56–73.
Zaguskin, V. [1960]. "Handbook of Numerical Methods for the Solution of Algebraic and Transcendental Equations." Fizmatgiz, Moscow; transl. by G. Harding, Pergamon Press, Oxford, 1961.
Zangwill, W. [1967]. Minimizing a function without calculating derivatives, *Comput. J.* 10, 293–296.

Zaplitna, G. [1965]. On an approximate method of a solution of nonlinear operator equations (Ukrainian), *Dopovidi Akad. Nauk Ukrain. RSR* **1965**, 1434–1437.
Zarantonello, E. [1960]. Solving functional equations by contractive averaging, Math. Res. Center Rept. 160, Univ. of Wisconsin, Madison, Wisconsin.
Zarantonello, E. [1964]. The closure of the numerical range contains the spectrum, *Bull. Amer. Math. Soc.* **70**, 781–787.
Zarantonello, E. [1967]. The closure of the numerical range contains the spectrum, *Pacific J. Math.* **22**, 575–595.
Zeleznik, F. [1968]. Quasi-Newton methods for nonlinear equations, *J. Assoc. Comput. Mach.* **15**, 265–271.
Zhidkov, E., and Puzynin, I. [1967]. A method of introducing a parameter in the solution of boundary value problems for second order nonlinear ordinary differential equations (Russian), *Ž. Vyčisl. Mat. i Mat. Fiz.* **7**, 1086–1095.
Zhidkov, E., and Puzynin, I. [1968]. An application of the continuous analog of Newton's method to the approximate solution of a certain nonlinear boundary value problem, *Dokl. Akad. Nauk SSSR* **180**, 18–21; *Soviet Math. Dokl.* **9**, 575–578.
Zhidkov, N. (see Berezin, I.)
Zielke, G. [1968]. Inversion of modified symmetric matrices, *J. Assoc. Comput. Mach.* **15**, 402–408.
Zinčenko, D. [1963a]. Some approximate methods of solving equations with nondifferentiable operators (Ukrainian), *Dopovidi Akad. Nauk Ukrain. RSR* **1963**, 156–161.
Zinčenko, D. [1963b]. A class of approximate methods of solving operator equations with non-differentiable operators (Ukrainian), *Dopovidi Akad. Nauk Ukrain. RSR* **1963**, 852–856.
Zuber, R. [1966]. A method of successive approximation, *Bull. Acad. Polon. Sci. Ser. Sci. Math. Astronom. Phys.* **14**, 559–563.
Zuev, A. [1966]. On an algorithm for solving nonlinear systems by the sweep method (Russian), *Trudy Mat. Inst. Steklov* **74**, 152–155.

AUTHOR INDEX

Numbers in italics refer to the pages on which the complete references are listed.

A

Aalto, S., *523*
Ablow, C., *523*
Adachi, R., *523*
Agaev, G., *523*
Ahamed, S., *523*
Akhieser, N., 26, *523*
Akilov, G., 65, 74, 82, 188, 390, 406, 414, 428, *522*, *538*
Albrecht, J., 446, 447, *523*
Alexandroff, P., 154, 155, *523*
Allen, B., 26, *523*
Altman, M., 164, 188, 258, 429, 493, 499, 500, *523*, *524*
Ambartsumanian, V., 18
Ames, W., *524*
Anderson, D., 200, 204, *524*
Anselone, P., 20, 37, 429, *521*, *524*
Antosiewicz, H., *525*
Apostol, T., 65, 73, 81, 82, 169, 170, *521*, *525*
Apslund, E., 88, *525*
Armijo, L., 493, 499, *525*
Avila, J., 341, *525*
Axelson, O., *525*

B

Baer, R., 265, *525*
Bailey, P., 12, *525*
Balakrishnan, A., *525*
Balázs, M., *537*
Baluev, A., 446, 454, *525*
Banach, S., 125, 390, *525*
Bard, Y., 249, *525*
Barnes, J., 200, 212, *525*
Bartis, M., *525*
Bartle, R., 406, *525*
Beale, E., *525*
Beightler, C., *556*
Bellman, R., 37, 139, 188, 201, 222, 429, *525*, *526*
Belluce, L., 125, *526*
Bel'tjukov, B., *526*
Ben-Israel, A., 270, 406, *526*
Bennett, A., 428, *526*
Berezin, I., 390, 393, 446, *526*
Berge, C., 88, 89, *526*
Berman, G., 276, *526*
Bers, L., 16, 165, 222, 440, *526*
Bezlyudnaya, L., *526*
Birkhoff, G., 32, 439, 471, *526*, *527*

Birkhoff, G. D., *526*
Bittner, E., *526*
Bittner, L., 200, 235, 378, 429, *527*
Bliss, G., 26, *527*
Block, H., *527*
Blum, E., 499, *527*
Blundell, P., *527*
Blutel, E., 316, *527*
Bohl, E., 439, 446, *527*
Bondarenko, P., *527*
Booth, A., 258, *527*
Booth, R., *527*
Bosarge, W., 235, 341, *527*
Box, M., 266, *527*, *528*
Braess, D., 499, 507, *528*
Bramble, J., 17, 58, 164, *528*
Brandler, F., *528*
Brannin, F., *528*
Brauer, F., *528*
Brouwer, L., 154, 155, 163, *528*
Browder, F., 125, 168, 407, *528*
Brown, K., 228, 229, *528*
Broyden, C., 212, 213, *529*
Brumberg, V., *529*
Bryan, C., 222, 431, *529*
Bryant, V., *529*
Budak, B., *529*
Bueckner, H., *529*
Buehler, R., 264, *552*
Bussmann, K., 428, *529*

C

Cacciopoli, R., 139, 390, *529*
Caspar, J., 145, 391, *529*
Cauchy, A., 97, 247, 258, 316, 390, 499, *529*
Cavanagh, R., 317, 353, 429, 430, *529*
Cesari, L., *529*
Chandrasekhar, S., 18, 21, *529*
Chen, K., 203, 363, 429, *529*
Chen, W., *529*
Cheney, E., 25, 406, *529*
Chernoff, H., 188, 247, 248, 258, 499, 507, *530*
Chu, S., 391, *529*, *530*
Ciarlet, P., 26, *530*
Coffman, C., 306, *530*
Collatz, L., 188, 201, 390, 399, 420, 429, 439, 446, 471, *521*, *530*

Concus, P., 31, *530*
Conte, S., 228, *528*
Crockett, J., 188, 247, 248, 258, 499, 507, *530*
Cronin, J., 155, 160, *530*
Cryer, C., *530*
Curry, H., 258, 493, 499, *530*

D

Daniel, J., 265, 512, *530*
Daugavet, I., *530*
Davidenko, D., 235, *530*, *531*
Davidon, W., 212, 248, *531*
Davis, J., 235, *531*
de Figueiredo, D., 168, *531*
Deist, F., 235, *531*
Dennis, J., 406, *528*, *531*
Derendjaev, I., *531*
Diaz, J., 390, 391, 407, *527*, *529*, *530*, *531*
Dieudonné, J., 65, 81, 82, 172, 175, *521*, *531*
D'jakonov, Ye., *531*
Dolph, C., 164, *531*
Douglas, J., Jr., 222, 391, *531*
Draper, N., 25, *531*
Dressel, H., *551*
Dreyfus, S., *531*
Dubovik, L., *531*
Dück, W., *532*
Duleau, J., *532*
Dulley, D., *532*
Dunford, N., 163, *532*
Durand, E., 307, *532*

E

Edelstein, M., 406, *532*
Edmunds, D., *532*
Eggleston, H., 88, *532*
Ehrmann, H., 139, 399, *532*
Elkin, R., 89, 101, 102, 103, 108, 265, 275, 478, 493, 494, 499, 500, 507, 508, 518, 519, 520, *532*
Emelina, L., *532*

F

Faddeev, D., 37, 44, *532*
Faddeeva, V., 37, 44, *532*
Falb, P., *527*

AUTHOR INDEX   561

Farforovskaya, J., *534*
Feder, D., *532*
Fel'dman, I., *532*
Feldstein, A., 286, *532*
Fenchel, W., 88, 101, *532*
Fenyo, I., *533*
Fiacco, A., *533*
Ficken, F., 139, 234, *533*
Filippi, S., *533*
Fine, H., 428, *533*
Firestone, R., 286, *532*
Fleming, W., *533*
Fletcher, R., 212, 213, 248, 265, 273, *533*
Flood, M., *533*
Fok, V., 235
Fomin, S., 390, *540*
Forster, P., *533*
Forsythe, G., 16, 222, 307, 343, *521*, *533*
Fox, L., *533*
Frank, T., 18, *533*
Fréchet, M., *533*
Freudenstein, F., 235, *533*
Frey, T., *533*
Fridman, V., 270, *533*
Fridrih, F., *533*
Fujii, M., *533*

G

Gagliardo, E., *534*
Galanov, B., *534*
Gale, D., 140, 145, *534*
Gantmacher, F., 37, *534*
Gardner, G., 399, *534*
Gâteaux, R., *534*
Gauss, K., 200
Gavurin, M., 235, *534*
Gendžojan, G., *534*
Geraščenko, S., *534*
Ghinea, M., *534*
Givens, J., 38
Glasmacher, W., *533*
Glazman, I., 499, *534*
Gleyzal, A., 188, 258, 507, *534*
Golab, S., *534*
Goldfeld, S., 242, *534*
Gol'dman, N., *529*
Goldstein, A., 110, 258, 363, 406, 408, 420, 478, 493, 499, 507, 518, 519, *521*, *529*, *534*, *535*

Gotusso, L., *535*
Graves, L., 131, *536*
Gray, J., *535*
Grebenjuk, V., *535*
Greenspan, D., 32, 222, 463, *521*, *535*
Greenstadt, J., *535*
Greub, W., 81, *535*
Groschaftová, Z., *535*
Gunn, J., 222, *535*
Gurr, S., *535*

H

Hadamard, J., 139, 154, *535*
Hadeler, K., *535*
Hahn, W., 353, *536*
Hajtman, B., *536*
Hanson, E., *536*
Hanson, M., 101, *536*
Hardaway, R., *536*
Hart, H., 227, *536*
Hartley, H., 269, 507, *536*
Haselgrove, C., *536*
Heinla, L., *554*
Heinrich, V., 200
Heinz, E., 155, 160, 163, *536*
Henrici, P., 13, 50, 200, 429, *536*
Hestenes, M., 263, 271, *536*
Hildebrandt, T., 131, *536*
Hill, W., *536*
Hirasawa, Y., 188, *536*
Holst, W., *556*
Holt, J., *536*
Homma, T., *536*
Homuth, H., *536*
Hooke, R., *536*
Hooker, W., *536*
Hopf, H., 154, 155, *523*
Horwitz, L., *536*
Householder, A., 37, 44, 50, 57, 307, *522*, *536*
Hrouda, J., *537*
Hubbard, B., 16, 17, 58, *528*, *537*
Hunter, W., *536*

I

Ichida, K., *554*
Isaev, V., *537*
Ivanov, V., *537*

J

Jain, P., *535*
Jakovlev, M., 235, 507, *537*
Jankó, B., 188, *537*
Jeeves, T., *536*
John, F., 16, *526*
Johnson, L., 363, 429, 430, *538*
Juncosa, M., *526*

K

Kaazik, Y., *538*
Kac, I., *538*
Kačurovskii, R., 89, 167, *538*
Kagiwada, H., 201, *526*
Kalaba, R., 188, 201, 222, 429, 454, *525*, *526*, *538*
Kalaida, O., *538*
Kannan, R., *538*
Kantorovich, L., 65, 74, 82, 188, 390, 406, 414, 420, 428, 439, 446, 499, *522*, *538*, *539*
Karlin, S., 163
Kas'janjuk, S., *555*
Kasriel, R., *539*
Katetov, M., *539*
Keller, H., 12, 30, 390, 429, *522*, *539*
Kellogg, O., *526*
Kellogg, R., 222, 332, 391, 471, *527*, *539*
Kempthorne, O., 264, *552*
Kenneth, P., *539*, *544*
Kerner, M., 97, *539*
Kerr, D., *539*
Kiefer, J., 276, *539*
Kincaid, W., *539*
Kirija, V., 235
Kirk, W., 125, *526*, *539*
Kitchen, J., 305, 306, *539*
Kivistik, L., 258, *539*, *540*
Kiyono, T., *554*
Kizner, W., 235, *540*
Kleinmichel, H., 235, 341, *540*
Knill, R., *540*
Kogan, T., *540*
Kolmogoroff, A., 390, *540*
Kolomy, J., 408, 414, *540*
Koppel', H., 203, 429, *540*
Korganoff, A., 201, 307, 363, *540*
Kosolev, A., *540*
Kotze, W., *540*
Kowalik, J., 266, *522*, *540*
Krasnoselskii, M., 20, 160, 164, 258, 407, *522*, *540*, *541*
Krawczyk, R., 378, *541*
Krein, M., 406
Krivonos, J., *541*
Kronecker, L., 154, 155, 163, *541*
Kuiken, H., *541*
Kulik, S., *541*
Kuntzmann, J., 307, *541*
Kuo, M., *541*
Kurpel', M., *541*
Kusakin, A., *541*
Kushner, H., *541*
Kuznečov, I., *544*
Kwan, C., *541*

L

Ladyzhenskaya, O., 16, 30, *541*
Lahaye, E., 235, *541*
Lahmann, H., *532*
Lancaster, P., 399, 429, *541*
Lance, G., *541*
Langer, R., *542*
Langlois, W., *542*
Laptinskii, V., *542*
Lastman, G., *542*
Lavi, A., *542*
Lavrent'ev, I., *542*
Leach, E., 131, *542*
Lee, E., 429, *542*
Lees, M., 12, 13, 145, 163, *542*
Leon, A., 266, *533*, *542*
Leray, J., 160, 163, *542*
Levenberg, K., 188, 242, 270, 507, *542*
Levin, A., *542*
Levy, M., 139, *542*
Lezanski, T., *542*
Li, T., 353, *542*
Lieberstein, H., 222, *542*
Liebl, P., 310, *542*
Lika, D., 188, *542*, *543*
Lin, C., *543*
Lin'kov, E., 408, 420, *543*
Liouville, J., 390
Ljaščenko, M., *543*
Ljubič, Y., 499, *543*
Lohr, L., 429, *543*
Lorentz, G., 172, 173, 175, *543*

AUTHOR INDEX

Lotkin, M., *543*
Luchka, A., 226, *543*
Ludwig, R., 200, *543*
Lumiste, Y., 499, *543*
Luxemburg, W., *543*
Lyubčenko, I., *543*

M

Maddison, R., 271, *543*
Madorskii, V., *543*
Maergoiz, M., 201, 363, *538*, *543*
Maistrov'skii, G., *543*
Mamedov, A., 429, *543*
Mancino, O., *543*
Mangasarian, O., 101, 102, 104, *544*
Marčuk, G., *544*
Marquardt, D., 259, 270, 507, *544*
Martos, B., 102, *544*
Maslova, N., *544*
Matveev, V., *544*
McAllister, G., 18, *544*
McCormick, G., 139, *533*
McGill, R., *539*, *544*
Mead, R., *546*
Meeter, D., *544*
Meinardus, G., 25, *544*
Melon, S., *544*
Mertvecova, M., 188, *544*
Metcalf, F., 407, *531*
Meyer, F., *553*
Meyer, G., 139, 235, 341, *544*
Meyers, P., *544*
Migovič, F., *541*
Mihklin, S., 26, *544*
Minkowski, H., 88, *544*
Minty, G., 89, 145, 164, 167, 168, *531*, *544*, *545*
Mirakov, V., *545*
Moore, J., *545*
Moore, R. H., 20, 429, *524*, *545*
Morozov, V., *545*
Morrison, D., *545*
Moser, J., 429, *545*
Moses, J., *545*
Moszyński, K., *545*
Motzkin, T., 17, 227, *536*, *545*
Muir, T., 170, *545*
Myers, G., *545*
Mysovskii, I., 414, *545*

N

Nagumo, M., 155, *545*
Nashed, M., 102, 248, 499, 506, *539*, *545*, *546*
Nečepurenko, M., 188, *546*
Nelder, J., *546*
Nemyckii, V., *546*
Neustadt, L., *525*
Newman, D., *546*
Nicolovius, R., *546*
Nicula, A., *546*
Nikaido, H., 139, 145, *534*
Noble, B., 201, 429, *546*

O

Opial, Z., 145, 168, 407, *546*
Ortega, J., 57, 222, 293, 305, 306, 331, 345, 363, 399, 400, 414, 420, 428, 440, 446, 454, 463, 470, 471, 512, *546*
Osborne, M., 266, *522*, *540*
Ostrowski, A., 37, 51, 187, 200, 259, 294, 305, 399, 428, 478, 493, 499, *522*, *546*, *547*
Overholt, K., 278, *547*

P

Panov, A., 306, *547*
Parmet, I., *547*
Parter, S., 222, 463, *535*, *547*
Peaceman, D., *547*
Penrose, R., 270, *547*
Pereyra, V., 406, 429, *547*
Perron, O., 305, *547*
Perry, C., *523*
Peteanu, V., *547*
Petry, W., *547*
Petryshyn, W., 125, 407, *528*, *547*
Picard, E., 390
Pietrzykowski, T., *548*
Pinsker, A., *539*
Pitteway, M., *532*
Poljak, B., 89, 101, 108, 235, 353, 414, 499, *548*
Poll', V., *548*
Ponstein, J., 101, 102, 103, 104, *548*
Porsching, T., 471, *548*
Powell, M., 212, 248, 264, 265, 271, 273, 478, *533*, *548*

Prager, M., 239, *548*
Price, H., 58, *548*
Price, J., 363, *535*
Pugačev, B., *548*
Puzynin, I., *557*

Q

Quandt, R., 242, *534*

R

Rachford, H., *547*
Rademacher, H., 88, *549*
Rado, T., 30, *549*
Rall, L., 37, 317, 429, *522*, *524*, *535*, *543*, *549*
Reeves, C., 265, *533*
Rehbock, F., 428, *549*
Reiss, E., *539*
Rheinboldt, W., 57, 139, 140, 222, 306, 331, 399, 400, 406, 414, 419, 420, 428, 440, 446, 454, 463, 470, 471, 512, *525*, *546*, *549*
Rice, J., *549*
Roberts, S., 235, *549*
Robinson, S., 203, 380, *549*
Rockafellar, R., *549*
Rockoff, M., 222, 293, 305, 306, 331, 345, *546*
Rosenbloom, P., 235, *549*
Rosenbrock, H., 265, 478, *549*
Roth, B., 235, *533*
Roth, C., *549*
Rothe, E., 165, *550*
Runge, C., 316, *550*
Rutickii, Y., 258, *541*

S

Saaty, T., *550*
Šafiev, R., 188, *550*
Saibel, E., *547*
Šamanskii, V., 188, 316, 365, 367, *550*
Samokis, B., *530*
Sapagovas, M., *550*
Sarachik, P., *536*
Sarbasov, K., *544*
Sard, A., 131, *550*
Šarkovskii, O., *550*
Savenko, S., *550*
Scarf, H., *550*

Schaefer, H., 407, *550*
Schauder, J., 160, 163, *542*, *551*
Schechter, E., *551*
Schechter, M., 16, *526*
Schechter, S., 32, 115, 222, 223, 518, 519, *551*
Schmetterer, L., *551*
Schmidt, J., 201, 242, 286, 316, 363, 364, 365, 399, 429, 430, 454, *551*
Schoenheinz, H., *551*
Scholz, D., 363, 429, 430, *538*
Schröder, E., 307, *551*
Schröder, J., 420, 439, 440, 446, 447, 471, 472, *551*, *552*
Schultz, M., 26, 32, 145, 163, *527*, *530*, *542*
Schwartz, J., 131, 139, 155, 160, 163, *522*, *532*, *552*
Schwetlick, H., 363, *551*
Seda, V., *552*
Sefor, L., 235, *531*
Sen, R., *552*
Senčuk, Ju., *534*
Sergeev, A., 203, *552*
Shah, B., 264, *552*
Shah, M., *552*
Shampine, L., 12, *525*
Shanno, D., *552*
Shinbrot, M., 165, 169, *552*
Shipman, J., 235, *549*
Sidlovskaya, N., 235, 341, *552*
Simeonov, S., *552*
Simpson, R., *552*
Sisler, M., *552*, *553*
Slugin, S., 446, *553*
Smith, H., 25, *531*
Smith, R., 353, *553*
Sonin, V., *537*
Southwell, R., *553*
Spang, H., *553*
Späth, H., 243, *553*
Spiridonov, V., *553*
Srinivasacharyulu, K., *553*
Stečenko, V., *553*
Steffensen, J., *553*
Stein, M., *553*
Stepleman, R., 18, 115, 116, 117, 164, 318, 454, 507, *553*
Sternberg, S., *528*
Stewart, G., 248, *553*
Stiefel, E., 263, *536*

AUTHOR INDEX

Strygin, V., *542*
Swann, W., 265, *553*
Sylvester, R., *553*
Syn, W., *552*

T

Takahashi, I., 265, *553*
Takota, M., *553*
Tamme, E., *554*
Tapia, R., *554*
Taylor, A., 50, *554*
Temple, G., 499, *554*
Thompson, G., *536*
Thurston, G., *554*
Todd, J., *554*
Tornheim, L., 200, 293, 294, 378, 507, *554*
Traub, J., 187, 188, 200, 286, 294, 307, 316, 317, 522, *554*
Trinkaus, H., 242, *551*
Trotter, H., 242, *534*
Tsuda, T., *554*
Turner, L., *554*

U

Ul'm, S., 188, 203, 363, 429, *554*
Urabe, M., 399, *554*, *555*
Ural'tseva, N., 16, 30, *541*
Ustaal, A., *540*

V

Vainberg, M., 65, 66, 74, 167, 499, *522*, *555*
Vandergraft, J., 57, 439, 454, 455, *555*
Varga, R., 26, 32, 50, 57, 222, 225, 293, 307, 332, 341, 342, 343, 344, *522*, *527*, *530*, *555*
Varjuhin, V., *555*
Vasudevan, R., *526*
Vertgeim, B., *555*
Vitasek, E., 239, *548*
Voevodin, V., *555*
Vogl, L., *542*
Voigt, R., 205, 295, 307, 308, 309, 334, 353, 365, 366, 369, 379, 380, *555*

Vulich, B., *539*
Vyhandu, L., *555*

W

Wachspress, E., 222, *555*
Wall, D., 293, *555*
Walsh, J., *555*
Waltman, P., 12, *525*
Wang, H., *528*
Warga, J., 399, *555*
Wasow, W., 16, 17, 222, 307, 343, *521*, *533*, *545*
Wasscher, E., *555*
Wegge, L., 201, 223, 363, *555*
Weinitschke, H., 391, *556*
Weinstock, R., 26, *556*
Weissinger, J., 390, 399, *556*
Wells, M., *556*
Wetterling, W., *556*
Whiteside, D., *556*
Wilde, D., 264, *556*
Wilkinson, J., 37, 44, *556*
Willers, F., 428, *556*
Winslow, A., *556*
Witte, B., *556*
Wolfe, P., 200, *556*
Wouk, A., 390, *556*

Y

Yamamuro, S., 165, *556*
Yohe, M., 222, *535*
Yoshiaki, M., 446, *556*

Z

Zaguskin, V., *556*
Zangwill, W., 265, *556*
Zaplitna, G., *556*
Zarantonello, E., 167, 407, 408, *557*
Zeleznik, F., 213, *557*
Zhidkov, E., *557*
Zhidkov, N., 390, 393, 446, *526*
Zielke, G., 50, *557*
Zinčenko, D., 414, 415, 431, *557*
Zuber, R., *557*
Zuev, A., *557*

SUBJECT INDEX

A

A-orthogonal vectors, 260, 266
Absolute value
  of matrices, 52
  of vectors, 52
Affine mapping, 34
Almost linear mapping, 11
Alternating direction (ADI) method, *see* Peaceman–Rachford method
Altman principle, 251, 258, 259, 483, 496–498
Antitone mapping, 52
Approximate contraction, 393–400
Asymptotically monotone on rays, 169

B

Ball, 2
Banach space, 50, 74, 82, 88, 89, 124, 125, 131, 139, 145, 160, 163, 164, 167, 187, 188, 201, 235, 286, 306, 307, 341, 363, 364, 390, 407, 414, 428, 493, 499
Basic types of matrices, 34–36
Berman algorithm, 276–278
Bernstein polynomial, 172
Bilinear mapping, 75
Block-tridiagonal matrix, 16
Boundary-value problem, 112, 141, 386, 471
  elliptic, 14–18, 112, 222, 225, 344, 440, 459, 463
  two-point, 9–14, 23, 111, 116, 145, 201, 235, 344, 392, 393, 454
Boundary value theorem, 157
Brouwer fixed point theorem, 161, 164

C

Cauchy–Schwarz inequality, 39
Chain rule, 62
Change of variable theorem, 169
Chaplygin method, 446
Characteristic polynomial, 35, 37
Chebyshev approximation, 25
Coercive mapping, 165–169
Coerciveness theorem, 166
Compact support, 149, 155
Comparable vectors, 51
Comparison principle for steplength algorithms, 484
Composite mapping, 62
Condition number, 42
Conjugate basis, 260
Conjugate-direction methods, 260–267, 509–513
Conjugate-gradient method, 261–263, 271–273, 509–513
Conjugate vectors, 260
Connected functional, 98–101, 105, 108

Continuous iterative process, 235
Consistent approximation, 355–369, 371, 375
Consistently ordered matrix, 344
Continuation method, 230–236, 334–341
Continuation property, 132–141
Continuous differentiability, 60, 71
Contraction, 119–125, 164, 408, 433, 439
　approximate, 393–400
　iterated, 400–401, 406–407
　P–, 433, 440
Contractions, sequence of, 393–400
Contraction mapping theorem, 120, 124, 125, 385, 410
　generalizations, 164, 383–408
Control problem, 12
Convergence factor, 281–298, see also Q-factor, R-factor
Convergence of iterative methods, see particular method
Convergence theorem, types of, 5, see also various theorems (global, local, monotone, semilocal convergence)
Convex functional, 82–90, 99–109, 142
Convex mapping, 448–450, 454, 455, see also Convex functional
Coordinate vector, 34
Critical point, 93–98, 100, 101, 473–520
Cube in R^n, 130
Cubic convergence, 286, see also local convergence of particular method
Curry principle, 251, 258, 483, 496–498

D

Damped methods, see particular method
Damping factor, 253
Daniel algorithm, 262, 509
　convergence, 509, 510
Davidon–Fletcher–Powell method, 212, 213, 246–249, 264, 266, 273
Degree integral, 149, 151
Degree of mapping, 147–160
δ^2 process, 200
Derivative, see also F-derivative, G-derivative
　higher order, 75, 81
Descent method, 243, 473
Diagonal dominance theorem, 48
Diagonal mapping, 11

Diagonally dominant matrix, 48
Diagonally isotone mapping, 464
Difference equation, 238, 294, 305, 306, 353, 409, 420
Differential, see Frechet differential, Gateaux differential
Differential correction method, see Gauss–Newton method
Differential inequalities for convexity, 84–88, 448–450
Differentiation with respect to a parameter, 235
Direction vector, 244
Dirichlet problem, 14
Discrete analog, 10, 15, 19
Discrete Newton method, see Newton method
Discrete Ritz method, 24, 26, 27
Discretization error, 12, 16
Divergence theorem, 170
Divided difference operator, 201–203, 213, 363
Domain-invariance theorem, 145, 160
Domain of iterative process, 236

E

Eigenvalue, 35
Eigenvector, 35
Elliptic boundary value problem, 14–18, 112, 222, 225, 344, 440, 454, 463
Elliptic norm, 245
Ellipticity condition, 17, 30, 31
Essentially periodic sequence, 513–516
Euclidean norm, 39
Euler equation, 22, 26, 30, 31
Euler's method, 233–235, 339
Excision theorem, 158

F

F-derivative, 61, 62, 65–67
　partial, 127, 128, 131, 132
　second, 75–81
　strong, 71, 72, 74, 126, 127, 131
F-function, see Forcing function
f-minimal solution, 96
Fibonaccian search, 276–278
First integral of difference equation, 416
Fixed point, 119

Fletcher–Reeves algorithm, 262, 510
  convergence, 510, 511
Forcing function, 479
Frechet differential, 66
Frechet derivative, *see* F-derivative
Free steering sequence or method, 223, 513, 517, 519
Function, *see* Mapping
Functional,
  connected, 98–101, 105, 108
  convex, 82–90, 99–109, 142
  hemivariate, 474, 475, 478, 479
  midpoint convex, 89
  pseudoconvex, 102–104, 518
  quasiconvex, 99–105, 108, 275, 479, 489
Functional averaging method, 226
Fundamental theorem of integral calculus, 70

G

G-derivative, 59–62, 65–67, 71, 74
  second, 75–81, 87
Gateaux derivative, *see* G-derivative
Gateaux differential, 65, 66
Gauss method, *see* Gauss–Newton method
Gauss–Newton method, 227, 267–270
  convergence, 320, 505, 506
  damped, 503
    convergence, 504
Gauss-secant method, 269
Gauss–Seidel method, *see* SOR method
Gauss–Seidel–Newton method, *see* SOR-Newton method
Gauss–Seidel principle, 224
Gaussian elimination, 227
General position, 191
Generalization of contraction mapping theorem, 164, 383–408
Generalized inverse, 270
Generalized linear methods, 214–230, 320–334, 456–472, 516–518
Gerschgorin circle theorem, 49
Global convergence theorem, 5, 385, 387, 392, 393, 434, 438, 440, 453, 454, 461, 462, 464, 469, 471, 503, 505, 506, 508, 510, 512, 516–518
Goldstein principle, 255, 256, 490, 496
Goldstein–Armijo algorithm, 491, 496, 503
Gradient equation, 33

Gradient mapping, 95, 142, 167
Gradient method, 245, 247, 248, 258
  convergence, 496
Gradient-Newton method, 252
Gradient-related methods, 474, 495
  convergence, 495–510, 519
Gradient-related sequence, 495
Gradient-vector, 60
Green's function, 19

H

H-equation, 18, 20, 21
Hadamard inequality, 380
Hadamard theorem, 137–139
Hammerstein equation, 19, 20
Hemicontinuous, 61, 67, 71, 79
Hemivariate functional, 474, 475, 478, 479
Hessian matrix, 76, 87
Higher order derivative, 75, 81
Hilbert space, 139, 164, 167, 168, 499, 512
Hölder continuity, 63, 64
Homeomorphism, 63, 107, 110, 122, 132, 137, 139, 141, 143, 146, 167, 169, 466, 467
  local, 124, 125, 132
Homotopy, 135, 231
Homotopy invariance theorem, 156
Hyperrectange, 130

I

Imbedding method, *see* Continuation method
Implicit function theorem, 128, 131
Initial points of iterative processes, 236
Inner product, 39
Integral equation, 18–20, 392, 429
Interval in R^n, 68
Inverse function theorem, 125, 131
Inverse isotone mapping, 466, 467, 471
Inverse mapping, 63
Irreducible matrix, 46
Isolated solution, 124
Isotone mapping, 52
Iterative process, 236
  continuous, 235
  domain, 236
  initial points, 236

SUBJECT INDEX 569

J

Jacobi method, 217
  nonlinear, 220, 225, 464, 500
    global convergence, 434, 440, 469, 471
    monotone convergence, 465, 467, 471
Jacobi–Newton method, 220, 222, 229
  local convergence, 332, 334, 343
Jacobi-regula-falsi method, 333
Jacobi-secant method, 221, 223, 368
  local convergence, 353
Jacobi–Steffensen method, 221, 223
Jacobian determinant, 170
Jacobian matrix, 60
Jordan block, 36
Jordan canonical form, 36, 37
  modified, 44

K

Kantorovich lemma, 417, 442, 446
Kronecker theorem, 161

L

l_p-norm, 39–41, 52, 57, 67
Least-squares problem, 21, 25
Least-squares solution, 96, 267
Leray–Schauder theorem, 162, 165
Level set, 98
  proper, 103
Linear convergence, 285, 291, 294, *see also* local convergence of particular method
Linear convergence theorem, 301
Linear interpolation in R^n, 192, 205, 206
Lipschitz-continuity, 63, 64, 70
Local convergence theorems, 299–334, 347–380
Local homeomorphism, 124, 125, 132

M

M-function, 468–472
M-matrix, 54
m-step method, 237
Majorization of iterative methods, 409, 414–416, 419, 420, 428, 429
Majorization principle for steplength, 253, 481
Majorizing sequence, 409, 414

Mapping, 2
  affine, 34
  almost-linear, 11
  antitone, 52
  asymptotically monotone on rays, 169
  bilinear, 75
  coercive, 165–169
  contractive, *see* Contraction
  convex, 448–450, 454, 455, *see also* Functional, convex
  diagonal, 11
  diagonally isotone, 464
  inverse, 63
  inverse isotone, 466, 467, 471
  isotone, 52
  monotone, 85, 141–146, 167–169, 387
  multilinear, 81
  nonexpansive, *see* Nonexpansion
  norm-coercive, 136, 139
  off-diagonally antitone, 465
  one-to-one, 63
  order-convex, 448–450, 454, 455
  weakly coercive, 165–168
Mappings, uniformly differentiable, 349
Matrices, uniformly nonsingular, 369, 370
Matrix, basic types, 34–36
  block tridiagonal, 16
  consistently ordered, 344
  diagonally dominant, 48
  irreducible, 46
  nonnegative, 53
  positive-definite, 35
  strictly lower (upper) triangular, 53
  two-cyclic, 344
Matrix-norm, 40, *see also* Norm
Maximal solution, 445
Maximizer, 93
Mean value theorems, 68–74, 78–82, 127
Measure zero, 130
Metric space, 124, 390, 398
Midpoint convex functional, 89
Mildly nonlinear equation, 31
Minimal solution, 445
Minimal surface problem, *see* Plateau problem
Minimization in R^1, 275–278
Minimization methods, 240–278
  convergence, 473–520
Minimization principle for steplength, 249, 258, 259, 487–490

Minimization problems, 21–26
Minimizer, 21, 93–118
Modified Jordan canonical form, 44
Modification methods, 206–214, 270
Modifications of Newton's method, *see* Newton's method
Modulus of continuity, 64, 67
Monotone convergence theorems, 441–447, 450–455, 457–460, 465, 467
Monotone mapping, 85, 141–146, 167–169, 387
Monotonic norm, 52, 58
Multilinear mapping, 81
Multistep method, 237, 347–380, 388–391, 435
Mysovskii theorem, 412, 428

N

Natural (component-wise) partial ordering, 51, 57
Network problem, 471
Neumann lemma, 45, 53, 122
Newton's method, 4, 183–185, 187, 201, 204, 214, 220, 228, 232–235, 238, 241, 244, 269
  damped, 186, 243, 244, 246–248, 258, 501
    global convergence, 503
  discrete, 186–189, 196, 201
    local convergence, 360, 363, 365
  global convergence, 453, 454, 464
  local convergence, 287, 312, 316–320, 336–339, 365
  modifications, 187, 241
    convergence, 313–316, 357, 358, 505
  monotone convergence, 451–455, 459
  for over-(under-)determined systems, 406
  semilocal convergence, 401–403, 407, 412–414, 421, 425, 428, 429, 506
  simplified, 182, 187, 316
    convergence, 318, 421, 455
Newton–Jacobi method, 219, 221, 229
  local convergence, 332
Newton–Kantorovich theorem, 421, 425, 428, 429
Newton–Peaceman–Rachford method, 219, 230
  local convergence, 328

Newton–SOR method, 215–219, 222–225, 229, 235, 237
  damped, 504
    convergence, 504
  global convergence, 462, 463
  local convergence, 321, 322, 327, 331, 332, 342, 343, 346, 350–352, 355
  monotone convergence, 458–460
  semilocal convergence, 427
Nonexpansion, 119–122, 403–408
Nonlinear discretization, 31, 116
Nonlinear iterative method, *see* particular method
Nonnegative matrix, 53
Nonnegative vector, 52
Nonstationary method, 237, 347–350, 395, 435
Norm, 38–45
  elliptic, 245
  Euclidean, 39
  l_p, 39–41, 52, 57, 67
  monotonic, 52, 58
  strictly and uniformly convex, 44, 45
Norm-coercive mapping, 136, 139
Norm-coerciveness theorem, 136, 139
Norm-equivalence theorem, 39, 42
Norm-reducing method, 186
Normal equations, 22

O

Off-diagonally antitone mapping, 465
One-to-one mapping, 63
Operator, *see* Mapping
  of monotone kind, 471
Operator norm, *see* matrix norm
Order convergence, 439
Order-convex mapping, 448–450, 454, 455
Order-interval, 441
Order of convergence, *see* Q-order, R-order
Ostrowski theorem, 300, 305, 349
Over-(under-)determined systems, 22, 25, 97, 129–131, 213, 267–271, 320, 406, 503, 504

P

P-contraction, 433, 440
P-matrix, 145
Paraboloid method, 240–243, 268

SUBJECT INDEX 571

Parallel chord method, 181
  convergence, 183, 418–420
Partan method, 264, 266
Partial derivative, see F-derivative, G-derivative
Partial ordering, 51, 52, 57, 432, 440
  natural (component-wise), 51, 57
Partially ordered space, 439, 446, 454, 471
Path, 135
Path-connected set, 99
Peaceman–Rachford method, 218
  nonlinear, 221, 222, 225, 227
    global convergence, 387, 391
    local convergence, 329–332
Peaceman–Rachford–Newton method, 221, 222
  local convergence, 329–332
Permutation matrix, 46
Perron-Frobenius theory, 57
Perturbation lemma, 45
Picard iteration, 182
  convergence, 386
Plateau equation, 17, 31
Plateau problem, 27, 31, 115, 513
Point of attraction, 299, 305, 306, 348, see also Local convergence theorems
Poincaré–Bohl theorem, 157
Polynomial system, 20
Potential mapping, see Gradient mapping
Positive cone, 52
Positive definite matrix, 35
Positive definite second derivative, 87–89, 94
Primary iteration, 215
Principle of sufficient decrease, 479
Property A, 344
Proper level set, 103
Pseudoconvex functional, 102–104, 518
Pseudometric space, 398, 399, 439

Q

Q-factor, 282–287, 295–298
Q-faster, 283, 285
Q-order, 284–286, 296–298
Quadratic convergence, 184, 286, 291, see also local convergence of particular method
Quasiconvex functional, 99–105, 108, 275, 479, 489

Quasilinear equation, 17, 30
Quasilinearization, 188, 454
Quasi-Newton method, 212
Quotient-convergence factor, see Q-factor

R

R-factor, 287–298
R-faster, 289, 290
R-order, 290, 291, 296–298
Rate of convergence, see Q-factor, R-factor, local convergence of particular method
Regula-falsi method, 189, 205
  convergence, 366, 379
Regular splitting, 56–58
Relaxation parameter, 215
Restriction of mapping, 63
Reverse modulus of continuity, 482
Ritz method, 23, 26, 27, 31
  discrete, 24, 26, 27
Root convergence factor, see R-factor
Rosenbrock method, 262, 265, 266
  convergence, 519
Runge–Kutta method, 235, 236

S

Sard theorem, 130, 131
Schauder fixed point theorem, 164
Schultz iteration, 446
Search method in R^1, 275–278
Secant approximation, see Secant method
Secant method, 189–198, 200, 201, 203–208, 212
  convergence, 361–365, 372, 373, 378, 379, 429
Secant-SOR method, 219
Second derivative, 74–82, see also F-derivative, G-derivative
  positive definite, 87–89, 94
Secondary iteration, 215
Seidel method, 223
Semilocal convergence, see particular method
Sequence of contractions, 393–400
Set notation, 2
Sherman–Morrison formula, 50
Sherman–Morrison–Woodbury formula, 50
Shooting method, 11, 12

Simplified Newton method, *see* Newton's method
Simultaneous displacement method, *see* Jacobi method
SOR method, 214, 215, 220, 222, 229, 341, 342, 344
  nonlinear, 219, 222–226, 238, 325, 441, 464
    global convergence, 438, 440, 469, 471, 516–519
    local convergence, 326–328, 331, 332, 342–346
    monotone convergence, 465, 467, 470
  SOR-Newton method, 220, 222, 223, 225, 229, 239, 252, 487, 519
    local convergence, 323, 327, 333, 343–346, 353
SOR-regula-falsi iteration, 334
SOR-secant method, 221, 353
SOR-Steffensen method, 221
Spectral radius, 43
Splitting, 56–58, 217, 225
Square root of matrix, 38
Stationary process, 237
Steepest descent method, 245–247, *see also* Gradient-related methods
Steffensen method, 198–201, 203
  convergence, 362, 363, 365, 373, 378, 379, 429
Steffensen–Peaceman–Rachford method, 219
Steplength, 244
Steplength algorithms, 249–260, 429–494
Steplength sets, 255
Stieltjes matrix, 54
Strictly and uniformly convex norm, 44, 45
Strictly lower (upper) triangular matrix, 53
Strong derivative, *see* F-derivative
Strongly consistent approximation, *see* Consistent approximation
Strongly downward sequence, 475
Subinverse, 53–56, 443
Sublinear convergence, 286, 291
Successive displacement method, *see* Gauss–Seidel method
Successive overrelaxation method, *see* SOR method
Superinverse, 53

Superlinear convergence, 285, 291, *see also* local convergence of particular method
Symmetry of higher derivative, 76, 77, 81, 82
Symmetry principle, 95

T

Tangent-hyperbola method, 188
Taylor formula, 79
Tensor product of matrices, 67, 310
Tietze–Urysohn extension theorem, 175
Toeplitz lemma, 399
Two-point boundary value problem, 9–14, 23, 111, 116, 145, 201, 235, 344, 392, 393, 454
Types of convergence theorems, 5, *see also* various theorems (global, local, monotone, semi-local convergence)
Two-cyclic matrix, 344

U

Underrelaxation factor, 253
Uniform approximation, 25
Uniform monotonicity theorem, 143, 167
Uniformly differentiable mappings, 349
Uniformly linearly independent sequence, 514, 518
Uniformly nonsingular matrices, 369, 370
Unimodal function, 275
Univariate relaxation method, 244
  convergence, 516–519
Urysohn equation, 19

V

Vector, 1, 34
  nonnegative, 52

W

Weak regular splitting, 56–58, 456
Weakly coercive mapping, 165–168
Weakly linear convergence, 294
Weierstrass approximation theorem, 171, 172

Z

Zangwill algorithm, 265, 519